Phytoremediation for Environmental Sustainability

Ram Prasad
Editor

Phytoremediation for Environmental Sustainability

Springer

Editor
Ram Prasad
Department of Botany
Mahatma Gandhi Central University
Motihari, Bihar, India

ISBN 978-981-16-5620-0 ISBN 978-981-16-5621-7 (eBook)
https://doi.org/10.1007/978-981-16-5621-7

This Springer imprint is published by the registered company Springer Nature Singapore Pte Ltd.
The registered company address is: 152 Beach Road, #21-01/04 Gateway East, Singapore 189721, Singapore

Contents

Editor and Contributors

About the Editor

Ram Prasad Ph.D., is associated with the Department of Botany, Mahatma Gandhi Central University, Motihari, Bihar, India. His research interest includes applied and environmental microbiology, plant-microbe interactions, sustainable agriculture, and nanobiotechnology. Dr. Prasad has more than 225 publications to his credit, including research papers, review articles, book chapters and six patents issued or pending, and edited or authored several books. Dr. Prasad has 12 years of teaching experience and has been awarded the Young Scientist Award and Prof. J.S. Datta Munshi Gold Medal by the International Society for Ecological Communications; FSAB fellowship by the Society for Applied Biotechnology; the American Cancer Society UICC International Fellowship for Beginning Investigators, USA; Outstanding Scientist Award in the field of Microbiology by Venus International Foundation; BRICPL Science Investigator Award; Research Excellence Award, etc. He has been serving as a board member of various national and international journals. Previously, Dr. Prasad served as Assistant Professor at Amity University Uttar Pradesh, India; Visiting Assistant Professor, Whiting School of Engineering, Department of Mechanical Engineering at Johns Hopkins University, Baltimore, United States; and Research Associate Professor at the School of Environmental Science and Engineering, Sun Yat-sen University, Guangzhou, China.

Contributors

Komal Agrawal Bioprocess and Bioenergy Laboratory, Department of Microbiology, Central University of Rajasthan, Bandarsindri, Kishangarh, Ajmer, Rajasthan, India

Meri Barbafieri Istituto di Ricerca sugli Ecosistemi Terrestri, CNR, Pisa, Italy

Sahana Basu Department of Biotechnology, Assam University, Silchar, Assam, India

Ankita Bhatt Bioprocess and Bioenergy Laboratory, Department of Microbiology, Central University of Rajasthan, Bandarsindri, Kishangarh, Ajmer, Rajasthan, India

Sylwia Budzyńska Department of Chemistry, Faculty of Forestry and Wood Technology, Poznań University of Life Sciences, Poznań, Poland

A. J. Palace Carvalho Chemistry Department, School of Sciences and Technology, University of Évora, Évora, Portugal
LAQV-REQUIMTE, University of Évora, Évora, Portugal

Disha B. Changela Department of Biochemistry, Saurashtra University, Rajkot, Gujarat, India

Chongrak Polprasert Department of Civil Engineering, Faculty of Engineering, Thammasat University, Bangkok, Thailand

Kruti G. Dangar Department of Biochemistry, Saurashtra University, Rajkot, Gujarat, India

Chirayu Desai Department of Biological Sciences, P. D. Patel Institute of Applied Sciences, Charotar University of Science and Technology, CHARUSAT Campus, Changa, Anand, Gujarat, India

Neha Dhingra Department of Zoology, University of Delhi, Delhi, India

Bhupinder Dhir School of Sciences, Indira Gandhi National Open University, New Delhi, India

Bhawna Dipta ICAR-Central Potato Research Institute, Shimla, India

Ana V. Dordio Chemistry Department, School of Sciences and Technology, University of Évora, Évora, Portugal
MARE-Marine and Environmental Science Centre, University of Évora, Évora, Portugal

Kinga Drzewiecka Department of Chemistry, Faculty of Forestry and Wood Technology, Poznań University of Life Sciences, Poznań, Poland

Elisabetta Franchi Eni S.P.A., Renewable Energy and Environmental Laboratories, S. Donato Milanese, Milan, Italy

Liliana Giraldo Departamento de Química, Facultad de Ciencias, Universidad Nacional de Colombia, Bogotá, Colombia

Piotr Goliński Department of Chemistry, Faculty of Forestry and Wood Technology, Poznań University of Life Sciences, Poznań, Poland

Martina Grifoni Istituto di Ricerca sugli Ecosistemi Terrestri, CNR, Pisa, Italy

Diana Hernández-Monje Departamento de Química, Facultad de Ciencias, Universidad Nacional de Colombia—Sede Bogotá, Bogotá, Colombia

Anna Ilek Department of Botany and Forest Habitats, Faculty of Forestry and Wood Technology, Poznań University of Life Sciences, Poznań, Poland

Kunal R. Jain Post Graduate Department of Biosciences, Sardar Patel University, Satellite Campus, Bakrol, Gujarat, India

Prakriti Jhilta Dr Yashwant Singh Parmar University of Horticulture and Forestry, Nauni, Solan, Himachal Pradesh, India

Suphiya Khan Department of Bioscience and Biotechnology, Banasthali Vidyapith, Banasthali Tonk, Rajasthan, India

Magdalena Krzesłowska Laboratory of General Botany, Faculty of Biology, Adam Mickiewicz University, Uniwersytetu Poznańskiego 6, Poznań, Poland

Anil Kumar Laboratory of Plant Physiology and Biochemistry, Department of Botany, University of Rajasthan, Jaipur, Rajasthan, India

Gautam Kumar Department of Life Science, Central University of South Bihar, Panchanpur, Gaya, Bihar, India

Indrajeet Kumar Department of Botany, Institute of Science, Banaras Hindu University, Varanasi, India

Umesh Kumar Department of Botany, Institute of Science, Banaras Hindu University, Varanasi, India

Fatma Kusur Department of Molecular Biology and Genetics, Faculty of Science and Letters, Usak University, Usak, Turkey

Datta Madamwar Department of Biological Sciences, P. D. Patel Institute of Applied Sciences, Charotar University of Science and Technology, CHARUSAT Campus, Changa, Anand, Gujarat, India

Abdulrezzak Memon Department of Molecular Biology and Genetics, Faculty of Science and letters, Usak University, Usak, Turkey

Muhammet Memon Izmir International Biomedicine and Genome Institute (IBG-Izmir), Dokuz Eylül University Saglik Yerleskesi, Izmir, Turkey

Mirosław Mleczek Department of Chemistry, Faculty of Forestry and Wood Technology, Poznań University of Life Sciences, Poznań, Poland

Patrycja Mleczek Department of Ecology and Environmental Protection, Faculty of Environmental Engineering and Mechanical Engineering, Poznań University of Life Sciences, Poznań, Poland

Agnieszka Mocek-Płóciniak Department of General and Environmental Microbiology, Poznań University of Life Science, Poznań, Poland

Astrid Roxanna Moreno-Marenco Departamento de Química, Facultad de Ciencias, Universidad Nacional de Colombia, Bogotá, Colombia

Juan Carlos Moreno-Piraján Departamento de Química, Facultad de Ciencias, Universidad de los Andes, Bogotá, Colombia

Hamid Mukhtar Institute of Industrial Biotechnology, Government College University (GCU), Lahore,, Pakistan

Mehmooda Munazir Department of Botany, Government College Women University, Sialkot, Punjab, Pakistan

Mubashrah Munir Department of Biological Sciences, University of Veterinary and Animal Sciences, Lahore, Punjab, Pakistan

Ulla Neumann Central Microscopy, Max Planck Institute for Plant Breeding Research, Cologne, Germany

Dishant Patel Department of Biological Sciences, P. D. Patel Institute of Applied Sciences, Charotar University of Science and Technology, CHARUSAT Campus, Changa, Anand, Gujarat, India

Francesca Pedron Istituto di Ricerca sugli Ecosistemi Terrestri, CNR, Pisa, Italy

Gianniantonio Petruzzelli Istituto di Ricerca sugli Ecosistemi Terrestri, CNR, Pisa, Italy

Aneta Piechalak Department of Genome Biology, Institute of Molecular Biology and Biotechnology, Adam Mickiewicz University in Poznań, Poznań, Poland

A. P. Pinto Chemistry Department, School of Sciences and Technology, University of Évora, Évora, Portugal
Institute of Mediterranean Agricultural and Environmental Sciences (ICAAM), University of Évora, Évora, Portugal

Ram Prasad Department of Botany, Mahatma Gandhi Central University, Motihari, Bihar, India

Rahmatullah Qureshi Department of Botany, Pir Mehr Ali Shah Arid Agriculture University, Rawalpindi, Punjab, Pakistan

Ketaki V. Ramani Department of Microbiology, School of Science, RK University, Rajkot, Gujarat, India

Anchal Rana CSIR-Institute of Himalayan Bioresources Technology, Palampur, Himachal Pradesh, India

Irene Rosellini Istituto di Ricerca sugli Ecosistemi Terrestri, CNR, Pisa, Italy

Sujoy Sarkar Department of Life Science, Central University of South Bihar, Panchanpur, Gaya, Bihar, India

Kunal Seth Department of Botany, Government Science College, Valsad, Gujarat, India

Rajesh K. Sharma Department of Botany, Institute of Science, Banaras Hindu University, Varanasi, India

Ranju Sharma Department of Zoology, University of Delhi, Delhi, India

Neha Singh Department of Bioscience and Biotechnology, Banasthali Vidyapith, Banasthali Tonk, Rajasthan, India

Ngangbam Sarat Singh Department of Zoology, Dr. SRK Government Arts College, Yanam, Puducherry, India

Prince K. Singh Department of Botany, Institute of Science, Banaras Hindu University, Varanasi, India

Neerja Srivastava Department of Biochemistry, IBSBT, CSJM University, Kanpur, India

Szymon Suski Laboratory of Electron Microscopy, Nencki Institute of Experimental Biology of Polish Academy of Sciences, Warszawa, Poland

Suthirat Kittipongvises Environmental Research Institute, Chulalongkorn University (ERIC), Bangkok, Thailand
Research Unit of Green Mining Management (GMM), Chulalongkorn University, Bangkok, Thailand
Center of Excellence for Environmental and Substance Management (HSM), Chulalongkorn University (ERIC), Bangkok, Thailand

Małgorzata Szostek College of Natural Sciences, Departament of Soil Sciences, Environmental Chemistry and Hydrology, University of Rzeszów, University of Rzeszów, Rzeszów, Poland

Antonius C. J. Timmers Central Microscopy, Max Planck Institute for Plant Breeding Research, Cologne, Germany
Laboratory of Plant Microorganism Interactions, INRA UMR 441, CNRS UMR2594, Castanet-Tolosan, France

Pradeep Verma Bioprocess and Bioenergy Laboratory, Department of Microbiology, Central University of Rajasthan, Bandarsindri, Kishangarh, Ajmer, Rajasthan, India

Adam Woźny Laboratory of General Botany, Faculty of Biology, Adam Mickiewicz University, Uniwersytetu Poznańskiego 6, Poznań, Poland

Jay S. Yadav Department of Botany, Institute of Science, Banaras Hindu University, Varanasi, India

Metal Hyperaccumulator Plants and Their Role in Phytoremediation

1

Abdulrezzak Memon, Fatma Kusur, and Muhammet Memon

Abstract

Several hundred plant species are documented as metal hyperaccumulators, and a majority of them are restricted to metalliferous soils and are known as obligate hyperaccumulators. However, some other plant species are widely spread in metalliferous and non-metalliferous soils, and hyperaccumulate metals when occurring in metalliferous habitats. These plant species are listed as facultative hyperaccumulators. This phenomenon of metal hyperaccumulation has profound implications in the field of phytoremediation.

Metal hyperaccumulator plants have developed a number of regulatory mechanisms, including heavy metal absorption, transportation, chelation, and detoxification, for their survival in the metal-contaminated environment. Several metalloproteins or metallochaperone-like proteins containing conserved heavy metal-associated (HMA) domains are involved in metal binding and transport. P1B-metal transporting ATPases are of particular interest for their role in metal transport at the cellular and subcellular levels in accumulator plants. The genomic data of accumulator plants in *the Brassicaceae* have shown many upregulated and downregulated genes in accumulator plants when encountering heavy metal stress. Nucleotide and protein sequences from different websites such as http://www.ncbi.nih.gov, http://www.tigr.org, http://www.brassica.info, etc. that encode heavy metal ATPases and transporter protein homologs were collected.

A. Memon (✉) · F. Kusur
Department of Molecular Biology and Genetics, Faculty of Science and Letters, Usak University, Usak, Turkey
e-mail: armemon@usak.edu.tr

M. Memon
Izmir International Biomedicine and Genome Institute (IBG-Izmir), Dokuz Eylül University Saglik Yerleskesi, Izmir, Turkey

R. Prasad (ed.), *Phytoremediation for Environmental Sustainability*,
https://doi.org/10.1007/978-981-16-5621-7_1

The functional and evolutionary similarities in the genes and proteins induced by heavy metals among different accumulator and non-accumulator species were analyzed. In the present communication, we have overviewed these findings and highlighted the role of transporter proteins in metal homeostasis in hyperaccumulator plants.

Keywords

Phytoremediation · Hyperaccumulator plants · Metal transporters · Heavy metal ATPases

1.1 Introduction

Heavy metal pollution is a serious global challenge that needs urgent attention. The high amount of heavy metals, especially toxic metals, reduces plant growth and negatively affects the physiological and metabolic processes, including the inhibition in respiration and photosynthesis, which could lead to plant death (Garbisu and Alkorta 2001; Schmidt 2003; Schwartz et al. 2003). In addition, metal contamination in the soil has a negative impact on the soil microbial population, and it alters the composition and structure of the soil (Giller et al. 1998; Kozdroj and van Elsas 2001; Kurek and Bollag 2004). In the USA and China major problem of land contamination by heavy metals have been reported and represent a great challenge for agriculturist and environmentalists (McKeehan and Kan 2000; Liu et al. 2007). Small industrial units are pouring their untreated effluents into surface drains that extend through agricultural fields in India, Pakistan, and Bangladesh, causing significant soil and water pollution (Lone et al. 2008). The plants absorb contaminants through the root system and transport them up in the shoots. Heavy metals such as Cu, Zn, Mn, Fe are essential micronutrients for plant growth but are potentially phytotoxic to plants when found in high amounts in the soil. As Cd, Pb, Cr, Ni, and Hg have been identified in polluted soils and water and most of these metal/metalloids are non-essential to plant growth and toxic to the plant both at a cellular and subcellular level (Memon et al. 2001). The toxicity of these metals alters or inhibits numerous metabolic processes at the cellular level, such as inhibiting enzymes required for cell functioning and disrupting the membrane integrity. The toxic amount of the metal increases the production of reactive oxygen species (ROS) (Pagliano et al. 2006). It induces oxidative stress, deteriorates membrane integrity, and damages the DNA (Quartacci et al. 2001). However, some unique plant species can grow and flourish on both the natural metalliferous soils and as well as on heavy metal polluted soils because of anthropogenic activities.

The European Union launched a comprehensive heavy metal survey program to estimate the heavy metal content of the topsoil of European Union Countries named LUCAST Top Soil Survey of the European Union (Tóth et al. 2013). This survey has opened up new possibilities to get detailed information on the soil cover in Europe, including the heavy metal content data of these soils (Tóth et al. 2016). This survey

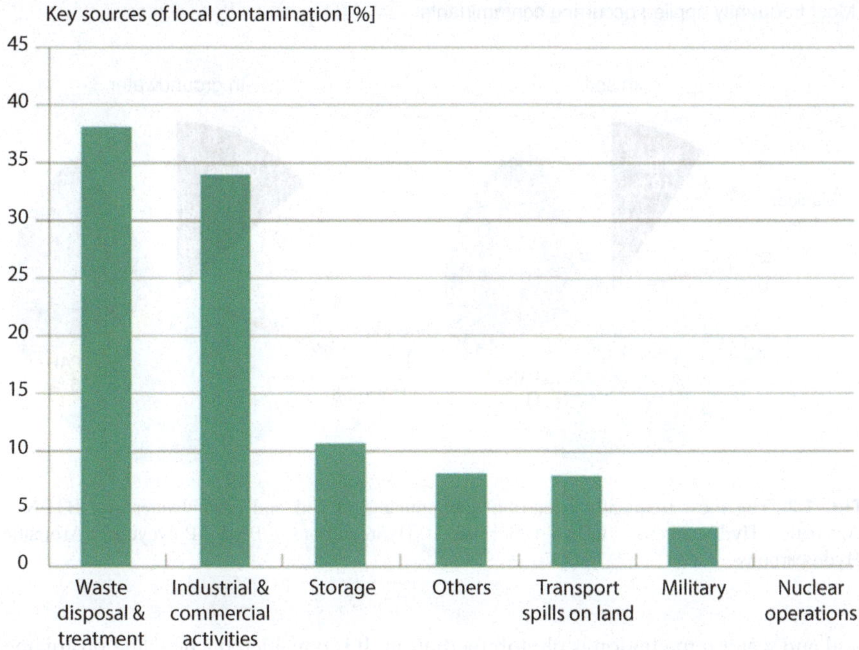

Fig. 1.1 Key sources of contamination reported in 2011 (Van Liedekerke et al. 2014)

is useful to identify the potential heavy metal-contaminated sites and will allow the environmentalists to monitor, control, and clean these contaminated sites for reuse.

The European research study called "Progress in the management of Contaminated Sites in Europe" reported about 2.5 million potentially contaminated sites, of which about 14% (340,000 sites) are estimated to be contaminated (Van Liedekerke et al. 2014). Among EU countries, Belgium, Finland, and Lithuania reported having the highest number of contaminated sites. The major sources of contamination that have the highest impact on soil and water pollution across Europe are shown in Fig. 1.1 (Van Liedekerke et al. 2014). The key contributing factors for soil and water pollution seem to be waste disposal and treatment and industrial and commercial activities (Fig. 1.1).

The most common contaminant in soils and groundwater across Europe is shown in Fig. 1.2. It is noticed that heavy metals are the major contaminants present in water and soils in Europe.

The estimated cost of managing contaminated areas in Europe is around €6.5 billion per year. It corresponds to an average annual national expenditure on managing contaminated sites in on average about €10 per capita (Van Liedekerke et al. 2014). Because of the high cost of the conventional management techniques, there is an urgent need to find out cheaper and more efficient remediation technologies that can be successfully applied to remediate polluted soil and water across Europe and the rest of the world. One of the most efficient biological approaches to contaminated

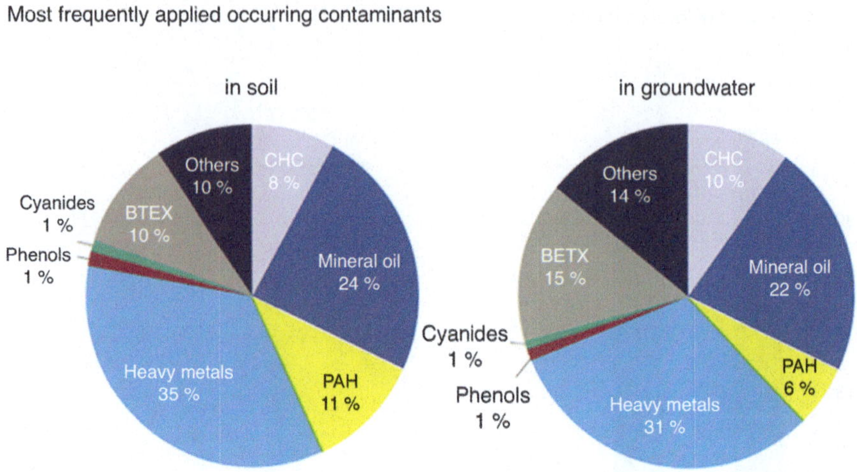

Fig. 1.2 The most frequently occurring contaminants in soil and groundwater are: BTEX—Aromatic Hydrocarbons; CHC—Chlorinated Hydrocarbons; PAH—Polycyclic Aromatic Hydrocarbons

soil and water remediation is phytoremediation. It is considered a new and promising technology for the reclamation of polluted sites and is less costly than other conventional methods like physicochemical approaches, etc. (Garbisu and Alkorta 2001; McGrath et al. 2001; Raskin et al. 1997).

1.2 Advancing Phytoremediation Potential to Clean up the Environmental Pollution

Phytoremediation is a biological process where plants extract, remove, stabilize, or degrade the pollutants from the soil and waters (Salt et al. 1998). Some specific plants can extract, immobilize or metabolize and accumulate organic and inorganic contaminants and remediate the polluted areas for reuse for either agriculture or social (recreational parks, gardens, etc.) purposes. It is considered economical and environmentally friendly biotechnology where plants and microorganisms are used to remove contaminants from polluted soils and industrial waste.

Currently, several physicochemical approaches are being used to clean up contaminated soils. However, physicochemical approaches are generally costly, used on a small scale because they have toxic effects when used on a large scale. Therefore, phytoremediation as a safe biological approach represents a perspective alternative and efficient solution for sustainable environmental cleanup (Salt et al. 1998; Peer et al. 2005; Golubev et al. 2011). Primarily, phytoremediation technology aims to remove or degrade or immobilize environmental pollutants, especially anthropogenic origin, to restore the contaminated sites for reuse in agriculture, forestry, and other public and private applications. Six different phytoremediation methods are briefly listed here; all are commonly used in the phytoremediation of metals and other organic contaminants from soil and water. They include phytoextraction, rhizofiltration, phytostabilization, phytovolatilization, and phytodegradation (Salt et al. 1998; Peer et al. 2005; Thakare et al. 2021; Sarma et al. 2021).

Phytoextraction technology is generally focused on the use of plants to extract and remove metals from soil and water and has been extensively developed by several academic and industrial groups in several countries. The major criterion of this technology is to extract and accumulate metals from the polluted sites and accumulate them in the aerial part of the plant, which can be removed to dispose of or burnt to recover metals (Chaney et al. 2018). We will be mainly discussing this technology in our review paper. Recently Chaney's group has introduced a new term, "agromining," which is possibly derived from this technology and encapsulates the entire series of processes involved in producing metals for commercial or industrial use (van der Ent et al. 2015; Chaney et al. 2018). Rhizofiltration uses plant roots or rhizomes for extracting metals from wastewaters. Phytostabilization is a technique that uses plant roots to absorb contaminants from the soil and make them harmless by preventing them from leaching. In Phytovolatilization, plants take up the elements like Se, As, and Hg and translocate and volatilize pollutants from their foliage. Phytodegradation technology uses plants and related microorganisms to degrade and remove organic contaminants from the soil and water (Salt et al. 1998; Garbisu and Alkorta 2001; Peer et al. 2005).

Phytoremediation offers many advantages over the other conventional physical and chemical methods like precipitation with lime, ion exchange, and precipitation with bio-sulfide, biosorption, etc., which are costly and difficult to handle at a large scale (Khalid et al. 2017). Phytoremediation efforts are mostly focused on using plants in combination with root rhizosphere microorganisms to eliminate toxic heavy metals from soils and water and speeding up the degradation of organic and inorganic contaminants (Silver and Phung 2005; Gerhardt et al. 2017; Sonowal et al. 2022). The advantages that phytoremediation offers are the low cost, minimization of the chemical and biological volume to be disposed of, high efficiency in detoxifying very dilute effluents, and the reuse of the collected heavy metals from contaminated areas. There are several factors that could be considered in developing effective and successful phytoremediation technology. One of the most important factors is identifying or developing (through molecular breeding) an ideal plant/or plant species for effective phytoextraction of toxic metals from the polluted soils or the environment (Suman et al. 2018). Other factors include the use of modern agronomical practices, optimizing crop and soil management practices, and developing cutting edge-technologies for extracting metals efficiently from biomass (Zhuang et al. 2007; Kidd et al. 2015). To develop a suitable plant for phytoextraction following parameters should be considered: rapid metal entry into root tissues needs to be accompanied by efficient metal transport into the shoots. Metal uptake efficiency primarily depends on the bioavailability of the metal in soil (Lu et al. 2018). Bioavailability of heavy metals is the primary factor for effective phytoextraction and describes the degree of availability of the pollutants which plant can take from the soil and sediment. However, metal bioavailability is a complex process and is dependent on many other factors related to the soil structure and chemical composition (McGrath and Zhao 2003). Rhizospheric microbes and root exudates such as siderophores and organic acids can alter the bioavailability of heavy metals in the soil (Thijs et al. 2017). Several elements in the soil and plant roots can mobilize the metals from the soil and enhance the metal uptake through the plant roots. For instance, initial metal uptake can be achieved by mobilizing the metal bound to soil particles through the secretion of organic exudates like mugenic and aveic acids from roots which cause the acidification of the soil and the chelation of metals (Muszyńska et al. 2015). After uptake, the metal is translocated from roots to shoots through xylem tissues. To enter the metal in xylem tissues, it must cross the endodermis through the transporters or channels in the membrane. Once the metal is loaded into the xylem (possibly through metal ATPases and other transporters), it is transported into the leaves and then can be stored in different cells, depending on the chemical form of the metal, since it can be converted into less toxic forms through different chemical mechanisms (conversion or complexation) (Peer et al. 2005).

There are many advantages in using phytoremediation technology for removing contaminants from the environment compared to traditional technologies. First of all, it is cheap and cost-effective, around 50% to 90% cheaper than other conventional chemical or engineering options (Salt et al. 1998; Peer et al. 2005). Secondly, it is easy to dispose of the plants, and it will cause limited disturbance to the landscape (Batty and Dolan 2013). The metals can be extracted easily from biomass to prevent the resulting plant material as hazardous waste. There are some disadvantages of using this technology; for example, it takes a longer time to remediate the soil for reuse. This can be addressed using plant species with a short growth cycle and high biomass (Pollard 2016; Suman et al. 2018).

1.3 Use of Hyperaccumulator Plants for Phytoremediation of Metals from the Polluted Soils

Plants can degrade organic and inorganic contaminants, mainly with the help of root rhizosphere microorganisms (Lone et al. 2008). The metal hyperaccumulator plants grow on metalliferous soils and accumulate extraordinarily high amounts of heavy metals in the aboveground parts, far above the levels found in most plant species, without suffering phytotoxic effects (van der Ent et al. 2013). Hyperaccumulators have three essential characteristics which are lacking in non-hyperaccumulator species; an increase in heavy metal uptake rate, high root-to-shoot translocation, and a greater ability to detoxify and sequester heavy metals in shoots. On the molecular level hyperaccumulator plants have different gene expression and regulation patterns than non-accumulator plants (Goolsby and Mason 2016). The hyperaccumulator plants efficiently absorb and translocate metals from the roots to the shoots and sequester them in the cell wall and vacuole (Memon and Schroder 2009; Memon 2016). Accumulator plants constitutively overexpress the genes encoding membrane transporter proteins, such as ZIPs, HMAs, MATE, YSL, and MTPs for metal transport in the cell (Rascio and Navari-Izzo 2011; Memon and Schroder 2009; Memon 2016). Hyperaccumulator plant species are an important economic source for removing the contaminants from the soil, and the metals can be harvested from the growing plants for marketing (Chaney et al. 2018).

Hyperaccumulator plants actively absorb and take up large amounts of one or more heavy metals from the soil and efficiently translocate to the shoot and accumulated in the aboveground parts of the plant, especially with the leaves, at concentrations 100–1000 fold higher than those found in non-hyperaccumulator species without showing any toxicity symptoms (Reeves and Brooks 1983; Bhargava et al. 2012). Although a distinct feature, hyperaccumulation also relies on hyper tolerance, a distinct feature of the hyperaccumulator plants essential for these plants to avoid heavy metal toxicity. Goolsby and Mason (2015) have highlighted several key issues related to defining the hyperaccumulation trait and proposed a more objective definition of hyper accumulation than the definition previously proposed by van der Ent et al. (2013). This redefined definition reflects both the genetic and physiological mechanisms underlying hyperaccumulation and the evolutionary aspects of this phenomenon. They suggested that hyperaccumulation and tolerance should be considered two distinct continuous traits mediated by genetically and physiologically distinct mechanisms. The plant phenotypes span a wide range of combinations of both traits producing four general categories: tolerant accumulator (traditional hyperaccumulators; e.g., *Astragalus bisculcatus* for Se), non-tolerant accumulator (excluded from the naturalistic definition of hyperaccumulation; e.g., *Thlaspi goesingense* for Zn), non-tolerant non-accumulator, and tolerant non-accumulator (Goolsby and Mason 2015). The two last categories are typically collapsed together as non-hyperaccumulators (for example, *Arabidopsis thaliana* for Cd and *Silene vulgaris* for Cu).

The heavy metal accumulation ability of the plant varies significantly and is dependent on the type of the species and cultivars within the species. The different mechanisms of ion uptake are operating in each species, based on their genetic, morphological, physiological, and anatomical characteristics. To date, there are more than 700 plant species known worldwide to accumulate metals in large amounts, and these accumulator species are of interest for their potential use in the phytoremediation of metal-contaminated soils (Reeves et al. 2018). For example, *Noccaea caerulescens* (*Thlaspi*) and *Arabidopsis hallari* are characterized as hyperaccumulator plants of Zn/Cd. Several crops *Brasssica* spp. such as *B. nigra* L., *B. juncea* L. Czern, *B. napus* L., and *B. rapa* L. exhibit enhanced accumulation of Cu, Zn, and Cd (Ebbs et al. 1997). A list of hyperaccumulator plants is given in Table 1.1. In this table, several plant species belong to different families accumulate metal both in roots and/or shoots. Hyperaccumulator plants have got a considerable interest in exploiting their accumulation traits for practical use, in particular, to develop cheap and clean technologies for phytoremediation of heavy metal from contaminated soils or for phytomining valuable metals from mineralized sites (Chaney et al. 2018).

However, there are many factors that could be considered for efficient phytoremediation and also for beneficial agromining, such as plant tolerance to pollutants, agronomic characteristics of the plant species, climatic conditions (rainfall, temperature), soil physicochemical properties, and the recent technologies

Table 1.1 A list of hyperaccumulator plants. Metal shoot/root ratio and the plant tissues where metal is highly accumulated is given (Memon, Kusur, and Memon unpublished data)

Plant name	Metal	TF (C_s/C_r)	Tissue	References
Arabidopsis halleri	Cd	0.23	Root	Bert et al. (2003)
Arabis paniculata		1.45	Root, shoot	Tang et al. (2009)
Arabis gemmifera		6.13	Shoot	Kubota and Takenaka (2003)
Thlaspi caerulescens		–	Shoot	Baker et al. (1994)
T. goesingense		0.5	Root	Lombi et al. (2000)
N. praecox (T. praecox)		–	Shoot (5960 ppm)	Vogel-Mikuš et al. (2008)
Sedum alfredii		1.05	Root,shoot	Xiong et al. (2004)
Tamarix smyrnensis		1.36	Root,shoot	Manousaki et al. (2008)
Rorippa globosa		2.21	Shoot	Sun et al. (2011)
Arabis gemmifera	Zn	6.48	Shoot	Kubota and Takenaka (2003)
A. paniculata		1.98	Shoot	Tang et al. (2009)
T. goesingense		–	Shoot	Baker et al. (1994)
Thlaspi caerulescens		–	Shoot	Reeves and Brooks (1983)
Arabidopsis halleri		0.16	Root	Küpper et al. (2000)
Sedum alfredii		0.43	Root	Sun et al. (2005)
Salix viminalis		–	Shoot	Schmidt (2003)
Brassica napus		4.02	Shoot	Brunetti et al. (2011)
Aeolanthus biformifolius	Cu	–	Shoot (13,700 ppm)	Brooks et al. (1978)
Crassula helmsii		–	Shoot (9200 ppm)	Küpper et al. (2009)
Elsholtzia splendens		0.033	Root	Weng et al. (2005)
Sorghum sudanense L.		3.41	Shoot	Wei et al. (2008)
Chrysanthemum coronarium L.		7.58	Shoot	Wei et al. (2008)
Brassica napus		2.13	Shoot	Brunetti et al. (2011)
Spartina argentinensis	Cr	5.1	Shoot	Redondo-Gómez (2013)
Brassica juncea		0.56	Root	Seth et al. (2012)
Brassica napus		5.04	Shoot	Brunetti et al. (2011)
Elodea canadensis		0.05	Root	Ranieri et al. (2013)
Arabis gemmifera		0.15	Root	Kubota and Takenaka (2003)
Hemidesmus indicus	Pb	0.66	Root	Sekhar et al. (2005)
Brassica oleracea		0.54	Root	Zhu et al. (2004)
B. campestris		0.62	Root	
Arabis paniculata		1.96	Shoot	Zeng et al. (2009)
Brassica juncea		0.2	Root	Seth et al. (2012)
T. caerulescens		–	Shoot (0.66 ppm)	Baker et al. (1994)

(continued)

Table 1.1 (continued)

Plant name	Metal	TF (C_s/ C_r)	Tissue	References
Sedum alfredii		0.003	Root	Sun et al. (2005)
Brassica napus		5.04	Shoot	Brunetti et al. (2011)
Sesbania drummondii		1.1	Root, shoot	Ruley et al. (2006)

TF, translocation factor = (Cs, concentration of metal in shoots/ Cr, concentration of metal in roots)

available for the recovery of metals from the harvested plant biomass. The naturally occurring heavy metal accumulator plants are good candidates for phytoextraction (Table 1.1) because they take metal from the soil in two or three orders of magnitude than non-accumulator plants growing on natural uncontaminated soils. Table 1.1 shows the TF (translocation factor) value of metals in plants. Several accumulator plant species had translocation factor (TF) of metals more than one, suggesting that plants remove the metals from the soil by phytoextraction and translocate them to shoots (Brunetti et al. 2011; Kubota and Takenaka 2003). On the contrary, non-accumulator plants have TF less than one and cannot accumulate metal in shoots.

It appears that both chemical and biological approaches are still not wholly efficient and need more efforts for their effective use in the future (Kidd et al. 2015). Some plants may accumulate one metal, whereas others can accumulate two or more metals at a time, which could be beneficial for phytoremediation and phytomining (see Table 1.1) (Chaney et al. 2007).

Although the annual biomass yield is an essential trait for phytoremediation, the ability to hyperaccumulates and hypertolerate metals is of greater importance than high biomass (Chaney et al. 1997). Hyperaccumulator plants absorb and transport many valuable metals from the contaminated soil and accumulate them in their shoots. These marketable metals could be recovered from the plant biomass for use in the metal industries (Brooks et al. 1998; Chaney et al. 2018). Commercial technologies have been developed for Ni phytomining using Alyssum Ni hyperaccumulator species (Broadhurst et al. 2004). However, other high price metals (Au, Tl, Co, and U) can be extracted using hyperaccumulator plants from the soil or mine tailing containing concentrations of the metals at a level uneconomic for conventional extraction techniques.

1.3.1 Selection of Plant Species for Phytoextraction

As mentioned above, one of the requirements for plants to be used in phytoremediation of soil is to take up the heavy metals from the contaminated soils efficiently. In other words, if a plant species accumulate and concentrate metals in their shoots at levels greater than those in the soil is an excellent candidate for remediation of the polluted soils. The plants that grow in their natural habitats and accumulate 100 µg/g for Cd, Se, and Tl; 300 µg/g for Co, Cu, and Cr; 1000 µg/g for Ni, Pb, and As; 3000 µg/g for Zn; and 10,000 µg/g for Mn in their dried foliage are proposed to be hyperaccumulators (Rascio and Navari-Izzo 2011; van der Ent et al. 2013). To find out the hyperaccumulator plants and their accumulation capacity and specificity to the metal accumulation, a global database (www.hyperaccumulators.org) was created in 2015 and is administered and maintained by the Center for Mined Land Rehabilitation, University of Queensland, Brisbane, Australia. The data about all known metal and metalloid accumulator plants are deposited, continuously updated, and is free to use (Reeves et al. 2017). This database currently contains more than 700 different metal hyperaccumulator species, and most of the plant species are Ni accumulators (523 spp.). Some plant species accumulate Cu (53 species.), Co (42 species), Mn (42 species), Se (41 species), Zn (20 species), Pb (8 species), Cd (7 species), and As (5 species). A very few plant species are accumulators of rare elements (Reeves et al. 2017). The most strongly represented hyperaccumulator plant species are in the Brassicaceae (83 species) and Phyllanthaceae (59 species) families.

1.3.2 Hyperaccumulator Plant Species in Brassicaceae

Brassicaceae comprises approximately 338 genera and 3700 species. In the Brassicaceae family, the Brassica genus contains about 100 species, including essential oilseed crops (for example, *Brassica napus*, *B. juncea*) and many common vegetable plants (Ozturk et al. 2012; Warwick and Black 1991). Among Brassica, *B. rapa* has the smallest genome, at ca. 529 Mb, and *B. napus* have the largest one, at ca 1132 Mb (Lysak et al. 2005; Nagaharu 1935). The genome of both species is sequenced, and the data is available in the public domain http://www.brassicagenome.net/databases.php; https://www.ncbi.nlm.nih.gov/genome/?term=brassica%20napus (Memon 2016; Liu et al. 2016). Around 80–90% homology between the exons of putative orthologous genes in Arabidopsis and Brassica is reported (Ozturk et al. 2012). Due to that, the annotated Arabidopsis genome sequence can be exploited for the comparative analysis of *Arabidopsis* and *Brassica* genomes. The plant species in Brassica (e.g., *B. juncea; B. napus; B. nigra*) produce high biomass and accumulate and tolerate high metals (including Cd, Cu, Ni, Pb, U, Zn) in their tissues (Anjum et al. 2012a, b; Kumar et al. 1995). *B. juncea* is considered a suitable candidate for phytoremediation of multiple heavy metals from the soil. It is highly metal tolerant and comparatively accumulates more metals in its shoots than other Brassica species Zn, Cd, and Pb. For example, this species accumulates a high amount of Cd in their shoots (1450 µg Cd/g dry wt), three times more than reported in *B. napus* (555 µg/g dry wt). In addition, it absorbs a huge amount of other metals such as Pb (28% reduction) and Se (reduced between 13 and 48%) (Szczygłowska et al. 2014). It also accumulates more Zn from the soil than *Noccaea caerulescens*, a known hyperaccumulator of zinc. It appears that *B. juncea* produces ten times more biomass than *N. caerulescens* (Anjum et al. 2012a, b; Szczygłowska et al. 2014). *B .nigra* Diyarbakir ecotype (Southern Anatolia), a diploid, is known as Cu accumulator (Memon and Zahirovic 2014) and accumulated around 20,000 μg g^{-1} DW Cu in their shoots (Ozturk et al. 2012; Cevher-Keskin et al. 2019). Because of the high Cu accumulation capacity of *B. nigra,* this plant could become a suitable candidate for phytoremediation of Cu-polluted soils (Cevher-Keskin et al. 2019; Dalyan et al. 2017; Kumar et al. 2012; Memon and Zahirovic 2014).

1.4 Subcellular Localization of Metals in Hyperaccumulator Plants

To understand the mechanism of metal hyperaccumulator, the detailed physiological knowledge of metal absorption by roots, translocation to the shoots, and the subcellular localization of the metals in the leaves are of great importance (Memon and Schroder 2009; Tangahu et al. 2011). Microarray analysis with Cu accumulator *B. nigra* Diyarbakar ecotype showed several hundredfold increases in metal transport ATPases and other genes related to metal transport and accumulation in plants treated with 500 μM Cu (Memon and Zahirovic 2014). Several other genes related to signal transduction, metabolism, and transport facilitation were highly expressed with high Cu. For example, the genes involved in the glutathione pathway (γ-ECS, PC, etc.) were also highly expressed in root and shoot tissues (Memon and Zahirovic 2014; Merakli and Memon, unpublished data). Because of its high growth both at low and high Cu, this plant was classified as a facultative metallophyte (Memon 2016).

It is interesting to know the mechanisms responsible for making these metals in an innocuous form in the plant cell. One of the primary mechanisms for detoxification in the plant cell is storing and depositing the metals in the vacuolar compartment (Memon et al. 2001; Reeves et al. 2018; Tangahu et al. 2011). Different organic acid chelators such as malate, citrate, histidine, and nicotinamide play a role in translocating metal cations through the xylem (Salt et al. 1995; Stephan et al. 1996; von Wirén et al. 1999). To maintain the metal homeostasis in the cell, hyperaccumulator plants efficiently absorb metal from the soil and transport it to shoots and sequestered them in the subcellular compartments (e.g., cell wall, vacuole, etc.) or secreted in the trichomes (Hanikenne and Nouet 2011; Memon and Schroder 2009; Memon and Yatazawa 1982; Ovečka and Takáč 2014). Previously, we carried out an electroprobe X-ray microprobe analysis to understand the subcellular localization of Mn in the leaves of Mn accumulator plant *Acanthopanas sciadophylloides* and tea. The micro-distribution pattern of Mn showed that a large portion of Mn was located in the cell wall and vacuolar compartment of epidermal cells (Fig. 1.3), and it was almost absent from the cytoplasm (Memon et al. 1981; Memon and Yatazawa 1984). One of the detoxification mechanisms proposed was Mn^{2+} complex with malate in the cytoplasm and then transported to the vacuole where it is dissociated from malate and forms a stable complex with oxalate. Under this condition, malate functions as a "transport vehicle" through the cytoplasm and oxalate as the "terminal acceptor" in the vacuole (Memon and Schroder 2009;

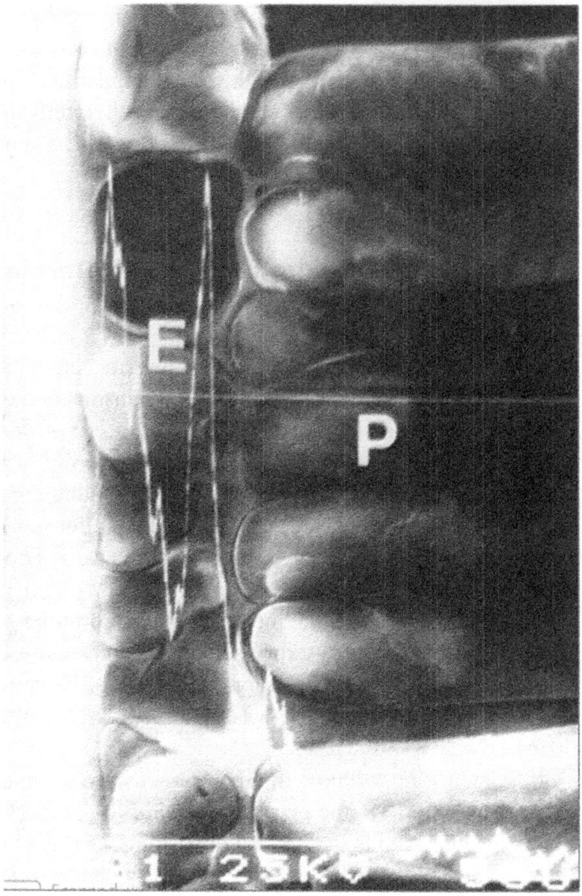

Fig. 1.3 Secondary electron images line scan profile of a leaf section of a tea plant with Mn (Ka radiation) peaks. It shows the localization of Mn at the subcellular level in the epidermis. *E* epidermis, *P* palisade parenchyma cells (Memon et al. 1981)

Memon and Yatazawa 1984). There are several other mechanisms involved in metal detoxification, e.g., production of superoxide dismutase, peroxidase, catalase, glutathione reductase, and nonenzymatic antioxidants (e.g., flavonoids, reduced glutathione, ascorbic acid), which play a significant role in neutralizing the free radicals caused by ROS and minimize the plant cell damage (Küpper et al. 1999; van de Mortel et al. 2006; Li et al. 2015).

Metal accumulation and compartmentalization patterns differ depending on plant species and element type. According to Küpper et al. (2000), *A. hallari* accumulates more Zn and Cd in the mesophyll cells than in the epidermis, but *N. caerulescens* accumulates six times more Zn and Cd in epidermis cells than in mesophyll cells. *B. juncea* (a metal tolerant and accumulator plant), on the other hand, accumulates 40 times more Cd in trichomes compared to leaves (Dalyan et al. 2017; Küpper et al. 1999). *Alyssum lesbiacum* also accumulates a significant amount of Zn and Ni in leaf trichomes (Reeves et al. 2018).

Various alternative detoxifying and accumulation mechanisms have been proposed (Hanikenne and Nouet 2011; Isaure et al. 2015; Memon 2016; Rascio and Navari-Izzo 2011) in which metals can be bound and sequestered by phytochelatins, metallothioneins, metalloenzymes, and metal-activated enzymes. Recent advancements in the next-generation sequencing technologies have opened up new possibilities to understand the metal detoxification mechanisms in plants at the cellular and molecular level (Verbruggen et al. 2013).

1.5 Metal Transporters and Their Function in the Plant Cell

Several genes and proteins related to metal absorption and transport have been identified and characterized in several accumulator plants. These metal transporters are subdivided into six main groups, including natural resistance macrophage protein (NRAMP), ZRT-like protein (ZIP), cation diffusion facilitator (CDF), Yellow-stripe-like (YSL), and heavy metal P1B-type ATPases (HMAs) (Guerinot 2000; Memon 2016; Merlot et al. 2018). Table 1.2 shows the genomic structure and protein length of different metal transporters, including metal ATPases, NRAMPs, and ZIP proteins identified from different plant species. To maintain the metal homeostasis in the cell, a metal accumulator plant can activate several transporters, which can function either in excluding metal at the root or sequestering them at the subcellular level in the vacuole, chloroplast, and some other cellular compartments. Analysis of the *A. thaliana* genome has shown the genes of several metal transporter families, including 15 members of zinc and iron transporters (ZIP), eight members of Cation Diffusion Facilitator (CDF), six members of copper transporters (CTR), six members of NRAMP homologous, and eight members of Cu, Zn/Cd transporting ATPases (Mäser et al. 2001; Merlot et al. 2018) (http://www.cbs.umn.edu/arabidopsis/). The role of some other transporter families, such as vacuolar cation proton exchanger (CAX) and ABC transporters in metal homeostasis, have been elucidated (Colangelo and Guerinot 2006; Hall and Williams 2003; Memon 2016; Memon and Schroder 2009; Sarma et al. 2018). Li et al. have identified 55 AtHMPs and 46 OsHMPs in dicot Arabidopsis and monocot rice, respectively (Li et al. 2020). These proteins are named metalloproteins or metallochaperone-like proteins containing heavy metal-associated (HMA) domains comprising a conserved HMA domain with around 30 amino acid residues. Several other proteins that transport or detoxify heavy metals have this conserved domain. Two cysteine residues in this domain bind with copper, zinc, cadmium, cobalt, and other heavy metals (Li et al. 2020). These HMA domain-containing proteins fall into four groups; HPPs (heavy metal-associated plant proteins), HIPPs (heavy metal-associated isoprenylated plant proteins), ATX1-like copper transport proteins, and heavy metal ATPases (HMAs) (Memon 2016).

Among the genes of transporter families described above, P1B-type ATPases, an ion pump, which utilizes the energy resulting from ATP hydrolysis to carry membrane transport of multiple metal ions in the subcellular level, is of particular importance. These ATPases maintain the homeostasis of the heavy metals in the

Table 1.2 Genomic structure, cDNA, and protein length of different transporters of different plant species (Memon, Kusur, and Memon unpublished data)

Plant name	Gene name	Genomic DNA base pairs (bp)[a]	cDNA base pairs (bp)	Exon	Intron	Protein length amino acids (aa)
A. thaliana	HMA1*	4776	2460	13	12	819
M. trunculata		9415	2490			829
B. napus		4359	2331			776
G. max		14,420	2454			817
S. tuberosum		9994	2454			817
A. lyrata		3448	2421	11	10	806
B. rapa		4207	2457	13	12	818
Z. mays	HMA2	6917	3726	10	9	1241
O. sativa		7771	3204			1067
B. napus		8845	2661	15	14	886
B. rapa		6062	2715	9	8	904
O. lucimarinus		2328		1	1	776
G. max		8194	1683	10	9	560
A. lyrata	HMA3	3369	2274	10	9	757
C. sativus		6642	2667			888
Z. mays		3484	2959	6	5	893
B. oleracea		4012	2292	8	7	763
B. rapa		8243	3864	10	9	1287
B. napus		3396	2291	9	8	763
G. max	HMA4	12,008	2865	8	7	954
A. lyrata		7886	3828	10	9	1275
S. oleracea		9964	2901			966
B. napus		8158	3585			1194
B. oleracea		7550	3588			1195
B. rapa		7723	3573			1190
*M. truncula*ta		9737	2991	5	4	996
A. thaliana	HMA5	3657	2988	6	5	995
B. napus		3604	2922	4	3	973
B. oleracea		5077	2922			973
B. rapa		3542	2934			977
A. thaliana	HMA6	7322				
C. sativa		7368	2856	19	18	951
A. thaliana	HMA7	7773		10	9	
C. sativa		5401	3021			1006
N. tabacum		4525	2667	3	9	888
G. max	HMA8	8496	2711	17	16	903
C. sativa		5735	2655			884
A. lyrata	NRAMP1	3274	1581	12	11	526
B. napus		5770	1599	13	12	532
B. oleracea		3319	1599	11	10	532

(continued)

Table 1.2 (continued)

Plant name	Gene name	Genomic DNA base pairs (bp)[a]	cDNA base pairs (bp)	Exon	Intron	Protein length amino acids (aa)
B. rapa		3344	1599			532
A. thaliana	NRAMP2	2703	1593	4	3	530
A. lyrata		2708	1599			532
B. napus		1969	1077	5	4	358
B. rapa		2755	1599			532
A. thaliana	NRAMP3	2630	1530	4	3	509
A. lyrata		2539	1524			507
B. napus		3735	1542			513
O. sativa		2367	1536	14	13	511
A. thaliana	NRAMP4	2632	1539	3	2	512
B. napus		1741	1539	6	5	512
B. oleracea		2465	1536	3	2	511
B. rapa		2350	1539			512
A. thaliana	NRAMP5	2321	1593	4	3	530
A. lyrata		2322	1590			529
B. napus		2590	1596			531
B. oleracea		2577	1596			531
A. thaliana	NRAMP6	4441	1584	13	12	527
B. napus		3338	867	7	8	288
B. oleracea		3220	1561	13	12	520
B. rapa		7237	1512			503
A. thaliana	ZIP1	1551	1068	2	1	355
G. max		3147	1065	3	2	354
A. thaliana	ZIP2	1696	1062	2	1	353
O. sativa		4301	1101	3	2	366
N. attenuata		1251	996			331
A. thaliana	ZIP3	2861	1020	3	2	339
O. sativa		259	1095			364
H. annaus	ZIP4	2524	1254	5	4	417
G. hirsutum		2531	1256	4	3	422
O. sativa	ZIP5	4301	1101	3	2	366
N. attenuata		3458	1032			343
A. lyrata	ZIP6	1639	1008	2	1	335
M. trunculata		612	306	1		101
C. sativus		612	306			101
A. thaliana	ZIP7	1613	1098	3	2	365
O. sativa		3353	1155	4	3	384
A. thaliana	ZIP8	1728		5	4	
O. sativa		3239	1173	3	2	390
A. lyrata	ZIP9	2516	1170	4	3	389
Q. suber			1059	1	1	187

(continued)

Table 1.2 (continued)

Plant name	Gene name	Genomic DNA base pairs (bp)[a]	cDNA base pairs (bp)	Exon	Intron	Protein length amino acids (aa)
A. thaliana	ZIP10	1804	1095	3	2	364
O. sativa		2621	1215	5	4	404
A. thaliana	ZIP11	1051	981	2	1	326
H. annaus		2932	981			326
A. thaliana	ZIP12	1758	1068	2	2	355
O. brachyantha		11,024	1776	20	19	591

HMA Heavy metal ATPase, *NRAMP* Natural resistance-associated macrophage protein, *ZIP* Zinc-regulated, iron-regulated transporter-like proteins

[a]Partial sequences (bp) are given for some of the genomic DNAs

cell and are present in prokaryotic and eukaryotic cells, including bacteria, plants, and mammals.

1.6 Function of Heavy Metal ATPases (HMAs) in Plants

There are three main pumps (ATPases) present in plant cells. The first Fo-F1 type ATPase is present in chloroplast and mitochondrial membrane and is involved in ATP synthesis. V-type ATPases are present in the tonoplast membrane and generate the H^+ gradient required for transport across the tonoplast membrane. The third one P-type ATPases are present in the plasma membrane and other organelle membranes and are involved in the active pumping of charged substrates across the cell membranes and form a phosphorylated intermediate during the reaction cycle (Palmgren and Nissen 2011). The P-type ATPases are classified into five major families (P1-P5) and divided into several subgroups (Axelsen and Palmgren 2001). Heavy metal ATPases (P1B ATPases) are a subclade of P1-ATPase and transport heavy metals such as Cu, Zn, Cd, Pb, and Co and are the main pumps required in metal detoxification and metal homeostasis in the cell (Østerberg and Palmgren 2018). P1B-ATPases contain six to eight transmembrane domains (TMs), an HP locus, and a CPx/SPC motif (Williams and Mills 2005), required for metal binding and transport. The majority of these ATPases possess conserved regions such as DKTGT, GDGxNDxP, PxxK, and S/TGE in their sequence necessary for their proper function (Williams and Mills 2005). Based on their substrate specificity, these ATPases are subdivided into two groups, Cu/Ag (Cu^+-ATPases) and Zn/Cd/Co/Pb transporters (Zn^{2+}-ATPases) (Axelsen and Palmgren 2001).

The plant genome contains many copies of P1B-ATPases, especially *Arabidopsis thaliana* has eight, rice has nine, and soybean has 25 genes in their genome (Fang et al. 2016; Williams and Mills 2005). Table 1.2 shows the genomic size and structure, cDNA, and protein length of different metal ATPases identified in the genome in various plant species. HMA2, HMA3, and HMA4 have high sequence homology among them and transport Zn and Cd. HMA2 and HMA4 are the plasma membrane transporters in pericycle cells and are involved in root-to-shoot transport of Zn/Cd. HMA3 is located in the tonoplast and has a detoxification function through vacuolar sequestration of Zn/Cd (Hanikenne et al. 2008; Hussain et al. 2004; Liu et al. 2017; Morel et al. 2009; Wong and Cobbett 2009). Table 1.2 shows that HMA4 protein has longer amino acid sequences than other metal transporters and has an essential function in Zn/Cd hypertolerance and hyperaccumulation in accumulator plants like *Arabidopsis hallari* and *Noccaea caerulescens*. Three copies of HMA4 have been identified in *A. hallari* and are highly conserved in coding sequences but diverge in promoter sequences (Nouet et al. 2015). Their complementation experiment with the *A. thaliana, hma2hma4* mutant (severe Zn-deficiency phenotype) showed that all three copies restored root-to-shoot translocation of Zn. Each copy had a different impact on the metal homeostasis in the *A. thaliana*. This observation indicates a functional difference among the three *A. halleri HMA4* copies, possibly due to the differences in expression levels rather than in expression profile (Nouet et al. 2015).

The C-terminus of the HMA4, one of the well-known ATPase transporter located in the plasma membrane, binds Zn, has considerably divergent amino acid motifs between *A. thaliana* (non- accumulator) and *A. hallari* (accumulator). The di-Cys motif in this region has a high affinity for Zn binding in accumulator plants (Lekeux et al. 2018). Similarly, BjHMA4 transporter protein in *B. juncea* showed a repeat region BjHMA4R in the C-terminus not far from the last transmembrane domain in the cytosol (Wang et al. 2019). It binds Cd^{2+} and improves Cd tolerance and accumulation in *B. juncea*. AtHMA1, a chloroplast membrane protein, transports Cu and Zn into and out of the chloroplast, respectively (Zhao et al. 2018). SpHMA1 in *S. plumbizincicola* leaves a chloroplast Cd exporter and protects photosynthesis by inhibiting the Cd accumulation in the chloroplast (Zhao et al. 2018). The RNA interference of chloroplast SpHMA1 and CRISPR/Cas9-induced HMA1 mutant lines significantly increased Cd accumulation in the chloroplasts than wild-type *Sedum plumbizincicola*. AtHMA5 is localized in the plasma membrane and contributes to the detoxification of excess Cu in roots by increasing Cu translocation from roots to shoots (Kobayashi et al. 2008). On the contrary, AtHMA6 (PAA1) and AtHMA8 (PAA2) are located in chloroplast envelope and thylakoids and transport Cu into the chloroplast (Abdel-Ghany et al. 2005; Shikanai et al. 2003). 20 *HMA* genes (*GmHMA1* to *GmHMA20*) in soybean are phylogenetically divided into 6 clusters (Fang et al. 2016). Six GmHMAs (5, 19,13,16,14, and 18) were classified as Zn^{2+} ATPases, while the remaining HMAs were identified as Cu^{+}-ATPases (Fang et al. 2016). 17 *HMA* genes in *Populus trichocarpa* were shown to be differentially regulated by high metal stress (Li et al. 2015).

Genomic analysis of metal accumulator species *A. hallari, N. caerulecence, B. juncea, B. napus, and B. nigra* have identified the specific role of several metal transporters, including metal ATPases in metal accumulation and tolerances in plants (see Table 1.2) (Cevher-Keskin et al. 2019; Dalyan et al. 2017; Memon 2016). HMA genes are identified both in model plants like *A. thaliana*, rice and in other crop plants like *B. napus, B. rapa, B. juncea, Glycine max,* and *P. trichocarpa* (see Table 1.3). The role of HMA1 to HMA4 in Cu, Zn, and Cd transport in the model plants has been extensively studied and well-characterized at the gene and protein level. Functional studies of these transporters in yeast have given helpful information related to the function of these transporter proteins in the eukaryotic cells (Fang et al. 2016; Wang et al. 2019).

In the last decade, many plant transporter genes involved in metal uptake and translocation are characterized. However, identification and functional analysis of many other transporter genes are still awaiting identification. More studies on the expression and function of these transporter genes at the cellular and subcellular levels coordinated with the structural analysis of the transporter proteins will reveal the fundamental role of these transporters in the detoxication mechanism in accumulator plants. Two different approaches could be taken related to the expression of transporter genes in the accumulator and non-accumulator plants. For accumulator plants, the overexpression of metal uptake and translocation transporters would increase the translocation of toxic metals to aerial parts, which would target phytoremediation. For non-accumulator edible crop plants, the low uptake transporters could be engineered or overexpressed to minimize the transport of toxic cation in edible crops.

1.7 Conclusion

The recent developments in phytoremediation have been summarized, and the role of obligate and facultative accumulator plant species in metal accumulation and detoxification has been discussed. X-ray microprobe analyzer data with frozen leaf tissues of accumulator plant shows the subcellular localization of metals in the cell, especially their localization in the cell wall and storage vacuole, and keeps the toxic amount of metal away cytoplasm. The recent genomic analysis of one diploid Brassica rapa and another tetraploid (amphidiploid) *Brassica napus* have shown the differential gene expression of metal transporters when encountering low and high metal concentrations in the soil. Recent progress in the genetic and molecular analysis of the metal transporters has elucidated the molecular mechanism of metal absorption, accumulation, and detoxification in hyperaccumulator plants and their role in phytoremediation. Phytoremediation is an environmentally friendly and green technology that holds great potential for environmental cleanup. In the future, it will become an established technology for removing hazardous pollutants from the environment. It will guarantee a greener and cleaner planet for all of us in the coming years.

Acknowledgement This work is supported by Usak University BAP project F010 to Prof. Dr. Abdul Razaque Memon.

References

Abdel-Ghany SE, Muller-Moule P, Niyogi KK, Pilon M, Shikanai T (2005) Two P-type ATPases are required for copper delivery in Arabidopsis thaliana chloroplasts. Plant Cell 17(4):1233–1251. https://doi.org/10.1105/tpc.104.030452

Anjum NA, Ahmad I, Pereira ME, Duarte AC, Umar S, Khan NA (2012a) The plant family Brassicaceae: contribution towards phytoremediation, vol 21. Springer

Anjum NA, Gill SS, Ahmad I, Duarte AC, Umar S, Khan NA, Pereira E (2012b) Metals and metalloids accumulation variability in Brassica species–a review. CRC Press/Taylor Francis Group, Boca Raton

Axelsen KB, Palmgren MG (2001) Inventory of the superfamily of P-type ion pumps in Arabidopsis. Plant Physiol 126(2):696–706. https://www.ncbi.nlm.nih.gov/pubmed/11402198

Baker A, Reeves R, Hajar A (1994) Heavy metal accumulation and tolerance in British populations of the metallophyte Thlaspi caerulescens J. & C. Presl (Brassicaceae). New Phytol 127(1):61–68

Batty LC, Dolan C (2013) The potential use of phytoremediation for sites with mixed organic and inorganic contamination. Crit Rev Environ Sci Technol 43(3):217–259

Bert V, Meerts P, Saumitou-Laprade P, Salis P, Gruber W, Verbruggen N (2003) Genetic basis of Cd tolerance and hyperaccumulation in Arabidopsis halleri. Plant Soil 249(1):9–18

Bhargava A, Carmona FF, Bhargava M, Srivastava S (2012) Approaches for enhanced phytoextraction of heavy metals. J Environ Manage 105:103–120. https://doi.org/10.1016/j.jenvman.2012.04.002

Broadhurst CL, Chaney RL, Angle JS, Maugel TK, Erbe EF, Murphy CA (2004) Simultaneous hyperaccumulation of nickel, manganese, and calcium in Alyssum leaf trichomes. Environ Sci Technol 38(21):5797–5802

Brooks R, Morrison R, Reeves R, Malaisse F (1978) Copper and cobalt in African species ofAeolanthus Mart.(Plectranthinae, Labiatae). Plant Soil 50(1–3):503–507

Brooks RR, Chambers MF, Nicks LJ, Robinson BH (1998) Phytomining. Trends Plant Sci 3(9):359–362

Brunetti G, Farrag K, Rovira PS, Nigro F, Senesi N (2011) Greenhouse and field studies on Cr, Cu, Pb and Zn phytoextraction by Brassica napus from contaminated soils in the Apulia region, Southern Italy. Geoderma 160(3–4):517–523

Cevher-Keskin B, Yildizhan Y, Yuksel B, Dalyan E, Memon AR (2019) Characterization of differentially expressed genes to Cu stress in Brassica nigra by Arabidopsis genome arrays. Environ Sci Pollut Res Int 26(1):299–311. https://doi.org/10.1007/s11356-018-3577-7

Chaney RL, Malik M, Li YM, Brown SL, Brewer EP, Angle JS, Baker AJ (1997) Phytoremediation of soil metals. Curr Opin Biotechnol 8(3):279–284

Chaney RL, Angle JS, Broadhurst CL, Peters CA, Tappero RV, Sparks DL (2007) Improved understanding of hyperaccumulation yields commercial phytoextraction and phytomining technologies. J Environ Qual 36(5):1429–1443

Chaney RL, Baker AJ, Morel JL (2018) The long road to developing agromining/phytomining. In: Agromining: farming for metals. Springer, pp 1–17

Colangelo EP, Guerinot ML (2006) Put the metal to the petal: metal uptake and transport throughout plants. Curr Opin Plant Biol 9(3):322–330. https://doi.org/10.1016/j.pbi.2006.03.015

Dalyan E, Yüzbaşıoğlu E, Keskin BC, Yıldızhan Y, Memon A, Ünal M, Yüksel B (2017) The identification of genes associated with Pb and Cd response mechanism in Brassica juncea L. by using Arabidopsis expression array. Environ Exp Bot 139:105–115

Ebbs S, Lasat M, Brady D, Cornish J, Gordon R, Kochian L (1997) Phytoextraction of cadmium and zinc from a contaminated soil. J Environ Qual 26(5):1424–1430

Fang X, Wang L, Deng X, Wang P, Ma Q, Nian H et al (2016) Genome-wide characterization of soybean P 1B-ATPases gene family provides functional implications in cadmium responses. BMC Genomics 17(1):376

Garbisu C, Alkorta I (2001) Phytoextraction: a cost-effective plant-based technology for the removal of metals from the environment. Bioresour Technol 77(3):229–236

Gerhardt KE, Gerwing PD, Greenberg BM (2017) Opinion: taking phytoremediation from proven technology to accepted practice. Plant Sci 256:170–185

Giller KE, Witter E, Mcgrath SP (1998) Toxicity of heavy metals to microorganisms and microbial processes in agricultural soils: a review. Soil Biol Biochem 30(10–11):1389–1414

Golubev SN, Muratova AY, Wittenmayer L, Bondarenkova AD, Hirche F, Matora LY et al (2011) Rhizosphere indole-3-acetic acid as a mediator in the Sorghum bicolor–phenanthrene–Sinorhizobium meliloti interactions. Plant Physiol Biochem 49(6):600–608

Goolsby EW, Mason CM (2015) Toward a more physiologically and evolutionarily relevant definition of metal hyperaccumulation in plants. Front Plant Sci 6:33

Goolsby EW, Mason CM (2016) Response: commentary: toward a more physiologically and evolutionarily relevant definition of metal hyperaccumulation in plants. Front Plant Sci 6:1252

Guerinot ML (2000) The ZIP family of metal transporters. Biochim Biophys Acta 1465 (1–2):190–198. https://www.ncbi.nlm.nih.gov/pubmed/10748254

Hall JL, Williams LE (2003) Transition metal transporters in plants. J Exp Bot 54(393):2601–2613. https://doi.org/10.1093/jxb/erg303

Hanikenne M, Nouet C (2011) Metal hyperaccumulation and hypertolerance: a model for plant evolutionary genomics. Curr Opin Plant Biol 14(3):252–259. https://doi.org/10.1016/j.pbi.2011.04.003

Hanikenne M, Talke IN, Haydon MJ, Lanz C, Nolte A, Motte P et al (2008) Evolution of metal hyperaccumulation required cis-regulatory changes and triplication of HMA4. Nature 453(7193):391–395. https://doi.org/10.1038/nature06877

Hussain D, Haydon MJ, Wang Y, Wong E, Sherson SM, Young J et al (2004) P-type ATPase heavy metal transporters with roles in essential zinc homeostasis in Arabidopsis. Plant Cell 16(5):1327–1339. https://doi.org/10.1105/tpc.020487

Isaure M-P, Huguet S, Meyer C-L, Castillo-Michel H, Testemale D, Vantelon D et al (2015) Evidence of various mechanisms of Cd sequestration in the hyperaccumulator Arabidopsis halleri, the non-accumulator Arabidopsis lyrata, and their progenies by combined synchrotron-based techniques. J Exp Bot 66(11):3201–3214

Khalid S, Shahid M, Dumat C, Niazi NK, Bibi I, Gul Bakhat HFS et al (2017) Influence of groundwater and wastewater irrigation on lead accumulation in soil and vegetables: implications for health risk assessment and phytoremediation. Int J Phytoremediation 19(11):1037–1046

Kidd P, Mench M, Álvarez-López V, Bert V, Dimitriou I, Friesl-Hanl W et al (2015) Agronomic practices for improving gentle remediation of trace element-contaminated soils. Int J Phytoremediation 17(11):1005–1037

Kobayashi Y, Kuroda K, Kimura K, Southron-Francis JL, Furuzawa A, Kimura K et al (2008) Amino acid polymorphisms in strictly conserved domains of a P-type ATPase HMA5 are involved in the mechanism of copper tolerance variation in Arabidopsis. Plant Physiol 148(2):969–980. https://doi.org/10.1104/pp.108.119933

Kozdroj J, van Elsas JD (2001) Structural diversity of microorganisms in chemically perturbed soil assessed by molecular and cytochemical approaches. J Microbiol Methods 43(3):197–212

Kubota H, Takenaka C (2003) Field Note: Arabis gemmifera is a hyperaccumulator of Cd and Zn. Int J Phytoremediation 5(3):197–201

Kumar PN, Dushenkov V, Motto H, Raskin I (1995) Phytoextraction: the use of plants to remove heavy metals from soils. Environ Sci Technol 29(5):1232–1238

Kumar V, Mahajan M, Yadav SK (2012) Toxic metals accumulation, tolerance and homeostasis in brassicaoilseed species: overview of physiological, biochemical and molecular mechanisms. In: The plant family Brassicaceae. Springer, pp 171–211

Küpper H, Zhao FJ, McGrath SP (1999) Cellular compartmentation of zinc in leaves of the hyperaccumulator Thlaspi caerulescens. Plant Physiol 119(1):305–312

Küpper H, Lombi E, Zhao F-J, McGrath SP (2000) Cellular compartmentation of cadmium and zinc in relation to other elements in the hyperaccumulator Arabidopsis halleri. Planta 212(1):75–84

Küpper H, Götz B, Mijovilovich A, Küpper FC, Meyer-Klaucke W (2009) Complexation and toxicity of copper in higher plants. I. Characterization of copper accumulation, speciation, and toxicity in Crassula helmsii as a new copper accumulator. Plant Physiol 151(2):702–714

Kurek E, Bollag J-M (2004) Microbial immobilization of cadmium released from CdO in the soil. Biogeochemistry 69(2):227–239

Lekeux G, Laurent C, Joris M, Jadoul A, Jiang D, Bosman B et al (2018) di-Cysteine motifs in the C-terminus of plant HMA4 proteins confer nanomolar affinity for zinc and are essential for HMA4 function in vivo. J Exp Bot 69(22):5547–5560

Li D, Xu X, Hu X, Liu Q, Wang Z, Zhang H et al (2015) Genome-wide analysis and heavy metal-induced expression profiling of the HMA gene family in Populus trichocarpa. Front Plant Sci 6: 1149

Li J et al (2020) Heavy metal stress-associated proteins in rice and Arabidopsis: genome-wide identification. Phylogenet Duplication Express Profiles Anal 11. https://doi.org/10.3389/fgene.2020.00477

Liu W-X, Shen L-F, Liu J-W, Wang Y-W, Li S-R (2007) Uptake of toxic heavy metals by rice (Oryza sativa L.) cultivated in the agricultural soil near Zhengzhou City, People's Republic of China. Bull Environ Contamin Toxicol 79(2):209–213

Liu J, Wang W, Mei D, Wang H, Fu L, Liu D et al (2016) Characterizing variation of branch angle and genomewide association mapping in rapeseed (Brassica napus L.). Front Plant Sci 7(21):1–10. https://doi.org/10.3389/fpls.2016.00021

Liu H, Zhao H, Wu L, Liu A, Zhao FJ, Xu W (2017) Heavy metal ATPase 3 (HMA3) confers cadmium hypertolerance on the cadmium/zinc hyperaccumulator Sedum plumbizincicola. New Phytol 215(2):687–698. https://doi.org/10.1111/nph.14622

Lombi E, Zhao F, Dunham S, McGrath S (2000) Cadmium accumulation in populations of Thlaspi caerulescens and Thlaspi goesingense. New Phytol 145(1):11–20

Lone MI, He ZL, Stoffella PJ, Yang XE (2008) Phytoremediation of heavy metal polluted soils and water: progresses and perspectives. J Zhejiang Univ Sci B 9(3):210–220. https://doi.org/10.1631/jzus.B0710633

Lu H, Li Z, Gascó G, Méndez A, Shen Y, Paz-Ferreiro J (2018) Use of magnetic biochars for the immobilization of heavy metals in a multi-contaminated soil. Sci Total Environ 622:892–899. https://doi.org/10.1016/j.scitotenv.2017.12.056

Lysak MA, Koch MA, Pecinka A, Schubert I (2005) Chromosome triplication found across the tribe Brassiceae. Genome Res 15(4):516–525

Manousaki E, Kadukova J, Papadantonakis N, Kalogerakis N (2008) Phytoextraction and phytoexcretion of Cd by the leaves of Tamarix smyrnensis growing on contaminated non-saline and saline soils. Environ Res 106(3):326–332

Mäser P, Thomine S, Schroeder JI, Ward JM, Hirschi K, Sze H et al (2001) Phylogenetic relationships within cation transporter families of Arabidopsis. Plant Physiol 126(4):1646–1667

McGrath SP, Zhao FJ (2003) Phytoextraction of metals and metalloids from contaminated soils. Curr Opin Biotechnol 14(3):277–282. https://www.ncbi.nlm.nih.gov/pubmed/12849780

McGrath S, Zhao F, Lombi E (2001) Plant and rhizosphere processes involved in phytoremediation of metal-contaminated soils. Plant Soil 232(1–2):207–214

McKeehan MD, Kan TC (2000) System and method for supporting mixed-phase transactions in an object-oriented environment. In: Google Patents

Memon AR (2016) Metal hyperaccumulators: mechanisms of hyperaccumulation and metal tolerance. In: Phytoremediation. Springer, pp 239–268

Memon AR, Schroder P (2009) Implications of metal accumulation mechanisms to phytoremediation. Environ Sci Pollut Res Int 16(2):162–175. https://doi.org/10.1007/s11356-008-0079-z

Memon AR, Yatazawa M (1982) Chemical nature of manganese in the leaves of manganese accumulator plants. Soil Sci Plant Nutr 28(3):401–412

Memon AR, Yatazawa M (1984) Nature of manganese complexes in manganese accumulator plant-Acanthopanax sciadophylloides. J Plant Nutr 7(6):961–974

Memon AR, Zahirovic E (2014) Genomics and transcriptomics analysis of Cu accumulator plant Brassica nigra L. J Appl Biol Sci 2:1–8
Memon AR, Chino M, Yatazawa M (1981) Microdistribution of aluminum and manganese in the tea leaf tissues as revealed by X-ray microanalyzer. Commun Soil Sci Plant Anal 12(5):441–452
Memon AR, Aktopraklıgil D, Özdemir A, Vertii A (2001) Heavy metal accumulation and detoxification mechanisms in plants. Turk J Bot 25(3):111–121
Merlot S, de la Torre VSG, Hanikenne M (2018) Physiology and molecular biology of trace element hyperaccumulation. In: Agromining: farming for metals. Springer, pp 93–116
Morel M, Crouzet J, Gravot A, Auroy P, Leonhardt N, Vavasseur A, Richaud P (2009) AtHMA3, a P1B-ATPase allowing Cd/Zn/Co/Pb vacuolar storage in Arabidopsis. Plant Physiol 149(2):894–904. https://doi.org/10.1104/pp.108.130294
Muszyńska E, Hanus-Fajerska E, Ciarkowska K (2015) Studies on Gypsophila fastigiata parameters verifying its suitability to reclamation of post-flotation Zn-Pb wastes. Geol Geophys Environ 41(1):17–24
Nagaharu UJJJB (1935) Genome analysis in Brassica with special reference to the experimental formation of B. napus and peculiar mode of fertilisation. Jpn J Bot 7:389–452
Nouet C, Charlier J-B, Carnol M, Bosman B, Farnir F, Motte P, Hanikenne M (2015) Functional analysis of the three HMA4 copies of the metal hyperaccumulator Arabidopsis halleri. J Exp Bot 66(19):5783–5795
Østerberg JT, Palmgren M (2018) Heavy metal pumps in plants: structure, function and origin. In: Advances in botanical research. Elsevier, pp 57–89
Ovečka M, Takáč T (2014) Managing heavy metal toxicity stress in plants: biological and biotechnological tools. Biotechnol Adv 32(1):73–86
Ozturk M, Memon AR, Gucel S, Sakcali MS (2012) Brassicas in Turkey and their potential role for degraded habitats' remediation. In: The plant family Brassicaceae. Springer, pp 265–287
Pagliano C, Raviolo M, Dalla Vecchia F, Gabbrielli R, Gonnelli C, Rascio N et al (2006) Evidence for PSII donor-side damage and photoinhibition induced by cadmium treatment on rice (Oryza sativa L.). J Photochem Photobiol B Biol 84(1):70–78
Palmgren MG, Nissen P (2011) P-type ATPases. Annual Rev Biophys 40:243–266
Peer WA, Baxter IR, Richards EL, Freeman JL, Murphy AS (2005) Phytoremediation and hyperaccumulator plants. In: Molecular biology of metal homeostasis and detoxification. Springer, pp 299–340
Pollard AJ (2016) Heavy metal tolerance and accumulation in plants of the southeastern United States. Castanea 81(4):257–269
Quartacci MF, Cosi E, Navari-Izzo F (2001) Lipids and NADPH-dependent superoxide production in plasma membrane vesicles from roots of wheat grown under copper deficiency or excess. J Exp Bot 52(354):77–84
Ranieri E, Fratino U, Petruzzelli D, Borges AC (2013) A comparison between Phragmites australis and Helianthus annuus in chromium phytoextraction. Water Air Soil Pollut 224(3):1465
Rascio N, Navari-Izzo F (2011) Heavy metal hyperaccumulating plants: how and why do they do it? And what makes them so interesting? Plant Scie 180(2):169–181
Raskin I, Smith RD, Salt DE (1997) Phytoremediation of metals: using plants to remove pollutants from the environment. Curr Opin Biotechnol 8(2):221–226
Redondo-Gómez S (2013) Bioaccumulation of heavy metals in Spartina. Funct Plant Biol 40(9):913–921
Reeves R, Brooks R (1983) European species of Thlaspi L.(Cruciferae) as indicators of nickel and zinc. J Geochem Explor 18(3):275–283
Reeves R, Baker A, Jaffré T, Erskine P, Echevarria G, van der Ent A (2017) A global database for hyperaccumulator plants of metal and metalloid trace elements. New Phytol 14907
Reeves RD, Baker AJ, Jaffré T, Erskine PD, Echevarria G, van der Ent A (2018) A global database for plants that hyperaccumulate metal and metalloid trace elements. New Phytol 218(2):407–411

Ruley AT, Sharma NC, Sahi SV, Singh SR, Sajwan KS (2006) Effects of lead and chelators on growth, photosynthetic activity and Pb uptake in Sesbania drummondii grown in soil. Environ Polluton 144(1):11–18

Salt DE, Blaylock M, Kumar NP, Dushenkov V, Ensley BD, Chet I, Raskin I (1995) Phytoremediation: a novel strategy for the removal of toxic metals from the environment using plants, vol 13

Salt DE, Smith R, Raskin I (1998) Phytoremediation. Annual Rev Plant Biol 49(1):643–668

Sarma RK, Gowtham I, Bharadwaj R, Hema J, Sathishkumar R (2018) Recent advances in metal induced stress tolerance in plants: possibilities and challenges. In: Plants under metal and metalloid stress. Springer, pp 1–28

Sarma H, Forid N, Prasad R, Prasad MNV, Ma LQ, Rinklebe J (2021) Enhancing phytoremediation of hazardous metal(loid)s using genome engineering CRISPR–Cas9 technology. J Hazard Mater. https://doi.org/10.1016/j.jhazmat.2021.125493

Schmidt U (2003) Enhancing phytoextraction. J Environ Qual 32(6):1939–1954

Schwartz C, Echevarria G, Morel JL (2003) Phytoextraction of cadmium with Thlaspi caerulescens. Plant Soil 249(1):27–35

Sekhar KC, Kamala C, Chary N, Balaram V, Garcia G (2005) Potential of Hemidesmus indicus for phytoextraction of lead from industrially contaminated soils. Chemosphere 58(4):507–514

Seth CS, Misra V, Chauhan L (2012) Accumulation, detoxification, and genotoxicity of heavy metals in indian mustard (Brassica juncea L.). Int J Phytoremediation 14(1):1–13

Shikanai T, Muller-Moule P, Munekage Y, Niyogi KK, Pilon M (2003) PAA1, a P-type ATPase of Arabidopsis, functions in copper transport in chloroplasts. Plant Cell 15(6):1333–1346. https://www.ncbi.nlm.nih.gov/pubmed/12782727

Silver S, Phung LT (2005) Genes and enzymes involved in bacterial oxidation and reduction of inorganic arsenic. Appl Environ Microbiol 71(2):599–608

Sonowal S, Nava AR, Joshi SJ, Borah SN, Islam NF, Pandit S, Prasad R, Sarma H (2022) Biosurfactants assisted heavy metals phytoremediation: green technology for the United Nations sustainable development goals. Pedosphere 2(1):198–210

Stephan UW, Schmidke I, Stephan VW, Scholz G (1996) The nicotianamine molecule is made-to-measure for complexation of metal micronutrients in plants. Biometals 9(1):84–90

Suman J, Uhlik O, Viktorova J, Macek T (2018) Phytoextraction of heavy metals: a promising tool for clean-up of polluted environment? Front Plant Sci 9:1476

Sun Q, Ye Z, Wang X, Wong MH (2005) Increase of glutathione in mine population of Sedum alfredii: a Zn hyperaccumulator and Pb accumulator. Phytochemistry 66(21):2549–2556

Sun Y, Zhou Q, Xu Y, Wang L, Liang X (2011) The role of EDTA on cadmium phytoextraction in a cadmium-hyperaccumulator Rorippa globosa. J Environ Chem Ecotoxicol 3(3):45–51

Szczygłowska M, Bodnar M, Namieśnik J, Konieczka P (2014) The use of vegetables in the biomonitoring of cadmium and lead pollution in the environment. Criti Rev Anal Chem 44(1):2–15

Tang Y-T, Qiu R-L, Zeng X-W, Ying R-R, Yu F-M, Zhou X-Y (2009) Lead, zinc, cadmium hyperaccumulation and growth stimulation in Arabis paniculata Franch. Environ Exp Bot 66(1):126–134

Tangahu BV, Abdullah S, Rozaimah S, Basri H, Idris M, Anuar N, Mukhlisin M (2011) A review on heavy metals (As, Pb, and Hg) uptake by plants through phytoremediation. Int J Chem Eng 2011

Thakare M, Sarma H, Datar S, Roy A, Pawar P, Gupta K, Pandit S, Prasad R (2021) Understanding the holistic approach to plant-microbe remediation technologies for removing heavy metals and radionuclides from soil. Curr Res Biotechnol. https://doi.org/10.1016/j.crbiot.2021.02.004

Thijs S, Sillen W, Weyens N, Vangronsveld J (2017) Phytoremediation: state-of-the-art and a key role for the plant microbiome in future trends and research prospects. Int J Phytoremediation 19(1):23–38

Tóth G, Jones A, Montanarella L (2013) The LUCAS topsoil database and derived information on the regional variability of cropland topsoil properties in the European Union. Environ Monit Assess 185(9):7409–7425

Tóth G, Hermann T, Da Silva M, Montanarella L (2016) Heavy metals in agricultural soils of the European Union with implications for food safety. Environ Int 88:299–309

van de Mortel JE, Villanueva LA, Schat H, Kwekkeboom J, Coughlan S, Moerland PD et al (2006) Large expression differences in genes for iron and zinc homeostasis, stress response, and lignin biosynthesis distinguish roots of Arabidopsis thaliana and the related metal hyperaccumulator Thlaspi caerulescens. Plant Physiol 142(3):1127–1147

Van der Ent A, Baker AJ, Reeves RD, Pollard AJ, Schat H (2013) Hyperaccumulators of metal and metalloid trace elements: facts and fiction. Plant Soil 362(1–2):319–334

van der Ent A, Baker AJ, Reeves RD, Chaney RL, Anderson CW, Meech JA et al (2015) Agromining: farming for metals in the future? ACS Publications

Van Liedekerke M, Prokop G, Rabl-Berger S, Kibblewhite M, Louwagie G (2014) Progress in the management of contaminated sites in Europe. Ref Rep Jt Res Centre Eur Commission 1(1):4–6

Verbruggen N, Hanikenne M, Clemens S (2013) A more complete picture of metal hyperaccumulation through next-generation sequencing technologies. Front Plant Sci 4:388

Vogel-Mikuš K, Simčič J, Pelicon P, Budnar M, Kump P, Nečemer M et al (2008) Comparison of essential and non-essential element distribution in leaves of the Cd/Zn hyperaccumulator Thlaspi praecox as revealed by micro-PIXE. Plant Cell Environ 31(10):1484–1496

von Wirén N, Klair S, Bansal S, Briat J-F, Khodr H, Shioiri T et al (1999) Nicotianamine chelates both FeIII and FeII. Implications for metal transport in plants. Plant Physiol 119(3):1107–1114

Wang J, Liang S, Xiang W, Dai H, Duan Y, Kang F, Chai T (2019) A repeat region from the Brassica juncea HMA4 gene BjHMA4R is specifically involved in Cd2+ binding in the cytosol under low heavy metal concentrations. BMC Plant Biol 19(1):89. https://doi.org/10.1186/s12870-019-

Warwick SI, Black LD (1991) Molecular systematics of Brassica and allied genera (subtribe Brassicinae, Brassiceae)—chloroplast genome and cytodeme congruence. Theor Appl Genet 82(1):81–92

Wei L, Luo C, Li X, Shen Z (2008) Copper accumulation and tolerance in Chrysanthemum coronarium L. and Sorghumsudanense L. Arch Environ Contam Toxicol 55(2):238–246

Weng G, Wu L, Wang Z, Luo Y, Christie P (2005) Copper uptake by four Elsholtzia ecotypes supplied with varying levels of copper in solution culture. Environ Int 31(6):880–884

Williams LE, Mills RF (2005) P1B-ATPases–an ancient family of transition metal pumps with diverse functions in plants. Trends Plant Sci 10(10):491–502

Wong CKE, Cobbett CS (2009) HMA P-type ATPases are the major mechanism for root-to-shoot Cd translocation in Arabidopsis thaliana. New Phytol 181(1):71–78. https://doi.org/10.1111/j.1469-8137.2008.02638.x

Xiong Y, Yang X, Ye Z, He Z (2004) Characteristics of cadmium uptake and accumulation by two contrasting ecotypes of Sedum alfredii Hance. J Environ Sci Health Part A 39 (11–12):2925–2940

Zeng X, Ma LQ, Qiu R, Tang Y (2009) Responses of non-protein thiols to Cd exposure in Cd hyperaccumulator Arabis paniculata Franch. Environ Exp Bot 66(2):242–248

Zhao H, Wang L, Zhao FJ, Wu L, Liu A, Xu W (2018) SpHMA1 is a chloroplast cadmium exporter protecting photochemical reactions in the Cd hyperaccumulator Sedum plumbizincicola. Plant Cell Environ

Zhu Y-G, Chen S-B, Yang J-C (2004) Effects of soil amendments on lead uptake by two vegetable crops from a lead-contaminated soil from Anhui, China. Environ Int 30(3):351–356

Zhuang P, Yang Q, Wang H, Shu W (2007) Phytoextraction of heavy metals by eight plant species in the field. Water Air Soil Pollut 184(1–4):235–242

Role of Soil Microflora in Phytoremediation of Heavy Metal Contaminated Soils

2

Kunal Seth and Anil Kumar

Abstract

Heavy metals are of great environmental concern as they are non-biodegradable, accumulate in the environment, enter into the food chain, and exert adverse effects on all living organisms including microorganisms, plants, and animals. Among different technologies, phytoremediation is a better option for reclamation of heavy metal polluted soils. Several plants including hyperaccumulators have been reported with significant remediation potential. The phytoremediation potential of these plants is also affected by microorganisms present in the plant rhizosphere. The potential role of microorganisms in phytoremediation of heavy metal contaminated sites is becoming apparent. The capability of soil microorganisms to promote the uptake and accumulation of heavy metals from soil is an important aspect of phytoremediation. The establishment of a microbiocenosis with potential to stimulate the uptake of heavy metals depends upon the microbial dynamics in the rhizospheres. Soil microorganism including plant growth-promoting rhizobacteria, P-solubilizing bacteria, mycorrhiza-helping bacteria, plant endophytic bacteria, arbuscular mycorrhizal fungi, and soil fungi have the potential to increase the phytoremediation potential of plants. Use of genetically engineered microorganisms in phytoremediation increases the plants heavy metal accumulation and widens the horizon of microbial use in the technique. The mechanism of soil microbes assisted phytoremediation include acceleration of metal mobility, immobilization, nutrient acquisition, metal detoxification, transformation, and mitigation of heavy metal stresses in plants. Soil

K. Seth
Department of Botany, Government Science College, Valsad, Gujarat, India

A. Kumar (✉)
Laboratory of Plant Physiology and Biochemistry, Department of Botany, University of Rajasthan, Jaipur, Rajasthan, India

R. Prasad (ed.), *Phytoremediation for Environmental Sustainability*,
https://doi.org/10.1007/978-981-16-5621-7_2

microbes involved in biogeochemical processes operating in the rhizosphere affect the mobility and availability of heavy metal to the plant through the release of chelating agents, biosurfactants and biomethylation, metal speciation, acidification, dissolution, phosphate solubilization, and redox changes. In addition, plant-associated bacteria can also increase plant resistance to the pathogen and ensure nitrogen fixation and the production of growth regulators. This chapter presents the recent advances and applications made hitherto in understanding the functional role of plant-microbe interactions in the phytoremediation.

Keywords

Heavy metals · Phytoremediation · Soil microorganisms · Plant growth-promoting rhizobacteria · Phosphate-solubilizing bacteria · Arbuscular mycorrhizal fungi

Abbreviations

ACC	1-Aminocyclopropane-1-carboxylate
ACCD	1-Amino-cyclopropane-1-carboxylic acid deaminase
AMF	Arbuscular mycorrhizal fungi
HMs	Heavy metals
IAA	Indole-3-acetic acid
MHB	Mycorrhiza helper bacteria
MTs	Metallothioneins
PAHs	Polycyclic aromatic hydrocarbons
PCs	Phytochelatins
PGPR	Plant growth-promoting rhizobacteria
PSB	Phosphate-solubilizing bacteria

2.1 Introduction

Phytoremediation is an inexpensive, easily applicable, environmentally safe strategy to treat contaminated sites. Phytoremediation can be applied in situ and include different approaches, namely phytoextraction, phytostabilization, phytodegradation, phytovolatilization, phytostimulation, and rhizofiltration. Several comprehensive reviews have been written to summarize important aspects of this plant-based novel technology (Salt et al. 1998; Pilon-Smits 2005; Harish and Sundaramoorthy 2008; Kumar and Aery 2016). Though the phytoremediation is economic and has minimal impacts on soil structure it is usually slow compared to the traditional physical and chemical remediation technologies and can even be inefficient in case of some long-lasting contaminants like polycyclic aromatic hydrocarbons (Yang et al. 2020). Moreover, the contaminant should be approachable to the roots in rhizosphere. The success of phytoremediation depends not only on the plant's accumulation capacity but also on the bioavailability of contaminants. To overcome these, the concept of using green plants in the remediation of contaminated soil is now assisted by an interdisciplinary approach, in which plant–microbe interaction is exploited to enhance the removal, immobilization, or degradation of certain metals from contaminated soils. The assisted phytoremediation methods not only restore the quality of contaminated soils but also reduce the need for artificial fertilization. Furthermore, the use of microbes associated with plants adds new dimensions to phytoremediation technology. Soil harbors a variety of microorganisms, viz. plant growth-promoting rhizobacteria (PGPR), endophytes, fungi, mycorrhiza, and algae also form an association with plants that directly or indirectly play important roles in growth and development of plants and physicochemical properties of soil (Prasad et al. 2017). Various metabolites produced by soil microflora including siderophores, organic acids, biosurfactants, etc. play an important role in numerous biogeochemical processes happening in the rhizosphere. The main purpose is to detoxify the contaminants and amelioration of abiotic stress in plants. Rhizospheric microorganisms also hasten the mobilization or immobilization of heavy metals. Moreover, organic and mineral substances having acidifying, chelating, and/or reducing effects which are of great importance in the metal absorption by plants (Kamal et al. 2010; Koptsik 2014). The complex inter-relationships between plants and microorganisms, on the whole, are capable of increasing the effectiveness of phytoremediation technology.

Soil contamination is an ever-increasing global problem. In context to serious and complicated cases of environmental pollution, the socioeconomic reforms and environmental policies are underway to minimize the release of pollutants in the environment. It becomes very important to develop effective bioremediation strategies. Therefore, it is critical to enrich the knowledge about microbial-assisted phytoremediation to identify the most appropriate phytoremediation strategy.

2.2 Heavy Metal Pollution and Microbe-Assisted Phytoremediation

A wide variety of soil contaminants such as heavy metals, organic pollutants including hydrocarbons, polyaromatic hydrocarbons, polychlorinated biphenyls, herbicides, and pesticides have been found to pose a critical concern to human health and the environment. Heavy metal contamination in soil is a serious environmental concern as large area of land is facing this problem. Moreover, anthropogenic activities such as over use of fertilizers and pesticides, sludge or municipal composts, emissions from municipal waste incinerators, automobiles, residues of metalliferous mines and smelting industries, etc. are responsible for heavy metal pollution in the environment (Aery 2016; Kumar and Aery 2016; Kumar 2020). Several heavy metals and organic contaminants can enter the food chain and can cause mutagenicity and carcinogenicity. Heavy metal pollution in agrarian soils has become a critical environmental anxiety due to their long-term persistent nature and potential toxic ecological effects. Heavy metal contamination affect the seed germination, growth

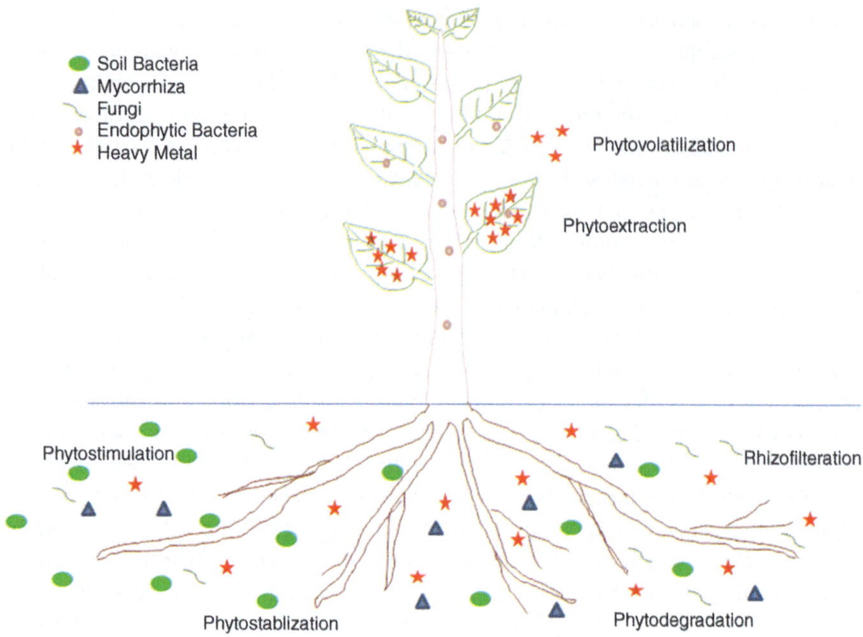

Fig. 2.1 Diversity of soil microorganism and their relation with different types of phytoremediation

and development of plants and cell membrane, cell organelles and enzymes involved in metabolism, detoxification in animals (Kumar and Aery 2010, 2011, 2012a, b, 2016; Kamal et al. 2010; Raikwar et al. 2008; Tchounwou et al. 2012).

Remediation of contaminated soils using plants has become an important method to deal with heavy metal contamination. But the absorption of heavy metals by the plants is a slow process (Göhre and Paszkowski 2006). Soil microorganisms have the capability to remove waste and convert composite waste into simple non-toxic compounds. Because of this they can improve growth and survival of plants under heavy metal stress. Interaction of plants with soil microbes can play a vital role in adaption of plants to metal contaminated environment and can be explored to increase microbe-assisted metal removal of heavy metal contaminated soils (Prasad 2021; Kumar 2020; Kumar et al. 2021). The relation between soil microorganisms and types of phytoremediation is summarized in Fig. 2.1.

2.2.1 Role of Plant Growth-Promoting Rhizobacteria in Phytoremediation

The beneficial free-living soil bacteria which exist in association with roots of different plants are generally referred as plant growth-promoting rhizobacteria (PGPR) (Prasad et al. 2015). PGPR include several diverse genera and enhance the phytoremediation capabilities of plants by increasing their growth and biomass. PGPR influence plant growth by producing growth-promoting compounds like phytohormone, vitamins, enzymes, and antibiotics. Certain PGPR strains can solubilize inorganic P or mineralize organic P by producing organic acids and thereby provide accessible P to the plant which is often growth-limiting element and is generally present in the soil in a quite unavailable form. They can produce iron chelators known as siderophores to facilitate iron to plants. PGPR are very effective in immobilization of heavy metals and reduce the deleterious effects caused by plant pathogens or disease. PGPR inoculation significantly enhances biomass production, organic matter content, nitrogen and phosphorus content of the soil with the rapid decomposition of heavy metal contaminated soil. As an example, the microbial-assisted phytoremediation involving the application of PGPR has been reported to increase organic N and bioavailable P in sewage sludge treated soil compared to the uninoculated soil (Grobelak et al. 2018). PGPR bioaugmentation on *Spartina maritima* in vivo has been reported to decrease antioxidant enzymatic activities, respiration rate, reduction in respiratory carbon consumption and alleviation of abiotic stress caused by heavy metal contamination (Mesa-Marín et al. 2018). PGPR alleviate toxic effects on plants by producing different substances like siderophores, organic acids, biosurfactants, and extra-cellular polymeric substances (Sessitsch et al. 2013). For successful bioinoculants PGPR strains must be able to rapidly establish, colonize plant root, and survive under stress environment. Ma et al. (2016) showed that plant growth-promoting bacteria endowed with abiotic stress-resistant contribute for efficient phytoremediation process. Inoculation of drought-resistant rhizobacterial strains (*Pseudomonas libanensis* TR1 and *Pseudomonas reactans* Ph3R3) to *Brassica oxyrrhina* imposed positive effects on plant development and metal phytoremediation under drought conditions. *Pseudomonas* strains TR1 and Ph3R3 significantly improve uptake and translocation of Cu and Zn (Ma et al. 2016). PGPR have been shown to improve chromium uptake by promoting extensive proliferation in roots secreting metal sequestering molecules in rhizosphere and upregulation of genes involved in stress alleviation (Ahemad 2015a). Franchi et al. (2017)) reported that indigenous metal-tolerant bacterial strains isolated from the metal contaminated site can show great potential in assisting phytoremediation and show a positive effect on the plant biomass. The selected bacterial consortium, when augmented in addition to thiosulfate, showed increased phytoaccumulation efficacy up to 85% for As and up to 45% for Hg. Similarly, rhizospheric bacteria isolated from the rhizosphere of halophyte *Arthrocnemum macrostachyumis* showed promising remediation capability, plant growth-promoting properties and found to be multiresistant to heavy metals (Navarro-Torre et al. 2016). PGPR are even found beneficial for remediation of sites co-contaminated with polycyclic aromatic hydrocarbons (PAHs) and heavy metals. Some species of soil microbes such as *Lysobacter, Kaistobacter,* and *Pontibacter* significantly increase heavy metal accumulation and are important to regulate plant growth and heavy metal uptake (Lin et al. 2021). Heavy metals often create difficulty in remediating these sites as high levels of heavy metals can significantly inhibit PAHs mineralization. However, the bioavailability of PAHs can be increase by organic acids present in root exudates and biosurfactants secreted by PGPR (Chen et al. 2017). The colonization of bacteria in rhizosphere depends upon the nutrient level present in the root exudates. Moreover, high concentration of contaminants also affects the microbial communities by reducing total microbial biomass or changing microbial community structure.

2.2.2 Role of P-Solubilizing Bacteria in Phytoremediation

Several phosphate-solubilizing bacteria (PSB) with high tolerance to heavy metals and plant growth-promoting activities have been explored for phytoremediation of metal contaminated soils. Inorganic phosphate solubilization is an important mechanism of plant growth promotion by rhizobacteria. Plant-associated bacteria release organic acids into the soil to solubilize the unavailable phosphate complexes converting them into ortho-phosphate for plant uptake and utilization. PSB not only increase P utilization in low P condition but also protect the plants from pathogens by the production of antibiotics, HCN, phenazines, and antifungal compounds (Ahemad and Kibret 2014). In synergistic association with plants, PSB remediates metalliferous soils largely through facilitating either phytostabilization or phytoextraction. PSB promote plant growth by producing organic acid, secreting siderophores, releasing IAA and activity of enzyme ACC deaminase (Ahemad 2015b). Low molecular weight organic acids, namely lactic, citric, 2-ketogluconic, malic, glycolic, oxalic, malonic, tartaric, valeric, piscidic, succinic and formic acid have chelating properties that increase the bioavailability of metal to plants (Chen et al. 2017). PSB can also enhance phytoextraction of metals by solubilizing insoluble and biologically unavailable metals which are strongly adhered to soil particles (Gamalero and Glick 2012; Aery 2016). Yang et al. (2018) reported that PSB strains *Pseudomonas fluorescens* gim-3 and *Bacillus cereus* qh-35 have the ability to dissolve $CdCO_3$ and solid Cd in soil. Gluconic acid produced by the peripheral direct oxidation pathway is responsible for Cd dissolution in high-Cd-mobilizing PSB. It has been reported that PSB exerts positive impact on the soil microflora, soil quality, increased growth of *Wedelia trilobata,* and increased absorption and translocation of Cu from Cu-contaminated soil (Lin et al. 2018).

2.2.3 Role of Mycorrhizal-Helping Bacteria in Phytoremediation

In soil ecosystems, numerous bacterial taxa colonize and form biofilm-like structures on the surface of extraradical hyphae and spores of arbuscular mycorrhizal fungi (AMF) mycelia (Scheublin et al. 2010; Lecomte et al. 2011; Cruz and Ishii 2012). Several bacterial taxa belonging to α-, β-, and γ-proteobacteria and firmicutes have been identified and isolated from mycelial surface of many AMF species (Scheublin et al. 2010; Lecomte et al. 2011). Some AMF taxa can harbor bacteria within the cytoplasm as endobacteria (Bonfante and Anca 2009). These endobacteria could be obligate biotrophs and unable to grow without AMF (Jargeat et al. 2004). Mycorrhizal symbiosis increases the tolerance of plants to heavy metals and other contaminants. Establishment of ectomycorrhizas on roots is depended on the bacterial communities present in the rhizosphere. Some bacteria show a helper effect with the fungi and known as "Mycorrhiza Helper Bacteria" (MHB). The term "MHB" was proposed by Garbaye in Garbaye 1994, who first suggested that rhizospheric bacteria can promote mycorrhization of plants. MHB can induce mycorrhization in the plant's root system, by increasing root–fungus interaction and colonization, by detoxification of mycorrhizal accumulated metabolites, by reducing the effects of unfavorable environmental factors, and by inhibiting the growth of competing microorganisms (Frey-Klett et al. 2007; Giri et al. 2005; Prasad et al. 2021). Mycelia growth capability of MHB can be attributed to the production of growth factors and neutralization of antagonistic substances. MHB improve the fungal adaptation to different soil types by producing more fungal reproductive propagules. MHB has also been found to mediate ectomycorrhiza influenced plant's elemental uptake. Hydroxamate siderophore produced by ectomycorrhiza increase elemental availability mediated by MHB siderophore to enhance nutrient uptake such as P, Cu, Zn, and Fe in plants (Dhawi 2016). Some MHB influence both plants-mycorrhizal association as well as PSB. Thus, tripartite associations of mycorrhizal fungi, bacteria, and plants result in a consortium that promotes plant growth and influence mycorrhizal symbiosis for improved phytoremediation of heavy metals. It has been observed that the effect of MHB is not plant-specific but it may be fungal selective (Garbaye et al. 1992; Garbaye 1994). Though the bacterial–mycorrhizal interaction seems to be species-specific, MHB provides carbon source to mycorrhiza in the form of malate and citrate which have a major role in mycorrhiza bacteria signaling of mycorrhiza establishment.

2.2.4 Role of Endophytic Bacteria in Phytoremediation

Very low concentrations of soil contaminants can inhibit plant growth and metabolic activities of soil-associated microbes. Microbes concede interactions between plant-soil-microbe systems and play major role in nutrient cycling and effective phytoremediation of contaminated soils. Rhizospheric bacteria colonize the vicinity of roots and get benefit from root exudates, but some bacteria are capable to enter in internal tissues of the plant as endophytes and can establish a mutualistic association. The endophytes enter in plant tissue through the root, germinating radicles and root hair cells; however, aerial route of entry through flowers, stems, and cotyledons have also been reported (Kobayashi and Palumbo 2000). Cell wall-degrading enzymes make penetration feasible for the entry of endophytic bacteria into plants (Reinhold-Hurek et al. 2006). Endophytic bacteria possess the plant growth-promoting abilities. The inoculation of endophytic bacteria improves the plant's adaptation and growth in contaminated soil and enhances the degradation of pollutants. Endophytic bacteria promote plant growth through mechanisms involving bacterial metabolites, such as indole-3-acetic acid (IAA), detoxifying the metals by siderophores, organic acids and methylation, solubilization of metal phosphates, and by the activity of enzyme bacterial 1-amino-cyclopropane-1-carboxylic acid deaminase (ACCD). The bacterial endophytes have several advantages over rhizospheric bacteria. Endophytic bacteria can interact more closely with their host plant as compared to rhizobacteria due to colonization in plant interior. They seem to be efficient colonizers of the rhizosphere as well as of the endosphere, therefore, they can degrade contaminants in both environments in a synergistic manner. While rhizospheric microorganisms are conditioned by adverse biotic and abiotic conditions (Seghers et al. 2004; Afzal et al. 2011), endophytic bacteria remain secure from biotic and abiotic stresses than rhizospheric bacteria (Rosenblueth and Martínez-Romero 2006). Considering their life strategies, endophytic bacteria can be either facultative or obligate in nature. Facultative endophytic bacteria are able to survive and colonize outside the plant body. However, the existence of obligate endophytic bacteria depends on the host plant and they may be transferred through seeds or vegetative part or via vectors (Hamilton et al. 2012). Once entered, a plant endophytic bacterium may either become localized at the point of entry or spread systematically in the plant body. These bacteria can reside within cells, or in the intercellular spaces, or within the components of the vascular system. Various types of pollutant degrading endophytic bacteria are found to proliferate in the intercellular spaces of the plant due to availability of high of nutrients, sugars, and amino acids (Bacon and Hinton 2007). The population and diversity of contaminant-degrading endophytic bacteria usually depend on the concentration of contaminants in the environment (Peng et al. 2013). Burges et al. (2017) reported that inoculation of the endophytes improved plant growth and Zn phytoextraction of *Noccaea caerulescens* and *Rumex acetosa* plants. It has been observed that the bacteria present on polluted areas are tolerant to higher concentrations of metals compared to those present in unpolluted areas. Studies revealed that endophytic bacteria that were not exposed to a polluted environment also have degradation genes. This indicates that endophytic bacteria may be inherently equipped with genes to destroy complex organic substances, which may be used as carbon source by the bacteria. The same enzymes are supposed to be involved in the degradation of contaminants. Plant growth-promoting endophytic bacterium *Burkholderia phytofirmans* PsJN has hydrocarbon-degrading alkane monooxygenase (alkB) and alkane hydroxylase (CYP450) genes which may contribute to its extraordinary ability to colonize in plant tissues (Mitter et al. 2013).

2.2.5 Role of Arbuscular Mycorrhizal Fungi in Phytoremediation

Arbuscular mycorrhizal fungi (AMF) are a type of fungi which found associated with roots of land plants and form a mutualistic relationship with them. They are obligate in nature and cannot complete their life cycle without association with plants (Ferrol et al. 2004; Prasad et al. 2017). Recently their role in the natural ecosystem has been recognized. AMF can extend the virtual roots system of heavy metal-accumulating plants. It is reported that the phytoremediation efficiency of legume plant *Robinia pseudoacacia* increased for Pb when associated with AMF. Association with AMF increases the nutrient uptake of legume and photosynthesis rate and subsequently increase the biomass of the plant thereby help in phytoremediation (Yang et al. 2016). Similar study was conducted with *Triticum aestivum* L. plants in Zn contaminated soils. Results showed that the association of plants with AMF help in tolerating higher heavy metal concentration in the soil and can help in the phytostabilization of Zn contaminated soil (Kanwal et al. 2016). In a recent study, the impact of AMF *Glomus constrictum* on various physiological parameters of the *Tagetes erecta* evaluated. Inoculation of AMF induces the growth rate and reduces the heavy metal content in the tissue of *T. erecta*. It is concluded that inoculation with AMF help the plant in tolerating heavy metal in the soil and thereby have a protective role (Elhindi et al. 2018). Inoculation of *Pseudomonas libanensis* and *Claroideoglomus claroideum* to plants significantly improved the heavy metal accumulation in heavy metal-saline contaminated soil. This proves the concept that bioaugmentation of plants with AMF improves the efficacy of phytoremediation in soil contaminated with heavy metals (Ma et al. 2019).

2.2.6 Role of Fungi in Phytoremediation

Many fungal endophytes which are heavy metal resistant have found to enhance the phytoremediation efficiency of plants by enhancing the plant growth, increasing tolerance to metals and by influencing the metal translocation and accumulation (Li et al. 2012). One important study revealed that *Solanum nigrum* inoculated with fungal endophyte (strain RSF-6L) has a higher growth rate, better tolerance to Cd under Cd contaminated soil (Khan et al. 2017). Sharma et al. (2019) investigate the role of fungal endophytes in metal tolerance accumulation of Pb-Zn hyperaccumulator *Arabis alpine*. Association of fungal endophytes helps the plant by decreasing the accumulation of Pb and Cd and thereby increases the host tolerance to metal contamination. Tong et al. (2017) investigated the grass species growing on copper tailing dam China. They found that frequency of endophyte infection is increased over a period of time and infection rates of *Bothriochloa ischaemum* and *Festuca rubra* were positively related to concentration of Cd. Furthermore, fungal endophytes associated with *Imperata cylindrical* and *Elymus dahuricus* help plants to develop Pb tolerance. In a similar study, Ali et al. (2019) reported the association of endophytic *Aureobasidium pullulans* BSS6 help in increase in phytoremediation potentials of *Cucumis sativus* under Cd and Pb stress. In a recent study, eight different Cd-tolerant endophytic fungal isolates with plant growth-promoting properties are characterized in *Trifolium repens* (Liu et al. 2019). It has been reported that endophytes produce IAA and extra-cellular enzymes like phosphatase, cellulase, and glucosidase, which significantly improve the stress tolerance by increasing antioxidant enzyme activity and by decreasing lipid peroxidation in the plant (Bilal et al. 2018; Ali et al. 2019). Overall fungal association increases metal tolerance of the plant by reducing metal uptake and boosting the antioxidant system.

2.2.7 Genetically Engineered Microorganisms and Phytoremediation

Despite its proven value phytoremediation is still underused as it is not considered to be highly efficient, predictable, and fast clean-up technology. Many studies have shown variable results at the field scale like slow and partial degradation and long clean-up period. Furthermore, its success depends upon the choice of plant species, its establishment, contaminant concentration, environmental factors, soil conditions, and soil microflora present in the rhizosphere. Although highly diverse microbial communities present in the soil can efficiently degrade many pollutants, still many of these pollutants are beyond the degradation capability of microorganisms. Bioaugmentation is the strategy to introduce exogenous microorganism at the contaminated site to augment the indigenous microorganisms. This strategy can also result in horizontal gene transfer between the exogenous to the indigenous microorganisms. Considering the reclamation of metalliferous sites, genetic engineering might be a better tool for increasing the efficiency of phytoremediation. Among the available genetic engineering tools, CRISPR-Cas9 and CRISPR-Cpf1 gene-editing tools have shown the potential to improve agro-ecosystems and improve phytoremediation efficiency by improving plant-microbe interactions (Basu et al. 2018; Sarma et al. 2021). Cas9/sgRNA system has been successfully implemented to form CRISPR crops; this technology can also be effectively utilized to develop custom made plant growth-promoting rhizobacteria to improve phytoremediation efficiency (Seth and Harish 2016). Many genes of microbial origin were discovered that encode enzymes involved in the detoxification of contaminants. These genes can be overexpressed by employing gene-editing tools in plant growth-promoting microorganisms. Similarly, rhizobacteria that produce phytohormone IAA facilitate plants to resist metal stress and improve nitrogen fixation as well. By utilizing gene-editing tools, these rhizobacteria can be designed to improve the IAA threshold in the rhizosphere (Basu et al. 2018). Plants respond to heavy metal toxicity in a variety of different ways. Most plants secrete metal-binding peptides, metallothioneins (MTs), and phytochelatins (PCs) when exposed to heavy metal toxicity. The production of these peptides has been found to be increased through genetically modified microorganisms. Recombinant rhizobia (*Mesorhizobium huakuii* subsp. *rengei* B3) carrying synthetic MTs (MTL4) and cDNA encoding phytochelatin synthase from *Arabidopsis thaliana* (AtPCS) has been reported to show up to 25 fold increase in Cd accumulation in *Astragalus sinicus* (Ike et al. 2007). In another study, Sriprang et al. (2003) successfully introduced genes from *Arabidopsis thaliana* into diazotrophic *Mesorhizobium huakuii* sub sp. *rengei* B3 to produce PCs and accumulate Cd under the control of bacterial specific promoter. Enhanced Cd resistance and accumulation has been reported in case of *Pseudomonas putida* KT2440 transformed with the phytochelatin synthase gene of *Schizosaccharomyces pombe* (Yong et al. 2014).

2.3 Mechanisms of Soil Microorganism Induced Phytoremediation

Soil microorganisms can assist plants in heavy metal uptake in both ways, i.e. directly and indirectly. In a direct manner, soil microbes increase the uptake of heavy metals by synthesizing siderophores, producing phytohormones, and fixing and solubilizing minerals. Whereas in an indirect manner soil microbe increases the uptake of heavy metals by increasing plant growth, improving plant health by inhibiting plant pathogen, transforming heavy metals, etc. Microbes have different protection strategies to combat heavy metals stress like compartmentalization, exclusion, and the synthesis of metal-binding proteins like phytochelatins, metallothioneins, Cd-binding peptides (CdBPs), cysteines (gcgcpcgcg) (CP), and histidines $(ghhphg)_2$ (HP) (Sharma et al. 2021). The mechanism of soil microorganisms induced phytoremediation is summarized in Fig. 2.2.

2.3.1 Microbial Secretion

Microbial secretion plays the main role among mechanisms of phytoremediation assisted by microbes. Soil microbes (bacteria and fungi) produce metal-chelating agents in low iron conditions known as siderophores. Siderophores are low molecular weight organic compounds having very high and specific affinity to chelate iron (Oswald 2010; Das et al. 2007). About 500 different siderophores are known (Illmer and Buttinger 2006). The most well-known siderophore is Fe (III) chelator

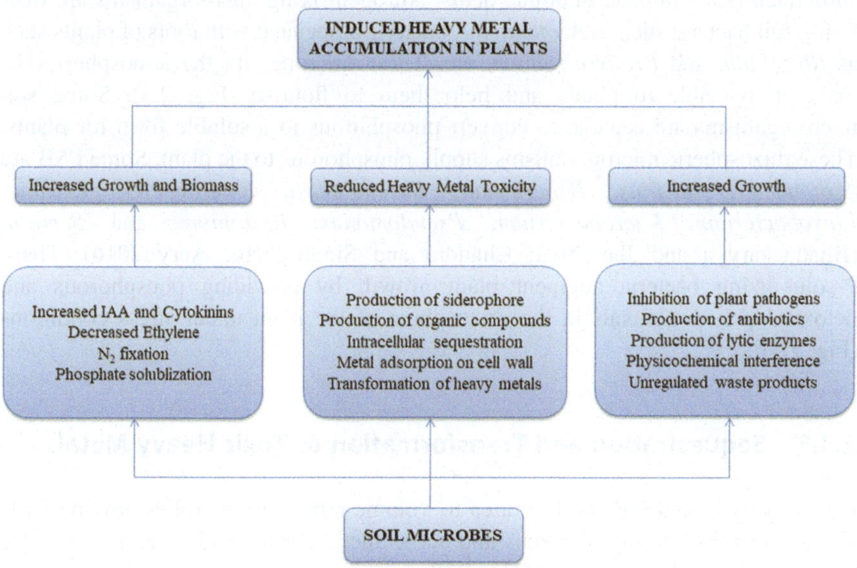

Fig. 2.2 Outline of soil microorganisms induced heavy metal accumulation in plant

pyoverdine (Sharma and Johri 2003). Siderophores plays an important role in the acquisition of several heavy metals such as Cd, Cu, Pb, Zn, and U (Leong 1986; Dimkpa et al. 2009; Rashmi et al. 2013; Gaonkar and Bhosle 2013) by forming complexes with heavy metals, reducing the toxicity of metals (Grobelak and Hiller 2017) and increasing their availability to plants (Jing et al. 2007). Soil microorganisms also regulate heavy metal solubility and mobilization of mineral compounds in the rhizosphere by producing low molecular weight organic acids (Rajkumar et al. 2012). These include citric, lactic, tartaric, malic, oxalic, succinic, malonic, formic and 5-ketogluconic acids (Panhwar et al. 2013). Plants may not have the capability to synthesize necessary endogenous growth hormones for optimal growth and development under heavy metal stress. Under the stress condition, the plant relies on the soil microorganisms for growth hormone such as IAA, cytokinins, etc. (Ahemad and Khan 2012; Ambawade and Pathade 2015). These hormones induce the growth and biomass production of plants and increase the rate of phytoremediation. A number of plant growth-promoting bacteria facilitate plant growth and development by depressing ethylene concentration by decreasing the quantity of 1-aminocyclopropane-1-carboxylate (ACC, the precursor of the plant hormone ethylene) through action of the enzyme ACCD (Glick 2004). At low ethylene concentrations increased rate of leaf elongation and enlargement of primary leaves in plants has been observed (Dubois et al. 2018).

2.3.2 Fixation and Solubilization of Minerals

Some soil microbes lead to the reduction of molecular nitrogen to ammonia and subsequently assimilated in amino acids. Nitrogen-fixing microorganisms are free-living soil bacteria such as *Azotobacter*, bacteria associated with roots of plants such as *Rhizobium* and *Bradyrhizobium*, etc. These microbes fix the atmospheric N_2, make it available to plants and help them to flourish (Fig. 2.3). Some soil microorganisms are capable to convert phosphorous to a soluble form for plants. These rhizospheric microorganisms supply phosphorous to the plant. Some PSB are *Azotobacter*, *Bacillus*, *Beijerinckia*, *Burkholderia*, *Enterobacter*, *Erwinia*, *Flavobacterium*, *Microbacterium*, *Pseudomonas*, *Rhizobium*, and *Serratia* (Bhattacharyya and Jha 2012; Chandra and Singh 2016; Aery 2016). These P-solubilizing bacteria augment plant growth by providing phosphorous and detoxifying heavy metals in the rhizosphere of the plant under stress conditions (Fig. 2.3).

2.3.3 Sequestration and Transformation of Toxic Heavy Metals

Due to activity and high surface area to volume ratio, soil microbes provide large surface area for binding of metals and act as metal chelators (Jing et al. 2007; He et al. 2012; Thakare et al. 2021). Soil microorganisms have the capability to accumulate heavy metals in biomass by intracellular sequestration or precipitation

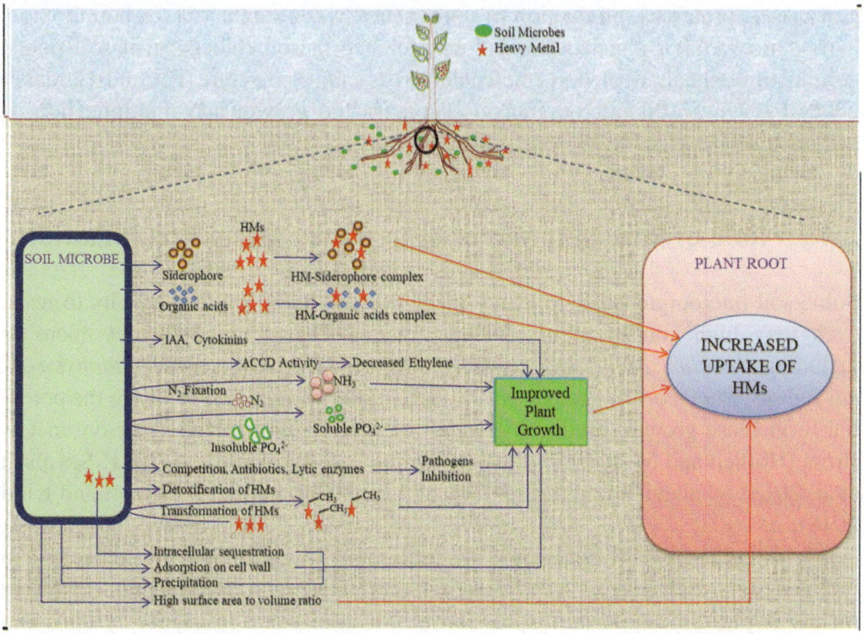

Fig. 2.3 Mechanisms of soil-microbial induced heavy metal accumulation in plant

or through metal adsorption onto cell walls (Gadd 2004). Soil microbes can transform toxic heavy metals that are more stable, less mobile or inert in innocuous or less toxic heavy metals. The process of heavy metal transformation includes several types of chemical reactions such as oxidation, reduction, methylation, and demethylation. Metal transformation is sometimes by-products of normal metabolism and confers no known advantage on the organism responsible (Silver and Misra 1984). Redox reactions are the main reaction involves in chemical transformation of harmful heavy metals (Tandon and Singh 2016) and plays a vital role in the detoxification of harmful heavy metals, especially As, Cr, Hg, and Se (Gadd 2010; Rajapaksha et al. 2013). This indicates that soil microorganisms can work on the bimodal way, i.e. immobilizing the metal in their biomass by sequestration on one hand and on the other transformation to less toxic forms, increasing solubility in the soil leading to higher bioavailability to the plants (Fig. 2.3).

2.3.4 Inhibition of Plant Pathogens

Soil microbes can suppress plant pathogens by several different mechanisms such as competition for nutrients, space, oxygen, and resources, producing antibiotics, viz., gliotoxin, pyrrolnitrin, pyocyanine, 2,4-diacetyl phloroglucinol and inducing resistance of host plants (Pal and Gardener 2006; Hibbing et al. 2010; Karim et al. 2018). Other important mechanisms are production of lytic enzymes like chitinases, glucanases, proteases, production of unregulated waste (ammonia, carbon dioxide, hydrogen cyanide), physicochemical interference in soil (blockage of soil pores, germination signals depletion, molecular cross-talk confusion) (Pal and Gardener 2006) by which soil microorganisms suppress the growth and multiplication of pathogen and induce the growth of host plants.

2.4 Role of Resistant Microbes in Heavy Metal Accumulation

Some soil microorganisms are heavy metal tolerant and have the capacity to resist even very high concentration of heavy metals. Heavy metal-tolerant microbes including *Bacillus*, *Pseudomonas*, *Streptomyces*, *Methylobacterium*, *Staphylococcus aureus*, *Pediococcus pentosaceus*, *Acidiphilium*, *Acidocella*, etc. have the potential to increase growth, productivity, and tolerance of host plants to heavy metals stress (Hoflich and Metz 1997; Castro-Silva et al. 2003; Ilias et al. 2011; Sessitsch et al. 2013) by alleviating metal toxicity and supplying nutrients (Benizri and Kidd 2018).

Metal resistance has been known as a requirement for plant-associated bacteria in polluted soils (Salt et al. 1999). It also affects uptake and accumulation of heavy metals in plants because the expression of bacterial metal resistance systems can change the bioavailability of heavy metals (van der Lelie et al. 1999). It has been suggested that most metal resistant soil microbes having plant growth-promoting characters could increase the bioavailability of heavy metals by solubilization or mobilization and can be successfully utilized for phytoextraction of heavy metals (Mishra et al. 2017).

Heavy metal-tolerant microbes can be isolated from contaminated soils, sewage sludge, and mining waste. These bacteria have high heavy metal tolerance to several metals such as Zn, Cu, Ni, and Co. Cd-resistant bacterial genus *Variovorax paradoxus*, *Rhodococcus* sp., *Flavobacterium* sp. *Bacillus subtilis, B. pumilus, Rhizobium, Ochrobactrum* sp. found to stimulate root elongation in presence of Cd, Co, Pb, and As by producing IAA and siderophores in plants (Belimov et al. 2005; Pandey et al. 2013; Yu et al. 2014). Cu-resistant bacteria *Pseudomonas putida* increases to the uptake and translocation of Cu in *Elsholtzia splendens* (Xu et al. 2015). It has been also reported that some rhizobacteria take part in the metal accumulation of hyperaccumulator plants and increase the uptake and tolerance of heavy metals (Thijs et al. 2017). Application of Cd- or Pb-resistant fungi (*Paecilomyces lilacinus* and *Hypocrea virens*) improved the ability of *Solanum nigrum* to accumulate heavy metals and increase plant yield (Gao et al. 2012). Furthermore, it has been reported that simultaneous addition of thiosulfate with metal-tolerant microorganism increase uptake of heavy metals by stimulating bioavailability to plants (Franchi et al. 2017).

However, in some instances, it has been reported that metal resistant soil microbes improve the plant growth by immobilization of heavy metals and reduce the uptake and translocation of heavy metals in plants via precipitation, complex formation, and adsorption. Plant growth-promoting soil bacteria have been reported to stimulate growth, reduce the bioavailability, decrease the accumulation of heavy metals in plants by immobilization and metal resistant attribute (Yuan et al. 2017; Wang et al. 2018; Han et al. 2018; Mallick et al. 2018). Cd-tolerant bacteria such as *Pseudomonas aeruginosa* and *Burkholderia gladioli* reduce the uptake of Cd and mitigate Cd stress of *Lycopersicon esculentum* (Khanna et al. 2019).

2.5 Conclusion and Future Perspective

Phytoremediation can provide an inexpensive and eco-friendly way to remediate the heavy metal contaminated site. Association between plants and soil microbes increase the efficiency of uptake of heavy metals. Soil microorganisms assist plants to remove heavy metals from the contaminated soils either by degrading the contaminants in the rhizosphere or by increasing metal-accumulating capability of plants. Use of genetically engineered microorganisms in phytoremediation increase the plants heavy metal accumulation and widen the horizon of microbial use in the technique. The mechanism of soil microbes assisted phytoremediation include acceleration of metal mobility, immobilization, nutrient acquisition, metal detoxification, transformation and mitigation of heavy metal stresses in plants. Soil microbes involved in biogeochemical processes operating in the rhizosphere, affect the mobility and availability of heavy metal to the plant through the release of chelating agents, biosurfactants and biomethylation, metal speciation, acidification, dissolution, phosphate solubilization, and redox changes. In addition, plant-associated bacteria can also increase plant resistance to the pathogen and ensure nitrogen fixation and the production of growth regulators. The role of soil microorganisms in phytoremediation of heavy metals is not fully explored. Despite its proven value there are certain challenges that needs to be overcome. There are many areas of poor understanding where more research is required. These are:

- A better understanding of the soil microbes–plant interaction can foster systemic improvements in phytoremediation.
- Soil microbes induced rhizospheric processes and their effect on solubility and bioavailability of heavy metals is needed to be explored.
- Rhizospheric microflora associated with plant and change in their diversity and population on contaminated lands is yet to be elucidated.
- Identification and isolation of metal resistant microflora with plant growth-promoting traits and directly or indirectly involved in the degradation of contaminants.
- Use of genetically modified microorganisms to improve the phytoremediation efficiency is still in its infancy.
- Use of genetic engineering to enhance phytoremediation capacities by inserting transgenes.
- Heavy metal accumulation capacity of hyperaccumulators induced by soil microorganisms.

- Use of artificial chelators in combination with soil microbes to improve phytoremediation efficiency.

The development of phytoremediation strategy requires an interdisciplinary approach to understand the complicated interactions between plants, soils, and contaminant under particular conditions. In order to achieve effective phytoremediation and to make the leap from lab to field it is necessary to build a greater understanding of the many and diverse process that are involved.

Acknowledgments This chapter is dedicated to our Ph.D. supervisor Professor Naresh Chander Aery, the multifacet personality and pioneer in India in the field of Geobotany and Biogeochemistry. Prof. N. C. Aery has studied various aspects of environment, viz., Geobotany and Biogeochemistry of Mineral Deposits, Phytoremediation, Phytostabilization, Biomonitoring, Ecotoxicology, and Biodiversity. He has studied the mineral deposits of Pb, Zn, Cd, W, and U. His work on Khetri Copper Deposits, Rajasthan, India has resulted in the development of concept of "Absorption Barriers in Plants." It is our salute to his stature.

References

Aery NC (2016) Phytoremediation of contaminated lands. Athena Academic, London

Afzal M, Yousaf S, Reichenauer TG et al (2011) Soil type affects plant colonization, activity and catabolic gene expression of inoculated bacterial strains during phytoremediation of diesel. J Hazard Mater 186(2–3):1568–1575. https://doi.org/10.1016/j.jhazmat.2010.12.040

Ahemad M (2015a) Enhancing phytoremediation of chromium-stressed soils through plant-growth-promoting bacteria. J Genet Eng Biotechnol 13(1):51–58. https://doi.org/10.1016/j.jgeb.2015.02.001

Ahemad M (2015b) Phosphate-solubilizing bacteria-assisted phytoremediation of metalliferous soils: a review. 3 Biotech 5(2):111–121. https://doi.org/10.1007/s13205-014-0206-0

Ahemad M, Khan MS (2012) Evaluation of plant-growth promoting activities of rhizobacterium *Pseudomonas putida* under herbicide stress. Ann Microbiol 62:1531–1540. https://doi.org/10.1007/s13213-011-0407-2

Ahemad M, Kibret M (2014) Mechanisms and applications of plant growth promoting rhizobacteria: current perspective. J King Saud Uni Sci 26(1):1–20. https://doi.org/10.1016/j.jksus.2013.05.001

Ali A, Bilal S, Khan AL (2019) Endophytic *Aureobasidium pullulans* BSS6 assisted developments in phytoremediation potentials of *Cucumis sativus* under Cd and Pb stress. J Plant Interact 14(1):303–313. https://doi.org/10.1080/17429145.2019.1633428

Ambawade MS, Pathade GR (2015) Production of indole acetic acid (IAA) by *Stenotrophomonas maltophilia* BE25 isolated from roots of Banana (*Musa* spp.). Int J Sci Res 4:2644–2650

Bacon CW, Hinton DM (2007) Bacterial endophytes: the endophytic niche, its occupants, and its utility. In: Plant-associated bacteria. Springer, Dordrecht, pp 155–194. https://doi.org/10.1007/978-1-4020-4538-7_5

Basu S, Rabara RC, Negi S, Shukla P (2018) Engineering PGPMOs through gene editing and systems biology: a solution for phytoremediation? Trends Biotechnol 36(5):499–510. https://doi.org/10.1016/j.tibtech.2018.01.011

Belimov AA, Hontzeas N, Safronova VI (2005) Cadmium-tolerant plant growth-promoting bacteria associated with the roots of Indian mustard (*Brassica juncea* L. Czern.). Soil Biol Biochem 37(2):241–250. https://doi.org/10.1016/j.soilbio.2004.07.033

Benizri E, Kidd PS (2018) The role of the rhizosphere and microbes associated with hyperaccumulator plants in metal accumulation. In: Agromining: farming for metals. Springer, Cham, pp 157–188. https://doi.org/10.1007/978-3-319-61899-9_9

Bhattacharyya P, Jha D (2012) Plant growth-promoting rhizobacteria (PGPR): emergence in agriculture. World J Microbiol Biotechnol 28:1327–1350. https://doi.org/10.1007/s11274-011-0979-9

Bilal S, Shahzad R, Khan AL et al (2018) Endophytic microbial consortia of phytohormones-producing fungus *Paecilomyces formosus* LHL10 and bacteria Sphingomonas sp. LK11 to Glycine max L. regulates physio-hormonal changes to attenuate aluminum and zinc stresses. Front Plant Sci 9:1273. https://doi.org/10.3389/fpls.2018.01273

Bonfante P, Anca IA (2009) Plants, mycorrhizal fungi, and bacteria: a network of interactions. Annu Rev Microbiol 63:363–383. https://doi.org/10.1146/annurev.micro.091208.073504

Burges A, Epelde L, Blanco F (2017) Ecosystem services and plant physiological status during endophyte-assisted phytoremediation of metal contaminated soil. Sci Total Environ 584:329–338. https://doi.org/10.1016/j.scitotenv.2016.12.146

Castro-Silva MA, Souza Lima AO, Gerchenski AV et al (2003) Heavy metal resistance of microorganisms isolated from coal mining environments of Santa Catarina. Braz J Microbiol 1:45–47. https://doi.org/10.1590/S1517-83822003000500015

Chandra P, Singh E (2016) Applications and mechanisms of plant growth-stimulating rhizobacteria. In: Plant-microbe interaction: an approach to sustainable agriculture. Springer, Singapore, pp 37–62. https://doi.org/10.1007/978-981-10-2854-0_3

Chen X, Liu X, Zhang X et al (2017) Phytoremediation effect of *Scirpus triqueter* inoculated plant-growth-promoting bacteria (PGPB) on different fractions of pyrene and Ni in co-contaminated soils. J Hazard Mater 325:319–326. https://doi.org/10.1016/j.jhazmat.2016.12.009

Cruz AF, Ishii T (2012) Arbuscular mycorrhizal fungal spores host bacteria that affect nutrient biodynamics and biocontrol of soil-borne plant pathogens. Biol Open 1(1):52–57. https://doi.org/10.1242/bio.2011014

Das A, Prasad R, Srivastava A, Giang PH, Bhatnagar K, Varma A (2007) Fungal siderophores: structure, functions and regulations. In: Varma A, Chincholkar SB (eds) Microbial siderophores, vol 12. Springer, Berlin, pp 1–42

Dhawi F (2016) Mycorrhiza, bacteria and plant an organized model of rhizoshere interaction. Int J Sci Eng Res 7:61–77

Dimkpa CO, Merten D, Svatoš A et al (2009) Siderophores mediate reduced and increased uptake of cadmium by *Streptomyces tendae* F4 and sunflower (*Helianthus annuus*), respectively. J Appl Microbiol 107(5):1687–1696. https://doi.org/10.1111/j.1365-2672.2009.04355.x

Dubois M, Van den Broeck L, Inzé D (2018) The pivotal role of ethylene in plant growth. Trends Plant Sci 23(4):311–323. https://doi.org/10.1016/j.tplants.2018.01.003

Elhindi KM, Al-Mana FA, El-Hendawy S et al (2018) Arbuscular mycorrhizal fungi mitigates heavy metal toxicity adverse effects in sewage water contaminated soil on *Tagetes erecta* L. Soil Sci Plant Nutr 64(5):662–668. https://doi.org/10.1080/00380768.2018.1490631

Ferrol N, Azcón-Aguilar C, Bago B (2004) Genomics of arbuscular mycorrhizal fungi. In: Applied mycology and biotechnology: fungal genomics. Elsevier, Amsterdam, pp 379–403

Franchi E, Rolli E, Marasco R et al (2017) Phytoremediation of a multi contaminated soil: mercury and arsenic phytoextraction assisted by mobilizing agent and plant growth promoting bacteria. J Soils Sediments 17(5):1224–1236. https://doi.org/10.1007/s11368-015-1346-5

Frey-Klett P, Garbaye JA, Tarkka M (2007) The mycorrhiza helper bacteria revisited. New Phytol 176(1):22–36. https://doi.org/10.1111/j.1469-8137.2007.02191.x

Gadd GM (2004) Microbial influence on metal mobility and application for bioremediation. Geoderma 122(2–4):109–119. https://doi.org/10.1016/j.geoderma.2004.01.002

Gadd GM (2010) Metals, minerals and microbes: Geomicrobiology and bioremediation. Microbiology 156:609–643. https://doi.org/10.1099/mic.0.037143-0

Gamalero E, Glick BR (2012) Plant growth-promoting bacteria and metals phytoremediation. In: Phytotechnologies: remediation of environmental contaminants. CRC Press, Boca Raton, pp 361–376. https://doi.org/10.1201/b12954-24

Gao Y, Miao C, Wang Y et al (2012) Metal-resistant microorganisms and metal chelators synergistically enhance the phytoremediation efficiency of *Solanum nigrum* L. in Cd-and Pb-contaminated soil. Environ Technol 33(12):1383–1389. https://doi.org/10.1080/09593330.2011.629006

Gaonkar T, Bhosle S (2013) Effect of metals on a siderophore producing bacterial isolate and its implications on microbial assisted bioremediation of metal contaminated soils. Chemosphere 93(9):1835–1843. https://doi.org/10.1016/j.chemosphere.2013.06.036

Garbaye J (1994) Mycorrhization helper bacteria: a new dimension to the mycorrhizal symbiosis. Acta Bot Gallica 141(4):517–521. https://doi.org/10.1111/j.1469-8137.1994.tb04003.x

Garbaye J, Churin JL, Duponnois R (1992) Effects of substrate sterilization, fungicide treatment, and mycorrhization helper bacteria on ectomycorrhizal formation of pedunculate oak (*Quercus robur*) inoculated with *Laccaria laccata* in two peat bare-root nurseries. Biol Fertil Soils 13(1):55–57. https://doi.org/10.1007/BF00337239

Giri B, Giang PH, Kumari R, Prasad R, Sachdev M, Garg AP, Oelmuller R, Varma A (2005) Mycorrhizosphere: strategies and functions. In: Buscot F, Varma A (eds) Microorganisms in soils: roles in genesis and functions, vol 3. Springer, Berlin, pp 213–252

Glick BR (2004) Bacterial ACC deaminase and the alleviation of plant stress. Adv Appl Microbiol 56:291–312. https://doi.org/10.1016/S0065-2164(04)56009-4

Göhre V, Paszkowski U (2006) Contribution of the arbuscular mycorrhizal symbiosis to heavy metal phytoremediation. Planta 223(6):1115–1122. https://doi.org/10.1007/s00425-006-0225-0

Grobelak A, Hiller J (2017) Bacterial siderophores promote plant growth: screening of catechol and hydroxamate siderophores. Int J Phytoremediation 19(9):825–833. https://doi.org/10.1080/15226514.2017.1290581

Grobelak A, Kokot P, Hutchison D et al (2018) Plant growth-promoting rhizobacteria as an alternative to mineral fertilizers in assisted bioremediation—sustainable land and waste management. J Environ Manag 227:1–9. https://doi.org/10.1016/j.jenvman.2018.08.075

Hamilton CE, Gundel PE, Helander M et al (2012) Endophytic mediation of reactive oxygen species and antioxidant activity in plants: a review. Fungal Divers 54(1). https://doi.org/10.1007/s13225-012-0158-9

Han H, Sheng X, Hu J et al (2018) Metal-immobilizing *Serratia liquefaciens* CL-1 and *Bacillus thuringiensis* X30 increase biomass and reduce heavy metal accumulation of radish under field conditions. Ecotoxicol Environ Saf 161:526–533. https://doi.org/10.1016/j.ecoenv.2018.06.033

Harish, Sundaramoorthy S (2008) Phytoremediation: a novel strategy to heal pollution problem in a natural way. In: Advances in applied biotechnology. Agrobios, Jodhpur, pp 303–326

He S, He Z, Yang X, Baligar VC (2012) Mechanisms of nickel uptake and hyperaccumulation by plants and implications for soil remediation. Adv Agron 117:117–189. https://doi.org/10.1016/B978-0-12-394278-4.00003-9

Hibbing ME, Fuqua C, Parsek MR, Peterson SB (2010) Bacterial competition: surviving and thriving in the microbial jungle. Nat Rev Microbiol 8(1):15. https://doi.org/10.1038/nrmicro2259

Hoflich G, Metz R (1997) Interaction of plant microorganism-association in heavy metal containing soils from sewage farms. Bodenkultur 48:238–247

Ike A, Sriprang R, Ono H et al (2007) Bioremediation of cadmium contaminated soil using symbiosis between leguminous plant and recombinant rhizobia with the MTL4 and the PCS genes. Chemosphere 66(9):1670–1676. https://doi.org/10.1016/j.chemosphere.2006.07.058

Ilias M, Rafiqullah IM, Debnath BC et al (2011) Isolation and characterization of chromium (VI)-reducing bacteria from tannery effluents. Indian J Microbiol 51(1):76–81. https://doi.org/10.1007/s12088-011-0095-4

Illmer P, Buttinger R (2006) Interactions between iron availability, aluminium toxicity and fungal siderophores. Biometals 19(4):367–377. https://doi.org/10.1007/s10534-005-3496-1

Jargeat P, Cosseau C, Ola'h B et al (2004) Isolation, free-living capacities, and genome structure of "*Candidatus* Glomeribacter gigasporarum," the endocellular bacterium of the mycorrhizal fungus *Gigaspora margarita*. J Bacteriol 186(20):6876–6884. https://doi.org/10.1128/JB.186.20.6876-6884.2004

Jing Y, He Z, Yang XJ (2007) Role of soil rhizobacteria in phytoremediation of heavy metal contaminated soils. Zhejiang Univ Sci B 8:192. https://doi.org/10.1631/jzus.2007.B0192

Kamal S, Prasad R, Varma A (2010) Soil microbial diversity in relation to heavy metals. In: Sherameti I, Varma A (eds) Soil heavy metals, vol 19. Springer, Berlin, pp 31–64

Kanwal S, Bano A, Malik RN (2016) Role of arbuscular mycorrhizal fungi in phytoremediation of heavy metals and effects on growth and biochemical activities of wheat (*Triticum aestivum* L.) plants in Zn contaminated soils. Afr J Biotechnol 15(20):872–883. https://doi.org/10.5897/AJB2016.15292

Karim H, Hamka L, Kurnia N, Junda M (2018) Effectivity of anatagonistic bacteria in controlling of *Fusarium* wilt diseases of banana (*Musa paradisiaca*) by *in vitro*. J Phys Conf Ser 1028(1):012014. https://doi.org/10.1088/1742-6596/1028/1/012014

Khan AR, Ullah I, Waqas M (2017) Host plant growth promotion and cadmium detoxification in *Solanum nigrum*, mediated by endophytic fungi. Ecotoxicol Environ Safe 136:18. https://doi.org/10.1016/j.ecoenv.2016.03.014

Khanna K, Jamwal VL, Gandhi SG et al (2019) Metal resistant PGPR lowered Cd uptake and expression of metal transporter genes with improved growth and photosynthetic pigments in *Lycopersicon esculentum* under metal toxicity. Sci Rep 9(1):5855. https://doi.org/10.1038/s41598-019-41899-3

Kobayashi DY, Palumbo JD (2000) Bacterial endophytes and their effects on plants and uses in agriculture. In: Microbial endophytes. CRC Press, Boca Raton, pp 213–250

Koptsik GN (2014) Problems and prospects concerning the phytoremediation of heavy metal polluted soils: a review. Eurasian Soil Sci 47(9):923–939. https://doi.org/10.1134/S1064229314090075

Kumar A (2020) Inorganic soil contaminants and their biological remediation. In: Singh P, Singh SK, Prasad SM (eds) Plant responses to soil pollution. Springer, Singapore. https://doi.org/10.1007/978-981-15-4964-9_8

Kumar A, Aery NC (2010) Studies on the effect of tungsten on seed germination and early seedling growth of cowpea. In: Masih et al (eds) Natural resource management for sustainable agriculture. Aavishkar Publishers, Jaipur, pp 274–280

Kumar A, Aery NC (2011) Effect of tungsten on growth, biochemical constituents, molybdenum and tungsten contents in wheat. Plant Soil Environ 57(11):519–525. https://doi.org/10.17221/345/2011-PSE

Kumar A, Aery NC (2012a) Effect of tungsten on the growth, dry-matter production, and biochemical constituents of cowpea. Commun Soil Sci Plant Anal 43(7):1098–1107. https://doi.org/10.1080/00103624.2012.656171

Kumar A, Aery NC (2012b) Influence of tungsten and molybdenum on seed germination and early seedling growth of wheat—a comparative study. NBU J Plant Sci 6(1):1–5

Kumar A, Aery NC (2016) Impact, metabolism, and toxicity of heavy metals in plants. In: Singh A, Prasad S, Singh R (eds) Plant responses to xenobiotics. Springer, Singapore, pp 141–176. https://doi.org/10.1007/978-981-10-2860-1_7

Kumar V, Prasad R, Kumar M (2021) Rhizobiont in bioremediation of hazardous waste. Springer, Singapore. https://www.springer.com/gp/book/9789811606014

Lecomte J, St-Arnaud M, Hijri M (2011) Isolation and identification of soil bacteria growing at the expense of arbuscular mycorrhizal fungi. FEMS Microbiol Lett 317(1):43–51. https://doi.org/10.1111/j.1574-6968.2011.02209.x

Leong J (1986) Siderophores: their biochemistry and possible role in the biocontrol of plant pathogens. Annu Rev Phytopathol 24(1):187–209. https://doi.org/10.1146/annurev.py.24.090186.001155

Li HY, Wei DQ, Shen M et al (2012) Endophytes and their role in phytoremediation. Fungal Divers 54(1):11–18. https://doi.org/10.1007/s13225-012-0165-x

Lin M, Jin M, Xu K et al (2018) Phosphate-solubilizing bacteria improve the phytoremediation efficiency of *Wedelia trilobata* for Cu-contaminated soil. Int J Phytoremediation 20(8):813–822. https://doi.org/10.1080/15226514.2018.1438351

Lin H, Liu C, Li B et al (2021) *Trifolium repens* L. regulated phytoremediation of heavy metal contaminated soil by promoting soil enzyme activities and beneficial rhizosphere associated microorganisms. J Hazard Mater 402:123829. https://doi.org/10.1016/j.jhazmat.2020.123829

Liu C, Lin H, Dong Y et al (2019) Identification and characterization of plant growth–promoting endophyte RE02 from *Trifolium repens* L. in mining smelter. Environ Sci Pollut R 1:1–2. https://doi.org/10.1007/s11356-019-04904-w

Ma Y, Rajkumar M, Zhang C, Freitas H (2016) Inoculation of *Brassica oxyrrhina* with plant growth promoting bacteria for the improvement of heavy metal phytoremediation under drought conditions. J Hazard Mater 320:36–44. https://doi.org/10.1016/j.jhazmat.2016.08.009

Ma Y, Rajkumar M, Oliveira RS (2019) Potential of plant beneficial bacteria and arbuscular mycorrhizal fungi in phytoremediation of metal-contaminated saline soils. J Hazard Mater 120813. https://doi.org/10.1016/j.jhazmat.2019.120813

Mallick I, Bhattacharyya C, Mukherji S et al (2018) Effective rhizoinoculation and biofilm formation by arsenic immobilizing halophilic plant growth promoting bacteria (PGPB) isolated from mangrove rhizosphere: a step towards arsenic rhizoremediation. Sci Total Environ 610: 1239–1250. https://doi.org/10.1016/j.scitotenv.2017.07.234

Mesa-Marín J, Del-Saz NF, Rodríguez-Llorente ID et al (2018) PGPR reduce root respiration and oxidative stress enhancing *Spartina maritima* root growth and heavy metal rhizoaccumulation. Front Plant Sci 9:1500. https://doi.org/10.3389/fpls.2018.01500

Mishra J, Singh R, Arora NK (2017) Alleviation of heavy metal stress in plants and remediation of soil by rhizosphere microorganisms. Front Microbiol 8:1706. https://doi.org/10.3389/fmicb.2017.01706

Mitter B, Petric A, Shin MW (2013) Comparative genome analysis of *Burkholderia phytofirmans* PsJN reveals a wide spectrum of endophytic lifestyles based on interaction strategies with host plants. Front Plant Sci 4:120. https://doi.org/10.3389/fpls.2013.00120

Navarro-Torre S, Mateos-Naranjo E, Caviedes MA et al (2016) Isolation of plant-growth-promoting and metal-resistant cultivable bacteria from *Arthrocnemum macrostachyum* in the Odiel marshes with potential use in phytoremediation. Mar Pollut Bull 110(1):133–342. https://doi.org/10.1016/j.marpolbul.2016.06.070

Oswald A (2010) Evaluating soil rhizobacteria for their ability to enhance plant growth and tuber yield in potato. Ann Appl Biol 157:259–271. https://doi.org/10.1111/j.1744-7348.2010.00421.x

Pal KK, Gardener BM (2006) Biological control of plant pathogens. Plant Health Instr 1–25. https://doi.org/10.1094/PHI-A-2006-1117-02

Pandey S, Ghosh PK, Ghosh S et al (2013) Role of heavy metal resistant *Ochrobactrum* sp. and *Bacillus* spp. strains in bioremediation of a rice cultivar and their PGPR like activities. J Microbiol 51(1):11–17. https://doi.org/10.1007/s12275-013-2330-7

Panhwar QA, Jusop S, Naher UA et al (2013) Application of potential phosphate-solubilizing bacteria and organic acids on phosphate solubilization from phosphate rock in aerobic rice. Sci World J 272409. https://doi.org/10.1155/2013/272409

Peng A, Liu J, Gao Y, Chen Z (2013) Distribution of endophytic bacteria in *Alopecurus aequalis* Sobol and *Oxalis corniculata* L. from soils contaminated by polycyclic aromatic hydrocarbons. PLoS One 8(12):e83054. https://doi.org/10.1371/journal.pone.0083054

Pilon-Smits E (2005) Phytoremediation. Annu Rev Plant Biol 56:15–39. https://doi.org/10.1146/annurev.arplant.56.032604.144214

Prasad R (2021) Environmental pollution and remediation. Springer. (ISBN: 978-981-15-5499-5) https://www.springer.com/gp/book/9789811554988

Prasad R, Kumar M, Varma A (2015) Role of PGPR in soil fertility and plant health. In: Egamberdieva D, Shrivastava S, Varma A (eds) Plant growth-promoting Rhizobacteria (PGPR) and medicinal plants. Springer International Publishing, Switzerland, pp 247–260

Prasad R, Bhola D, Akdi K, Cruz C, Sairam KVSS, Tuteja N, Varma A (2017) Introduction to mycorrhiza: historical development. In: Varma A, Prasad R, Tuteja N (eds) Mycorrhiza. Springer, pp 1–7

Prasad R, Nayak SC, Kharwar RN, Dubey NK (2021) Mycoremediation and environmental sustainability, vol 3. Springer. (ISBN: 978-3-030-54421-8) https://www.springer.com/gp/book/9783030544218

Raikwar MK, Kumar P, Singh M, Singh A (2008) Toxic effect of heavy metals in livestock health. Vet World 1(1):28

Rajapaksha AU, Vithanage M, Ok YS, Oze C (2013) Cr (VI) formation related to Cr (III)-muscovite and birnessite interactions in ultramafic environments. Environ Sci Technol 47:9722–9729. https://doi.org/10.1021/es4015025

Rajkumar M, Sandhya S, Prasad MNV, Freitas H (2012) Perspectives of plant-associated microbes in heavy metal phytoremediation. Biotechnol Adv 30:1562–1574. https://doi.org/10.1016/j.biotechadv.2012.04.011

Rashmi V, ShylajaNaciyar M, Rajalakshmi R et al (2013) Siderophore mediated uranium sequestration by marine cyanobacterium *Synechococcus elongatus* BDU 130911. Bioresour Technol 130:204–210. https://doi.org/10.1016/j.biortech.2012.12.016

Reinhold-Hurek B, Maes T, Gemmer S et al (2006) An endoglucanase is involved in infection of rice roots by the not-cellulose-metabolizing endophyte *Azoarcus* sp. strain BH72. Mol Plant-Microbe Interact 19(2):181–188. https://doi.org/10.1094/MPMI-19-0181

Rosenblueth M, Martínez-Romero E (2006) Bacterial endophytes and their interactions with hosts. Mol Plant-Microbe Interact 19(8):827–837. https://doi.org/10.1094/MPMI-19-0827

Salt DE, Smith RD, Raskin I (1998) Phytoremediation. Annu Rev Plant Biol 49(1):643–668. https://doi.org/10.1146/annurev.arplant.49.1.643

Salt DE, Benhamou N, Leszczyniecka M et al (1999) A possible role for rhizobacteria in water treatment by plant roots. Int J Phytoremediation 1(1):67–79. https://doi.org/10.1080/15226519908500005

Sarma H, Forid N, Prasad R, Prasad MNV, Ma LQ, Rinklebe J (2021) Enhancing phytoremediation of hazardous metal(loid)s using genome engineering CRISPR–Cas9 technology. J Hazard Mater. https://doi.org/10.1016/j.jhazmat.2021.125493

Scheublin TR, Sanders IR, Keel C, Van Der Meer JR (2010) Characterisation of microbial communities colonising the hyphal surfaces of arbuscular mycorrhizal fungi. ISME J 4(6):752. https://doi.org/10.1038/ismej.2010.5

Seghers D, Wittebolle L, Top EM et al (2004) Impact of agricultural practices on the *Zea mays* L. endophytic community. Appl Environ Microbiol 70(3):1475–1482. https://doi.org/10.1128/AEM.70.3.1475-1482.2004

Sessitsch A, Kuffner M, Kidd P et al (2013) The role of plant-associated bacteria in the mobilization and phytoextraction of trace elements in contaminated soils. Soil Biol Biochem 60:182–194. https://doi.org/10.1016/j.soilbio.2013.01.012

Seth K, Harish (2016) Current status of potential applications of repurposed Cas9 for structural and functional genomics of plants. Biochem Biophys Res Commun 480(4):499–507. https://doi.org/10.1016/j.bbrc.2016.10.130

Sharma A, Johri BN (2003) Growth promoting influence of siderophore-producing *Pseudomonas* strains GRP3A and PRS9 in maize (*Zea mays* L.) under iron limiting conditions. Microbiol Res 158(3):243–248. https://doi.org/10.1078/0944-5013-00197

Sharma VK, Li XY, Wu GL et al (2019) Endophytic community of Pb-Zn hyperaccumulator *Arabis alpina* and its role in host plants metal tolerance. Plant Soil 437(1–2):397–411. https://doi.org/10.1007/s11104-019-03988-0

Sharma P, Pandey AK, Udayan A, Kumar S (2021) Role of microbial community and metal-binding proteins in phytoremediation of heavy metals from industrial wastewater. Bioresour Technol 22:124750. https://doi.org/10.1016/j.biortech.2021.124750

Silver S, Misra TK (1984) Bacterial transformations of and resistances to heavy metals. In: Genetic control of environmental pollutants. Springer, Boston, MA, pp 23–46. https://doi.org/10.1007/978-1-4684-4715-6_3

Sriprang R, Hayashi M, Ono H (2003) Enhanced accumulation of Cd^{2+} by a *Mesorhizobium* sp. transformed with a gene from *Arabidopsis thaliana* coding for phytochelatin synthase. Appl Environ Microbiol 69(3):1791–1796. https://doi.org/10.1128/AEM.69.3.1791-1796.2003

Tandon PK, Singh SB (2016) Redox processes in water remediation. Environ Chem Lett 14:15–25. https://doi.org/10.1007/s10311-015-0540-4

Tchounwou PB, Yedjou CG, Patlolla AK, Sutton DJ (2012) Heavy metal toxicity and the environment. In: Molecular, clinical and environmental toxicology. Springer, Basel, pp 133–164. https://doi.org/10.1007/978-3-7643-8340-4_6

Thakare M, Sarma H, Datar S, Roy A, Pawar P, Gupta K, Pandit S, Prasad R (2021) Understanding the holistic approach to plant-microbe remediation technologies for removing heavy metals and radionuclides from soil. Curr Res Biotechnol. https://doi.org/10.1016/j.crbiot.2021.02.004

Thijs S, Sillen W, Weyens N, Vangronsveld J (2017) Phytoremediation: state-of-the-art and a key role for the plant microbiome in future trends and research prospects. Int J Phytoremediation 19:23–38. https://doi.org/10.1080/15226514.2016.1216076

Tong J, Miaowen C, Juhui J (2017) Endophytic fungi and soil microbial community characteristics over different years of phytoremediation in a copper tailings dam of Shanxi, China. Sci Total Environ 574:881–888. https://doi.org/10.1016/j.scitotenv.2016.09.161

van der Lelie D, Corbisier P, Diels L et al (1999) The role of bacteria in the phytoremediation of heavy metals. In: Phytoremediation of contaminated soil and water. Lewis Publishers, London, pp 265–281. https://doi.org/10.1201/9781439822654.ch15

Wang Q, Zhang WJ, He LY, Sheng XF (2018) Increased biomass and quality and reduced heavy metal accumulation of edible tissues of vegetables in the presence of Cd-tolerant and immobilizing *Bacillus megaterium* H3. Ecotoxicol Environ Saf 148:269–274. https://doi.org/10.1016/j.ecoenv.2017.10.036

Xu C, Chen X, Duan D et al (2015) Effect of heavy-metal-resistant bacteria on enhanced metal uptake and translocation of the Cu-tolerant plant, *Elsholtzia splendens*. Environ Sci Pollut Res 22(7):5070–5081. https://doi.org/10.1007/s11356-014-3931-3

Yang Y, Liang Y, Han X et al (2016) The roles of arbuscular mycorrhizal fungi (AMF) in phytoremediation and tree-herb interactions in Pb contaminated soil. Sci Rep 6:20469. https://doi.org/10.1038/srep20469

Yang P, Zhou XF, Wang LL et al (2018) Effect of phosphate-solubilizing bacteria on the mobility of insoluble cadmium and metabolic analysis. Int J Environ Res Public Health 15(7):1330. https://doi.org/10.3390/ijerph15071330

Yang Y, Liu Y, Li Z et al (2020) Significance of soil microbe in microbial-assisted phytoremediation: an effective way to enhance phytoremediation of contaminated soil. Int J Environ Sci Technol 17(4):2477–2484. https://doi.org/10.1007/s13762-020-02668-2

Yong X, Chen Y, Liu W (2014) Enhanced cadmium resistance and accumulation in *Pseudomonas putida* KT 2440 expressing the phytochelatin synthase gene of *Schizosaccharomyces pombe*. Lett Appl Microbiol 58(3):255–261. https://doi.org/10.1111/lam.12185

Yu X, Li Y, Zhang C et al (2014) Culturable heavy metal-resistant and plant growth promoting bacteria in V-Ti magnetite mine tailing soil from Panzhihua, China. PLoS One 9(9):e106618. https://doi.org/10.1371/journal.pone.0106618

Yuan Z, Yi H, Wang T et al (2017) Application of phosphate solubilizing bacteria in immobilization of Pb and Cd in soil. Environ Sci Pollut Res 24:21877–21884. https://doi.org/10.1007/s11356-017-9832-5

Phytoremediation of Heavy Metal Contaminated Soil and Water

3

Neha Dhingra, Ranju Sharma, and Ngangbam Sarat Singh

Abstract

Heavy metals among the other contaminants present in the environment pose a great threat. Natural activities as well as many human activities have contributed to alarming levels of heavy metals contamination in the environment. These contaminants migrate into non-contaminated areas by the process of leaching through the soil or by spreading through the sewage sludge. Several methodologies are being used in order to clean up the environment from these contaminants, but most of the methodologies are costly as well as do not give their best results. Various physical and chemical methodologies tend to generate sludge, thus increasing the costs. These physico-chemical technologies tend to render the land usage as they remove all the nutrients from the soil. Currently, phytoremediation is the most preferred technology for an effective as well as affordable solution which can be used to extract or remove the inactive metals and metal pollutants from contaminated soil and water. Phytoremediation is an eco-friendly as well as a cost-effective technology. In this chapter we would discuss about phytoremediation technology, including the heavy metal uptake mechanisms and various studies related to phytoremediation. In this chapter we also review the advantages of this technology used in order for reducing them, along with heavy metal uptake mechanisms in phytoremediation technology as well as various factors affecting these uptake mechanisms. Also plants capable of phytoremediation along with their capabilities to reduce the contaminants have also been discussed.

N. Dhingra · R. Sharma (✉)
Department of Zoology, University of Delhi, Delhi, India

N. S. Singh
Department of Zoology, Dr. SRK Government Arts College, Yanam, Puducherry, India

R. Prasad (ed.), *Phytoremediation for Environmental Sustainability*,
https://doi.org/10.1007/978-981-16-5621-7_3

Keywords

Bioremediation · Phytoremediation · Heavy metal · Toxic compounds · Eco-friendly

3.1 Introduction

Environmental pollution is a serious health concern because it acts as a major source of health risk and leads to several diseases worldwide (Briggs 2003). One of the major contaminants bothering the environment these days. Although the ill effects of heavy metals have been known, but due to increased and continuous exposure to heavy metals can even lead to death (Jarup 2003; Ahmed et al. 2017). Increasing industrial development has led to increasing heavy metal disposal causing environmental pollution (Suvaryan et al. 2011; Adesuyi et al. 2015; Jiao et al. 2015). Heavy metals are our primary concern as they cannot be remediated by degradation. Several methods have being used for removing the pollutants from the contaminated environments. Soils that are contaminated with heavy metals can be treated by acid leaching, soil washing, physical or mechanical separation of the contaminant, electro-chemical treatment, electrokinetics, chemical treatment, thermal or pyrometallurgical separation, and biochemical processes (Mulligan et al. 2001; Tangahu et al. 2011; Agnello et al. 2016; Behera and Prasad 2020a).

Remediation techniques such as treatment by activated carbon adsorption, usage of microbes, air stripping (Susarla et al. 2002; Atanes et al. 2019), chemical, biological, biochemical, and biosorptive treatment (Mulligan et al. 2001; Agnello et al. 2016) (Fig. 3.1) are being used in order to remove heavy metals from contaminated sites (Behera and Prasad 2020b).

Usage of some of these remediation methodologies requires a high cost (Raskin et al. 1997; Tangahu et al. 2011), takes long time (Susarla et al. 2002; Chen and Achal 2019), logistical problems (Vangronsveld et al. 2009; Garcia-Sanchez et al. 2018) and involves various technical complexity (Ali et al. 2013; Guo and Zhou 2020). Therefore, we need an alternative solution in order to remediate heavy metal contaminants from the environment. Bioremediation is an innovative and promising methodology available for removal and recovery of the heavy metals from the polluted water and lands (Dixit et al. 2015). Phytoremediation is one of the branches of bioremediation techniques that can be used as an alternative solution for heavy metal remediation (Ali et al. 2013; Etteieb et al. 2020). The objective of this review is to give the information about phytoremediation of heavy metals from the environment.

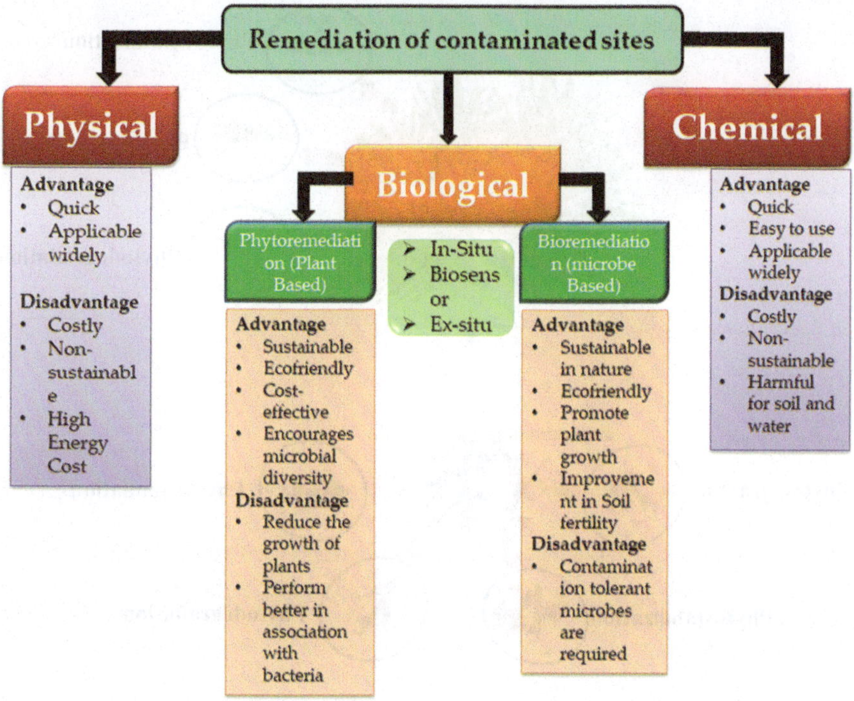

Fig. 3.1 Different types of remediation techniques

3.2 Phytoremediation and Its Mechanisms

The term of phytoremediation is relatively new term, it consists of the Greek prefix *phyto* which is means 'plant' and the Latin root *remedium* which is means 'to correct or remove evil. Basic information for phytoremediation comes from a variety of research areas including constructed wetlands, oil spills, and agricultural plant accumulation of heavy metals. Phytoremediation can be defined as an emerging technology using the desired plants to clean up the contaminated environment from hazardous contaminant so as to improve the quality of environment (U.S. Environmental Protection Agency 2000; Tangahu et al. 2011). Phytoremediation has received attention due to it being an innovative, cost-effective alternative to the more established chemical and physical treatment methods used at contaminated sites (Ali et al. 2013; Goncalves et al. 2017; Burakov et al. 2018). Phytoremediation is known as "green technology" owing to its advantages such as it being a cost-effective, efficient, environment- and eco-friendly technology (Sarma et al. 2021; Sonowal et al. 2022). There are various mechanisms that a plant undergoes for remediating heavy metal contaminants from the environment, such as phytoextraction, phytofiltration, phytostabilization, phytodegradation,

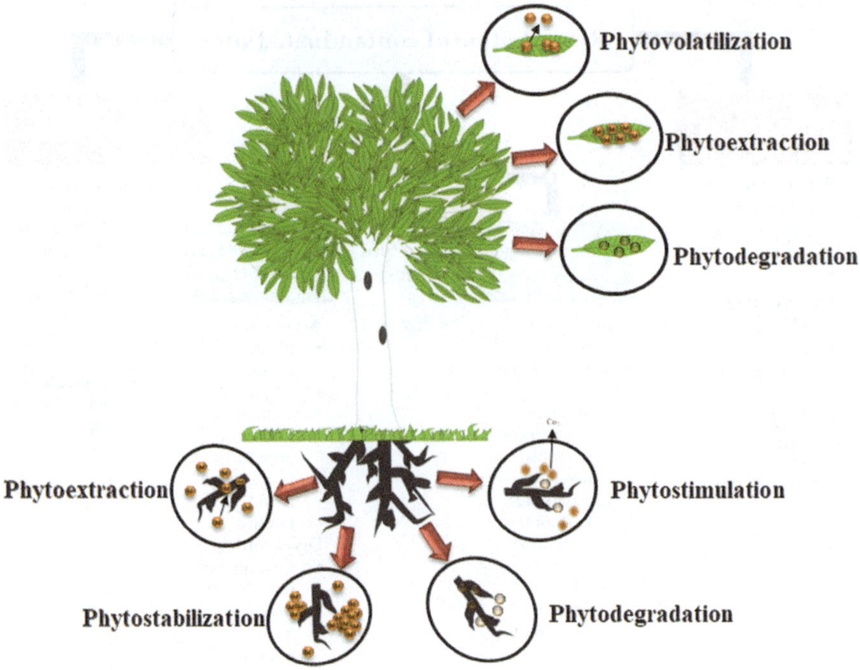

Fig. 3.2 Different types of phytoremediation mechanism

phytovolatilation, rhizodegradation (U.S. Environmental Protection Agency 2000; Ali et al. 2013; Dixit et al. 2015; Kulkarni et al. 2018; Thakare et al. 2021) (Fig. 3.2).

3.3 Different Strategies of Phytoremediation Mechanism

3.3.1 Rhizofiltration

It is defined as the technique that involves use of plants, both terrestrial as well as aquatic; in order to absorb, concentrate, and precipitate the contaminants from polluted aqueous sources having low contaminant concentration in their roots zone. Rhizofiltration can be used to partially treat the industrial discharge, agricultural runoff, or acid mine drainage. It can be useful in case of certain metals such as lead, cadmium, copper, nickel, zinc, and chromium, which are most primarily retained within the roots of the plants (U.S. Environmental Protection Agency 2000; Goland-Goldhirsh 2006; Kushwaha et al. 2018). The advantages of using rhizofiltration technique include it ability to be used both as in-situ or ex-situ application methods and plant species other than hyperaccumulators can also be used for this technique. Certain plants like sunflower, Indian mustard, tobacco, rye, spinach, and corn have been studied for their capability to remove metals such as lead from effluent, with sunflower showing the highest capability. Indian mustard has seen to be capable of effectively removing a wide concentration range (4–500 mg/l) of metal such as lead (Dasgupta et al. 2011). This technology has been tried in the field with contamination of uranium (U) in the water having the concentrations of 21–874 μg/l; the amount of treated U concentration was reported by Dushenkov was found to be <20 μg/l before their discharge into the environment (Goland-Goldhirsh 2006).

3.3.2 Phytostabilization

This technique is mostly preferred for the remediation of soil, sediments, and sludges (Ali et al. 2013) and it also depends on roots capability to contain the mobility of contaminants and their bioavailability in the soil. Phytostabilization can occur through the processes of sorption, precipitation, complex action, or metal valence reduction. The primary responsibility of the plants is to reduce the amount of water percolating through the soil matrix, which may tend to form hazardous leachates and thus prevent soil erosion and the distribution of toxic metal contaminants to other areas. Presence of dense root system can help stabilize the soil and thus prevents soil erosion (Vangronsveld et al. 2009; Reeves et al. 2018). It is a very effective technique whenever rapid immobilization of the contaminants is to be done in order to preserve the ground and surface water and this technique does not demand the disposal of biomass. However, this technique has a major disadvantage that is, the contaminant is present in soil as it is, and therefore this requires regular monitoring.

3.3.3 Phytoextraction

It is one of the best techniques that can be used to remove the contamination from soil and isolate it, without hampering the soil structure and its fertility. It is also called as phytoaccumulation (Ali et al. 2013). As the plant can absorb, concentrate as well as precipitate the toxic metal contaminants and radionuclide from contaminated soils into the plant biomass, as it is best suited technique for the remediating of dispersed polluted areas, where the contaminants occurred only at relatively low concentration and are present superficially (Henry 2000).

Several methods have been tested but the two basic strategies of phytoextraction that are being used are (1) Chelate assisted phytoextraction or the induced phytoextraction, where the artificial chelates are added to the soil in order to increase the mobility as well as uptake of metal contaminants. (2) Continuous phytoextraction where the removal of metal contaminates depends on the natural capability of the plant for remediation (Baker et al. 1999). Usage of hyperaccumulator plant species has further boosted this technique and has supported it for maximum benefits. For this technology to work effectively and a greater scale, the plants must be able to extract large concentrations of heavy metal contaminants into their roots, translocate these heavy metals to the surface biomass, and thus produces a large quantity of plant biomass. The heavy metal that has been removed from the contaminated soil can be recycled from the contaminated plant biomass (Burges et al. 2018; Reeves et al. 2018).

Physio-chemical factors also play an important role in the process of remediation, such as growth rate, element selectivity, resistance to disease, method of harvesting; tend to influence the process of remediation (Arslan et al. 2017; Reeves et al. 2018). There is certain limitation in this technique of remediation such as slow growth, shallow root system, small biomass production, final disposal of the contaminants causes a hindrance in the action of these hyperaccumulator species (Burges et al. 2018; Reeves et al. 2018).

3.3.4 Phytovolatilization

This technique involves the use of plants where they take up contaminants from the soil, then transforming them into volatile form and transpires them into the atmosphere. This technique occurs as the growing trees and various other plants take up water along with the organic and inorganic contaminants. Some of the contaminants are capable of passing through the plants into their leaves and then volatilise into the atmosphere at lower concentrations (Kanwar et al. 2020). This technique has mostly been used in the removal of mercury metal ions; the mercuric ion is thus transformed into less toxic elemental mercury. The disadvantage of this technique is that, mercury ions are released back into the atmosphere, thus they are likely to be recycled back by precipitation and then again redeposit back into the ecosystem (Mwegoha 2008; Kanwar et al. 2020).

3.3.5 Phytodegradation

Using the process of phytoremediation for organic contaminants, the metabolism of the plant contributes to the remediation of the contaminant by the process of transformation, break down, stabilization or volatilising the contaminant compounds from soil and groundwater resources. This process involves the breakdown of organic compounds, taken up by the plants to a simpler molecule, and then they are incorporated into the plant tissues (Berti and Cunningham 2000). Plants tend to have certain enzymes that can help in the breakdown as well as conversion of ammunition wastes, chlorinated solvents such as trichloroethylene and other herbicides. The enzymes present in the plant are usually dehalogenases, oxygenases, and reductases (Erakhrumen 2017). Rhizodegradation method basically involves the breakdown of organic contaminants present in the soil through the microbial activity present in the root zone (rhizosphere) and is a slower process as compared to phytodegradation. Yeast, fungi, bacteria, and other microorganisms consume and tend to digest various organic substances like fuels.

All phytoremediation techniques are not exclusive in nature and might be used simultaneously, but the process of metal extraction depends entirely on its bio available fraction in the soil.

3.4 Plant Response to Heavy Metals

Plants have implied basic three strategies for growing on metal contaminated soil (Erakhrumen 2017).

3.4.1 Metal Excluders

These plants prevent metal ions from entering their aerial parts (leaves) or by maintaining low along with constant metal concentration in soil that is they mainly restrict the presence of metal ions in their roots. The plants tend to alter its membrane permeability, in order to alter the metal binding capacity of cell walls or exude more chelating substances (Raskin and Ensley 2000).

3.4.2 Metal Indicators

The plant species are capable of actively accumulating the metal ions in their aerial tissues and generally reflect the concentration of metal in the soil. They are tolerating to the existing concentration level of metals by producing chelators (intracellular metal binding compounds) or can change the metal compartmentalisation pattern by storing the metal ions in non-sensitive parts of the plants (Kanwar et al. 2020).

3.4.3 Metal Accumulator Plant Species

These plants are capable of concentrating the metal ions in their aerial parts, to higher levels as compared to the soil. Hyperaccumulators are the plants capable of absorbing high levels of contaminants concentrated either in their roots, shoots, and/or leaves (Zhao et al. 2003; Zha et al. 2004). Baker and Brooks have discussed that these metal hyperaccumulator plants contain more than or up to 0.1%, i.e. more than (1000 mg/g) of copper, cadmium, chromium, lead, nickel cobalt or 1% (>10,000 mg/g) of zinc or manganese in their dry matter. The amount of metal concentration in case of cadmium and other rare metals, it is >0.01% by dry weight (United States Protection Agency Reports 2000). Scientists have studied these hyperaccumulator species by collecting these plants from the areas where soil has higher than usual concentration of metals, in the case of polluted areas or geographically rich in a particular element (Ghosh and Singh 2005). The *Brassicaceae* family is known to have a large number of hyperaccumulating plant species capable of accumulating with widest range of metals, these include 87 species from 11 genera (Lureysens et al. 2004).

3.5 Plants Heavy Metals Uptake and Responses

Various studies have discussed the potential of plant as heavy metals bioaccumulator as well as remediator from contaminated soil and water. The studies have suggested the use of phytoremediation technology as an alternative solution to remediate heavy metal contaminated areas.

Each plant has displays different responses to heavy metals stress. Various plants are sensitive towards it while some of the plants have a high tolerance limit to several heavy metal contaminations. As a result of plant–metal interaction (Sreelal and Jayanthi 2017; Galal et al. 2018), many plants tend to accumulate heavy metals from soil thus resulting in decreased growth and development. However, some of the plants show a high tolerance range as well as can maintain the growth and development under heavy metals stress (Hussain et al. 2018).

Plants show different responses to heavy metals exposure depends entirely on its level of tolerance to the heavy metal contamination. For examples, the process wilting, yellowing and growth inhibition of it was seen in Chives plants (*Allium schoenoprasum*) on exposing them to stress of Ni, Co and Cd at 0.25 mM concentrations (Mwegoha 2008). On chickpea (*Cicer arietinum*) plants, when exposed to increasing metal concentrations of Pb and Cr along with different time intervals, displayed an inhibition in the seed germination and decreased dry weight of plants. Cd stress of 20 μM concentration was observed to not affect the root dry weight, shoot height, shoot dry weight, leaf number and total chlorophyll concentration (a and b) of pea plant significantly (Mwegoha 2008). The dry weight of maize plant (*Zea mays*) was seen to decrease extremely on Zn-amended soil along with increasing concentration of Zn.

The plant selection to be used as accumulator is the most important factor affecting the rate of heavy metal removal in phytoremediation. There are certain parameters to be kept in mind in order to select remediating plants: plant biomass (Ma et al. 2016; Muthusaravanan et al. 2018), as the metal removal rate totally depends on the plant biomass harvested and presence of heavy metal concentration in the harvested biomass. In order to protect the ecosystem native species are preferred as compared to exotic species. Ecosystem protection, native species are preferred to exotic plants, as exotic species can be invasive in nature and disturbs the balance of the ecosystem.

Physical characteristics of soil contamination, in order to remediate the surface-contaminated soils, shallow rooted plant species should be used, whereas deep-rooted plants would be an apt choice for contamination at deeper levels.

3.5.1 Ex Situ Method

This method works on removing the contaminants from the soil for treatment. This technique requires the removal of contaminated soil for the treatment on or off site as well as returning the treated soil to the restored site. The conventional methods (Muthusaravanan et al. 2018) that are applied for remediating the contaminated soils relies mainly on excavation, detoxification, and/or destruction of contaminant physically or chemically, as a result of these methods the contaminant undergo stabilization, solidification, immobilization, incineration or destruction.

3.5.2 In Situ Method

It is remediation method without excavation of polluted site. Wang and Greger (2006) gave the definition of in situ remediation technologies as remediation of the contaminants, along with immobilization of the contaminants to reduce their bioavailability and separation of the contaminants from the soil (Wang and Greger 2006). In situ methods are preferred over the ex situ methods due to its cost effectiveness and reduced impact on the environment.

Basically, the ex situ method involves the excavation of the contaminated soil with heavy metals and burial of the contaminated soil at the landfill site (Dasgupta et al. 2013). But the offsite burial is not considered to be a preferable option as it merely shifting the contamination problem to a different place (Salido et al. 2003) and also due to the hazards that are associated with the transport of contaminated soil (Rakhshaee et al. 2009). An on-site management where the heavily contaminated soil is diluted to a safer level by importing the clean soil and mixing with the contaminated soil (Van Ginneken et al. 2007). Barriers and on-site containment provide an alternative methodology, where covering the contaminated soil with an inert material (Rahman et al. 2016). The technique involving immobilization of inorganic contaminant can also be used as a remedial strategy for soils contaminated with heavy metal (Van Ginneken et al. 2007). The contaminant can be achieved by complexing the contaminants (Heavy metal) or by increasing the pH of the soil by using chelating compounds such as lime (Liu et al. 2000). Increased pH helps in decreasing the solubility of heavy metals like Cd, Cu, Ni, and Zn in the soil. Although by this method the risk of potential exposure of these contaminants to plants is reduced, but their concentration in the soil remains unchanged. Most of these conventional remediation technologies are not cheap in terms of implementation as well as causes further disturbance to the already damaged ecosystem (Ginneken et al. 2000). Plant based bioremediation strategies are termed as phytoremediation, this involves the use of green plants along with the micro biota associated for the in situ treatment of contaminated soil and ground water (Rakhshaee et al. 2009). The concept of using plants capable of accumulating heavy metal was first introduced in the year 1983, but it has been practised since last 300 years irrespectively (Rodriguez et al. 2005). The physico-chemical techniques that are used for soil remediation render the land unfertile for plant growth as they tend to eliminate all biological activities, including microbes that are useful such as nitrogen fixing bacteria, mycorrhiza, fungi, as well as fauna during the process of decontamination of the contaminated soil (Mwegoha 2008). The cost of conventional methods used for the process of remediation may cost from $10 to 1000 per cubic metre. The costs for phytoextraction techniques are estimated to be as low as $0.05 per cubic metre (Mwegoha 2008).

3.6 Advantages of Phytoremediation

Phytoremediation techniques are more acceptable publicly, are aesthetically pleasing, and lead to less damages as compared to current techniques of physical and chemical techniques (Huang et al. 2019). Advantages of this technology lie in its effectiveness in reduction of the contaminant, bearing low-cost, being applicable for a wide range of contaminants present, and overall, it is an eco-friendly method.

The major advantage of using heavy metal adsorption technology by plant biomass are in its effectiveness in decreasing the concentration of heavy metal ions to low levels of concentration and involves use of inexpensive biosorbent materials. Phytoremediation is possibly the cleanest as well as the cheapest technology that can be employed in the process of remediating hazardous sites (Huang et al. 2019). The process of phytoremediation encompasses different methodologies that lead to degradation of the contaminants (Hu et al. 2020a, b).

This technology has been receiving attention due to its innovative, cost-effective nature as compared to the more established treatment methods used at hazardous waste sites (Huang et al. 2019). Phytoremediation potentially offers low-cost solutions for soil contamination (Dinake et al. 2019; Häder et al. 2020). It is inexpensive in nature as compared to the conventional physico-chemical methods, since it does not require expensive equipment or the involvement of highly specialized personnel. It is cost-effective for large volumes of water having low concentrations of contaminants and for large areas having low to moderately contaminated surface soils (Dinake et al. 2019). This technique is applicable to a wide range of contaminants (organic and inorganic), metals and radionuclides (Huang et al. 2019). Phytoremediation is considered as a new approach for the clean-up of polluted soils, water, and air (Häder et al. 2020; Behera and Prasad 2020c). Phytoremediation research can also contribute to the improvement of poor soils such as those with high aluminium or salt levels (Dinake et al. 2019). Phytoextraction is considered as an environmentally friendly method to remove metals from contaminated soils in situ. This method can be used in much larger-scale clean-up operations and has been applied for other heavy metals (Diderjean et al. 2002). In Situ applications decrease the amount of soil disturbance compared to conventional methods. It can be performed with minimal environmental disturbance with topsoil left in a usable condition and may be reclaimed for agricultural use. The organic pollutants may be degraded to CO_2 and H_2O, removing environmental toxicity (Diderjean et al. 2002).

Phytoremediation can be an alternative to the much harsher remediation technologies of incineration, thermal vaporization, solvent washing, or other soil washing techniques, which essentially destroy the biological component of the soil and can majorly affect its chemical and physical characteristics as well as creating a nonviable solid waste. Phytoremediation is the most ecological clean-up technology that can be used for contaminated soils and is also known as the green technology. Phytoremediation could be a feasible option in order to decontaminate the heavy-metal-polluted soils, as the biomass produced during the phytoremediation process could be economically utilized in the form of bioenergy.

3.7 Limitations of Phytoremediation Technology

On the other hand, there are limitations in the phytoremediation system. Among them are it is a time-consuming process, the amount of biomass produced, the root depth, chemistry of soil and the level of contamination present at the site, the age of plant, the impacts of contaminated vegetation on health, and climatic condition affecting the process.

The process of phytoremediation is a time-consuming process; it may take several seasons to clean up a contaminated site. The by-products formed at the end of remediation processes from the organic and inorganic contaminants may act as cytotoxic to the plants (Lureysens et al. 2004). The process is limited by the growth rate of the plants as they may require more time to phytoremediate a site as compared to other traditional clean-up technologies. Processes like Excavation and disposal or incineration takes weeks to month's time on order to accomplish, while the process of phytoextraction may take several years. Therefore, for sites that are extremely contaminated and pose acute risks for human, phytoremediation cannot be the choice of remediation technique (Bañuelos 2000; Gerth and Hartmut 2004; Bhat et al. 2019). This methodology might be best suited for places where human contact is limited or where the soil contamination does not need an immediate remediation (Interstate Technology and Regulatory Cooperation (ITRC) 2001; Dinake et al. 2019; Häder et al. 2020).

The success of this process is limited by factors such as growing time, climate, root depth, soil chemistry, and level of contamination (Loutre et al. 2003; Cameselle and Gouveia 2019). Root contact is a primary limitation on applicability of this process. Remediation process involving plants requires that the contaminants must be in contact with the root zone of the plants. Either the plants must extend their roots to the contaminants or the contaminated media must be in the range of the plants (Bañuelos 2000). Plant age affects the physiological activity of the plant, especially its roots, thus roots of a young plant show greater ability to absorb ions as compared to those of an old plant. Thus, it is important to utilize young and healthy plants as remediators for the process of remediation.

High concentrations of contaminants hinder the plant growth (Bhat et al. 2019), thus it may limit the application of this technique on some sites. The phytotoxicity can be dealt with a remedial approach where the presence of high-concentration waste is eliminated with ex situ application that can quickly eliminate acute risk, whereas in situ phytoremediation can be used for longer period of time in order to clean the lower contaminant concentrations present at the site (Bañuelos 2000). A major hindrance in the process of remediation of toxic contaminants is the maximum amount of contaminant that can be accumulated by plants (Ali et al. 2019a, b).

Plants capable of accumulating highest levels of toxic metal contents are known as "hyperaccumulators", they are measured on dry weight basis, ranging about 2000 ppm (0.2%) for higher toxic elements (Cd, Pb) to above 2% for the lesser toxic ones (Zn, Ni, Cu) (Marmiroli et al. 2005). This technique is restricted to sites with low contaminant concentrations, the treatment is limited to soils from 1 m to within a few metres of the surface (McCutcheon and Schnoor 2003; Al-Thani and Yasseen 2020). While the plants accumulating metal needs to be harvested and either recycled or disposed of in accordance with proper regulations, most of the phytoremediative plants does not require further treatments or disposal measures (Marmiroli et al. 2004). The harvested biomass from plants can be classified as a hazardous waste; hence, their recycle as well as disposal should be done properly. Consumption of contaminated plant biomass is one of the major concerns; contaminants might enter the food chain through animals that feed on the plant material containing contaminants (McCutcheon and Schnoor 2003).

3.8 Implementation of Knowledge for Application

3.8.1 Constructed Wetlands

Requisition from claiming constructed wetlands for medicine of defiled waters will be getting developing enthusiasm toward Europe. A few cases have been depicted in distinctive reason the gatherings. Modern effluents holding aniline, nitrobenzene and sulfapyridine (Ramos et al. 2004).

3.8.2 Short Rotation Coppice Forestry

Short rotation coppice is a plantation of trees, poplars or willows, which needs to be kept under 15 years and generate plant biomass for few other purposes in the paper and pulp industry. In particular, coppicing comprises over cutting those trunks toward the base at intervals from claiming 2–3 years. Also, new shoots develop starting with the stump. This sort from claiming ranger service likewise speaks to a sourball of renewable energy, constituting during those same periods a sink for climatic carbon. Using plants which could consume overwhelming metals, expend CO_2, handling biomass, combines ranger service for phytotechnologies. A few creators would be considering metal uptake in willow and poplar, so as will evaluate the biodiversity existing "around cultivars, clones and accessions. Concerning illustration, a late example of poplar clones were broke down to uptake from claiming a few metals, demonstrating to how Cd, Zn, al were consumed with the secondary effectiveness (Lureysens et al. 2004).

3.8.3 Interactions with Microorganisms

It is well known that plant–microorganisms cooperation's assume imperative parts in phytoremediation. In recent years discovery a long-time revelation of the part of endophytic microscopic organisms in phytoremediation need prompted a few fascinating considerations. Building endophytic microscopic organisms of the species *Burkholderia cepacia* with plasmid pTOM increased degradation of toluene in yellow lupine plants (Barac et al. 2004), In the same run through bringing down poisonous quality of the plant.

3.8.4 Atmosphere Contaminants

Phytotechnologies have traditionally been restricted to contaminants receptive through plant roots, possibly in soil sediments or in water. However, contaminants might enter those plants additionally from the atmosphere, and a new application of the phytotechnologies might be the removal of pollutants from the troposphere (Morikawa et al. 2003). Nitrogen dioxide is a pollutant which might be consumed through stomata also consolidated into organic compounds. There is extensive variability among plant taxa in this regard, and a survey of about 300 species showed that the most efficient plant is *Eucalyptus viminalis*, 657 times more efficient than Tillandsia, the less efficient taxon. These plants could be used to assembly "green walls", covering the vertical surfaces of building where plants are able to assimilate NO_2 in great quantities. Recently the authors have depicted a positive effect of NO_2 on plant biomass growth, characterizing it as a "plant vitalisation signal" (Morikawa et al. 2005), but this still waits for a confirmation.

3.9 Recent Trends

3.9.1 Natural Remediation

A new trend in use of phytotechnologies is encountered by the supposed "Assisted Natural Remediation", clearly a non-ordinary application (Adriano et al. 2004). In assisted natural remediation, corrections are added to the soil in order to speed up natural processes of remediation. In the case of metals, alterations contribute to immobilization with complexation, adsorption, precipitation, and chemical reactions: the main purpose is to lower the bioavailability of the metal, and not its total concentration (Fig. 3.3). The literature reviewed by Adriano et al. (2004) includes successful examples to which assisted natural remediation has been applied.

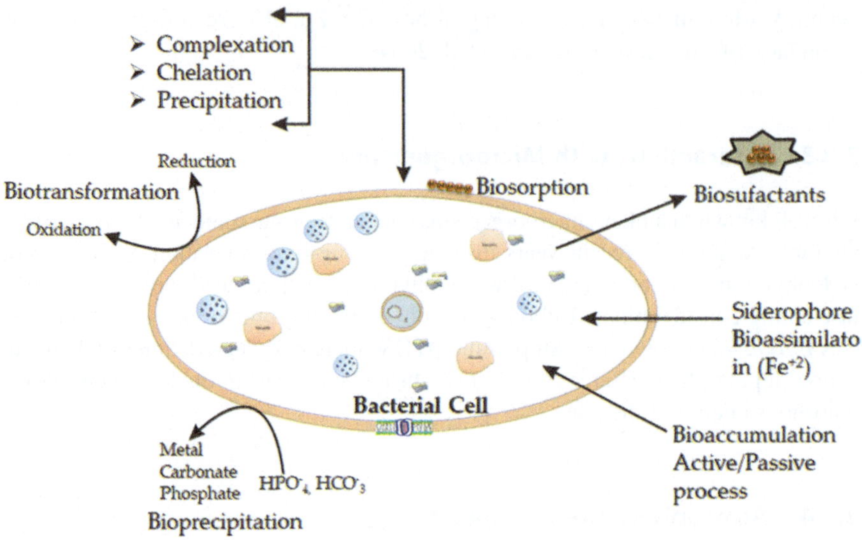

Fig. 3.3 Various ways of natural remediation

3.9.2 Biofortification

Studies on the communication between plant tissues, heavy metals, and/or trace elements have prompted the idea of biofortification, in which plants enriched in micronutrient content are seen as an aid against malnutrition. Uniquely in contrast to phytoaccumulation of metals, which is considered as a danger for the natural pecking order, biofortification of harvests with explicit components may get invaluable (Welch and Graham 2004). Invigorated yields are appropriate for development on micronutrient-poor in light of the fact that their bio concentration limit will prompt higher substance of micronutrients in edible tissues. Knowledge of mechanisms controlling metal accumulation is a prerequisite for elucidating the biochemical basis of these phenomena. Information is also requested for those anti-nutrients that decrease element availability: examples include phytic acid, fibres, and polyphenols.

3.9.3 Glucosinolates and Biofumigation

Biofumigation is a recent application of the properties of plant chemicals. In particular, several Brassicaceae are exploited in the fight towards pests and pathogens in agriculture due to the production of specific secondary metabolites called glucosinolates (Mithen 2001). These sulphur-containing compounds have an anticarcinogenic activity in man, they contribute to the characteristic flavour of cruciferous plants, and their degradation products can deter herbivores and inhibit microorganisms (Häder et al. 2020). From this we can derive that the use of

Brassicaceae as "green manure" to be added to the soil during preparation, in order to decrease the load due to pathogens and pests. This is a sustainable substitute to the use of chemical fumigants. A feature to be explored is the possible connection between production of specific glucosinolates in the plant and the presence of heavy metals in the environment which may act as inducers or repressors. Since glucosinolates contain sulphur, like metallothioneins and phytochelatins, they could have an impact in the sulphur metabolism of these heavy metal sequestering peptides. This in turn may determine an interlock in the process of pest resistance and heavy metal resistance (Häder et al. 2020).

3.9.4 Uptake and Transport

Interaction among plants and metals starts in the root environment. All phytotechnologies can be applied only if the contaminant is in contact with roots, and most of them rely on contaminant uptake by roots. This is the reason why plasma membrane transporters are a subject of research for phytotechnology implementation. Heavy metals uptake involves the same kind of transporters which otherwise provide macro- and micronutrients entrance. Recently, Perfus-Barbeoch et al. (2002) have demonstrated the involvement of Ca channels in Cd uptake in *Arabidopsis thaliana*. The possibility of Cd mimicking Ca in plant cells can also justify its toxicity with perturbation of metabolism and homeostasis of this vital element. Studies performed with plant cell protoplasts have tried to ascertain if differences in transport in sink tissues could explain the different behaviour of hyperaccumulator plants (Cosio et al. 2004; Häder et al. 2020). The results obtained with *A. halleri* and *Thlaspi caerulescens* show that plasma membranes of leaf cells do not account for differences in transport. Therefore, it has been hypothesized that other mechanisms may be active to direct the metals to their subcellular compartments, where they are stored: vacuoles and lignocellulosic material such as cell wall may be among these. Studies of metal transport, and especially in the case of radionuclides, can benefit from autoradiographic techniques, as shown by Soudek et al. (2004) with Cs. Imaging techniques allowed comparison between different species for uptake efficiency, but they also revealed potential sink tissues, providing useful information for implementation of phytoextraction.

3.9.5 Accumulation and Sequestration

During recent years, several scientists have tried to explain the differences, which are present in between the hyperaccumulator taxa and non-accumulator congeners by exploring the analytical techniques which help us to extract the information on speciation along with localisation of certain heavy metal ions present in the plant tissues. In order to understand the molecular basis of the hyperaccumulation capacity as well as to define the storage strategies that are important in order to develop as well as implement the process of phytoextraction. There are various analytical techniques that are based on X-ray emission (scanning electron microscopy and microanalysis) that have been used to show Ni accumulation in the leaf trichomes of *Alyssum bertolonii*, in comparison with the non-accumulator *Alyssum montanum* which stores the Nickel in the roots of the plants (Marmiroli et al. 2004; Häder et al. 2020). *Arabidopsis halleri* which is a zinc and cadmium hyperaccumulator that has been studied by Sarret et al. (2002). The zinc metal ion gets mainly sequestered in the vacuoles of leaf trichomes as well as the mesophyll cells. The researchers have determined it through EXAFS that there are two main forms of zinc metal ions that are present in the plant roots were in the form of malate and phytate (or possibly phosphate), whereas in the trichomes of the leaves it is coordinated along with the C atoms, presumably the C atoms do belong to the organic acids. The careful computations thus suggested that zinc ions in the leaf trichomes, even if it is highly concentrated, that cannot constitute the part of the major sink.

There is another interesting fact that concerns that zinc has binding in the non-accumulator *Arabidopsis lyrata*, in which the phosphate species were predominantly involved in them. Having a similar approach that is based on the EXAFS that it has been possible to show that lead metal ions can be accumulated in the roots of the walnut trees by the coordination with the carbon atoms of cellulose and lignin molecules (Marmiroli et al. 2005).

In order to unravel the molecular and biochemical mechanisms of the hyperaccumulation, the search for the genes and the proteins is being carried on with genomics and proteomics approaches. The genomic efforts that have been promoted by the scientists in various research projects that include the studying various properties of the plant hyperaccumulators, and the results are thus demonstrating that how different genes are being induced by various metal ions in these plants and their congeners (Van de Mortel et al. 2004; Häder et al. 2020).

The knowledge of comparative genomics and proteomics also adds on to the information of orthologous genes, novel sequences, and molecular markers that are necessary (Tuomainen et al. 2004).

Bernard et al. (2004) have studies showing that how the ectopic expression of *Thlaspi* genes present in the yeast led to the isolation of a new gene function that was seen to be involved in cadmium metal ion transport and also probably is responsible for hyperaccumulation, a P-type ATPase.

The method that was taken under considerations is genetic mapping as the method of choice in the case of quantitative traits, and some scientists are trying to build maps of quantitative trait Loci (QTLs) for the process of hyperaccumulation and tolerance in model plants and in hyperaccumulators.

Considering the phylogenetic relationships that the known hyperaccumulator plants present in the family Brassicaceae, *Arabidopsis thaliana* is one of the best model plants available for studies, owing to the complete information of the genomic sequence of the plant along with the genetic knowledge. The research group that is led by Martin Broadley has recently studied and mapped QTL involved in metal ions accumulation of *A. thaliana* (Payne et al. 2004). There are various accessions that were being analysed for heavy metal accumulation that leads to the complete description of a twofold variation in heavy metal concentration. The various crosses among the contrasting phenotypes as well as the analysis in segregating the progenies have led to the mapping of putative QTLs present on several chromosomes; therefore the existence of the two QTLs present on the chromosomes I and V was then confirmed using the analysis of segregating populations from the independent crosses. Mapping of the candidate genes present in these regions will then lead to a new hypothesis about there make up and working of these QTLs.

3.9.6 Genetic Bases of Tolerance

Classical genetical studies have been explored for the sole purpose of addressing the issue of genetical bases of stress tolerance along with the addition to various genomic and proteomic approaches involved towards gene identification. For the purpose, there was a selection of model plants of choice that have been traditionally been *Arabidopsis halleri* and *Thlaspi caerulescens*, these two hyperaccumulators that can be crossed with the non-hyper accumulator ecotypes that can be used for further studies of traits segregation for further usage.

Genecological observations of the plants have demonstrated that the tolerance to one heavy metal is a trait which is independent in nature from accumulation of the same metal ions, and also that the tolerance is controlled by a few major genes that are responsible for remediation. The interspecific crosses that have been done between *A. halleri* and *A. lyrata* ssp. *petraea* have contributed the required information about the tolerance and hyperaccumulation to Cadmium (Bert et al. 2003; Häder et al. 2020) tolerance and hyperaccumulation tends to segregate the responsible independent characters, whereas the Cadmium tolerance co-segregates along with the zinc tolerance. Moreover, the cadmium and zinc hyperaccumulation then seems to be co-regulated or controlled by the same set of genes.

The same genetical approach has been pursued in the case of *Thlaspi* with their various ecotypes that differ in their accumulation capacity (Zha et al. 2004). The segregation results reveals that there are at least two genes that are responsible for the zinc accumulation, whereas in the case of cadmium accumulation it was seen that there is more than one gene could be that could be involved in there remediation. The correlation that is present between the accumulation of zinc, cadmium and manganese that is they are consistent with a multiple transporter facility along with simultaneously there is specificity for all the three metals. Also, in case of *Thlaspi*, the cadmium tolerance As well as the accumulation segregate as the independent characters.

3.9.7 New Contaminants

Phytotechnologies have basically been applicable in case of heavy metal contaminants, nutrients, and radionuclides. There are various new contaminants of interest that may include arsenic: only recently that the hyperaccumulator plants for this contaminant been described: the fern *Pteris vittata* (Zhao et al. 2003) and other species of the same genus (Zhao et al. 2002) have reported to be of great use in order to treat the contaminants and other heavy metal ions as well. As tolerance, as reported by in (Zhao et al. 2003), but in case of *Pteris* fern As is present mainly in an inorganic form as the arsenite is found to be present in the vacuole of leaf cells. Various scientists have suggested that phytochelatins may play a role in binding of the small quantities of as that are found in the cytoplasm. The contaminants of mercury ions are also attracting new interest, and there are studies suggesting that plants may be capable of volatilizing the mercury ions as metallic mercury (Ernst et al. 2005).

3.10 Conclusion and Perspectives

The excessive contamination of the heavy metal contaminants in the environment is of great concern owing to its potential impact on the health of human and animal. There are cheaper and effective technologies that are needed in order to protect the natural resources and biological lives. Many substantial efforts are being made in order to identify the plant species along with their mechanisms of uptake and hyperaccumulation of heavy metals during the last decade. There are many genetic variations that were studied among various plant species. The remediation mechanisms of heavy metal uptake, accumulation, exclusion, translocation, osmoregulation, and compartmentation differ along with the plant species and also help to determine its specific role in phytoremediation. There are variations that exist for the purpose of hyperaccumulation of various heavy metal ions among different plant species and within plant populations. The variations present in the plant species do not correlate with either the presence of metal concentration in the soil or with the degree of metal tolerance in the plant. So as to develop the new crop species/plants that are having the ability of metal extraction from the contaminated environment, using the traditional breeding techniques along with hybrid generation through radiation and chemicals are all in progress to achieve these. Along with the development of biotechnology stream, the capabilities of plant hyperaccumulators can be enhanced greatly through using some specific metal gene identification techniques and thus it can be transferred in the promising plant species.

Thus, this technique can play a significant role which is responsible in extracting of these heavy metals from the contaminated soils. The usage of cleaning technologies is due to its site-specific nature owing to its spatial and climatic variations and it is not economically feasible to be applicable everywhere. Therefore, the need of cheaper technologies is being looked for more practical use. The recent advancements in the field of plant biotechnology have recently created a new hope in the development of plant hyperaccumulating species. However, there is dearth of research work which is required in the respect of heavy metal uptake studies at cellular level including efflux and influx of different metal ions by different cell organelles and membranes. Various rhizospheric studies that are done under the control and field conditions are also studied in order to examine the antagonistic and synergistic effects of various metal ions present in the soil solution along with the polluted waters. There is an urgent need of in-depth soil microbial studies that are required in order to identify the microorganisms that are highly associated along with metal solubility or precipitations. Till date the availability of these methods used for the recovery of heavy metal ions from the plant biomass of hyperaccumulators are still studied. The use of traditional disposal approaches such as burning and ashing are not applicable in order to remediate volatile metal ions; therefore, ongoing investigations are still needed so as to develop various new methods that are effective in order to be used for the recovery of metals from the hyperaccumulator plant biomass.

References

Adesuyi AA, Hjoku KL, Akinola MO (2015) Assessment of heavy metals pollution in soil and vegetation around selected industries in Lagos State, Nigeria. J Geosci Env Prot 3:11–19

Adriano DC, Wenzel WW, Vangronsveld J, Bolan NS (2004) Role of assisted natural remediation in environmental cleanup. Geoderma 122(2–4):121–142

Agnello AC, Bagard M, Van Hullebusch ED, Esposito G, Huguenot D (2016) Comparative bioremediation of heavy metals and petroleum hydrocarbons co-contaminated soil by natural attenuation, phytoremediation, bioaugmentation and bioaugmentation-assisted phytoremediation. Sci Total Environ 563–564:693–703

Ahmed MB, Zhou JL, Ngo HH, Guo W, Thomaidis NS, Xu J (2017) Progress in the biological and chemical treatment technologies for emerging contaminant removal from wastewater: a critical review. J Hazard Mater 323:274–298

Ali H, Khan E, Sajad MA (2013) Phytoremediation of heavy metals—concepts and applications. Chemosphere 91:869–881

Ali H, Khan E, Ilahi I (2019a) Environmental chemistry and ecotoxicology of hazardous heavy metals: environmental persistence, toxicity, and bioaccumulation. J Chem 2019:1–14

Ali I, Basheer AA, Mbianda XY, Burakov A, Galunin E, Burakova I, Mkrtchyan E, Tkachev A, Grachev V (2019b) Review article graphene based adsorbents for remediation of noxious pollutants from wastewater. Environ Int 127:160–180

Al-Thani RF, Yasseen BT (2020) Phytoremediation of polluted soils and waters by native Qatari plants: future perspectives. Environ Pollut 259:113694

Arslan M, Imran A, Khan QM, Afzal M (2017) Plant–bacteria partnerships for the remediation of persistent organic pollutants. Environ Sci Pollut Res 24:4322–4336

Atanes E, Cuesta-García B, Nieto-Márquez A, FernándezMartínez F (2019) A mixed separation-immobilization method for soluble salts removal and stabilization of heavy metals in municipal solid waste incineration fly ash. J Environ Manag 240:359–367

Baker AJM, McGrath SP, Reeves RD, Smith JAC (1999) Metal hyperaccumulator plants: a review of the ecology and physiology of a biological resource for phytoremediation of metal-polluted soils. In: Terry N, Vangronsveld J, Banuelos G (eds) Phytoremediation of contaminated soils. CRC Press, Boca Raton, FL, pp 85–107

Bañuelos GS (2000) Phytoextraction of selenium from soils irrigated with selenium-laden effluent. Plant Soil 224:251–258

Barac T, Taghavi S, Borremans B, Provoost A, Oeyen L, Colpaert JV, Vangronsveld J, van der Lelie D (2004) Engineered endophytic bacteria improve phytoremediation of water-soluble, volatile, organic pollutants. Nat Biotechnol 22(5):583–588

Behera BK, Prasad R (2020a) Strategies for soil management. In: Behera BK, Prasad R (eds) Environmental technology and sustainability. Elsevier, pp 143–167

Behera BK, Prasad R (2020b) Environmental technology and sustainability. Elsevier. ISBN: 9780128191033. https://www.elsevier.com/books/environmental-technology-and-sustainability/behera/978-0-12-819103-3

Behera BK, Prasad R (2020c) Conceptual development for a clean environment. In: Behera BK, Prasad R (eds) Environmental technology and sustainability. Elsevier, pp 1–40

Bert V, Meerts P, Saumitou-Laprade P, Salis P, Gruber W, Verbruggen N (2003) Genetic basis of Cd tolerance and hyperaccumulation in *Arabidopsis halleri*. Plant Soil 249:9–18

Berti WR, Cunningham SD (2000) In: Raskin I (ed) Phytoremediation of toxic metals: using plants to clean up the environment. Wiley-Interscience, John Wiley and Sons, Inc., New York, pp 71–88

Bernard C, Roosens N, Czernic P, Lebrun M, Verbruggen N (2004) A novel CPx-ATPase from the cadmium hyperaccumulator *Thlaspi caerulescens*. FEBS Lett 569:140–148

Bhat SA, Hassan T, Majid S (2019) Heavy metal toxicity and their harmful effects on living organisms—a review. Int J Med Sci Diagno Res 3(1):106–122

Briggs D (2003) Environmental pollution and the global burden disease. Br Med Bull 68:1–24

Burakov AE, Galunin EV, Burakova IV, Kucherova AE, Agarwal S, Tkachev AG, Gupta VK (2018) Adsorption of heavy metals on conventional and nanostructured materials for wastewater treatment purposes: a review. Ecotoxicol Environ Saf 148:702–712

Burges A, Alkorta I, Epelde L, Garbisu C (2018) From phytoremediation of soil contaminants to phytomanagement of ecosystem services in metal contaminated sites. Int J Phytoremediation 20: 384–397

Cameselle C, Gouveia S (2019) Phytoremediation of mixed contaminated soil enhanced with electric current. J Hazard Mater 361:95–102

Cosio C, Martinoia E, Keller C (2004) Hyperaccumulation of Cadmium and Zinc in *Thlaspi caerulescens* and *Arabidopsis halleri* at the leaf cellular level. Plant Physiol 134:716–725

Chen X, Achal V (2019) Biostimulation of carbonate precipitation process in soil for copper immobilization. J Hazard Mater 368:705–713

Dasgupta S, Satvat PS, Mahinrakar AB (2011) Ability of *Cicer arientinum* (L.) for bioremoval of lead and chromium from soil. IJTES 2:338–341

Dasgupta D, Ghosh R, Sengupta TK (2013) Biofilm-mediated enhanced crude oil degradation by newly isolated Pseudomonas species. ISRN Biotechnol. https://doi.org/10.5402/2013/250749

Diderjean L, Gondet L, Perkins R, Lau S-MC, Schaller H, O'Keefe DP, Werck-Reickhart D (2002) Engineering herbicide metabolism in tobacco and Arabidopsis with CYP76B1, a cytochrome P450 enzyme from Jerusalem artichoke. Plant Physiol 130:179–189

Dinake P, Kelebemang R, Sehube N (2019) A comprehensive approach to speciation of lead and its contamination of firing range soils: a review. Soil Sediment Contam 28:1–29

Dixit R, Wasiulah, Malaviya D, Pandiyan K, Singh UB, Sahu A, Shukla R, Singh BP, Rai JP, Sharma PK, Lade H, Paul D (2015) Bioremediation of heavy metals from soil and aquatic environment: an overview of principles and criteria of fundamental processes. Sustainability 7: 2189–2212

Erakhrumen AA (2017) Phytoremediation: an environmentally sound technology for pollution prevention, control and remediation in developing countries. Educ Res Rev 2:151–156

Ernst D, Battke F, Halbach S (2005) Ascorbate promotes emission of mercury vapour from plants. In: Abstracts of the 1st scientific workshop and management committee meeting of COST action 859, p 26

Etteieb S, Magdouli S, Zolfaghari M, Brar S (2020) Monitoring and analysis of selenium as an emerging contaminant in mining industry: a critical review. Sci Total Environ 698:134339

Galal TM, Eid EM, Dakhil MA, Hassan LM (2018) Bioaccumulation and rhizofiltration potential of Pistia stratiotes L. for mitigating water pollution in the Egyptian wetlands. Int J Phytoremediation 20:440–447

Garcia-Sanchez M, Kosnar Z, Mercl F, Aranda E, Tlustos P (2018) A comparative study to evaluate natural attenuation, mycoaugmentation, phytoremediation, and microbialassisted phytoremediation strategies for the bioremediation of an aged PAH-polluted soil. Ecotoxicol Environ Saf 147:165–174

Gerth A, Hartmut T (2004) The role of plants in TNT-phytoremediation. In: Abstracts of COST action 859 1st WG2 workshop, 2004—exploiting "-omics" approaches in phytotechnologies, p 22

Ghosh, Singh (2005) A review on phytoremediation of heavy metals and utilization of its byproduct. Appl Ecol Env Res 3:1–18

Goland-Goldhirsh A (2006) Plant tolerance to heavy metals, a risk for food toxicity or a means for food fortification with essential metals: the *Allium schoenoprasum* model. In: Twardowska I, Allen HE, Haggblom MM (eds) Soil and water pollution monitoring, protection and remediation. Springer, Amsterdam, pp 479–486

Goncalves AL, Pires JC, Simões M (2017) A review on the use of microalgal consortia for wastewater treatment. Algal Res 24:403–415

Guo J, Zhou Y (2020) Transformation of heavy metals and dewaterability of waste activated sludge during the conditioning by Fe^{2+}-activated peroxymonosulfate oxidation combined with rice straw biochar as skeleton builder. Chemosphere 238:124628

Häder D-P, Banaszak AT, Villafañe VE, Narvarte MA, González RA, Helbling EW (2020) Anthropogenic pollution of aquatic ecosystems: emerging problems with global implications. Sci Total Environ 713:136586

Henry JR (2000) An overview of phytoremediation of lead and mercury. NNEMS Report. Washington, DC, pp 3–9

Hu B, Xue J, Zhou Y, Shao S, Fu Z, Li Y, Chen S, Qi L, Shi Z (2020a) Modeling bioaccumulation of heavy metals in soil-crop ecosystems and identifying its controlling factors using machine learning. Environ Pollut 262:114308

Hu H, Zhao J, Wang L, Shang L, Cui L, Gao Y, Li B, Li Y-F (2020b) Synchrotron-based techniques for studying the environmental health effects of heavy metals: current status and future perspectives. Trends Anal Chem 122:115721

Huang L, Liu C, Liu X, Chen Z (2019) Immobilization of heavy metals in e-waste contaminated soils by combined application of biochar and phosphate fertilizer. Water Air Soil Pollut 230:26

Hussain Z, Arslan M, Malik MH, Mohsin M, Iqbal S, Afzal M (2018) Integrated perspectives on the use of bacterial endophytes in horizontal flow constructed wetlands for the treatment of liquid textile effluent: phytoremediation advances in the field. J Environ Manag 224:387–395

Interstate Technology and Regulatory Cooperation (ITRC) (2001) Phytotechnology technical and regulatory guidance document. www.itrcweb.org

Jarup L (2003) Hazards of heavy metal contamination. Br Med Bull 68:167–182

Jiao X, Teng Y, Zhan Y, Wu J, Lin X (2015) Soil heavy metal pollution and risk assessment in Shenyang industrial district, Northeast China. PLoS One 10:1–9

Kanwar VS, Sharma A, Srivastav AL (2020) Phytoremediation of toxic metals present in soil and water environment: a critical review. Environ Sci Pollut Res 27:44835–44860

Kulkarni P, Olson ND, Paulson JN, Pop M, Maddox C, Claye E, Mongodin EF (2018) Conventional wastewater treatment and reuse site practices modify bacterial community structure but do not eliminate some opportunistic pathogens in reclaimed water. Sci Total Environ 639:1126–1137

Kushwaha A, Hans N, Kumar S, Rani R (2018) A critical review on speciation, mobilization and toxicity of lead in soil-microbe-plant system and bioremediation strategies. Ecotoxicol Environ Saf 147:1035–1045

Liu D, Jiang W, Liu C, Xin C, Hou W (2000) Uptake and accumulation of lead by roots, hypocotyls and shoots of Indian mustard [Brassica juncea (L.)]. Bioresour Technol 71:273–277

Loutre C, Dizon DP, Brazier M, Slater M, Cole DJ, Edwards R (2003) Isolation of a glucosyltransferase from *Arabidopsis thaliana* active in the metabolism of the persistent pollutant 3,4-dichloroaniline. Plant J 34:485–493

Lureysens I, Blust R, De Temmerman L, Lemmens C, Ceulemans R (2004) Clonal variation in heavy metal accumulation and biomass production in a poplar coppice culture: I. Seasonal variation in leaf, wood and bark concentrations. Environ Pollut 131:485–494

Ma Y, Rajkumar M, Zhang C, Freitas H (2016) Beneficial role of bacterial endophytes in heavy metal phytoremediation. J Environ Manag 174:14–25

Marmiroli M, Gonnelli C, Maestri E, Gabbrielli R, Marmiroli N (2004) Localisation of nickel and mineral nutrients Ca, K, Fe, Mg with scanning electron microscopy microanalysis in tissues of the nickel-hyperaccumulator *Alyssum bertolonii* Desv. and the non-accumulator *Alyssum montanum* L. Plant Biosyst 138:231–243

Marmiroli M, Antonioli G, Maestri E, Marmiroli N (2005) Evidence of the involvement of plant ligno-cellulosic structure in the sequestration of Pb: an X-ray spectroscopy-based analysis. Environ Pollut 134:217–227

McCutcheon SC, Schnoor JL (eds) (2003) *Phytoremediation:* transformation and control of contaminants. Wiley, New York

Mithen R (2001) Glucosinolates—biochemistry, genetics and biological activity. Plant Growth Regul 34:91–103

Morikawa H, Takahashi M, Hakata M, Sakamoto A (2003) Screening and genetimanipulation of plants for decontamination of pollutants from the environments. Biotechnol Adv 22:9–15

Morikawa H, Takahashi M, Nakagawa M, Sakamoto A, Ohsumi C, Matsubara T (2005) Atmospheric nitrogen dioxide as a plant vitalization signal to improve size and cell constituent. In: Abstracts of the 1st scientific workshop and management committee meeting of COST action 859, p 189

Mulligan CN, Yong RN, Gibbs BF (2001) Remediation technologies for metalcontaminated soils and groundwater: an evaluation. Eng Geol 60:193–207

Muthusaravanan S, Sivarajasekar N, Vivek J, Paramasivan T, Naushad M, Prakashmaran J, Al-Duaij OK (2018) Phytoremediation of heavy metals: mechanisms, methods and enhancements. Environ Chem Lett 16:1339–1359

Mwegoha SJW (2008) The use of phytoremediation technology for abatement soil and groundwater pollution in Tanzania: opportunities and challenges. J Sustain Dev Afr 10:140–156

Payne KA, Bowen HC, Hammond JP, Hampton CR, Lynn JR, Mead A, Swarup K, Bennett MJ, White PJ, Broadley MR (2004) Natural genetic variation in caesium (Cs) accumulation by *Arabidopsis thaliana*. New Phytol 162:535–548

Perfus-Barbeoch L, Leonhardt N, Vavasseur A, Forestier C (2002) Heavy metal toxicity:cadmium permeates through calcium channels and disturbs the plant water status. Plant J 32:539–548

Rahman MF, Islam M, Begum MC, Kabir AH, Alam MF (2016) Genentic variation in cadmium tolerance is related to transport and antioxidant activities in field peas (*Pisum sativum* L.). Arch Agron Soil Sci 63:578–585s

Rakhshaee R, Giahi M, Pourahmad A (2009) Studying effect of cell wall's carboxyl-carboxylate ratio change of *Lemna minor* to remove heavy metals from aqueous solution. J Hazard Mater 163:165–173

Ramos H, Davies L, Carias C, Novais J, Martins-Dias S (2004) Phytoremediation of organic compounds in constructed wetlands. In: Abstracts of COST action 859 1st WG4 workshop, 2004—integration and application of phytotechnologies, p 8

Raskin I, Ensley BD (2000) Phytoremediation of toxic metals: using plants to clean up the environment. Wiley, New York, pp 53–70

Raskin I, Smith RD, Salt DE (1997) Phytoremediation of metals: using plants to remove pollutants from the environment. Curr Opin Biotechnol 8:221–226

Reeves RD, Baker AJ, Jaffré T, Erskine PD, Echevarria G, van der Ent A (2018) A global database for plants that hyperaccumulate metal and metalloid trace elements. New Phytol 218:407–411

Rodriguez L, Lopez-Bellido JF, Carnicer A, Recreo F, Tallos A, Monteagudo MJ (2005) Mercury recovery from soils by phytoremediation. In: Environmental chemistry. Springer, Berlin, pp 197–204

Salido LA, Hasty LK, Lim JM, Butcher JD (2003) Phytoremediation of arsenic and lead in contaminated soil using Chinese Brake ferns (*Pteris vittata*) and Indian mustard (*Brassica juncea*). Int J Phytoremediation 5:89–103

Sarma H, Forid N, Prasad R, Prasad MNV, Ma LQ, Rinklebe J (2021) Enhancing phytoremediation of hazardous metal(loid)s using genome engineering CRISPR–Cas9 technology. J Hazard Mater. https://doi.org/10.1016/j.jhazmat.2021.125493

Sarret G, Saumitou-Laprade P, Bert V, Proux O, Hazemann J-L, Traverse A, Marcus MA, Manceau A (2002) Forms of zinc accumulated in the hyperaccumulator *Arabidopsis helleri*. Plant Physiol 130:1815–1826

Sonowal S, Nava AR, Joshi SJ, Borah SN, Islam NF, Pandit S, Prasad R, Sarma H (2022) Biosurfactants assisted heavy metals phytoremediation: green technology for the United Nations sustainable development goals. Pedosphere 2(1):198–210. https://doi.org/10.1016/S1002-0160(21)60067-X

Soudek P, Tykva R, Vanek T (2004) Laboratory analyses of ^{137}Cs uptake by sunflower, reed and poplar. Chemosphere 55:1081–1087

Sreelal G, Jayanthi R (2017) Review on phytoremediation technology for removal of soil contaminant. Indian J Sci Res 14:127–130

Susarla S, Medina VF, McCutheon SC (2002) Phytoremediation: an ecological solution to organic chemical contamination. Ecol Eng 18:647–658

Suvaryan Y, Sargsyan V, Sargsyan A (2011) The problem of heavy metal pollution in The Republic of Armenia: overview and strategies of balancing socioeconomic and ecological development. In: Simeonov LI, Kochubovski MV, Simeonova BG (eds) Environmental heavy metal pollution and effects on child mental development: risk and prevention strategies. Springer, pp 309–315

Tangahu BV, Abdullah SRS, Basri H, Idris M, Anuar N, Mukhlisin M (2011) A review on heavy metals (As, Pb, and Hg) uptake by plants through phytoremediation. Int J Chem Eng 31:939161

Thakare M, Sarma H, Datar S, Roy A, Pawar P, Gupta K, Pandit S, Prasad R (2021) Understanding the holistic approach to plant-microbe remediation technologies for removing heavy metals and radionuclides from soil. Curr Res Biotechnol. https://doi.org/10.1016/j.crbiot.2021.02.004

Tuomainen M, Nunan N, Lehesranta S, Tervahauta A, Hassinen V, Schat H, Auriola S, McNicol J, Karenlampi S (2004) Proteomics and metal-responsive proteins in *Thlaspi*. In: Abstracts of COST action 859 1st WG2 workshop, 2004—exploiting "-omics" approaches in phytotechnologies, p 42

U.S. Environmental Protection Agency (2000) Introduction to phytoremediation. National Risk Management Research Laboratory, EPA/600/R-99/107. http://www.clu-in.org/download/remed/introphyto.pdf

United States Protection Agency Reports (2000) Introduction to phytoremediation. EPA 600/R-99/107

Van de Mortel J, Rigola D, Talukdar S, Fiers M, Pieper B, Moeller B, Schat H, Aarts M (2004) Comparative genomics of Zn accumulation in *Thlaspi caerulescens* and *Arabidopsis thaliana*. In: Abstracts of COST action 859 1st WG2 workshop, 2004—exploiting "-omics" approaches in phytotechnologies, p 8

Van Ginneken L, Meers E, Guisson R et al (2007) Phytoremediation for heavy metal-contaminated soils combined with bioenergy production. J Environ Eng Landsc 15:227–236

Vangronsveld J, Herzig R, Weyens N, Boulet J, Andriaensen K, Ruttens A, Thewys T, Vassilev A, Meers E, Nehevajova E, van der Lelie D, Mench M (2009) Phytoremediation of contaminated soils and groundwater: lesson from the field. Environ Sci Pollut Res 16(7):765–794

Wang Y, Greger M (2006) Use of iodide to enhance the phytoextraction of mercury-contaminated soil. Sci Total Environ 368:30–39

Welch RM, Graham RD (2004) Breeding for micronutrients in staple food crops from a human nutrition perspective. J Exp Bot 55:353–364

Zha HG, Jiang RF, Zhao F-J, Vooijs R, Schat H, Barker JHA, McGrath SP (2004) Co-segregation analysis of cadmium and zinc accumulation in *Thlaspi caerulescens* interecotypic crosses. New Phytol 164:299–312

Zhao F-J, Dunham SJ, McGrath SP (2002) Arsenic hyperaccumulation by different fern species. New Phytol 156:27–31

Zhao FJ, Wang JR, Barker JHA, Schat H, Bleeker PM, McGrath SP (2003) The role of phytochelatins in arsenic tolerance in the hyperaccumulator *Pteris vittata*. New Phytol 159: 403–410

Effective Removal of Radioactive Waste from Environment Using Plants

4

Bhupinder Dhir

Abstract

Radionuclides released into the environment pose threat to human health and the environment, hence their remediation becomes mandatory. Phytoremediation technology assists in the removal of radioactive contaminants from the environment in an effective way. Plant surfaces take up radionuclides present in the atmosphere by leaves and those immersed in water or bound to soil through roots. Terrestrial and aquatic plants both remove radionuclides by mechanisms such as accumulation and rhizofiltration. The plasma membrane transporters present on the surface of cells facilitate the transfer of radioactive elements in plants. The potential of plants can be exploited for developing eco-friendly technologies for large scale remediation of harmful radioactive elements from the environment.

Keywords

Phytoremediation · Rhizofiltration · Radionuclides · Phytoextraction · Radioactive elements · Phytovolatilization

4.1 Introduction

Radionuclides get added to the environment through releases from various sources such as nuclear tests, nuclear weapons, accidental spills, and discharges from nuclear facilities or operations (Table 4.1). Radionuclide emissions in the atmosphere include fallout from atmospheric bomb, nuclear accidents, emissions from reprocessing plants, nuclear power stations (isotopes are ^{3}H, ^{14}C, ^{35}S), and waste disposal sites (^{14}C, ^{36}Cl, ^{99}Tc, ^{129}I, ^{135}Cs, ^{237}Np and ^{239}Pu, ^{240}Pu, ^{42}Pu) (Hattink

B. Dhir (✉)
School of Sciences, Indira Gandhi National Open University, New Delhi, India

R. Prasad (ed.), *Phytoremediation for Environmental Sustainability*,
https://doi.org/10.1007/978-981-16-5621-7_4

Table 4.1 Major sources of radioactive contamination in the environment

Source	Radionuclides
Nuclear weapon testing	^{14}C, ^{137}Cs, ^{90}Sr, ^{95}Zr
Nuclear weapon production	^{137}Cs, ^{106}Ru, ^{95}Zr
Discharges in mining	^{226}Ra, ^{210}Pb, ^{210}Po, ^{232}Th, ^{222}Rn
Mining and milling sites	^{238}U
Nuclear accidents	^{137}Cs, ^{90}Sr, ^{131}I, ^{210}Po, ^{95}Zr, ^{144}Ce

et al. 2000; Ould-Dada et al. 2001; Peles et al. 2002). In addition, natural sources such as parent material, organic/mineral component, soil solution, leakage from unnatural sources such as buried radioactive materials in the soil also contribute to radioactive pollution. The presence of radionuclides in soil and water affects ecosystem stability and poses a threat to human health.

A variety of physicochemical methods such as washing, ion exchange, leaching with chelating agents, flocculation, and reverse osmosis have been used for the treatment of radionuclide contamination. These methods remove radioactive elements from soil and water but the efficiency of radionuclide removal varies depending upon the chemistry of the element, rate of deposition, and decay. Limitations of these methodologies generated the need for framing environmentally safe and affordable technologies. Phytoremediation proved to be an effective alternate to remove a variety of contaminants including radioactive elements from the environment (Zhu and Shaw 2000; Dhir 2013; Malhotra et al. 2014; Yan et al. 2021).

Phytoextraction, rhizofiltration, phytovolatilization, and phytostabilization are the major phytoremediation techniques involved in the remediation of radionuclides. Phytoextraction is a process in which plant biomass accumulates radionuclides and radionuclides get transported from soil into the aboveground parts of the plant which are harvested. It depends on the natural ability of vascular plants to take chemical elements through the roots, deliver them to the vascular tissue, transport and compartmentalize radioactive elements in the aboveground biomass. The phytoextraction technology has proven effective in treating areas having low-level of radionuclide contamination. Rhizofiltration utilizes plant roots to precipitate and concentrate radionuclides from polluted sites while in phytovolatilization plants remove radionuclides (such as ^{3}H) from the leaves/ foliage via volatilization (Prasad 2007). In phytostabilization process plants stabilize radionuclides in soils rendering them harmless. The features such as absorption, translocation, bioaccumulation, and contaminant degradation help plants to remove contaminants such as radionuclides from the environment (Yadav and Kumar 2019).

The present chapter highlights the role of plants in remediation/removal of the radioactive contamination present in the soil or water. The potential of the plants for removing radionuclides from the environment has been assessed.

4.2 Removal of Radionuclides by Plants

Radionuclides get deposited on the external plant surface directly from the atmosphere or by wet/dry deposition via resuspension from soil (Bell et al. 1988; Tagami 2012). High level of radionuclide accumulation has been reported in plants growing in soils having huge radioactive deposits and mine tailings.

4.2.1 Accumulation and Uptake of Radionuclides by Plants

Uptake and accumulation of radionuclides by various crop plants and tree species has been well documented (Planinšek et al. 2018; Duong et al. 2021) (Tables 4.2 and 4.3). Phytoextraction technique has proven to be the major mechanism involved in the uptake of radionuclides such as ^{90}Sr (strontium), ^{95}Nb (niobium), ^{99}Tc (techtenium), ^{106}Ru (rubidium), ^{144}Ce (cesium), $^{226,228}Ra$ (radon), $^{239,240}Pu$ (plutonium), ^{241}Am (americium), $^{228,230,232}Th$ (thorium), ^{244}Cm (curium), ^{237}Np (neptunium) (Kabata-Pendias and Pendias 1996; Dushenkov 2003; Hattink et al. 2004). Phytoextraction of radionuclides depends on the bioavailability of radionuclides in soil, rate of uptake by plant roots and efficiency of radionuclide transport through the vascular system. The rate of transfer uptake of radionuclide from soil/water to plant is related to transfer factors (TF) which gives a measure of ratio of concentration of element in the plant to that present in the mine tailings or soil. TF can be used as an index for the growth of a target element in the plant and its transfer from the medium to the plant. This factor varies for each plant. The difference in TF values for the plants tissues may be due to change in metabolic rate.

Table 4.2 Radionuclide accumulation in crop plants

Element	Plant species
Co	Broccoli, tomato
Rb	Tomato, chard, sunflower, cucumber
Sr	Cucumber, sunflower, turnip
Cs	Tomato, chard, cucumber, sweet potato
U	Sunflower

Table 4.3 Radioactive elements taken up by tree species

Tree species	Radioactive elements
Paxistima myrsinites	^{210}Pb, ^{210}Po
Ribes lacustre	^{210}Pb, ^{210}Po
Abies lasiocarpa	^{226}Ra
Pinus contorta	^{226}Ra
Salix scouleriana *S. scouleriana*	^{226}Ra
Alnus incana	^{226}Ra
Larix occidentalis	^{226}Ra
Pseudotsuga menziesii	^{226}Ra
Acacia auriculiformis	^{226}Ra, ^{238}U, ^{137}Cs, ^{228}Ra, ^{40}K

Phytovolatilization uses the ability of plant to transpire enormous amounts of water. It is used for remediation of ^{3}H (Tritium), a radioactive isotope of hydrogen. A small portion of ^{3}H is absorbed by plant roots and most of it remains in the plant tissues in the form of easily exchangeable hydroxyl ions or is incorporated into organic molecules through photosynthesis. The roots of terrestrial plants efficiently remove radionuclides such as uranium (U) from aqueous streams. The process is referred as rhizofiltration. High concentrations of uranium get accumulated in the roots and bioaccumulation coefficients above 30,000 have been noted for the element.

Besides, phytostabilization, phytostimulation, phytotransformation, phytofiltration, and phytoextraction, combination of plant–microbe interaction also help in removal/remediation of radionuclides from soil. Microbes are known possess great potential to bio-transform, biosorb, and biomineralize radionuclides through their inherent catabolic process. Microbe assisted phytoremediation technique has proved beneficial in improving the mobility/removal of radionuclides from soil surfaces (Sarma and Prasad 2018; Thakare et al. 2021). The technique has played a role in restore balance of the soil (Lajayer et al. 2019). Transporters like NRAMP, ZIP families CDF, ATPases (HMAs) family like P1B-ATPases play an important role in phytoremediation of radioactive elements.

Radionuclide accumulation varies according to plant species and radioactive element. A high uptake rate has been found for many elements while low rate has been noted for others. The mobility and uptake of radionuclides in plants is controlled by external factors such as chemical composition of the soil, pH, and temperature and several plant physiological factors. The degree of radionuclide accumulation also depends on competition between the elements present in the soil, at the uptake site or within the plant tissue. Transfer of tritium into vegetation occurs mainly through stomata on the leaf surface and uptake of soil water. It gets incorporated in organic matter within plant tissues in both exchangeable and non-exchangeable form. Soluble forms of lithium (Li) present in soils get readily absorbed by the plants.

Accumulation of radionuclides by plants is determined by the content of exchangeable and mobile forms of radionuclide. The availability of radionuclides in soil depends on several factors including pH (Ebbs et al. 1998; Echevarria et al. 2001). It is suggested that pH regulates the solubility of Rb^{+} in soils. Addition of chelators (Huang et al. 1998) and soil supplements stimulates radionuclide removal from soil (Dushenkov et al. 1999). The presence of organic matter, inorganic colloids (clay), and competing elements strongly affect the uptake of radionuclides. Radionuclides such as ^{137}Cs and ^{134}Cs get firmly bound to clay fraction of the soil. Clay strongly binds Cs and restricts the uptake by root. Addition of monovalent cations similar to Cs causes removal of ^{137}Cs from the soil (Dushenkov et al. 1999). Addition of organic matter increases the uptake of Cs by plants. Nutrient elements such as Ca^{+}, Mg^{2+} depress the uptake of Cs. The availability of Cs decreases with increasing soil moisture and percentage of fine sand and silt.

High U concentration has been noted in roots of plants raised in citric acid-treated soil (@5000 mg kg^{-1}) (Huang et al. 1998). Addition of fertilizers increases the retention of radioactive elements such as ^{137}Cs in roots. Fertilizer (potassium sulfate) treatment lowers the concentrations of ^{241}Am, ^{244}Cm, ^{232}Th, and ^{238}U. Some radioactive elements act similar to nutrients. Cesium (Cs) and rubidium (Rb) act similar to K. Similarly Sr act similar to Ca and Se is similar to S. They follow the same path as the nutrient. Uptake of radionuclide is related to essential elements such as K. The presence of potassium in the medium strongly disturbs the uptake of Cs (Bystrzejewska-Piotrowska and Urban 2003). Cesium uptake capacity decreased when 1.3 mM potassium was present in the medium (Marčiulionienė et al. 2015). The potassium concentrations of 0.5–3 mg l^{-1} cause inhibition of Cs uptake. These elements are interrelated in a complex, concentration-dependent, manner. Excessive addition of monovalent cations results in strong competition between the ions for uptake by plants. This affects the levels of radionuclide (such as ^{137}Cs) accumulation in plants (Lasat et al. 1998). The presence of potassium depresses the uptake of Cs. Rubidium and cesium follow the similar uptake route as for potassium. The size of the hydrated Rb^+ molecule is similar to that of hydrated K^+, thus the binding site at the plasma membrane of root cells cannot distinguish between the two cations. Hence, studies established that addition of potassium fertilizers decreases uptake of Cs uptake while addition of nitrogen increases uptake of ^{137}Cs by plants (Seel et al. 1995). The influx of Cs and Li into cells occurs via potassium transporters.

4.2.2 Mechanism of Radionuclide Accumulation in Plants

Radionuclides enter the plant by two ways. First is through direct deposition on the leaf surface, i.e. through foliar absorption and second is through root uptake from the soil (Fig. 4.1). The interaction of radioactive elements in the plant occurs either in the aerial portion of the plant, i.e., or in the soil-root zone of the plant (rhizosphere).

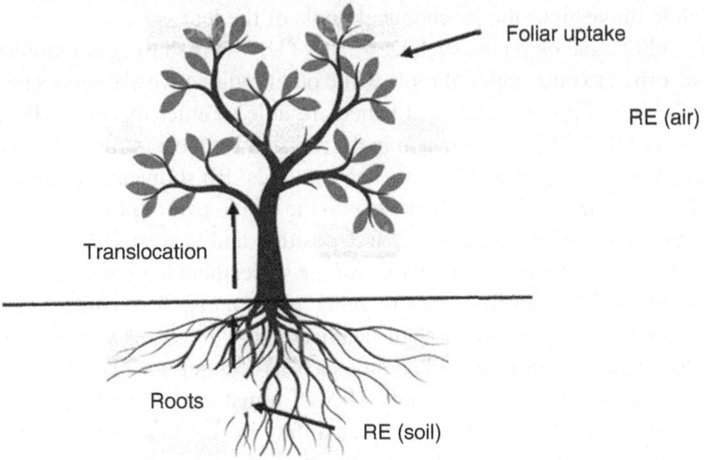

Fig. 4.1 Major routes of uptake of radioactive elements (RE)/radionuclides by plants

The root uptake has been considered as the main radionuclide transfer pathway in plants. The adsorption of radionuclides such ^{210}Pb by leaves and uptake by roots has been reported in many plant species (Chandrashekara et al. 2015). The absorption of radionuclides is followed by translocation to other plant parts particularly leaf or other organs such as stems or storage organs within the plant. Radionuclides also get distributed throughout the reproductive structures such as seeds and fruit.

Foliar uptake of radioactive particles has been commonly reported in plants. Leaf axils, leaf sheaths, grasses, and vegetation have shown accumulation of radioactive elements in the range of 0.5–5.0. After uptake, the radionuclides get transported across the cuticle and epidermis of plant leaves. The absorption of radionuclides by leaves may take place through cuticle and epidermis. The epidermal features help in retention of radioactive particles. The absorbed element get in the cuticle or cell wall of the outer cell or translocated via phloem and further to other plant parts. The phenomenon of uptake of radioactive elements is different for upper and lower leaf surfaces. The cuticle is a selectively permeable membrane, cationic, negatively charged, and hydrophobic. The cuticle is made up of waxy substances or cutin. The cuticular wax is a non-cellular and insoluble substance composed of long-chain fatty acids, alcohols, and esters. Being hydrophobic, it reduces the contact of the surface contaminant with the leaf. Fine, dense pubescence entraps radioactive particles. Hence the structural components of the cuticle and the epidermis of the leaf act as barriers to ionic movement toward the interior of the leaf. The radioactive ion cross the plasmalemma of the epidermal cell and reach the protoplasmic pool. Uptake of radionuclides through the epidermis and across the plasmalemma of the epidermal cell occurs by active and passive transport mechanism (Tagami 2012). Active transport moves the elements into the protoplasm while in passive diffusion elements move into the apparent free space (AFS) of the cell walls (Greger 2004). Active transport is associated with biosynthetic processes such as oxidative phosphorylation. Passive transport of radionuclides occurs by leaf and is affected by other cations that compete for sites on exchange compounds (Ambe et al. 1999). The radionuclide moves into the biochemical pools of the leaf.

Some radioactive elements such as ^{234}U, ^{238}U, ^{238}Pu show less mobility, hence remain adsorbed to outer layer of roots while other radioactive elements such as ^{85}Sr, ^{90}Sr, ^{137}Cs show high mobility and hence are able to enter the plant (Baeza et al. 1999). Most of the ^{137}Cs gets retained in the roots and some part (only 25%) of it gets translocated to shoots. Some elements such as^{137}Cs, Rb show accumulation in fruit and seeds. The translocation of elements into the edible part of plants depends on the element, the plant and the time between deposition and harvest. The rate of translocation varies for each species. *Calluna vulgaris* (heather) have shown high rate of translocation of ^{134}Cs from leaves to other plant parts. In contrast members of Ericaceae such as *Erica tetralix* (bell heather) and *Vaccinium myrtillus* (bilberry) showed low translocation rate. Elements such as technetium (Tc), tellurium (Te), iodine (I), and cesium (Cs) show only 10% of translocation to the grain of cereal crops (Echevarria et al. 1997). Plants differ in their ability to accumulate radionuclides. Elements such as Cs are absorbed by leaves primarily via metabolic processes linked to developmental of the plants (Carvalho et al. 2006). About 5–30%

of the Cs is absorbed by plant leaves and a substantial portion is translocated to other plant parts. Apart from this, Cs is also absorbed by roots. The ability of plant species to accumulate ^{137}Cs in the aboveground parts differs (Zhu and Smolders 2000). High level of ^{137}Cs level has been found in the wood xylem of tree trunks as well as storage roots and leaves. The accumulation of ^{137}Cs varied from 2- to 4-fold within cereals and about 27-fold in field crops (Sanzharova et al. 1997). *Amaranthus* species *A. cruentus, A. retro-flexus,* and *A. caudatus* have reported high level of ^{137}Cs accumulation in the aboveground parts (Dushenkov et al. 1999). In *Picea abies* high accumulation of ^{134}Cs has been found in roots. The sunflower has been reported to absorb 150 μg Cs in 100 h whereas a vetiver (*Vetiveria zizanioides*) absorbed 61% of ^{137}Cs in 168 h (Singh et al. 2009). Cs is adsorbed onto the cell surface. The sunflower plants showed potential to absorb radionuclides ^{134}Cs and ^{60}Co from hydroponic media. Most of the ^{134}Cs accumulated on the leaves, then inside the stem and the lowest at the root (Achmad and Hadiyanto 2018).

Strontium (Sr), barium (Ba), and radium (Ra) are the elements considered to be analogous to calcium (Ca). Calcium and Sr exist largely as immobile complexes with glutauronic acids and pectate in the plant tissue. The root uptake of Sr from soil includes mass-flow and exchange diffusion. ^{90}Sr is found to be more concentrated in leaves than in storage roots. In *Picea abies*, ^{85}Sr is predominantly accumulated in fine-roots. Accumulation of Sr in the range of 10–1500 μg DW^{-1} has been noted in plants. After foliar deposition Sr becomes moderately mobile in plants. *Vaccinium myrtillus* has been found to be a hyperaccumulator of beryllium (Be). High concentrations of Ba (barium) (up to 10,000 μg g DW^{-1}) have been reported for different trees and shrubs (Carini et al. 2016). Radium (^{226}Ra) accumulation has been found to be more in fruit than that of leaves and roots. Selective uptake of Gallium (Ga) by plants has been reported. Plants take up indium (I) and concentrations up to 100 μg g DW^{-1} has been reported from pine trees. Around 17,000 μg g DW^{-1} has been reported from the flowers of *Galium* sp.

Tellurium (Te) is rare radioactive element. Bacteria methylate Te and reduction of tellurite to Te is also influenced by micro-organisms. In onion and garlic high concentrations of Te (300 μg g DW^{-1}) have been reported. Woody seed plants can accumulate high levels of yttrium(Y) up to 700 μg g DW^{-1}. The translocation of Cerium (Ce) is very low, i.e. after foliar application and after root uptake. Higher concentration of ^{144}Ce has been found in the shoots than in roots. Ce applied to foliage is also absorbed to a lesser extent, which is probably the reason for the low translocation. Niobium (Nb) (^{95}Nb) is generally complexed with organic agents and is relative mobile and therefore available for uptake by plants. Accumulation ranging from 1 to 10 μg g DW^{-1} has been reported in plants such as *Rubus arcticus*. Vanadium (V) is easily taken up by plant roots and absorption is passive. Uptake of vanadium depends upon pH. A high pH decreases the uptake. It is more rapidly absorbed by roots as VO^{3-} and HVO_4^{2-} species under neutral and alkaline soil solutions. Biotransformation of V from vanadate (VO_3^-) to vanadyl (VO^{2+}) occurs during plant uptake. Rhenium is found in anionic form as ReO_4^-. It is taken up by plants and concentrations up to 70–300 μg g DW^{-1} have been reported. After foliar deposition ^{183}Re shows medium mobility in the plant body. Radioactive cobalt

(^{58}Co) is easily taken up through the cuticle. Concentrations reported in terrestrial plants range from 0.4 to 200 ng g DW^{-1}. Iridium (Ir) uptake by plants has also been reported. Terrestrial plants contain concentration of 20 ng g DW^{-1} and accumulate it in the leaf margins. After foliar deposition the element has been found to immobile in the plant.

Uptake of technetium (Tc) in plants occurs in the form of TcO^{4-}. It is transported as TcO_4^- across plasma membrane into the cytosol. The Tc uptake occurs via active mechanism involving transport of TcO_4^- across the cell membrane and the reduction of Tc^{7+}. Reduction from $TcO_4^-(Tc^{7+})$ to Tc^{5+} is probably mediated by ferredoxin within the chloroplast, and up to 10 bioorganic complexes with Tc were found in leaves.

In greenhouse experiments sunflower (*Helianthus annuus*) showed U removal capacity of more than 95% from contaminated water in 24 h (Dushenkov et al. 1997; Sorochinsky et al. 1998). In pilot rhizofiltration system sunflower plants showed U accumulation of more than 1% in roots (Dushenkov et al. 1997). The sunflower, vetiver, and purple guinea grass showed ability to absorb U from water. All three plants could accumulate U in their roots. Sunflower showed the best capacity for removal of U. The non-addition of plant nutrients to the culture solution prevent competition between U and nutrient absorption. The amount of U in the plants increased with the length of the exposure period (Roongtanakiat et al. 2010). Sunflower plants also showed capacity to accumulate ^{134}Cs and ^{60}Co mostly in the leaves and roots (Achmad and Hadiyanto 2018). Dushenkov et al. (1997) reported similar results where the plants grown in a radioactive solution for long showed more radionuclides were absorption. Pilot scale studies also proved that aquatic plants possess high capacity to treat radionuclide-contaminated water. Greenhouse experiments demonstrated successful removal of Cs, Sr, and U from contaminated water (Dushenkov et al. 1997). Aquatic plants accumulates significant amounts of radionuclides depicting a high bioconcentration factor for ^{90}Sr, ^{137}Cs in case of *Cladophora glomerata,* and ^{90}Sr and ^{137}Cs for *Elodea canadensis*. *Lemna aoukikusa*, a floating vascular plant showed successful elimination of Cs and I from contaminated water. Cyanobacterium *Stigonema ocellatum* (NIES-2131) show high capacity to remove elements like Sr and I. A large number of radionuclide such as ^3H, U, Pu, ^{137}Cs and ^{90}Sr has been treated using plants (Negri and Hinchman 2000).

Hydroponic studies demonstrated that U uptake occurs at pH 5. At this pH, uranium occurs as uranyl (UO_2^{2+}) cation which is readily taken up and translocated by plants (Ebbs et al. 1998). Uranium forms stable uranium-phosphate complexes in roots and this prevents translocation of uranium to aboveground plant parts. In contrast, elements such as ^{90}Sr show about 80% of localization in the shoots. The cultivars of the same species show variation in accumulation of radiocesium (Cs). Rubidium (Rb) generally concentrates in flowers and young leaves. Studies have demonstrated accumulation of ^{86}Rb within reproductive structures and young tissues. *Phaseolus vulgaris* (bush bean) showed ten times higher absorption of ^{89}Sr than *Zea mays* (maize) after 72 h of treatment and two times higher than *Raphanus sativus* (radish) and *Lactuca sativa* (lettuce). Pine and aster exhibit ability to accumulate U. Grouseberry, larch, fireweed, and grass accumulate high levels of ^{226}Ra from the soil.

4.2.3 Effect of Radionuclide Accumulation on Plant Growth

The growth of plants gets affected after accumulation at high concentration of radionuclides in the plant tissues (Markovic and Stevovic 2019). Morphological changes, reduction of stem growth and root biomass are some of the changes noted in plants in response to high radionuclide accumulation. The decrease in growth rate in plants depends upon the rate of translocation from root to shoot. Uranium accumulation reduced plant biomass (ash weight) in plants. This may be due to detrimental effects of U on plant growth.

Cesium accumulation in cress, i.e. *Lepidium sativum* plants grown in hydroponic culture lead to high accumulation in leaves after both root and foliar treatments. Exposure to high concentration of cesium (3 mM) affected water uptake, tissue hydration (FW/DW), and production of biomass (DW). The gas exchange parameters such as stomatal conductance (C) and transpiration rate (E) also showed strong inhibition while the rate of photosynthesis did not get altered significantly. Changes in photochemistry of photosystem II (PSII) related to the alteration in photosynthetic potential. Decreased stomatal opening affected the rate of transpiration and uptake of water. The decrease in tissue hydration decreases photosynthetic CO_2 assimilation, synthesis of organic matter, and affects light reactions of photosynthesis (Bystrzejewska-Piotrowska and Urban 2003).

A greenhouse experiment showed that growth attributes such as relative growth rate, net assimilation rate, leaf are index, specific leaf area, dry matter allocation and production of reproductive organs showed a decrease as the radionuclide content in the plant increased. The decrease has been noted in plants such as *Cakile maritima*, *Senecio glaucus* and *Rumex pictus* at different stages of growth (seedling, juvenile, flowering, fruiting and senescing). High level of radionuclide accumulation in these plants affected dry matter allocation and root to shoot weight ratio (Hegazy et al. 2011).

4.3 Conclusions

Plants possess capacity to treat radioactive particles. The removal of radionuclides in plants occurs via adsorption by leaves or absorption by roots. The uptake and removal of radionuclides from the environment is regulated by various physical and environmental factors. The radionuclide removal capacity also varies for each plant species depending upon its affinity for taking up radioactive elements followed by translocation and/or accumulation in the plant tissues. The radionuclide accumulation capacity of the plants can be exploited for developing large scale technologies for treatment of water and soil contaminated with radioactive waste provided that we have a better understanding of the physiological and molecular mechanisms related to radionuclide accumulation in plants.

References

Achmad CAA, Hadiyanto (2018) Root uptake and distribution of radionuclides ^{134}Cs and ^{60}Co in sunflower plants (*Helianthus annuus. L*). E3S Web of Conferences 73:05027. https://doi.org/10.1051/e3sconf/2018730

Ambe S, Shinonaga T, Ozaki T, Enomoto S, Yasuda H, Uchida S (1999) Ion competition effects on the selective absorption of radionuclides by komatsuna (*Brassica rapa* var. perviridis). Environ Exp Bot 41:185–194. https://doi.org/10.1016/S0098-8472(99)00003-9

Baeza A, Paniagua JM, Rufo M, Sterling A, Barandica J (1999) Radiocaesium and radiostrontium uptake by turnips and broad beans via leaf and root absorption. Appl Radiat Isot 50:467–474. https://doi.org/10.1016/S0969-8043(98)00078-5

Bell JNB, Minski MJ, Grogan HA (1988) Plant uptake of radionuclides. Soil Use Manag 4:76–84. https://doi.org/10.1111/j.1475-2743.1988.tb00740.x

Bystrzejewska-Piotrowska G, Urban PL (2003) Accumulation of cesium in leaves of *Lepidium sativum* and its influence on photosynthesis and transpiration. Acta Biol Cracov Ser Bot 45:131–137

Carini F, Brambilla M, Mitchell NG, Tsukada H (2016) Radionuclides behavior in fruit plants. In: Takahashi T (ed) Radiological issues for Fukushima's revitalized future. pp 159–172. https://doi.org/10.1007/978-4-431-55848-4_14

Carvalho C, Mosquera B, Anjos RM, Sanches N, Bastos J, Macario K, Veiga R (2006) Accumulation and long-term behavior of radiocaesium in tropical plants. Braz J 36:1345–1348. https://doi.org/10.1590/S0103-97332006000800002

Chandrashekara K, Karunakara N, Somashekarappa HM (2015) Activity distribution and uptake of radionuclides in medicinal plants of coastal Karnataka, India. Res J Recent Sci 4:101–109

Dhir B (2013) Phytoremediation: role of aquatic plants in environmental clean-up. Springer, India. https://doi.org/10.1007/978-81-322-1307-9

Duong VH, Nguyen TD, Kocsis E, Csordas A, Hegedus M, Kovacs T (2021) Transfer of radionuclides from soil to *Acacia auriculiformis* trees in high radioactive background areas in North Vietnam. J Environ Radioact 229–230:106530

Dushenkov S (2003) Trends in phytoremediation of radionuclides. Plant Soil 249:167–175. https://www.jstor.org/stable/24129337

Dushenkov S, Vasudev D, Kapulnik Y, Gleba D, Fleisher D, Ting KC, Ensley B (1997) Removal of uranium from water using terrestrial plants. Environ Sci Technol 311:23468–23474. https://doi.org/10.1021/es970220l

Dushenkov S, Mikheev A, Prokhnevsky A, Ruchko M, Soroch-insky B (1999) Phytoremediation of radiocesium-contaminated soil in the vicinity of Chernobyl, Ukraine. Environ Sci Technol 33:469–475. https://doi.org/10.1021/es980788+

Ebbs SD, Brady DJ, Kochian LV (1998) Role of uranium speciation in the uptake and translocation of uranium by plants. J Exp Bot 49:1183–1190. https://www.jstor.org/stable/23696367

Echevarria G, Vong PC, Morel JL (1997) Bioavailability of Technetium-99 as affected by plant species and growth, application form, and soil incubation. J Environ Qual 26:947–956. https://doi.org/10.2134/jeq1997.00472425002600040004x

Echevarria G, Sheppard NI, Morel J (2001) Effect of pH on sorption of uranium in soils. J Environ Radioact 53:257–264. https://doi.org/10.1016/s0265-931x(00)00116-8

Greger M (2004) Uptake of nuclides by plants. Technical ReportTR-04-14. Swedish Nuclear Fueland Waste Management Co.

Hattink J, de Goeij JJM, Wolterbeek HTH (2000) Uptake kinetics of ^{99}Tc in common duckweed. Environ Exp Bot 44:9–22. https://doi.org/10.1016/S0098-8472(00)00045-9

Hattink J, Weltje L, Wolterbeek HT, de Goeij JJM (2004) Accumulation, elimination and retention of ^{99}Tc by duckweed. J Radioanal Chem 259:135–139. https://doi.org/10.1016/S0048-9697(02)00679-4

Hegazy AK, Emam MH, Alatar AA (2011) Growth and reproductive attributes of radionuclide phytoremediators in the Mediterranean coastal black sands. Afr J Biotechnol 10:16781–16792. https://doi.org/10.5897/AJB11.1257

Huang JW, Blaylock MJ, Kapulnik Y, Ensley BD (1998) Phytoremediation of uranium-contaminated soils: role of organic acids in triggering uranium hyperaccumulation in plants. Environ Sci Technol 32:2004–2008. https://doi.org/10.1021/es971027u

Kabata-Pendias A, Pendias H (1996) Trace elements in soils and plants. CRC Press, Boca Raton, FL, 365 pp

Lajayer BA, Moghadam NK, Maghsoodi MR, Ghorbanpour M, Kariman K (2019) Phytoextraction of HMs from contaminated soil, water and atmosphere using ornamental plants: mechanisms and efficiency improvement strategies. Environ Sci Pollut Res 26:8468–8484

Lasat MM, Ebbs SD, Kochian LV (1998) Phytoremediation of a radiocaesium-contaminated soil: evaluation of caesium-137 bioaccumulation in shoots of three plant species. J Environ Qual 27:165–169. https://doi.org/10.2134/jeq1998.00472425002700010023x

Malhotra R, Agarwal S, Gauba JP (2014) Phytoremediation of radioactive metals. J Civil Eng Environ Technol 1:75–79

Marčiulionienė D, Lukšienė B, Jefanova O (2015) Accumulation and translocation peculiarities of ^{137}Cs and ^{40}K in the soil—plant system. J Environ Radioact 150:86–92. https://doi.org/10.1016/j.jenvrad.2015.07.012

Markovic J, Stevovic S (2019) Radioactive isotopes in soils and their impact on plant growth. https://doi.org/10.5772/intechopen.81881

Negri CM, Hinchman RR (2000) The use of plants for the treatment of radionuclides. In: Raskin I (ed) Phytoremediation of toxic metals: using plants to clean up the environment. Wiley-Interscience, John Wiley and Sons, Inc., New York, pp 107–132

Ould-Dada Z, Fairlie I, Read C (2001) Transfer of radioactivity to fruit: significant radionuclides and speciation. J Environ Radioact 52:159–174. https://doi.org/10.1016/s0265-931x(00)00031-x

Peles JD, Smith MH, Brisbin IH Jr (2002) Ecological half-life of ^{137}Cs in plants associated with a contaminated stream. J Environ Radioact 59:169–178. https://doi.org/10.1016/s0265-931x(01)00069-8

Planinšek P, Smodiš B, Benedik L (2018) Uptake of natural radionuclides from contaminated soil into vegetables and consequent dose assessment. J Radioanal Nuclear Chem 318:2373–2379

Prasad MNV (2007) Aquatic plants in phytotechnology. In: Singh SN, Tripathi RD (eds) Environmental bioremediation technologies. Springer, pp v259–v274. https://doi.org/10.1007/978-3-540-34793-4_11

Roongtanakiat N, Sudsawad P, Ngernvijit N (2010) Uranium absorption ability of sunflower, vetiver and purple Guinea Grass Kasetsart. J Nat Sci 44:182–190

Sanzharova NI, Fesenko SF, Lisyanskii KB, Kuznetsov VK, Ab-ramova TN, Kotik VA (1997) Forms and accumulation dynamics of ^{137}Cs in crops after the accident at the Chernobyl Nuclear Power Plant. Pochvodenie 2:159–164

Sarma H, Prasad MNV (2018) Metabolic engineering of rhizobacteria associated with plants for remediation of toxic metals and metalloids. In: Prasad MNV (ed) Transgenic plant technology. Elsevier. eBook ISBN: 9780128143902. https://doi.org/10.1016/B978-0-12-814389-6.00014-6

Seel JF, Whicker FW, Adriano D (1995) Uptake of ^{137}Cs in vegetable crops grown on a contaminated lake bed. Health Phys 68:793–799. https://doi.org/10.1097/00004032-199506000-00005

Singh S, Thorat V, Kaushik CP, Raj K, Susan Eapen S, D'Souza SF (2009) Potential of *Chromolaena odorata* for phytoremediation of ^{137}Cs from solution and low level nuclear waste. Ecotoxicol Environ 69:743–745. https://doi.org/10.1016/j.jhazmat.2008.05.097

Sorochinsky BV, Mikheev AN, Kuchko MV, Prokhrevsky AT (1998) Decontamination of small water reservoirs of the 10-km zone of Chernobyl by rhizofiltration. In: Problems of Chernobyl exclusion zone, pp 97–102

Tagami K (2012) Transfer of radionuclides to the higher plants through direct deposition and root uptake pathways. J Open Access 61:267–279. https://doi.org/10.3769/radioisotopes.61.267

Thakare M, Sarma H, Datar S, Roy A, Pawar P, Gupta K, Pandit S, Prasad R (2021) Understanding the holistic approach to plant-microbe remediation technologies for removing heavy metals and radionuclides from soil. Curr Res Biotechnol 3:84–98. https://doi.org/10.1016/j.crbiot.2021.02.004

Yadav D, Kumar P (2019) Phytoremediaton of hazardous radioactive waste. In: Saleh HE-D (ed) Assessment and management of radioactive and electronic wastes. IntechOpen. https://doi.org/10.5772/intechopen.88055

Yan L, Van Le Q, Sonne C, Yang Y, Yang H, Gu H, Ma NL, Lam SS, Peng W (2021) Phytoremediation of radionuclides in soil, sediments and water. J Hazard Mater 407:124771. https://doi.org/10.1016/j.jhazmat.2020.124771

Zhu YG, Shaw G (2000) Soil contamination with radionuclides and potential remediation. Chemosphere 41:121–128. https://doi.org/10.1016/s0045-6535(99)00398-7

Zhu YG, Smolders E (2000) Plant uptake of radiocaesium: a review of mechanisms, regulation and application. J Exp Bot 51:1635–1645. https://doi.org/10.1093/jexbot/51.351.1635

Phytoremediation of Heavy Metals and Radionuclides: Sustainable Approach to Environmental Management

5

Prakriti Jhilta, Bhawna Dipta, and Anchal Rana

Abstract

Heavy metals (HMs) and radionuclides pose a serious threat to human health because of their ubiquity, non-biodegradability, and long-term persistence in the environment. The presence of high amounts of these pollutants in soil has a detrimental influence on soil fertility, agricultural productivity, and yield. HMs and radionuclides contamination have been remediated using a variety of conventional approaches. However, these technologies have limitations, such as excessive cost, intensive labor, and alteration of the soil native microflora by affecting soil properties with the potential to pollute the environment with the release of secondary pollutants. As a result, switching to a more cost-effective and eco-friendly method is very desirable. Phytoremediation technology for HMs and radionuclides decontamination has been recognized as a novel, low-cost, and ecologically acceptable solution. The present chapter explains the major processes of phytoremediation, as well as the function of transgenic plants in increasing plant efficacy for HMs and radionuclides decontamination. The role of plant growth regulators (PGRs), beneficial microorganisms, arbuscular mycorrhizal fungi (AMF), and nanoparticles (NPs) in phytoremediation is also discussed.

P. Jhilta
Dr Yashwant Singh Parmar University of Horticulture and Forestry, Nauni, Solan, Himachal Pradesh, India

B. Dipta
ICAR-Central Potato Research Institute, Shimla, Himachal Pradesh, India

A. Rana (✉)
CSIR-Institute of Himalayan Bioresource Technology, Palampur, Himachal Pradesh, India

R. Prasad (ed.), *Phytoremediation for Environmental Sustainability*,
https://doi.org/10.1007/978-981-16-5621-7_5

Keywords

Arsenic · Gibberellic acid · Metallothionein · Phytostabilization · Nanoparticles · Heavy metals · Radionuclides · Phytoremediation

5.1 Introduction

Pollutants are substances that are found at higher concentrations in the environment than their natural abundance and have a negative impact on the ecosystem. Organic pollutants include benzene, toluene, polychlorinated biphenyls (PCBs), polyaromatic hydrocarbons (PAHs), dioxins, nitro-aromatics, dyes, polymers, pesticides, and chlorinated organics. Inorganic pollutants, on the other hand, comprise a variety of toxic heavy metals (HMs) and radionuclides. HMs are highly notorious contaminants because of their abundance, non-biodegradability, and long-term persistence in the environment. They include copper (Cu), cadmium (Cd), chromium (Cr), cobalt (Co), zinc (Zn), iron (Fe), nickel (Ni), mercury (Hg), lead (Pb), arsenic (As), aluminum (Al), silver (Ag), and platinum (Pt). HMs pollute the soil and water and have toxic, genotoxic, teratogenic, and mutagenic impacts on living organisms. Once accumulated in soils, these metals have an inverse effect on soil fertility and diminish agricultural production. Furthermore, even at low concentrations, they induce endocrine disruption and neurological problems. They are classified as priority pollutants by environmental protection agencies across the globe because they can pose serious health risks. Like HMs, radionuclides cannot be naturally or synthetically degraded. In addition, numerous studies have reported that cesium (^{37}Cs) and strontium (^{90}Sr) are not removed from the top 0.4 meters of soil even under high rainfall, and the migration rate from the top few centimeters of soil is slow. Therefore, radionuclides have become a threat to public health when exposed and/or deposited in the soil and water. Moreover, exposure to radioactivity is a common and natural phenomenon. For instance, exposure to cosmic radiation, radon (Rn) gas from rocks and soil, or potassium (^{40}K) through food.

Furthermore, elevated levels of these pollutants in soils have a negative impact on crop development, and yields by dissolving cell organelles and disrupting membranes, acting as genotoxic substances, disrupting physiological processes like photosynthesis, or inactivating respiration, protein synthesis, and carbohydrate metabolism. Hence, remediation of these pollutants has become a necessity to sustain a stable environment. Several traditional remediation approaches have been explored to remediate HMs and radionuclides contamination. However, these technologies are costly and hazardous with the potential to release secondary pollutants into the environment. Therefore, adaptation to an alternative, cost-effective, eco-friendly technology having high removal efficiency is highly desirable. Phytoremediation has been identified as an emerging, low-cost, and eco-sustainable approach to HMs and radionuclides decontamination (Sarma et al. 2021). Phytoremediation uses plants to remove, degrade, or detoxify toxic metals (Nedjimi 2021; Thakare et al. 2021). Phytoextraction, phytostabilization, phytovolatilization, phytodegradation, and rhizodegradation are types of phytoremediation techniques that have been utilized for soil decontamination. The present chapter discusses various sources and toxic effects of HMs and radionuclides, plant strategies for avoiding and/or tolerating hazardous metals, as well as the importance of genetic engineering (GE) in improving efficiency of phytoremediation. The role of plant growth regulators (PGRs), beneficial microorganisms, arbuscular mycorrhizal fungi (AMF), and nanoparticles (NPs) in assisting phytoremediation is also highlighted.

5.2 Heavy Metals (HMs) and Radionuclides

Heavy metals (HMs) are defined as elements with numerous metallic properties, i.e., ductility, conductivity, stability, ligand specificity, etc., an atomic number >20, and a density >5 g/cm^3. They are generally present in the environment at a trace level (<1 g/kg/ppb). HMs can also be classified into essential and non-essential HMs. Essential HMs consist of Co, Cr, Cu, Fe, Mn, Ni, and Zn and non-essential includes Pb, Cd, and Hg. In addition, according to their level of toxicity, they can also be grouped as extremely poisonous, moderately poisonous, and relatively less poisonous. Radionuclides, on the other hand, are a class of chemicals where the nucleus of the atom is unstable. Radionuclides achieve stability through changes in the nucleus (spontaneous fission, emission of alpha particles, or conversion of neutrons to protons or the reverse). The emission of radionuclides from nuclear power plants, as well as their subsequent mobility in the environment, is a subject of intense public concern. HMs and radionuclides are emitted from both natural as well as anthropogenic sources, such as automobile exhaust, smelting, warfare, electronic industries, agrochemical use, irrigation, waste disposal, fossil fuel consumption, nuclear plants, and nuclear weapons testing as shown in Table 5.1.

The accumulation of HMs in the soil causes severe health problems for plants, animals, and humans. According to the United States Environmental Protection Agency (USEPA), soil HM contamination has caused health issues for about ten million humans all over the world. As a result, HM accumulation in plants via the soil–root interface is a serious threat (Sakizadeha and Ghorbani 2017). The most well-known case of Hg poisoning is the Minamata disease in Japan. Another example of HM poisoning is the disaster in the Spanish national reserve. The water in the reservoir was polluted with traces of HMs, mineral sediment, and acidic chemicals. In addition, Hinckley water contamination is another most common example of Cr contamination in the world. Lead poisoning is also not uncommon and is probably the best example of an HM poisoning. It has been reported that 890,000 children aged 1–5 have elevated blood lead levels in the USA (Pirkle et al. 1998). The Kyshtym disaster (1957), Stationary Low-Power Reactor Number One, also known as SL-1 accident (1961), Three Mile Island accident (1979), Chernobyl accident (1986), and Fukushima Daiichi disaster (2011) are a few major nuclear disasters in history. The Chernobyl disaster in Ukraine is a common example. The Chernobyl accident happened in a dangerously constructed nuclear power reactor

Table 5.1 Sources of heavy metals (HMs) and radionuclides in the environment

Contaminant	Sources
Heavy metals	
Zinc (Zn)	Electroplating and smelting
Cadmium (Cd)	Smelting, incineration, fuel combustion, waste batteries, e-waste, and paint sludge
Copper (Cu)	Mining, electroplating, and smelting operations
Mercury (Hg)	Chlor-alkali plants, thermal power plants, electrical appliances, fluorescent lamps, and hospital waste
Chromium (Cr)	Mining, leather tanning, industrial coolants, and chromium salt manufacturers
Lead (Pb)	Lead-acid batteries, e-waste, coal-based thermal power plants, bangle industry, ceramics, paints, and smelting operations
Arsenic (As)	Geogenic/natural processes, smelting operations, thermal power plants, and fuel-burning
Cobalt (Co)	Volcanic emissions, weathering of rocks, and decomposition of plant waste
Nickel (Ni)	Smelting operations, battery industry, and thermal power plants
Manganese (Mn)	Mining, alloy production, goods processing, iron-manganese operations, welding, and agrochemical production
Iron (Fe)	Geogenic, industrial, agricultural, pharmaceutical, domestic effluents, and atmospheric sources
Aluminum (Al)	Mining and processing of aluminum ores or the production of aluminum metal, alloys, and compounds, coal-fired power plants and incinerators
Radionuclides	
Uranium ($^{235, 238}U$)	Mining/milling of uranium ores, geological repositories of nuclear waste, testing of nuclear weapons, and natural sources
Thorium (^{232}Th)	Natural, mining, milling and processing, phosphate fertilizer production, tin processing, industrial boilers, and military operations
Strontium ($^{89, 90}Sr$)	Spent nuclear fuel, nuclear accidents, nuclear fallout, nuclear fission, nuclear weapons testing, geological repository of nuclear waste, and radioactive storage leaking
Radium ($^{226, 228}Ra$)	Decay product of U and Th from mill tailing and production of phosphate fertilizers
Cobalt (^{60}Co)	Car, truck, and airplane exhausts, burning coal and oil, industrial processes, and nuclear medicines
Iodine (^{131}I)	Nuclear tests, fuel reprocessing, and spent nuclear fuel
Cesium (^{137}Cs)	Nuclear accidents and weapons testing
Carbon (^{14}C)	Natural and nuclear weapons explosions
Tritium (^{3}H)	Nuclear accidents and testing of nuclear weapons
Potassium (^{40}K)	Natural
Plutonium (^{239}Pu)	Geological repositories of nuclear waste, nuclear accidents, testing of nuclear weapons, and fuel reprocessing
Radon ($^{220,226}Rn$)	Decay product of U and Th from mill tailing

with a total meltdown of the core and 10 days of free emission of radionuclides into the atmosphere. In addition, nuclear disasters, such as Fukushima, have contaminated coastal ecosystems by dispersing radionuclides. Several amendments' applications, independently of their type and concentration, reduced their concentrations in the soil available fraction and the soil leachates. Any change in the concentration of these metals will either cause deficiency or will interfere with cellular functions, ultimately adversely affecting the growth of plants, as presented in Table 5.2.

5.3 Phytoremediation: An Environmental Tool for the Reclamation of Contaminated Sites

Phytoremediation is a broad concept that refers to a variety of processes involving plant–soil–atmosphere interactions. It is an emerging technology that involves the use of plants to extract, sequester, degrade, or immobilize pollutants from the soil and water. Potential plants for phytoremediation usually possess four important characteristics; (1) rapid growth and high biomass, (2) abstruse root system, (3) harvestable, and (4) accumulation of excessive concentration of pollutants in the shoots. The ability of plants to remove HMs and radionuclides from soils has been reported by many researchers. *Eichhornia crassipes* roots removed 54% of the initial U within 4 min of contact time (Bhainsa and D'Souza 2001). Entry et al. (2001) compared the potential of bahiagrass, Johnson grass, and switchgrass to accumulate ^{137}Cs and ^{90}Sr from contaminated soils in the presence and absence of either sphagnum peat or poultry litter amendments. Johnson grass growing on soil treated with chicken litter showed the highest accumulation of these radionuclides. Among three plants, viz., Indian mustard, redroot pigweed, and tepary bean, redroot pigweed showed the highest accumulation of ^{137}Cs and ^{90}Sr (Fuhrmann et al. 2002). Bystrzejewska-Piotrowska and Urban (2004) reported that onion plants (*Allium cepa*) may play an important role in the ^{137}Cs recycling by facilitating the transfer of fallout ^{137}Cs to the soil. Eapen et al. (2006) reported that *Calotropis gigantea* plants accumulated ^{90}Sr and ^{137}Cs more in their roots than in their shoots. Sasmaz and Sasmaz (2009) reported that *Astragalus gummifer* can be utilized to rehabilitate the soil contaminated by Sr. In another study, *Ocimum basilicum* seeds showed significant uptake of both ^{137}Cs and ^{90}Sr. The maximum adsorption capacity was 160 mg Cs g^{-1} and 247 mg Sr g^{-1} seed dry weight (Chakraborty et al. 2007). *Melastoma malabathricum* L. was reported to accumulate a relatively high range of Pb and As concentration (Selamat et al. 2014). In comparison to other plants, *Miscanthus floridulus* and *Cyperus iria* are reported to have the potential for phytoremediation of radionuclide ^{232}Th in the soil (Yan 2016). In another study, Bhat et al. (2016) reported that *Centella asiatica* can uptake and accumulate Fe significantly in the aerial parts. Silva et al. (2018) suggested that *Cassia alata* plants can be used for the phytoremediation of Cd. *Hypnum plumaeforme* has been described as a possible Rn pollution accumulator plant, as well as a possible indicator plant for Rn pollution monitoring (Zhang et al. 2019). Phytoremediation

Table 5.2 Effects of heavy metals (HMs) and radionuclides on plants

Contaminant	Harmful effects
Zn	Excessive concentration of Zn hampers growth and development, metabolism and causes oxidative damage in plants. It also affects the catalytic efficiency of enzymes, which results in retarded growth and ultimately causes senescence
Cd	Elevated levels of Cd show symptoms of injury, i.e., chlorosis, inhibition of growth, root tips browning, and finally death. It might also reduce the absorption of nitrate and its transport from roots to shoots by inhibiting nitrate reductase activity. It can also induce lipid peroxidation, inhibit chlorophyll biosynthesis, and reduce the activity of enzymes that are involved in the fixation of CO_2
Cu	Cu in the soil is cytotoxic. Elevated concentrations of Cu cause oxidative stress and the development of reactive oxygen species (ROS). It can disturb metabolic pathways and also damage macromolecules. Cu causes leaf chlorosis and plant growth retardation
Hg	A high level of Hg^{2+} inhibits mitochondrial function and causes oxidative stress by activating the development of ROS. This ultimately leads to the disruption of biomembrane lipids and plant cellular metabolism
Cr	A high concentration of Cr affects the germination of seeds. It can also interfere with the process of photosynthesis, i.e., CO_2 fixation, electron transport, photo-phosphorylation, and enzyme activities
Pb	The toxic concentration adversely affects growth and photosynthetic processes by inhibiting the activity of carboxylation enzymes. It also inhibits elongation of roots and stems and expansion of leaves. Pb poisoning also impairs mineral nutrition by inhibiting enzyme activity, creating a water imbalance, altering membrane permeability, and disrupting mineral nutrition
As	Roots are generally the first tissue to be exposed to As, where the metalloid inhibits root extension and proliferation. As interferes with critical metabolic processes, which can lead to death. Antioxidant resistance systems are triggered by As exposure
Co	Crop dropping, suppression of greening, discolored veins, premature leaf closure, and decreased shoot weight are the toxic effects of Co
Ni	Excessive Ni^{2+} in the soil induces a variety of physiological changes and toxicity in plants, including chlorosis and necrosis. A high Ni^{2+} environment causes nutrient imbalance, which leads to cell membrane dysfunction
Mn	The accumulation of too much Mn in the leaves reduces the photosynthetic rate. Mn toxicity is characterized by necrotic brown spotting on leaves, petioles, and stems. The symptom is commonly known as "crinkle leaf." It is also linked to browning and chlorosis in these tissues. Excess Mn is said to block a Fe-related mechanism, preventing chlorophyll synthesis
Fe	The excess Fe^{2+} produces free radicals, which irreversibly destroy cellular structure and damage membranes, DNA, and proteins
Al	An elevated concentration of Al causes a reduction in plant growth, thickening of roots, root tip dieback, yellowing and purpling, wilting, loss of apical dominance, and sometimes loss of geotropism occurs. Al also reduces the performance of several enzymes such as ATPases
Radionuclides	An elevated concentration of U can cause macroscopic effects such as stunted growth and reduced biomass production. U can interact with macromolecules and can affect enzyme capacities and membrane permeability, inducing oxidative stress-related responses in plants. The higher concentration of Sr damages various processes of photosynthesis, such as energy absorption, energy transfer, and photosynthetic carbon assimilation, and induces oxidative stress

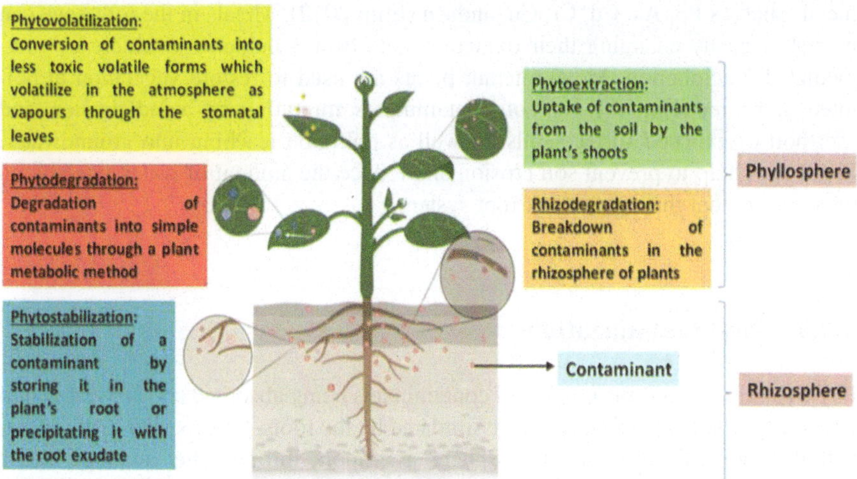

Fig. 5.1 An overview of the soil contaminant cleanup mechanisms

employs various techniques, such as phytoextraction, phytostabilization, phytovolatilization, phytodegradation, and rhizodegradation for the remediation of polluted soil as present in Fig. 5.1.

5.3.1 Phytoextraction

Plants can absorb nutrients from the soil naturally. The absorption of chemicals through the plant's root system and the accumulation of metal and radioactive contaminants from the soil in their shoots is known as phytoextraction. Phytoextraction is also called phytoaccumulation. The contaminants, as well as plant biomass containing metals and radionuclides, have been extracted during the post-harvest process (Raskin and Ensley 2000). Phytoextraction has been applied to many contaminants such as metals-Ag, Cd, Co, Cr, Cu, Hg, Mn, Mo, Ni, Pb, and Zn (Salt et al. 1995), metalloids-As and Se (Kumar et al. 1995), and radionuclides- ^{90}Sr, ^{95}Nb, ^{99}Tc, ^{106}Ru, ^{144}Ce, ^{226,228}Ra, ^{239,240}Pu, ^{241}Am, ^{228,230,232}Th, ^{244}Cm, and ^{237}Np (Nisbet and Shaw 1994; Kabata-Pendias and Pendias 1996). The ability of a plant to translocate and accumulate contaminants varies, depending on the plant species (Susarla et al. 2002).

5.3.2 Phytostabilization

The alternative approach is to slow down contaminant movement and stabilize the contaminant by storing it in the plant roots or precipitating it with root exudate. This method is best for dealing with radionuclides with short half-lives (Lee 2013) and metals such as Pb, As, Cd, Cr, Cu, and Zn (Etim 2012). Metals in the root zone can be stabilized by changing their oxidation state from soluble to insoluble by root-mediated precipitation. Metal-tolerant plants are used to restore vegetation at polluted sites, reducing the risk of contaminants migrating by wind erosion and transport of exposed surface soils, as well as pollution leaching into groundwater. Plants also help to prevent soil erosion and reduce the amount of water available in the environment through a thick root system.

5.3.3 Phytovolatilization

Phytovolatilization is the process of contaminants being absorbed and converted into less toxic volatile forms which are assimilated by the roots, translocated to the shoot, and then volatilized in the atmosphere as vapors through the stomatal leaves (Tollsten and Muller 1996; Raskin and Ensley 2000). Phytovolatilization can occur with contaminants present in the soil, sediment, or water. This approach has the benefit of converting the contaminant, mercuric ion, into a less toxic material. The downside is that Hg emitted into the atmosphere is likely to be recycled by precipitation and then redeposited in lakes and oceans. Phytovolatilization of radionuclides, which takes advantage of a plant's ability to transpire massive quantities of water, is currently being used for tritium (^{3}H) remediation. Tritium, a radioactive hydrogen isotope with a half-life of around 12 years, decays to stable He. However, since phytovolatilization requires the release of pollutants into the environment, a risk assessment of the effects on the ecosystem and human health may be required.

5.3.4 Phytodegradation

Phytodegradation, also known as phytotransformation, is the degradation of organic contaminants into simple molecules through a plant metabolic method. Contaminant-metabolizing enzymes formed by plants can be released into the rhizosphere, where they may continue to work in contaminant transformation. Dehydrogenase, nitrogenase, laccase, and nitrilase are examples of plant-formed enzymes in plant sediments and soils and released by roots (Schnoor et al. 1995). The plant degrades the organic contaminant and uses it for its own purposes. Plants can pick up nitrate and integrate it into proteins or other nitrogen-containing compounds or it can be converted to nitrogen gas. Some organic contaminants, such as chlorinated solvents, herbicides, 2,4,6-trinitrotoluene, and trichloroethylene, are remedied through phytodegradation.

5.3.5 Rhizodegradation

Rhizodegradation is a term that describes the breakdown of pollutants in the rhizosphere of plants. Plants provide habitats for bacteria and mycorrhizal fungi to work together to degrade pollutants. The bacteria flourished in the rhizosphere, causes the contaminant to degrade. Plant exudates, such as sugar, amino acids, enzymes, and other components increase the microbial population (Shahzad et al. 2015). The rate of rhizodegradation can be accelerated by soil characteristics such as aeration and moisture content (Kirk et al. 2005). Organic chemicals such as petroleum hydrocarbons, polycyclic aromatic hydrocarbons (PAHs), chlorinated solvents, pesticides, polychlorinated biphenyls (PCBs), benzene, toluene, ethylbenzene, and xylenes are removed through rhizodegradation. Table 5.3 presents various phytoremediation techniques employed for the remediation of HMs and radionuclides.

5.4 Plants Strategies Towards Metals

Plants tolerant to the presence of high concentrations of metals in the soil are classified as metallophytes. To cope with the toxicity of high amounts of elements in the soil, metallophytes exhibit two major strategies, viz., exclusion and accumulation (Baker 1981). In exclusion, plants resist the translocation of metals to their tissues. The metal excluder restricts the amount of metal translocated from roots to shoots, thus maintaining low levels of metal concentration in the aerial sections of the plants. Exclusion involves modification of the pH in the rhizosphere by secretion of organic acids from roots which bind to the metals and decrease their bioavailability. Other mechanisms involve the accumulation of metals in cell walls. However, beyond a certain threshold dose, this mechanism usually breaks down and the metal is taken up by the roots. In *Silene paradoxa*, the generation of metal-excluding root cell walls was suggested to be one of the factors contributing to low Cu accumulation and thus limiting the Cu uptake by the root cells by decreasing their pectin concentration in the cell wall and increasing pectin methylation, thus preventing the binding of Cu (Colzi et al. 2012). Seregin et al. (2003) reported maize as an excluder plant, with its root system acting as a barrier, restricting Ni uptake by above-ground organs. In another study, Wei et al. (2005) reported *Oenothera biennis* and *Commelina communis* as Cd excluders and *Taraxacum mongolicum* as a Zn excluder.

In accumulation, metals are accumulated in a non-toxic form in the upper plant parts at both low and high concentrations. Plants can be distinguished as indicators, accumulators, and hyperaccumulators based on their ability to accumulate metals in their tissues. Indicator plants sequester metals in the above-ground aerial tissue, but the level of metal within their tissue reflects those in the surrounding soil. These plants are of biological and ecological importance since they are pollution indicators. Accumulator plant species can accumulate greater metal concentrations in the aerial portions of the plant with a shoot/root ratio of >1 (Baker 1981). Hyperaccumulator plants can accumulate extraordinarily high amounts of metals in the aerial organs, far

Table 5.3 Remediation of heavy metals (HMs) and radionuclides by different types of phytoremediation

Type	Plant species	HMs/ radionuclides	References
Phytoextraction	*Brassica juncea* and *Brassica chinensis*	U	Huang et al. (1998)
Phytovolatilization	*Arabidopsis thaliana*	Hg	Rugh et al. (1996)
Phytovolatilization	*Liriodendron tulipifera*	Hg	Rugh et al. (1998)
Phytovolatilization	*Arabidopsis* and *Brassica juncea*	Se	LeDuc et al. (2004)
Phytoextraction	*Nyssa sylvatica* and *Liquidambar styraciflua*	^{238}U and ^{232}Th	Hinton et al. (2005)
Phytoextraction	*Rumex crispus*	Zn and Cd	Zhuang et al. (2007)
Phytovolatilization	*Brassica juncea*	Hg	Moreno et al. (2008)
Phytostabilization	*Atriplex halimus subsp. schweinfurthii*	Cd	Nedjimi and Daoud (2009)
Phytoextraction	*Calotropis procera*	Pb and Cd	D'Souza et al. (2010)
Phytoextraction	*Catharanthus roseus*	^{137}Cs	Fulekar et al. (2010)
Phytoextraction	*Salix* spp. and *Helianthus annuus*	U	Mihalik et al. (2010)
Phytoextraction	*Raphanus sativus*	^{88}Sr and ^{133}Cs	Wang et al. (2012)
Phytostabilization	*Solanum nigrum*	Cd	Khan et al. (2014)
Phytostabilization	*Vigna radiata*	Cd	Prapagdee et al. (2014)
Phytoextraction	*Pteris vittata*	As	Lei et al. (2016)
Phytostabilization	*Hibiscus cannabinus*	Cd	Chen et al. (2017)
Phytostabilization	*Leptochloa fusca*	U and Pb	Ahsan et al. (2017)
Phytostabilization	*Helianthus annuus* cv. Zaria	Cd	Shahabivand et al. (2017)
Phytostabilization	*Canavalia ensiformis*	Cu	Santana et al. (2018)
Phytoextraction	*Chlorophytum laxum* R. Br	Cd	Chuaphasuk and Prapagdee (2019)
Phytoextraction	*Lepidium sativum*	Hg	Smolinska (2019)
Phytoextraction	*Rhizophora apiculata*	Mn	Khan et al. (2020)
Phytoextraction	*Vetiveria zizanioides*	U	Pentyala and Eapen (2020)

above the levels found in the majority of species, without experiencing phytotoxic effects. These plants have a high rate of metal uptake, a faster root-to-shoot translocation, and a better ability to detoxify and sequester toxic metal in their leaves (Rascio and Navari-Izzo 2011). The criterion for hyperaccumulators of Co, Cu, Cr, Pb, and Ni are plants containing over 1000 μg/g of any of these elements in the dry matter; for Mn and Zn, the criterion is 10,000 μg/g (Baker and Brooks 1989). The fate of hyperaccumulation depends on the plant species, soil physicochemical properties such as pH, cation exchange capacity (CEC), organic matter content, electrical conductivity (EC), and metal concentration in the soil. Hyperaccumulators are excellent models for studying metal control, including the physiology of metal intake, transport, and sequestration, as well as evolution and adaption in harsh settings. Above-ground parts assimilate high amounts of metal as compared to ground parts in hyperaccumulator plants. Species belonging to the family Brassicaceae, Asteraceae, Amaranthaceae, Cyperaceae, Fabaceae, Lamiaceae, Poaceae, and Euphorbiaceae have been qualified as hyperaccumulators (Table 5.4). However, high metal specificity, lower biomass production with specific ecology and requirements in terms of climate, soil characteristics, water regime, are some of the obstacles to hyperaccumulator plants based remediation technology.

HMs/radionuclides can be transferred by apoplastic and symplastic channels through the roots, stems, and leaves (Song et al. 2017). Acidification of the rhizosphere via plasma membrane proton pumps and release of ligands capable of chelating the metal allows plants to desorb metals from the soil matrix. The metal ion can be transferred through the root in a radial fashion. Before reaching the xylem for transport to the shoot, the metal passes through the epidermis, cortex, casparian strip in the endodermis, and the pericycle of the roots. Once metal reaches the xylem, it is transported to the leaves by the flow of xylem sap, where it crosses a membrane to enter the leaf tissues. Once metal penetrates the leaf tissues, it can be sequestered in numerous subcellular compartments, such as the cell wall, cytosol, and vacuole, or volatilized through the stomata. Cellular compartmentation of metals in leaves varies between hyperaccumulator species. Kupper et al. (1999) demonstrated that Zn was sequestered predominantly in the epidermal vacuoles in *Thlaspi caerulescens* leaves instead of its mesophyll cells. However, in another study, *Arabidopsis halleri* preferentially accumulated Zn in its mesophyll cells as compared to epidermal cells (Kupper et al. 2000).

5.5 Phytoremediation by Transgenic Plants

Genetic engineering (GE) is used as an efficient method for evaluation and a better understanding of various important steps at the molecular level for improving plant tolerance to various environmental stresses and metal toxicity. A gene from a foreign source, such as a plant species, bacteria, or animals, is transferred and incorporated into the genome of a target plant. The foreign gene inherited after DNA recombination confers unique traits to the plants. GE can significantly improve metal absorption, transport, oxidation, and sequestration. Important crop plants like maize, rice, and sorghum are frequently grown in acidic soils where Al toxicity is a major issue. Overproduction of citrate resulted in Al tolerance in transgenic *Nicotiana tabacum* and *Carica papaya* plants. This study demonstrates that organic acid excretion is a mechanism of Al tolerance in higher plants (de la Fuente et al. 1997). *Arabidopsis thaliana* expressing *merBpe* that encodes for organomercurial lyase (MerB) grew vigorously at a wide range of concentrations of monomethylmercuric chloride and phenylmercuric acetate (Bizily et al. 1999). *Arabidopsis thaliana* expressing *merBpe*

Table 5.4 Potential hyperaccumulator species

HMs	Hyperaccumulator species	References
Zn	*Sedum alfredii*	Yang et al. (2002)
	Potentilla griffithii	Qiu et al. (2006)
	Thlaspi caerulescens	Banasova et al. (2008)
	Justicia procumbens	Phaenark et al. (2009)
Cd	*Sedum alfredii*	Ni and Wei (2003)
	Viola baoshanensis	Liu et al. (2004)
	Thlaspi caerulescens	Banasova et al. (2008)
	Chromolaena odoratum, *Gynura pseudochina*, *Impatiens violaeflora*, and *Justicia procumbens*	Phaenark et al. (2009)
	Lonicera japonica	Liu et al. (2009)
	Prosopis laevigata	Buendia-Gongalez et al. (2010)
	Coronopus didymus	Sidhu et al. (2017)
	Vetiveria zizanioides	Kumar et al. (2018)
Cu	*Helianthus annuus* and *Hydrangea paniculata*	Forte and Mutiti (2017)
	Lactuca sativa	Shams et al. (2019)
Hg	*Mentha arvensis*	Manikandan et al. (2015)
Cr	*Leersia hexandra*	Zhang et al. (2007)
	Prosopis laevigata	Buendia-Gongalez et al. (2010)
	Iris ensata	Usman et al. (2012)
	Nopalea cochenillifera	Adki et al. (2013)
Pb	*Sesbania drummondii*	Sahi et al. (2002)
	Helianthus annuus	Boonyapookana et al. (2005)
	Colocasia esculenta	Islam et al. (2016)
	Hydrangea paniculata	Forte and Mutiti (2017)
As	*Pteris vittata*	Ma et al. (2001)
	Pteris cretica, *Pteris longifolia*, and *Pteris umbrosa*	Zhao et al. (2002)
	Pityrogramma calomelanos	Francesconi et al. (2002)
	Lemma gibba	Mkandawire and Dudel (2005)
Co	*Haumaniastrum robertii* and *Haumaniastrum katangense*	Kabeya et al. (2018)
Ni	*Sebertia acuminata*	Jaffre et al. (1976)
	Berkheya coddii	Robinson et al. (1997a)
	Alyssum bertolonii	Robinson et al. (1997b)
	Streptanthus polygaloides	Reeves et al. (1981)
Mn	*Austromyrtus bidwillii*	Bidwell et al. (2002)
	Phytolacca acinosa	Xue et al. (2004)
	Schima superba	Yang et al. (2008)
	Phytolacca americana	Pollard et al. (2009)
Fe	*Imperata cylindrica*	Rodriguez et al. (2005)
	Centella asiatica	Bhat et al. (2016)

may be used to degrade methylmercury at polluted sites and sequester Hg(II). Expression of *CAX2* (calcium exchanger 2) in *Nicotiana tabacum* accumulated more Ca^{2+}, Cd^{2+}, and Mn^{2+} and was more tolerant to elevated Mn^{2+} levels. The expression of *CAX2* also increased Cd^{2+} and Mn^{2+} transport in isolated root tonoplast vesicles. These findings imply that *CAX2* has a broad substrate range and maybe a key component in improving plant ion tolerance (Hirschi et al. 2000). *Arabidopsis thaliana* plants expressing *Escherichia coli* arsenate reductase (*arsC*) and γ-glutamylcysteine synthetase (γ-ECS) genes enhanced As tolerance and hyperaccumulation of As in above-ground parts (Dhankher et al. 2002). Pilon et al. (2003) expressed a mouse (*Mus musculus*) Se-Cys lyase (SL) in the cytosol or chloroplasts of *Arabidopsis* to direct Se flow away from incorporation into proteins. SL specifically catalyzes the decomposition of Se-Cys into elemental Se and alanine. The transgenics showed SL activities up to two-fold in cytosolic lines and six-fold in chloroplastic lines compared to wild-type plants. Se incorporation into proteins was reduced two-fold in both types of SL transgenics, indicating that the approach successfully redirected Se flow in the plant. Enhanced shoot Se concentrations up to 1.5-fold were shown in both the cytosolic as well as chloroplastic lines.

Eapen et al. (2003) developed hairy root cultures of *Brassica juncea* and *Chenopodium amaranticolor* by *Agrobacterium rhizogenes* mediated genetic transformation. The stable, transformed root systems of *B. juncea* and *C. amaranticolor* uptake 20–23% and 13% of U from the solution containing up to 5000 mM concentration, respectively. Wangeline et al. (2004) reported that Indian mustard [*Brassica juncea* (L.) Czern.] transgenics overexpressing ATP sulfurylase were more tolerant to As(III), As(V), Cd, Cu, Hg, and Zn, but less tolerant to Mo and V than the wild-type. LeDuc et al. (2004) overexpressed the gene encoding selenocysteine methyltransferase (SMT) from *Astragalus bisulcatus* in *Arabidopsis* and *B. juncea*. SMT transgenic seedlings tolerated Se, particularly selenite, producing three- to seven-fold greater biomass and three-fold longer root lengths. A significant increase in Se accumulation and volatilization was also observed in SMT plants. To enhance the phytoextraction capacity of *Linum usitatissimum* L., the linseed breeding line AGT 917 was engineered to constitutively express the genetic fusion of the α-domain of mammalian metallothionein 1a (αMT1a) and the β-glucuronidase *gus* gene. The stem of the *αMT1/2* line contained an average of 3.3 and 1.9 times higher levels of Cd than stems of the corresponding AGT 917 when grown in soils amended with Cd at 20 and 360 mg kg^{-1} (Vrbova et al. 2013). In another study, expression of the bacterial Hg transporter *MerE* promoted the transport and accumulation of methylmercury in transgenic *Arabidopsis*, which may be a useful method for improving the efficacy of plants to facilitate the phytoremediation of methylmercury pollution (Sone et al. 2013). Transgenic plants enhancing the phytoremediation of HMs are depicted in Table 5.5.

Table 5.5 Genes introduced into plants for improved phytoremediation of heavy metals (HMs)

Target plant species	Gene introduced	Effect	References
Brassica napus L. and *Nicotiana tabacum* L.	MT-II (human metallothionein-II)	The growth of transformed seedlings was unaffected by up to 100 μM $CdCl_2$	Misra and Gedamu (1989)
Nicotiana tabacum cv. NC89	MT-I (mouse metallothionein-I)	The growth of transformed plants was unaffected by up to 200 μM Cd	Pan et al. (1994)
Arabidopsis thaliana	*merApe9* (mercuric ion reductase)	Transgenic *merApe9* seedlings evolved considerable amounts of Hg^0 relative to control plants. Plants were also resistant to toxic levels of Au^{3+}	Rugh et al. (1996)
Liriodendron tulipifera	*merA18* (mercuric ion reductase)	*merA18* plants conferred resistance to toxic, ionic Hg and released elemental Hg ten times the rate of wild-type plantlets	Rugh et al. (1998)
Brassica juncea	*gshI* (γ-glutamylcysteine synthetase)	Transgenic plants accumulated more Cd and showed increased tolerance than wild-type plants	Zhu et al. (1999)
Nicotiana tabacum	*RCS1* (cysteine synthase)	Transgenics showed up to three-fold higher activity of cysteine synthase and exhibited enhanced Cd tolerance	Harada et al. (2001)
Populus deltoides	*merA9* and *merA18* (mercuric ion reductase)	Transgenic *merA9* and *merA18* plants evolved two- to four-fold the amount of elemental Hg and accumulated significantly higher biomass than wild-type plantlets	Che et al. (2003)
Nicotiana glauca R. Graham	*TaPCS1* (phytochelatin synthase)	Transformed seedlings showed increased tolerance for Pb and Cd	Gisbert et al. (2003)
Arabidopsis thaliana	*YCF1* (yeast cadmium factor 1)	Enhanced tolerance and accumulation of Pb and Cd	Song et al. (2003)
Arabidopsis thaliana	SMT (selenocysteine methyltransferase)	Selenite tolerance and foliar Se accumulation were significantly improved	Ellis et al. (2004)
Arabidopsis thaliana	*AsMT2b* (metallothionein)	Showed stronger Cd tolerance and higher Cd accumulation	Zhang et al. (2006)
Nicotiana glauca	*TaPCS1* (phytochelatin synthase)	Overexpressed gene showed a greater accumulation of Zn, Pb, and Cd	Martinez et al. (2006)
Brassica juncea	*AtPCS1* (phytochelatin synthase)	Significantly higher tolerance to Cd and As was exhibited by transgenic plants	Gasic and Korban (2007)

Arabidopsis thaliana	*GSH1* and *AsPCS1* (γ-glutamylcysteine synthetase and phytochelatin synthase)	Increased tolerance and accumulation of Cd and As in dual-gene transgenic lines were observed	Guo et al. (2008)
Nicotiana tabacum	*tzn*1 (zinc transporter)	Transgenic plants showed enhanced accumulation of Zn (up to 11 times) compared to control plants	Dixit et al. (2010)
Agrostis stolonifera	*PaGCS* (γ-glutamylcysteine synthetase)	Transgenic lines accumulated more Cd^{2+} and phytochelatins (PCs) than the wild-type line	Zhao et al. (2010)
Nicotiana tabacum	*AtPCS1* and *CePCS* (phytochelatin synthase)	Transformants accumulated more As both in roots and leaves	Wojas et al. (2010)
Brassica juncea	*YCF1* (yeast cadmium factor 1)	Transformed seedlings were found to be 1.3–1.6 times more resistant to Cd stress and 1.2–1.4 times more resistant to Pb stress than wild-type plants	Bhuiyan et al. (2011)
Nicotiana tabacum	*tcu*-1 (Cu transporter)	Exhibited higher acquisition of Cu (up to 3.1 times)	Singh et al. (2011)
Nicotiana tabacum	Met(GluCys)6Gly (phytochelatin synthase)	Significantly enhanced Cd accumulation in shoots than control	Postrigan et al. (2012)
Populus alba × *Populus tremula* var. *glandulosa*	*ScYCF1* (yeast cadmium factor 1)	Transgenic poplar plants exhibited enhanced growth, reduced toxicity symptoms, and increased Cd content in the aerial tissue compared to the non-transgenic lines. Plants also accumulated increased amounts of Cd, Zn, and Pb in the roots	Shim et al. (2013)
Nicotiana tabacum	*ScMTII* (metallothionein)	Transgenic tobacco plants accumulated 3.5–4.5-fold more Cd above the threshold level of 100 mg Cd kg^{-1}	Daghan et al. (2013)
Beta vulgaris L.	*StGCS-GS* (γ-glutamylcysteine synthetase-glutathione synthetase)	Transgenic sugar beets accumulated more Cd, Zn, and Cu ions in the shoots and exhibited increased biomass, root length, and relative growth compared to the wild-type	Liu et al. (2015)
Populus tremula × *Populus alba*	*γ-ECS* (γ-glutamylcysteine synthetase)	Exhibited greater Cd accumulation in the aerial parts than wild-type plants in response to Cd^{2+} exposure	He et al. (2015)
Nicotiana tabacum var. Sumsun	*AtACR2* (arsenic reductase 2)	Accumulated higher amount of As in the roots as compared to the wild-type	Nahar et al. (2017)

5.6 Plant Growth Regulators (PGRs) Facilitated Phytoremediation

Plant growth regulators (PGRs) are organic substances that regulate increased plant tolerance to abiotic stress by stimulating expression of the genes associated with antioxidant activity, modulation of cellular redox homeostasis, and alteration in transcription element activities. PGRs include auxins, gibberellins, cytokinins, ethylene, abscisic acid, salicylic acid, jasmonates, brassinosteroids, and strigolactones (Bulak et al. 2014). The exogenous application of indole acetic acid alleviated the negative effect of Cr on growth, protein, nitrogen content, and nitrogen metabolism, and led to a decrease in oxidative injuries caused by Cr (Gangwar and Singh 2011). In *A. thaliana*, 5 μM of gibberellic acid was reported to alleviate Cd toxicity by reducing Cd uptake and lipid peroxidation (Zhu et al. 2012). Ali et al. (2015) reported that the application of gibberellic acid-3 enhanced the length, fresh and dry weight of shoots and roots as well as grain yield of mungbean in the Ni contaminated soils. Application of gibberellic acid-3 significantly increased the biomass of *Solanum nigrum* by 56% and increased Cd concentrations in the shoot by 16% at 1000 mg L^{-1} (Ji et al. 2015). The exogenous abscisic acid can decrease Zn concentrations in *Populus* x *canescens* tissues by modulating the transcript levels of key genes involved in Zn uptake and detoxification, and by activating the antioxidative defense system (Shi et al. 2015). In another study, the addition of exogenous abscisic acid enhanced the tolerance of grapevine (*Vitis vinifera* L.) to excess Zn due to the expression of both *VviZIP* genes and detoxification-related genes (Song et al. 2019). The application of different PGRs (indole-3-acetic acid, indole-3-butyric acid, diethyl aminoethyl hexanoate, 6-benzylaminopurine, 1-naphthylacetic acid, abscisic acid, 2,4-dichlorophenoxyacetic acid, ethrel, brassinolide, gibberellic acid-3, and compound sodium nitrophenolate) enhanced the growth of *Amaranthus hypochondriacus* L. and the phytoextraction efficiency of Cd. However, the application of indole-3-butyric acid or diethyl aminoethyl hexanoate was reported to fix more Cd in upper and lower epidermal cells (Sun et al. 2019). Zhang et al. (2020) reported increased tolerance of tea plants to Cd stress on exogenous application of indole acetic acid (10 μM). Gong et al. (2020) reported that the exogenous application of indole-3-acetic acid reduced the malondialdehyde (MDA) concentrations in Cu stressed seedlings and increased biomass, proline content, and the activities of antioxidant enzymes. Thus, indole-3-acetic acid alleviated Cu toxicity and enhanced Cu tolerance in spinach seedlings.

5.7 Microbial Facilitated Phytoremediation

Beneficial microorganisms associated with plants enhance the efficiency of the phytoremediation process either by altering the metal accumulation in plant tissues or by conferring plant metal tolerance and/or enhancing plant biomass production. These microorganisms influence metal uptake through translocation, transformation, chelation, immobilization, solubilization, precipitation, volatilization, and complexation of metal, ultimately facilitating phytoremediation. Siderophores producing microorganisms inhabiting the rhizosphere are believed to play an important role in HM and radionuclide phytoextraction. Siderophores solubilize unavailable forms of HM and radionuclide bearing minerals by complexation reaction. The production of siderophores by *Streptomyces tendae* F4 has been reported to enhance uptake of Cd in sunflower (Dimkpa et al. 2009). Microbial production of other metabolites such as organic acids (Sayer et al. 1999; Saravanan et al. 2007), biosurfactants (Juwarkar et al. 2007; Sonowal et al. 2022), hormones (Ma et al. 2008), and extracellular polymeric substances such as exopolysaccharides and lipopolysaccharides (Joshi and Juwarkar 2009) also contribute to phytoremediation. Through oxidation or reduction reactions, several plant-associated microorganisms can alter HM and radionuclide mobility. A significant increase in the mobility of Cu, Cd, Hg, and Zn by >90% was reported when co-inoculated with Fe-reducing bacteria and the Fe/S oxidizing bacteria (Beolchini et al. 2009).

Many researchers have reported improved phytoremediation efficiency with plant-associated microorganisms. Chen et al. (2013) suggested that two metal-resistant and plant growth-promoting bacteria, viz., *Burkholderia* sp. J62 and *Pseudomonas thivervalensis* Y-1-3-9, promoted the growth and Cd uptake of *Brassica napus*. The study indicates there might be potential for developing an effective plant–microbe partnership for phytoextraction of Cd from heavily Cd contaminated soils. In another study, inoculation of *Pseudomonas* sp. Lk9 significantly increased shoot dry biomass (14%) and accumulated Cd (46.6%), Zn (16.4%), and Cu (16.0%) in aerial parts of *Solanum nigrum* L. compared to uninoculated plants. This symbiotic association between *S. nigrum* L. and *Pseudomonas* sp. Lk9 also resulted in a significant increase in the soil microbial biomass C (39.2%) and acid phosphatase activity (28.6%.) (Chen et al. 2014). Soil inoculation with *Arthrobacter* sp. TISTR 2220 enhanced Cd accumulation in the roots, above-ground tissues, and whole plant of *Ocimum gratissimum* L. by 1.2-fold, 1.4-fold, and 1.1-fold, respectively. This synergistic use of *Arthrobacter* sp. with *O. gratissimum* L. could be a feasible economic and environmental option for the reclamation of Cd polluted areas (Prapagdee and Khonsue 2015). Szuba et al. (2017) reported that *Paxillus involutus* accumulated Pb in the roots and stems of *Populus* × *canescens* trees, thus improving the host plant growth. Inoculation of *Leptochloa fusca* (L.) Kunth with endophytic bacterial consortia (*Pantoea stewartii* ASI11, *Enterobacter* sp. HU38, and *Microbacterium arborescens* HU33) resulted in a 22–51% increase in root length, 25–62% increase in shoot height, 10–21% increase in chlorophyll content, and 17–59% more plant biomass in U and Pb contaminated soils. Enhanced metal uptake capacity by 53–88% for U and 58–97% for Pb was also observed (Ahsan et al. 2017).

Piriformospora indica enhanced growth, Chl a, Chl b, proline content and showed the ability to immobilize Cd in the root and reduce Cd concentrations in the stem and leaves. This alleviated metal toxicity in the *Helianthus annuus* cv. Zaria plants, and also resulted in phytostabilization of Cd polluted soils (Shahabivand et al. 2017). Inoculation of three metallotolerant siderophore-producing *Streptomyces* sp. B1-B3 strains significantly stimulated plant biomass, reduced oxidative stress, and enhanced the uptake and bioaccumulation of Zn, Cd, and Pb in *Salix dasyclados* L (Zloch et al. 2017). *Bacillus cereus* (T1B3) removed Cr^{6+} (82%), Fe (92%), Mn (67%), Zn (36%), Cd (31%), Cu (25%), and Ni (43%) in the HM amended extract medium. Results indicated that inoculating the native hyperaccumulator *Vetiveria zizanioides* with the T1B3 strain improves the phytoremediation efficiency of *V. zizanioides* (Nayak et al. 2018). Jin et al. (2019) reported that *Simplicillium chinense* QD10 significantly enhanced the phytoextraction of Cd and Pb by *Phragmites communis*. Irshad et al. (2019) reported higher As uptake and removal efficiency by *Vallisneria denseserrulata* and the indigenous *Bacillus* sp. XZM partnership. *Enterobacter* sp. FM-1, a potent bioaugmentation agent, facilitated Mn and Cd phytoextraction in *Polygonum hydropiper* L. and *Polygonum lapathifolium* L. (Li et al. 2020). *Cupriavidus basilensis* strain r507 showed excellent As tolerance, rapid arsenite oxidation ability, and strong colonization of *Pteris vittata*. Inoculation of *P. vittata* with strain r507 accumulated As (up to 171%), suggesting the feasibility of co-cultivating hyperaccumulators with facilitator bacteria for practical As phytoremediation (Yang et al. 2020).

5.8 Arbuscular Mycorrhizal Fungi (AMF) Facilitated Phytoremediation

Arbuscular Mycorrhizal Fungi (AMF) increases tolerance to HMs and radionuclides, improves acquisition of water and nutrients, and results in the establishment of plants in contaminated soil (Thakare et al. 2021). AMF improves phytoremediation via chelation/complexation, compartmentation in vacuoles, metal retention in vesicules, arbuscules, spore and cell walls, and production of glomalin (Cabral et al. 2015). Entry et al. (1999) reported the accumulation of ^{137}Cs and ^{90}Sr from the contaminated soil by bahiagrass (*Paspalum notatum*), Johnson grass (*Sorghum halpense*), and switchgrass (*Panicum virgatum*) after inoculation with *Glomus mosseae* and *Glomus intraradices*. Arriagada et al. (2004) reported that combined inoculation of *Glomus deserticola* and *Trichoderma koningii* resulted in the highest accumulation of Cd in the stem of the eucalyptus plant.

The AM association helped *Phyllanthus niruri* and *Paspalum vaginatum* plants to survive in a disturbed ecosystem by enhancing the uptake and recycling of radionuclides, particularly ^{137}Cs and ^{90}Sr (Selvaraj et al. 2004). Wang et al. (2005) reported that inoculation of an AM fungal consortium consisting of *Gigaspora margarita* ZJ37, *Gigaspora decipiens* ZJ38, *Scutellospora gilmori* ZJ39, *Acaulospora* spp., and *Glomus* spp., increased not only the shoot biomass but also the uptake of Cu, Zn, Pb, and Cd into the shoots of *Elsholtzia splendens* Nakai ex F. Maekawa. Hashem et al. (2016) reported that AMF inoculation mitigated the Cd stress tolerance of *Cassia italica* Mill by reducing lipid peroxidation and enhancing the antioxidant activity. *Trigonella foenum graecum* plants accumulated high concentrations of Cd in their root systems from AMF inoculation. Furthermore, AMF colonization diminished the negative effects of Cd on plant development by increasing antioxidant enzyme activity, soluble protein content, and decreasing malondialdehyde (MDA) content (Abdelhameed and Metwally 2019). Thus, AMF presents a viable strategy to remediate and reclaim sites contaminated with HMs and radionuclides.

5.9 Nanoparticles (NPs) Facilitated Phytoremediation

The use of nanotechnology in conjunction with phytoremediation is progressing rapidly. A nanoparticle's size typically falls between 1 and 100 nanometers. The ability of nanoparticles (NPs) to penetrate within plants and translocate from roots to other areas of the plants is largely determined by their size. Owing to their small size and large surface area, NPs can penetrate the contaminant zone where other particles are unable to, enabling NPs to have a wider range of applications. NPs cause physiological and morphological changes in plants and the plants' response strongly depend on the NPs type, dose, and speciation as well as on the plant species involved. NPs raise the pH of the soil and adsorb metal, reducing its mobility and bioavailability. NPs also boost the plant defense system by regulating the metal transport genes, promoting the synthesis of protective agents, and scavenging reactive oxygen species (ROS) (Zhou et al. 2020; Prasad et al. 2017).

Singh and Lee (2016) observed that an application of 300 mg/kg of nano-titanium dioxide (TiO_2) particles significantly enhanced the Cd uptake (507.6 μg/g) by soybean plants (*Glycine max*) from contaminated soil. In another study, the application of 5 g/kg nano-hydroxyapatite (NHAP) to Pb contaminated soils significantly increased the ryegrass biomass (Jin et al. 2016). Souri et al. (2017) reported a significant beneficial effect of salicylic acid nanoparticles (SANPs) on the growth and phytoremediation efficiency of *Isatis cappadocica* against As toxicity. The maximum As accumulation in the shoots and roots reached 705 mg/kg and 1188 mg/kg, respectively. Gong et al. (2017) studied the effect of 100, 500, and 1000 mg/kg starch stabilized nZVI (S-nZVI) particles on the Cd accumulation in *Boehmeria nivea* (L.) Gaudich (ramie). The addition of S-nZVI particles increased the Cd accumulation in the roots, stems, and leaves by 16–50%, 29–52%, and 31–73%, respectively. Huang et al. (2018) observed maximum accumulation of Pb (1175.40 μg/pot) in ryegrass (*Lolium perenne*) with the treatment of 100 mg/kg nZVI. However, decreased Pb accumulation was reported in high concentrations of nZVI (1000 and 2000 mg/kg). In another study, thidiazuron (TDZ) growth regulator and magnesium oxide (MgO) nanoparticles increased plant growth, phenolic and flavonoid contents, free radical scavenging activity, and Pb phytoaccumulation by radish (*Raphanus sativus* L.) (Hussain et al. 2019).

5.10 Conclusion

Phytoremediation is a cost-effective plant-based approach for the reclamation of HM and radionuclide polluted sites that has a high level of public acceptance. Plants can also be genetically modified to achieve desirable traits such as rapid growth, high biomass output, high metal tolerance and accumulation, and strong adaptation to a variety of climatic and geological settings. The prospect of using transgenic plants to clean up contaminated sites has been thoroughly investigated and many plant species harboring transgenes of various origins and presumptive functions have been surveyed. Furthermore, PGRs, plant-associated microorganisms, AMF, and NPs also boost phytoremediation efficiency.

References

Abdelhameed RE, Metwally RA (2019) Alleviation of cadmium stress by arbuscular mycorrhizal symbiosis. Int J Phytoremediation 28:1–9. https://doi.org/10.1080/15226514.2018.1556584

Adki VS, Jadhav JP, Bapat VA (2013) *Nopalea cochenillifera*, a potential chromium (VI) hyperaccumulator plant. Environ Sci Pollut Res 20:1173–1180. https://doi.org/10.1007/s11356-012-1125-4

Ahsan MT, Najam-ul-haq M, Idrees M et al (2017) Bacterial endophytes enhance phytostabilization in soils contaminated with uranium and lead. Int J Phytoremediation 19:937–946. https://doi.org/10.1080/15226514.2017.1303813

Ali MA, Asghar HN, Khan MY et al (2015) Alleviation of nickel-induced stress in mungbean through application of gibberellic acid. Int J Agric Biol 17:990–994

Arriagada CA, Herrera MA, Garcia-Romera I, Ocampo JA (2004) Tolerance to Cd of soybean (*Glycine max*) and eucalyptus (*Eucalyptus globulus*) inoculated with arbuscular mycorrhizal and saprobe fungi. Symbiosis 36:1–15

Baker AJM (1981) Accumulators and excluders-strategies in the response of plants to heavy metals. J Plant Nutr 3:643–654. https://doi.org/10.1080/01904168109362867

Baker AJM, Brooks RR (1989) Terrestrial higher plants which hyperaccumulate metallic elements: a review of their distribution, ecology and phytochemistry. Biorecovery 1:81–126

Banasova V, Horak O, Nadubinska M et al (2008) Heavy metal content in *Thlaspi caerulescens* J. et C. Presl growing on metalliferous and non-metalliferous soils in Central Slovakia. Int J Environ Pollut 33:133–145. https://doi.org/10.1504/IJEP.2008.019388

Beolchini F, Dell'Anno A, Propris LD et al (2009) Auto- and heterotrophic acidophilic bacteria enhance the bioremediation efficiency of sediments contaminated by heavy metals. Chemosphere 74:1321–1326. https://doi.org/10.1016/j.chemosphere.2008.11.057

Bhainsa KC, D'Souza SF (2001) Uranium (VI) biosorption by dried roots of *Eichhornia crassipes* (water hyacinth). J Environ Sci Health A Tox Hazard Subst Environ Eng 36:1621–1631. https://doi.org/10.1081/ESE-100106247

Bhat IUH, Mauris EN, Khanam Z (2016) Phytoremediation of iron from red soil of tropical region by using *Centella asiatica*. Int J Phytoremediation 18:918–923. https://doi.org/10.1080/15226514.2016.1156637

Bhuiyan MSU, Min SR, Jeong WJ et al (2011) Overexpression of a yeast cadmium factor 1 (*YCF1*) enhances heavy metal tolerance and accumulation in *Brassica juncea*. Plant Cell Tissue Organ Cult 105:85–91. https://doi.org/10.1007/s11240-010-9845-y

Bidwell SD, Woodrow IE, Batianoff GN, Sommer-Knudsen J (2002) Hyperaccumulation of manganese in the rainforest tree *Austromyrtus bidwillii* (Myrtaceae) from Queensland, Australia. Funct Plant Biol 29:899–905. https://doi.org/10.1071/PP01192

Bizily SP, Rugh CL, Summers AO, Meagher RB (1999) Phytoremediation of methylmercury pollution: *merB* expression in *Arabidopsis thaliana* confers resistance to organomercurials. Proc Natl Acad Sci U S A 96:6808–6813. https://doi.org/10.1073/pnas.96.12.6808

Boonyapookana B, Parkplan P, Techapinyawat S, DeLaune RD, Jugsujinda A (2005) Phytoaccumulation of lead by sunflower (*Helianthus annuus*), tobacco (*Nicotiana tabacum*), and vetiver (*Vetiveria zizanioides*). J Environ Sci Health A Tox Hazard Subst Environ Eng 40: 117–137. https://doi.org/10.1081/ese-200033621

Buendia-Gongalez L, Orozco-Villafuerte J, Cruz-Sosa F et al (2010) *Prosopis laevigata* a potential chromium (VI) and cadmium (II) hyperaccumulator desert plant. Bioresour Technol 101:5862–5867. https://doi.org/10.1016/j.biortech.2010.03.027

Bulak P, Walkiewicz A, Brzezinska M (2014) Plant growth regulators-assisted phytoextraction. Biol Plant 58:1–8. https://doi.org/10.1007/s10535-013-0382-5

Bystrzejewska-Piotrowska G, Urban PL (2004) Accumulation and translocation of cesium-137 in onion plants (*Allium cepa*). Environ Exp Bot 51:3–7. https://doi.org/10.1016/S0098-8472(03)00039-X

Cabral L, Soares CRFS, Giachini AJ, Siqueira JO (2015) Arbuscular mycorrhizal fungi in phytoremediation of contaminated areas by trace elements: mechanisms and major benefits of their applications. World J Microbiol Biotechnol 31:1655–1664. https://doi.org/. https://doi.org/10.1007/s11274-015-1918-y

Chakraborty D, Maji S, Bandyopadhyay A, Basu S (2007) Biosorption of cesium-137 and strontium-90 by mucilaginous seeds of *Ocimum basilicum*. Bioresour Technol 98:2949–2952. https://doi.org/10.1016/j.biortech.2006.09.035

Che D, Meagher RB, Heaton ACP et al (2003) Expression of mercuric ion reductase in Eastern cottonwood (*Populus deltoides*) confers mercuric ion reduction and resistance. Plant Biotech J 1:311–319. https://doi.org/10.1046/j.1467-7652.2003.00031.x

Chen ZJ, Sheng XF, He LY et al (2013) Effects of root inoculation with bacteria on the growth, Cd uptake and bacterial communities associated with rape grown in Cd-contaminated soil. J Hazard Mater 244–245:709–717. https://doi.org/10.1016/j.jhazmat.2012.10.063

Chen L, Luo S, Li X et al (2014) Interaction of Cd-hyperaccumulator *Solanum nigrum* L. and functional endophyte *Pseudomonas* sp. Lk9 on soil heavy metals uptake. Soil Biol Biochem 68: 300–308. https://doi.org/10.1016/j.soilbio.2013.10.021

Chen Y, Yang W, Chao Y et al (2017) Metal-tolerant *Enterobacter* sp. strain EG16 enhanced phytoremediation using *Hibiscus cannabinus* via siderophore-mediated plant growth promotion under metal contamination. Plant Soil 413:203–216. https://doi.org/10.1007/s11104-016-3091-y

Chuaphasuk C, Prapagdee B (2019) Effects of biochar-immobilized bacteria on phytoremediation of cadmium-polluted soil. Environ Sci Pollut Res 26:23679–23688. https://doi.org/10.1007/s11356-019-05661-6

Colzi I, Arnetoli M, Gallo A et al (2012) Copper tolerance strategies involving the root cell wall pectins in *Silene paradoxa* L. Environ Exp Bot 78:91–98. https://doi.org/10.1016/j.envexpbot.2011.12.028

D'Souza RJ, Varun M, Masih J, Paul MS (2010) Identification of *Calotropis procera* L. as a potential phytoaccumulator of heavy metals from contaminated soils in Urban North Central India. J Hazard Mater 184:457–464. https://doi.org/10.1016/j.jhazmat.2010.08.056

Daghan H, Arslan M, Uygur V, Loleli N (2013) Transformation of tobacco with *ScMTII* gene-enhanced cadmium and zinc accumulation. CLEAN Soil Air Water 41:503–509. https://doi.org/10.1002/clen.201200298

de la Fuente JM, Ramirez-Rodrıguez V, Cabrera-Ponce JL, Herrera-Estrella L (1997) Aluminum tolerance in transgenic plants by alteration of citrate synthesis. Science 276:1566–1568. https://doi.org/10.1126/science.276.5318.1566

Dhankher OP, Li Y, Rosen BP et al (2002) Engineering tolerance and hyperaccumulation of arsenic in plants by combining arsenate reductase and γ-glutamylcysteine synthetase expression. Nat Biotechnol 20:1140–1145. https://doi.org/10.1038/nbt747

Dimkpa CO, Merten D, Svatos A et al (2009) Siderophores mediate reduced and increased uptake of cadmium by *Streptomyces tendae* F4 and sunflower (*Helianthus annuus*), respectively. J Appl Microbiol 107:1687–1696. https://doi.org/10.1111/j.1365-2672.2009.04355.x

Dixit P, Singh S, Vancheeswaran R et al (2010) Expression of a *Neurospora crassa* zinc transporter gene in transgenic *Nicotiana tabacum* enhances plant zinc accumulation without co-transport of cadmium. Plant Cell Environ 33:1696–1707. https://doi.org/10.1111/j.1365-3040.2010.02174.x

Eapen S, Suseelan KN, Tivarekar S et al (2003) Potential for rhizofiltration of uranium using hairy root cultures of *Brassica juncea* and *Chenopodium amaranticolor*. Environ Res 91:127–133. https://doi.org/10.1016/s0013-9351(02)00018-x

Eapen S, Singh S, Thorat V et al (2006) Phytoremediation of radiostrontium (^{90}Sr) and radiocesium (^{137}Cs) using giant milky weed (*Calotropis gigantea* R.Br.) plants. Chemosphere 65:2071–2073. https://doi.org/10.1016/j.chemosphere.2006.06.049

Ellis DR, Sors TG, Brunk DG et al (2004) Production of Se-methylselenocysteine in transgenic plants expressing selenocysteine methyltransferase. BMC Plant Biol 4:1. https://doi.org/10.1186/1471-2229-4-1

Entry JA, Watrud LS, Reeves M (1999) Accumulation of 137Cs and 90Sr from contaminated soil by three grass species inoculated with mycorrhizal fungi. Environ Pollut 104:449–457. https://doi.org/10.1016/S0269-7491(98)00163-8

Entry JA, Watrud LS, Reeves M (2001) Influence of organic amendments on the accumulation of ^{137}Cs and ^{90}Sr from contaminated soil by three grass species. Water Air Soil Pollut 126:385–398. https://doi.org/10.1023/A:1005201220596

Etim EE (2012) Phytoremediation and its mechanisms: a review. Int J Environ Bioenerg 2:120–136

Forte J, Mutiti S (2017) Phytoremediation potential of *Helianthus annuus* and *Hydrangea paniculata* in copper and leadcontaminated soil. Water Air Soil Pollut 228:77

Francesconi K, Visoottiviseth P, Sridokchan W, Goessler W (2002) Arsenic species in an arsenic hyperaccumulating fern, *Pityrogramma calomelanos*: a potential phytoremediator of arsenic-contaminated soils. Sci Total Environ 284:27–35

Fuhrmann M, Lasat MM, Ebbs SD et al (2002) Uptake of cesium-137 and strontium-90 from contaminated soil by three plant species; application to phytoremediation. J Environ Qual 31: 904–909. https://doi.org/10.2134/jeq2002.9040

Fulekar MH, Singh A, Thorat N et al (2010) Phytoremediation ^{137}Cs from low level nuclear waste using *Catharanthus roseus*. Indian J Pure Appl Phys 48:516–519

Gangwar S, Singh VP (2011) Indole acetic acid differently changes growth and nitrogen metabolism in *Pisum sativum* L. seedlings under chromium (VI) phytotoxicity: implication of oxidative stress. Sci Hortic 129:321–328

Gasic K, Korban SS (2007) Transgenic Indian mustard (*Brassica juncea*) plants expressing an *Arabidopsis* phytochelatin synthase (*AtPCS1*) exhibit enhanced As and Cd tolerance. Plant Mol Biol 64:361–369. https://doi.org/10.1007/s11103-007-9158-7

Gisbert C, Ros R, De Haro A et al (2003) A plant genetically modified that accumulates Pb is especially promising for phytoremediation. Biochem Biophys Res Commun 303:440–445. https://doi.org/10.1016/S0006-291X(03)00349-8

Gong X, Huang D, Liu Y et al (2017) Stabilized nanoscale zerovalent iron mediated cadmium accumulation and oxidative damage of *Boehmeria nivea* (L.) Gaudich cultivated in cadmium contaminated sediments. Environ Sci Technol 51:11308–11316. https://doi.org/10.1021/acs.est.7b03164

Gong Q, Li Z, Wang L et al (2020) Exogenous of indole-3-acetic acid application alleviates copper toxicity in spinach seedlings by enhancing antioxidant systems and nitrogen metabolism. Toxics 8:1. https://doi.org/10.3390/toxics8010001

Guo J, Dai X, Xu W, Ma M (2008) Overexpressing *GSH1* and *AsPCS1* simultaneously increases the tolerance and accumulation of cadmium and arsenic in *Arabidopsis thaliana*. Chemosphere 72:1020–1026. https://doi.org/10.1016/j.chemosphere.2008.04.018

Harada E, Choi YE, Tsuchisaka A et al (2001) Transgenic tobacco plants expressing a rice cysteine synthase gene are tolerant to toxic levels of cadmium. J Plant Physiol 158:655–661. https://doi.org/10.1078/0176-1617-00314

Hashem A, AbdAllah EF, Alqarawi AA, Egamberdieva D (2016) Bioremediation of adverse impact of cadmium toxicity on *Cassia italica* Mill by arbuscular mycorrhizal fungi. Saudi J Biol Sci 23: 39–47. https://doi.org/10.1016/j.sjbs.2015.11.007

He J, Li H, Ma C et al (2015) Overexpression of bacterial γ-glutamylcysteine synthetase mediates changes in cadmium flux, allocation and detoxification in poplar. New Phytol 205:240–254. https://doi.org/10.1111/nph.13013

Hinton TG, Knox AS, Kaplan DI, Sharitz R (2005) Phytoextraction of uranium and thorium by native trees in a contaminated wetland. J Radioanal Nucl Chem 264:417–422. https://doi.org/10.1007/s10967-005-0731-8

Hirschi KD, Korenkov VD, Wilganowski NL, Wagner GJ (2000) Expression of *Arabidopsis* CAX2 in tobacco. Altered metal accumulation and increased manganese tolerance. Plant Physiol 124: 125–133. https://doi.org/10.1104/pp.124.1.125

Huang JW, Blaylock MJ, Kapulnik Y, Ensley BD (1998) Phytoremediation of uranium-contaminated soils: role of organic acids in triggering uranium hyperaccumulation in plants. Environ Sci Technol 32:2004–2008. https://doi.org/10.1021/es971027u

Huang D, Qin X, Peng Z et al (2018) Nanoscale zero-valent iron assisted phytoremediation of Pb in sediment: impacts on metal accumulation and antioxidative system of *Lolium perenne*. Ecotoxicol Environ Saf 153:229–237. https://doi.org/10.1016/j.ecoenv.2018.01.060

Hussain F, Hadi F, Akbar F (2019) Magnesium oxide nanoparticles and thidiazuron enhance lead phytoaccumulation and antioxidative response in *Raphanus sativus* L. Environ Sci Pollut Res 26:30333–30347. https://doi.org/10.1007/s11356-019-06206-7

Irshad S, Xie Z, Wang J et al (2019) Indigenous strain *Bacillus* XZM assisted phytoremediation and detoxification of arsenic in *Vallisneria denseserrulata*. J Hazard Mater 381:120903. https://doi.org/10.1016/j.jhazmat.2019.120903

Islam MS, Kashem MA, Khan TO (2016) Phytoextraction efficiency of lead by arum (*Colocasia esculenta* L.) grown in soil. Int J Soil Sci 11:130–136. https://doi.org/10.3923/ijss.2016.130.136

Jaffre T, Brooks RR, Lee J, Reeves RD (1976) *Sebertia acuminata*: a hyperaccumulator of nickel from New Caledonia. Science 193:579–580

Ji P, Tang X, Jiang Y et al (2015) Potential of gibberellic acid 3 (GA3) for enhancing the phytoremediation efficiency of *Solanum nigrum* L. Bull Environ Contam Toxicol 95:810–814. https://doi.org/10.1007/s00128-015-1670-x

Jin Y, Liu W, Li XL et al (2016) Nano-hydroxyapatite immobilized lead and enhanced plant growth of ryegrass in a contaminated soil. Ecol Eng 95:25–29. https://doi.org/10.1016/j.ecoleng.2016.06.071

Jin Z, Deng S, Wena Y et al (2019) Application of *Simplicillium chinense* for Cd and Pb biosorption and enhancing heavy metal phytoremediation of soils. Sci Total Environ 697:134148. https://doi.org/10.1016/j.scitotenv.2019.134148

Joshi PM, Juwarkar AA (2009) *In vivo* studies to elucidate the role of extracellular polymeric substances from *Azotobacter* in immobilization of heavy metals. Environ Sci Technol 43:5884–5889. https://doi.org/10.1021/es900063b

Juwarkar AA, Nair A, Dubey KV et al (2007) Biosurfactant technology for remediation of cadmium and lead contaminated soils. Chemosphere 68:1996–2002. https://doi.org/10.1016/j.chemosphere.2007.02.027

Kabata-Pendias A, Pendias H (1996) Trace elements in soils and plants. CRC Press, Boca Raton, FL, 365 pp

Kabeya FI, Pongrac P, Lange B et al (2018) Tolerance and accumulation of cobalt in three species of *Haumaniastrum* and the influence of copper. Environ Exp Bot 149:27–33. https://doi.org/10.1016/j.envexpbot.2018.01.018

Khan AR, Ullah I, Khan AL et al (2014) Phytostabilization and physicochemical responses of Korean ecotype *Solanum nigrum* L. to cadmium contamination. Water Air Soil Pollut 225:2147. https://doi.org/10.1007/s11270-014-2147-y

Khan WR, Rasheed F, Zulkifli SZ et al (2020) Phytoextraction potential of *Rhizophora apiculata*: a case study in matang mangrove forest reserve, Malaysia. Trop Conserv Sci. https://doi.org/10.1177/1940082920947344

Kirk JL, Klironomos JN, Lee H, Trevors JT (2005) The effects of perennial ryegrass and alfalfa on microbial abundance and diversity in petroleum contaminated soil. Environ Pollut 133:455–465. https://doi.org/10.1016/j.envpol.2004.06.002

Kumar PN, Dushenkov V, Motto H, Raskin IJE (1995) Phytoextraction: the use of plants to remove heavy metals from soils. Environ Sci Technol 29:1232–1238. https://doi.org/10.1021/es00005a014

Kumar D, Bharti SK, Anand S et al (2018) Bioaccumulation and biochemical responses of *Vetiveria zizanioides* grown under cadmium and copper stresses. Environ Sustain 1:133–139. https://doi.org/10.1007/s42398-018-0009-z

Kupper H, Zhao FJ, McGrath SP (1999) Cellular compartmentation of zinc in leaves of the hyperaccumulator *Thlaspi caerulescens*. Plant Physiol 119:305–311. https://doi.org/10.1104/pp.119.1.305

Kupper H, Lombi E, Zhao FJ, McGrath SP (2000) Cellular compartmentation of cadmium and zinc in relation to other elements in the hyperaccumulator *Arabidopsis halleri*. Planta 212:75–84. https://doi.org/10.1007/s004250000366

LeDuc DL, Tarun AS, Montes-Bayon M et al (2004) Overexpression of selenocysteinemethyltransferase in *Arabidopsis* and Indian mustard increases selenium tolerance and accumulation. Plant Physiol 135:377–383. https://doi.org/10.1104/pp.103.026989

Lee JH (2013) An overview of phytoremediation as a potentially promising technology for environmental pollution control. Biotechnol Bioprocess Eng 18:431–439. https://doi.org/10.1007/s12257-013-0193-8

Lei M, Wan X, Guo G et al (2016) Phytoextraction of arsenic-contaminated soil with *Pteris vittata* in Henan Province, China: comprehensive evaluation of remediation efficiency correcting for atmospheric depositions. Environ Sci Pollut Res. https://doi.org/10.1007/s11356-016-8184-x

Li Y, Lin J, Huang Y et al (2020) Bioaugmentation-assisted phytoremediation of manganese and cadmium co-contaminated soil by Polygonaceae plants (*Polygonum hydropiper* L. and *Polygonum lapathifolium* L.) and *Enterobacter* sp. FM-1. Plant Soil 448:439–453. https://doi.org/10.1007/s11104-020-04447-x

Liu W, Shu W, Lan C (2004) *Viola baoshanensis*, a plant that hyperaccumulates cadmium. Chin Sci Bull 49:29–32. https://doi.org/10.1007/BF02901739

Liu Z, He X, Chen W et al (2009) Accumulation and tolerance characteristics of cadmium in a potential hyperaccumulator-*Lonicera japonica* Thunb. J Hazard Mater 169:170–175. https://doi.org/10.1016/j.jhazmat.2009.03.090

Liu D, An Z, Mao Z et al (2015) Enhanced heavy metal tolerance and accumulation by transgenic sugar beets expressing *Streptococcus thermophilus* StGCS-GS in the presence of Cd, Zn and Cu alone or in combination. PLoS One 10:e0128824. https://doi.org/10.1371/journal.pone.0128824

Ma L, Komar K, Tu C et al (2001) A fern that hyperaccumulates arsenic. Nature 409:579. https://doi.org/10.1038/35054664

Ma Y, Rajkumar M, Fritas H (2008) Inoculation of plant growth promoting bacterium *Achromobacter xylosoxidans* strain Ax10 for the improvement of copper phytoextraction by *Brassica juncea*. J Environ Manag 90:831–837. https://doi.org/10.1016/j.jenvman.2008.01.014

Manikandan R, Sahi SV, Venkatachalam P (2015) Impact assessment of mercury accumulation and biochemical and molecular response of *Mentha arvensis*: a potential hyperaccumulator plant. Sci World J 715217. https://doi.org/10.1155/2015/715217

Martinez M, Bernal P, Almela C et al (2006) An engineered plant that accumulates higher levels of heavy metals than *Thlaspi caerulescens*, with yields of 100 times more biomass in mine soils. Chemosphere 64:478–485. https://doi.org/10.1016/j.chemosphere.2005.10.044

Mihalik J, Tlustos P, Szakova J (2010) Comparison of willow and sunflower for uranium phytoextraction induced by citric acid. J Radioanal Nucl Chem 285:279–285. https://doi.org/10.1007/s10967-010-0538-0

Misra S, Gedamu L (1989) Heavy metal tolerant transgenic *Brassica napus* L. and *Nicotianatabacum* L. plants. Theor Appl Genet 78:161–168. https://doi.org/10.1007/BF00288793

Mkandawire M, Dudel G (2005) Accumulation of arsenic in *Lemna gibba* L. (duckweed) in tailing waters of two abandoned uranium mining sites in Saxony, Germany. Sci Total Environ 336:81–89. https://doi.org/10.1016/j.scitotenv.2004.06.002

Moreno FN, Anderson CW, Stewart RB, Robinson BH (2008) Phytofiltration of mercury-contaminated water: volatilisation and plant-accumulation aspects. Environ Exp Bot 62:78–85. https://doi.org/10.1016/j.envexpbot.2007.07.007

Nahar N, Rahman A, Nawani NN et al (2017) Phytoremediation of arsenic from the contaminated soil using transgenic tobacco plants expressing *ACR2* gene of *Arabidopsis thaliana*. J Plant Physiol 218:121–126. https://doi.org/10.1016/j.jplph.2017.08.001

Nayak AK, Panda SS, Basu A, Dhal NK (2018) Enhancement of toxic Cr(VI), Fe, and other heavy metals phytoremediation by the synergistic combination of native *Bacillus cereus* strain and *Vetiveria zizanioides* L. Int J Phytoremediation 20:682–691. https://doi.org/10.1080/15226514.2017.1413332

Nedjimi B (2021) Phytoremediation: a sustainable environmental technology for heavy metals decontamination. SN Appl Sci 3:286. https://doi.org/10.1007/s42452-021-04301-4

Nedjimi B, Daoud Y (2009) Cadmium accumulation in *Atriplex halimus* subsp. *schweinfurthii* and its influence on growth, proline, root hydraulic conductivity and nutrient uptake. Flora 204:316–324. https://doi.org/10.1016/j.flora.2008.03.004

Ni TH, Wei YZ (2003) Subcellular distribution of cadmium in mining ecotype *Sedum alfredii*. Acta Bot Sin 45:925–928

Nisbet A, Shaw S (1994) Summary of a five-year lysimeter study on the time dependent transfer of ^{137}Cs, $_{90}$Sr, ^{239,240}Pu and ^{241}Am to crops from three contrasting soil types. Distribution between different plant parts. J Environ Radioact 23:171–187. https://doi.org/10.1016/0265-931X(94)90059-0

Pan A, Yang M, Tie F et al (1994) Expression of mouse metallothionein-I gene confers cadmium resistance in transgenic tobacco plants. Plant Mol Biol 24:341–351. https://doi.org/10.1007/BF00020172

Pentyala VB, Eapen S (2020) High efficiency phytoextraction of uranium using *Vetiveria zizanioides* L. Nash. Int J Phytoremediation 22:1137–1146. https://doi.org/10.1080/15226514.2020.1741506

Phaenark C, Pokethitiyook P, Kruatrachue M, Ngernsansaruay C (2009) Cd and Zn accumulation in plants from the padaeng zinc mine area. Int J Phytoremediation 11:479–495. https://doi.org/10.1080/15226510802656243

Pilon M, Owen JD, Garifullina GF et al (2003) Enhanced selenium tolerance and accumulation in transgenic *Arabidopsis* expressing a mouse selenocysteinelyase. Plant Physiol 131:1250–1257. https://doi.org/10.1104/pp.102.014639

Pirkle JL, Kaufmann RB, Brody DJ et al (1998) Exposure of the U.S. population to lead, 1991-1994. Environ Health Perspect 106:745–750. https://doi.org/10.1289/ehp.98106745

Pollard AJ, Stewart HL, Roberson CB (2009) Manganese hyperaccumulation in *Phytolacca americana* L. from the Southeastern United States. Northeast Nat 16:155–162

Postrigan BN, Knyazev AB, Kuluev BR et al (2012) Expression of the synthetic phytochelatin gene in tobacco. Russ J Plant Physiol 59:275–280. https://doi.org/10.1134/S1021443712020136

Prapagdee B, Khonsue N (2015) Bacterial-assisted cadmium phytoremediation by *Ocimum gratissimum* L. in polluted agricultural soil: a field trial experiment. Int J Environ Sci Technol 12:3843–3852. https://doi.org/10.1007/s13762-015-0816-z

Prapagdee S, Piyatiratitivorakul S, Petsom A, Tawinteung N (2014) Application of biochar for enhancing cadmium and zinc phytostabilization in *Vigna radiata* L. cultivation. Water Air Soil Pollut 225:2233. https://doi.org/10.1007/s11270-014-2233-1

Prasad R, Gupta N, Kumar M, Kumar V, Wang S, Abd-Elsalam KA (2017) Nanomaterials act as plant defense mechanism. In: Prasad R, Kumar M, Kumar V (eds) Nanotechnology. Springer, Singapore, pp 253–269

Qiu R, Fang X, Tang Y, Du S, Zeng X (2006) Zinc hyperaccumulation and uptake by *Potentilla griffithii* Hook. Int J Phytoremediation 8:299–310. https://doi.org/10.1080/15226510600992865

Rascio N, Navari-Izzo F (2011) Heavy metal hyperaccumulating plants: how and why do they do it? And what makes them so interesting? Plant Sci 180:169–181. https://doi.org/10.1016/j.plantsci.2010.08.016

Raskin I, Ensley BD (2000) Recent developments for in situ treatment of metal contaminated soils. In: Phytoremediation of toxic metals: using plants to clean up the environment. Wiley, New York

Reeves RD, Brooks RR, Macfarlane RM (1981) Nickel uptake by californian *Streptanthus* and *Caulanthus* with particular reference to the hyperaccumulator *S. polyoaloides* gray (Brassicaceae). Am J Bot 68:708–712

Robinson BH, Brooks RR, Howes AW et al (1997a) The potential of the high biomass nickel hyperaccumulator *Berkheya coddii* for phytoremediation and phytomining. J Geochem Explor 60:115–126

Robinson BH, Chiarucci A, Brooks RR et al (1997b) The nickel hyperaccumulator plant *Alyssum bertolonii* as a potential agent for phytoremediation and phytomining of nickel. J Geochem Explor 59:75–86

Rodriguez N, Menendez N, Tornero J, Amils R, De La Fuente V (2005) Internal iron biomineralization in *Imperata cylindrica*, a perennial grass: chemical composition, speciation and plant localization. New Phytol 165:781–789. https://doi.org/10.1111/j.1469-8137.2004.01264.x

Rugh CL, Wilde HD, Stack NM et al (1996) Mercuric ion reduction and resistance in transgenic *Arabidopsis thaliana* plants expressing a modified bacterial *merA* gene. Proc Natl Acad Sci U S A 93:3182–3187. https://doi.org/10.1073/pnas.93.8.3182

Rugh CL, Senecoff JF, Meagher RB, Merkle SA (1998) Development of transgenic yellow poplar for mercury phytoremediation. Nat Biotechnol 16:925–928. https://doi.org/10.1038/nbt1098-925

Sahi SV, Bryant NL, Sharma NC, Singh SR (2002) Characterization of a lead hyperaccumulator shrub, *Sesbania drummondii*. Environ Sci Technol 36:4676–4680. https://doi.org/10.1021/es020675x

Sakizadeha M, Ghorbani H (2017) Concentration of heavy metals in soil and staple crops and the associated health risk. Arch Hyg Sci 6:303–313

Salt D, Blaylock M, Kumar N et al (1995) Phytoremediation: a novel strategy for the removal of toxic metals from the environment using plants. Nat Biotechnol 13:468–474. https://doi.org/10.1038/nbt0595-468

Santana NA, Rabuscke CM, Soares VB et al (2018) Vermicompost dose and mycorrhization determine the efficiency of copper phytoremediation by *Canavalia ensiformis*. Environ Sci Pollut Res 25:12663–12677. https://doi.org/10.1007/s11356-018-1533-1

Saravanan VS, Madhaiyan M, Thangaraju M (2007) Solubilization of zinc compounds by the diazotrophic, plant growth promoting bacterium *Gluconacetobacter diazotrophicus*. Chemosphere 66:1794–1798. https://doi.org/10.1016/j.chemosphere.2006.07.067

Sarma H, Islam NF, Prasad R, Prasad MNV, Ma LQ et al (2021) Enhancing phytoremediation of hazardous metal(loid)s using genome engineering CRISPR–Cas9 technology. J Hazard Mater 414:125493. https://doi.org/10.1016/j.jhazmat.2021.125493

Sasmaz A, Sasmaz M (2009) The phytoremediation potential for strontium of indigenous plants growing in a mining area. Environ Exp Bot 67:139–144. https://doi.org/10.1016/j.envexpbot.2009.06.014

Sayer JA, Cotter-Howells JD, Watson C et al (1999) Lead mineral transformation by fungi. Curr Biol 9:691–694. https://doi.org/10.1016/s0960-9822(99)80309-1

Schnoor JL, Light LA, McCutcheon SC et al (1995) Phytoremediation of organic and nutrient contaminants. Environ Sci Technol 29:318–323. https://doi.org/10.1021/es00007a002
Selamat SN, Abdullah SRS, Idris M (2014) Phytoremediation of lead (Pb) and arsenic (As) by *Melastoma malabathricum* L. from contaminated soil in separate exposure. Int J Phytoremediation 16:694–703. https://doi.org/10.1080/15226514.2013.856843
Selvaraj T, Padmanabhan C, Jeong YJ, Kim H (2004) Occurrence of vesicular-arbuscular mycorrhizal (VAM) fungi and their effect on plant growth in endangered vegetations. J Microbiol Biotechnol 14:885–890
Seregin IV, Kozhevnikova AD, Kazyumina EM, Ivanov VB (2003) Nickel toxicity and distribution in maize roots. Russ J Plant Physiol 50:711–717. https://doi.org/10.1023/A:1025660712475
Shahabivand S, Parvaneh A, Aliloo AA (2017) Root endophytic fungus *Piriformospora indica* affected growth, cadmium partitioning and chlorophyll fluorescence of sunflower under cadmium toxicity. Ecotoxicol Environ Saf 145:496–502. https://doi.org/10.1016/j.ecoenv.2017.07.064
Shahzad T, Chenu C, Genet P et al (2015) Contribution of exudates, arbuscular mycorrhizal fungi and litter depositions to the rhizosphere priming effect induced by grassland species. Soil Biol Biochem 80:146–155. https://doi.org/10.1016/j.soilbio.2014.09.023
Shams M, Ekinci M, Turan M et al (2019) Growth, nutrient uptake and enzyme activity response of lettuce (*Lactuca sativa* L.) to excess copper. Environ Sustain 2:67–73. https://doi.org/10.1007/s42398-019-00051-7
Shi WG, Li H, Liu TX et al (2015) Exogenous abscisic acid alleviates zinc uptake and accumulation in *Populus* × *canescens* exposed to excess zinc. Plant Cell Environ 38:207–223. https://doi.org/10.1111/pce.12434
Shim D, Kim S, Choi YI et al (2013) Transgenic poplar trees expressing yeast cadmium factor 1 exhibit the characteristics necessary for the phytoremediation of mine tailing soil. Chemosphere 90:1478–1486. https://doi.org/10.1016/j.chemosphere.2012.09.044
Sidhu GPS, Singh HP, Batish DR, Kohli RK (2017) Tolerance and hyperaccumulation of cadmium by a wild, unpalatable herb Coronopus didymus (L.) Sm (Brassicaceae). Ecotoxicol Environ Safe 135:209–215. https://doi.org/10.1016/j.ecoenv.2016.10.001
Silva JRR, Fernandes AR, Silva Junior ML et al (2018) Tolerance mechanisms in *Cassia alata* exposed to cadmium toxicity: potential use for phytoremediation. Photosynthetica 56:495–504. https://doi.org/10.1007/s11099-017-0698-z
Singh J, Lee BK (2016) Influence of nano-TiO_2 particles on the bioaccumulation of Cd in soybean plants (*Glycine max*): a possible mechanism for the removal of Cd from the contaminated soil. J Environ Manag 170:88–96. https://doi.org/10.1016/j.jenvman.2016.01.015
Singh S, Korripally P, Vancheeswaran R, Eapen S (2011) Transgenic *Nicotiana tabacum* plants expressing a fungal copper transporter gene show enhanced acquisition of copper. Plant Cell Rep 30:1929–1938. https://doi.org/10.1007/s00299-011-1101-3
Smolinska B (2019) The influence of compost and nitrilotriacetic acid on mercury phytoextraction by *Lepidium sativum* L. J Chem Technol Biotechnol. https://doi.org/10.1002/jctb.5970
Sone Y, Nakamura R, Pan-Hou H et al (2013) Increase methylmercury accumulation in *Arabidopsis thaliana* expressing bacterial broad-spectrum mercury transporter MerE. AMB Express 3:52. https://doi.org/10.1186/2191-0855-3-52
Song WY, Sohn EU, Martinoia E et al (2003) Engineering tolerance and accumulation of lead and cadmium in transgenic plants. Nat Biotechnol 21:914–919. https://doi.org/10.1038/nbt850
Song Y, Jin L, Wang X (2017) Cadmium absorption and transportation pathways in plants. Int J Phytoremediation 19:133–141. https://doi.org/10.1080/15226514.2016.1207598
Song C, Yan Y, Rosado AR, Zhang Z, Castellarin SD (2019) ABA alleviates uptake and accumulation of zinc in grapevine (*Vitis vinifera* L.) by inducing expression of ZIP and detoxificationrelated genes. Front Plant Sci 10:872
Sonowal S, Nava AR, Joshi SJ, Borah SN, Islam NF, Pandit S, Prasad R, Sarma H (2022) Biosurfactants assisted heavy metals phytoremediation: Green technology for the United

Nations sustainable development goals. Pedosphere 2(1):198–210. https://doi.org/10.1016/S1002-0160(21)60067-X

Souri Z, Karimi N, Sarmadi M, Rostami E (2017) Salicylic acid nanoparticles (SANPs) improve growth and phytoremediation efficiency of *Isatis cappadocica* Desv., under As stress. IET Nanobiotechnol 11:650–655. https://doi.org/10.1049/iet-nbt.2016.0202

Sun S, Zhou X, Cui X et al (2019) Exogenous plant growth regulators improved phytoextraction efficiency by *Amaranths hypochondriacus* L. in cadmium contaminated soil. Plant Growth Regul. https://doi.org/10.1007/s10725-019-00548-5

Susarla S, Medina VF, McCutcheon SC (2002) Phytoremediation: an ecological solution to organic chemical contamination. Ecol Eng 18:647–658. https://doi.org/10.1016/S0925-8574(02)00026-5

Szuba A, Karlinski L, Krzeslowska M, Hazubska-Przybyl T (2017) Inoculation with a Pb-tolerant strain of *Paxillus involutus* improves growth and Pb tolerance of *Populus* × *canescens* under in vitro conditions. Plant Soil 412:253–266. https://doi.org/10.1007/s11104-016-3062-3

Thakare M, Sarma H, Datar S, Roy A, Pawar P, Gupta K, Pandit S, Prasad R (2021) Understanding the holistic approach to plant-microbe remediation technologies for removing heavy metals and radionuclides from soil. Curr Res Biotechnol. https://doi.org/10.1016/j.crbiot.2021.02.004

Tollsten L, Muller P (1996) Volatile organic compounds emitted from beech leaves. Phytochemistry 43:759–762. https://doi.org/10.1029/1999JD900144

Usman ARA, Lee SS, Awad YM et al (2012) Soil pollution assessment and identification of hyperaccumulating plants in chromated copper arsenate (CCA) contaminated sites, Korea. Chemosphere 87:872–878. https://doi.org/10.1016/j.chemosphere.2012.01.028

Vrbova M, Kotrba P, Horacek J et al (2013) Enhanced accumulation of cadmium in *Linum usitatissimum* L. plants due to overproduction of metallothionein α-domain as a fusion to β-glucuronidase protein. Plant Cell Tissue Organ Cult 112:321–330. https://doi.org/10.1007/s11240-012-0239-1

Wang F, Lin X, Yin R (2005) Heavy metal uptake by arbuscular mycorrhizas of *Elsholtzia splendens* and the potential for phytoremediation of contaminated soil. Plant Soil 269:225–232. https://doi.org/10.1007/s11104-004-0517-8

Wang D, Wen F, Xu C et al (2012) The uptake of Cs and Sr from soil to radish (*Raphanus sativus* L.)—potential for phytoextraction and remediation of contaminated soils. J Environ Radioact 110:78–83. https://doi.org/10.1016/j.jenvrad.2012.01.028

Wangeline AL, Burkhead JL, Hale KL et al (2004) Overexpression of ATP sulfurylase in Indian mustard: effects on tolerance and accumulation of twelve metals. J Environ Qual 33:54–60. https://doi.org/10.2134/jeq2004.5400

Wei S, Zhou Q, Wang X (2005) Identification of weed plants excluding the uptake of heavy metals. Environ Int 31:829–834. https://doi.org/10.1016/j.envint.2005.05.045

Wojas S, Clemens S, Sklodowska A, Antosiewicz DM (2010) Arsenic response of *AtPCS1*- and *CePCS*-expressing plants-effects of external As(V) concentration on As-accumulation pattern and NPT metabolism. J Plant Physiol 167:169–175. https://doi.org/10.1016/j.jplph.2009.07.017

Xue SG, Chen YX, Reeves RD (2004) Manganese uptake and accumulation by the hyperaccumulator plant *Phytolacca acinosa* Roxb. (Phytolaccaceae). Environ Pollut 131:393–399. https://doi.org/10.1016/j.envpol.2004.03.011

Yan X (2016) Uptake of radionuclide thorium by twelve native plants grown in uranium mill tailings soils from south part of China. Nucl Eng Des 304:80–83. https://doi.org/10.1016/j.nucengdes.2016.04.019

Yang X, Long X, Ni W et al (2002) *Sedum alfredii* H: a new Zn hyperaccumulating plant first found in China. Chin Sci Bull 47:1634–1637. https://doi.org/10.1007/BF03184113

Yang SX, Deng H, Li MS (2008) Manganese uptake and accumulation in a woody hyperaccumulator, *Schima superba*. Plant Soil Environ 54:441–446

Yang C, Ho YN, Makita R et al (2020) *Cupriavidus basilensis* strain r507, a toxic arsenic phytoextraction facilitator, potentiates the arsenic accumulation by *Pteris vittata*. Ecotoxicol Environ Saf 190:110075. https://doi.org/10.1016/j.ecoenv.2019.110075

Zhang H, Xu W, Dai W et al (2006) Functional characterization of cadmium-responsive garlic gene *AsMT2b*: a new member of metallothionein family. Chin Sci Bull 51:409–416. https://doi.org/10.1007/s11434-006-0409-9

Zhang XH, Liu J, Huang HT et al (2007) Chromium accumulation by the hyperaccumulator plant *Leersia hexandra* Swartz. Chemosphere 67:1138–1143. https://doi.org/10.1016/j.chemosphere.2006.11.014

Zhang Y, Li C, Luo X (2019) Enrichment effect of *Hypnum plumaeforme* on ^{210}Po and ^{210}Pb. Int J Phytoremediation. https://doi.org/10.1080/15226514.2019.1652560

Zhang C, He Q, Wang M et al (2020) Exogenous indole acetic acid alleviates Cd toxicity in tea (*Camellia sinensis*). Ecotoxicol Environ Saf 190:110090. https://doi.org/10.1016/j.ecoenv.2019.110090

Zhao FJ, Dunham SJ, McGrath SP (2002) Arsenic hyperaccumulation by different fern species. New Phytol 156:27–31

Zhao C, Qiao M, Yu Y et al (2010) The effect of the heterologous expression of *Phragmites australis* γ-glutamylcysteine synthetase on the Cd^{2+} accumulation of *Agrostis palustris*. Plant Cell Environ 33:877–887. https://doi.org/10.1111/j.1365-3040.2009.02113.x

Zhou P, Adeel M, Shakoor N et al (2020) Application of nanoparticles alleviates heavy metals stress and promotes plant growth: an overview. Nanomaterials 11:26. https://doi.org/10.3390/nano11010026

Zhu YL, Pilon-Smits EAH, Tarun AS et al (1999) Cadmium tolerance and accumulation in Indian mustard is enhanced by overexpressing γ-glutamylcysteine synthetase. Plant Physiol 121:1169–1177. https://doi.org/10.1104/pp.121.4.1169

Zhu XF, Jiang T, Wang ZW et al (2012) Gibberellic acid alleviates cadmium toxicity by reducing nitric oxide accumulation and expression of IRT1 in *Arabidopsis thaliana*. J Hazard Mater 239–240:302–307. https://doi.org/10.1016/j.jhazmat.2012.08.077

Zhuang P, Yang QW, Wang HB, Shu WS (2007) Phytoextraction of heavy metals by eight plant species in the field. Water Air Soil Pollut 184:235–242. https://doi.org/10.1007/s11270-007-9412-2

Zloch M, Kowalkowski T, Tyburski J, Hrynkiewicz K (2017) Modeling of phytoextraction efficiency of microbially stimulated *Salix dasyclados* L. in the soils with different speciation of heavy metals. Int J Phytoremediation 19:1150–1164. https://doi.org/10.1080/15226514.2017.1328396

Remediation Technologies, from Incineration to Phytoremediation: The Rediscovery of the Essential Role of Soil Quality

6

Gianniantonio Petruzzelli, Francesca Pedron, Meri Barbafieri, Irene Rosellini, Martina Grifoni, and Elisabetta Franchi

Abstract

The remediation of contaminated soils started years ago using consolidates technologies (incineration, inertization, etc.) usually employed in the waste treatment. This has contributed to consider a contaminated soil as a hazardous waste. This approximation was unfortunately transferred in many European legislations and on this basis soil quality have been used only marginally considered in the clean-up procedures. For many years, soil quality has been identified by the concentration values of contaminant and excavation and landfill disposal of soil has been largely used.

In recent years, the knowledge of technologies has rapidly grown and soil remediation is now based on innovative technologies, which are largely dependent on soil properties. The new environmental policies are increasingly promoting "Green remediation" and "Natural Based Solution" strategies: which consider all environmental effects of remedy and incorporate all the options to maximize environmental benefit. These remediation strategies restore contaminated sites to productive use with a great attention to the global environmental quality, including the preservation of soil functionality by the use of minimally invasive technologies such as bioremediation and phytoremediation.

Moving from the definition of remedial targets based on contaminant concentrations, it is essential to select technologies with low environmental

G. Petruzzelli (✉) · F. Pedron · M. Barbafieri · I. Rosellini · M. Grifoni
Istituto di Ricerca sugli Ecosistemi Terrestri, CNR, Pisa, Italy
e-mail: gianniantonio.petruzzelli@cnr.it; francesca.pedron@cnr.it; meri.barbafieri@cnr.it; irene.rosellini@cnr.it; martina.grifoni@iret.cnr.it

E. Franchi
Eni S.P.A., Renewable Energy and Environmental Laboratories, S. Donato Milanese, Milan, Italy
e-mail: elisabetta.franchi@eni.com

R. Prasad (ed.), *Phytoremediation for Environmental Sustainability*,
https://doi.org/10.1007/978-981-16-5621-7_6

impact to avoid the destruction in a very short time of an essential non-renewable resource, such as the soil.

Keywords

Soil contamination · Soil quality · Remediation technologies · Natural based solution · Green remediation · Phytoremediation · Biomass valorization

6.1 Introduction

Contamination is an important component of soil degradation processes as soil pollution greatly influences the quality of water, food, and human health. About 342,000 contaminated sites in Europe (Panagos et al. 2013; EEA 2014) and 450,000 sites in the USA (Marcomini et al. 2009) have been identified. Soil contamination has been recognized as an important issue for action in European soil protection strategies. Contaminated soils have mainly been found in industrial sites, landfills, and energy production plants, but accumulations of heavy metals and organic compounds sometimes also occur in agricultural land.

Conventional approach for the remediation of contaminated sites is mainly based on reducing the concentration of contaminants in the soil to reach the remediation targets, which are often determined from a risk analysis. This approach only considers the problem of the contaminated matrix and neglects the environmental effects of the remediation activities. In this context, the choice of the remediation technology is typically dictated by the intervention cost and by the time required for realization (Reddy and Kumar 2018). If the environmental impacts of such remediation activities are not considered, then significant consequences, such as the emission of toxic substances, the production of greenhouse gases, and the energy costs of transporting materials and the production of waste, may be neglected.

The destruction of the soil is one of the most significant negative outcomes derived from remediation, as it has been regarded as another type of waste and often placed in landfills. As we must currently face new challenges related to environmental sustainability, we can no longer delay implementing technologies that reduce the dangers derived from soil contamination, thus balancing the economic and social environmental effects. A critical review of the traditional remediation approach, which is exclusively aimed at achieving lower levels of contaminant concentration in the contaminated matrix, is therefore necessary. Instead, a more holistic approach that considers all the effects of reclamation activities can help improve the quality of the environment and attempt to balance the environmental, economic, and social issues. In this context, legislation based on the generic concentration values of contaminants, or risk analyses that are too cautious and not site specific, have led to the destruction of large quantities of soil. If this soil is recovered and reused productively, the environmental impact is much lower than if it is destroyed by placing it in landfills.

The discovery of soil contamination must be followed by remedial action. Remediation technologies have rapidly advanced in recent years; at present many treatment processes appear to be feasible at the field scale, and soil remediation activity is now based on site-specific risk assessment procedures. The effects of innovative technologies largely dependent on soil properties have been successfully applied. Hazardous organic compounds, in particular, are commonly treated by biological technologies, bioremediation and also phytoremediation, which can be partially applied to metals.

The technology selection should be based on reducing exposure to contaminants and acting on the pathways leading to the receptors, not just on the exclusive elimination of the pollution source. Clean-up approaches that restore soil quality after remediation treatments are also of increasing interest (Behera and Prasad 2020a). This issue is of major importance in the USA, where from 2009 the EPA promoted innovative clean-up strategies (Green remediation). Green remediation is defined as the practice of considering all environmental effects of remedy and ensuring the environmental benefits of clean-up actions are maximized. Also, the European Commission developed within the Horizon 2020 Framework Program new research and innovation policies in which remediation should be based on "Natural Based Solutions", NBS (EU-European Commission 2015). These NBS and Green Remediation strategies restore contaminated sites to productive use with close attention to global environmental quality, including the preservation of soil functionality. They are based on the following principles: using minimally invasive technologies and passive energy technologies such as bioremediation and phytoremediation as primary remedies minimizing soil disturbance and reducing bioavailability of contaminants through the adequate control of contaminant sources and plumes.

This chapter presents a brief overview of the emergence of the importance of soil in remediation technologies, from the consolidated technologies related to hazardous waste treatments to green and sustainable remediation. Many reviews of remediation technologies have been produced, which can provide further details and information (Khan et al. 2004; Kuppusamy et al. 2016). The focus of this chapter is on soil as a fundamental environmental matrix, which for many years has been neglected in remediation interventions but is of primary importance as it is the basis of many environmental equilibria.

6.2 Remediation Technologies

The remediation processes can be divided schematically into the general categories of physical, chemical, biological, thermal, and inertization. However, this is a drastic simplification, as the same technology may operate under different processes. For example, the vitrification process can be considered both as an inertization and a thermal treatment. Furthermore, remediation technologies are distinguished into "ex situ" and "in situ" techniques. The first techniques are based on the removal of the contaminated environmental matrix and the relative treatment in appropriate mobile plants located on the site itself (on site techniques) or in fixed installations in locations that can be at a considerable distance (off-site). Alternative techniques have been developed that do not include the phase of removal of the contaminated environmental matrix but enable it to be treated "in situ" by means of chemical, physical, and biological technologies, etc. Each treatment methodology has its own particular characteristics and limitations, therefore each technology is able to interact only with certain classes of pollutants, while the chosen treatment can be ineffective for other classes.

6.2.1 Thermal Technologies

At the beginning, the technologies that have been applied for the remediation of contaminated soils are derived directly from the methods used for the treatment of toxic and harmful wastes. Among the most widely used there were thermal technologies including incineration. Incineration using high temperatures (900–1200 °C) is capable of destroying all organic contaminants. However, this technology has significant environmental limitations because the soil, before being incinerated, is excavated and the energy required is very high, as are the costs. Of course, incineration has no effect on heavy metals, which can be dispersed in the environment during the process (Sabbas et al. 2003). In soil contaminated by these elements, the only usable thermal treatment has been soil vitrification (Timmerman and Peterson 1990; Thompson et al. 1992). With this procedure, it is possible to immobilize heavy metals, but the energy consumption is huge and involves very high costs (Mulligan et al. 2001).

These incineration processes have been updated and are now used with energy recovery (Arafat et al. 2015), however, contaminated soils are treated as hazardous wastes and the soil is destroyed. In terms of the quality of the soil, low-temperature thermal desorption technologies (LTTD) represent an evolution of thermal technologies (George et al. 1995; de Percin 1995). The two processes are distinguished by the temperatures, as in the case of LTTD the range is between 90 and 600 °C. The two technologies are also conceptually different because with incineration the contaminants are destroyed, while in thermal desorption they are subject only to a phase change (Guemiza et al. 2020). In thermal desorption, heat is applied to the soil to volatilize the contaminants, which are transferred to the gas phase. Heat and mass transfer are based on the specificity of the used thermal desorption system.

The two processes leave the treated soil in very different quality conditions. In the incineration process, irreversible changes are induced into the structure of the soil itself, which cannot be relocated on site but only landfilled. In the thermal desorption there is only a partial loss of the specific characteristics of the soil due to the destruction of the organic substance that occurs at 450 °C and a structural rearrangement of the oxides, and after the treatment the soil can be relocated in the site.

While in the incineration the characteristics of the soil have no importance as soil is considered simply as a waste, in the processes of thermal desorption the characteristics of the soil can play a very important role. The main factors that determine the process efficiency of thermal desorption include the characteristics of the soil to be treated (humidity, concentration of pollutants, organic substance content, ashes etc.), in addition to other process parameters such as the temperature at which it operates, the oxygen demand and the transfer heat. A further fundamental factor is the length of time the contaminated soil remains at the selected temperature; in general, it is possible to treat 50 t/h of soil. The efficiency of the thermal desorption treatment of polluted soil is also related to the porosity and texture of the soil, and sand particles will desorb contaminants easier than clays and silts. Also, the type of clays present is of considerable importance, as, for example, soils with higher kaolinite content than bentonite, is more easily treated with this technology. In fact, bentonite can form strong hydrogen bonds between the hydroxyl groups of the clay and the contaminants in the water of the pores in the soil, thus forming a barrier to the movement of gas through the soil particles (Zivdar et al. 2019). Theoretically, thermal treatments can be applied on-site and off-site, however, as in most off-site treatments, there are environmental problems associated with transporting the soil, so on-site treatments are receiving increasing interest, also for economic reasons. In Europe, mobile thermal treatment units are often used, which are particularly useful when the quantity of contaminated soil to be treated is very high and the costs for transport to a treatment plant would therefore be unsustainable.

As previously mentioned, thermal treatment processes for the soil are applicable in the case of volatile and semi-volatile organic compound contaminants, such as PCBs, dioxins, chlorinated phenols, pesticides, and herbicides (Kaštánek and Kaštánek 2005; Khaitan et al. 2006; Pavel and Gavrilescu 2008).

LTTD is a full-scale commercially available technology. The full-scale performance must be based on the knowledge of the soil characteristics, to define the working conditions for the equipment and to obtain the highest performance of thermal treatment systems. This technique has been used to clean-up many sites polluted by several organic contaminants. Successful applications have been achieved for remediating volatile organic contaminants in many kinds of soils (Merino and Bucalá 2007; Kastanek et al. 2016), petroleum hydrocarbon (Chien 2012), and PAH (Falciglia et al. 2016). In the last few years, "in situ" thermal desorption techniques (ISTD) have also been used. ISTD is effective for separating PAHs (by volatilization or destruction) from contaminated soils (Kuppusamy et al. 2017). Here, heat is used to separate the PAHs from the soil through a physical process but the costs are lower because no excavation of the soil is required. It is considered a fairly safe technique and produces little or no emissions of PAHs into the atmosphere. This is obtained by a carrier gas or a vacuum system, which carry the volatile PAHs compounds in the gas retention system for disposal.

6.2.1.1 Effects on Soil Functions

The passage from incineration to thermal desorption represents an important step towards the consideration of the soil as an environmental matrix rather than as a waste. Using a thermal desorption treatment, the contaminated soil is not considered as waste, and its characteristics are evaluated with a view to a possible reuse. This new point of view is of substantial importance as over time, the technology continues to evolve to achieve greater energy efficiency, through finding innovative methods to achieve low-temperature desorption, low emissions with considerable energy savings, and lower impacts on soil quality (Hou et al. 2016). The magnitude of the effects that this technology has on the functionality of the soil is determined by the process temperature and the duration of the heating (O'Brien et al. 2016). The temperature reduction from 600 °C to values below 400 °C can keep 70% and 90% of the soil's organic substance intact by using short-term treatment cycles (Sierra et al. 2016). The maintenance of low desorption temperatures reduces the variation in texture that occurs with heating at high temperatures, which is a major consideration. At high temperatures, the clay proportion is reduced in relation to the sandy proportion, with a consequent reduction in the water and nutrient retention capacity of the soil. This also implies a greater susceptibility of the soil to erosion and wind transport phenomena (Yi et al. 2016).

6.2.2 Physical Treatments

Remediation techniques based on physical processes are used to isolate or concentrate pollutants. Physical treatments do not destroy contaminants. They are often used as a first stage in remediation processes and followed by other decontamination technologies.

The main advantages are the speed of treatment, the applicability to many categories of contaminants and the relatively low costs. These technologies do not require a very thorough characterization of the site, but the residues they generate require further treatments, and the characteristics of the soil may limit the applicability.

6.2.2.1 Soil Washing

Soils in contaminated sites can have very different particle sizes. The treatment methods are applicable when within the various size classes, the dimensions of the contaminated particles distributed in a very narrow range, i.e. in some granulometric classes, there is no contamination.

The separation technologies based on the particle dimensions exploit the characteristic that in many cases the contaminants preferentially adsorb on the finer soil materials. If there is a distribution of different levels of contamination according to the dimensions of soil particles, the separation of different granulometric classes makes the treatment operations and the final use of the contaminated soil much more effective. These technologies enable the immediate reuse of specific parts of the soil itself, and a significant reduction in the volume of the soil that is contaminated to be further treated or landfilled. In optimal conditions the "physical separation" of the polluting substances can provide very high yields, but different factors influence this type of process, including:

- the composition of the contaminated soil: this technology is particularly suitable for sandy soils;

- the chemical form of the contaminant: for example, a metal in the form of a mineral can be successfully separated from the constituents of the soil according to the different densities.

Washing treatments are interesting examples of these physical treatments.

Soil washing is an "ex situ" technology for removing both organic and inorganic contaminants, based on physical and chemical principles, by means of a dimensional separation of the particles (Williford and Bricka 2000). With this and other similar procedures, inorganic contaminants such as heavy metals, and organics such as halogenated hydrocarbons, PAH, and benzene, can theoretically be treated (Saponaro et al. 2002; Urum et al. 2003).

This is one of the few efficient technologies for the clean-up of heavy metals from contaminated soils. The technology is characterized by its broad applicability, which derived from its simple operability, the ability to be effective for any contaminant concentration level, and the relatively low cost. This dimensional separation is achieved through different processes:

- crushing of the soil, if the contaminants are in the larger aggregates;
- mechanical disintegration with high-pressure water jets to break up the aggregates;
- energetic rubbing of soil aggregates to remove contaminants that adhere to the surface.

The process releases the contaminants from the larger soil particles, concentrating them in a small portion of 5–30% of the original soil. This objective is achieved mainly by an intense mixing of the soil with the washing liquid (water, with or without additives). Here, the contaminants that are bound to the coarser particles by adhesion and compaction forces are released in the processes of abrasion and wet rubbing. A separation of the solid from the liquid phase occur, while the soil is recovered in two distinct fractions: a decontaminated reusable fraction, and a smaller fraction with dimensions generally lower than 63 μm, consisting of the fine clay and silty fraction that contains all the contaminants. Naturally, after the washing process, the decontaminated fraction can be subjected to another cycle to eliminate any contaminants that were not completely separated. Chemical reagents, such as surfactants, can also be added to facilitate these processes. Decontaminated soil can be relocated on site or used as a filling material. With either outcome, the soil is considered and treated as non-hazardous waste (Chu and Chan 2003).

The ability of this process to remove contaminants depends on the properties of each class of substance, such as polarity, volatility, water solubility, and the characteristics of the contaminated soil such as pH, organic substance, cation exchange capacity, surface area, and on that from the characteristics of the washing method and the contact time.

The best results are obtained with recently contaminated soils, as the phases of adhesion and compaction, which bind the most polluted particles to the other constituents of the soil, increase over time.

This technology consists of a series of consolidated process units and machinery, which are very similar to numerous longstanding mining processes. In addition to a separation process based on the particle size to concentrate the contaminants in a small portion of the original contaminated soil, this technology can also be used to transfer pollutants to the liquid phase used for the treatment (Isoyama and Wada 2007). In this case, the process is similar to that of "solvent extraction" and with this term is defined by USEPA since the contaminants are dissolved or suspended in the liquid phase and removed for subsequent treatment.

This technology is particularly attractive when it is used exclusively as a physical treatment capable of concentrating contaminants (without solubilizing them) in a fine contaminated fraction, allowing a considerable amount of soil to be recovered.

Apart from the different configurations, a soil washing process involves the following phases:

- excavation of the soil;
- screening to remove the coarser materials;
- washing with water only, to separate the contaminants;
- recovery of a proportion of the decontaminated soil that can be relocated in the site or elsewhere without restrictions;
- management of the contaminated fine fraction;
- treatment of the washing liquid.

A knowledge of soil characteristics is fundamental to this technology. First, the texture of the soil must be established, because soils with clay percentages above 30% are difficult to treat (Mulligan et al. 2001). In case of heavy metal contamination, knowledge of the chemical forms in which they are present in the soil is of primary importance, as metals bound to solid surfaces with chemical bonds are not removable with this technology (Petruzzelli et al. 2004; Grifoni et al. 2017).

Soil washing is primarily applicable to industrial soils generally characterized by mixture of wastes, where metals are present as discrete particles, while it is less appropriate for contamination of agricultural soils unless these soils have a high content of sand.

The technology can be effectively used in combination with other technologies (Elgh-Dalgren et al. 2009; Villa et al. 2010; Jeon et al. 2010).

6.2.2.2 Effects on Soil Functions

Like any "ex situ" technology, soil washing allows for better control of the process parameters, but also leads to the total destruction of the soil. Nevertheless, with this technology a considerable proportion of the excavated material can be relocated on site, in line with the administrative procedures of remediation legislation.

Soil washing removes the fine particles of the soil, which are mainly of a clayey nature, so for the part of the soil that can be recovered the reactive surfaces for the transport of water and nutrients are reduced (Yi and Sung 2015). The loss of clay material modifies the pore system with a reduction in water retention and cation exchange capacities. This in turn also alters the soil's chemical properties with a reduction of the amount of exchangeable Ca, Mg, and K, which can adversely affect microorganism and plant growth.

Despite these negative aspects and the fact that the technology results in the total deconstruction of the soil, the possibility of replacing a large part of the material on site enables a new soil to be recreated, which has pre-determined chemical-physical properties capable of recovering the typical functions of the soil in a reasonably rapid time. Soil reconstitution is not simply a re-aggregation of the different materials that have separated during the treatment of soil washing, but through the correct processing of the mixture of these materials, it recombines them as a new soil with specific properties. These soils have been defined as "anthropogenic", as their pedogenesis largely derived from the mixture of different components obtained by mechanical treatment (IUSS 2007).

The washing process removes the soluble organic matter and particulate organic matter, thus decreasing the humus content of soil (Ko et al. 2005). The essential point of the soil reconstitution after soil washing is therefore the incorporation of organic matter in the mineral proportion of the uncontaminated materials, to reactivate the microbial activity that has an essential role in organic matter and nutrient cycling in the soil (Mukhopadhyay et al. 2014). Thus, after the soil has been relocated on site, it may be necessary to accelerate the recovery processes of the soil functionality by adding humic-like materials (compost, biochar) and fertilizers (NPK).

Soil washing is a stand-alone technology but a final finishing phase can be added, using plant cover. Thus, the use of a train technology can speed up the reuse of the soil.

6.3 Electrokinetic Remediation

Electrokinetic remediation (EKRT) is an "in situ" technology, which requires low levels of direct electric current applied to a contaminated soil through electrodes distributed inside the soil to form an electric field across the contaminated site, which promote the contaminants' migration towards the electrodes where they are collected and treated. The positive ions are attracted to the cathode, which is negatively charged, while the negative ones move towards the anode (positive). Non-ionic species are transported together with the electro-osmosis-induced water flow. The flow direction and the amount of contaminants moved are influenced by the type and structure of the soil and the mobility of contaminants. This technology applied "in situ" is strictly dependent on the soil properties, such as pH, texture, and organic matter content. It is effective in highly clayey soils, which are difficult to treat with other methods.

The technology is characterized by three basic processes. The main transport mechanisms responsible for the migration of pollutants to the cathode or anode are electromigration, electrophoresis and electro-osmosis (Acar and Alshawabkeh 1993; Acar et al. 1995; Reddy and Cameselle 2009).

- Electromigration is the transport of ions in the soil solution determined by the strength of the applied electrical field, in which the cations (positively charged) migrate towards the cathode and the anions towards the anode.
- Electrophoresis is the process of the migration of charged colloidal particles (including bacteria) due to the applied electrical field. This phenomenon is more complex than the migration of ions in solution as the charge of the colloids can change according to local conditions (for example, at low pH this may be positive, and negative at high pH), therefore it can change the direction of particle migration. Mass transport by electrophoresis is negligible in soil of low permeability but is important for the transport of particles of colloidal size and micelles.
- Electro-osmosis is the migration of the interstitial aqueous solution induced by the electrical field, and therefore the solution transports all solutes present including non-ionized compounds (organic contaminants). The transport is governed by complex mechanisms, dependent on the characteristics of soil solid surfaces and their interactions with contaminants in the soil solution (Probstein and Hicks 1993; Pamukcu 2009).

The application of an electrical field to soil produces several chemical reactions (dissolution–precipitation, redox reaction, adsorption–desorption) that greatly affect the transport and speciation of the contaminants and the removal efficiency (Yeung 2009).

The mechanism of the migration of the interstitial solution by the electric field is determined by the surface charge of the solid matrix. This is normally negative and determines a "concentration" of positive ions in a layer close to the solid surfaces (double layer). The ions closest to the solid surface are immobile while those beyond a certain distance migrate towards the cathode; their migration also drags the interstitial solution towards the cathode. As the process is determined by the surface charge of the soil solid phase, it is also influenced by the local conditions, of which the pH is extremely important. Acidity conditions, which can also arise during the process due to the electrolysis of water, can determine the change of the sign of the surface charge and therefore the inversion of the electro-osmotic flow. As the electro-osmotic process determines the displacement of the interstitial solution, the effectiveness in removing contaminants strongly adsorbed on the soil surfaces is, for obvious reasons, very limited.

The electrolysis of water is the predominant process in wet soils. This leads to the formation of H^+ ions and oxygen at the anode and OH^- ions and hydrogen at the cathode. The immediate result is a local pH change in the soil solution with an increase over time of anode acidity and cathode alkalinity. The electric current therefore creates an acid front that starts from the anode and a basic front from the cathode. The acid front moves faster than the basic front due to the greater mobility of H^+ ions compared to OH^-, and thus the electro-osmotic flow direction is generally towards the cathode. The generation of the acid front is fundamental. In fact, the low pH conditions in the acid front can promote the solubilization of metals, facilitating their collection at the cathode (the H^+ ions replace the adsorbed ions onto the soil particles, releasing them in solution and making them available for migration).

Where the two fronts meet, water is formed with a sharp change in the pH value, which influences the solubility of the contaminants and their adsorption to the solid surfaces of the soil. For heavy metals decontamination, it is necessary to maintain a high solubility of metal ions throughout the soil with a low pH and to avoid the conditions that lead to precipitation.

Soil properties are of paramount importance for the applicability of this technology. The pH value of the soil largely determines the efficiency of the remediation, as the metals that can be removed are exclusively those present in the liquid phase of the soil. Metals are usually present as cations and are therefore mainly transported by electromigration to the cathode. However, certain elements such as Cr and As could be present as oxy-anions and therefore the transport will be towards the anode. Electroremediation can also be used to remove other anions such as sulphates and cyanides.

Some organic substances such as phenols, chlorophenols, toluene, can also be treated (Cameselle et al. 2013; Gill et al. 2014). For organic compounds, the predominant transport mechanism is electro-osmosis, although electromigration may also contribute due to partial dissociation of the compounds. The electroremediation process may be less effective in the case of immiscible non-polar organic compounds, although they can be transported by electro-osmosis and electrophoresis if they are bound to colloidal particles.

The major advantage of this technique is the accurate control of the flow direction of water and dissolved contaminants, even when moving through heterogeneous soils, with the retention of contaminants in a restricted area. This technology does, however, have some limits. For metal contaminated soil, the whole process depends on the soil acidity, which increases in the soil during current application, which is a condition that favours the release of heavy metals in the soil solution. If the buffer capacity of the soil is high, it could be difficult to reach appropriate acidity conditions. Here, it may be necessary to use an additive to mobilize contaminants in the soil solution. Moreover, we must consider that soil acidification may produce negative environmental effects due to the increased potential leaching of contaminants. The efficiency of the EKRT may be drastically reduced by soil heterogeneity and anomalies, and the presence of high concentration of non-target ions.

Before full-scale application, it is essential to evaluate the characteristics of the soil, in particular the pH, soil texture, and organic matter content. The quantity of metals in the soil solution, which is the proportion that can be removed with this technology, can be evaluated by sequential extraction procedures that select the extractants on the basis of the metals present in the contaminated soil (Petruzzelli et al. 2015). This procedure is the basis for defining the additives (for example, complexing agents) required to increase metal solubility.

A feasibility test in the laboratory can be performed with a small cell, but this may be irrelevant because the soil that is placed in the cell is typically homogeneous and not representative of the heterogeneity of the field. Thus, it is generally preferable to carry out the feasibility test directly in the field on a small scale (Gill et al. 2014).

6.3.1 Effects on Soil Functions

During the electroremediation, a pH gradient is created between the anode and the cathode, which causes the electrolysis of the water with the production of H^+ and OH^- ions that cluster near the electrodes. This phenomenon causes a decrease in the pH of the soil near the anode and a shift towards basic values near the cathode. The extent of this change depends on the difference in potential applied and the duration of the process (Cang et al. 2012; Zhou et al. 2015). This pH change has a significant influence on the solubility of metals in the soil, which tend to accumulate at the cathode or anode depending on the specific chemical properties. The process is also accompanied by a change in the distribution of the elements of fertility in the soil. Relative increases of N and P were reported to the anode and relative decreases at the cathode, while K increased in the area around the cathode (Chen et al. 2006; Zhou et al. 2015).

The formation of a pH gradient creates a change in the functionality of the soil that can be differentiated along the area affected by the electroremediation treatment (Zhou et al. 2015). The main negative effects are evident in the microbiological properties of the soil, with a decrease in microbial respiration and in bacterial and fungal populations (Lear et al. 2007). The most significant effects are found near the anode where the soil has lower pH values. However, the diminishing effects of bacterial abundance are often found throughout the treated area (Kim et al. 2010). The reduction in the number of bacteria is sometimes accompanied by a reduction in enzymatic activity (Cang et al. 2012).

The electrokinetic remediation can also induce the phenomena of deterioration of the soil's humic substance and can inhibit or reduce plant growth. In fact, in some areas there may be deficiencies of fertility elements due to the effects of the treatment (O'Brien et al. 2017).

The basic condition required to remove contaminants from the soil with this technology is that their mobility is high: that is, the contaminants must be present in the liquid phase of the soil. This has some similarities to biological technology (bioremediation, phytoextraction), which are based on the bioavailable fraction of contaminants, i.e., the amount in the soil solution.

6.4 The New Vision of Soil in Remediation

Industrial and urban development has had a significant impact on the entire ecosystem since its inception. However, from the twentieth century onwards the general increase in environmental pollution prompted leaders and global institutions to adopt new strategies to reduce the real and potential associated risks. Therefore, new environmental policies aim to address the current environmental challenges (climate change, food security, or natural disasters) and to actively encourage an increasing use of solutions that are efficient and sustainable, such as "nature based solutions (NBS)" (Faivre et al. 2017). NBS are actions inspired or copied from nature and based on the efficient use of energy and resources, to provide solutions to climate mitigation and adaptation challenges while simultaneously offering economic, social, and environmental benefits (EU-European Commission 2015; IUCN 2016). Each NBS should follow the concept of sustainability (Keesstra et al. 2018) by guaranteeing environmental and social protection in an economically efficient way, thus stimulating innovation for a green economy.

Initial NBS actions were mainly focused on urban resilience, through multifunctional "green" interventions such as green infrastructure, and to solutions for climate change mitigation and adaptation (Eggermont et al. 2015). However, after the European Commission also assessed the NBS potential from an economic point of view, new research and innovation policies were developed in the issue of contaminated sites (EU-European Commission 2015).

Soil contamination is currently one of the most widespread and serious environmental problems linked to industrial and urban development.

Thus, NBS approaches to the remediation of contaminated soils have been developed and promoted (EU-European Commission 2015; Bourguignon 2017), as they can offer more advantages than traditional approaches. An adequate implementation of NBS for soil remediation can allow efficient soil cleaning to be realized, while simultaneously reducing energy and resource consumption (Song et al. 2019). Among the NBS technologies for remediation, phytoremediation is an option that most closely meets the criteria of environmental, social and economic sustainability.

As this is an "in situ" technology, the costs and energy consumption are reduced, and in addition it can provide aesthetic benefits, minimize the disturbance of the surrounding environment, and has the possibility of the more efficient use of by-products (Pandey and Souza-Alonso 2019; Grifoni et al. 2020). The latter represents a strength point of phytoremediation, as it can make the technology highly competitive within circular and green economy schemes.

Phytoremediation is a low environmental impact and low-cost technology based on biological processes. The evaluation of the technology through the Life Cycle Assessment (LCA) tool provides a method for evaluating the total environmental impact throughout the remediation. For example, from the phytoextraction of contaminants to the production of biomass, including its use and its disposal, and accounting for the energy and resource inputs. Recent paper based on LCA comparison of different technologies clearly shows the major advantage of phytoremediation in terms of environmental impact, and ecological footprint with respect to consolidated technologies or excavation and landfill disposal (Vocciante et al. 2019).

The use of plants specifically selected for the remediation of contaminated sites is an essential tool for achieving sustainable development. Plant-based remediation represents one of the best strategies for maintaining healthy ecosystems, providing a low-expense method of site restoration for future use. Phytotechnologies fall fully within the green remediation category (Misra and Misra 2019; Ashraf et al. 2019). These technologies are extremely environmentally-friendly and achieve clean-up targets of reducing energy use, while protecting ecosystem services during site remediation (Behera and Prasad 2020b).

The full-scale application of technology requires both a deep understanding of the physiological mechanisms of plants that determine phytoextraction or phytostabilization processes and of the effects on the process of site-specific environmental variables, in particular of the characteristics of the soil that determine the bioavailability of contaminants, which are the bases of all the processes of phytoremediation.

6.4.1 Bioavailability

The evaluation of the bioavailability of contaminants for plants is essential for the applicability of the technology. This aspect is often neglected in the phytoremediation strategies on a real scale. In the soil–plant system, bioavailability is determined by a set of reactions which are regulated by the properties of the contaminants, the soil, and the plants (Petruzzelli et al. 2015) (Fig. 6.1).

The contaminants in soil are retained by solid phases (clays, oxides, humic substances) with bonds of different nature and strength through adsorption processes, which depend on the specific characteristics of contaminant and soil (Petruzzelli et al. 2013b).

The influence of specific soil characteristics on the potential transfer of the contaminant to plants can be explained through the bioavailability processes that regulate the uptake of heavy metals. The bioavailability process can be divided into several phases. Initially (step A) an element passes from the solid phase, in which it is not available for environmental processing, to the liquid phase. In the liquid phase, the contaminant becomes potentially available for plant uptake and following transport processes (diffusion, dispersion, etc.) it can be transferred to the plant

Fig. 6.1 Bioavailability processes in the soil–plant system

root system (step B). During these transport processes, additional reactions (oxidation–reduction, hydrolysis, photolysis, degradation, etc.) able to modify the properties of the contaminants may occur. The same mechanisms can also transport substances that are bound to very small solid particles, such as those that are colloidal in nature (step C). Finally, the contaminant is taken up by the plant and accumulates in the different plant tissues (Step D).

Due to the high persistence of heavy metals in soil they are very dangerous to human health (Abrahams 2002).

In the case of phytoextraction the technology is based on the transfer of contaminants from the soil to the root system, and this fundamental step is governed by the specific characteristics of the soil, which determine the release of contaminants from the solid phase to the soil solution where the contaminants are available to plant uptake. The sorption/desorption reactions of metals on solid phase closely depend on the soil properties, such as pH, cation exchange capacity (CEC), redox potential (Eh), etc.

pH is the most important parameter that governs the concentration of inorganic elements in the soil solution (Li et al. 2003; Chaney et al. 2005). This parameter can provide the largest amount of information related to the chemical-physical characteristics of the soil. The pH depends on the pedogenetic processes that lead to the formation of the soil and on the reactions established between the solid, liquid, and gaseous phases of the pedosphere. pH is the most important parameter that governs the concentration of inorganic elements in the soil solution, regulating the phenomena of precipitation and solubilization. The solubility of most metals tends to decrease with increasing pH. In alkaline conditions, the precipitation processes reduce the concentration of these metal ions in solution, while the opposite phenomenon occurs in an acid environment. The adsorption of the metals is closely related to the pH due to the competition of the H^+ ions for the same adsorption sites on the surfaces of the soil. With phytoremediation, the rhizosphere activity can modify the pH due to the preferential uptake of anions or cations and the consequent release of H^+ or OH^- ions, and the release of organic compounds. Modification of pH can also be derived by the activation of plants' adaptive responses to stress abiotic that are present in contaminated soils (Fraire-Velazquez and Emmanuel 2013).

Cation exchange capacity (CEC) is a soil parameter representing the amount of exchangeable cations that a soil can retain at exchange sites. The cations retained on the surfaces by forces of electrostatic attraction are in dynamic equilibrium with the cations present in the liquid phase of the soil and are easily exchangeable with those in solution. Therefore, a soil with a high CEC has a greater capacity to reintegrate in the liquid phase the cations absorbed by the roots of plants. The exchange sites are mainly present on the surfaces of clay minerals and soil organic matter. Clay minerals have permanent electrical charges due to isomorphic substitutions in the crystal lattice. The clays, as primary constituents of the soil, retain the metals through specific adsorption and ion exchange reactions through a mechanism of interaction with the hydroxyl ions to which the metals are subsequently bound. The various types of clays have considerable differences in their adsorbent capacities. The adsorption and retention processes are higher in the expandable clays, in which the adsorption takes place in the interlayer spaces. The influence of clay content on bioavailability and plant uptake has been reported for some species (Abdullah and Sarem 2010). The organic matter, the iron and aluminium oxides, have instead electrical charges variable with the pH due to the dissociation of acid groups on surface sites. Surfaces with a dependent pH charge tend to develop a higher CEC as the pH increases. Soil organic matter is the material that derives from the decomposition of plant and animal residues that are structured in complex chemical forms. Due to the ability of humic substances to form complexes with metals and to act as adsorbent surfaces for organic compounds, their content in the soil is of paramount importance for contaminant bioavailability. The effects on bioavailability are variable according to the solubility of complexes, metal—fulvic acids complexes are soluble thus increase the content of metals in the soil solution. In contrast, humic acids of higher molecular weight form very stable complexes with metals, drastically reducing their bioavailability in relation to the strength of the bonds formed in the complexation reactions (Wang et al. 2010).

Under relatively oxidizing conditions, the metal solubility is particularly affected by the hydrous oxides of Fe and Mn, since the latter can reduce metal concentrations in soil solution through specific adsorption and precipitation reactions. Due to the strength of the linkages, the specific adsorption of metals by oxides drastically reduces their bioavailability.

Another important soil characteristic is its redox potential. The redox potential of a soil derives from the numerous oxidation–reduction reactions, which occur in the soil environment. These reactions are controlled by the activity of free electrons in solution expressed as the redox potential, which is a measure of the electron availability in the soil environment. High levels of Eh are characteristics of dry and aerated soils, while soils that are submerged or particularly rich in organic substances tend to have low Eh values (Charlatchka and Cambier 2000). The redox potential influences the solubility and therefore the bioavailability for the plants of all the elements with more oxidation states. The redox processes can induce pH changes as the modification of the oxidation state of an element involves the consumption or release of protons.

Several other factors may affect the solubility and bioavailability of metals in soils. An increase in ionic strength in the soil solution tends to reduce the adsorption of metals by surfaces due to competition of alkaline ions (Petruzzelli et al. 2015). The soil temperature influences the decomposition processes of humic materials, changing the stability of the complexes with the metals and their solubility. Microbial biomass can promote the precipitation of metals as sulphides and reduce their content in the soil solution by also offering new surfaces for adsorption processes, while root systems can increase solubility for the release of complexing substances (radical exudates) or protons that increase the acidity in the area of the rhizosphere.

An important aspect that governs the interactions between the solid phase of the soil and the contaminants is the time factor; with the permanence in the soil a contaminant is subject to transformations that lead it to be more restrained by the solid phase and to become less available for environmental processes. This aspect is particularly important for organic compounds whose bioavailability is governed by the characteristics of the soil synthetically described above, and also by processes, such as volatilization and biodegradation, which reduce over time their presence in the soil. Over time, due to ageing processes, there is a reduction in the bioavailability of organic compounds in the soil with the formation of stable, non-bioavailable residues. The ageing process is greatly influenced by the content and nature of humic substances, by the initial concentration of the compounds, by the microbial biomass and also by some physical properties of the soil, in particular from the pore system, both in shape and size (Alexander 2000). In fact, organic compounds can be subjected not only to irreversible adsorption phenomena but also to physical entrapment processes within the micropores of the soil. In the soil there are numerous natural organic compounds synthesized by plants, animals, and microorganisms. These compounds are continuously destroyed by the biodegradation processes, and consequently over a fairly long time scale their concentration in the soil remains substantially constant. Among these natural organic compounds, humic substances have an effective absorbent effect against contaminants, reducing their bioavailability. However, anthropogenic chemical substances, although subject to the same biodegradation processes as natural ones, often have significantly different molecular structures that microorganisms and do not have the ability to degrade in a reasonably short time. Importantly, in the absence of enzymatic activity they will tend to persist for long periods in the soil, and depending on the intrinsic toxicity (Doick et al. 2005) and bioavailability may give rise to significant environmental and health hazards.

Bioavailability is the key to assess the risks from pollution, however there are no official methodologies to measure it. For years, several extractants for assessing the bioavailable fraction of a substance or an element have been explored in the field of soil chemistry. However, bioavailability is determined by a complex series of processes, involving chemistry, biology, and ecotoxicology (Harmsen et al. 2005). Thus, only the combination of chemical and biological tests can provide information on bioavailability, which are essential for obtaining the greatest efficiency from phytotechnologies (Petruzzelli et al. 2015). By considering the "soil–plant" system from the chemical point of view, it is possible to determine the concentration of metals present in the soil solution and/or the most easily releasable from the solid phase. This can be accomplished by either the direct analysis of the soil solution or by using mild extractants (water or alkaline salts solutions), which provide a good indication of the metals present in a soluble form or easily releasable in the liquid phase of the soil (Rayment and Lyons 2012). Biological tests should be carried out by growing plant species on the specific contaminated soil, which enables an evaluation of the physiological response of plants to bioavailable metals. At the end of the growth, the analysis of the metal content in the plants will, in addition to the chemical test, provide a further indication of the bioavailable quantity involved in phytoremediation strategies. A preliminary study of the specific characteristics of the soil, of the chemical forms of the metal and of the possibilities of growth of the selected plant species is essential to make reliable predictions on the efficiency of the technology on a specific contaminate site (Pedron et al. 2009).

6.4.2 Phytoextraction Based on Bioavailability Processes

Phytoremediation technologies have had considerable interest from the scientific community and stakeholders since they were first developed. After years of phytoremediation application, further efforts to achieve a development of technology that makes it more applicable at the field scale are required.

At a real scale, positive results have often been obtained from the remediation of soils contaminated by organic compounds, as the plants support and stimulate soil microbial activity to enhance the degradation of contaminants, but in the case of heavy metal pollution optimal results at the field scale are yet to be achieved in most cases.

An essential step for the implementation of phytoremediation is to consider that plants exclusively act on the bioavailable proportion of the contaminants the only one really dangerous. The target of phytoremediation should be therefore to reduce or eliminate this bioavailable proportion and not to decrease the total concentration of metals in the soil.

For example, in most contaminated sites a substantial proportion of total metals is irreversibly linked to the surfaces of the soil. The phytoextraction is not able to remove this fraction, so the aim of the technology should be to act only on the proportions of metals that are or can be made bioavailable. This strategy is defined as "bioavailable contaminant stripping" and has been implemented (Pedron and Petruzzelli 2011; Petruzzelli et al. 2013a) as "enhanced bioavailable contaminant stripping (EBCS)". This new approach also considers the capacity of soil to replenish with time the liquid phase with metals from lower available pools. This phytoextraction approach is integrated into the new regulations based on site-specific risk assessments and no longer on generic contaminant concentration values.

The phytoextraction efficiency can be increased by enhancing the productivity of selected plants with the fertilizers use. Amendments such as amino-poly-carboxylic acids (APCAs) consisting of several carboxyl groups bonded to one or more nitrogen atoms can be added to soil to facilitate the desorption of metals from the solid phase and thus to increase their solubility (assisted phytoextraction). This is a widely used procedure (Doumett et al. 2011), but the use of chelators to form stable and water-soluble complexes with metals can increase their long-term concentrations in the soil solution, in excess of the translocation capacity of plants (Luo et al. 2005; Santos et al. 2006; Cao et al. 2007). This feature has other potential risks, as soluble chelated metals can easily percolate into the soil spreading contamination into the subsoil or into ground or surface waters. A successfully used option is to add the chelator in several doses separated over time, which in addition to increasing the amount of metal absorbed by the plant reduces the risk of leaching (Tassi et al. 2004; Barbafieri et al. 2017).

An alternative option to chelating agents involves the use of natural low-molecular-weight organic acids (including citric, malic, oxalic, and tartaric acids) with higher biodegradability, and thus lower persistence, and much lower toxicity (Evangelou et al. 2006; Doumett et al. 2008). Due to their rapid biodegradability, these ligands only persist in soil for a short time (Evangelou et al. 2008). Repeated applications may therefore be required to maintain high enough metal bioavailability in soils to support plant metal uptake. If additives have been previously used for several years, there are additional ways to improve phytoremediation, and one innovative method of increasing technology efficiency is to use plant growth regulators.

6.4.3 Further Phytoremediation Improvement

In recent years, many efforts have been made to increase the efficiency of phytotechnologies through laboratory and field studies, looking for new potential strategies to make the technology more and more applicable. There are many innovations of great relevance reported in specific reviews (Ansari et al. 2018). Among these, the support of exogenous application of plant growth regulators (PGRs) and the use of plant growth-promoting bacteria (PGPB) to improve the effectiveness of phytoextraction processes are of particular interest.

6.4.3.1 Plant Growth Regulators (PGRs)

Limitations due contaminant toxicity and or reduced fertility of contaminated areas deeply affect the efficiency of phytoremediation reducing its applicability. Therefore, new studies are devoted to improving the technology performance, trying to overcome its intrinsic limitations.

The application of PGR has been tested and demonstrated in different trials (Barbafieri 2016; Barbafieri et al. 2012, 2018; Bulak et al. 2014). The PGRs are small organic compounds that have the same role of natural phytohormones in regulation of several plant growth and developmental processes. In phytoremediation, they can be supplied, alone or in combination with other treatments, to promote the plant growth and the metal uptake by plants, since they are able to alleviate metal toxicity. Among most widely PGRs, there are the cytokines, ethylene, abscisic acid, salicylic acid, auxin, gibberellin.

Research activities on phytoextraction are above all focused on overcoming the two main drawbacks: the survival of plants in unfavourable environmental conditions and the long time needed to reduce contaminants to the requested level. The first case is due to contaminant toxicity and to low fertility, common conditions in contaminated sites; instead, the second case is due to physiological plant limitations in contaminants uptake and translocation or also low and slow plant biomass production.

Thus, the new investigations on plant growth regulators, integrating expertise on soil chemistry with plant biology have the objective to explore the potentiality of phytohormones exogenously supplied in overcoming limitations of phytoremediation.

In several studies (Barbafieri et al. 2012; Barbafieri 2016), the results have clearly shown the interesting advantages on improving phytoremediation technologies when plant growth regulators are employed:

- Increased plant biomass production;
- Increased contaminant uptake and translocation in plant biomass;
- Increased organic contamination degradation;
- Increased plant resistant to abiotic stress (metal contamination).

For example, Cassina et al. (2012) have carried out a laboratory tests with crop plants (sunflower and mustard) on microcosms (small pots and controlled growth condition in growth chamber), to study mercury phytoextraction from a contaminated soil, combining a phytohormone (Cytokinin, CK) and thioligand (thiosulphate, TS) treatments. Results have shown that the combination of exogenous CK and TS treatments improved the technology performance. The treatment showed the synergistic effects on increasing the Hg phytoextraction up to 450%. In one growing cycle the plants reduced labile-Hg pools of about 40% from the contaminated soil that represent the most dangerous fraction as it can enter in environmental process and/or in food chain. Also, Barbafieri (2014) reported the synergistic effect of combination of CK treatment with nitrogen fertilization (modulated application, MA), on morphological, physiological parameters and biomass production of sunflower using in boron phytoextraction test. The MA treatment helped overcome the stress caused by boron phytotoxicity with an increasing of plant biomass, so improving its phytoextraction ability by five times.

Other investigations have also shown the benefic effects of phytohormones application on hyperaccumulator plants. In Cassina et al. (2011), a plant nickel hyperaccumulator, *Alyssum murale* shown a higher biomass production without reduction of metal hyperaccumulation when treated with cytokinin. The obtained net improvement of the Ni phytoextraction was of about 75%.

Extended studies are necessary to elucidate the role of the PGR in signaling pathways, defence mechanisms, alleviation of contaminant toxicity as well as uptake or degradation of contaminants.

6.4.3.2 Plant Growth-Promoting Bacteria (PGPB)

Phytoremediation assisted by plant growth-promoting bacteria (PGPB) is a winning strategy. These plant beneficial bacteria living in close association with roots have several positive effects on plant growth and development (Prasad et al. 2015). They can act as biofertilizers by increasing available mineral nutrients, modulate phytohormone levels and carry out a protective action against potential phytopathogens. Usually, the beneficial effects of PGPB are generally more evident when plants are grown under stress conditions such as salinity, presence of heavy metals or toxic organic molecules (Glick 2020). The plants are able to attract and feed the bacteria living in the rhizosphere by selecting their secreted molecules (radical exudates such as organic acids, vitamins, sugars) in function of which microorganisms will be more useful to the plant in that particular environmental condition. This mechanism allows to create a specific and suitable microbiome at the root level. This mechanism reveals that plants and their related microbiota symbolize a mutual and characteristic ecological element known as a holobiont (Vandenkoornhuyse et al. 2015). Plant growth-promoting bacteria can influence plant growth and development with various direct and indirect mechanisms. Among direct mechanisms there are the production of bacterial molecules directly affecting plant health such as auxins (indole-3-acetic acid—IAA—the most important), cytokinins, gibberellins, ACCD (1-aminocyclopropane-1-carboxylate deaminase) which lowers the level of ethylene produced by the plant in stress conditions, siderophores (small peptide molecules with high affinity for iron), nitrogen fixation, and inorganic phosphate solubilization. Among the indirect mechanisms we can mention the production of antibiotic molecules and cell wall degrading enzymes against phytopathogens (mainly fungi). Another important indirect mechanism is the so-called induced systemic resistance (ISR) alleviating the negative effects produced by pathogenic agents. Several PGPB are effective to activate ISR by a network of synchronized signals (Lucas et al. 2014). Thus, PGPB may protect plants against the effects of negative pressures including drought, flooding, high salinity, metal and organic contaminants, fungal and bacterial phytopathogens (Olanrewaju et al. 2017).

Many plants are able to grow on metal-rich soils without showing any particular symptoms of suffering. Tolerance may depend on the fact that roots put in place a mechanism of exclusion of metals or conversely, are able to accumulate them in their tissues (Baker 1989).

PGPB can significantly contribute to increase metal uptake by plants and therefore the efficiency and the rate of phytoextraction (Rajkumar et al. 2009; Glick 2010; Franchi et al. 2019). The increase in uptake is often associated with the simultaneous addition of mobilizers. Phytoremediation is a strictly site-specific technology that depends on the type of contamination. The nature of the rhizosphere microbial community is also deeply influenced by the specificity of the contaminants. The microorganisms grown in the presence of the specific target contaminant are exactly those potentially most useful. For this reason, several rhizospheric and endophytic strains have been isolated from metal-tolerant plants. Many of these microbial strains have been shown to possess peculiar plant growth promotion properties and have produced beneficial effects to host plants when used as inoculum (Ma et al. 2011).

When the contaminants are organic molecules the degradation is carried out in close synergy between microflora and plant enzymes. The organic molecules are used as energy and carbon source and, often, in the presence of these contaminants the microbial diversity, generally high in the soil, decreases since microorganisms able to degrade those specific molecules overcome. This condition facilitates the selection of microorganisms that have naturally been selected to degrade specific organic contaminants. As already described, radical exudates play a key role in biodegradation promoting plant–microbe and microbe–microbe interactions stimulating the selection of beneficial specific PGPBs providing them carbon source and energy (Phillips et al. 2012; Huang et al. 2014).

When a soil is contaminated with oil residues, one of the most effective biological techniques is landfarming and bioaugmentation with microbial indigenous consortia. In the case of particularly recalcitrant contaminants or complex mixtures the association of phytoremediation is very useful. The landfarming increases the soil oxygenation promoting and stimulating the biodegradation by the microbial flora and the inoculum of selected indigenous hydrocarbon-oxidizing microorganisms allows to significantly accelerate the biodegradation process. The next step through the use of plant species effectively complete this strategy and since frequently the hydrocarbon-oxidizing microbial strains also own several plant growth-promoting properties and can therefore be considered PGPB, this approach leads to excellent results (Franchi et al. 2017).

It is important to point out that endophytes are certainly more beneficial to the plant as they are in close contact with plant tissues and can therefore exert a direct beneficial effect even if endophytic bacteria may also be found free-living in the soil (Santoyo et al. 2016).

Rhizospheric and endophytic PGPB develop comparable mechanisms to promote plant growth with the difference that once established within the tissues of the host plant, endophytic PGPB are much less subject to fluctuations of soil conditions (i.e. temperature, pH, and water content) and also to the competition with other rhizospheric bacteria for binding sites on host plant root surfaces (Glick 2012). The privileged position of endophytes, within plant tissues, makes them less susceptible to environmental variations and to the complex interactions of the rhizosphere. This particular situation makes them a very important tool for the study of plant growth-promoting mechanisms, of which many aspects are still not well defined.

6.4.3.3 Biomass Valorization

Phytoremediation is a green remediation technique that uses plants to clean-up contaminated soils, removing or degrading inorganic and organic pollutants. As this is often used as an "in situ" technology, the costs and energy consumption are reduced, and in addition it can provide aesthetic benefits and minimize the disturbance of the surrounding environment (Pandey and Souza-Alonso 2019; Grifoni et al. 2020; Sarma et al. 2021). The sustainability assessment of phytoremediation also focuses on the secondary life cycle (Hou et al. 2018) and on the potentially more efficient use of its by-products. Therefore, confirming that the biomass is a resource represents a strength point of phytoremediation, as it makes the technology highly competitive within circular and green economy schemes, and more economical. Indeed, the emergence of the concepts of green and bio-economies also implies the sustainability, efficiency, and economy of biomass (Scarlat et al. 2015). The initial aim of phytoremediation was only the efficiency of contaminant treatment, by reducing their bioavailable amount in polluted soil, but the emergence of the NBS concept shifts the focus to optimizing sustainability and synergies between nature and society (O'Connor and Hou 2018; Song et al. 2019). Potential products of high ecological and economic value can be generated (oil, biochemical, pulp-paper biomass, aromatic essential oils, biochar, biodiesel, biosurfactant, bioplastics, etc.) from biomass derived from phytoremediation (phytobiomass) (Pandey and Souza-Alonso 2019).

The research on this topic is still in the early stages and needs to be deepened. This can be achieved through phytomanagement, namely by cost effectively combining phytoremediation with other technologies (Evangelou et al. 2015). One attractive strategy is an integrated approach combining phytoremediation with bioenergy production (Licht and Isebrands 2005; Andersson-Sköld et al. 2014), thus developing sustainable energy option through plant biomass enhancement and valorization obtained from repeated phytoremediation cycles. Until a few years ago phytobiomass was only considered as a type of waste for disposal, as the potential of phytoremediation had not been fully realized. However, several recent studies have demonstrated the potential use of biomass as a renewable energy source and a bioenergetics resource (Van Ginneken et al. 2007; Prabha et al. 2021).

The term "biomass" refers to organic matter generated by plants and includes several biological materials with high calorific values, from which is possible to produce different types of renewable energy. The main processes to convert biomass into bioenergy are thermochemical conversion (combustion, pyrolysis, gasification, and liquefaction), biochemical conversion (anaerobic digestion, fermentation), and mechanical extraction (with esterification) (McKendry 2002a). These mainly produce bioenergy in the form of electricity and heat (by direct combustion), gaseous energy (biomethane and biohydrogen), and liquid biofuels (bioethanol and biodiesel). However, the end use of biomass depends on its composition and source (Guldhe et al. 2017). In phytoremediation several types of biomass can be obtained, depending on the phytotechnology used, and contaminated or not. However, only some phytotechnologies (i.e., phytoextraction and phytodegradation) involve the absorption and transfer of contaminants within the vegetal tissues, resulting in contaminated biomass. However, energy production from biomass enriched with contaminants can also be economically viable, as several studies have confirmed (Witters et al. 2012b; Gomes 2012; Vigil et al. 2015; Tian and Zhang 2016), but the risk management is essential as the contaminants in phytobiomass may be of concern, particularly in the processing of biofuels (Gomes 2012).

Sustainable phytoremediation studies are often based on life cycle assessment methodology. LCA is a tool to evaluate all the environmental impacts associated with the entire life cycle of a product, which in case of phytoremediation are from sowing to biomass or waste disposal. The resources consumed, the emissions and wastes produced and released into the environment are also evaluated and quantified (Onwubuya et al. 2009). For the bio-reuse of by-products, the conversion process is fully evaluated together with the precautionary measures necessary to avoid any secondary risks presented by the pollutants in terms of environment and human health, and to ensure that these do not affect the energy conversion process. Vigil et al. (2015) confirmed the valorization of biomass associated with its conversion into energy, which is the key factor when considering phytoremediation as a sustainable technology compared to traditional techniques. Witters et al. (2012a, b) also demonstrated the environmental and economic benefits of phytoremediation over to traditional remediation using the LCA. The authors found that the best CO_2 abatement and production of energy by combustion were derived from the contaminated digestate of *Zea mays* L.

In addition, phytobiomass valorization overcomes the serious issue of the accumulation of large quantities of dangerous biomass (Sas-Nowosielska et al. 2004). Thus, the phytobiomass conversion in renewable energy can be a solution to its volume reduction, simultaneously satisfying the increased global demand for bioenergy (Yadav et al. 2018). To date, plant biomass is one of the main sources

Table 6.1 An illustrative list of the main energy crops studied for a sustainable soil remediation

Bioenergy crop	Soil pollutants	Sustainable bioenergy production
Jatropha curcas (perennial shrub species)	Heavy metals	Biodiesel (seed oil)
Populus spp. (arboreal woody plants)	Organics, heavy metals	Bioethanol (biomass)
Salix spp. (arboreal species)	Organics, heavy metals	Bioethanol (biomass)
Arundo donax (perennial herbaceous plant)	Organics, heavy metals	Bioenergy, bioethanol (biomass)
Miscanthus (perennial herbaceous species)	Organics, heavy metals	Bioethanol (biomass)
Ricinus communis (annual or perennial herbaceous or arborescent plant)	Organics, heavy metals	Biodiesel (biomass and seed oil)
Zea mays (annual herbaceous plant)	Heavy metals	Bioenergy (biomass)
Helianthus annuus (annual herbaceous plant)	Heavy metals	Bioenergy, bioethanol (biomass and seed oil)
Brassica spp. (biennial or perennial herbaceous plants)	Heavy metals	Biofuel, biodiesel (seed oil)
Cannabis sativa (annual herbaceous plant)	Heavy metals	Bioenergy (biomass)

of energy, contributing 14% of the global energy demand (McKendry 2002b; Shrinkhal 2017). To fulfil this demand, extensive areas are needed for energy crop cultivation, which would entail a reduction of important arable land for food and forage production. This competition could be avoided by using non-edible species with low nutrient content but with high yield, or by exploiting large contaminated areas that are not suitable for food crop cultivation (Bardos et al. 2011; Schreurs et al. 2011). Thus, the agricultural lands would not be affected.

Although many different plant species have been recognized and used for the remediation of contaminated soils, research into suitable energy crops for sustainable phytoremediation programmes is ongoing (Pandey and Souza-Alonso 2019). One factor that significantly influences phytoremediation efficiency, in addition to type and bioavailability of contaminants and soil properties, is the vegetal species (Petruzzelli et al. 2019). The selection of energy crops for phytoremediation should be based on the following characteristics: rapid growth, high biomass production, deep roots, contaminant tolerance, ease of harvesting, being non-edible, and the ability to provide phytoproducts with high economic value (Vangronsveld et al. 2009; Tripathi et al. 2016; Singh et al. 2017). In phytoremediation programmes, annual herbaceous or perennial and woody energy crops of high density and short rotation (up to 4 years) and short plant cycles are preferred (Pandey et al. 2016; Jha et al. 2017; Pandey and Souza-Alonso 2019). Widely tested bioenergy crops for the remediation of contaminated soils are reported in Table 6.1.

However, in addition to biomass availability, the environmental impact must also be considered in the species selection. Low inputs of fertilizers, the potential carbon sequestration in roots and soil, the potential CO_2 abatement and the contribution to GHG (greenhouse gas) reduction should be evaluated (Fiorese and Guariso 2010; Witters et al. 2012a). Thus, in recent years *Cannabis sativa* (hemp) has gained much interest. This species has been considered an important food and non-food source since historical times (Linger et al. 2002), and was traditionally used in several agro-industrial fields for the production of textiles, paper pulp, materials for building cosmetics and in the pharmaceutical industry (Salentijn et al. 2015). Due to its useful characteristics, new sustainable applications for industrial purposes are being researched, in the fields of both phytoremediation and bioenergy production. For phytoremediation purposes, these traits include high biomass yield, an extensive root system, high tolerance to soil contaminants, microbial resistance, high land use efficiency, improving of soil health, the short growing season, use in organic crop rotation, and the ability to adapt to various climatic conditions and different types of soil (Linger et al. 2002; Citterio et al. 2003; Kumar et al. 2017). Hemp is also considered low-input and low-impact crops, as it has low management and feedstock costs, which make it a promising species for the phytomanagement of contaminated sites (Li et al. 2010; Rehman et al. 2013; Kumar et al. 2017). In particular, *C. sativa* exhibits a low nutrient and fertilizer demand, low emissions associated with cultivation activities (fertilizer application, weeding, irrigation, harvesting) and transportation to transformation sites have been assessed (Van Der Werf 2004; Casas and Rieradevalli Pons 2005). The study by Finnan and Styles (2013) demonstrated the considerable net GHG abatement potential and positive net energy balance of hemp, using the LCA method. The authors hypothesized an increase in net GHG abatement by up to 21 Mt./CO_2eq./year, if traditional annual energy crops were replaced with hemp plantations.

The various uses of *C. sativa* have been reported in several studies. Its ability to decontaminate polluted soils by heavy metals or petroleum hydrocarbons has been reported, and Linger et al. (2005) showed the ability of hemp to remove a considerable amount of Cd from soil. The authors found various Cd amounts in vegetal tissues of plants, with concentration up to 800 mg kg^{-1} in roots, and in the range of 50–100 mg kg^{-1} in stems and leaves. Similarly, the study by Shi and Cai (2009) confirmed that the highest Cd concentration was absorbed by roots, reaching a concentration of about 4000 mg kg^{-1}, and only partially translocated to the above-ground tissues. With regard to the organic compounds, Campbell et al. (2002) reported a reduction of about 50% and 13% of concentration of benzo[α]pyrene and chrysene, respectively, spiked in a soil in which hemp was grown.

The potential of hemp as an energy crop for the sustainable management of contaminated soils has also been explored, as several bioenergy options can be obtained from its biomass (Fig. 6.2): bioethanol (Sipos et al. 2010; Kreuger et al. 2011; Kuglarz et al. 2016), biodiesel (Li et al. 2010; Ahmad et al. 2011), biogas (Kreuger et al. 2011; Pakarinen et al. 2011; Prade et al. 2011), solid fuel (Prade et al. 2011), and biohydrogen (Agbor et al. 2014).

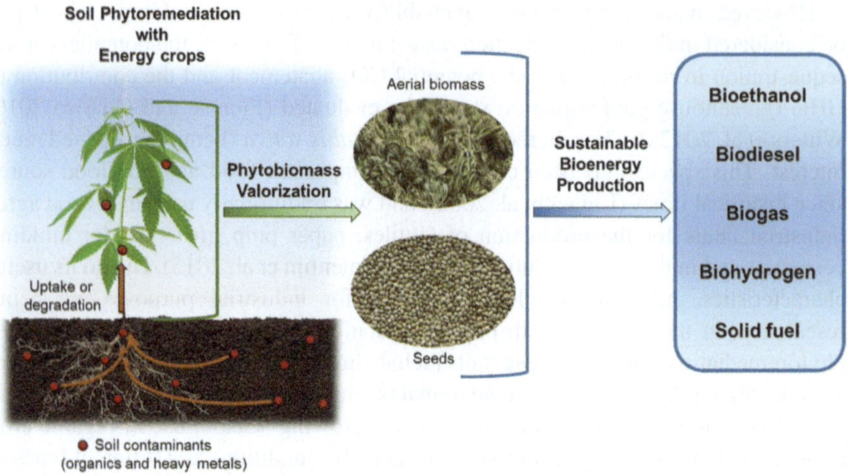

Fig. 6.2 Hemp as an example of biomass valorization from phytoremediation

However, the possibility of a synergy between the energy production and phytoremediation of *C. sativa* has not yet been investigated in depth, so the feasibility of this new sustainable strategy requires further efforts both from the scientific community and in the regulatory and planning field, to overcome the intrinsic limitations and to tackle major environmental challenges such as pollution, the energy crisis, and climate change.

6.4.3.4 Effects on Soil Functions

Phytoremediation through plants uses the functionality of the soil for the remediation of contaminated sites. Like other bioremediation options, the technology has some limits in terms of the long time required to complete the remediation and the difficulty of growing plants if soil contamination levels are particularly high. Furthermore, as the action of the plants is particularly effective in the areas explored by the roots residual concentrations of contaminants can remain in some areas of the contaminated site if they have not been reached by the root system of the plants.

In any case, plant growth promotes an improvement in both physical and chemical soil properties. From the point of view of physical fertility at the end of a phytoremediation intervention, an improvement in soil porosity occurs, and radical exudates can stimulate the formation of stable aggregates (Pedron and Petruzzelli 2011). This in turn improves the transportation of water, air, and nutrients in the soil (Xu et al. 2019).

Phytoremediation usually also involves a progressive increase in organic matter during the period of the treatment application. This effect is particularly evident with phytostabilization, in which the bioavailability of the contaminants is also drastically reduced. The improvement of the chemical-physical characteristics of the soil implies an additional enhancement in the biological properties, with a significant growth of microorganisms stimulated by the release of radical exudates by the plants

(Hamdi et al. 2012). This growth, which often occurs relatively quickly, is accompanied by an increase in enzymatic activities (Mikkonen et al. 2011).

The combined and synergistic action of plants and microorganisms is able to improve the functionality of the soil, and to restore fertility levels to suitable conditions for plant growth following the reduction of the contaminants. The technology is, therefore, able to establish a virtuous cycle, because the increase in biomass production leads to a higher quantity of contaminants being removed and improves the quality of the soil for the subsequent growth cycle. Furthermore, the improvement of the physical quality of the soil, in particular the porosity and stability of the aggregates, improves water retention and the transport of nutrients, favouring the development of the root system and increasing the microbial biomass (Shahsavari et al. 2015). The effects are particularly evident in the case of contamination by organic compounds. However, the possible presence of toxic compounds deriving from the incomplete degradation of organic contaminants must also be considered (Mikkonen et al. 2012).

This aspect may also result in the possible migration of these compounds along the soil profile, depending on the specific characteristics of the contaminants and the soils involved. The problem of possible contaminant migration is of primary importance when assisted phytoextraction is used. Here, the addition of mobilizing agent doses must be extremely accurate, as previously described.

6.5 Conclusion and Perspectives

In traditional soil remediation (incineration, inertization, etc.), contaminated soils were considered and treated as hazardous waste, and therefore soil excavation and landfilling were widely used. This classification of soil has led many legislatures to undervalue soil as a resource, identifying soil quality only by the concentration value of a contaminant.

Instead, soil is a natural resource that requires an extremely long time to renew. Soil management, even in terms of reclamation, must be sustainable and remediation interventions must attempt to consider and recover soil quality, in accordance with economic and social sustainability. This is a major challenge in an increasingly industrialized world, where the soil resource is highly susceptible to environmental changes and anthropic pressure. It therefore becomes crucial to identify and develop remediation strategies aimed at minimizing soil losses and maintaining the high level of functionality of this important environmental matrix.

The sustainable approach requires the clear definition of restoration objectives in addition to soil decontamination, and therefore how any remediation activities can influence the final soil use should be evaluated. Understanding the impacts of each technology on the functionality of the soil is thus essential. Any quantitative assessment of overall soil quality is difficult to describe in simple terms, as soil is an extremely complex matrix that performs essential functions and is a keystone of multiple environmental equilibria. Due to its importance, it is necessary to understand the role of soil and to adequately consider it in remediation and restoration activities (Callaham et al. 2008; Heneghan et al. 2008).

Thus, in the remediation of contaminated sites, the most effective technologies in terms of environmental sustainability are those that can help restore a site to productive use while drastically reducing any potential environmental impact.

References

Abdullah WS, Sarem SM (2010) The potential of *Chrysanthemum* and *Pelargonium* for phytoextraction of lead-contaminated Soils. Jordan J Civil Eng 4:409–417

Abrahams PW (2002) Soils: their implications to human health. Sci Total Environ 291:1–32. https://doi.org/10.1016/S0048-9697(01)01102-0

Acar YB, Alshawabkeh AN (1993) Principles of electrokinetic remediation. Environ Sci Technol 27:2638–2647. https://doi.org/10.1021/es00049a002

Acar YB, Gale RJ, Alshawabkeh AN, Marks RE, Puppala S, Bricka M, Parker R (1995) Electrokinetic remediation: Basics and technology status. J Hazard Mater 40:117–137. https://doi.org/10.1016/0304-3894(94)00066-P

Agbor V, Zurzolo F, Blunt W, Dartiailh C, Cicek N, Sparling R, Levin DB (2014) Single-step fermentation of agricultural hemp residues for hydrogen and ethanol production. Biomass Bioenergy 64:62–69. https://doi.org/10.1016/j.biombioe.2014.03.027

Ahmad M, Ullah K, Khan MA, Zafar M, Tariq M, Ali S, Sultana S (2011) Physicochemical analysis of hemp oil biodiesel: a promising non edible new source for bioenergy. Energy Sources Part A Recover Util Environ Eff 33:1365–1374. https://doi.org/10.1080/15567036.2010.499420

Alexander M (2000) Aging, bioavailability, and overestimation of risk from environmental pollutants. Environ Sci Technol 34:4259–4256. https://doi.org/10.1021/ES001069+

Andersson-Sköld Y, Hagelqvist A, Crutu G, Blom S (2014) Bioenergy grown on contaminated land—a sustainable bioenergy contributor? Biofuels 5:487–498. https://doi.org/10.1080/17597269.2014.996728

Ansari AA, Gill SS, Gill R, Lanza G, Newman L (2018) Phytoremediation. Management of environmental contaminants, 6th edn. Springer

Arafat HA, Jijakli K, Ahsan A (2015) Environmental performance and energy recovery potential of five processes for municipal solid waste treatment. J Clean Prod 105:233–240. https://doi.org/10.1016/j.jclepro.2013.11.071

Ashraf SS, Ali Q, Zahir ZA, Ashraf SS, Asghar HN (2019) Phytoremediation: environmentally sustainable way for reclamation of heavy metal polluted soils. Ecotoxicol Environ Saf 174:714–727

Baker AJM (1989) Terrestrial higher plants which accumulate metallic elements: a review of their distribution, ecology and phytochemistry. Biorecovery:81–126

Barbafieri M (2014) A method for the phytoextraction of boron in contaminated soil, and related kit of products for the use in this method

Barbafieri M (2016) Use of phytohormones for strengthening metal(loid) phytoextraction: limitations and a case study. In: Ahammed GJ, Yu J-Q (eds) Plant hormones under challenging environmental factors. Springer, Dordrecht, pp 157–179

Barbafieri M, Peralta-Videa J, Pedron F, Gardea-Torresdey J (2012) Plant growth regulators and improvements in phytoremediation process efficiency: studies on metal contaminated soils. In: Anjum N, Pereira M, Ahmad I, Duarte A, Umar S, Khan N (eds) Phytotechnologies: remediation of environmental contaminants. CRC Press, pp 375–388

Barbafieri M, Pedron F, Petruzzelli G, Rosellini I, Franchi E, Bagatin R, Vocciante M (2017) Assisted phytoremediation of a multi-contaminated soil: investigation on arsenic and lead combined mobilization and removal. J Environ Manage 203:316–329

Barbafieri M, Morelli E, Tassi E, Pedron F, Remorini D, Petruzzelli G (2018) Overcoming limitation of "recalcitrant areas" to phytoextraction process: the synergistic effects of exogenous cytokinins and nitrogen treatments. Sci Total Environ 639:1520–1529. https://doi.org/10.1016/J.SCITOTENV.2018.05.175

Bardos RP, Bone B, Andersson-Sköld Y, Suer P, Track T, Wagelmans M (2011) Crop-based systems for sustainable risk-based land management for economically marginal damaged land. Remediat J 21:11–33. https://doi.org/10.1002/rem.20297

Behera BK, Prasad R (2020a) Conceptual development for a clean environment. In: Behera BK, Prasad R (eds) Environmental technology and sustainability. Elsevier, pp 1–40

Behera BK, Prasad R (2020b) Environmental technology and sustainability. Elsevier. ISBN: 9780128191033. https://www.elsevier.com/books/environmental-technology-and-sustainability/behera/978-0-12-819103-3

Bourguignon D (2017) Nature-based solutions. concept, opportunities and challenges

Bulak P, Walkiewicz A, Brzezińska M (2014) Plant growth regulators-assisted phytoextraction. Biol Plantarium 58:1–8. https://doi.org/10.1007/s10535-013-0382-5

Callaham MA, Rhoades CC, Heneghan L (2008) A striking profile: soil ecological knowledge in restoration management and science. Restor Ecol 16:604–607. https://doi.org/10.1111/j.1526-100X.2008.00490.x

Cameselle C, Gouveia S, Eddine D, Belhadj B (2013) Advances in electrokinetic remediation for the removal of organic contaminants in soils. In: Rashed N (ed) Organic pollutants—monitoring, risk and treatment. InTech, pp 209–229

Campbell S, Paquin D, Awaya JD, Li QX (2002) Remediation of benzo[a]pyrene and chrysene-contaminated soil with industrial hemp (*Cannabis sativa*). Int J Phytoremediation 4:157–168. https://doi.org/10.1080/15226510208500080

Cang L, Zhou DM, Wang QY, Fan GP (2012) Impact of electrokinetic-assisted phytoremediation of heavy metal contaminated soil on its physicochemical properties, enzymatic and microbial activities. In: Electrochimica Acta Pergamon, pp 41–48

Cao A, Carucci A, Lai T, La Colla P, Tamburini E (2007) Effect of biodegradable chelating agents on heavy metals phytoextraction with *Mirabilis jalapa* and on its associated bacteria. Eur J Soil Biol 43:200–206. https://doi.org/10.1016/j.ejsobi.2007.02.002

Casas XA, Rieradevalli Pons J (2005) Environmental analysis of the energy use of hemp—analysis of the comparative life cycle: diesel oil vs. hemp-diesel. Int J Agric Resour Gov Ecol 4:133–139. https://doi.org/10.1504/ijarge.2005.007195

Cassina L, Tassi E, Morelli E, Giorgetti L, Remorini D, Chaney RL, Barbafieri M (2011) Exogenous cytokinin treatments of an NI hyper-accumulator, *Alyssum Murale*, grown in a serpentine soil: implications for phytoextraction. Int J Phytoremediation 13:90–101. https://doi.org/10.1080/15226514.2011.568538

Cassina L, Tassi E, Pedron F, Petruzzelli G, Ambrosini P, Barbafieri M (2012) Using a plant hormone and a thioligand to improve phytoremediation of Hg-contaminated soil from a petrochemical plant. J Hazard Mater 231–232:36–42. https://doi.org/10.1016/j.jhazmat.2012.06.031

Chaney R, Angle J, McIntosh M, Reeves R, Li Y, Brewer E, Chen K, Roseberg R, Perner H, Synkowski EC, Broadhurst CL, Wang S, Bakerg AJM (2005) Using hyperaccumulator plants to phytoextract soil Ni and Cd Agromining-Phytomining View project soil quality and health View project.

Charlatchka R, Cambier P (2000) Influence of reducing conditions on solubility of trace metals in contaminated soils. Water Air Soil Pollut 118:143–167. https://doi.org/10.1023/A:1005195920876

Chen X, Shen Z, Lei Y, Zheng S, Ju B, Wang W (2006) Effects of electrokinetics on bioavailability of soil nutrients. Soil Sci 171:638–647. https://doi.org/10.1097/01.ss.0000228038.57400.a8

Chien YC (2012) Field study of in situ remediation of petroleum hydrocarbon contaminated soil on site using microwave energy. J Hazard Mater 199–200:457–461. https://doi.org/10.1016/j.jhazmat.2011.11.012

Chu W, Chan KH (2003) The mechanism of the surfactant-aided soil washing system for hydrophobic and partial hydrophobic organics. Sci Total Environ 307:83–92. https://doi.org/10.1016/S0048-9697(02)00461-8

Citterio S, Santagostino A, Fumagalli P, Prato N, Ranalli P, Sgorbati S (2003) Heavy metal tolerance and accumulation of Cd, Cr and Ni by *Cannabis sativa* L. Plant Soil 256:243–252. https://doi.org/10.1023/A:1026113905129

de Percin PR (1995) Application of thermal desorption technologies to hazardous waste sites. J Hazard Mater 40:203–209. https://doi.org/10.1016/0304-3894(94)00085-U

Doick KJ, Burauel P, Jones KC, Semple KT (2005) Distribution of aged 14C– PCB and 14C–PAH residues in particle-size and humic fractions of an agricultural soil. Environ Sci Technol 39:6575–6583

Doumett S, Lamperi L, Checchini L, Azzarello E, Mugnai S, Mancuso S, Petruzzelli G, Del Bubba M (2008) Heavy metal distribution between contaminated soil and *Paulownia tomentosa*, in a pilot-scale assisted phytoremediation study: Influence of different complexing agents. Chemosphere 72:1481–1490. https://doi.org/10.1016/j.chemosphere.2008.04.083

Doumett S, Fibbi D, Azzarello E, Mancuso S, Mugnai S, Petruzzelli G, del Bubba M (2011) Influence of the application renewal of glutamate and tartrate on Cd, Cu, Pb and Zn distribution between contaminated soil and paulownia tomentosa in a pilot-scale assisted phytoremediation study. Int J Phytoremediation 13:1–17. https://doi.org/10.1080/15226510903567455

EEA-European Environment Agency (2014) Progress in management of contaminated sites

Eggermont H, Balian E, Azevedo JMN, Beumer V, Brodin T, Claudet J, Fady B, Grube M, Keune H, Lamarque P, Reuter K, Smith M, Van Ham C, Weisser WW, Le Roux X (2015) Nature-based solutions: new influence for environmental management and research in Europe. In: GAIA—Ecological Perspectives for Science and Society (ed) GAIA. Oekom Verlag, pp 243–248

Elgh-Dalgren K, Arwidsson Z, Camdzija A, Sjöberg R, Ribé V, Waara S, Allard B, von Kronhelm T, van Hees PAW (2009) Laboratory and pilot scale soil washing of PAH and arsenic from a wood preservation site: changes in concentration and toxicity. J Hazard Mater 172:1033–1040. https://doi.org/10.1016/j.jhazmat.2009.07.092

EU-European Commission (2015) Towards an EU research and innovation policy agenda for Nature-Based Solutions & Re-Naturing Cities: Final report of the Horizon 2020 expert group on "Nature-based solutions and re-naturing cities." Publications Office of the European Unio

Evangelou MWH, Ebel M, Schaeffer A (2006) Evaluation of the effect of small organic acids on phytoextraction of Cu and Pb from soil with tobacco *Nicotiana tabacum*. Chemosphere 63:996–1004. https://doi.org/10.1016/j.chemosphere.2005.08.042

Evangelou MWH, Ebel M, Hommes G, Schaeffer A (2008) Biodegradation: the reason for the inefficiency of small organic acids in chelant-assisted phytoextraction. Water Air Soil Pollut 195:177–188. https://doi.org/10.1007/s11270-008-9738-4

Evangelou MWH, Papazoglou EG, Robinson BH, Schulin R (2015) Phytomanagement: phytoremediation and the production of biomass for economic revenue on contaminated land. In: Ansari AA, Gill SS, Gill R, Lanza GR, Newman L (eds) Phytoremediation: management of environmental contaminants, vol 1. Springer, pp 115–132

Faivre N, Fritz M, Freitas T, de Boissezon B, Vandewoestijne S (2017) Nature-based solutions in the EU: innovating with nature to address social, economic and environmental challenges. Environ Res 159:509–518. https://doi.org/10.1016/j.envres.2017.08.032

Falciglia PP, De Guidi G, Catalfo A, Vagliasindi FGA (2016) Remediation of soils contaminated with PAHs and nitro-PAHs using microwave irradiation. Chem Eng J 296:162–172. https://doi.org/10.1016/j.cej.2016.03.099

Finnan J, Styles D (2013) Hemp: a more sustainable annual energy crop for climate and energy policy. Energy Policy 58:152–162. https://doi.org/10.1016/j.enpol.2013.02.046

Fiorese G, Guariso G (2010) A GIS-based approach to evaluate biomass potential from energy crops at regional scale. Environ Model Softw 25:702–711. https://doi.org/10.1016/j.envsoft.2009.11.008

Fraire-Velazquez S, Emmanuel V (2013) Abiotic stress in plants and metabolic responses. In: Vahdati K, Leslie C (eds) Abiotic stress—plant responses and applications in agriculture. InTech, Rijeka, pp 25–48

Franchi E, Rolli E, Marasco R, Agazzi G, Borin S, Cosmina P, Pedron F, Rosellini I, Barbafieri M, Petruzzelli G (2017) Phytoremediation of a multi contaminated soil: mercury and arsenic phytoextraction assisted by mobilizing agent and plant growth promoting bacteria. J Soils Sediments 17:1224–1236. https://doi.org/10.1007/s11368-015-1346-5

Franchi E, Cosmina P, Pedron F, Rosellini I, Barbafieri M, Petruzzelli G, Vocciante M (2019) Improved arsenic phytoextraction by combined use of mobilizing chemicals and autochthonous soil bacteria. Sci Total Environ 655:328–336. https://doi.org/10.1016/j.scitotenv.2018.11.242

George CE, Azwell DE, Adams PA, Rao GVN, Averett DE (1995) Evaluation of steam as a sweep gas in low temperature thermal desorption processes used for contaminated soil clean up. Waste Manag 15:407–416. https://doi.org/10.1016/S0956-053X(99)80029-4

Gill RT, Harbottle MJ, Smith JWN, Thornton SF (2014) Electrokinetic-enhanced bioremediation of organic contaminants: a review of processes and environmental applications. Chemosphere 107: 31–42

Glick BR (2010) Using soil bacteria to facilitate phytoremediation. Biotechnol Adv 28:367–374

Glick BR (2012) Plant growth-promoting bacteria: mechanisms and applications. Scientifica (Cairo) 2012:1–15. https://doi.org/10.6064/2012/963401

Glick BR (2020) Beneficial plant-bacterial interactions, 2nd edn. Springer

Gomes HI (2012) Phytoremediation for bioenergy: challenges and opportunities. Environ Technol Rev 1:59–66. https://doi.org/10.1080/09593330.2012.696715

Grifoni M, Petruzzelli G, Barbafieri M, Rosellini I, Pedron F (2017) Soil quality protection at heavy metal-contaminated manufactured gas plant sites: role of biological remediation

Grifoni M, Pedron F, Barbafieri M, Rosellini I, Petruzzelli G, Franchi E (2020) Sustainable valorization of biomass: from assisted phytoremediation to green energy production. In: Prasad MNV (ed) Handbook on assisted and amendments enhanced sustainable remediation technology. Wiley, UK

Guemiza K, Saffar T, Gmar M, Magdouli S, Foudhaili T (2020) Thermal desorption and incineration. In: Hussain C (ed) The handbook of environmental remediation: classic and modern techniques. Royal Society of Chemistry, pp 229–253

Guldhe A, Singh B, Renuka N, Singh P, Misra R, Bux F (2017) Bioenergy: a sustainable approach for cleaner environment. In: Phytoremediation potential of bioenergy plants. Springer, Singapore, pp 47–62

Hamdi H, Benzarti S, Aoyama I, Jedidi N (2012) Rehabilitation of degraded soils containing aged PAHs based on phytoremediation with alfalfa (*Medicago sativa* L.). Int Biodeterior Biodegrad 67:40–47. https://doi.org/10.1016/j.ibiod.2011.10.009

Harmsen J, Rulkens W, Eijsackers H (2005) Bioavailability: concept for understanding or tool for predicting? In: Land contamination & reclamation. EPP Publications Ltd, pp 161–172

Heneghan L, Miller SP, Baer S, Callaham MA, Montgomery J, Pavao-Zuckerman M, Rhoades CC, Richardson S (2008) Integrating soil ecological knowledge into restoration management. Restor Ecol 16:608–617. https://doi.org/10.1111/j.1526-100X.2008.00477.x

Hou D, Gu Q, Ma F, O'Connell S (2016) Life cycle assessment comparison of thermal desorption and stabilization/solidification of mercury contaminated soil on agricultural land. J Clean Prod 139:949–956. https://doi.org/10.1016/j.jclepro.2016.08.108

Hou D, Ding Z, Li G, Wu L, Hu P, Guo G, Wang X, Ma Y, O'Connor D, Wang X (2018) A sustainability assessment framework for agricultural land remediation in China. L Degrad Dev 29:1005–1018. https://doi.org/10.1002/ldr.2748

Huang XF, Chaparro JM, Reardon KF, Zhang R, Shen Q, Vivanco JM (2014) Rhizosphere interactions: root exudates, microbes, and microbial communities1. Botany 92:267–275

Isoyama M, Wada SI (2007) Remediation of Pb-contaminated soils by washing with hydrochloric acid and subsequent immobilization with calcite and allophanic soil. J Hazard Mater 143:636–642. https://doi.org/10.1016/j.jhazmat.2007.01.008

IUCN-International Union for Conservation of Nature (2016) Nature-based solutions to address global societal challenges. IUCN, Gland

IUSS Working Group WRB (2007) World reference base for soil resources 2006. A framework for international classification, correlation and communication. FAO, Rome

Jeon C-S, Yang J-S, Kim K-J, Baek K (2010) Electrokinetic removal of petroleum hydrocarbon from residual clayey soil following a washing process. CLEAN Soil Air Water 38:189–193. https://doi.org/10.1002/clen.200900190

Jha AB, Misra AN, Sharma P (2017) Phytoremediation of heavy metal-contaminated soil using bioenergy crops. In: Phytoremediation potential of bioenergy plants. Springer, Singapore, pp 63–95

Kaštánek F, Kaštánek P (2005) Combined decontamination processes for wastes containing PCBs. J Hazard Mater 117:185–205. https://doi.org/10.1016/j.jhazmat.2004.09.026

Kastanek F, Topka P, Soukup K, Maleterova Y, Demnerova K, Kastanek P, Solcova O (2016) Remediation of contaminated soils by thermal desorption; effect of benzoyl peroxide addition. J Clean Prod 125:309–313. https://doi.org/10.1016/j.jclepro.2016.03.134

Keesstra S, Nunes J, Novara A, Finger D, Avelar D, Kalantari Z, Cerdà A (2018) The superior effect of nature based solutions in land management for enhancing ecosystem services. Sci Total Environ 610–611:997–1009

Khaitan S, Kalainesan S, Erickson LE, Kulakow P, Martin S, Karthikeyan R, Hutchinson SLL, Davis LC, Illangasekare TH, Ng'oma C (2006) Remediation of sites contaminated by oil refinery operations. Environ Prog 25:20–31. https://doi.org/10.1002/ep.10083

Khan FI, Husain T, Hejazi R (2004) An overview and analysis of site remediation technologies. J Environ Manage 71:95–122. https://doi.org/10.1016/j.jenvman.2004.02.003

Kim SH, Han HY, Lee YJ, Kim CW, Yang JW (2010) Effect of electrokinetic remediation on indigenous microbial activity and community within diesel contaminated soil. Sci Total Environ 408:3162–3168. https://doi.org/10.1016/j.scitotenv.2010.03.038

Ko I, Chang YY, Lee CH, Kim KW (2005) Assessment of pilot-scale acid washing of soil contaminated with As, Zn and Ni using the BCR three-step sequential extraction. J Hazard Mater 127:1–13. https://doi.org/10.1016/j.jhazmat.2005.06.041

Kreuger E, Prade T, Escobar F, Svensson SE, Englund JE, Björnsson L (2011) Anaerobic digestion of industrial hemp-effect of harvest time on methane energy yield per hectare. Biomass Bioenergy 35:893–900. https://doi.org/10.1016/j.biombioe.2010.11.005

Kuglarz M, Alvarado-Morales M, Karakashev D, Angelidaki I (2016) Integrated production of cellulosic bioethanol and succinic acid from industrial hemp in a biorefinery concept. Bioresour Technol 200:639–647. https://doi.org/10.1016/j.biortech.2015.10.081

Kumar S, Singh R, Kumar V, Rani A, Jain R (2017) *Cannabis sativa*: A plant suitable for phytoremediation and bioenergy production. In: Bauddh K, Singh B, Korstad J (eds) Phytoremediation potential of bioenergy plants. Springer, Cham, pp 269–285

Kuppusamy S, Palanisami T, Megharaj M, Venkateswarlu K, Naidu R (2016) Ex-situ remediation technologies for environmental pollutants: a critical perspective. In: Reviews of environmental contamination and toxicology. Springer, New York, pp 117–192

Kuppusamy S, Thavamani P, Venkateswarlu K, Lee YB, Naidu R, Megharaj M (2017) Remediation approaches for polycyclic aromatic hydrocarbons (PAHs) contaminated soils: technological constraints, emerging trends and future directions. Chemosphere 168:944–968

Lear G, Harbottle MJ, Sills G, Knowles CJ, Semple KT, Thompson IP (2007) Impact of electrokinetic remediation on microbial communities within PCP contaminated soil. Environ Pollut 146: 139–146. https://doi.org/10.1016/j.envpol.2006.06.037

Li YM, Chaney R, Brewer E, Roseberg R, Angle JS, Baker A, Reeves R, Nelkin J (2003) Development of a technology for commercial phytoextraction of nickel: economic and technical considerations. In: Plant and soil. Springer, pp 107–115

Li SY, Stuart JD, Li Y, Parnas RS (2010) The feasibility of converting *Cannabis sativa* L. oil into biodiesel. Bioresour Technol 101:8457–8460. https://doi.org/10.1016/j.biortech.2010.05.064

Licht LA, Isebrands JG (2005) Linking phytoremediated pollutant removal to biomass economic opportunities. In: Biomass and bioenergy. Pergamon, pp 203–218

Linger P, Müssig J, Fischer H, Kobert J (2002) Industrial hemp (*Cannabis sativa* L.) growing on heavy metal contaminated soil: fibre quality and phytoremediation potential. Ind Crops Prod 16: 33–42. https://doi.org/10.1016/S0926-6690(02)00005-5

Linger P, Ostwald A, Haensler J (2005) *Cannabis sativa* L. growing on heavy metal contaminated soil: growth, cadmium uptake and photosynthesis. Biol Plant 49:567–576. https://doi.org/10.1007/s10535-005-0051-4

Lucas JA, García-Cristobal J, Bonilla A, Ramos B, Gutierrez-Mañero J (2014) Beneficial rhizobacteria from rice rhizosphere confers high protection against biotic and abiotic stress inducing systemic resistance in rice seedlings. Plant Physiol Biochem 82:44–53. https://doi.org/10.1016/j.plaphy.2014.05.007

Luo C, Shen Z, Li X (2005) Enhanced phytoextraction of Cu, Pb, Zn and Cd with EDTA and EDDS. Chemosphere 59:1–11. https://doi.org/10.1016/j.chemosphere.2004.09.100

Ma Y, Prasad MNV, Rajkumar M, Freitas H (2011) Plant growth promoting rhizobacteria and endophytes accelerate phytoremediation of metalliferous soils. Biotechnol Adv 29:248–258

Marcomini A, Suter GW II, Critto A (2009) Decision support systems for risk-based management of contaminated sites. Springer, New York

McKendry P (2002a) Energy production from biomass (part 1): overview of biomass. Bioresour Technol 83:37–46. https://doi.org/10.1016/S0960-8524(01)00118-3

McKendry P (2002b) Energy production from biomass (part 2): Conversion technologies. Bioresour Technol 83:47–54. https://doi.org/10.1016/S0960-8524(01)00119-5

Merino J, Bucalá V (2007) Effect of temperature on the release of hexadecane from soil by thermal treatment. J Hazard Mater 143:455–461. https://doi.org/10.1016/j.jhazmat.2006.09.050

Mikkonen A, Kondo E, Lappi K, Wallenius K, Lindström K, Hartikainen H, Suominen L (2011) Contaminant and plant-derived changes in soil chemical and microbiological indicators during fuel oil rhizoremediation with *Galega orientalis*. Geoderma 160:336–346. https://doi.org/10.1016/j.geoderma.2010.10.001

Mikkonen A, Hakala KP, Lappi K, Kondo E, Vaalama A, Suominen L (2012) Changes in hydrocarbon groups, soil ecotoxicity and microbiology along horizontal and vertical contamination gradients in an old landfarming field for oil refinery waste. Environ Pollut 162:374–380. https://doi.org/10.1016/j.envpol.2011.12.012

Misra S, Misra KG (2019) Phytoremediation: an alternative tool towards clean and green environment. In: Sustainable green technologies for environmental management. Springer, Singapore, pp 87–109

Mukhopadhyay S, Maiti SK, Masto RE (2014) Development of mine soil quality index (MSQI) for evaluation of reclamation success: a chronosequence study. Ecol Eng 71:10–20. https://doi.org/10.1016/j.ecoleng.2014.07.001

Mulligan CN, Yong RN, Gibbs BF (2001) An evaluation of technologies for the heavy metal remediation of dredged sediments. J Hazard Mater 85:145–163. https://doi.org/10.1016/S0304-3894(01)00226-6

O'Brien PL, DeSutter TM, Casey FXM, Derby NE, Wick AF (2016) Implications of using thermal desorption to remediate contaminated agricultural soil: physical characteristics and hydraulic processes. J Environ Qual 45:1430–1436. https://doi.org/10.2134/jeq2015.12.0607

O'Brien PL, DeSutter TM, Casey FXM, Wick AF, Khan E (2017) Evaluation of soil function following remediation of petroleum hydrocarbons—a review of current remediation techniques. Curr Pollut Rep 3:192–205. https://doi.org/10.1007/s40726-017-0063-7

O'Connor D, Hou D (2018) Targeting cleanups towards a more sustainable future. Environ Sci Process Impacts 20:266–269

Olanrewaju OS, Glick BR, Babalola OO (2017) Mechanisms of action of plant growth promoting bacteria. World J Microbiol Biotechnol 33:197

Onwubuya K, Cundy A, Puschenreiter M, Kumpiene J, Bone B, Greaves J, Teasdale P, Mench M, Tlustos P, Mikhalovsky S, Waite S, Friesl-Hanl W, Marschner B, Müller I (2009) Developing decision support tools for the selection of "gentle" remediation approaches. Sci Total Environ 407:6132–6142. https://doi.org/10.1016/j.scitotenv.2009.08.017

Pakarinen A, Maijala P, Stoddard FL, Santanen A, Tuomainen P, Kymäläinen M, Viikari L (2011) Evaluation of annual bioenergy crops in the boreal zone for biogas and ethanol production. Biomass Bioenergy 35:3071–3078. https://doi.org/10.1016/j.biombioe.2011.04.022

Pamukcu S (2009) Electrochemical transport and transformations. In: Reddy KR, Cameselle C (eds) Electrochemical remediation technologies for polluted soils, sediments and groundwater. Wiley, p 554

Panagos P, Van Liedekerke M, Yigini Y, Montanarella L (2013) Contaminated sites in Europe: review of the current situation based on data collected through a European network. J Environ Public Health 2013. https://doi.org/10.1155/2013/158764

Pandey VC, Souza-Alonso P (2019) Market opportunities: in sustainable phytoremediation. In: Phytomanagement of polluted sites: market opportunities in sustainable phytoremediation. Elsevier, pp 51–82

Pandey VC, Bajpai O, Singh N (2016) Energy crops in sustainable phytoremediation. Renew Sustain Energy Rev 54:58–73

Pavel LV, Gavrilescu M (2008) Overview of ex-situ decontamination technologies for soil cleanup. Environ Eng Manag J EcoAdvertising Offer 7:815–834

Pedron F, Petruzzelli G (2011) Green remediation strategies to improve the quality of contaminated soils. Chem Ecol 27:89–95. https://doi.org/10.1080/02757540.2010.534086

Pedron F, Petruzzelli G, Barbafieri M, Tassi E (2009) Strategies to use phytoextraction in very acidic soil contaminated by heavy metals. Chemosphere 75:808–814. https://doi.org/10.1016/j.chemosphere.2009.01.044

Petruzzelli G, Barbafieri M, Bonomo L, Saponaro S, Milani A, Pedron F (2004) Bench scale evaluation of soil washing for heavy metal contaminated soil at a former manufactured gas plant site. Bullettin Environ Contam Toxicol 73:38–44. https://doi.org/10.1007/s00128-004-0390-4

Petruzzelli G, Pedron F, Gorini G, Pezzarossa B, Tassi E, Barbafieri M (2013a) Enhanced bioavailable contaminant stripping (EBCS): metal bioavailability for evaluation of phytoextraction success. E3S Web Conf 1:31001. https://doi.org/10.1051/e3sconf/20130131001

Petruzzelli G, Pedron F, Rosellini I, Barbafieri M (2013b) Phytoremediation towards the future: focus on bioavailable contaminants. In: Gupta DK (ed) Plant-based remediation processes, 1st edn. Springer, Berlin, pp 273–289

Petruzzelli G, Pedron F, Rosellini I, Barbafieri M (2015) The bioavailability processes as a key to evaluate phytoremediation efficiency. In: Ansari AA, Gill SS, Gill R, Lanza GRNL (eds) Phytoremediation. Springer, Cham, pp 31–43

Petruzzelli G, Grifoni M, Barbafieri M, Rosellini I, Pedron F (2019) Sorption: release processes in soil-the basis of phytoremediation efficiency

Phillips LA, Greer CW, Farrell RE, Germida JJ (2012) Plant root exudates impact the hydrocarbon degradation potential of a weathered-hydrocarbon contaminated soil. Appl Soil Ecol 52:56–64. https://doi.org/10.1016/j.apsoil.2011.10.009

Prabha J, Kumar M, Tripathi R (2021) Opportunities and challenges of utilizing energy crops in phytoremediation of environmental pollutants: a review. In: Bioremediation for environmental sustainability. Elsevier, pp 383–396

Prade T, Svensson SE, Andersson A, Mattsson JE (2011) Biomass and energy yield of industrial hemp grown for biogas and solid fuel. Biomass Bioenergy 35:3040–3049. https://doi.org/10.1016/j.biombioe.2011.04.006

Prasad R, Kumar M, Varma A (2015) Role of PGPR in soil fertility and plant health. In: Plant growth-promoting rhizobacteria (PGPR) and medicinal plants (eds. Egamberdieva D, Shrivastava S, Varma A), Springer 47–260.

Probstein RF, Hicks RE (1993) Removal of contaminants from soils by electric fields. Science 260: 498–503. https://doi.org/10.1126/science.260.5107.498

Rajkumar M, Prasad MNV, Freitas H, Ae N (2009) Biotechnological applications of serpentine soil bacteria for phytoremediation of trace metals. Crit Rev Biotechnol 29:120–130

Rayment GE, Lyons DJ (2012) New, comprehensive soil chemical methods book for Australasia. Commun Soil Sci Plant Anal 43:412–418. https://doi.org/10.1080/00103624.2012.641802

Reddy KR, Cameselle C (2009) Overview of electrochemical remediation technologies. In: Reddy KR, Cameselle C (eds) Electrochemical remediation technologies for polluted soils, sediments and groundwater. Wiley, pp 1–28

Reddy K, Kumar G (2018) Green and sustainable remediation of polluted sites: new concept, assessment tools, and challenges. ce/papers 2:83–92. https://doi.org/10.1002/cepa.663

Rehman MSU, Rashid N, Saif A, Mahmood T, Han JI (2013) Potential of bioenergy production from industrial hemp (Cannabis sativa): Pakistan perspective. Renew Sustain Energy Rev 18: 154–164

Sabbas T, Polettini A, Pomi R, Astrup T, Hjelmar O, Mostbauer P, Cappai G, Magel G, Salhofer S, Speiser C, Heuss-Assbichler S, Klein R, Lechner P (2003) Management of municipal solid waste incineration residues. Waste Manag 23:61–88. https://doi.org/10.1016/S0956-053X(02) 00161-7

Salentijn EMJ, Zhang Q, Amaducci S, Yang M, Trindade LM (2015) New developments in fiber hemp (*Cannabis sativa* L.) breeding. Ind Crops Prod 68:32–41. https://doi.org/10.1016/j.indcrop.2014.08.011

Santos FS, Hernández-Allica J, Becerril JM, Amaral-Sobrinho N, Mazur N, Garbisu C (2006) Chelate-induced phytoextraction of metal polluted soils with *Brachiaria decumbens*. Chemosphere 65:43–50. https://doi.org/10.1016/j.chemosphere.2006.03.012

Santoyo G, Moreno-Hagelsieb G, del Carmen Orozco-Mosqueda M, Glick BR (2016) Plant growth-promoting bacterial endophytes. Microbiol Res 183:92–99

Saponaro S, Bonomo L, Barbafieri M, Petruzzelli G (2002) Soil washing feasibility at a manufacturing gas plant site. Soil Sediment Contam 11:751–767. https://doi.org/10.1080/20025891107078

Sarma H, Forid N, Prasad R, Prasad MNV, Ma LQ, Rinklebe J (2021) Enhancing phytoremediation of hazardous metal(loid)s using genome engineering CRISPR–Cas9 technology. J Hazard Mater. https://doi.org/10.1016/j.jhazmat.2021.125493

Sas-Nowosielska A, Kucharski R, Małkowski E, Pogrzeba M, Kuperberg JM, Kryński K (2004) Phytoextraction crop disposal—an unsolved problem. Environ Pollut 128:373–379. https://doi.org/10.1016/j.envpol.2003.09.012

Scarlat N, Dallemand JF, Monforti-Ferrario F, Nita V (2015) The role of biomass and bioenergy in a future bioeconomy: policies and facts. Environ Dev 15:3–34. https://doi.org/10.1016/j.envdev.2015.03.006

Schreurs E, Voets T, Thewys T (2011) GIS-based assessment of the biomass potential from phytoremediation of contaminated agricultural land in the Campine region in Belgium. Biomass Bioenergy 35:4469–4480. https://doi.org/10.1016/j.biombioe.2011.09.005

Shahsavari E, Adetutu EM, Taha M, Ball AS (2015) Rhizoremediation of phenanthrene and pyrene contaminated soil using wheat. J Environ Manage 155:171–176. https://doi.org/10.1016/j.jenvman.2015.03.027

Shi G, Cai Q (2009) Cadmium tolerance and accumulation in eight potential energy crops. Biotechnol Adv 27:555–561. https://doi.org/10.1016/j.biotechadv.2009.04.006

Shrinkhal R (2017) Coupling phytoremediation appositeness with bioenergy plants: a sociolegal perspective. In: Phytoremediation potential of bioenergy plants. Springer, Singapore, pp 457–472

Sierra MJ, Millán R, López FA, Alguacil FJ, Cañadas I (2016) Sustainable remediation of mercury contaminated soils by thermal desorption. Env Sci Pollut Res 23:4898–4907. https://doi.org/10.1007/s11356-015-5688-8

Singh S, Sounderajan S, Kumar K, Fulzele DP (2017) Investigation of arsenic accumulation and biochemical response of in vitro developed *Vetiveria zizanoides* plants. Ecotoxicol Environ Saf 145:50–56. https://doi.org/10.1016/J.ECOENV.2017.07.013

Sipos B, Kreuger E, Svensson SE, Réczey K, Björnsson L, Zacchi G (2010) Steam pretreatment of dry and ensiled industrial hemp for ethanol production. Biomass Bioenergy 34:1721–1731. https://doi.org/10.1016/j.biombioe.2010.07.003

Song Y, Kirkwood N, Maksimović Č, Zhen X, O'Connor D, Jin Y, Hou D (2019) Nature based solutions for contaminated land remediation and brownfield redevelopment in cities: a review. Sci Total Environ 663:568–579. https://doi.org/10.1016/j.scitotenv.2019.01.347

Tassi E, Pedron F, Barbafieri M, Petruzzelli G (2004) Phosphate-assisted phytoextraction in As-contaminated soil. Eng Life Sci 4:341–346. https://doi.org/10.1002/elsc.200420037

Thompson LE, Bates SO, Hansen JE (1992) Technology status report: in situ vitrification applied to buried wastes. Richland, WA (United States)

Tian Y, Zhang H (2016) Producing biogas from agricultural residues generated during phytoremediation process: possibility, threshold, and challenges. Int J Green Energy 13:1556–1563

Timmerman CL, Peterson ME (1990) Pilot-scale testing of in situ vitrification of Arnold engineering development center site 10 contaminated soils. Richland, WA (United States)

Tripathi V, Edrisi SA, Abhilash PC (2016) Towards the coupling of phytoremediation with bioenergy production. Renew. Sustain. Energy Rev. 57:1386–1389

Urum K, Pekdemir T, Gopur M (2003) Optimum conditions for washing of crude oil-contaminated soil with biosurfactant solutions. Process Saf Environ Prot Trans Inst Chem Eng Part B 81:203–209. https://doi.org/10.1205/095758203765639906

Van Der Werf HMG (2004) Life cycle analysis of field production of fibre hemp, the effect of production practices on environmental impacts. In: Euphytica. Kluwer Academic Publishers, pp 13–23

Van Ginneken L, Meers E, Guisson R, Ruttens A, Elst K, Tack FMG, Vangronsveld J, Diels L, Dejonghe W (2007) Phytoremediation for heavy metal-contaminated soils combined with bioenergy production. J Environ Eng Landsc Manag 15:227–236. https://doi.org/10.1080/16486897.2007.9636935

Vandenkoornhuyse P, Quaiser A, Duhamel M, Le Van A, Dufresne A (2015) The importance of the microbiome of the plant holobiont. New Phytol 206:1196–1206. https://doi.org/10.1111/nph.13312

Vangronsveld J, Herzig R, Weyens N, Boulet J, Adriaensen K, Ruttens A, Thewys T, Vassilev A, Meers E, Nehnevajova E, van der Lelie D, Mench M (2009) Phytoremediation of contaminated soils and groundwater: lessons from the field. Environ Sci Pollut Res 16:765–794. https://doi.org/10.1007/s11356-009-0213-6

Vigil M, Marey-Pérez MF, Martinez Huerta G, Álvarez Cabal V (2015) Is phytoremediation without biomass valorization sustainable? Comparative LCA of landfilling vs. anaerobic co-digestion. Sci Total Environ 505:844–850. https://doi.org/10.1016/j.scitotenv.2014.10.047

Villa RD, Trovó AG, Nogueira RFP (2010) Soil remediation using a coupled process: soil washing with surfactant followed by photo-Fenton oxidation. J Hazard Mater 174:770–775. https://doi.org/10.1016/j.jhazmat.2009.09.118

Vocciante M, Caretta A, Bua L, Bagatin R, Franchi E, Petruzzelli G, Ferro S (2019) Enhancements in phytoremediation technology: environmental assessment including different options of biomass disposal and comparison with a consolidated approach. J Environ Manage 237:560–568. https://doi.org/10.1016/j.jenvman.2019.02.104

Wang Q, Li Z, Cheng S, Wu Z (2010) Effects of humic acids on phytoextraction of Cu and Cd from sediment by Elodea nuttallii. Chemosphere 78:604–608. https://doi.org/10.1016/j.chemosphere.2009.11.011

Williford CW, Bricka RM (2000) Physical separation of metal-contaminated soils. In: Iskandar IK (ed) Environmental restoration of metals-contaminated soils. Lewis Publishers, pp 121–165

Witters N, Mendelsohn R, Van Passel S, Van Slycken S, Weyens N, Schreurs E, Meers E, Tack F, Vanheusden B, Vangronsveld J (2012a) Phytoremediation, a sustainable remediation technology? II: economic assessment of CO_2 abatement through the use of phytoremediation crops for renewable energy production. Biomass Bioenergy 39:470–477. https://doi.org/10.1016/j.biombioe.2011.11.017

Witters N, Mendelsohn RO, Van Slycken S, Weyens N, Schreurs E, Meers E, Tack F, Carleer R, Vangronsveld J (2012b) Phytoremediation, a sustainable remediation technology? Conclusions from a case study. I: energy production and carbon dioxide abatement. Biomass Bioenergy 39: 454–469. https://doi.org/10.1016/j.biombioe.2011.08.016

Xu L, Xing X, Liang J, Peng J, Zhou J (2019) In situ phytoremediation of copper and cadmium in a co-contaminated soil and its biological and physical effects. RSC Adv 9:993–1003. https://doi.org/10.1039/C8RA07645F

Yadav KK, Gupta N, Kumar A, Reece LM, Singh N, Rezania S, Ahmad Khan S (2018) Mechanistic understanding and holistic approach of phytoremediation: a review on application and future prospects. Ecol Eng 120:274–298. https://doi.org/10.1016/j.ecoleng.2018.05.039

Yeung AT (2009) Geochemical processes affecting electrochemical remediation. In: Reddy KR, Cameselle C (eds) Electrochemical remediation technologies for polluted soils, sediments and groundwater. Wiley, Hoboken, pp 65–94

Yi YM, Sung K (2015) Influence of washing treatment on the qualities of heavy metal-contaminated soil. Ecol Eng 81:89–92. https://doi.org/10.1016/j.ecoleng.2015.04.034

Yi MY, Park S, Munster C, Kim G, Sung K (2016) Changes in ecological properties of petroleum oil-contaminated soil after low-temperature thermal desorption treatment. Water Air Soil Pollut 277:1–10. https://doi.org/10.1007/s11270-016-2804-4

Zhou M, Wang H, Zhu S, Liu Y, Xu J (2015) Electrokinetic remediation of fluorine-contaminated soil and its impact on soil fertility. Environ Sci Pollut Res 22:16907–16913. https://doi.org/10.1007/s11356-015-4909-5

Zivdar Z, Heidarzadeh N, Asadollahfardi G (2019) Remediation of diesel-contaminated soil by low-temperature thermal desorption. Int J Environ Sci Technol 16:6113–6124. https://doi.org/10.1007/s13762-018-2020-4

Morphology and Physiology of Plants Growing on Highly Polluted Mining Wastes

7

Magdalena Krzesłowska, Piotr Goliński, Małgorzata Szostek, Agnieszka Mocek-Płóciniak, Kinga Drzewiecka, Aneta Piechalak, Anna Ilek, Ulla Neumann, Antonius C. J. Timmers, Sylwia Budzyńska, Patrycja Mleczek, Szymon Suski, Adam Woźny, and Mirosław Mleczek

Abstract

Biological methods have been described and suggested as a useful tool in studies on plant growth, development, and phytoremediation abilities in heavily polluted soil. Pot experiments are easier in practice, although field studies show a clearer picture of plant response to the stressors present in a polluted environment. Speciation (a form of toxic element), as well as mycorrhiza in the soil, play a role which is hard to overestimate. The enzymatic activity involved in this

M. Krzesłowska · A. Woźny
Laboratory of General Botany, Faculty of Biology, Adam Mickiewicz University, Uniwersytetu Poznańskiego 6, Poznań, Poland

P. Goliński · K. Drzewiecka · S. Budzyńska · M. Mleczek (✉)
Department of Chemistry, Faculty of Forestry and Wood Technology, Poznań University of Life Sciences, Poznań, Poland
e-mail: miroslaw.mleczek@up.poznan.pl

M. Szostek
College of Natural Sciences, Departament of Soil Sciences, Environmental Chemistry and Hydrology, University of Rzeszów, University of Rzeszów, Rzeszów, Poland

A. Mocek-Płóciniak
Department of General and Environmental Microbiology, Poznań University of Life Science, Poznań, Poland

A. Piechalak
Department of Genome Biology Institute of Molecular Biology and Biotechnology, Adam Mickiewicz University in Poznań, Poznań, Poland

A. Ilek
Department of Botany and Forest Habitats, Faculty of Forestry and Wood Technology, Poznań University of Life Sciences, Poznań, Poland

U. Neumann
Central Microscopy, Max Planck Institute for Plant Breeding Research, Cologne, Germany

R. Prasad (ed.), *Phytoremediation for Environmental Sustainability*,
https://doi.org/10.1007/978-981-16-5621-7_7

process will be discussed in this chapter. Plants used for recultivation should show a wide ecological tolerance to stressors. Plant resistance to trace elements (TE), induced systematic resistance (ISR), as well as root architecture alterations indicate a defence strategy in response to TE and possible TE accumulation in the tissue, with successful phytoremediation. Long-term exposure of plants to extremely high concentration levels of TE damages both their roots and functioning.

Keywords

Plant physiology · Polluted substrates · Root architecture · Trace elements · Wastes

7.1 Introduction

The level of environmental contamination with industrial debris (waste); not only its concentration but also the form (speciation) of the elements there, plays an important role. Physical and/or chemical methods are used in the reclamation of ecosystems through stabilization of elements in water and wind erosion. Biological methods—in the case of plants called phytoremediation—reduce the bioavailability and mobility of toxic elements. Pot experiments, which are easy to perform and flexible in correction, are often applied in such methods. Field studies, although much more difficult to modify, show real and well-documented plant responses to the stressors present in a polluted environment.

Although the following relationship appears to be obvious: the higher level of element concentration, the higher metal/metalloid uptake is observed in experimental plants; this is not always the case as there are several factors and soil properties that can influence phytoextraction. Microorganisms present in ecosystems play a particularly important role in this process. Enzymatic activity is a key parameter as it indicates the biological status (ecotoxicity) of the contaminated environment (waste). Increased enzymatic biosynthesis of phytochelatins is coupled with the heightened activity of the plant antioxidative system. Contaminating metal

A. C. J. Timmers
Central Microscopy, Max Planck Institute for Plant Breeding Research, Cologne, Germany

Laboratory of Plant Microorganism Interactions, INRA UMR 441, CNRS UMR2594, Castanet-Tolosan, France

P. Mleczek
Department of Ecology and Environmental Protection, Faculty of Environmental Engineering and Mechanical Engineering, Poznań University of Life Sciences, Poznań, Poland

S. Suski
Laboratory of Electron Microscopy, Nencki Institute of Experimental Biology of Polish Academy of Sciences, Warszawa, Poland

accumulation in the tissue is associated with unspecific plant response (pigment content, lipid peroxidation, or antioxidant activity).

Plant resistance to toxic elements takes the form of plant-fungi associations (mycorrhizal interactions), which lead to induced systematic resistance (ISR). Plant roots are in direct touch (contact) with all the nutrients and toxic components of an ecosystem and usually accumulate trace elements (TE). Their architecture alterations, influenced by the TE, is supported by mycorrhizal fungi in a defence strategy designed to cope with TE toxicity and the phytoremediation capability of plants. The above is accompanied by a cell wall (CW) and vacuole compartmentalization of TE. Binding TE with CWs, especially cross-linking by low-methyl esterified pectin's and enlargement of the CW capacity for sequestration, simultaneously increases the rigidity of the structure and therefore inhibits root elongation. Among the harmful effects of TE on root cell architecture, the most serious are alterations in the nucleus and nucleolus ultrastructure, while long-term plant exposure to extremely high concentrations of TE irreversibly damages root cells.

7.2 Plants Growing on Highly Polluted Substrates

Economic development promotes an increased demand for mineral raw materials, e.g. metal ores or coal extracted with the use of mining technologies. Highly contaminated industrial waste is becoming one of the most serious ecological problems worldwide, not only because of the high concentration of toxic elements but their diverse structure and considerable amount (Lottermoser 2010; Candeias et al. 2014; Behera and Prasad 2020). In 2014, the total amount of waste generated in the EU by all industrial activities and households was estimated to be 2503 million tonnes, the highest ever quantity recorded in the EU in the period 2004–2014. Over 64% of this amount of waste in 2014 was represented by mineral wastes from mining and extraction activities (EUROSTAT 2018). Nevertheless, there remain high concentrations of precious elements whose recovery from waste to pure form would be highly beneficial for industry (Aghaei et al. 2017). The concentrations of these elements are usually too high for the environment but too low for applied technologies to retrieve. The use of expensive technical methods for the thousands or millions of tons of waste deposited in settlers has long been an unrealistic idea. Conventional technologies used for reclamation of mining wastes are based on physical and chemical stabilization processes. Physical stabilization is targeted at a reduction of wind and water erosion; it consists of covering mine wastes with harmless materials, i.e. usually mining waste rocks, gravel, or topsoil collected from neighbouring areas. These remain provisional solutions and usually fail to produce long-lasting effects (Mohanty et al. 2010). Chemical reclamation consists of soil leaching/acid extraction and soil washing. The processes are usually based on the use of organic reagents for the removal of trace elements. They are considerably more efficient than physical methods and offer good results, especially in small objects with a high contamination level (Wang et al. 2017).

However, biological methods also seem to be a promising solution to mine waste remediation (Sangeeta and Maiti 2010; He and Kappler 2017). Particular attention has been paid in this area to phytoremediation (Wang et al. 2017). The goal of phytoremediation is to reduce the bioavailability and mobility of harmful pollutants from mining wastes, protect groundwaters, and prevent the spread of harmful substances into the consecutive elements of the food chain (Mohanty et al. 2010; Kuppusamy et al. 2016; Saha et al. 2017; Wang et al. 2017). Some plants have evolved several mechanisms for accumulation of excessive amounts of trace elements in tissues, even in extremely adverse conditions (Conesa et al. 2009; Mleczek et al. 2018). The knowledge of the application of phytoremediation techniques to neutralize contaminants originating from various mining wastes is still insufficient; therefore, further research is indispensable to find plant species that will be the most effective in solving the problem of different mining wastes (Mohanty et al. 2010). Studies on the possibility of the using plants for remediation of industrial wastes, including mining tailings were undertaken both as pot and field experiments.

7.2.1 Pot Experiments

Pot experiments using mining tailings offer the possibility to quickly modify the substrate (mixing, homogenization, enrichment) due to the low mass of used wastes. An example can be seen in the studies of Gupta and Sinha (2006), where different additions of tannery sludge were used in a pot experiment. Thanks to the relatively easy preparation of model systems, these studies indicated the optimal amount of tannery sludge (25%) for the practical use of phytoremediation by *Sesamum indicum* (L.). Too high a concentration of heavy metals and/or metalloids in wastes contribute to the death of plants and serious changes in their structure, which precludes their practical use (Krzesłowska et al. 2019; see also Sect. 7.6.2). It is worth emphasizing that such experiments have usually been limited to selected plant species such as *Pennisetum sinese* Roxb (Napier grass, HGN) as described by Ma et al. (2016a). The authors characterized the uptake of arsenic (As), cadmium (Cd), chromium (Cr), copper (Cu), manganese (Mn), lead (Pb), and zinc (Zn) in plants growing on tailing waste from Dabaoshan, Shaoguan (China) and showed that the efficiency of phytoextraction ranged from 12 to 26% with the highest content in total biomass of shoots and the highest concentration in fibrous roots. Such studies have highlighted the important role of a well-developed root system for further plant growth and heavy metal accumulation. In spite of the clear inhibition of *P. sinese* growth, its well-developed root system qualifies HGN as a suitable plant for effective phytoextraction of heavy metals from tailing wastes.

Studies on trees or shrubs and their use in phytoremediation (dendroremediation), owing to their extensive root system, have become more widespread and wide-ranging in recent years, yet the limitation imposed by the age and size of plants remains (Mleczek et al. 2017, 2019). Shi et al. (2017) analysed *Salix integra* growing on mine tailings from Hangzhou (China). This willow species was characterized by diverse tolerance and metal uptake with respect to the percentage of wastes in the mixed substrate. The results obtained in this study, like those—described in numerous other papers—have pointed to the potential of the plant in phytostabilization or phytoextraction of mine tailings. Unfortunately, the use of tree species that seem promising in practice in pot experiments is generally highly problematic and gives only part responses to numerous questions since the influence of environmental factors is limited (Budzyńska et al. 2017, 2021). Therefore, the best solution would appear to be to look (compare) for plants whose uptake of metals most efficient and then conduct field experiments, indicating at the same time differences in plant responses depending on their origin. An example of this approach can be found in the studies of Wang et al. 2007.

7.2.2 Field Experiments

Generally, field experiments with mine tailings do not allow us to implement such daring modifications of substrates as in pot experiments, but the information gained shows the real plant response to a polluted environment (García-Salgado et al. 2012). To date, these experiments have probably been conducted on each kind of polluted waste (Dudka et al. 1996; Yanqun et al. 2004; Ang et al. 2010; Fernández-Martínez et al. 2015). These mining wastes were characterized by different concentrations of toxic elements which greatly reduced plant growth and development (Dudka et al. 1996; Mleczek et al. 2017; Krzesłowska et al. 2019 see Sect. 7.6). Limitations in plant growth were rather more an effect of a possible nutrient imbalance than the phytotoxicity of metals, as described, e.g. by Dudka et al. (1996) who studied *Hordeum vulgare* L., *Poa pratensis* L., *Solanum tuberosum* L., and *Trifolium pratense* L. exposed to extremely high concentrations of Cd, Cu, Pb, and Zn (4000; 106; 3452 and 11,375 mg kg^{-1}, respectively). Mining wastes usually contain an extremely high concentration of toxic elements such as As (Santos-Jallath et al. 2012); mercury (Hg) (Fernández-Martínez et al. 2015), Pb (Freitas et al. 2004; Čudić et al. 2016), or antimony (Sb) (Okkenhaug et al. 2011). The high concentrations of these elements and their great quantity deposited in the environment have created a pressing need to select the most effective plants for phytoextraction of toxic elements (Anawar et al. 2011; García-Salgado et al. 2012). For this reason, numerous papers have focused on a comparison of several plant species growing on the same wastes (Zhao et al. 2014). However, only some of them have compared a wide range of plant species, e.g. Baroni et al. (2004) or Karimi et al. (2013) studies (64 and 49 species, respectively).

The most popular plants researched were trees and bushes (Madejón et al. 2005; Domínguez et al. 2008), but also grasses, mosses, native ferns or rushes (Craw et al. 2007; Čudić et al. 2016). Tree and bush species were especially interesting as regards their traits such as long life or an extensive root system capable of deep penetration into the waste (Pratas et al. 2005; Migeon et al. 2009). Moreover, trees are characterized by selective uptake of elements described, e.g. by Zhao et al. (2014), who studied 18 Chinese tree species. The authors showed that particular tree species are capable of *i*) effective phytoextraction of a single metal; *ii*) hyperaccumulation; or *iii*) multi-metal accumulation. There is a clear and interesting response of selected plants (also trees) growing on mining sludge. The majority of available literature data describes the following relationship: the greater the concentration of toxic metals in mining wastes the higher their concentration in plant organs (Martínez-Sánchez et al. 2011; Alagić et al. 2013; Karimi et al. 2013), however, the metals/metalloids uptake is not always dependent on its concentration in substrate (Baroni et al. 2004; Santos-Jallath et al. 2012).

The problem of mining waste decontamination is an effect of two distinct stress factors: the high concentration of toxic metals and high salinity. Additionally, mining wastes are often substrates with alkali pH, high salinity, a heavily concise structure (percentage participation of loam over 10), low content of bioavailable nutrients, and low water retention capacity (Ssenku et al. 2014; Kuppusamy et al. 2016). The significance of pH, waste chemistry, and the mechanism used by plants in these unfavourable conditions, for example, of *Betula* sp. and *Tilia* sp., was clearly described by Alagić et al. (2013). Tree species from highly polluted areas (Bor Region in Serbia—mining and metallurgy) were characterized by different mechanisms of As and Cd uptake: a high ability of assimilation through leaves in the case of *Tilia* sp. and clear transport from the soil for *Betula* sp. Similarly, Kalinovic et al. (2016) compared tree species (*Pinus nigra* Arn., *Sambucus nigra* L., and *T. grandifolia*) growing in areas polluted by emissions from a Cu smelter and by the tailing ponds of open pit mines. They found that rates of metal uptake were dependent on plant species and that element concentration differed in particular plants. However, conducting studies on extremely polluted wastes and interpretation of obtained results was in many cases, less effective concerning the analysis of total metal concentration only. To describe the real response of a plant, data on the concentration of bioavailable metal forms is necessary (Fernández-Martínez et al. 2015). This factor alone, especially in case of mining wastes highly contaminated with trace elements, can be crucial to plant survival, adaptation, and element uptake.

For this reason, low bioavailability was 'increased' by the addition of numerous additions, e.g. sewage sludge to combine waste into two different processes in one remedy technique (Forsberg and Ledin 2006). A high concentration of elements in mining wastes with a simultaneous high concentration of their bioavailable forms were, in the majority of studies, the cause of high phytoextraction of elements to plant roots or even to above-ground plant parts (Jung et al. 2002; Okkenhaug et al. 2011). A distinct relationship between As concentration in polluted substrate and plants was described by Martínez-Sánchez et al. (2011). It is necessary to state that despite the close correlation between the As concentration in the substrate and plant, the efficiency of As will depend on the plant species. A confirmation of this can be found in the results described by Chang et al. (2005) who analysed *Pteridium aquilinum* growing on areas polluted by two gold (Au) mines (Duckum mine and Myoungbong mine in the Republic of Korea). The high potential of the selected *Polypodiopsida Cronquist* species growing on highly polluted post-industrial wastes was also widely discussed (Visoottiviseth et al. 2002; Wang et al. 2007).

A high concentration of metals in wastes is not unequivocal with its effective transport to plants, as described by Craw et al. (2007) who studied two historic gold mining sites in north Westland (New Zealand) and the growth of diversified plant species with clear phytostabilization. Similar observations were also described by Fernández-Martínez et al. (2015), who showed that selected plant species can be considered as excluders only and by Santos-Jallath et al. (2012), who found a low correlation between As in soil and plants in spite of a high concentration of this metalloid in wastes of 183–14,660 mg kg^{-1}. The authors of the works conducted their studies mainly using plants naturally growing in situ but also plants cultivated in these areas (Domínguez et al. 2008; Álvarez-Ayuso et al. 2012; García-Salgado et al. 2012). An efficient phytoremediation strategy for plants cultivated on mining wastes was generally the result of the specific traits of a growing plant and the impact of environmental factors directly acting on this (Otones et al. 2011). Parraga-Aguado et al. (2014) analysed *P. halepensis* growing on a mine tailing disposal site in Spain observed the immobilization of Cd, Cu, Pb, and Sb in woody stems as well as the effective transport of As, Cd, Sb, Pb, and Zn to leaf litterfall with a possible return to the environment after the vegetation period.

7.3 Physicochemical Properties of Mining Wastes: Implication for Phytoextraction

Mining wastes usually include waste rocks, overburdens, slags, and tailings on land surfaces. Mine wasteland generally comprises a bare stripped area, loose soil piles, waste rock and overburdened surfaces, subsided land areas, tailings dams, and other land degraded by mining facilities, among which the tailings dams and waste rock surfaces often pose extremely stressful conditions for restoration (Li 2006; Gautam et al. 2016a). The largest amount of waste is generated by the mining and processing of coal, followed by non-ferrous and ferrous ores and industrial minerals. It is estimated that approximately 2–12 tonnes of overburden are removed with every tonne of metal extracted from ores (Mohanty et al. 2010). Storage of these wastes in heaps results in degradation of soils, contamination of water resources, pollution of air in adjacent areas, and a consequent reduction of biodiversity (Álvarez-Valero et al. 2008; Candeias et al. 2014; Ssenku et al. 2014; Saha et al. 2017). The high variability of the physicochemical properties of these wastes, as well as local hydrological, topographical, and meteorological conditions, contribute to a varied rate of release of various types of pollutants, including toxic trace elements. Therefore, detailed characterisation of these wastes is required to assess their potential toxicity and the possibility of further disposal thereof (Lottermoser 2010; Candeias et al. 2014; Cappuyns et al. 2014; Azarova et al. 2019).

Mining wastes contain numerous contaminants, e.g. salts, metalloids, metals, radionuclides, and others (Conesa et al. 2009; Lottermoser 2010; Ssenku et al. 2014; Kuppusamy et al. 2016), which are dispersed in the environment on a local and regional scale through the process of weathering or wind and water erosion (Conesa et al. 2009; Candeias et al. 2014). The content of Zn, Cu, Pb, nickel (Ni), Cd, and Cr in soils is usually in the range from 0.0001 to 0.065%, whereas iron (Fe) and Mn account for 10% and 0.002%, respectively (Kabata-Pendias 2004; Ernst 2006). Except for Fe, all elements present at a level higher than 0.1% become toxic to plants (Bothe and Słomka 2017). Macro and micronutrients, such as sodium (Na), potassium (K), calcium (Ca), magnesium (Mg), Fe, Zn, Mn, cobalt (Co), Cu, Ni, and molybdenum (Mo), are important for plant growth and development. However, their high concentrations in the mining waste substrate disturb physiological processes, causing abnormal plant development (Gautam et al. 2016a). The concentrations of As, Cd, Cu, Mn, Pb, and Zn are as low as 1 g kg^{-1} in modern mining waste and can be higher than 50 g kg^{-1} in historical wastes (Mendez and Maier 2008). A prerequisite for the application of phytoextraction techniques is the presence of trace elements in soluble and exchangeable fractions, which determines their bioavailability. Trace elements contained in mining wastes exhibit substantially higher bioavailability than those contained in non-contaminated soils (Ssenku et al. 2014). The mobility and bioavailability of trace elements in mining wastes depends on many physical and chemical properties, e.g. particle size distribution, structure, bulk density, water holding capacity, nutrient content, pH, salinity, and others (Sheoran et al. 2016; Jović et al. 2017; Kushwaha et al. 2018). Plants used in the phytoextraction process can influence the mobility and bioavailability of trace elements, thereby contributing to changes in the properties of the substrate. This is achieved by compounds secreted by plant roots such as organic acids, which induce changes in pH (Gerhardt et al. 2017). Therefore, prediction of the effectiveness of mining waste reclamation with phytoextraction techniques requires determination of the physicochemical properties of these wastes and their interactions (Sessitsch et al. 2013; Baldantoni et al. 2014; Sheoran et al. 2016).

The physical and chemical properties of mining wastes differ depending on their mineralogy, geochemistry, particle size in the extracted material, and moisture content. The basic sources of these materials are rocks, soil, and sediments from surface mining operations, especially from outcrops (Lottermoser 2010). Mining wastes are non-homogeneous geological materials that can consist of sedimentary, metamorphic, or magmatic rocks, soils, and loose sediments. The mineral composition of mining wastes is dominated by silica (70.43%), followed by Al_2O_3 (7.32%), Na_2O (2.32%), K_2O (0.08%), CaO (0.69%), FeO (2.14%), and Fe_2O_3 (2.8%) (Flores Badillo et al. 2015; Almeida et al. 2018). The mineral composition of mining wastes, e.g. calcite, plays an important role in controlling the level and potential bioavailability of trace elements in the environment (Kim et al. 2014).

Many mining wastes are characterized by unfavourable physical conditions, e.g. the absence of structure, tendency to crust, or low water retention capacity. As in the case of soil, the physical properties of mining wastes determine their usefulness in plant production and, hence, in the effectiveness of phytoremediation techniques. The physical properties of mining wastes are associated with their chemical composition and can significantly modify the availability of nutrients and the influence absorption of trace elements by the roots of plants used in phytoextraction (Almeida et al. 2018). The water holding capacity and availability for plants depend on the physical properties. The water content in the substrate significantly influences the efficiency of the phytoextraction process, as it determines, e.g. the bioavailability of contaminants. Low water capacity and/or low nutrient (phosphorus (P), nitrogen (N)) content increase the toxicity of trace elements in the substrate (Bothe and Słomka 2017). At higher humidity, plants take up more trace elements and produce a greater amount of biomass, which is directly reflected in a more efficient phytoextraction process (Dineshkumar et al. 2019). The water holding capacity of mining wastes is strictly dependent on their particle size distribution, which fundamentally determines the further management thereof and their impact on the environment. The particle size in these materials can vary depending on the parent rock, and their diameter usually represents that of the sand to clay fraction (Sun et al. 2018; Festin et al. 2019). The particle size may range from exceedingly fine (e.g., phosphate slimes, <0.01 mm) to very fine, e.g., most tailings that result from metal ore processing, typically <0.1 mm, to very coarse, e.g., typical blasted overburden, where the particle size exceeds many centimetres and a large part may have a size of the order of 1 m and greater. Very small particles of mining wastes can sometimes be transferred over long distances, thus contaminating large areas (Daemen and Akgün 2012; Gautam et al. 2016a). Larger fractions have no ability to accumulate organic matter and are poor in nutrients, devoid of structure, and vulnerable to crusting. They are also characterized by low water holding capacity (Rivera-Becerril et al. 2013). The high potential evapotranspiration and low water retention capacity of these fractions suggest that the water deficit limits the possibility of introducing plants, especially in arid regions (Sun et al. 2018). The finer fractions of these wastes usually exhibit greater water retention capacity, but water infiltration is often limited by the poor structural properties; hence, water often accumulates on the surface of these wastes (Hossner and Hons 1992; Sun et al. 2018). Finer fractions are also associated with the levels of toxic elements such as As, Pb, or Cd (Acosta et al. 2011; Silva et al. 2014). The predominance of fine fractions in these metals may lead to their excessive contents (Festin et al. 2019). Most mining wastes are characterized by high bulk density (Gautam et al. 2016a), often exceeding 1.6 g cm^{-3} (Saha et al. 2017). Such a high density not only inhibits the development and growth of plant roots but also causes a reduction in general porosity and a deterioration of the air-water status of the substrate, which often determines the potential applicability of individual phytoextraction techniques. The most optimal system has a total porosity value of approximately 50%. Importantly, in the adverse air-water conditions of the substrate related to its high density and low porosity, plants may show nutrient deficiencies, even at an excess of elements in the substrate.

The reaction of mining wastes varies from very acidic to alkaline. The pH value depends on the content of carbonates, the potential release of acids by these wastes, and the parent rock (e.g. dolomites and limestones are alkaline) (Mendez and Maier 2008; Conesa et al. 2009; Rivera-Becerril et al. 2013; Cappuyns et al. 2014; Gautam et al. 2016a; Almeida et al. 2018; Festin et al. 2019). Extremely low or high pH values make mining wastes an unfavourable environment for the development of microorganisms and higher plants (Ssenku et al. 2014). The growth of most plants is hampered at soil pH < 4, which results from, e.g. reduced solubility and assimilation of essential nutrients by plants such as Ca and Mg (Dineshkumar et al. 2019). Additionally, low pH leads to the activation of Al^{3+} ions, thus inducing their toxicity. At the substrate pH value of <5.5, this element is released from minerals containing aluminium (Al) oxides and hydroxides, which results in the generation of the Al^{3+} ion. Plant roots are especially sensitive to the Al^{3+} ion, which is highly toxic and often limits plant growth and development (Sun et al. 2018). At low pH, the solubility and biotoxicity of trace elements contained in mining wastes increases. The pH of the substrate influences not only bioavailability but also the accumulation of metal ions in roots. At lower pH, cationic forms of trace elements become more soluble, while anion forms are better soluble at higher pH values. Reduction in the pH value increases Zn, Mn, and Co absorption and diminishes, e.g. Ni uptake (Dineshkumar et al. 2019). In turn, in alkaline conditions, the absorption of nutrients is low as they are present as insoluble salts and cannot be taken up by plant roots. At pH > 7, the absorption of P is significantly reduced. At such an extreme reaction value it is impossible to achieve the expected effects of the phytoextraction process as most plants develop normally in the range of slightly acidic and neutral reactions; therefore, raising or reducing the pH value in such wastes to the range of 5.5–7.0 or 5.5–6.5 is crucial for the achievement of good results. The highest absorption of plant nutrients is observed at slightly acidic and neutral reactions; it is directly reflected in better plant growth and development as well as increased efficiency of the phytoextraction process (Ssenku et al. 2014).

The content of plant nutrients, in particular nitrogen, organic matter, and available phosphorus forms in mining wastes is an important determinant of the efficiency of the phytoextraction process (Rivera-Becerril et al. 2013; Ssenku et al. 2014; Almeida et al. 2018). Normal plant growth and development requires the presence of such elements as N, P, and K in the substrate. In comparison with non-contaminated soils, many mining wastes are characterized by lower levels of organic matter and available forms of essential plant nutrients (P, N, and K) (Ssenku et al. 2014).

A low N mineralization rate, low P availability and low content of organic matter are typical of mining wastes; hence, the efficiency of the phytoextraction process is low due to the poor nutritional status of plants. Management of such wastes should consist in the improvement of their condition, e.g. by application of exogenous sources of organic matter. Furthermore, some elements that play the role of micronutrients required for normal plant growth and development are often present in these wastes in unavailable forms or in inadequate proportions (Sun et al. 2018). To increase the efficiency of phytoextraction, it is necessary to identify thoroughly which nutrients are deficient and to apply additional fertilization with these components. At sufficient nutrient content, the adverse effect of abundant contaminants in mining wastes is neutralized. An example is Ca, which is not only required for normal plant growth but also neutralizes the excessive acidification of mining wastes. It has been emphasized that the pH and Ca content in the substrate are more important for normal plant growth and development than the total content of toxic trace elements contained therein (Sun et al. 2018).

Substrate salinity influences almost all aspects of plant development, including germination, vegetative growth, and reproductive development. It limits normal plant growth and development by its significant effect on water and nutrient uptake by plant roots. The salinity of mining wastes is associated with the excessive accumulation of sodium ions used in technological processes (Sun et al. 2018). The impact of excessive substrate salinity on plant development is mainly related to its influence on water and ion metabolism in plants, which is modified, and thus, normal development is affected. Substrate salinity significantly modifies ionic ratios Na^+/Ca^{2+}, Na^+/K^+, Ca^{2+}/Mg^{2+}, and Cl^-/NO_3^-, which leads to disturbances in the uptake of nutrients, in particular, K^+, Ca^{2+}, and NO_3^-, and plant growth inhibition (Bano and Fatima 2009). Sodium is a necessary element for normal plant growth and development and plays an important role in the maintenance of proper turgor in plant cells. However, its elevated concentration in the substrate exerts a toxic effect on plants. Increased uptake of Na^+ ions has an impact on the uptake of K^+; hence, plants may exhibit a deficiency of the former element as well. Substrate salinity exerts a significant effect on water availability for plants. At excessive salinity, the ability of plants to take up water from the substrate is impaired, which leads to osmotic stress and growth retardation (Munns 2002; Nouri et al. 2017). The direct impact of salt on plants can result in ionic stress, mainly caused by Na^+ and Cl^- ions. Increasing concentrations of these ions in the plant organs causes leaf chlorosis, necrosis, and premature fall and can lead to the early death of individual organs and, consequently, entire plants. Moreover, salinity significantly determines the mobility of such elements as Pb, Cd, Zn, and Cu and the effectiveness of phytoextraction through competition with calcium for sorption sites, complexation with chlorides, complexation with sulphates, competition with Mg and/or Ca, etc. (Acosta et al. 2011; Filipović et al. 2018).

7.4 The Role of Microorganisms in the Disposal of Energy Waste (Furnace Waste)

The dynamic economic development of many countries, mainly in the second half of the twentieth century, resulted in the production of huge amounts of waste, which became a very serious problem in environmental protection. Waste is classified according to different criteria, depending on production (Rosik-Dulewska 2015). Industrial waste is a special group whose parameters strictly depend on the type of industry and production technology. In countries where the production of energy is based on coal or lignite, large amounts of energy waste are produced as a result of the combustion of ground coal. Non-combustible mineral parts in the form of dust are transported into smokestacks, where they are captured by filters and fly ash is formed. Apart from fly ash, coal combustion in boilers also generates slag. This waste falls to the bottom of the furnace or remains on the grate and then it is discharged outside. The vast majority of ash and slag hereinafter referred to as energy waste or furnace waste is collected in over ground landfills or in pits left after excavated minerals.

There are differences in the chemical composition of energy waste, depending on the type of coal burned, combustion technology, the type of transport to the landfill and storage method. Depending on the chemical composition of ashes and slags, they are classified into the following three types according to the so-called Oxide Module (OM):

$$\frac{SiO_2 + Al_2O_3}{CaO + MgO + Fe_2O_3}$$

- carbon silicate (OM $\leq$ 2.0),
- silicate (OM 2.1–5.9),
- aluminium silicate (OM $\geq$ 6.0).

Fresh energy waste is almost completely devoid of biological life due to its particularly unfavourable physicochemical properties. The following physical parameters are unfavourable: inadequate structure, low specific density, low bulk density, and high porosity. As far as the grain size is concerned, fresh energy waste usually has a sandy, or less frequently, a loamy structure. Therefore, it has a low capacity to retain water, which should be easily accessible, especially to plants and microorganisms. As far as the chemical parameters of energy waste are concerned, its high alkalinity (pH 8–12) is noteworthy. It results from the considerable content of alkali metal hydroxides, and a small amount of organic matter, i.e. available forms of N and P. Apart from that, the solubility of these ashes is also low—it generally ranges from about 2–10%. Energy waste usually contains trace amounts of organic xenobiotics, including polycyclic aromatic hydrocarbons (PAHs) and basic heavy metals (Cd, Cr, Cu, Ni, Pb, and Zn).

Energy waste is stored in landfills. As it generates high amounts of dust, it should undergo reclamation as soon as possible, which is very difficult and tedious. For successful land reclamation, it is first necessary to repair the chemism of energy waste by adequate NPK fertilization and by lowering its pH. It is also necessary to improve its physical properties – loosen or crush the waste mass and select adequate vegetation, depending on the plans to restore an agricultural or forest ecosystem in the future. Successful technical and biological reclamation initiates the transformation of these barren and almost 'inanimate' parent rocks into living forms, which are classified as technosols (WRB 2014).

A few years after successful reclamation organic matter available to microorganisms begins to accumulate in shallow surface levels, i.e. accumulation and humus levels because energy waste contains only small amounts (2–8%) of incompletely burnt pieces of coal (black carbon), which is a low-activity form in the soil environment (Gustafsson and Gschwend 1997; Zikeli et al. 2002, 2004; Strzyszcz 2004; Cornelissen and Gustafsson 2006). The investigations which were carried out 20 years after the initiation of land reclamation in a landfill with ashes and slag from a lignite-based power plant showed relatively small populations of basic groups of microorganisms, i.e. heterotrophic bacteria and bacteria of the *Azotobacter* sp. genus, actinobacteria, and moulds (Mocek-Płóciniak 2018).

Samples were collected from newly formed forest soils classified into the Technosols group, Spolic Technosols subgroup (WRB 2014). During the 4-year study (2012–2015) the counts of heterotrophic bacteria in the upper soil layers (0–15 cm) were fairly diversified and usually ranged from 10 to 90 $\times$ 10^5 CFU g^{-1} d.w. of soil. These values were several folds greater than the counts of these bacteria

Table 7.1 Ranges of number and biomass of the most important soil organisms (Martyniuk 2017)

Organisms	Number in 1 g	Biomass (kg ha^{-1})
Bacteria	10^7–10^9	300–3000
Actinobacteria	10^6–10^8	300–3000
Moulds	10^5–10^6	500–5000
Algae	10^3–10^6	10–1500
Protozoa	10^3–10^5	5–200
Nematodes	10^1–10^2	1–100
Earthworms	30–300 per m^2	10–1000

in the samples collected from a depth of 80–100 cm (usually 2–12 × 10^5 CFU g^{-1} dw of soil) (Mocek-Płóciniak 2018). The mean count of these bacteria in the samples amounted to about 22.54 × 10^5 CFU g^{-1} d.w. of soil. There were even smaller counts of bacteria of the *Azotobacter* sp. The count of these bacteria in the topsoil usually ranged from 10 to 30 CFU g^{-1} dw of soil, whereas in the bottom levels (80–100 cm) it amounted to a few CFU g^{-1} dw of soil. The average count of *Azotobacter* in the entire mass of ashes forming the technosols was 19.85 CFU g^{-1} dw of soil. The count of actinobacteria in energy waste was also strongly diversified. In the consecutive years of the research, the count of these microorganisms did not always tend to decrease with the depth. The average count of actinobacteria in the whole research material (72 samples) was low, i.e. 13.19 × 10^5 CFU g^{-1} dw of soil. The samples collected from the landfill also contained small amounts of moulds which varied in individual years. The upper soil layers, especially at a depth of 0–5 cm, contained much more moulds (from about 1 to 170 × 10^5 CFU g^{-1} dw of soil) than the bottom levels at a depth of 80–100 cm (from 0 to 9.47 × 10^3 CFU g^{-1} dw of soil). The average count of moulds amounted to 26.15 × 10^3 CFU g^{-1} dw of soil. This was low in comparison with the count of these microorganisms in various soils formed from natural (postglacial) parent rocks (Table 7.1).

Enzymatic activity is another very important parameter illustrating the state of the biological environment of soils formed from industrial waste. It can be treated as a function of the activity of basic populations of the groups mentioned above of soil microorganisms and the root secretions of plants living on these soils. The activity of enzymes depends on their absolute amount, the size of the group of other reacting compounds than enzymes, and the catalytic efficiency (Murray et al. 1995). The catalytic efficiency in the emerging soil environment is affected by additional biotic and abiotic factors such as: the content of mineral and organic colloids, temperature, the properties and pH of water and air, the content and availability of biogenic elements as well as the count and species of microorganisms (Kobus 1995; Kucharski 1997). The value of enzyme activity reflects the in situ state. It is determined not only by the current soil conditions but also due to the accumulation of enzymes in the form of humus complexes, it is to a large extent determined by the history of events preceding the measurement, such as the climatic conditions and treatments applied to soil (Januszek 1999; Bielińska et al. 2016). The technosols formed from energy waste were considerably diversified not only in the count microorganisms but also in the content of the enzymes analysed in this study.

Dehydrogenase activity is an indicator of the intensity of the respiratory metabolism of soil microorganisms, mainly bacteria and actinobacteria (Praveen-Kumar and Tarafdar 2003). For this reason, dehydrogenase activity is regarded as a measure of the total microbial activity of soils and an index of ecotoxicity. During the studies (2012–2015) the dehydrogenase activity in the landfill soils was very low, i.e. 0.15–16.89 mg TPF kg^{-1} d.w. of soil 24 h^{-1} at a depth of 0–5 cm and 0.12–4.56 mg TPF kg^{-1} d.w. of soil 24 h^{-1} at a depth of 5–15 cm. The activity of these enzymes was almost unnoticeable at deeper levels (80–100 cm). This may have been caused by the supply of nutrients in the form of root secretions or the biomass of microorganisms (Yang et al. 2007; Futa 2017). The presence of carbon substrates induces and stimulates the biomass of enzymes by soil microorganisms (Renella et al. 2006; Fierer et al. 2003). The low dehydrogenase activity in the samples under analysis, which indicated the low overall microbial activity of the environment, was mostly related with the initial phase of the formation of biological balance (homeostasis) in the furnace waste soils and with the particular sensitivity of this group of intracellular enzymes to environmental factors (Januszek 1999; Bielińska et al. 2014).

Alkaline phosphatase—during the studies—exhibited the highest activity of all the enzymes analysed in the samples of technosols collected from the energy ash landfill. It was similar to the activity of this enzyme observed in soils formed from natural parent rocks. The alkaline phosphatase activity in most of the upper soil levels was more intense than at the lower levels. It was predictable due to the higher content of organic matter in the top soil levels, better air and water conditions and greater biomass of plant roots. In general, the higher alkaline phosphatase activity could also be attributed to low amounts of available forms of phosphorus at all levels of these soils (Kucharski et al. 2015). Urease is an enzyme that perfectly adapts to any environment, regardless of temperature, humidity, and pH. The only factor limiting urease activity is the availability of urea because the presence of this substrate is necessary for the synthesis of urease, which is an extracellular enzyme (Carbrera et al. 1994). The urease activity in the soils under study was low. This may have been the result of the nearly trace amount of easily available forms of N in this soil material. Proteases are a large group of enzymes due to the high diversity of proteins as substrates. They hydrolyse peptide bonds almost anywhere in the protein chain (Dahm and Rydlak 1997). The protease activity in energy waste forms was low. Bielińska and Futa (2009) observed a similar activity of this group of enzymes in other ash and slag substrates. It is noteworthy that there were significant correlations between the four enzymes in the technosols. However, the average activity of all the enzymes under analysis at a depth of 0–5 cm was significantly higher than at deeper levels.

To sum up, we can say that energy waste in the form of fly ash and furnace slags is a difficult parent material for the reclamation of gradually developing technosols. They are characterized by numerous unfavourable physicochemical, microbiological, and biochemical properties. The soil-forming processes, which were initiated by humans as the anthropogenic factor, are much slower here than in natural parent rocks. The period of more than 20 years of the formation of technosols proves that they have not yet achieved an optimal biological balance. This means that the composition of this microbiocenosis is not stable, and the count of microorganisms still varies. The enzymatic parameters also proved to be useful for the monitoring of the transformation of energy waste into soil. Although for many years the landfill has been remediated technically and biologically, as manifested by a beautiful oasis of plant and animal communities, so far neither explicit epipedons nor endopedons have appeared in the soil profile. Nor has biological life developed, which would be reflected by a stabile count of microorganisms and enzymatic activity. It seems that these processes must continue for a 50 years or more at least.

7.5 Physiological Aspects of Plant Survival on Heavily Polluted Sites

7.5.1 Plant Selection for Phytoremediation of Mine Tailings

Mine tailings, being waste disposal sites, pose a serious threat to surrounding ecosystems and local communities due to the considerable ease with which pollution can spread to the environment with wind and runoff water (Salas-Luévano et al. 2017). A recent approach to deal with contaminated mine tailings—having both environmental and social acceptance—is phytostabilization with tolerant plant species to create a permanent vegetation cover preventing the spread of the pollution via erosion, runoff, and percolation (Pulford and Watson 2003). The solidification of toxic elements (metals and metalloids) reduces their availability and enables sustainable revegetation of disturbed lands. The selection of pioneer plants for phytostabilization is crucial. However, it remains the subject of an on-going debate since there are no standard approaches for the management of multi-contaminated sites (Barbafieri et al. 2017). Recent studies have shown that selection of plants for recultivation of mine tailings should consider indigenous species with wide ecological tolerance instead of newly introduced exotic ones (Pratas et al. 2013; Kumar et al. 2017). Native species will most likely develop a fully functional ecosystem on degraded land by gradual alternation of the waste properties, mainly by improving organic carbon (C), N fixation, water storage, reduction of acidity and nutrient effluent, and also by improving the diversity of the soil microbiom (Shi et al. 2016; Demková et al. 2017a, b).

The selection of appropriate species may be achieved via different experimental setups, such as (1) investigations on specimens collected from mine tailings showing high tolerance to the pollution (Boojar and Goodarzi 2007; Barbafieri et al. 2017; Abreu et al. 2012; Pistelli et al. 2017); (2) cultivation of native plant species under controlled conditions using waste material as a substrate (da Silva et al. 2018); (3) in situ cultivation of plant species naturally occurring at mine tailings.

7.5.2 Physiological Determinants of Plant Tolerance to Mining Waste Materials

Metal ion toxicity triggers physiological and developmental changes that lead to adaptation and defence reactions in the plant. However, non-essential and excess of essential metals also cause irreversible damage. Heavy metal ions present in the soil can be taken up alongside nutrients with water and incorporated into plant tissues. Plants have to continuously maintain physiological concentrations of both essential and non-essential metal ions to achieve ionic homeostasis. Moreover, this homeostasis must be maintained in a cell-tissue and organ-specific manner (Sharma and Agrawal 2005; Hu et al. 2013). Plants growing in certain areas are exposed to high localized concentrations of metal ions. When exposed to excess metals, the vast majority of plant species adopt an excluder strategy which involves avoidance of exposure, minimizing their uptake, and intracellular sequestering in the cell wall and vacuoles to prevent their harmful effects in cells (DalCorso et al. 2013; Hossain and Komatsu 2013). Reduced growth is one of the most common physiological consequences of heavy metal exposure in plants (Hu et al. 2013; Tamás et al. 2008). Metal ions can have devastating effects on basic metabolism, transport processes, membranes, and cellular structure. There are reports about metal induced disturbances in the structural and physiological integrity of leaves which impact the rates of photosynthesis and respiration, and consequently energy provision. Heavy metal toxicity also affects the ability to take up water and nutrients and transport processes between various organs (Ying et al. 2010; Barceló and Poschenrieder 2004). Major changes in the functioning of the organs will affect developmental processes such as flowering, embryogenesis, and seed formation. Exposure to toxic metal ions or high concentrations of non-toxic ions, therefore, triggers stress reactions and necessitates adaptation at all levels: physiological, structural, and molecular (Gautam et al. 2016b; Tamás et al. 2008; Hall 2002; Hirayama and Shinozaki 2010).

The mechanisms of metal tolerance exhibited by some plant species are a unique and very interesting feature of plants in stress condition. Besides exclusion strategies, they include the extracellular chelation of metal ions, the restriction of ions in the apoplast, and the detoxification and compartmentalization of metal ions inside the plant tissues (Hossain and Komatsu 2013; Dickinson et al. 1991; Hall 2002). The most advanced strategies used by plants are based on hypertolerance and the hyperaccumulation of metal ions without any negative effects on growth and yield (DalCorso et al. 2013; Dickinson et al. 1991; Van der Ent et al. 2013). It has been observed that a high concentration of metal in hyperaccumulators may secondarily protect them against herbivores and pathogens (Boyd 2012; Cabot et al. 2013). Hyperaccumulators are a small number of plant species that grow on naturally or anthropogenically metal-contaminated soils and possess the ability to accumulate and tolerate extraordinarily high metal concentrations in above-ground tissues (e.g., $>1\%$ Zn, 0.1% Ni or 0.01% Cd in leaf DM) (Baker et al. 2000). These species are classified as either absolute metallophytes (occurring only on metalliferous soils) or pseudometallophytes (present at both metalliferous and non-metalliferous sites).

Tolerant species are favoured by natural selection in contaminated environments due to their ability to survive or else to competitively exclude non-tolerant plants.

Plant roots are the organs that are directly exposed to the heavy metal content of contaminated soils. The availability of metals to plants is strongly dependent on the chemical and physiological conditions in the rhizosphere. The availability of metal ions for plant roots increases in slightly acidic conditions and decreases in alkaline soils (McGrath et al. 1988). Acidic conditions may significantly reduce plant growth by the secretion of different root exudates like organic acids, peptides, amino acids; plant enzymes can increase the pH of the rhizosphere and counteract this effect (Pavlovkin et al. 2009). This mechanism greatly increases the extent of metal ion precipitation and complexation in the vicinity of the roots and thereby helps to reduce the impact of heavy metal toxicity (Reichman 2002). The ability of plants to buffer the rhizosphere is dependent on the type of soil and level of organic matter content, the availability of phosphorus, nitrogen, and iron. All of these factors have significant effects on the accessibility and uptake of zinc, cadmium, and other heavy metal ions (Dickinson et al. 1991; Hirayama and Shinozaki 2010; Broadley et al. 2007). Strategies for modifying an acidic rhizosphere are particularly important for soil rich in Al and Zn. Aluminium is an important growth-limiting factor in acidic soils (Horst et al. 2010). Aluminium binds primarily to cell surface components such as mucilage. In the cell walls of rhizodermal cells, Al interacts mainly with pectins and hemicellulose (Gautam et al. 2016a, b; Schmohl and Horst 2000). In the presence of silicon (Si) and boron (B), Al is mostly bound in the cell wall matrix and thus halted in an extracellular space that limits Al-induced changes. The cell wall, therefore, represents an important physical and physiological barrier against the symplastic entry of metal ions, moreover the properties of the cell wall help to determine the anatomical characteristics of the root in terms of its growth rate (see Sect. 7.6.3). The plasma membranes of the root cells are the first physiological barriers to the entrance of heavy metals into the symplast. The metal ions affect the plasma membrane like induction of lipid peroxidation in the plasma membrane and the loss of highly mobile essential ions, leading to serious ion imbalances in the cytoplasm (Horst et al. 2010). However, a significant part of the metal is bound at the plasma membrane interface, and it has been suggested that this could be one of the factors responsible for metal tolerance. Iwasaki et al. (1990) showed that 60% of Cu in the roots of both *Lolium multiflorum* (Italian ryegrass) and *T. pratense* was bound by the cell wall and plasma membrane. Additionally, in *Minuartia verna ssp. hercynica* growing on heavy metal-contaminated medieval mine dumps, high concentrations of Fe, Cu, Zn, and Pb have been found associated with cell walls and membranes. In comparison, no accumulation of heavy metal was detected in the cytoplasm suggesting a determined use of exclusion by the metal adapted subspecies (Liptáková et al. 2013). When some of the heavy metal (HM) ions overcome biophysical barriers and enter the cytoplasm, it triggers the initiation of several cellular defence mechanisms to nullify and attenuate their toxic effects. The primary strategy includes biosynthesis of diverse cellular biomolecules which play the role of ligands and chelators such as low-molecular-weight protein, nicotianamine, putrescine, spermine, mugineic acids, organic acids, glutathione, phytochelatins, and metallothioneins or cellular exudates such as flavonoid and phenolic compounds, protons, heat shock proteins, and specific amino acids, such as proline and histidine, and hormones such as salicylic acid, jasmonic acid, and ethylene (Hossain and Komatsu 2013; Hall 2002; Broadley et al. 2007; Boyd 2012). With an elevated level of metal ions, the balance of cellular redox systems is disturbed, which leads to the increased induction of reactive oxygen species (ROS) (Boyd 2012). To mitigate the harmful effects of free radicals an antioxidant defence mechanism is activated in cells, composed of enzymatic antioxidants like superoxide dismutase (SOD), catalase, (CAT), ascorbate peroxidase (APX), guaiacol peroxidase (GPX), and glutathione reductase (GR) and non-enzymatic antioxidants such as ascorbate (AsA), glutathione (GSH), carotenoids, alkaloids, tocopherols, proline, and phenolic compounds (flavonoids, tannins, and lignin) that act as the scavengers of free radicals (Pavlovkin et al. 2009; Boyd 2012; Sharma and Agrawal 2005). Some of the biological molecules involved in cellular metal detoxification can be multifunctional and have antiradical, chelating, or antioxidant roles. Exploitation and upregulation of any of these mechanisms and biomolecules may depend on plant species, the level of their metal tolerance (Hall 2002; Cabot et al. 2013), plant growth stage, and metal type. One of the molecules that performs a key function in response to metal stress is glutathione. Glutathione (GSH), a sulphur-containing tripeptide, is considered to be the most important cellular antioxidant involved in cellular defence (Sharma and Agrawal 2005) and functions directly as a free radical scavenger. Glutathione levels in plants are known to change under metal stress due to the role of GSH as an antioxidant, metal-ligand, and also the precursor for the biosynthesis of phytochelatins (PCs) (Barałkiewicz et al. 2009). Phytochelatins are short-chain thiol-rich repetitions of peptides of low-molecular-weight synthesized by the enzyme phytochelatin synthase (PCS) with the general structure of (γ-glutamyl-cysteinyl) -glycine that have a high affinity to bind to HMs (Barałkiewicz et al. 2009). Phytochelatins, as a key player in processes of metal homeostasis and detoxification, have been identified in organisms from yeast and fungi to many different species of animals (Rodrigo et al. 2016; Emamverdian et al. 2015). PCs are reported to have been used as biomarkers for the early detection of HM stress in plants (Emamverdian et al. 2015). In the cytosol, PCs bind HM, and metal-phytochelatin complexes are actively transported to the vacuole, the transport is probably mediated by an Mg ATP-dependent carrier or an ATP-binding cassette (ABC) transporter (Manara 2012). It has been shown that Cd significantly enhances the synthesis of phytochelatins (PCS) in plants. However, Sun et al. (2010) reported that the variation in phytochelatin production in the roots and shoots of two Cd-treated species, viz., *Rorippa globosa* and *R. islandica* might be used as a biomarker for Cd hyperaccumulation, and the synthesis of PCS may be related to an increase in the uptake of Cd ions into the cytoplasm. However, the authors suggest that PC biosynthesis is not the primary mechanism for Cd tolerance. Similarly, the uptake and accumulation of Cd have influenced the biosynthesis of PCs in *Brassica napus,* and in the shoot, the concentration of PC3 and PC4 was higher than the PC2 irrespective of the quantity of Cd uptake (Selvam and Wong 2008). This result suggests that in the detoxification of Cd, higher molecular weight thiol complexes are involved in the shoot. PCs types and chain lengths show variation among plant species as well as HM types. In legumes, it is reported that PCs with longer chains bind more strongly to Pb in comparison to shorter PCs (Sharma and Dietz 2006). Phytochelatins, along with other stress resistance factors can form a synergistic defence in plants under HM stress which, in turn, can strengthen plant's resistance to metal. Chen et al. (2008) demonstrated that the increased enzymatic biosynthesis of PCs coupled with the heightened activity of the antioxidative system in *B. chinensis* L. led to effective detoxification of Cd. But there is no conclusive study to show whether the number of chains can have any impact on the effectiveness of the PC or of the role of PCs in metal tolerance.

Beside PCs in the plant cell metallothioneins (MTs) are also present; they are a group of low molecular mass, cysteine-rich, metal-binding proteins (Sharma and Dietz 2006). It has been suggested that metallothioneins may have a role in trace metal metabolism and cell homeostasis rather than metal tolerance per se.

Different plant parts, species, and metals appear to elicit different responses and possibly more than one response. However, there are certain mechanisms which appear to hold promise as being more widespread than others. There are reports suggesting that plants tolerant to Zn, Fe, or Al exclude organic acids. Plants tolerant to Cd, and possibly Zn, synthesize phytochelatins although it is not clear whether this is a tolerance mechanism or a transport system to sequester metals away in vacuoles. High cellular concentrations of organic acids may have a role in metal tolerance, especially as the complexing agent in vacuoles.

Only scattered studies have been conducted to determine the physiological background of elevated tolerance of pioneer plants naturally inhabiting heavily polluted sites such as mine tailings. Among them selected species of grasses, perennial plants, shrubs, and arboreal plants can be pronominally found, indicating their ability to survive under stressed conditions such as high salinity and extremely elevated concentrations of metal/loids. Along with their accumulation abilities, plants collected at contaminated sites are most often analysed for unspecific responses such as pigment content, lipid peroxidation using the TBARS assay, total antioxidant activity using a DPPH radical, total phenolic compounds, total glutathione, free amino acids, proline accumulation, and the activity of antioxidative enzymes such as SOD, GPX, and CAT. Despite highly adverse growing conditions, for the majority of investigated species, a physiological shift to an oxidative stress state was barely observed. As reported, pioneer species collected at former mining sites or mine tailings did not suffer any significant oxidation of membrane lipids or enhanced activation of antioxidant systems in comparison to plants of the same species derived from ecologically clean areas (Pistelli et al. 2017) or other sites with different pollution levels (Boojar and Goodarzi 2007; da Silva et al. 2018). This indicates the existence of highly effective mechanisms developed in pioneer species enabling intense accumulation of heavy metals in selected organs without the induction of oxidative stress.

Other studies have attempted to determine the potential of native non-pioneer species for revegetation of disturbed areas characterized by multi-metal pollution. A recent study by Drzewiecka et al. (2019) revealed a diversified uptake of elements by the common tree species *Acer platanoides* and *T. cordata* accompanied by a species-specific pattern of physiological reaction to cultivation in mining sludge characterized by high salinity, pH, TOC and highly elevated concentrations of trace elements, including arsenic. Both species were assigned as metal excluders due to their low bioconcentration and translocation abilities for the majority of detected elements. Among several secondary metabolites investigated in photosynthetic tissue, biosynthesis of glutathione and low-molecular-weight organic acids, both showing chelating abilities towards metal ions, was greatly reduced. However, salicylic acid accumulation appeared to serve a critical role in the tolerance mechanisms of *A. platanoides,* determining lower retardation of foliar growth compared to plants cultivated in unpolluted soil than *T. cordata.*

7.5.3 Influence of Arbuscular Mycorrhiza on Plant Condition During Phytoremediation

According to Smith and Smith (2012), in natural environments, a non-mycorrhizal condition should be considered as abnormal for the majority of plant species. Consequently, experiments aimed at evaluating plant resistance to toxic elements (including survival rate, biomass production, and phytoextraction efficiency) or the physiological mechanisms underlying the elevated tolerance of some species should consider the effect of plant–fungi associations. Among mycorrhizal interactions of plants, arbuscular mycorrhiza (AM) with Glomeromycota with the largest genus *Glomus* is the most abundant symbiosis for vascular land plants and epiphytes. In general, Glomeromycota form a close and highly beneficial symbiosis with the roots of 70–90% of land plant species (Smith and Read 2008; Prasad et al. 2017). As reported by Okiobé et al. (2015), root colonization with Arbuscular Mycorrhizal Fungi (AMF) can increase the yield of plants from 50 to 200%, mainly via a facilitated influx of water, P and N to plant roots, as well as other mineral nutrients elevating plant nutritional status. Compared to sterile plants, mycorrhizal associations may lead to an increase of up to 80% of P, 60% of Cu, 25% of N, 25% of Zn, and 10% of K uptake (Soares and Siqueira 2008). In exchange, plants transfer some of their soluble carbohydrates to the fungus mycelium to be utilized as carbon sources in order to maintain the mycorrhizal symbiosis (Bonfante and Genre 2010).

AMF spores and mycelium are extremely resistant to high concentrations of heavy metals. Up to 40% AM fungal colonization of plant roots was reported for plants growing in multi-metal polluted soils despite high levels of Cd and Pb concentrations (1220 and 895 mg kg^{-1}, respectively) (Weissenhorn et al. 1994). As recently assumed by Schneider et al. (2016), arbuscular mycorrhiza is highly beneficial in fungi-assisted phytoremediation by influencing both the availability of metals for plants and the overall plant condition. AMF action is based on heavy metal dilution in plant tissue based on increased plant growth, reduced uptake by precipitation or metal chelation in the rhizosphere, and as a result of metal retention and immobilization in fungal structures, with a consequent reduction of their translocation to shoots (Schneider et al. 2016). Model studies have confirmed the efficient metal-binding capacities of *Glomus* mycelium towards Zn with concentrations exceeding 1200 and 600 mg kg^{-1} for *G. mosseae* and *G. versiforme*, respectively (Chen et al. 2001). Furthermore, symbiotic plant/fungus interactions lead to the phenomenon of induced systemic resistance (ISR). Previous or simultaneous colonization of roots enhances plant ability to fight biotic challenges, i.e. to cope with necrotizing pathogens or parasites (Burketova et al. 2015). Combined systemic acquired resistance (SAR) and ISR lead to an overproduction of plant hormones and signalling compounds, such as salicylic, jasmonic acids, and ethylene, resulting in the induction of pathogenesis-related (PR) proteins, phytoalexins, and intensified cell wall lignification (Choudhary et al. 2008; see Sect. 7.6.2). As a consequence, AMF colonization of plant roots may elicit a significant reduction in the incidence or severity of various diseases on a diversity of plant hosts also employed in phytoremediation processes. Considering the sustainable ecosystem created during revegetation of disturbed lands, management of the rhizosphere microbiome, including fostering of indigenous AMF communities or root inoculation with AMF enhances nutrient uptake, improves plant health, pest resistance, and drought tolerance (Bender et al. 2016; Varma et al. 2017a, b, c). Recent studies have confirmed the critical role of symbiotic fungi in phytoremediation strategies using enhanced species diversity (legume tree species co-cultured with grasses and N-fixing herbs) based on the reduction of nutrient loss, elevated availability of dissolved organics and mineral nutrients, and soil erosion resistance (Yang et al. 2016).

7.6 Alterations in Root Architecture as an Indicator of Plant Ability to Cope with Toxic Trace Elements

Plant roots are in direct contact with many elements, including toxic trace elements (TE) present in the contaminated substrate and are generally the main plant organs which accumulate TE (Baker 1981; Verbelen et al. 2006). The exception to this rule are hyperaccumulating plants (see also Sect. 7.5.2)—species, often endemic to naturally mineralized soils, which accumulate high concentrations of metals and metalloids in their above-ground tissues without developing any toxicity symptoms (Baker 1981; Baker et al. 2000; Suman et al. 2018; Ashraf et al. 2019). Moreover, plant roots, due to their relatively simple and predictable structural organization and developmental zonation, are considered to be an ideal model system to study the various responses of plants to TE (Verbelen et al. 2006). Structural alterations in plant roots are often well visible and relatively easy to estimate. Therefore, these traits could be useful as indicator symptoms for the assessment of the plant ability to cope with stress conditions. They could be beneficial in the selection of examined plant species for reclamation of TE contaminated soils or as bioindicators for the scale of pollution (Schneider et al. 2013; Pita-Barbosa et al. 2015; Krzesłowska et al. 2019).

This chapter includes a short characterization of root architectural alterations in response to TE, such as trace metals (Pb, Cd, Cu, Zn, Al) and metalloids (As). We selected the most common alterations in root architecture at the different levels of root organization that have been previously described for several vascular plant species: morphological, anatomical, and cellular modifications, considered as the symptoms of a plant defence strategy and abnormalities caused by TE which lead to deleterious effects.

We excluded from this chapter alterations of root architecture in hyperaccumulating plants, which are able to accumulate large amounts of TE in their aerial parts in general with no effect on yield as compared to agronomic crops or non-accumulator plants species which belong predominantly to the genera *Alyssum* and *Brassicaceae* (Baker et al. 2000; Prasad and Freitas 2003; Verbruggen et al. 2009; Suman et al. 2018; Ashraf et al. 2019) including such species as *Pteris vittata,* a hyperaccumulator of As (Ma et al. 2001; Danh et al. 2014), *Arabidopsis halleri*, a hyperaccumulator of Cd and Zn (Verbruggen et al. 2013; Schvartzman et al. 2018), *Sedum alfredii,* a hyperaccumulator of Cd (Zhou and Qiu 2005) and several *Noccaea* (*Thlaspi)* species which are hyperaccumulators, e.g. *N. caerulescence* Cd, Pb, Ni, *N. goesingense* Ni and Zn, *N. ochroleucum* Ni and Zn, *N. rotundifolium* Ni, Pb, Zn (Baker et al. 2000; Prasad and Freitas 2003). Moreover, most hyperaccumulators are not considered as suitable plant species for phytoremediation, in particular, phytoextraction, because of their small biomass (Suman et al. 2018; Ashraf et al. 2019).

7.6.1 Morphological Alterations in Root Architecture

Uptake and accumulation of TE by plant roots result in the occurrence of a range of alterations. On the one hand, they include modifications considered as symptoms of plant resistance strategy to TE, on the other hand, the deleterious effects of TE toxicity.

Plants are sessile organisms, and in general, they use two main strategies to cope with TE: exclusion and accumulation (by either sequestration or compartmentalization) (Baker 1981; Viehweger 2014). The alterations in root morphology are predominantly involved in the exclusion strategy, which protects plant roots from TE influx. One of the most common is an increase of root exudate secretion and formation of a mucilage barrier limiting the entry of TE into the root. It has been demonstrated that exclusion is the main defence strategy of plants to cope with Al (Barceló and Poschenrieder 2002; Cai et al. 2013) but it also functions in response to other TE, such as Pb, Cd, Cu (Morel et al. 1986; Seregin and Kozhevnikova 2008; Colzi et al. 2015). Root exudates predominantly contain organic acids (e.g. Barceló and Poschenrieder 2002), and/or polysaccharides such as pectins (Seregin and Kozhevnikova 2008; Yang et al. 2008; Cai et al. 2013)—components able to bind and immobilize TE ions (Barceló and Poschenrieder 2002; Seregin and Kozhevnikova 2008; Yang et al. 2008; Cai et al. 2013). Hence, the mucilage layer which appears on the root surface, in particular in root apex regions, protects the root from an influx of TE ions (Barceló and Poschenrieder 2002; Seregin and Kozhevnikova 2008; Cai et al. 2013; Colzi et al. 2015). The increase of the amount of mucilage, detectable even by light microscopy, has been demonstrated, e.g. in response to Al in *Pachyrhizus ahipa* (Poschenrieder et al. 2008), *Oryza sativa* (Cai et al. 2011), tolerant populations of *Glycine max* (Cai et al. 2013), and *Camelia sinensis* (Li et al. 2017) as well as in response to Cu in tolerant populations of *Silene paradoxa* (Colzi et al. 2015). It is worth emphasizing that binding TE by root exudates within the rhizosphere is considered to be one of the main defence strategies for TE, thanks to which many plant species belonging to so-called excluders or metallophytes (Baker 1981), are able to grow on highly TE polluted substrate, e.g. mine tailings (Baker 1981; Seregin and Kozhevnikova 2008; Colzi et al. 2015).

Other modifications considered as symptoms of a defence strategy visible in the root morphology is an increase in the number of living root border cells (RBC; cells of the root cap that have undergone cellular separation but are still attached to the root via a soluble polysaccharide matrix; Driouich et al. 2007). It was found that in *Vigna unguiculata* 'Red Caloona' exposed to As, RCB were separated from the root tip as layers (Kopittke et al. 2012). It was suggested that RCB possibly contributes to the plant's ability to withstand an excess of TE in two main ways: (1) accumulating high levels of TE and (2) secretion of mucilage where TE can be bound and retained (Cai et al. 2011, 2013; Kopittke et al. 2012). Both processes resulted in the limitation of TE entrance into the root (Cai et al. 2011, 2013). Such a mechanism has been demonstrated in several other plant species coping with TE, e.g. in *S. armeria* growing on Cu polluted mine tailings (Llugany et al. 2003), in *O. sativa* and *Glycine max*, exposed to Al (Cai et al. 2011, 2013) and in *Vigna unguiculata* 'Red Caloona' exposed to As (Kopittke et al. 2012).

Another morphological alteration regarded as a symptom of plant defence against TE is the appearance of Fe plaque on the root surface. This response was predominantly observed in rice exposed to As(V) and As(III) (Farooq et al. 2016). The release of O_2 into the rhizosphere of waterlogged soils (anaerobic conditions) can result in the formation of an Fe-rich plaque (ferrihydrite) surrounding the root system, with the concomitant oxidation of As(III) to As(V), which then adsorbs strongly to the Fe plaque (Farooq et al. 2016; Kopittke et al. 2017). The experiments with rice showed that Fe plaques adsorb both As(III) and As(V), minimizing As uptake by roots and consequently its toxic effects on root anatomy and subsequent As translocation to shoots (Deng et al. 2010).

An interesting morphological alteration of the root also considered as a symptom of plant defence strategy against TE was the increase of lateral root length. This occurred under field conditions where extension of lateral roots into less toxic surface soil was an adaptative growth response which can avoid TE toxicity (Poschenrieder et al. 2008).

To sum up, most alterations in the root morphological architecture are evidently to restrict the amount of TE entry into the root. It is worth noting that both a marked decrease of TE influx into the root, and a beneficial influence for plant adaptation to areas polluted with TE, can result from cooperation with symbiotic organisms, such as symbiotic bacteria, mycorrhizal (see also Sect. 7.5.3), and endophytic fungi (Rajkumar et al. 2012; Cabral et al. 2015; Ma et al. 2016b; Mishra et al. 2017; Domka et al. 2019). For example, mycorrhizal and endophytic fungi can play a role as a barrier, effectively immobilizing TE and reducing their uptake by host plants via binding metal ions to hyphal CWs and sequestration in vacuoles (Rajkumar et al. 2012; Cabral et al. 2015) as well as by the secretion of extracellular metal-chelating molecules, such as glycoprotein glomalin produced by arbuscular mycorrhizal fungi (Cabral et al. 2015; Sharma et al. 2017; Domka et al. 2019) or organic acids, siderophores, exopolysaccharides, and phenolic compounds produced by fungal endophytes and other mycorrhizal fungi (Mishra et al. 2017; Domka et al. 2019). The alterations caused by TE in the relationship of symbiotic organisms with plant roots and subsequently the alterations in plant root architecture, as well as the role of symbiotic organisms in the defence strategy of plants to cope with TE toxicity and their beneficial role in phytoremediation capability of plants is a very broad topic, widely studied and reviewed (Rajkumar et al. 2012; Cabral et al. 2015; Ma et al. 2016b; Mishra et al. 2017) and would require an entirely separate chapter.

In addition to modifications considered as symptoms of resistance strategies, morphological alterations in root architecture may also have harmful effects. One of the most common is the reduction of root length (Čiamporová 2002; Poschenrieder et al. 2008; Lux et al. 2015; Fahr et al. 2013), usually the result of the inhibition of both root elongation because of less turgor, higher rigidity of cell walls (see Sect. 7.6.3) and mitotic activity of meristematic cells (Samardakiewicz et al. 2009; Fahr et al. 2013; Gzyl et al. 2015). It is worth noting that together with a decrease of root length, a decrease of root biomass was often observed. For example, in soybean seedlings, the deleterious effect of As was even more evident in the root biomass than in the total root length for all As treatments compared to control plants, even for the lowest As concentration tested (25 μM; Armendariz et al. 2016). A reduction of root biomass was also detected in four tree species *A. platanoides*, *A. pseudoplatanus*, *T. cordata,* and *Ulmus laevis* growing on mining sludge containing extremely high levels of TE, e.g. As, Cd, Cu, Pb, Zn (Mleczek et al. 2017).

An obvious morphological alteration in response to TE is a change in the root colour. In contrast to the whitish colour of roots in plants not treated with TE (Fig. 7.1A), those of plants exposed to TE become brownish, e.g. roots of *Cajanus cajan* in response to As or even dark brown, lime trees growing on mining sludge (Fig.7.1B; Krzesłowska et al. 2019). Moreover, the shape of the taproot and lateral roots are often irregular in response to most TE such as Al (Čiamporová 2002), As (Pita-Barbosa et al. 2015), Cd, and Zn (Sofo et al. 2017) and to many TE present in mining sludge (Fig. 7.1B; Krzesłowska et al. 2019). Furthermore, in many plant species, lateral roots become markedly shorter and thicker. This trait, together with the reduction of the tap root length, and the increase of its thickness, results in the formation of a stunted root system, observed in many plant species, e.g. in wheat and maize exposed to Al (Čiamporová 2002; Doncheva et al. 2005), several plant species exposed to Pb (Fahr et al. 2013), in *C. cajan* (Pita-Barbosa et al. 2015) and soybean (Armendariz et al. 2016) exposed to As, as well as and in Arabidopsis exposed to Cd,

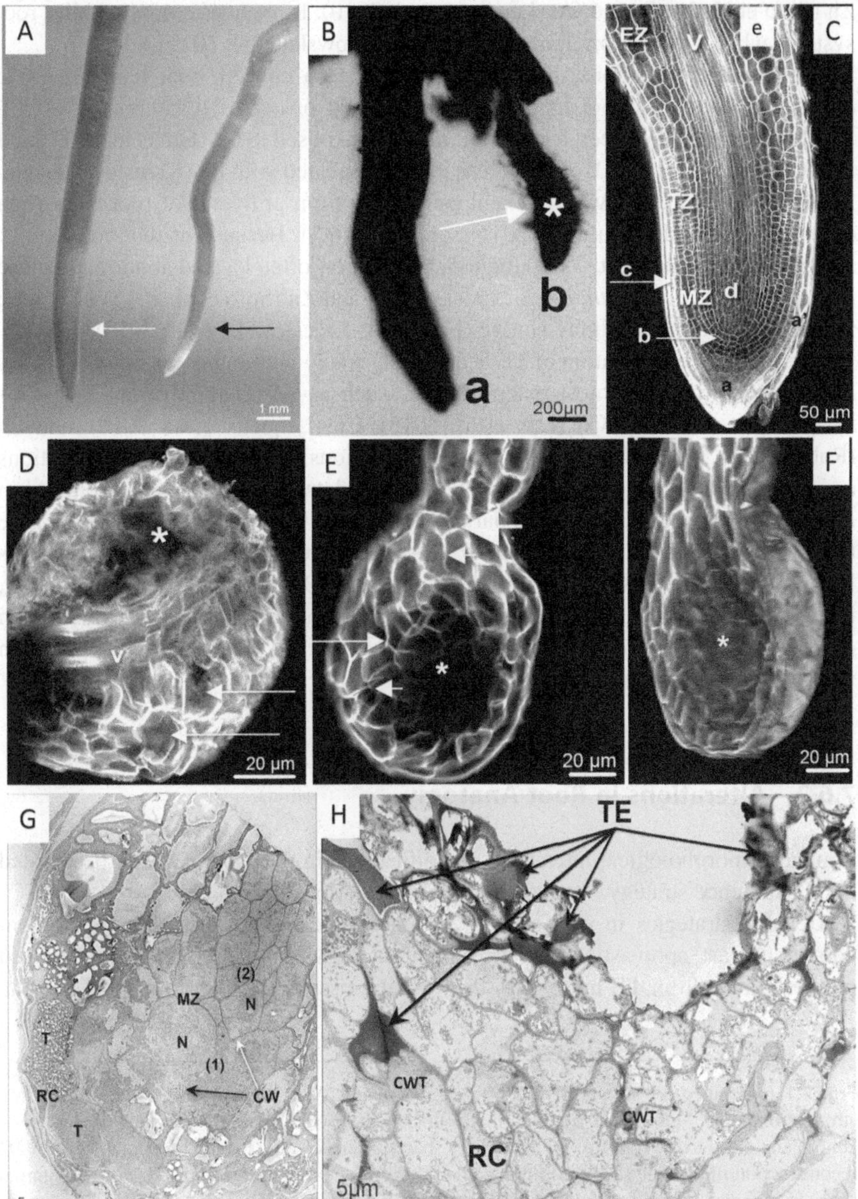

Fig. 7.1 *Tilia cordata* (**A**) Typical morphology of roots, in particular root apex, in control. Light grey—young root (black arrow), older light brown (white arrow). (**B**) Plants exposed to mining sludge extremely contaminated with, e.g. As, Cd, Cu, Cd, and Zn. Two root apices (**a** and **b**), characterized by irregular thickness, almost black probably because of the deposited TE. Root apex '**b**' markedly swollen (asterisk) at some distance from the tip. The root hair zone is located close to the root tip (white arrow). (**C**) Control—typical root architecture, all root apex zones and tissues are easy to distinguish (clearing technique). (**D–F**) The architecture of the root apex in the tree exposed to mining sludge—markedly but unevenly swollen. (**D–E**) Optical sections of cleared roots demonstrating (1) various thickness and shape, (2) the diversity of the arrangement of cells,

Zn (Sofo et al. 2017), and Cu (Lequeux et al. 2010). Lime trees exhibit similar root systems when growing on mining sludge (Krzesłowska et al. 2019).

Well visible alterations in root morphology also concern root hairs. In toxic concentrations of Al, root hair growth is reduced or ceases (Čiamporová 2002). Similar reactions have been found in *A. thaliana* exposed to Cd (Fan et al. 2011) and Pb (Krzesłowska et al. 2016) as well as in poplar treated with Pb (Krzesłowska et al. 2016). In addition, cell wall thickenings (CWT) form at the tip of root hairs (Fan et al. 2011; Krzesłowska et al. 2016; see Sect. 7.6.3). Because of the reduction of root apex zones (see Sect. 7.6.2) the root hair zone is often located abnormally close to the root tip, e.g. in *A. thaliana* exposed to Cd and Zn (Sofo et al. 2017) or in lime tree roots growing on mining sludge (Fig. 7.1B; Krzesłowska et al. 2019).

The highest concentration of TE is detected usually within the root apex. This has been demonstrated for most examined TE, such as Al (Garzón et al. 2011), As (Kopittke et al. 2012), Cd (Lux et al. 2015), Cu (Lequeux et al. 2010), and Pb (Rabęda et al. 2015). This could explain why serious morphological malformations are observed in this region, such as swollen tips detected, e.g. in response to Al in maize (Doncheva et al. 2005) and barley (Zelinová et al. 2011) as well as in the lime trees growing on mining sludge (Fig. 7.1B–G; Krzesłowska et al. 2019), curved tips in response, e.g. to Cd (Lux et al. 2015) and As (Pita-Barbosa et al. 2015). Moreover, in the root apex, a reduction or complete absence of the root cap has often been observed, e.g. in response to As (Pita-Barbosa et al. 2015) and a mixture of TE in mining sludge (Krzesłowska et al. 2019).

7.6.2 Alterations in Root Anatomy

Similar to morphological alterations, modifications in root anatomy can be divided into (1) defence strategy symptoms and (2) deleterious effects of TE.

Defence strategies in root anatomy mainly concern the formation of efficient barriers for fast, apoplastic radial movement of TE within the root. This mechanism protects plants from TE influx into the vascular tissues and their transport into the

Fig. 7.1 (continued) (3) diversity of cell sizes and shapes. Many cells in the external layers are strongly enlarged, surrounded by abnormally thick CWs (white arrows). Vascular tissues (v)—abnormally close to the root tip (**D**) and empty space (asterisk) (**E**). (**F**) The micrographs from 3D reconstruction illustrating the uniformly swollen shape of the root apex and the size of the internal space lacking any tissues and cells. (**G, H**) Different ultrastructure of the root apices, (**G**) root apex showing strong reduction of the root cap (RC) and the meristematic zone (MZ). Cells in these two regions vary in CW thickness (1) thin CWs and (2) thickened CWs. (H) Root apex where only several layers of root cap cells are preserved and the interior is empty. Numerous CW thickenings (CWT) and TE deposits, different in size and shape (arrows) are visible). Abbreviations: meristematic zone (MZ), transition zone (TZ), elongation zone (EZ), **a**—root cap, **a′**—lateral root cap, **b**—meristem, **c**—protoderm, **d**—ground meristem, **e**—cortex, and **v**—vascular tissues. (Krzesłowska et al. 2019—with permission)

stem and leaves (Seregin and Kozhevnikova 2008; Lux et al. 2011). Therefore, one of the most important alterations in plant root anatomy is the acceleration of endodermis maturation. Such a reaction was observed in several plant species exposed mainly to Cd and Pb (Seregin and Kozhevnikova 2008; Lux et al. 2011). Interestingly, acceleration of maturation has also been recently demonstrated for the exodermis (hypodermis), in maize exposed to Cd (Liška et al. 2016). Maturation of exo—and endodermis is associated with the appearance of suberin and lignin within their CWs (Esau 1977). The occurrence of suberin and lignin makes the CWs impermeable to aqueous solutions and consequently for the transport of both essential elements (Esau 1977) and TE (Seregin and Kozhevnikova 2008; Krzesłowska 2011; Lux et al. 2011). Therefore, mature exodermis and endodermis function as real barriers limiting radial, especially apoplastic, transport of TE within the root (Lux et al. 2011; Kopittke et al. 2012; Liška et al. 2016). However, it is worth noting that the endodermis layer is still permeable for water solution and many elements via symplastic transport (Esau 1977; Seregin and Kozhevnikova 2008).

Importantly, accelerated maturation, in response to TE, results in the formation of suberized endodermis closer to the root apex, as demonstrated in several plant species in response to Cd, e.g. *A. thaliana*, *S. dioica*, *Karwinskia humboldtiana*, *Cucurbita pepo* (Lux et al. 2011). A similar reaction also concerns the exodermis (Liška et al. 2016). Hence, accelerated maturation of both cell layers, exo- and endodermis, in plant roots exposed to TE leads to the appearance of both barrier tissues closer to the root apex than in roots of plants not exposed to TE. This extends the area of the root where the radial transport of TE is markedly limited, including normally unprotected regions close to the root apex.

The barrier role of the endodermis was commonly demonstrated for Pb transport in many plant species, e.g. in *Raphanus sativus* (Lane and Martin 1977), *Allium cepa* (Wierzbicka 1987), poplar (Książek and Woźny 1990), the aquatic plant *Lemna minor* (Kocjan et al. 1996). Moreover, such a function was clearly demonstrated for As (V) and As (III) transport in *Vigna unguiculata* 'Red Caloona' by using synchrotron-based X-ray fluorescence techniques (Kopittke et al. 2012). Hence, the limitation of radial apoplastic transport is a widespread defence strategy of plants against TE.

Cells with lignified cell walls play similar roles to that of the exo- and endodermis. It was observed that in maize exposed to Cd, the lignification was accelerated and concerned protoxylem cells and xylem parenchyma cells (Lux et al. 2015). An increase of lignification was also observed in tobacco exposed to Cd (Siemianowski et al. 2014) as well as in *Arabidopsis* (Lequeux et al. 2010) and tolerant populations of the metallophyte *S. paradoxa* (Colzi et al. 2015) treated with Cu. In addition, it was demonstrated that xylem elements occurred closer to the root apex, similarly to the accelerated maturation of the endodermis and exodermis described above. For instance, in pine trees exposed to Cd, premature xylogenesis was observed as well as in roots of barley (Lux et al. 2015) and soybean (Gzyl et al. 2015) exposed to the same metal. The occurrence of lignified vascular tissues abnormally close to the root tip was also detected in *T. cordata* growing on mining sludge (Fig. 7.1D; Krzesłowska et al. 2019). As mentioned above, lignified CWs—similarly to CWs containing suberin—are not permeable to water solutions transporting TE and thus form a barrier for radial toxic TE movement and their entry into the vascular tissues (Lequeux et al. 2010; Lux et al. 2015; Colzi et al. 2015).

Taking all these facts into consideration it can be concluded that alterations in root anatomical architecture involving higher suberification and lignification of root CWs as well as their occurrence closer to the root apex in plants exposed to TE, are the symptoms of plant defence strategies against TE (Seregin and Kozhevnikova 2008; Lux et al. 2015).

However, besides the symptoms of defence strategy, many other alterations, signs of the detrimental effects of TE toxicity, have been demonstrated in root anatomy architecture. Some of them relate to alterations in root morphology. The reduction of root length and the occurrence of swollen or curved tips were probably the result of the marked reduction of root apex zones. This predominantly concerned the reduction of the elongation zone (EZ), as observed in plants exposed to Al (Čiamporová 2002) and in *C. cajan* treated with As (Pita-Barbosa et al. 2015), but also the reduction of the meristematic zone (MZ) and the transition zone (TZ) (Fig. 7.1D, G), as in lime trees growing on mining sludge extremely contaminated with As, Cd, Cu, Pb (Krzesłowska et al. 2019). Together with TZ and MZ reduction, an irregular arrangement of the cells building the root apex tissues and a large diversity of disorders in their size and shape, e.g. the occurrence of many abnormally large cells, were well visible (Fig. 7.1D–H). The alterations in lime trees were analysed using confocal laser scanning microscopy combined with a clearing technique (Krzesłowska et al. 2019). Since plant tissues are not transparent, application of the clearing technique allowed the imaging of the whole root apex by confocal fluorescence microscopy and its subsequent 3D reconstruction (Timmers 2016). Thanks to this technique it was also possible to demonstrate one of the most dramatic disorders in root anatomy which occurred in lime trees growing on mining sludge, i.e. a lack of internal tissues in the root apex (Fig. 7.1E, F; Krzesłowska et al. 2019). Interestingly the external cell layers of the root apex, mainly root cap (if it was present) and ground meristem in such root apices were preserved (Fig. 7.1E, F, H; Krzesłowska et al. 2019).

In many plant species, the main abnormalities caused by TE were observed predominantly in the rhizodermis and cortex. For example, disintegration of cortical cells, a reduction of the cortex area, broken, collapsed cells, and larger intercellular spaces in this tissue occurred in many plant species, e.g. in response to Al (Čiamporová 2002), in willow and poplar exposed to Cd (Lux et al. 2015) as well as in response to As in *C. cajan* (Pita-Barbosa et al. 2015), several species of *Brassicaceae* (de Freitas-Silva et al. 2016) and in soybean (Armendariz et al. 2016).

Toxic effects of TE on root anatomy architecture were also visible in vascular tissues. For example, the secondary xylem vessel elements were reduced in diameter in plants exposed to As, such as *C. cajan* (Pita-Barbosa et al. 2015) and several species of *Brassicaceae* (de Freitas-Silva et al. 2016).

Serious alterations in root anatomy also concerned the lateral root zone. For example, in *C. cajan* treated with As the division orientation of phellogen and cambium cells and disintegration of the parenchyma cells adjacent to lateral roots were observed (Pita-Barbosa et al. 2015). Moreover, e.g. in response to both As and Cd, the primordia of lateral roots often did not develop and were retained within the cortex (Pita-Barbosa et al. 2015; Fattorini et al. 2017). In experiments examining maize root response to Cd applied only from one side of the root, it was demonstrated that primordia of the lateral roots developed only on the side which was not exposed to Cd. Thus Cd inhibited lateral root development (Lux et al. 2015).

7.6.3 Alterations in Root Architecture at the Cellular Level

Since the beginning of the research into plant cell reactions to TE, many alterations in cell ultrastructure have been described. It is a very broad topic widely studied and reviewed (Čiamporová 2002; Horst et al. 2010; Krzesłowska 2011; Lux et al. 2015; Fahr et al. 2013; Parrotta et al. 2015; Horiunova et al. 2016). Therefore, in this chapter, we have decided to focus mainly on the alterations in root cell architecture which can be considered as symptoms of defence strategies to TE such as CW remodelling, an increase of vacuolization and activity of vesicular transport. The most common detrimental effects of TE, modifying root architecture at the cellular level, are briefly described.

As alterations in morphology and anatomy were involved in exclusion strategies—many modifications in root structure at the cellular level are predominantly involved in compartmentalization strategies. Two cell compartments play a key role in compartmentalization: the cell wall and vacuole. These two compartments sequester TE ions protecting more sensitive sites in the protoplast from their toxicity (Krzesłowska 2011; Ovečka and Takáč 2014). The occurrence of TE, such as Pb, Cu, Cd, Al, within the CW has been reported since the earliest studies of plant cell reactions to these elements (e.g. Malone et al. 1974; Woźny et al. 1982; Wierzbicka 1998; Neumann and zur Nieden 2001; Sousa et al. 2008; Małecka et al. 2008; Krzesłowska 2011) and also recently (e.g. Colzi et al. 2012, 2015; Parrotta et al. 2015; Krzesłowska et al. 2016; Li et al. 2017; Krzesłowska et al. 2019).

Plant CWs accumulate large amounts of TE because this compartment is abundant in components able to bind divalent and trivalent metal cations, such as pectins, cellulose, hemicellulose, proteins (Krzesłowska 2011), and organic acids (Kopittke et al. 2017). However, pectins in particular play a crucial role in the binding and immobilization of TE within CW, especially low-methylesterified pectins (up to 40%), which are cross-linked by TE ions (Krzesłowska 2011; Inoue et al. 2013; Rabęda et al. 2015). Interestingly, alterations in root structure at the cellular level concerning plant CW are involved mainly in the increase of CW capacity for TE sequestration (Krzesłowska 2011; Le Gall et al. 2015). Because the CW capacity for TE binding and sequestration depends mainly on the amount of low-methylesterified pectins (Krzesłowska 2011; Inoue et al. 2013; Rabęda et al. 2015)—as expected—the level of this pectin fraction often increased in response to many TE (Krzesłowska 2011). The increase of low-methylesterified pectin levels has been clearly demonstrated in recent years, e.g. in wheat (Sun et al. 2016) and *Camellia sinensis*

(Li et al. 2017) exposed to Al, in sensitive populations of *S. paradoxa* in response to Cu (Colzi et al. 2012) as well as in Douglas fir trees treated with Cd (Astier et al. 2014).

Augmentation of CW capacity for TE sequestration also leads to the increase of CW thickness—one of the most widespread alterations observed at the cellular level in many plant species, e.g. in response to Al (Čiamporová 2002; Horst et al. 2010), As (Schneider et al. 2013; Armendariz et al. 2016), Cu (Colzi et al. 2015) or in plants growing on mining sludge containing a mixture of TE (Probst et al. 2009; Krzesłowska et al. 2019).

Furthermore, besides the general increase of the CW thickness local CW thickenings were also observed in roots of plants exposed to TE, in particular, Pb (Krzesłowska 2011; Le Gall et al. 2015). It is worth emphasizing that generally CW thickenings were characterized by high levels of low-methylesterified pectins and the occurrence of callose, which physically limit TE movement. It was detected that CW thickenings accumulated large amounts of TE. Hence, the formation of local CW thickenings also increased low-methylesterified pectin levels and the apoplast capacity for TE accumulation in plant roots (Krzesłowska 2011). Recent results have demonstrated, moreover, that formation of local CW thickenings is a widespread defence strategy of plants to cope with TE. Such alterations of CWs occurred in the root apices of poplar and Arabidopsis in response to Pb (Krzesłowska et al. 2016) and in lime trees exposed to a mixture of TE (e.g. As, Pb, Cd, Cu, and Zn) present in mining sludge (Fig. 7.1H; Krzesłowska et al. 2019). Furthermore, CW thickenings abundant in low-methylesterified pectins, accumulating high levels of Pb were also found in the apical zone of tip growing root hairs in Arabidopsis and poplar (Krzesłowska et al. 2016). Similar reactions were demonstrated for the root hair tips of Arabidopsis exposed to Cd (Fan et al. 2011). Hence, alterations in root structure at the cellular level concerning the formation of local CW thickenings, detected in diverse plant species and cell types differing in the type of growth, anisotropic (diffuse) and tip growing cells, demonstrated that this alteration in root architecture can be considered as a really widespread defence strategy of plants for coping with TE.

On the other hand, it should be remembered that binding TE within CWs, especially cross-linking by low-methylesterified pectins, and the enlargement of CW capacity for TE sequestration simultaneously increases the rigidity of this structure and therefore inhibits root elongation (Krzesłowska 2011). Thus, the increase of low-methylesterified pectin levels is a symptom of the defence strategy characteristic for plants which are not constitutively adapted to elevated amounts of TE in the substrate, e.g., in response to Al (Eticha et al. 2005; Amenós et al. 2009; Tolrà et al. 2009) or Cu (Colzi et al. 2012), Cd (Meyer et al. 2015) because in metallophytes, such as tolerant populations of *S. paradoxa,* rather a decrease of low-methylesterified pectins level was observed in their CWs (Colzi et al. 2012).

As mentioned above (see Sect. 7.6.2), remodelling of root CWs also includes the appearance of lignin and suberin that results in impermeable barrier formation for water solutions transporting TE. Such CWs also restrict the entrance of TE into the protoplast. The barrier role of CW for TE penetration is also underlined by callose deposition, known as a component impermeable to TE which protects the PM and protoplast from TM penetration or at least limits the amount of TE that is able to enter the interior of the cell. Callose is quickly synthetized in response to TE, as demonstrated mainly in response to Al and Pb (Krzesłowska 2011). However, the callose barrier is often not sufficient, and TE can penetrate the protoplasts, was demonstrated, e.g. for *Lemna minor* exposed to Pb (Samardakiewicz et al. 2012).

The alterations in root architecture at the cellular level also involve the increase of cell vacuolization, e.g. in *Arabidopsis* (Fan et al. 2011; Lux et al. 2015) and soybean (Gzyl et al. 2015) exposed to Cd, in *Arabidopsis* exposed to Pb (Fig. 7.2B; Krzesłowska and Neumann, unpublished data) as well as in *A. sativum* treated with Pb (Jiang and Liu 2010) and Cu (Colzi et al. 2015). Similar to the increase of CW capacity for TE sequestration, higher vacuolization of cells, predominantly meristematic cells, in plant roots exposed to TE, can also be considered as a defence strategy against TE. Vacuoles, besides CW, are the crucial plant cell compartment for TE sequestration (Verbruggen et al. 2009; Lux et al. 2015). High accumulation of TE in vacuoles was found for most TE. However, some TE are preferentially accumulated in these organelles, e.g. As(III) and Cd (Verbruggen et al. 2009; Farooq et al. 2016). As(III) and Cd ions are bound in the cytosol with glutathione or phytochelatins and these complexes are translocated via the tonoplast into the vacuole (Verbruggen et al. 2009; Farooq et al. 2016). However, the vacuole is also an important plant cell compartment for the accumulation of other TE, such as Pb (Samardakiewicz and Woźny 2000; Jiang and Liu 2010; Glińska and Gapińska 2013), Al (Poschenrieder et al. 2008), and Cu (Lequeux et al. 2010).

Besides high protoplast vacuolization, the alterations of root architecture at the cellular level exposed to TE include a noticeably higher number of vesicles indicating more intensive vesicular transport. For example, in *A. thaliana* exposed to Pb evidently more vesicles occurred in root apex cells (Fig. 7.2B) in comparison to the control (Fig. 7.2A; Krzesłowska and Neumann, unpublished data). Interestingly, the cargo of many vesicles were large Pb deposits, evidenced by X-ray microanalysis (Fig. 7.2C–C″; Krzesłowska and Neumann unpublished data). Pb deposits were located in the vesicle lumen and/or within membranes surrounding the vesicles (Fig. 7.2C; Krzesłowska and Neumann, unpublished data). Interestingly, many vesicles carrying Pb deposits were located in the vicinity of CWs accumulating Pb, in particular near CW junctions (Fig. 7.2C; Krzesłowska and Neumann unpublished data)—the regions of CWs where the highest Pb accumulation was detected (Rabęda et al. 2015). Furthermore, in *A. thaliana* and poplar roots, many vesicles containing Pb deposits were also visible in the vicinity of CW thickenings (Krzesłowska, unpublished data). It is generally known that TE, in particular, Pb, have been commonly detected within structures of the endomembrane system such as plasma membrane invaginations, vesicles, Golgi apparatus, the trans-Golgi network (TGN), the endoplasmic reticulum (ER) and vacuoles (Krzesłowska 2011). Hence, TE, already present within the endomembrane system, can be easily removed from the protoplast by the secretion pathway and sequestered in the CW and their thickenings (e.g. Malone et al. 1974; Woźny et al. 1982; Wierzbicka et al. 2007; Meyers et al. 2009; Krzesłowska et al. 2010). Therefore, a higher number of

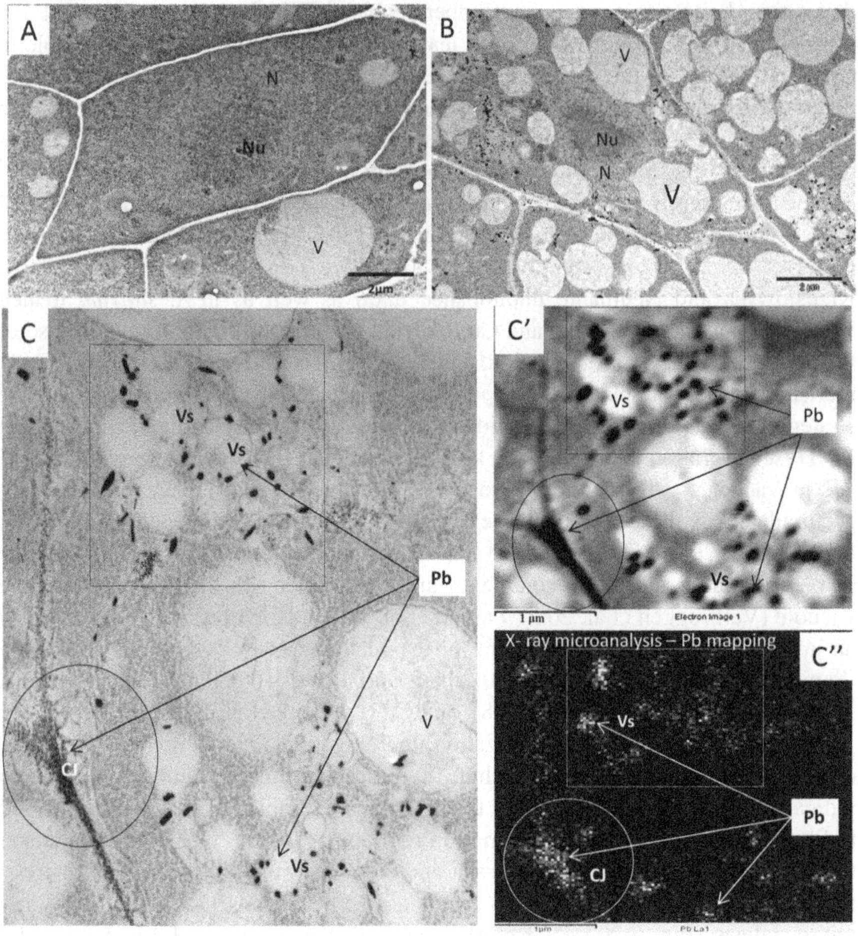

Fig. 7.2 Ultrastructure of *Arabidopsis thaliana* (L.) root apex (micrographs comes from JEM 1400 (JEOL Co., Japan). Pb identity and distribution mapping was confirmed by EDS X-ray microanalysis using transmission electron microscope JEM 1400 (JEOL Co., Japan)JEM1400 JEOL Co., Japan, equipped with a—Energy Dispersive X-ray Spectroscopy (EDS, INCA Energy TEM, Oxford Instruments, Great Britain). (**A**) Control cells surrounded by thin CWs. Nucleus (N) with nucleolus (Nu) located in the centre of the cell—typical for meristematic cells. (**B**–**C**″) Root apex cells of plants exposed to Pb (1 mM; 4 h). (**B**) Meristematic cells containing high number of relatively small vacuoles (V). Nucleus (N) with nucleolus (Nu) located in the centre of the cell typical for meristematic cells. (**C**) Cell showing high numbers of transport vesicles (Vs). Numerous electron dense Pb deposits (black arrows) located both on the membranes surrounding the Vs as well as in the Vs lumen. Vesicles located in the vicinity of CWs and CW junctions (CJ) both contain many Pb deposits. (**C**′) Detail of the boxed area in (**C**) used for the determination of Pb distribution by EDS—X-ray microanalysis. (**C**″) Mapping of Pb distribution examined by EDS X-ray microanalysis (eclipse and rectangular on the micrographs **C**, **C**′, and **C**″ include the same regions of interest; **C**′ and **C**″ comes from Energy Dispersive X-ray Spectroscopy (INCA Energy TEM, Oxford Instruments, Great Britain—therefore squeezed in comparison to **C**)

vesicles transporting TE can be considered as a symptom of the defence strategy involved in TE removal from the protoplast and their sequestration in CW and CW thickenings.

Among the harmful effects of TE on root cell architecture the most serious ones concern alterations in nucleus and nucleolus ultrastructure, e.g. Pb caused an increase of chromatin condensation in *Lemna minor* root apices (Samardakiewicz and Woźny 2005) Cr, Cd, and Pb caused formation of binucleate cells, micronuclei, 'budding' nuclei and nucleoli partly outside nuclei in *A. cepa* roots (Glińska et al. 2007). Moreover, TE, such as Pb, Cd, commonly caused inhibition of mitotic activity including a marked decrease in the number of dividing cells, e.g. in *Lemna minor* (Samardakiewicz and Woźny 2005), *Pisum sativum* (Fusconi et al. 2006), *A. cepa* (Wierzbicka 1988; Glińska et al. 2007) and even an absence of dividing cells as in the root apex of *T. cordata* exposed to mining sludge (Krzesłowska—unpublished data). Moreover, many alterations in the mitosis process were detected—including prolongation of prophase and metaphases, reduction of metaphase and anaphase and disorders of chromosomes, such as: c-metaphases, sticky and lagging chromosomes, chromosome bridges, induced, e.g. in the presence of Pb and Cd (Wierzbicka 1988; Samardakiewicz and Woźny 2005; Fusconi et al. 2006; Glińska et al. 2007; Samardakiewicz et al. 2009; Jiang et al. 2014; Gzyl et al. 2015).

One of the main targets of TE is actin and tubulin cytoskeleton where severe alterations in the arrangement have often been observed (Fusconi et al. 2007; Amenós et al. 2009; Samardakiewicz et al. 2009; Liu et al. 2009; Gzyl et al. 2015; Horiunova et al. 2016). The most harmful effects of TE on the cytoskeleton were observed in the root transition zone, concerning both microfilaments (Amenós et al. 2009) and microtubules (Samardakiewicz et al. 2009). For example, Al caused an assembly of dense but disorganized actin filaments at the cross walls and depolymerization of F-actin just beneath the plasma membrane in a sensitive variety of *Zea mays* (Amenós et al. 2009). It is worth emphasizing that disorders of F-actin in this zone, where meristematic cells exit the division phase and prepare for filamentous actin (F-actin)-dependent rapid cell elongation (Verbelen et al. 2006), besides increasing cell wall stiffness, resulting from binding TE ions mainly to low-methylesterified pectins described above, could be one of the most important reasons for the inhibition of root elongation caused by TE (Amenós et al. 2009; Horiunova et al. 2016).

Trace elements also affect microtubules, both cortical microtubules and microtubules involved in nucleus and cell division. It was demonstrated that TE could alter the 3-dimentional (3D) orientation of cortical microtubules and their dynamic instability (alteration of polymerization and depolymerization process). Moreover, it was demonstrated for soybean roots that Cd affected also the microtubule of the preprophase band and phragmoplast, e.g. disorders of microtubule array and their depolymerization (Gzyl et al. 2015; Horiunova et al. 2016) resulted in the formation of an incomplete cell plate and subsequently incomplete cell walls (Samardakiewicz et al. 2009; Krzesłowska et al. 2019). During mitosis TE, in particular, Pb, caused disorders of the mitotic spindle, often similar to colchicine

(Samardakiewicz et al. 2009). It is likely that most of the disorders in nucleus division described above are the result of alterations in the mitotic spindle caused by TE (Liu et al. 2009; Gzyl et al. 2015; Horiunova et al. 2016).

Root plastids seem to be a primary target of TE excess. Unfortunately, there are almost no data about the effect of TE deficiency or excess on nongreen plastids. The detailed review of Barceló and Poschenrieder (2006) came to the conclusion that except for Cd (decreased starch content and reduced internal membrane system in root plastids), no visible ultrastructural damage is observed in the organelles of the roots, but the metals rather disturb the polar zonation of the organelles within the cells. Changes in the amyloplasts and their arrangement in root columella cells may directly influence root gravitropism and growth direction and seem to be associated with, for example, Al stress-induced root growth defects.

Moreover, ultrastructural and morphological damage after Pb treatment was observed in the root meristematic cells of *A. sativum* during a long exposure (48–72 h), revealing mitochondrial swelling and loss of cristae. Plasmolysis occurred in some cells (Jiang and Liu 2010).

In general, extremely high concentrations of TE and long-term plant exposure could lead to complete damage of root cell protoplasts (e.g. Fig. 7.1H; Jiang and Liu 2010; Schneider et al. 2013; Armendariz et al. 2016; Krzesłowska et al. 2019) (Table 7.2).

7.7 Conclusions

The effectiveness of phytoextraction techniques in the remediation of mining wastes is mainly influenced by their physical and chemical properties. A prerequisite for the application of the phytoextraction process is at least low solubility of trace element ions, which determines their bioavailability. The solubility of elements is mainly influenced by pH (high in acidic conditions, while the elements are immobilisedin alkaline conditions). Mining wastes are an extreme environment for the development of most plants due to poor physical conditions, e.g. insufficient/excessive humidity, high salinity, extreme pH values, low content of available nutrient forms, low organic matter content, low biological activity, etc. Therefore, prior to the application of phytoremediation techniques, additional enrichment of mining wastes should be considered, e.g. with exogenous sources of organic matter. Otherwise, the phytoextraction process may not yield satisfactory results in such an extremely difficult environment for the life of most plants.

Microorganisms present in the ecosystem, inoculation of plants with microorganisms resistant to TE as well as the introduction of such microorganisms to the ecosystems enriched with TE can increase plant phytoremediation potential. Symbiotic microorganisms together with plant defence strategies such as the increase of enzymatic activity, e.g. involved in the antioxidative system, accompanied by TE compartmentalization by a cell wall and vacuole play an important role in plant resistance to TE stress.

Table 7.2 Alterations in root architecture—at the morphological, anatomical, and cellular level of organization

Morphological alterations	Anatomical alterations	Cellular alterations
Modifications considered as defence strategy symptoms		
Higher number of border cells accumulating TE	Suberized and lignified endodermis located close to the root tip	Increase of CW thickness
Increased amount of mucilage immobilizing TE on root apex surface	Increase and/or acceleration of root cell lignification including cortex, endodermis, xylem	Formation of local CW thickenings abundant in low-methylesterified pectins binding TE
Increase of Fe plaque thickness absorbing As		Increase of vesicular transport activity
More numerous and longer lateral roots		Higher vacuolization of protoplast
Disorders by TE		
Reduction of root length and root biomass	Reduction of root apex zones	Malformation of nucleus structure and shape
Brownish or even dark brown colour of the root apex and/or the whole root	Lack of root cap	Inhibition of mitotic activity, mitotic process disorders and alterations of chromosomes (chromosome bridges, c-metaphases, sticky chromosomes)
Root apex, swollen, curved	Disorders in rhizodermis and cortex tissue arrangement, occurrence of abnormal size and shape of cells, occurrence of cells with destroyed protoplasts	Alterations in microtubule and actin cytoskeleton arrangement
Root hairs developed closer to the root tip	Primordia of lateral roots retained in the cortex	Depolymerization of microtubules and F-actin
Increase in the length of the lateral root zone		Disturbance of the polar zonation of amyloplasts, within the cells, swelling of mitochondria and reduction of mitochondria cristae
Short lateral roots or lack of them		Plasmolysis and destruction or lack of protoplasts

The nature and the scale of alterations in plant root architecture, the organ which is in direct contact with TE in the substrate, could be considered/ used as the markers/indicators of plant phytoremediation ability in certain environment. However, we have to accept that long-time exposure of plants to extremely high concentration levels of TE, exceeding a certain dose of the stress factor—results in severe damages of root structure, their functions and the consequence death of plants.

References

Abreu MM, Santos ES, Magalhaes MCF, Fernandes E (2012) Trace elements tolerance, accumulation and translocation in *Cistus populifolius*, *Cistus salviifolius* and their hybrid growing in polymetallic contaminated mine areas. J Geochem Explot 123:52–60. https://doi.org/10.1016/j.gexplo.2012.05.001

Acosta JA, Jansen B, Kalbitz K, Faz A, Martinez-Martinez S (2011) Salinity increases mobility of heavy metals in soils. Chemosphere 85:1318–1324. https://doi.org/10.1016/j.chemosphere.2011.07.046

Aghaei E, Diaz Alorro R, Encila AN, Yoo K (2017) Magnetic adsorbents for the recovery of precious metals from leach solutions and wastewater. Metals 7:529. https://doi.org/10.3390/met7120529

Alagić SČ, Šerbula SS, Tošić SB, Pavlović AN, Petrović JV (2013) Bioaccumulation of arsenic and cadmium in birch and lime from the Bor region. Arch Environ Contam Toxicol 65:671–682. https://doi.org/10.1007/s00244-013-9948-7

Almeida CA, Oliveira AF, Pacheco AA, Lopes RP, Neves AA, Lopes Ribeiro de Queiroz ME (2018) Characterization and evaluation of sorption potential of the iron mine waste after Samarco dam disaster in Doce River basin—Brazil. Chemosphere 209:411–420. https://doi.org/10.1016/j.chemosphere.2018.06.071

Álvarez-Ayuso E, Otones V, Murciego A, García-Sánchez A, Santa Regina I (2012) Antimony, arsenic and lead distribution in soils and plants of an agricultural area impacted by former mining activities. Sci Total Environ 439:35–43. https://doi.org/10.1016/j.scitotenv.2012.09.023

Álvarez-Valero AM, Pérez-López R, Matos J et al (2008) Potential environmental impact at São Domingos mining district (Iberian Pyrite Belt, SW Iberian Peninsula): evidence from a chemical and mineralogical characterization. Environ Geol 55:1797. https://doi.org/10.1007/s00254-007-1131-x

Amenós M, Corrales I, Poschenrieder C, Illeš P, Baluška F, Barceló J (2009) Different effects of aluminum on the actin cytoskeleton and Brefeldin A-sensitive vesicle recycling in root apex cells of two maize varieties differing in root elongation rate and aluminum tolerance. Plant Cell Physiol 50:528–540. https://doi.org/10.1093/pcp/pcp013

Anawar HM, Freitas MC, Canha N, Santa Regina I (2011) Arsenic, antimony, and other trace element contamination in a mine tailings affected area and uptake by tolerant plant species. Environ Geochem Health 33:353–362. https://doi.org/10.1007/s10653-011-9378-2

Ang LH, Tang LK, Ho WM, Hui TF, Theseira GW (2010) Phytoremediation of Cd and Pb by four tropical timber species grown on an Ex-tin mine in peninsular Malaysia. Environ Ecol Eng 4: 2010. https://doi.org/10.5281/zenodo.1061521

Armendariz AL, Talano MA, Travaglia C, Reinoso H, Wevar Oller AL, Agostini E (2016) Arsenic toxicity in soybean seedlings and their attenuation mechanisms. Plant Physiol Biochem 98:119–127. https://doi.org/10.1016/j.plaphy.2015.11.021

Ashraf S, Ali Q, Zahir ZA, Ashraf S, Asghar HN (2019) Phytoremediation: environmentally sustainable way for reclamation of heavy metal polluted soils. Ecotoxicol Environ Saf 174: 714–727. https://doi.org/10.1016/j.ecoenv.2019.02.068

Astier C, Gloaguen V, Faugeron C (2014) Phytoremediation of cadmium-contaminated soils by young Douglas fir trees: effects of cadmium exposure on cell wall composition. Int J Phytoremediation 16:790–803. https://doi.org/10.1080/15226514.2013.856849

Azarova SV, Usmanova TV, Mezhibor AM (2019) Environmental problems of mining waste disposal sites in Russia. Rev Environ Contam Toxicol 247:59–84. https://doi.org/10.1007/398_2018_17

Baker AJM (1981) Accumulators and excluders—strategies in the response of plants to heavy metals. J Plant Nutr 3:643–654. https://doi.org/10.1080/01904168109362867

Baker AJM, McGrath SP, Reeves RD, Smith JAC (2000) Metal hyperaccumulator plants: a review of the ecology and physiology of a biological resource for phytoremediation of metal polluted soils. In: Terry N, Banuelos G (eds) Phytoremediation of contaminated soil and water. Lewis Publishers, Boca Raton, pp 85–107

Baldantoni D, Cicatelli A, Bellino A, Castiglione S (2014) Different behaviours in phytoremediation capacity of two heavy metal tolerant poplar clones in relation to iron and other trace elements. J Environ Manag 146:94–99. https://doi.org/10.1016/j.jenvman.2014.07.045

Bano A, Fatima M (2009) Salt tolerance in *Zea mays* (L.) following inoculation with *Rhizobium* and *Pseudomonas*. Biol Fertil Soil 45:405–413. https://doi.org/10.1007/s00374-008-0344-9

Barałkiewicz D, Kózka M, Piechalak A, Tomaszewska B, Sobczak P (2009) Determination of cadmium and lead species and phytochelatins in pea (*Pisum sativum*) by HPLC–ICP-MS and HPLC–ESI-MSn. Talanta 79:493–498. https://doi.org/10.1016/j.talanta.2009.04.026

Barbafieri M, Pedron F, Petruzzelli G, Rosellini I, Franchi E, Bagatin R, Vocciante M (2017) Assisted phytoremediation of a multi-contaminated soil: investigation on arsenic and lead combined mobilization and removal. J Environ Manag 203:316–329. https://doi.org/10.1016/j.jenvman.2017.07.078

Barceló J, Poschenrieder C (2002) Fast root growth responses, root exudates, and internal detoxification as clues to the mechanisms of aluminium toxicity and resistance. Environ Exp Bot 48:75–92. https://doi.org/10.1016/S0098-8472(02)00013-8

Barceló J, Poschenrieder C (2004) Structural and ultrastructural changes in heavy metal exposed plants. In: Prasad MNV (ed) Heavy metal stress in plants. Springer, Berlin, pp 223–248

Barceló J, Poschenrieder C (2006) Structural and ultrastructural changes in heavy metal exposed plants. In: Prasad MNV (ed) Heavy metal stress in plants: from biomolecules to ecosystems. Springer, Berlin, pp 223–248

Baroni F, Boscagli A, Di Lella LA, Protano G, Riccobono F (2004) Arsenic in soil and vegetation of contaminated areas in southern Tuscany (Italy). J Geochem Explor 81:1–14. https://doi.org/10.1016/S0375-6742(03)00208-5

Behera BK, Prasad R (2020) Environmental technology and sustainability. Elsevier. ISBN: 9780128191033. https://www.elsevier.com/books/environmental-technology-and-sustainability/behera/978-0-12-819103-3

Bender F, Wagg C, van der Heijden MGA (2016) An underground revolution: biodiversity and soil ecological engineering for agricultural sustainability. Trends Ecol Evol 31:440–452. https://doi.org/10.1016/j.tree.2016.02.016

Bielińska EJ, Futa B (2009) The influence of organic matter on biochemical transformations in anthropogenic soils on a landfill with ash from a power plant. Rocz Glebozn 60:18–26. (in Polish)

Bielińska EJ, Futa B, Mocek-Płóciniak A (2014) Soil enzymes as bioindicators of soil quality and health. Scientific monograph. Libropolis, Lublin. (in Polish)

Bielińska EJ, Futa B, Kołodziej B, Mocek-Płóciniak A, Kaczmarek Z (2016) Long-term ecochemical transformations in post-fire forest soils. Scientific monograph. Wydawnictwo Libropolis, Lublin. (in Polish)

Bonfante P, Genre A (2010) Mechanisms underlying beneficial plant–fungus interactions in mycorrhizal symbiosis. Nat Commun 1:48. https://doi.org/10.1038/ncomms1046

Boojar MMA, Goodarzi F (2007) The copper tolerance strategies and the role of antioxidative enzymes un three plant species grown on copper mine. Chemosphere 67:2138–2147. https://doi.org/10.1016/j.chemosphere.2006.12.071

Bothe H, Słomka A (2017) Divergent biology of facultative heavy metals plants. J Plant Physiol 219:45–61. https://doi.org/10.1016/j.jplph.2017.08.014

Boyd RS (2012) Plant defense using toxic inorganic ions: conceptual models of the defensive enhancement and joint effects hypotheses. Plant Sci 195:88–95. https://doi.org/10.1016/j.plantsci.2012.06.012

Broadley MR, White PJ, Hammond JP, Zelko I, Lux A (2007) Zinc in plants. New Phytol 173:677–702. https://doi.org/10.1111/j.1469-8137.2007.01996.x

Budzyńska S, Mleczek M, Goliński P, Rutkowski P, Niedzielski P (2017) The influence of As forms in substrate on the phytoextraction of this metalloid in *Ulmus laevis* Pall organs – Pot experiment. Microchem J 132:333–340. https://doi.org/10.1016/j.microc.2017.01.030

Budzyńska S, Niedzielski P, Mleczek M (2021) Time-dependent changes of arsenic and its selected forms in a hydroponic experiment with *Quercus robur* L. J Hazard Mater 405:124244. https://doi.org/10.1016/j.jhazmat.2020.124244

Burketova L, Trda L, Ott PG, Valentova O (2015) Bio-based resistance inducers for sustainable plant protection against pathogens. Biotechnol Adv 33:994–1004. https://doi.org/10.1016/j.biotechadv.2015.01.004

Cabot C, Gallego B, Martos S, Barceló J, Poschenrieder C (2013) Signal cross talk in *Arabidopsis* exposed to cadmium, silicon, and *Botrytis cinerea*. Planta 237:337–349. https://doi.org/10.1007/s00425-012-1779-7

Cabral L, Soares CRFS, Giachini AJ, Siqueira JO (2015) Arbuscular mycorrhizal fungi in phytoremediation of contaminated areas by trace elements: mechanisms and major benefits of their applications. World J Microbiol Biotechnol 31:1655–1664. https://doi.org/10.1007/s11274-015-1918-y

Cai MZ, Wang FM, Li RF, Zhang SN, Wang N, Xu GD (2011) Response and tolerance of root border cells to aluminum toxicity in soybean seedlings. J Inorg Biochem 105:966–971. https://doi.org/10.1016/j.jinorgbio.2011.04.004

Cai M, Wang N, Xing C, Wang F, Wu K, Du X (2013) Immobilization of aluminum with mucilage secreted by root cap and root border cells is related to aluminum resistance in *Glycine max* L. Environ Sci Pollut Res 20:8924–8933. https://doi.org/10.1007/s11356-013-1815-6

Candeias C, Melo R, Avila PF, Ferreira da Silva E, Salgueiro AR, Teixeira JP (2014) Heavy metal pollution in mine–soil–plant system in S. Francisco de Assis—Panasqueira mine (Portugal). Appl Geochem 44:12–26. https://doi.org/10.1016/j.apgeochem.2013.07.009

Cappuyns V, Alian V, Vassilieva E et al (2014) pH dependent leaching behavior of Zn, Cd, Pb, Cu and As from mining wastes and slags: kinetics and mineralogical control. Waste Biomass Valor 5:355–368. https://doi.org/10.1007/s12649-013-9274-3

Carbrera ML, Kissel DL, Bock BR (1994) Urea hydrolysis in soil. Effect of urea concentration and soil pH. Soil Biol Biochem 23:1121–1124. https://doi.org/10.1016/0038-0717(91)90023-D

Chang P, Kim JY, Kim KW (2005) Concentrations of arsenic and heavy metals in vegetation at two abandoned mine tailings in South Korea. Environ Geochem Healt 27:109–119. https://doi.org/10.1007/s10653-005-0130-7

Chen B, Christie P, Li L (2001) A modified glass bead compartment cultivation system for studies on nutrient and trace metal uptake by arbuscular mycorrhiza. Chemosphere 42:185–192. https://doi.org/10.1016/S0045-6535(00)00124-7

Chen L, Guo Y, Yang L, Wang Q (2008) Synergistic defensive mechanism of phytochelatins and antioxidative enzymes in *Brassica chinensis* L. against Cd stress. Chinese Sci Bull 53:1503. https://doi.org/10.1007/s11434-008-0062-6

Choudhary DK, Prakash A, Johri BN (2008) Induced systemic resistance (ISR) in plants: mechanism of action. Indian J Microbiol 47:289–297. https://doi.org/10.1007/s12088-007-0054-2

Čiamporová M (2002) Morphological and structural responses of plant roots to aluminium at organ, tissue and cellular level. Biol Plantarum 45:161–171. https://doi.org/10.1023/A:1015159601881

Colzi I, Arnetoli M, Gallo A, Doumett S, Del Bubba M, Pignattelli S, Gabbrielli R, Gonnelli C (2012) Copper tolerance strategies involving the root cell wall pectins in *Silene paradoxa* L. Environ Exp Bot 78:91–98. https://doi.org/10.1016/j.envexpbot.2011.12.028

Colzi I, Pignattelli S, Giorni E, Papini A, Gonnelli C (2015) Linking root traits to copper exclusion mechanisms in *Silene paradoxa* L. (Caryophyllaceae). Plant Soil 390:1–15. https://doi.org/10.1007/s11104-014-2375-3

Conesa HM, Moradi AB, Robinson BH et al (2009) Effects of increasing dosages of acid mining wastes in metal uptake by *Lygeum spartum* and soil metal extractability. Water Air Soil Pollut 202:379. https://doi.org/10.1007/s11270-009-9984-0

Cornelissen G, Gustafsson Ö (2006) Effects of added PAHs and precipitated humid acid coatings on phenanthrene sorption on environmental black carbon. Environ Pollut 141:526–531. https://doi.org/10.1016/j.envpol.2005.08.053

Craw D, Rufaut C, Haffert L, Paterson L (2007) Plant colonization and arsenic uptake on high arsenic mine wastes, New Zealand. Water Air Soil Pollut 179:351–364. https://doi.org/10.1007/s11270-006-9238-3

Čudić V, Stojiljković D, Jovović A (2016) Phytoremediation potential of wild plants growing on soil contaminated with heavy metals. Arh Hig Rada Toksikol 67:229–239. https://doi.org/10.1515/aiht-2016-67-2829

da Silva M, de Andrade SAL, De-Campos AB (2018) Phytoremediation potential of Jack bean plant for multi-element contaminated soils from Ribeira Valley, Brazil. Clean Soil Air Water 46: 1700321. https://doi.org/10.1002/clen.201700321

Daemen J, Akgün H (2012) Mining solid wastes. In: Meyers RA (ed) Encyclopedia of sustainability science and technology: SpringerReference. Springer, New York, pp 6665–6688

Dahm H, Rydlak K (1997) Some physical properties of bacteria of the *Pseudomonas*, accompanying the *Catharellus cibarius* mycorrhiza. In: Microorganisms vs soil life—materials of the national microbiological symposium, Cracow-Muszyna, 4–6 September. (in Polish)

DalCorso G, Fasani E, Furini A (2013) Recent advances in the analysis of metal hyperaccumulation and hypertolerance in plants using proteomics. Front Plant Sci 4:280. https://doi.org/10.3389/fpls.2013.00280

Danh LT, Truong P, Mammucardi R, Foster N (2014) A critical review of arsenic uptake mechanisms and phytoremediation potential of *Pteris Vittata*. Int J Phytoremediation 16:429–453. https://doi.org/10.1080/15226514.2013.798613

De Freitas-Silva L, de Araújo TO, da Silva LC, de Oliveira JA, de Araujo JM (2016) Arsenic accumulation in *Brassicaceae* seedlings and its effects on growth and plant anatomy. Ecotoxicol Environ Saf 124:1–9. https://doi.org/10.1016/j.ecoenv.2015.09.028

Demková L, Árvay J, Bobuľská L, Tomáš J, Stanovič R, Lošák T, Harangozo L, Vollmannová A, Bystrická J, Musilová J, Jobbágy J (2017a) Accumulation and environmental risk assessment of heavy metals in soil and plants of four different ecosystems in a former polymetallic ores mining and smelting area (Slovakia). J Environ Sci Health A 52:479–490. https://doi.org/10.1080/10934529.2016.1274169

Demková L, Bobul'ská L, Árvay J, Jezný T, Ducsay L (2017b) Biomonitoring of heavy metals contamination by mosses and lichens around Slovinky tailing pond (Slovakia). J Environ Sci Health A 52:30–36. https://doi.org/10.1080/10934529.2016.1221220

Deng D, Wu S-H, Wu F-Y, Deng H, Wong M-H (2010) Effects of root anatomy and Fe plaque on arsenic uptake by rice seedlings grown in solution culture. Environ Pollut 158:589–2595. https://doi.org/10.1016/j.envpol.2010.05.015

Dickinson N, Turner A, Lepp N (1991) How do trees and other long-lived plants survive in polluted environments? Funct Ecol 5:5–11. https://doi.org/10.2307/2389550

Dineshkumar M, Seenuvasan M, Sarojini G (2019) Phytoremediation strategies on heavy metal removal. In: Bui XT, Chiemchaisri C, Fujioka T, Varjani S (eds) Water and wastewater treatment technologies. Energy, environment, and sustainability. Springer, Singapore, pp 81–101

Domínguez MT, Maranón T, Murillo JM, Schulin R, Robinson BH (2008) Trace element accumulation in woody plants of the Guadiamar Valley, SW Spain: a large-scale phytomanagement case study. Environ Pollut 152:50–59. https://doi.org/10.1016/j.envpol.2007.05.021

Domka AM, Rozpądek P, Turnau K (2019) Are fungal endophytes merely mycorrhizal copycats? The role of fungal endophytes in the adaptation of plants to metal toxicity. Front Microbiol 10: 371. https://doi.org/10.3389/fmicb.2019.00371

Doncheva S, Amenós M, Poschenrieder C, Barceló J (2005) Root cell patterning: a primary target for aluminium toxicity in maize. J Exp Bot 56:1213–1220. https://doi.org/10.1093/jxb/eri115

Driouich A, Durand C, Vicré-Gibouin M (2007) Formation and separation of root border cells. Trends Plant Sci 12:14–19. https://doi.org/10.1016/j.tplants.2006.11.003

Drzewiecka K, Piechalak A, Goliński P, Gąsecka M, Magdziak Z, Szostek M, Budzyńska S, Niedzielski P, Mleczek M (2019) Differences of *Acer platanoides* L. and *Tilia cordata* Mill. response patterns/survival strategies during cultivation in extremely polluted mining sludge—a pot trial. Chemosphere. https://doi.org/10.1016/j.chemosphere.2019.05.051

Dudka S, Piotrowska M, Terelak H (1996) Transfer of cadmium, lead, and zinc from industrially contaminated soil to crop plants: a field study. Environ Pollut 94:181–188. https://doi.org/10.1016/S0269-7491(96)00069-3

Emamverdian A, Ding Y, Mokhberdoran F, Xie Y (2015) Heavy metal stress and some mechanisms of plant defense response. Sci World J 756120. https://doi.org/10.1155/2015/756120

Ernst WHO (2006) Evolution of metal tolerance in higher plants. For Snow Landsc Res 80:251–274

Esau K (1977) Anatomy of seed plants. Wiley, New York

Eticha D, Stass DA, Horst JW (2005) Cell-wall pectin and its degree of methylation in the maize root-apex: significance for genotypic differences in aluminium resistance. Plant Cell Environ 28:1410–1420. https://doi.org/10.1111/j.1365-3040.2005.01375.x

EUROSTAT (2018) Statistical book. Energy, transport and environment indicators. European Union 2018

Fahr M, Laplaze L, Bendaou N, Hocher V, El Mzibri M, Bogusz D, Smouni A (2013) Effect of lead on root growth. Front Plant Sci 4:175. https://doi.org/10.3389/fpls.2013.00175

Fan J-L, Wei X-Z, Wan L-C, Zhang L-Y, Zhao XQ, Liu W-Z, Hao H-Q, Zhang H-Y (2011) Disarrangement of actin filaments and Ca^{2+} gradient by $CdCl_2$ alters cell wall construction in *Arabidopsis thaliana* root hairs by inhibiting vesicular trafficking. J Plant Physiol 168:1157–1167. https://doi.org/10.1016/j.jplph.2011.01.031

Farooq FA, Islam F, Ali B, Najeeb U, Mao B, Gill RA, Yan G, Siddique KHM, Zhou W (2016) Arsenic toxicity in plants: cellular and molecular mechanisms of its transport and metabolism. Environ Exp Bot 132:42–52. https://doi.org/10.1016/j.envexpbot.2016.08.004

Fattorini L, Ronzan M, Piacentini D, Della Rovere F, De Virgilio C, Sofo A, Altamura MM, Falasca G (2017) Cadmium and arsenic affect quiescent centre formation and maintenance in *Arabidopsis thaliana* post-embryonic roots disrupting auxin biosynthesis and transport. Environ Exp Bot 144:37–48. https://doi.org/10.1016/j.envexpbot.2017.10.005

Fernández-Martínez R, Larios R, Gómez-Pinilla I, Gómez-Mancebo B, López-Andrés S, Loredo J, Ordóñez A, Rucandio I (2015) Mercury accumulation and speciation in plants and soils from abandoned cinnabar mines. Geoderma 253:30–38. https://doi.org/10.1016/j.geoderma.2015.04.005

Festin ES, Tigabu M, Chileshe MN et al (2019) Progresses in restoration of post-mining landscape in Africa. J For Res 30:381–396. https://doi.org/10.1007/s11676-018-0621-x

Fierer N, Schimel JP, Holden P (2003) Variations in microbial community composition through two soil depth profiles. Soil Biol Biochem 35:167–176. https://doi.org/10.1016/S0038-0717(02)00251-1

Filipović L, Romić M, Romić D, Filipović V, Ondrašek G (2018) Organic matter and salinity modify cadmium soil (phyto)availability. Ecotoxicol Environ Saf 147:824–831. https://doi.org/10.1016/j.ecoenv.2017.09.041

Flores Badillo J, Hernández Ávila J, Pérez Labra M, Patiño Cardona F, Ostos Santos JA, Trápala Pineda NY (2015) Developing alternative industrial materials from mining waste. In: Fergus JW, Mishra B, Anderson D, Sarver EA, Neelameggham NR (eds) Engineering solutions for sustainability. Springer, Cham, pp 119–125

Forsberg LS, Ledin S (2006) Effects of sewage sludge on pH and plant availability of metals in oxidising sulphide mine tailings. Sci Total Environ 358:21–35. https://doi.org/10.1016/j.scitotenv.2005.05.038

Freitas H, Prasad MNV, Pratas J (2004) Plant community tolerant to trace elements growing on the degraded soils of Sao Domingos mine in the south east of Portugal: environmental implications. Environ Int 30:65–72. https://doi.org/10.1016/S0160-4120(03)00149-1

Fusconi A, Repetto O, Bona E, Massa N, Gallo C, Dumas-Gaudot E, Berta G (2006) Effects of cadmium on meristem activity and nucleus ploidy in roots of *Pisum sativum* L. cv. Frisson seedlings. Environ Exp Bot 58:253–260. https://doi.org/10.1016/j.envexpbot.2005.09.008

Fusconi A, Gallo C, Camusso W (2007) Effects of cadmium on rootapical meristems of *Pisum sativum* L.: cell viability, cell proliferation and microtubule pattern as suitable markers for assessment of stress pollution. Mutat Res Genet Toxicol Environ Mutagen 632:9–19. https://doi.org/10.1016/j.mrgentox.2007.03.012

Futa B (2017) Enzymatic indicators of anthropogenic transformations of soil—a summary of professional accomplishments—typescript. Lublin University of Life Sciences, 32. (in Polish)

García-Salgado S, García-Casillas D, Quijano-Nieto MA, Bonilla-Simón MM (2012) Arsenic and heavy metal uptake and accumulation in native plant species from soils polluted by mining activities. Water Air Soil Pollut 223:559–572. https://doi.org/10.1007/s11270-011-0882-x

Gautam M, Pandey D, Agrawal SB, Agrawal M (2016a) Metals from mining and metallurgical industries and their toxicological impacts on plants. In: Singh A, Prasad S, Singh R (eds) Plant responses to xenobiotics. Springer, Singapore, pp 231–272

Gautam PK, Gautam RK, Banerjee S, Chattopadhyaya MC, Pandey JD (2016b) Heavy metals in the environment: fate, transport, toxicity and remediation technologies. Nova Sci Publishers 60: 101–130

Gerhardt KE, Gerwing PD, Greenberg BM (2017) Opinion: taking phytoremediation from proven technology to accepted practice. Plant Sci 256:170–185. https://doi.org/10.1016/j.plantsci.2016.11.016

Glińska S, Gapińska M (2013) The effect of pre-incubation of *A. cepa* L. roots in the ATH rich extract on Pb uptake and localization. Protoplasma 250:601–611. https://doi.org/10.1007/s00709-012-0445-z

Glińska S, Bartczak M, Oleksiak S, Wolska A, Gabara B, Posmyk M, Janas K (2007) Effects of anthocyanin-rich extract from red cabbage leaves on meristematic cells of *A. cepa* L. roots treated with heavy metals. Ecotoxicol Environ Saf 68:343–350. https://doi.org/10.1016/j.ecoenv.2007.02.004

Gupta AK, Sinha S (2006) Chemical fractionation and heavy metal accumulation in the plant of *Sesamum indicum* (L.) var. T55 grown on soil amended with tannery sludge: selection of single extractants. Chemosphere 64:161–173. https://doi.org/10.1016/j.chemosphere.2005.10.016

Gustafsson Ö, Gschwend PM (1997) Soot asa strongpartition medium for polycyclic aromatic hydrocarbons in aquatic systems. In: Eganhouse RP (ed) Molecular markers in environmental geochemistry. ACS Symposium Series 671, American Chemical Society, Washington, pp 365–381

Gzyl J, Chmielowska-Bąk J, Przymusiński R, Gwóźdź E (2015) Cadmium affects microtubule organization and posttranslational modifications of tubulin in seedlings of soybean (*Glycine max* L.). Front Plant Sci 6:937. https://doi.org/10.3389/fpls.2015.00937

Hall JL (2002) Cellular mechanisms for heavy metal detoxification and tolerance. J Exp Bot 53:1–11. https://doi.org/10.1093/jexbot/53.366.1

He JA, Kappler A (2017) Recovery of precious metals from waste streams. Microb Biotechnol 10: 1194–1198. https://doi.org/10.1111/1751-7915.12759

Hirayama T, Shinozaki K (2010) Research on plant abiotic stress responses in the post-genome era: past, present and future. Plant J 61:1041–1052. https://doi.org/10.1111/j.1365-313X.2010.04124.x

Horiunova II, Krasylenko YA, Yemets AI, Blume YB (2016) Involvement of plant cytoskeleton in cellular mechanisms of metal toxicity. Cytol Genet 50:47–59. https://doi.org/10.3103/S0095452716010060

Horst WJ, Wang Y, Eticha D (2010) The role of the root apoplast in aluminium induced inhibition of root elongation and in aluminium resistance of plants: a review. Ann Bot 106:185–197. https://doi.org/10.1093/aob/mcq053

Hossain Z, Komatsu S (2013) Contribution of proteomic studies towards understanding plant heavy metal stress response. Front Plant Sci 3:310. https://doi.org/10.3389/fpls.2012.00310

Hossner LR, Hons FM (1992) Reclamation of mine tailings. In: Lal R, Stewart BA (eds) Soil restoration. Advances in soil science. Springer, New York

Hu YF, Zhou G, Na XF, Yang L, Nan WB, Liu X (2013) Cadmium interferes with maintenance of auxin homeostasis in *Arabidopsis* seedlings. J Plant Physiol 170:965–975. https://doi.org/10.1016/j.jplph.2013.02.008

Inoue H, Fukuoka D, Tatai Y, Kamachi H, Hayatsu M, Ono M, Suzuki S (2013) Properties of lead deposits in cell walls of radish (*Raphanus sativus*) roots. J Plant Res 126:51–61. https://doi.org/10.1007/s10265-012-0494-6

Iwasaki K, Sakurai K, Takahashi E (1990) Copper binding by the root cell walls of Italian ryegrass and red clover. Soil Sci Plant Nutr 36:431–440. https://doi.org/10.1080/00380768.1990.10416911

Januszek K (1999) The enzymatic activity of selected forest soils in Southern Poland in the light of field and laboratory investigations. Zeszyty Naukowe AR w Krakowie, Rozprawy, 250. (in Polish)

Jiang W, Liu D (2010) Pb-induced cellular defense system in the root meristematic cells of *Allium sativum* L. BMC Plant Biol 10:40. https://doi.org/10.1186/1471-2229-10-40

Jiang Z, Zhang H, Qin R, Zou J, Wang J, Shi Q, Jiang W, Liu D (2014) Effects of lead on the morphology and structure of the nucleolus in the root tip meristematic cells of *Allium cepa* L. Int J Mol Sci 15:13406–13423. https://doi.org/10.3390/ijms150813406

Jović M, Šljivić-Ivanović M, Dimović S, Marković J, Smičiklas I (2017) Sorption and mobility of Co(II) in relation to soil properties. Geoderma 297:38–47. https://doi.org/10.1016/j.geoderma.2017.03.006

Jung MC, Thornton I, Chon HT (2002) Arsenic, Sb and Bi contamination of soils, plants, waters and sediments in the vicinity of the Dalsung Cu–W mine in Korea. Sci Total Environ 295:81–89. https://doi.org/10.1016/S0048-9697(02)00042-6

Kabata-Pendias A (2004) Soil-plant transfer of trace elelments—an environmental issue. Geoderma 122:143–149. https://doi.org/10.2016/j.geoderma.2004.01.004

Kalinovic TS, Serbula SM, Radojevic AA, Kalinovic JV, Steharnik MM, Petrovic JV (2016) Elder, linden and pine biomonitoring ability of pollution emitted from the copper smelter and the tailings ponds. Geoderma 262:266–275. https://doi.org/10.1016/j.geoderma.2015.08.027

Karimi N, Ghaderian SM, Schat H (2013) Arsenic in soil and vegetation of a contaminated area. Int J Environ Sci Technol 10:743–752. https://doi.org/10.1007/s13762-013-0227-y

Kim HJ, Kim Y, Choo CO (2014) The effect of mineralogy on the mobility of heavy metals in mine tailings: a case study in the Samsanjeil mine, Korea. Environ Earth Sci 71:3429–3441. https://doi.org/10.1007/s12665-013-2732-1

Kobus J (1995) Biological processes vs the development of soil fertility. Zesz Prob Post Nauk Rol 421a:209–219. (in Polish)

Kocjan G, Samardakiewicz S, Wozny A (1996) Regions of lead uptake in *Lemna minor* plants and localization of this metal within selected parts of the root. Biol Plant 38:107–117. https://doi.org/10.1007/BF02879643

Kopittke PM, De Jonge MD, Menzies NW, Wang P, Donner E, McKenna BA, Paterson D, Howard DL, Lombi E (2012) Examination of the distribution of arsenic in hydrated and fresh cowpea roots using two- and three-dimensional techniques. Plant Physiol 159:1149–1158. https://doi.org/10.1104/pp.112.197277

Kopittke PM, McKenna BA, Karunakaran C, Dynes JJ, Arthur Z, Gianoncelli A, Kourousias G, Menzies NW, Ryan PR, Wang P, Green K, Blamey FPC (2017) Aluminum complexation with malate within the root apoplast differs between aluminum resistant and sensitive wheat lines. Front Plant Sci 8:1377. https://doi.org/10.3389/fpls.2017.01377

Krzesłowska M (2011) The cell wall in plant cell response to trace metals: polysaccharide remodeling and its role in defense strategy. Acta Physiol Plant 33:35–51

Krzesłowska M, Lenartowska M, Samardakiewicz S, Bilski H, Woźny A (2010) Lead deposited in the cell wall of *Funaria hygrometrica* protonemata is not stable—a remobilization can occur. Environ Pollut 158:325–338. https://doi.org/10.1016/j.envpol.2009.06.035

Krzesłowska M, Rabęda I, Basińska A, Lewandowski M, Mellerowicz EJ, Napieralska A, Samardakiewicz S, Woźny A (2016) Pectinous cell wall thickenings formation: a common defense strategy of plants to cope with Pb. Environ Pollut 214:354–361. https://doi.org/10.1016/j.envpol.2016.04.019

Krzesłowska M, Timmers ACJ, Mleczek M, Niedzielski P, Rabęda I, Woźny A, Goliński P (2019) Alterations of root architecture and cell wall modifications in *Tilia cordata* Miller (Linden) growing on mining sludge. Environ Pollut 14(248):247–259. https://doi.org/10.1016/j.envpol.2019.02.019

Książek M, Woźny A (1990) Lead movement in poplar adventitious roots. Biol Plant 32:54–57. https://doi.org/10.1007/BF02897344

Kucharski J (1997) Enzymatic activity vs soil fertility. In: Barabasz W (ed) Microorganisms in the environment—occurrence, activity and significance. AR, Kraków, pp 327–348. (in Polish)

Kucharski J, Barabasz W, Bielińska EJ, Wyszkowska J (2015) The biological and biochemical properties of soils. In: Mocek A (ed) Gleboznawstwo. PWN, Warsaw, pp 232–281. (in Polish)

Kumar D, Tripathi DK, Liu S, Singh VK, Sharma S, Dubey NK, Prasad SM, Chauhan DK (2017) *Pongamia pinnata* (L.) Pierre tree seedlings offer a model species for arsenic phytoremediation. Plant Gene 11:238–246. https://doi.org/10.1016/j.plgene.2017.06.002

Kuppusamy S, Palanisami T, Megharaj M, Venkateswarlu K, Naidu R (2016) In-situ remediation approaches for the management of contaminated sites: a comprehensive overview. Rev Environ Contam Toxicol 236:1–115. https://doi.org/10.1007/978-3-319-20013-2_1

Kushwaha A, Hans N, Kumar S, Rani R (2018) A critical review on speciation, mobilization and toxicity of lead in soilmicrobe-plant system and bioremediation strategies. Ecotoxicol Environ Saf 147:1035–1045. https://doi.org/10.1016/j.ecoenv.2017.09.049

Lane SD, Martin ES (1977) A histochemical investigation of lead uptake in *Raphanus sativus*. New Phytol 79:281–286

Le Gall H, Philippe F, Domon J-M, Gillet F, Pelloux J, Rayon C (2015) Cell wall metabolism in response to abiotic stress. Plan Theory 4:112–166. https://doi.org/10.3390/plants4010112

Lequeux H, Hermans C, Lutts S, Verbruggen N (2010) Response to copper excess in *Arabidopsis thaliana*: impact on the root system architecture, hormone distribution, lignin accumulation and mineral profile. Plant Physiol Biochem 48:673–682. https://doi.org/10.1016/j.plaphy.2010.05.005

Li MS (2006) Ecological restoration of mineland with particular reference to the metalliferous mine wasteland in China: a review of research and practice. Sci Total Environ 357:38–53. https://doi.org/10.1016/j.scitotenv.2005.05.003

Li D, Shu Z, Ye X, Zhu J, Pan J, Wang W, Chang P, Cui C, Shen J, Fang W, Zhu X, Wang Y (2017) Cell wall pectin methyl-esterification and organic acids of root tips involve in aluminum tolerance in *Camellia sinensis*. Plant Physiol Biochem 119:265–274. https://doi.org/10.1016/j.plaphy.2017.09.002

Liptáková Ľ, Huttová J, Mistrík I, Tamás L (2013) Enhanced lipoxygenase activity is involved in the stress response but not in the harmful lipid peroxidation and cell death of short-term cadmium-treated barley root tip. J Plant Physiol 170:646–652. https://doi.org/10.1016/j.jplph.2012.12.007

Liška D, Martinka M, Kohanová J, Lux A (2016) Asymmetrical development of root endodermis and exodermis in reaction to abiotic stresses. Ann Bot 118:667–674. https://doi.org/10.1093/aob/mcw047

Liu D, Xue P, Meng Q, Zou J, Gu J, Jiang W (2009) Pb/Cu effects on the organization of microtubule cytoskeleton in interphase and mitotic cells of *Allium sativum* L. Plant Cell Rep 28:695–702. https://doi.org/10.1007/s00299-009-0669-3

Llugany M, Lombini A, Poschenrieder C, Dinelli E, Barceló J (2003) Different mechanisms account for enhanced copper resistance in *Silene armeria* ecotypes from mine soil and serpentine sites. Plant Soil 251:55–63. https://doi.org/10.1023/A:1022990525632

Lottermoser BG (2010) Introduction to mine wastes. In: Mine wastes. Springer, Berlin. https://doi.org/10.1016/j.scitotenv.2005.05.003

Lux A, Martinka M, Vaculík M, White PJ (2011) Root responses to cadmium in the rhizosphere: a review. J Exp Bot 62:21–37. https://doi.org/10.1093/jxb/erq281

Lux A, Lackovič A, Van Staden J, Lišková D, Kohanová J, Martinka M (2015) Cadmium translocation by contractile roots differs from that in regular, non-contractile roots. Ann Bot 115:1149–1154. https://doi.org/10.1093/aob/mcv051

Ma LQ, Komar KM, Tu C, Zhang W, Cai Y, Kennelley ED (2001) A fern that hyperaccumulates arsenic. Nature 409:579. https://doi.org/10.1038/35054664

Ma C, Ming H, Lin C, Naidu R, Bolan N (2016a) Phytoextraction of heavy metal from tailing waste using Napier grass. Catena 136:74–83. https://doi.org/10.1016/j.catena.2015.08.001

Ma Y, Rajkumar M, Zhang C, Freitas H (2016b) Beneficial role of bacterial endophytes in heavy metal phytoremediation. J Environ Manag 174:14–25. https://doi.org/10.1016/j.jenvman.2016.02.047

Madejón P, Marañón T, Murillo Carpio JM, Cabrera F (2005) Evolution of arsenic, lead, iron and manganese in evergreen trees affected by the Aznalcóllar mine spill (SW, Spain). In: Del Valls TA, Blasco J (eds) UNESCO. Unitwin- Wicop, Puerto Real, Cádiz, pp 91–98

Małecka A, Piechalak A, Morkunas I, Tomaszewska B (2008) Accumulation of lead in root cells of *Pisum sativum*. Acta Physiol Plant 30:629–637. https://doi.org/10.1007/s11738-008-0159-1

Malone C, Koeppe DE, Miller J (1974) Localization of lead accumulated by corn plants. Plant Physiol 53:388–394

Manara A (2012) Plant responses to heavy metal toxicity. In: Furini A (ed) Plants and heavy metals. SpringerBriefs in molecular science. Springer, Dordrecht, pp 27–53

Martínez-Sánchez MJ, Martínez-López S, García-Lorenzo ML, Martínez-Martínez LB, Pérez-Sirvent C (2011) Evaluation of arsenic in soils and plant uptake using various chemical extraction methods in soils affected by old mining activities. Geoderma 160:535–541. https://doi.org/10.1016/j.geoderma.2010.11.001

Martyniuk S (2017) Living soil. Farmer 4:116–120. (in Polish)

McGrath SP, Brookes PC, Giller KE (1988) Effects of potentially toxic metals in soil derived from past applications of sewage sludge on nitrogen fixation by *Trifolium repens*. Soil Biol Biochem 20:415–424. https://doi.org/10.1016/0038-0717(88)90052-1

Mendez MO, Maier RM (2008) Phytoremediation of mine tailings in temperate and arid environments. Rev Environ Sci Biotechnol 7:47–59. https://doi.org/10.1007/s11157-007-9125-4

Meyer CL, Juraniec M, Huguet S, Chaves-Rodriguez E, Salis P, Isaure MP, Goormaghtigh E, Verbruggen N (2015) Intraspecific variability of cadmium tolerance and accumulation, and cadmium-induced cell wall modifications in the metal hyperaccumulator *Arabidopsis halleri*. J Exp Bot 62:3215–3227. https://doi.org/10.1093/jxb/erv144

Meyers DE, Kopittke PM, Auchterlonie GJ, Webb RI (2009) Characterization of lead precipitate following uptake by roots of *Brassica juncea*. Environ Toxicol Chem 28:250–255. https://doi.org/10.1897/09-131.1

Migeon A, Richaud P, Guinet F, Chalot M, Blaudez D (2009) Metal accumulation by woody species on contaminated sites in the north of France. Water Air Soil Pollut 204:89. https://doi.org/10.1007/s11270-009-0029-5

Mishra J, Singh R, Arora NK (2017) Alleviation of heavy metal stress in plants and remediation of soil by rhizosphere microorganisms. Front Microbiol 8:1706. https://doi.org/10.3389/fmicb.2017.01706

Mleczek M, Goliński P, Krzesłowska M, Gąsecka M, Magdziak Z, Rutkowski P, Budzyńska S, Waliszewska B, Kozubik T, Karolewski Z, Niedzielski P (2017) Phytoextraction of potentially toxic elements by six tree species growing on hazardous mining sludge. Environ Sci Pollut Res 24:22183–22195. https://doi.org/10.1007/s11356-017-9842-3

Mleczek M, Gąsecka M, Waliszewska B, Magdziak Z, Szostek M, Rutkowski P, Kaniuczak J, Zborowska M, Budzyńska S, Mleczek P, Niedzielski P (2018) *Salix viminalis* L.—a highly effective plant in phytoextraction of elements. Chemosphere 212:67–78. https://doi.org/10.1016/j.chemosphere.2018.08.055

Mleczek M, Rutkowski P, Kaniuczak J, Szostek M, Budka A, Magdziak Z, Budzyńska S, Kuczyńska-Kippen N, Niedzielski P (2019) The signifcance of selected tree species age in their efciency in elements phytoextraction from wastes mixture. Int J Environ Sci Technol 16: 3579–3594. https://doi.org/10.1007/s13762-018-1996-0

Mocek-Płóciniak A (2018) The physicochemical, microbiological and biochemical properties of soils formed in landfills with ash and slag from power plants. Rozprawy Naukowe No. 499. Wydawnictwo Uniwersytetu Przyrodniczego w Poznaniu, Poznań (in Polish)

Mohanty M, Dhal NK, Patra P, Das B, Reddy PSR (2010) Phytoremediation: a novel approach for utilization of iron-ore wastes. Rev Environ Contam Toxicol 206:29–47. https://doi.org/10.1007/978-1-4419-6260-7_2

Morel JL, Mench M, Guckert A (1986) Measurement of Pb^{2+}, Cu^{2+} and Cd^{2+} binding with mucilage exudates from maize (*Zea mays* L.) roots. Biol Fertil Soils 2:29–34. https://doi.org/10.1007/BF00638958

Munns R (2002) Comparative physiology of salt and water stress. Plant Cell Environ 25:239–250. https://doi.org/10.1046/j.0016-8025.2001.00808.x

Murray RK, Granner DK, Mayes PA (1995) Harper's biochemystry. PZWL, Warsaw. (in Polish)

Neumann D, zur Nieden U (2001) Silicon and heavy metal tolerance in higher plants. Phytochemistry 56:685–692

Nouri H, Chavoshi Borujeni S, Nirola R, Hassanli A, Beecham S, Alaghmand S, Saint C, Mulcahy D (2017) Application of green remediation on soil salinity treatment: a review on halophytoremediation. Process Saf Environ 107:94–107. https://doi.org/10.1016/j.psep.2017.01.021

Okiobé ST, Abossolo Angue M, Bougnom BP et al (2015) Improvement of arbuscular mycorrhizal fungi inoculum production by nutrient solution concentration and soil texture variation. Int J Agron Agric Res 6:7–20

Okkenhaug G, Zhu YG, Luo L, Lei M, Li X, Mulder J (2011) Distribution, speciation and availability of antimony (Sb) in soils and terrestrial plants from an active Sb mining area. Environ Pollut 159:2427–2434. https://doi.org/10.1016/j.envpol.2011.06.028

Otones V, Álvarez-Ayuso E, García-Sánchez A, Santa Regina I, Murciego A (2011) Arsenic distribution in soils and plants of an arsenic impacted former mining area. Environ Pollut 159:2637–2647. https://doi.org/10.1016/j.envpol.2011.05.027

Ovečka M, Takáč T (2014) Managing heavy metal toxicity stress in plants: biological and biotechnological tools. Biotechnol Adv 32:73–86

Parraga-Aguado I, Querejeta JI, González-Alcaraz MN, Conesa HM (2014) Metal(loid) allocation and nutrient retranslocation in *Pinus halepensis* trees growing on semiarid mine tailings. Sci Total Environ 485:406–414. https://doi.org/10.1016/j.scitotenv.2014.03.116

Parrotta L, Guerriero G, Sergeant K, Cai G, Hausman J-F (2015) Target or barrier? The cell wall of early- and later-diverging plants vs cadmium toxicity: differences in the response mechanisms. Front Plant Sci 6:133. https://doi.org/10.3389/fpls.2015.00133

Pavlovkin J, Pal'ove-Balang P, Kolarovič L, Zelinová V (2009) Growth and functional responses of different cultivars of *Lotus corniculatus* to aluminum and low pH stress. J Plant Physiol 166: 1479–1487. https://doi.org/10.1016/j.jplph.2009.03.005

Pistelli L, D'Angiolillo F, Morelli E, Basso B, Rosellini I, Posarelli M, Barbafieri M (2017) Response of spontaneous plants from an ex-mining site of Elba island (Tuscany, Italy) to metal(loid) contamination. Environ Sci Pollut Res 24:7809–7820. https://doi.org/10.1007/s11356-017-8488-5

Pita-Barbosa A, Gonçalves EC, Azevedo AA (2015) Morpho-anatomical and growth alterations induced by arsenic in *Cajanus cajan* (L.) DC (Fabaceae). Environ Sci Pollut Res 22:11265–11274. https://doi.org/10.1007/s11356-015-4342-9

Poschenrieder C, Gunse B, Corrales I, Barceló J (2008) A glance into aluminium toxicity and resistance in plants. Sci Total Environ 400:356–368. https://doi.org/10.1016/j.scitotenv.2008.06.003

Prasad MNV, Freitas HMO (2003) Metal hyperaccumulation in plants biodiversity prospecting for phytoremediation technology. Electron J Biotechnol 6:285–321. https://doi.org/10.2225/vol6-issue3-fulltext-6
Prasad R, Bhola D, Akdi K, Cruz C, Sairam KVSS, Tuteja N, Varma A (2017) Introduction to mycorrhiza: historical development. In: Varma A, Prasad R, Tuteja N (eds) Mycorrhiza. Springer, pp 1–7
Pratas J, Prasad MNV, Freitas H, Conde L (2005) Plants growing in abandoned mines of Portugal are useful for biogeochemical exploration of arsenic, antimony, tungsten and mine reclamation. J Geochem Exp 85:99–107. https://doi.org/10.1016/j.gexplo.2004.11.003
Pratas J, Favas PJ, D'Souza R, Varun M, Paul MS (2013) Phytoremedial assessment of flora tolerant to heavy metals in the contaminated soils of an abandoned Pb mine in Central Portugal. Chemosphere 90:2216–2225. https://doi.org/10.1016/j.chemosphere.2012.09.079
Praveen-Kumar, Tarafdar JC (2003) 2,3,5-Triphenyltetrazolium chloride (TTC) as a electron acceptor of culturable soil bacteria, fungi and actinomycetes. Biol Fertil Soils 38:186–189. https://doi.org/10.1007/s00374-003-0600-y
Probst A, Liu H, Fanjul M, Liao B, Hollande E (2009) Response of *Vicia faba* L. to metal toxicity on mine tailing substrate: geochemical and morphological changes in leaf and root. Environ Exp Bot 66:297–308. https://doi.org/10.1016/j.envexpbot.2009.02.003
Pulford ID, Watson C (2003) Phytoremediation of heavy metal-contaminated land by trees—a review. Environ Int 29:529–540. https://doi.org/10.1016/S0160-4120(02)00152-6
Rabęda I, Bilski H, Mellerowicz EJ, Napieralska A, Suski S, Woźny A, Krzesłowska M (2015) Colocalization of low-methylesterified pectins and Pb deposits in the apoplast of aspen roots exposed to lead. Environ Pollut 205:315–326. https://doi.org/10.1016/j.envpol.2015.05.048
Rajkumar M, Sandhya S, Prasad MNV, Freitas H (2012) Perspectives of plant-associated microbes in heavy metal phytoremediation. Biotechnol Adv 30:1562–1574. https://doi.org/10.1016/j.biotechadv.2012.04.011
Reichman SM (2002) The responses of plants to metal toxicity: a review focusing on copper, manganese and zinc. Australian Minerals and Energy Environment Foundation, pp 1–40
Renella G, Egmaberiyeva D, Landi L, Mench M, Nannipieri P (2006) Microbial activity and hydrolase activities during decomposition of root exudates released by an artificial root surface in Cd-contaminated soils. Soil Biol Biochem 38:702–708. https://doi.org/10.1016/j.soilbio.2005.06.021
Rivera-Becerril F, Juárez-Vázquez LV, Hernández-Cervantes SC et al (2013) Impacts of manganese mining activity on the environment: interactions among soil, plants, and arbuscular mycorrhiza. Arch Environ Contam Toxicol 64:219–227. https://doi.org/10.1007/s00244-012-9827-7
Rodrigo MAM, Anjum NA, Heger Z, Zitka O, Vojtech A, Pereira E, Kizek R (2016) Role of phytochelatins in redox caused stress in plants and animals. In: Shanker AK, Shanker C (eds) Abiotic and biotic stress in plants—recent advances and future perspectives. IntechOpen
Rosik-Dulewska CZ (2015) Waste management essentials. PWN, 389. (in Polish)
Saha JK, Selladurai R, Coumar MV, Dotaniya ML, Kundu S, Patra AK (2017) Remediation and management of polluted sites. In: Soil pollution—an emerging threat to agriculture. Environmental chemistry for a sustainable world, vol 10. Springer, Singapore, pp 317–372
Salas-Luévano MA, Mauricio-Castillo JA, González-Rivera ML, Vega-Carrillo HR, Salas-Muñoz S (2017) Accumulation and phytostabilization of As, Pb and Cd in plants growing inside mine tailings reforested in Zacatecas, Mexico. Environ Earth Sci 76:806. https://doi.org/10.1007/s12665-017-7139-y
Samardakiewicz S, Woźny A (2000) The distribution of lead in duckweed (*Lemna minor* L.) root tip. Plant Soil 226:107–111. https://doi.org/10.1023/A:1026440730839
Samardakiewicz S, Woźny A (2005) Cell division in *Lemna minor* roots treated with lead. Aquat Bot 83:289–295. https://doi.org/10.1016/j.aquabot.2005.06.007

Samardakiewicz S, Krzesłowska M, Woźny A (2009) Tubulin cytoskeleton in plant cell response to trace metals. In: Maksymiec W (ed) Compartmentation of responses to stresses in higher plants, true or false. Kerala, Transworld Research Network, pp 149–162

Samardakiewicz S, Krzesłowska M, Bilski H, Bartosiewicz R, Woźny A (2012) Is callose a barrier for lead ions entering *Lemna minor* L. root cells? Protoplasma 249:347–351. https://doi.org/10.1007/s00709-011-0285-2

Sangeeta M, Maiti SK (2010) Phytoremediation of metal enriched mine waste: a review. Am Eurasian J Agric Environ Sci 9:560–575. http://www.idosi.org/.../17.pdf

Santos-Jallath J, Castro-Rodríguez A, Huezo-Casillas J, Torres-Bustillos L (2012) Arsenic and heavy metals in native plants at tailings impoundments in Queretaro, Mexico. Phys Chem Earth PT A/B/C 37:10–17. https://doi.org/10.1016/j.pce.2011.12.002

Schmohl N, Horst WJ (2000) Cell wall pectin content modulates aluminium sensitivity of *Zea mays* (L.) cells grown in suspension culture. Plant Cell Environ 23:735–742. https://doi.org/10.1046/j.1365-3040.2000.00591.x

Schneider J, Labory CRG, Rangel WM, Alves E, Guilherme LRG (2013) Anatomy and ultrastructure alterations of *Leucaena leucocephala* (Lam.) inoculated with mycorrhizal fung in response to arsenic-contaminated soil. J Hazard Mater 262:1245–1258. https://doi.org/10.1016/j.jhazmat.2012.05.091

Schneider J, Bundschuh J, do Nascimento CW (2016) Arbuscular mycorrhizal fungi-assisted phytoremediation of a lead-contaminated site. Sci Total Environ 572:86–97. https://doi.org/10.1016/j.scitotenv.2016.07.185

Schvartzman MS, Corso M, Fataftah N, Scheepers M, Nouet C, Bosman B, Carnol M, Motte P, Verbruggen N, Hanikenne M (2018) Adaptation to high zinc depends on distinct mechanisms in metallicolous populations of *Arabidopsis halleri*. New Phytol 218:269–282. https://doi.org/10.1111/nph.14949

Selvam A, Wong JWC (2008) Phytochelatin systhesis and cadmium uptake of *Brassica napus*. Environ Technol 29:765–773. https://doi.org/10.1080/09593330801987079

Seregin IV, Kozhevnikova AD (2008) Roles of root and shoot tissues in transport and accumulation of cadmium, lead, nickel, and strontium. Russ J Plant Physiol 55:1–22. https://doi.org/10.1134/S1021443708010019

Sessitsch A, Kuffner M, Kidd P, Vangronsveld J, Wenzel WW, Fallmann K, Puschemreiter M (2013) The role of plant-associated bacteria in the mobilization and phytoextraction of trace elements in contaminated soils. Soil Biol Biochem 60:182–194. https://doi.org/10.1016/j.soilbio.2013.01.012

Sharma RK, Agrawal M (2005) Biological effects of heavy metals: an overview. J Environ Biol 26: 301–313

Sharma SS, Dietz KJ (2006) The significance of amino acids and amino acid-derived molecules in plant responses and adaptation to heavy metal stress. J Exp Bot 57:711–726. https://doi.org/10.1093/jxb/erj073

Sharma S, Prasad R, Varma A, Sharma AK (2017) Glycoprotein associated with *Funneliformis coronatum, Gigaspora margarita* and *Acaulospora scrobiculata* suppress the plant pathogens in vitro. Asian J Plant Pathol. https://doi.org/10.3923/ajppaj.2017

Sheoran V, Sheoran AS, Poonia P (2016) Factors affecting phytoextraction: a review. Pedosphere 26:148–166. https://doi.org/10.1016/S1002-0160(15)60032-7

Shi X, Wang S, Sun H, Chen Y, Wang D, Pan H, Zou Y, Liu J, Zheng L, Zhao X, Jiang Z (2016) Comparative of *Quercus* spp. and *Salix* spp. for phytoremediation of Pb/Zn mine tailings. Environ Sci Pollut Res 24:3400–3411. https://doi.org/10.1007/s11356-016-7979-0

Shi X, Wang S, Sun H, Chen Y, Wang D, Pan H, Zou Y, Liu J, Zheng L, Zhao X, Jiang Z (2017) Comparative of *Quercus* spp. and *Salix* spp. for phytoremediation of Pb/Zn mine tailings. Environ Sci Pollut Res 24:3400–3411. https://doi.org/10.1007/s11356-016-7979-0

Siemianowski O, Barabasz A, Kendziorek M, Ruszczyńska A, Bulska E, Williams LE, Antosiewicz DM (2014) AtHMA4 expression in tobacco reduces Cd accumulation due to the induction of the apoplastic barrier. J Exp Bot 65:1125–1139. https://doi.org/10.1093/jxb/ert471

Silva V, Loredo J, Fernández-Martínez R et al (2014) Arsenic partitioning among particle-size fractions of mine wastes and stream sediments from cinnabar mining districts. Environ Geochem Health 36:831–843. https://doi.org/10.1007/s10653-014-9602-y

Smith SE, Read DJ (2008) Mycorrhizal symbiosis, 3rd edn. Academic Press, New York

Smith SE, Smith FA (2012) Fresh perspectives on the roles of arbuscular mycorrhizal fungi in plant nutrition and growth. Mycologia 104:1–13. https://doi.org/10.3852/11-229

Soares CRFS, Siqueira JO (2008) Mycorrhiza and phosphate protection of tropical grass species against heavy metal toxicity in multi-contaminated soil. Biol Fertil Soils 44:833–841. https://doi.org/10.1007/s00374-007-0265-z

Sofo A, Bochicchio R, Amato M, Rendina N, Vitti A, Nuzzaci M, Altamura MM, Falasca G, Rovere FD, Scopa A (2017) Plant architecture, auxin homeostasis and phenol content in *Arabidopsis thaliana* grown in cadmium- and zinc-enriched media. J Plant Physiol 216:174–180. https://doi.org/10.1016/j.jplph.2017.06.008

Sousa AI, Caçador I, Lillebø AI, Pardal MA (2008) Heavy metal accumulation in *Halimione portulacoides*: intra- and extra-cellular metal binding sites. Chemosphere 70:850–857. https://doi.org/10.1016/j.chemosphere.2007.07.012

Ssenku JE, Ntale M, Origa HO (2014) Physico-chemical characteristic of copper tailings and pyrite soils in Western Uganda: implication for phytoremediation. Int J Environ Monit Anal 2:191–198. https://doi.org/10.11648/j.ijema.20140204.12

Strzyszcz Z (2004) An assessment of the usefulness and the rules concerning the use of various waste for the rehabilitation of dumping grounds and areas degraded by industrial activity. IPIŚ PAN Work Stud 60:82. (in Polish)

Suman J, Uhlik O, Viktorova J, Macek T (2018) Phytoextraction of heavy metals: a promising tool for clean-up of polluted environment? Front Plant Sci 9:1476. https://doi.org/10.3389/fpls.2018.01476

Sun R, Jin C, Zhou Q (2010) Characteristics of cadmium accumulation and tolerance in *Rorippa globosa* (Turcz.) Thell., a species with some characteristics of cadmium hyperaccumulation. Plant Growth Regul 61:67–74. https://doi.org/10.1007/s10725-010-9451-3

Sun C, Lu L, Yu Y, Liu L, Hu Y, Ye Y, Jin C, Lin X (2016) Decreasing methylation of pectin caused by nitric oxide leads to higher aluminium binding in cell walls and greater aluminium sensitivity of wheat roots. J Exp Bot 67:979–989. https://doi.org/10.1093/jxb/erv514

Sun W, Ji B, Khoso SA, Tang H, Liu R, Wang L, Hu Y (2018) An extensive review on restoration technologies for mining tailings. Environ Sci Pollut Res 25:33911–33925. https://doi.org/10.1007/s11356-018-3423-y

Tamás L, Dudíková J, Ďurčeková K, Huttová J, Mistrík I, Zelinová V (2008) The impact of heavy metals on the activity of some enzymes along the barley root. Environ Exp Bot 62:86–91. https://doi.org/10.1016/j.envexpbot.2007.07.009

Timmers ACJ (2016) Light microscopy of whole plant organs. J Microsc-Oxford 263:165–170. https://doi.org/10.1111/jmi.12394

Tolrà R, Barcelò J, Poschenrieder CH (2009) Constitutive and aluminium-induced patterns of phenolic compounds in two maize varieties differing in aluminium tolerance. J Inorg Biochem 103:1486–1490. https://doi.org/10.1016/j.jinorgbio.2009.06.013

Van der Ent A, Baker AJ, Reeves RD, Pollard AJ, Schat H (2013) Hyperaccumulators of metal and metalloid trace elements: facts and fiction. Plant Soil 362:319–334. https://doi.org/10.1007/s11104-012-1287-3

Varma A, Prasad R, Tuteja N (2017a) Mycorrhiza: function, diversity and state-of-art. Springer. (ISBN: 978-3-319-53064-2). http://www.springer.com/us/book/9783319530635

Varma A, Prasad R, Tuteja N (2017b) Mycorrhiza: eco-physiology, secondary metabolites, nanomaterials. Springer. (ISBN: 978-3-319-57849-1). http://www.springer.com/us/book/9783319578484

Varma A, Prasad R, Tuteja N (2017c) Mycorrhiza: nutrient uptake, biocontrol, ecorestoration. Springer. (ISBN: 978-3-319-68867-1). http://www.springer.com/us/book/9783319688664

Verbelen J-P, De Cnodder T, Le J, Vissenberg K, Baluška F (2006) The root apex of *Arabidopsis thaliana* consists of four distinct zones of cellular activities: meristematic zone, transition zone, fast elongation zone, and growth terminating zone. Plant Signal Behav 1:296–304

Verbruggen N, Hermans C, Schat H (2009) Mechanisms to cope with arsenic or cadmium excess in plants. Curr Opin Plant Biol 12:364–372. https://doi.org/10.1016/j.pbi.2009.05.001

Verbruggen N, Juraniec M, Baliardini C, Meyer CL (2013) Tolerance to cadmium in plants: the special case of hyperaccumulators. Biometals 26:633–638. https://doi.org/10.1007/s10534-013-9659-6

Viehweger K (2014) How plants cope with heavy metals. Bot Stud 55:35. https://doi.org/10.1186/1999-3110-55-35

Visoottiviseth P, Francesconi K, Sridokchan W (2002) The potential of Thai indigenous plant species for the phytoremediation of arsenic contaminated land. Environ Pollut 118:453–461. https://doi.org/10.1016/S0269-7491(01)00293-7

Wang HB, Wong MH, Lan CY, Baker AJM, Qin YR, Shu WS, Chen GZ, Ye ZH (2007) Uptake and accumulation of arsenic by 11 *Pteris* taxa from southern China. Environ Pollut 145:225–233. https://doi.org/10.1016/j.envpol.2006.03.015

Wang L, Ji B, Hu Y, Liu R, Sun W (2017) A review on in situ phytoremediation of mine tailings. Chemosphere 184:584–600. https://doi.org/10.1016/j.chemosphere.2017.06.025

Weissenhorn I, Glashoff A, Leyval C et al (1994) Differential tolerance to Cd and Zn of arbuscular mycorrhizal (AM) fungal spores isolated from heavy metal-polluted and unpolluted soils. Plant Soil 167:189–196. https://doi.org/10.1007/BF00007944

Wierzbicka M (1987) Lead accumulation and its translocation barriers in roots of *Allium cepa* L.—autoradiographic and ultrastructural studies. Plant Cell Environ 10:17–26. https://doi.org/10.1111/j.1365-3040.1987.tb02075.x

Wierzbicka M (1988) Mitotic disturbances induced by low doses of inorganic lead. Caryologia 41:143–160. https://doi.org/10.1080/00087114.1988.10797856

Wierzbicka M (1998) Lead in the apoplast of *Allium cepa* L. root tips: ultrastructural studies. Plant Sci 133:105–119. https://doi.org/10.1016/S0168-9452(98)00023-5

Wierzbicka MH, Przedpełska E, Ruzik R, Ouerdane L, Połeć-Pawlak K, Jarosz M, Szpunar J, Szakiel A (2007) Comparison of the toxicity and distribution of cadmium and Pb in plant cells. Protoplasma 231:99–111. https://doi.org/10.1007/s00709-006-0227-6

Woźny A, Zatorska B, Młodzianowski F (1982) Influence of Pb on the development of lupin seedlings and ultrastructural localization of this metal in the roots. Acta Soc Bot Pol 5:345–351. https://doi.org/10.5586/asbp.1982.031

WRB (2014) (World Reference Base for Soil Resources). International soil classification system for naming soils and creating legends for soil maps. World Soil Resources Reports No. 106. FAO, Rome, 181

Yang ZX, Tang J, Chen X, Hu S (2007) Effect of coexist in plant species on soil microbes and soil enzymes in metal lead contaminated soils. Appl Soil Ecol 37:240–246. https://doi.org/10.1016/j.apsoil.2007.07.004

Yang JL, Li YY, Zhang YJ, Zhang SS, Wu YR, Wu P, Zheng SJ (2008) Cell wall polysaccharides are specifically involved in the exclusion of aluminium from the rice root apex. Plant Physiol 146:602–611. https://doi.org/10.1104/pp.107.111989

Yang Y, Liang Y, Han X et al (2016) The roles of arbuscular mycorrhizal fungi (AMF) in phytoremediation and tree-herb interactions in Pb contaminated soil. Sci Rep 6:20469. https://doi.org/10.1038/srep20469

Yanqun Z, Yuan L, Schvartz C, Langlade L, Fan L (2004) Accumulation of Pb, Cd, Cu and Zn in plants and hyperaccumulator choice in Lanping lead–zinc mine area, China. Environ Int 30:567–576. https://doi.org/10.1016/j.envint.2003.10.012

Ying RR, Qiu RL, Tang YT, Hu PJ, Qiu H, Chen HR, Morel JL (2010) Cadmium tolerance of carbon assimilation enzymes and chloroplast in Zn/Cd hyperaccumulator *Picris divaricata*. J Plant Physiol 167:81–87. https://doi.org/10.1016/j.jplph.2009.07.005

Zelinová V, Halušková L, Huttová J, Illéš P, Mistrík I, Valentovičová K, Tamás L (2011) Short-term aluminium-induced changes in barley root tips. Protoplasma 248:523–530. https://doi.org/10.1007/s00709-010-0199-4

Zhao X, Liu J, Xia X, Chu J, Wei Y, Shi S, Chang E, Yin W, Jiang Z (2014) The evaluation of heavy metal accumulation and application of a comprehensive bio-concentration index for woody species on contaminated sites in Hunan, China. Environ Sci Pollut Res 21:5076–5085. https://doi.org/10.1007/s11356-013-2393-3

Zhou W, Qiu B (2005) Effects of cadmium hyperaccumulation on physiological characteristics of *Sedum alfredii* Hance (Crassulaceae). Plant Sci 169:737–745. https://doi.org/10.1016/j.plantsci.2005.05.030

Zikeli S, Jahn R, Kastler M (2002) Initial soil development in lignite ash landfills and settling ponds in Saxony-Anhalt, Germany. J Plant Nutr Soil Sci 165:530–536. https://doi.org/10.1002/1522-2624(200208)165:4

Zikeli S, Kastler M, Jahn R (2004) Cation exchange properties of soils derived from lignite ashes. J Plant Nutr Soil Sci 167:439–444. https://doi.org/10.1002/jpln.200421361

Potential Impacts of Climatic Stress on the Performance of Phyto-bioremediation Techniques

8

Suthirat Kittipongvises and Chongrak Polprasert

Abstract

In this era, climate change is considered one of the greatest sustainability challenges and environmental threats facing our global society. Atmospheric warming from anthropogenic greenhouse gases (GHGs) will persist many centuries and continue to change in the global climate system. Evidently, both frequency and intensity of extreme climate events and natural disasters have been observed at the regional, continental, and global scales over comparable time periods, especially the additional warming of 1.5–2.0 °C. In terms of terrestrial biological systems, all microbial mechanisms have caused several changes in the global climate system (i.e. soil carbon and nitrogen cycling, terrestrial biogenic fluxes of GHGs). Although much attention has been paid to the linkage between microbial population and soil GHG fluxes, the significance of microorganism responses to climate-related environmental stress has remained neglected. The overall aim of this chapter is, therefore, to highlight the consequences of global climate change on the performance of bioremediation treatment processes. All potential effects of climatic parameters, such as increased atmospheric temperature and elevated CO_2 levels, extreme precipitations, soil moisture, soil warming,

S. Kittipongvises (✉)
Environmental Research Institute, Chulalongkorn University (ERIC), Bangkok, Thailand

Research Unit of Green Mining Management (GMM), Chulalongkorn University, Bangkok, Thailand

Center of Excellence on Hazardous Substance Management (HSM), Chulalongkorn University, Bangkok, Thailand
e-mail: suthirat.k@chula.ac.th

C. Polprasert
Department of Civil Engineering, Thammasat School of Engineering, Thammasat University, Pathumthani, Thailand

R. Prasad (ed.), *Phytoremediation for Environmental Sustainability*,
https://doi.org/10.1007/978-981-16-5621-7_8

water stress (drought) on the microbiological mechanisms (i.e. microbial diversity, structure, physiological change, etc.), fate and behavior of contaminants, the ability of plant to uptake the toxic contaminants the environment, and also the efficiency of bioremediation were critically addressed. Some bioremediation techniques (i.e. phytoremediation) were also emphasized and considered for the impacts of combined climatic stress on soil microbe–plant interactions (i.e. bioavailability and potential mobility of contaminants, etc.). Some sustainable bioremediation options in the climate change era and issues for future research were further discussed.

Keywords

Bioremediation · Contaminants · Global climate change · Impacts · Plants · Microorganisms · Soil

8.1 Introduction

As stated in the global Sustainable Development Goals (SDG 13), climate change is considered one of the biggest environmental threats facing our global society that does not respect national boarder. The consequences of changing climate patterns are now affecting every country on every continent. Anthropogenic greenhouse gases (GHGs) emissions are the principal cause of the warming observed since the mid-twentieth century (Venkatramanan et al. 2021a). Atmospheric warming from man-made carbon dioxide (CO_2) will persist for decades and continue to change the climate system, as illustrated in Fig. 8.1. The atmospheric CO_2 concentrations are projected to reach 500–1000 parts per million (ppm) by the year 2100 (IPCC 2007).

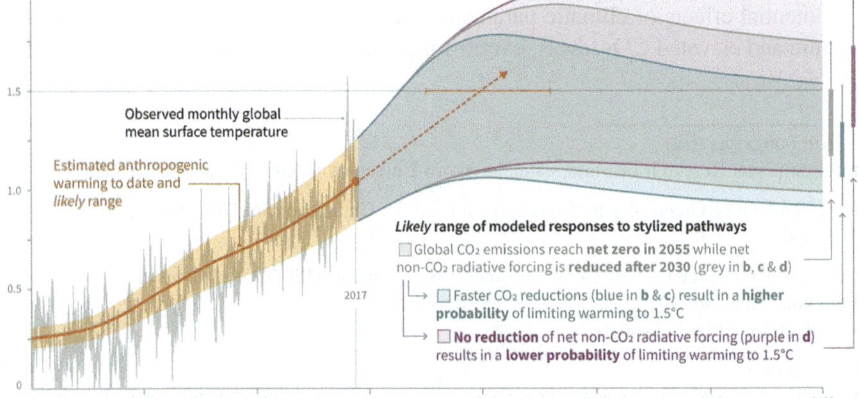

Fig. 8.1 Projected climate change and the probability of limiting global temperature rise to 1.5 °C. (Figure taken from IPCC (2018))

Table 8.1 Observed changes in the global climate system

Observed changes	
Temperature	Global mean surface temperature (GMST) has increased, showing a warming of 0.87 °C (0.75–0.99 °C), over the period of 2006–2015 higher than the average over the decade 1850–1900. For the temperature projections by the CMIP5 Model, the GMST change will likely reach the range of 2.6–4.8 °C (RCP8.5) for the period 2081–2100 relative to 1986–2005
Precipitation and water cycle	Anthropogenic influences have affected both the global water cycle and the precipitation patterns since 1960. Extreme precipitation events will likely be significant and increase over the twenty-first century. Changes in precipitation at 2 °C of atmospheric warming are expected to be higher than 1.5 °C scenario. With the global effects, the El Niño–southern oscillation (ENSO) related precipitation will consequently remain the dominant mode of both interannual and intraseasonal variability in the tropical region
Snow cover	At the end of the twenty-first century, the near-surface permafrost is expected to decrease approximately 37% and 81% according to RCP2.6 and RCP8.5, respectively
Sea level rise	During the period of 1901–2010, the rate of global mean sea level has risen by about 0.19 m (0.17–0.21 m)

Further, evidences for human influence on extreme weather and climate events have strengthened since the nineteenth century (IPCC 2007; Van der Putten 2012) (Table 8.1). Both frequency and intensity of climate change and natural disasters, such as droughts, floods, and storms have been continuously detected over time, especially the additional warming of 1.5 °C (IPCC 2018). Without any actions taken, the global surface temperature is projected to rise to 3 °C by the end of twenty-first century (United Nations 2019). Extreme weather and climate related events can cause the loss of life and significant impacts on ecosystem functioning, water supply, food security, livelihood, well-being, and socioeconomic development. Adverse impacts of global climate change on the coupled human-natural system have evidently been observed in our era. This means that, for example, an increase in average global temperature from approximately1.5–2 °C is likely to increase the exposure to the risks associated with sea level rise to both human and ecological systems (i.e. increased seawater intrusions, floods, and also damage to infrastructure). On land, the impacts of climate change on biodiversity and ecosystems (i.e. terrestrial, freshwater, and coastal systems) are likely to be higher at about 2 °C of global warming compared to 1.5 °C scenario (Venkatramanan et al. 2020, 2021b) (Fig. 8.2).

Climate-related environmental stress can have large impacts on the biotic processes in the terrestrial ecosystem where there is a huge pool of dynamic carbon. A change in the pattern of global climate is altering the distributions of species and simultaneously impacting interactions among organisms (Van der Putten 2012). Naturally, photosynthetic microorganisms consume atmospheric CO_2, while the heterotrophic microorganisms degrade organic compounds to emit atmospheric

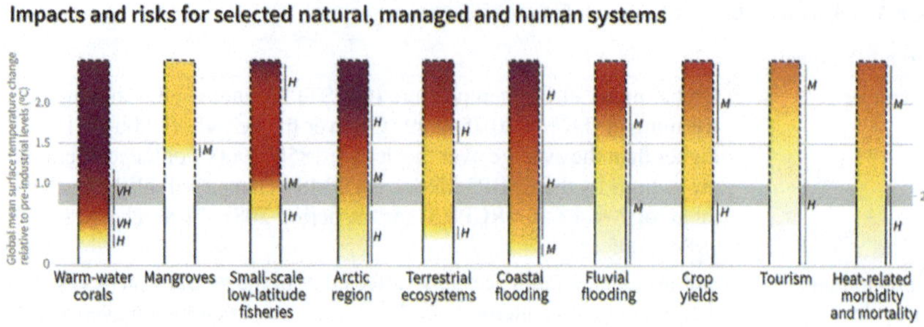

Fig. 8.2 Potential impacts and associated risks of climate change for selected natural, managed, and human systems (where *Red* represents widespread and severe impacts; *Yellow* represents moderate risks or the impacts of climate change are detectable with at least medium confidence; *White* represents undetectable risks or no impacts are attributable to the change in climate system). (Figure taken from IPCC (2018))

GHGs. The estimation conducted by Singh et al. (2010) found that approximately 120 billion tons of carbon at 2 °C of atmospheric warming is directly consumed by autotrophic microorganisms in soil, while heterotrophic microbials contribute 119 billion tons of carbon emission. The balance between biological processes and their metabolic pathways is considered as the key contributor to the net global carbon flux, depending on the climate conditions (i.e. temperature and precipitation patterns). This could mean that soil microbial communities are mutually linked to the global ecosystem functioning and biological processes as an important role in terrestrial carbon and nitrogen cycling. The terrestrial surface flux of 5000–7500 kg of carbon per year is evidently considered as a main component of the large-scale global carbon cycling (Raich and Schlesinger 1992). As the atmospheric GHGs (i.e. CO_2, CH_4, and N_2O) predominantly originate from microbial activities, all their mechanisms have caused several changes which have also influenced them (Zimmer 2010). In other words, the microbial world plays an enormous role in the global carbon and other biogeochemical cycles. Recently, a great deal of attention has been paid to microbial community and the dynamics of soil-atmosphere net GHGs fluxes. However, the significance of microbial response to environmental changes is mostly absent in the global climate change debate.

To address this issue, the major aim of this chapter was to highlight the consequences of climatic stress on the bioremediation performance. All impacts and consequences of global climate change and potential parameters, such as temperature, atmospheric CO_2 levels, precipitations, soil moisture, and soil warming on the microbiological mechanisms (i.e. microbial diversity, structure, physiological change, etc.), fate, behavior, the biodegradability, and ability of plants to uptake the toxic contaminant in the environment were addressed. In addition, by emphasizing phytoremediation and constructed wetlands bioremediation techniques, this chapter addressed how climatic change directly and indirectly affects both soil microbes and soil microbe–plant interactions (i.e. bioavailability and potential mobility of contaminants, etc.) and also discussed some emerging and sustainable bioremediation options in the climate change era and issues for future research.

8.2 Bioremediation Techniques

8.2.1 Overview

Bioremediation is referred to as any biological treatment process that mainly uses microbiota or natural microorganisms (typically heterotrophic bacteria, fungi, algae, or yeast) to degrade toxic contaminants into either less toxic or non-toxic compounds (i.e. carbon dioxide (CO_2), methane (CH_4), water (H_2O), inorganic salts, microbial biomass, and other byproducts). In this context, three main strategies for bioremediation technique are the following: natural attenuation, biostimulation, and bioaugmentation. Firstly, the term "natural attenuation" or intrinsic bioremediation refers to the natural degradation of the pollutant without any direct intervention. In this process, the environmental contaminants are directly transformed to less toxic condition by a variety of biodegradation mechanisms, such as dilution, sorption, volatilization, and so on. Secondly, biostimulation means the adjustment process of surrounding conditions (i.e. aeration, pH, and temperature) and the addition of nutrients, oxygen, or other electron acceptors to stimulate the microbial degradation rate of specific contaminants. Lastly, bioaugmentation focuses on the addition of actively growing and specialized microbial strains, nutrients, and electron donors and acceptors into the treatment system (USEPA 1991). The main treatment process involves reducing the solubility of the pollutants by altering pH value, the redox reactions, and also the adsorption of the pollutants in contaminated site. This could be further explained by the fact that this biological process commonly requires a mechanism for stimulating microbial activity by providing one or more of the

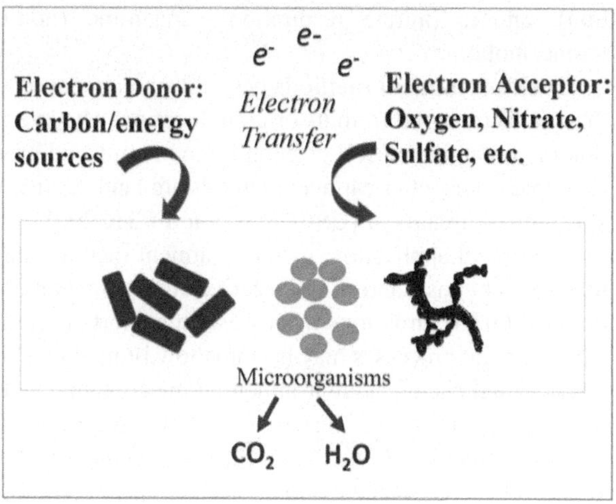

Fig. 8.3 Mechanisms of electron exchange by microorganisms during bioremediation

followings: electron acceptor (oxygen, nitrate), energy source (carbon), and nutrients (nitrogen, phosphorus, and trace elements). Microorganisms then degrade a wide variety of all organic contaminants (carbon-containing) found in the environment to obtain energy for their growth (Fig. 8.3).

Bioremediation is considered as an environmentally sound and cost-effective treatment method (USEPA 2005) that can be used in various applications such as the oil spill cleanup operation, the rehabilitation of the contaminated sites (i.e. soils, sediments, sludges, surface, and groundwater), and the remediation of petroleum hydrocarbons and hazardous organic compounds in soil (Thakare et al. 2021). This treatment technique can also enable appropriate reuse of the treated soil and consequently minimize the disposal of contaminated soil to landfill sites and often be completed where the contamination problem is located. Bioremediation method, suitable for treatment of a variety of organic contaminants under either anaerobic or aerobic conditions (Table 8.2), remains an active area of research and technology development. The following pollutants have been successfully remediated at many contaminated sites: halogenated and semi-volatile organic compounds (VOCs/SVOCs). Pollutants with a more limited treatment efficiency include: trinitrotoluene (TNT), dense nonaqueous phase liquids (DNAPLs), and also hexahydro-1,3,5-trinitro-1,3,5-triazine (RDX).

8.2.2 Types of Bioremediation

There are theoretically two different common types of bioremediation, namely in situ and ex situ bioremediation (USEPA 2016) (Table 8.3).

In situ treatment methods mainly involve managing or treating the contaminated material in the location in which it was found, without removing the contaminants from its original site. The examples of in situ are source remediation technologies such as slurry-phase lagoon aeration, bioventing, and also groundwater treatment technologies such as biosparging and aerobic bioremediation (O_2 respiration), anoxic (nitrate respiration), anaerobic (non-O_2 respiration), and even co-metabolic.

Ex situ treatment methods describe a treatment process where the contaminated material is removed from its original location. The overall processes require excavation of contaminated soil or pumping contaminated groundwater prior to treatment. Examples of ex situ treatment methods are land treatment, biopiles, composting, and slurry-phase treatment (USEPA 2000a; FRTR 2001).

In terms of application, in situ treatment methods may be advantageous since all management costs and some negative environmental impacts (i.e. energy consumption and GHGs emissions from waste transportation) may be reduced. However, in situ treatment processes may be somehow limited by ability to manipulate all related physical and chemical environment at the contaminated site. Comparatively, ex situ methods can be faster, easier to control, and employed to treat a wider range of pollutants than in situ methods. Ex situ treatment methods, however, require excavation of contaminated soils, sludge, and groundwater, leading to increased

Table 8.2 Classes of pollutants potentially suitable for bioremediation technique

Class	Specific example	Aerobic biodegradation process	Anaerobic biodegradation process
Chlorinated solvents			
Alkanes	Chloroform	+	+
Alkenes	Trichloroethylene	+	+
Aromatic compounds	Benzene, ethylbenzene, xylene, toluene	+	+
Nonhalogenated phenolics and cresols	2-Methylphenol	+	+
Chlorinated phenyls	Pentachlorophenol	+	+
Monochlorinated aromatic compounds	Chlorobenzene	+	
	Perchloroethylene		+
Polychlorinated biphenyls	4-Chlorobiphenyls 4,4-Dichlorobiphenyls		+
	Trichlorobiphenyl	+	+
Nitrogen heterocyclic compounds	Pyridine	+	
Polyaromatic hydrocarbon (PAHs)	Anthracene Pyrene Benzo(a)pyrene Fluorene	+	
Pesticides	Atrazine Carbaryl Carbofuran Parathion	+	+
Crude oil	Petroleum hydrocarbons (TPHs)	+	
Pharmaceuticals and personal care products (PPCPs)	Acetaminophen Sulfamethoxazole Ibuprofen Nonsteroidal anti-inflammatory drug	+	+

(Remark: + Suitable)

operations costs. In practice, they usually require additional treatment processes of the contaminants before, and sometimes, after the actual remediation step. Therefore, more uncertainty and risk of contamination can further arise. Overall, it could be noted that the effectiveness of both in situ and ex situ bioremediation techniques depends on several factors such as the degree of the contaminants, the nature and the metabolic potential of the microorganisms utilized, and also the prevailing environmental factors at the contaminated sites (Ojuederie and Babalola 2017).

Table 8.3 Examples of in situ and ex situ bioremediation methods

In situ treatment methods	
Source treatment technologies	
Bioventing	Bioventing is designed primarily to treat contaminants by providing O_2 to existing soil microbes (either extraction or injection of air) for enhancing the activities of indigenous bacteria and stimulate the degradation rate of in situ treatment
Slurry-phase lagoon aeration	Slurry-phase lagoon aeration commonly uses a lagoon as a place to combine both air and soil to promote in situ biological degradation of the contaminants
Groundwater technologies	
Biosparging/in situ air sparging (IAS)	Biosparging technique mainly involves the injection of air/O_2 and also gas-phase nutrients under pressure below the saturated zone in order to activate the rate of aerobic biodegradation by naturally occurring bacteria. This biosparging/in situ air sparging method also increases the mixing efficiency in the submerged zone and consequently increases contact between soil and groundwater
Aerobic	In the aerobic in situ treatment, O_2 sources are directly injected into groundwater in order to enhance the biodegradation rate
Anaerobic/anoxic	In anaerobic condition, carbon sources are directly injected into the contaminated groundwater zone to promote the degradation rate of specific contaminants. In anoxic process, there is no molecular oxygen but nitrite/nitrate is present. Nitrate then acts as an electron acceptor, while electron donor is organic compounds in the heterotrophic process
Ex situ treatment methods	
Biopiling/cells or mounds	Biopiling is an ex situ treatment method in which excavated contaminated soils are directly mixed with soil amendments and then bioremediated using forced aeration into piles to enhance biodegradation
Landfarming	Landfarming (ex situ land treatment) is a treatment method that removes all contaminants from soil, where excavated soils or sediments are spread over a much thinner layer over the ground surface (biocell) and then periodically tilled or turned over in order to aerate to the contaminated media
Composting	Ex situ composting method is a controlled biological treatment process that mainly treats organic pollutants by promoting the ability of microorganisms to degrade the contaminants under thermophilic conditions (40–50 °C). The process involves mixing the contaminated soils, sediments, or sludges in order to obtain the optimum levels of O_2 and H_2O for the biodegradation mechanisms. The selection of organic amendments depends mainly on the soil porosity and the balance of carbon and nitrogen needed to enhance the microorganism activities. In practice, the following three designs are commonly applied: (1) aerated static piles, (2) mechanically agitated in-vessel composting, and (3) windrow composting
Slurry-phase treatment or bioreactor	Slurry-phase treatment method is defined as a bioremediation treatment that mainly involves the excavation of toxic soils or sediments, mixing with H_2O and consequently placing it in a bioreactor to keep microorganisms in contact with the pollutants

(continued)

Table 8.3 (continued)

Soil washing	Soil washing is an ex situ water-based method to treat contaminated soils by concentrating them into a smaller volume of soil through gravity separation and attrition scrubbing techniques
Ex situ soil vapor extraction (SVE)	SVE is a full-scale treatment method where soils, sledges, or sediments are excavated and soil piles are placed over a network of aboveground piping and vacuum is applied to encourage volatilization of the organic contaminants

Table 8.4 Phytoremediation treatment processes and related contaminants and plant species

Phytoremediation	Contaminants	Plant species
Phytostabilization	Inorganic compounds (As, Cd, Cu, Cr, Pb, Zn)	*Brassica juncea*, hybrid poplars grasses
Photodegradation	Organic compounds, herbicides, chlorinated solvents	Hybrid poplars, algae
Rhizofiltration	Organic/inorganic compounds (i.e. Cd, Cu, Ni, Cr, Ni)	*Brassica juncea*, *Helianthus annuus*, spinach, and corn
Photovolatilization	Organic/inorganic compounds, chlorinated solvents	*Arabidopsis thaliana,* Poplars, Alfalfa, *Brassica juncea*
Constructed wetlands	Organic/inorganic compounds, metal-contaminated urban stormwater runoff, PPCPs, acid mine drainage	Cattail, *Typha* spp.; Common reed, *Phragmites communis*; Rush, *Juncus* spp.; Bulrush, *Scirpus* spp.

8.2.3 Plant–Microbe-Based Bioremediation

As outlined earlier, bioremediation treatment could be performed both in situ or ex situ technique. Altogether, the treatment process can be further conducted by "green plants" which simultaneously extract and remediate the contaminants from the contaminated sites. The classic examples of the microbial and plant-assisted bioremediation are phytoremediation and constructed wetland. Table 8.4 also presents the examples of plants and microorganisms which have been applied to degrade the environmental contaminants in different phytoremediation processes and constructed wetland systems.

8.2.3.1 Phytoremediation

Phytoremediation method ("*Phyto*" is defined as plant and "*Remedium*" represents restoring balance) basically refers to a cleanup green technology by the direct use of living green plants and associated soil microbes in the in situ process to remove, detoxify, and also immobilize both organic and inorganic pollutants in the environments (i.e. soils, sediments, sludges, surface and groundwater, etc.) (Sarma et al. 2021; Sonowal et al. 2022) (Fig. 8.4a). Several advantages of phytoremediation are ecofriendly, energy efficient, and cost-effective manner than other treatment methods like ex situ excavation and soil washing. Through the natural physical and bio-chemical activities of the green plants and microorganisms,

Fig. 8.4 (**a**) In situ phytoremediation technology for treatment of contaminated soil, Nakhon Ratchasima, Thailand; and (**b**) CWs treating a hospital wastewater containing pharmaceuticals and personal care products (PPCPs)

phytoremediation processes can be categorized into phytostabilization, phytodegradation, phytoextraction, rhizofiltration, and phytovolatilization (Fig. 8.5).

Phytostabilization (also known as phytosequestration or phytodeposition) involves the reduction of mobility and bioavailability of the contaminants in the environment through fixing or sequestering the toxic pollutants in the vadose zone directly through the accumulation of contaminants by plant roots or sorption within the rhizosphere. All microbial activities associated with the plant roots can also simultaneously accelerate the contaminant degradation. Additionally, the

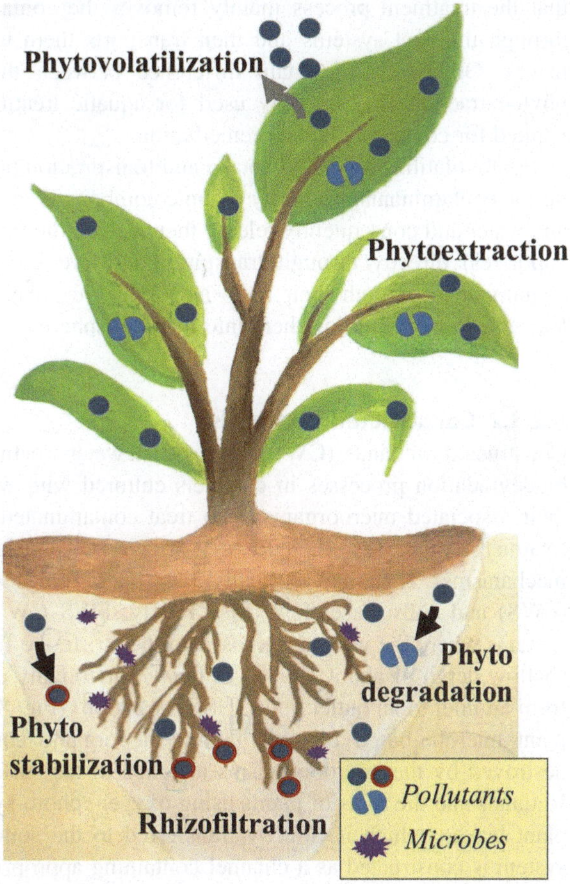

Fig. 8.5 Phytoremediation method. (Figure modified from Ghori et al. (2016))

remediation processes can be enhanced by adding organic soil amendments that immobilize the contaminants combined with plant species that are tolerant with high concentrations of toxic compounds.

Phytodegradation (also called phytotransformation) is the breakdown process of toxic contaminants taken up by plants by either plant metabolism or the effects of certain enzymes (i.e. oxygenase or dehalogenase) produced by the plants. Complex organic contaminants are consequently degraded into less toxic compounds and also incorporated into the plant tissues where they are metabolized. The efficiency of the transformation process depends mainly on plant types and can occur in roots, stems, or leaves of the plants.

Phytoextraction (also known as phytoaccumulation) involves the uptake of contaminants from soils, sediments, sludges, and water by plant roots into above-ground portions of plants (i.e. root-to-shoot transport processes). The mechanisms of phytoextraction mainly involve the processes of translocation of pollutants into both shoots and leaves of the plants. Rhizofiltration is also similar to phytoextraction in that the treatment process mainly removes the contaminants by absorbing them through the root systems and then transports them up into either their stems or leaves. Obviously, the main difference between the two mechanisms is that phytoextraction is commonly used for aquatic treatment, while rhizofiltration is applied for contaminated soil remediation.

Photovolatilization is the uptake and transpiration processes in which plants take up the contaminants, primarily organic compounds, from soils, sediments, sludges, and water and consequently release them as volatile form into the air (evaporates or vaporizes) directly through transpiration. More specifically, plants take up the contaminants through their roots and transpire volatile compounds through their leaves, and then emitting them into the atmosphere.

8.2.3.2 Constructed Wetlands

Constructed wetlands (CW) are the wastewater treatment system that use natural biodegradation processes in channels cultured with wetland vegetation, soils and their associated microorganisms to treat contaminated wastewater (Fig.8.4b). The treatment processes include a variety of physicochemical and biological mechanisms. There are basically two major types of CW: free water surface (FWS) and subsurface flow (SF). Firstly, FWS CW is defined as the treatment systems where the water is exposed to the air. In the FWS system, wastewater at a shallow depth of about 10–50 cm flows horizontally over a vegetated soil surface from an inlet to an outlet point (USEPA 2000b) (Fig. 8.6a). For the mechanisms of plant–microbe-based bioremediation, the organic contaminants are biologically destroyed by the microorganisms attached to the surface of media in constructed wetlands and the roots of plants using oxygen, photo-synthetically generated by the plant leaves, which is directly transferred to the soil–water matrix. The SF CW system is constructed as a channel containing appropriate media (e.g. coarse rock, gravel, sand, and soils) that supports the growth of emergent plants to treat wastewater (Fig. 8.6b). The level of wastewater is maintained to be below the bed of CW to promote the ability of microorganisms attached to the submerged substrate. Consequently, wastewater flows either horizontally or vertically through the medium and is also purified during the contact with the roots of plants growing in the surfaces of the media themselves based mainly on physicochemical and biological reactions similar to the FWS system. Depending on the climate conditions, the harvested aquatic-plant biomass can be utilized as animal feed or directly converted to soil conditioner. Since the CW units provide effective biological treatment in a passive manner and also minimize all energy and skilled operator attention, the operation and maintenance costs are cheaper than other conventional treatment technologies (Polprasert and Kittipongvises 2010).

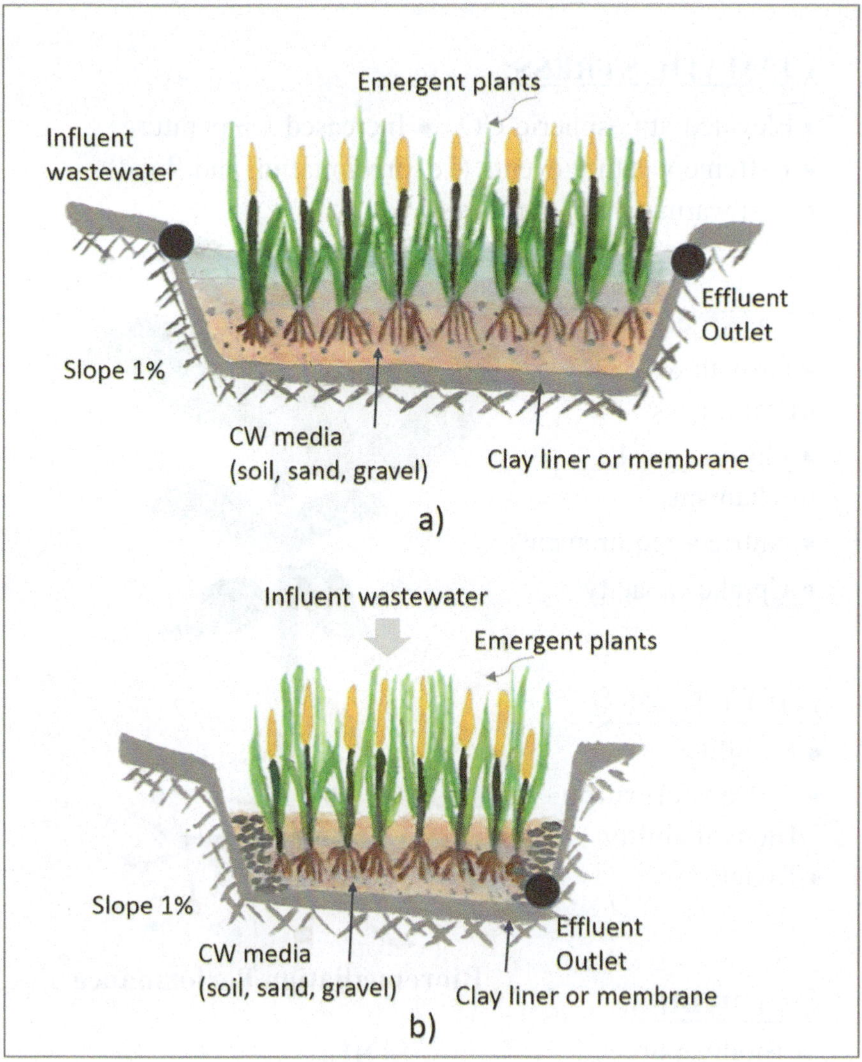

Fig. 8.6 Schematic diagrams of (**a**) FWS and (**b**) SF CW

8.3 Impacts of Climatic Change on the Bioremediation Performance

Global climate change has far reaching effects on all aspects of biogeochemical cycles. The consequences of climate stress can be observed in the alteration of the biological diversity. Changes in the intensity of climate events, such as high temperatures, drought, extreme precipitation, and storms are bound to have

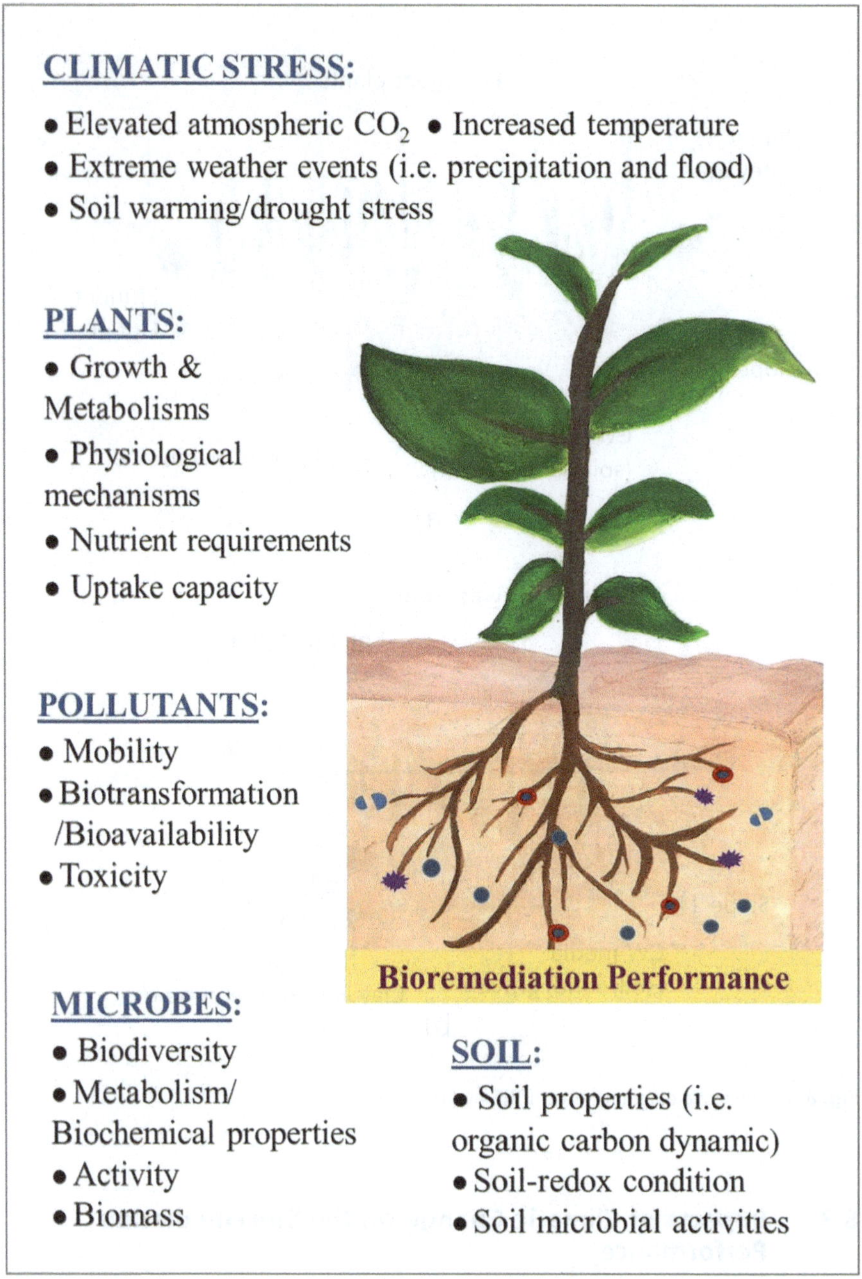

Fig. 8.7 Potential impacts of climatic stress on microbial and plant-assisted bioremediation

significant impacts on the bioremediation performance since the biodegradability of contaminants by soil microbial is dependent on the range of environmental and climate change impacts (Fig. 8.7). Also, the optimization of bioremediation treatment processes is a complex system involving several environmental factors, the important ones include: the existence of microorganisms capable to biodegrade the toxic contaminants, the presence of contaminants, and all related environmental and climate factors, such as temperature, moisture, humidity, pH, nutrients, and also the presence of O_2 or other electron acceptors in the treatment system. The details are as follows:

8.3.1 Effects on Fate and Behavior of Pollutants

Climate change can potentially affect the uptake, fate, long-range transport, biotransformation, bioavailability, and toxicity of numerous chemical toxicants by altering all physicochemical and biological mechanisms. In other words, due to changes in partitioning, bioaccumulation, carbon pathways, and biodegradation rates, altered environmental and climate conditions have a much greater potential on environmental distribution, transport, and exposure pathways of contaminants (Noyes et al. 2009; Rohr et al. 2013). For instance, rising temperature and precipitation are expected to have the biggest influence on the partitioning of toxic contaminants. Other processes, such as melting snow and ice sheet, biota dynamics, and organic carbon cycling, will be potentially altered by global climate change, leading to a significant increase in fugacity and concentrations of the contaminants (MacDonald et al. 2003). Also, higher soil surface and ambient temperature significantly enhance the volatilization and dispersion of both volatile organic and inorganic pollutants (i.e. methane, ammonia, sulfides, and nitrous oxide) from the soil to air. The speciation of elemental contaminants in environment depends mainly on biogeochemical cycle of the elements and temperature-mediated biotic interactions. As a case example illustrates, the elevated temperature can result in higher methylation rate of inorganic mercury in soil, water, and also sediment (Verta et al. 2010; Yang et al. 2016). In case of elemental mercury (Hg^0), the concentrations of Hg^0 in the surface soil (0–5 cm) will increase by about 9.4–40% under increased soil warming of 1–3.7 °C. The transformation to a more bioavailable form of Hg was detected due to the warming climate (Booth and Zeller 2005; Biswas et al. 2018). Besides this, with a climatic condition of increased rainfall variability, a high level of soil moisture content also resulted in an increase in the accumulation, exposure, and toxicity of inorganic mercury, such as Hg (II) in soils (Buch et al. 2017).

Additionally, based on the toxicity assessment of pesticides to earthworms under different temperatures conducted by Vilki and Ečimović (2015), the elevated ambient temperature mostly caused increase in pesticide toxicity (i.e. imidacloprid, chlorpyrifos and cypermethrin, lambda-cyhalothrin, difenoconazole and propiconazole, azoxystrobin and cyproconazole, tembotrione, and fluazifop-*p*-butyl) and also caused a higher level of the oxidative stress to the soil organisms. Furthermore, soil warming induced by climate change significantly influences the increase of contaminants transportation in soils. In the case of removal of organic contaminants, the geomembrane-assisted petroleum hydrocarbon remediation was applied at an Artic landfill by McWatters et al. (2016). Results found that the mobility rate of benzene, toluene, ethylbenzene, and xylene (BTEX) increased at elevated temperature of about 2 °C in the soil surface. For the persistent organic pollutants (POPs), toxico-kinetics of POPs could be altered as a direct result of temperature changes. As such, due to the increase in the rate of biotransformation, increased temperature may decrease the persistence of POPs in the environment. This climate-induced condition may be directly linked with an increased long-range transport of POPs to the remote regions (UNEP/AMAP 2011).

Drought stress, a meteorological term which means less water or rainfall, is commonly a prevalent environmental restraint that directly affects plant growth and productivity. In terms of drought effect on metal availability in soil and heavy metal uptake by the plant, water-deficit or drought condition may affect plant mineral nutrition through a reduced ion transport in plant roots and changes in physiological capacity of nutrient uptake by root (Thomas 1997). Obviously, the main effects of water stress on the plant uptake capacity of the contaminants were mainly observed in the root zone. An experiment conducted by Pascual et al. (2004) reported that, after the addition of sludge and mineral fertilizer, the uptake and transfer rate of the following metals, namely zinc (Zn), manganese (Mn), nickel (Ni), chromium (Cr), and lead (Pb), from contaminated soil to the plant root (Ryegrass; *Lolium multiflorum* Lam.) were significantly decreased under the water-deficit condition. The uptake behavior of ryegrass in the removal of metals under dry condition depends on size of the root plant, because water stress more drastically reduced the rate of root growth than those of the well-water plants. Moreover, Sardans et al. (2008a) investigated impacts of climatic changes on trace element accumulation in aboveground biomass and also soil and found that soil warming increased both concentrations and aboveground accumulation of the following trace elements: Al, As, Cr, Cu, and partially Pb. Due to the greater retranslocation and photosynthetic capacity in *Erica multiflora*, the relationship between temperature and the accumulation of toxic pollutants was significant in *E. multiflora* and *Globularia alypum*. More specifically, drought increased As and Cd (40–55%) in *E. multiflora* stems while decreased about 50% of Cu in leaves, 28% of Ni in stems, and 32% of Pb in leaf litter of *G. alypum*.

8.3.2 Effects on Microorganism Communities

Microorganism communities are considered to be a crucial important factor in both the biogeochemical cycles and climate feedbacks. Although there is no direct evidence linking climate change impacts on the bioremediation performance, some researches on microbial communities in soil and global warming (Castro et al. 2010) argued that changes in the physicochemical and biological properties of microorganisms may alter their metabolic processes and thereby the efficiency of bioremediation. Factors associated with climate change such as elevated atmospheric CO_2 concentration, altered temperature, precipitation, soil humidity, and seasonal patterns have both positive and negative consequences on plants, soil carbon balance (Classen et al. 2015), soil microorganism communities (Castro et al. 2010), and their physiological mechanisms. However, different kinds of microorganisms grown under different surrounding conditions might respond differently, by either accelerating or alleviating, to the impacts of climate change. Temperature is considered to be one of the most important determinants of both microbial activities and its growth rate. Goiun et al. (2013) reported that a higher temperature was directly linked to higher levels of microbial growth and its activity that potentially increasing the rate of metabolic reactions in the degradation of organic contaminants. However, prolonged warming directly affects multiple biomechanisms, ranging from shifts in the physiology of individual microorganism to changes in species composition (Bradford 2013). For instance, because of a reduction in microbial biomass and activity over time, some studies indicated that the initial acceleration rate of soil carbon decomposition was declined with long-term warming (Frey et al. 2008).

8.3.2.1 Temperature

Plant and microorganism communities are coupled through mutual responses to environmental changes such as precipitation and temperature. The effects of warming on several functional genes mainly involved in the degradation of soil carbon have been also reported (Wu et al. 2011). Under such situations, soil carbon and microbial activities may be imbalanced as the respiration rates respond more positively to rising temperature rather than the photosynthesis rates. Changes in physicochemical and microbial characteristics may alter the metabolic pathways and thereby the bioremediation. Since the bioavailability and toxicity of some toxic contaminants can increase with the higher temperature, the contaminant-induced degradation mechanisms may be impacted by climatic change, leading to the lower soil quality by the time of soil remediation commences. A study on the effect of temperature on the diesel oil biodegradation activity of psychrotrophic yeast (*Yarrowia lipolytica*) in the bioaugmentation process carried out by Margesin and Schinner (1997) found that the maximum degradation activity of the indigenous soil microorganisms (after 30 days) was observed at 15 °C than other temperatures tested (20–30 °C).

8.3.2.2 Elevated Ambient CO_2 Levels

The increased presence of CO_2 in the atmosphere will directly change the richness and structure of the rhizosphere and soil microbial community (Veresoglou et al. 2016). An experiment conducted by He et al. (2012) found that the bacterial richness (45 phyla) under elevated CO_2 (560 ppm) was decreased significantly compared to those under ambient CO_2 conditions (380 ppm). A significant reduction of the signal intensities of dominant microbial phyla, such as *Proteobacteria*, *Actinobacteria*, *Firmicutes*, *Acidobacteria*, and *Bacteroidetes* was detected at elevated atmospheric CO_2. Similarly, Chen et al. (2016) reported that increasing CO_2 concentrations in the atmosphere significantly reduced the diversity of soil bacterial communities and also the relative abundance of *Acidobacteria* and *Chloroflexi*. Overall, this could mean that the richness, composition of microbial communities in soil were shifted to respond to eCO_2. Moreover, soil microbial activity is directly linked to their extracellular enzymes production and the soil physicochemical properties that are influenced by climatic stress conditions. High levels of CO_2, for instance, can affect cell morphology of microorganisms. A study carried out by Wu et al. (2010) investigated the increased release of extracellular protein from the *E.coli* after exposure to CO_2. Changes in microbial morphology and its cell wall structure were also detected with progressively longer the exposure times to CO_2. As an indicator of cellular damage, the rate of lactate dehydrogenase release from microorganisms tends to increase, particularly when CO_2 concentrations reached 30,000 ppm (Wan et al. 2016). Generally, high CO_2 condition will cause the intracellular substances leakage, which could be the possible reason that high carbon dioxide levels inhibit most microbial growth (Yu and Chen 2019). Under such situations, as a result of physiological defense mechanism, the high CO_2 concentration will cause an increase in the formation of extracellular polymeric substances (EPS) that may consequently cause changes in the structure of microbial communities. Accumulation of CO_2 in the air might also inhibit the growth rate of soil microorganisms by displacing a partial or all of O_2 available for their metabolic processes. High CO_2 levels can inhibit the activity of functional enzymes in microbial communities and also modify the transcriptional regulation and protein expression by affecting the electron transport system and consequently influence the microbial physiology. More importantly, there are effects related to atmospheric CO_2 enrichment on plants and microbial interactions, thereby slowing decomposition of microorganisms in soils (Jastrow et al. 2005).

8.3.2.3 Changes in Soil Moisture Content

Change in climate is potentially leading to the occurrence of more heavy flooding and also extreme droughts which directly affect the moisture content in soil. Silva et al. (2008) reported that microbial respiration depends mainly on moisture content in soil than on temperature. It was predicted that if soil temperature increases, especially during warm and dry conditions, several microbial metabolisms and their functions are likely to be affected which will consequently disturb their ability to degrade the hazardous chemicals. Bioavailability of pollutant can be also altered by wetting-and-drying cycles. Generally, low water content in the soil ecosystem can limit all activities of soil microorganisms and the rate of biodegradation of contaminants, while a high content of water can limit oxygen availability of microbial and also negatively affect aerobic degradation of contaminants (i.e. hydrocarbons).

8.3.3 Effects on Soil

8.3.3.1 Temperature

Human-induced warming significantly influences the properties of soil, particularly the organic carbon dynamics. In a recent study, Crowther et al. (2016) revealed that up to about 1 °C of the warming of soil surface would consequently result in the loss of carbon from the upper soil horizons by approximately 30 ± 30 petagrams to 203 ± 161 petagrams of carbon. Soil warming induced by climate change, in general, can both directly and indirectly affect the availability of nutrient and transport pathways of heavy metals in soil by altering metal bioavailability and its distribution in the plant tissues. This is evident by the fact that an increase of soil organic matter (SOM) due to higher temperature could lead to a reduction of capacity of cation-exchange and ability of soils to stabilize nutrients and toxic metals (Sardans et al. 2008b). In other words, the decomposition of SOM due to soil warming can promote the increase in the available fraction of heavy metals in the soil resulting in greater amount of metal contaminant uptake by the plants. Besides this, the negative impacts of soil warming may be also found on the microbial enzymatic functions. A study conducted by Tan et al. (2018) found that higher soil temperature, particularly in alkaline conditions, may interrupt soil microorganism activity. Through changes in soil enzymatic activities, increased soil temperature can enhance trace elements mobility and transform from metal organic to exchangeable complex form, resulting in an increase of plant metal uptake.

8.3.3.2 Extreme Precipitation

An increase in precipitation and changes in both intensity and frequency of rainfall associated with climatic change can absolutely affect both soil properties and processes. Flood immersion triggered by extreme precipitation events, for instance, can alter the soil moisture regimes featured by the O_2 deprivation and the lowering of soil-redox potential (LeMonte et al. 2017). In a similar study, Fronhe et al. (2014) observed that the cycle of drying-wetting of soils leads to fluctuating both hydrological and soil-redox conditions, exerting influence on the metal mobility dynamics in soil through altering the pH, DOC, and the properties of Fe, Mn, and S minerals. The release of several toxic trace elements is a concern for floodplain areas and also for quality of the water body. The solubility of trace elements in flooded soils is commonly controlled by the following important factors: redox potential (E_H), DOC, sulfate, and Fe/Mn oxides. Shaheen et al. (2014) examined the impact of different flood-dry-cycles on the concentrations of toxic trace elements, namely As, Cr, Mo, and V in contaminated floodplain soil. Their results found that flooding condition caused a significant reduction in E_H and pH values. Overall, concentrations of soluble As, Cr, Fe, Mn, Mo, and also DOC were found to increase under reducing conditions in the long-term flooding. Under oxidizing conditions, both As and Cr elements tended to be mobilized during the short-term flooding. It is generally recognized that the trace element dynamics can be determined by the duration of the flooded period or the length of time that soil is exposed to flooding.

8.3.4 Effects on Plants

8.3.4.1 Elevated Ambient CO_2 Levels

Under elevated CO_2 levels in the atmosphere, these climatic effects may affect soil acidity, directly through inputs of carbonic acid released from plant roots and respiration of soil microbial that may consequently enhance the heavy metals mobility and their availability to uptake by plants (Öborn et al. 1995). Simultaneously, microbial communities in soil are affected by plant responses to a high CO_2 concentration, including increased rhizodeposition and faster uptake of nutrients. Further to this, in the context of plant defense mechanisms, elevated ambient atmospheric CO_2 may have stimulatory effect on plant growth and the chemical composition of plant tissue. Norby and Iversen (2006) reported that increased fine root production under an elevated CO_2 condition might allow green plants to match increased assimilation of carbon with a higher uptake of soil-derived elements. Carbon dioxide enrichment can also be a function of both the nutritional requirements of plant and also their uptake capacity. Natali et al. (2009) assessed the effects of carbon dioxide on the biological storage and stoichiometry of non-essential trace metals, such as aluminum (Al), lead (Pb), and vanadium (V) which are important contaminants in the environment. For example, there were significant effects of an elevated carbon dioxide on concentrations of heavy metals (i.e. Co, V, Cu, Fe, Mn, Ni, and Zn) in *P. taeda* foliage. As such, all metal concentrations found in the *P. taeda* leaf were significantly greater under ambient condition than under CO_2 enrichment condition. Jia et al. (2010) investigated the combined effects of Cd and enrichment of carbon dioxide on the growth, physiochemical characteristics, elemental composition, and antioxidant level in *Lolium mutiforum* and *Lolium perenne* grown in the condition of soils amended with cadmium. The results revealed that elevated CO_2 decreased concentrations of Cd in both roots and shoots parts of both *Lolium species*. In this context, carbon dioxide enrichment condition may ameliorate the Cd toxicity by increasing photosynthesis rate and altering the activity of antioxidant enzymes including superoxide dismutase (SOD) and catalase (CAT) of the *Lolium species*. Gelareh et al. (2018) also investigated the effects of CO_2 on physiology of wheat (*Triticum aestivum L.*) and sorghum (*Sorghum bicolor* (*L.*) *Moench.*) under Cd enrichment and found that, with increasing Cd concentrations, the activities of antioxidants and SOD and glutathione peroxidase (GSH-px) increased in wheat. In addition, elevated CO_2 then decreased the concentrations of Cd in both shoots and roots of wheat, but a reverse relationship was found in sorghum. For Pb uptake, Kim and Kang (2011) analyzed the effects of CO_2 on phytoextraction of Pb by pine seedlings and found that the elevation of carbon dioxide significantly increased both total biomass and Pb accumulation in the root of pine seedlings by increasing Pb bioavailability. However, the authors have pointed out that rising atmospheric CO_2 levels in the future might interfere the phytoremediation efficiency by influencing DOC quantity and DOC–metal complexation reaction in the treatment system. Taken together, the ability of a plant to uptake the metal under the elevated CO_2 may be differently affected as by different types of plant, its photosynthesis capacity and biomass production. Above all, further phytoremediation research is needed to understand how specific heavy metals influence plants in response to the condition of elevated atmospheric CO_2 levels. Other parameters including all physicochemical, biological environment, and also plant type can directly and indirectly influence the functions of soil including mobilization and immobilization of heavy metals in contaminated soils, organic matter decomposition, metal–microbe–plant interaction, and also performance of phytoremediation technique. From a toxicological perspective, the food crops grown in the contaminated site under the elevated carbon dioxide levels pose a major environmental health concern. For example, an experiment conducted by Rodriguez et al. (2011) found that elevated CO_2 concentrations can increase the total accumulation of Br, Cu, Co, Fe, Pb, Mn, Ni, and Zn in roots, stems, and seeds of *Glycine Max* (soybean) partially due to the effects of CO_2 on photosynthesis and also root growth and morphology. Guo et al. (2011) assessed the impacts of elevated carbon dioxide on the concentrations of heavy metals (i.e. Cu and Cd) in both rice and wheat grown in contaminated site. Their results found that high levels of CO_2 led to higher concentrations of Cd and lower concentrations of Cu in shoots and grain of both plant species. In similar studies, rice grains cultivated under CO_2 enrichment in metal contaminated soils had higher concentrations than the maximum permissible value in food based on the Codex Committee on Food Additives and Contaminants (CCFAC) European Union Regulation (CCFAC 2002; European Union 2006; Li et al. 2010). Rajkumar et al. (2013) suggested that further research on both individual and combined effects of climatic change on food crop productivities should be investigated.

8.3.4.2 Temperature/Drought

Drought condition and elevated temperature can change in the plasma membrane lipid composition, its fluidity as well as both passive and active metal flux. A study conducted by Li et al. (2012) investigated the impact of elevated temperature on heavy metal accumulation in *Solanum tuberosum L.* and found that a temperature rise of 3 °C significantly increased the concentrations of Fe, Cu, Zn leaf of about 24%, 25%, and 27%, respectively, while decreased the concentrations of Cu, Cd, Zn, Fe, and Pb in tuber of about 23%, 27%, 29%, 41%, and 55%, respectively. Because a higher temperature induced the tuber growth rate that exceeds its metal uptake rate, the decreasing concentrations in tubers are related to the dilution effect. Additionally, as a consequence of soil warming, the increased toxic contaminants accumulation in the plant tissues can then reduce the plant growth through the alteration in photosynthesis rate and also other metabolic activities. For instance, Li et al. (2011) studied the temperature effect on Cd-induced phytotoxicity in wheat roots and found that higher temperature in the surrounding environment increased the Cd accumulation in roots and reduced the root elongation. The SOD and CAT activities in wheat roots were decreased at a higher temperature condition, resulting in increased oxidative stress in plant. Altogether, Nuccio et al. (2016) found that soil warming significantly posed the additional stresses on the rhizobial microbial community. Particularly, changes in the diversity and the growth of arbuscular mycorrhizal fungi (AMF) communities in soil induced by climate warming could also reduce the capacity of plant to resist toxic contaminants. In extreme climate conditions, altered rainfall pattern or reduced soil moisture is considered as a primary constraint to the plant growth. There is evident that the drought stresses and heavy metal are more likely to co-occur because of poor water-holding capacity in the polluted soils. These combined stresses might cause a change in plant transpiration and metabolite accumulation. Whiting et al. (2003) found that the cultivation of the non-hyperaccumulator species, namely *A. montanum* and *L. heterophyllum* under Zn and Ni contamination together with water stress condition showed the decreases of both root growth and the survival rate. Angle et al. (2003) investigated the effects of soil moisture on the uptake of Ni by *Alyssum murale* and *Berkheya coddii* and Zn by *Thlaspi caerulescens* and found that hyperaccumulator plants grow well on soil with high moisture levels and could continue to hyperaccumulate these heavy metals in contaminated soil. In other words, under the drought conditions or low moisture content in soil, hyperaccumulating species showed a negative response to metal uptake and plant growth. Disante et al. (2011) assessed the impacts of Zn supply rate on the response of *Quercus suber* L. seedlings at severe drought condition and found that high Zn supply rates on seeding traits may exhibit a synergistic interaction with drought effects and water stress through decreased transpiratory losses and also slow decrease of carbon fixation in the treatment process. Tang et al. (2019) also studied the effects of regulated deficit irrigation on phytoremediation efficiency of Chard (*Beta vulgaris* L. var. *cicla*) in Cd contaminated soil using the following three different irrigation levels (T1 300 L, T2 200 L, and T3 100 L of water per block for irrigation during the organogenesis stage) and found that the regulated deficit irrigation condition decreased the shoot biomass of *B. vulgaris* L. var. *cicla* by about 15.8%, while increased the shoot Cd concentrations by approximately 23% and also maintained the constant ratio of root-shoot The Cd remediation potential efficiency of regulated deficit irrigation condition (T2) was 39.7% and 61.8% higher than that of T1 and T3, respectively. Further, Bhatia et al. (2005) conducted an experiment to examine a role of Ni in osmotic adjustment in the context of Ni hyperaccumulator *Stackhousia tryonii* by varying the following five levels of water stress: 20, 40, 60, 80, and 100% field capacity. Their results revealed that water stress (drought) had significant influence on both the growth of biomass and also Ni concentrations (dry weight basis) in *Stackhousia tryonii* shoots which significantly increased with a decrease in soil water content from 100 to 20%. Regarding to the osmoregulatory role, Ni hyperaccumulation plays an important role in protecting the plants against drought stress. Lastly, for the heavy metal and drought combined stresses, Santala and Ryser (2009) examined the response of white birch (*Betula papyrifera*) seedlings to simultaneous water and heavy metal stresses. The white birch seedlings were grown on a substrate with the following three levels of Cu-Ni containing slag (0%, 0.5%, and 2.5%) mixed with sand exposed to well watered and drought. Under combined stresses, results found that both low substrate moisture levels and the slag addition reduced the total dry mass, stem diameter, and cell size of the white birch. Overall, Table 8.5 presents some reports on the effects of environmental and climate changes on the bioremediation efficiencies.

Table 8.5 Impacts of environmental and climate change on the uptake of pollutants by plants in contaminated soils

Climate factors	Plants	Effects on the uptake of pollutants	References
Elevated CO_2			
Co, Al, Pb, V, Cu, Mn, Fe, Zn, Mo, and Ni	*P. taeda*, *L. styraciflua*, *Q. chapmanii*, *Q. geminate*, *Q. myrtifolia*	There was significant relationship between the enrichment of CO_2 on Co, V, Cu, Fe, Mn, Ni, and Zn concentrations in *P. taeda* leaves. Also, there was a significant effect of CO_2 enrichment on the concentrations of Mo in *L. styraciflua* leaves. In addition, Mn had significantly higher concentrations in leaves of all *Quercus species* under elevated CO_2 condition	Natali et al. (2009)
Br, Co, Cu, Fe, Mn, Ni, Pb, and Zn	*G. max*	Elevated CO_2 concentrations increased the accumulation of Br, Fe, Co, Ni, Cu, Mn, Zn, and Pb in roots, stems, and seeds of *G. max*	Rodriguez et al. (2011)
Cd	*L. mutiforum*, *L. perenne*	CO_2 enrichment condition decreased concentrations of Cd in both roots and shoots of both *L. mutiforum* and *L. Perenne*	Jia et al. (2010)
Cd	*T. aestivum* L., *S. bicolor* L.	CO_2 enrichment decreased the concentrations of Cd in both shoots and roots parts of *T. aestivum L*, but a reverse relationship was found in *S. bicolor* L.	Gelareh et al. (2018)
Cu and Cd	Wheat, rice	Enrichment of CO_2 led to lower Cu concentrations, but higher concentrations of Cd in shoots and grain parts of both plant species	Guo et al. (2011)
Pb	Pine seedlings	Elevated carbon dioxide concentrations significantly increased the Pb accumulation in the root of pine seedlings	Kim and Kang (2011)
Warming			
Cd, Cu, Fe, Zn, and Pb	*S. tuberosum* L.	Warming temperature increased the concentrations of Cu, Zn, and Fe in leaves, but decreased the concentrations of Cd, Cu, Fe, Zn, and Pb tuber in tubers of *S. tuberosum L.*	Li et al. (2012)
Cd	Wheat	Warming temperature increased the concentrations of Cd in roots of wheat while reduced wheat root cell elongation	Li et al. (2011)
Drought			
Ni and Zn	*A. murale*, *B. coddii*, *T. caerulescens*	Drought condition decreased the accumulation of Ni and Zn by *A. murale, B. Coddii, and T. caerulescens*	Angle et al. (2003)

(continued)

Table 8.5 (continued)

Climate factors	Plants	Effects on the uptake of pollutants	References
Ni and Zn	*A. montanum*, *L. heterophyllum*	Combined stresses of Ni/Zn contamination and water deficit decreased both root growth and the survival rate of both *A. montanum and L. heterophyllum*	Whiting et al. (2003)
Zn	*Q. suber* L.	Zn accumulation alleviated the effects of severe water stress and drought conditions	Disante et al. (2011)
Cd	*B. vulgaris var. cicla* L.	Under drought stress, the efficiency of Cd phytoremediation of regulated deficit irrigation condition (T2: 200 L of irrigation water) was 61.8% higher than that of T1 (100 L of irrigation water)	Tang et al. (2019)
Ni	*S. tryonii*	Water stress decreased both the growth of biomass and Ni concentrations in shoots of *S. tryonii*	Bhatia et al. (2005)
Zn	*B. papyrifera*	Combined drought and contamination of Zn showed the effects on the reduction of the total dry mass, the average length of the radial file of cells, stem diameter, and cell size of the white birch	Santala and Ryser (2009)

8.4 Conclusion and Implications

As one of the most pressing and globally recognized challenges of our times, climate change and atmospheric warming are crucial issues of various nations which have caused negative consequences on the ecology, environment, and socio-economic conditions. A latest report by the IPPC indicated that the additional warming of 1.5–2.0 °C above pre-industrial levels will have impacts on a coupled human–environment system including ecosystem functioning, terrestrial systems, and so on. With respect to bioremediation techniques, environmental and climate conditions can be paramount for the degradation performance since the abilities of microorganisms in soil are dependent mainly on environmental factors that are themselves being altered by climatic stresses, such as temperature, elevated CO_2, soil moisture, water-deficit conditions, etc. Specifically, climate variability and changes basically have effects on the characteristics of contaminants, the associated microbial activities, the physicochemical properties of soil, and also the ability of plants to uptake the toxic pollutants. Both direct and indirect effects of global climate change are expected to have large influence on the bioremediation processes. In support of this, impacts on the fate of environmental contaminants have been observed, especially on the dynamics, distribution, toxicity, mobility rate, and transportation of toxic contaminants in different environmental compartments.

Also, observed changes in climate are bound to affect microbial community structure, their activities and metabolic processes and thereby the efficiency of bioremediation. In the phyto-assisted bioremediation, changes in the interaction between soil microorganisms and plants induced by climatic stress could have great impact on the plant growth, its metabolism, physiological functions, and thereby bioremediation performance. Plant–microbe phytoremediation is an extremely complex system that can be affected by climate and other environmental changes in many ways depending on the types of plants, microorganisms, and also properties of soil. Further, the impacts of climate change (i.e. water stress) on the rhizobial microbial community, photosynthetic characteristics and antioxidant enzyme activities, and also bioremediation capacity should be given more consideration. Above all, for effective treatment results, bioremediation research or implementation should consider the following important factors:

- Different kinds of soil microorganisms may respond differently to changing climate parameters by either accelerating or alleviating the rate of bioremediation. Further research is therefore needed to better understand the patterns, processes, and mechanisms of soil microbes, as well as the relationship between their biogeochemical cycles and climate feedbacks. Many climate-induced impacts, not only on soil microbe–plant but also soil microbe–microbe, should also be taken into consideration.
- Further in-depth studies are required to explore physiological translocation mechanisms involved in the uptake of toxic contaminants, especially inside food crops. Management strategies and guidelines to reduce health risks from exposure to contaminated food crops and also all related environmental and health implications of soil–food systems are urgently needed. Further to this, it may be useful to consider using potential energy crops to generate new bioenergy resources along with the bioremediation techniques.
- A better understanding on the combined effects of various extreme climate events and climatic stresses (i.e. drought and heavy metal stresses) on plant growth and its ability to uptake or remove heavy metal from contaminated soil is prerequisite for effective remediation technologies. It is very important to select suitable types and the best-promising crop plants that would be able to tolerate multi-stress conditions, especially climatic change, without accumulating toxic substances in the food chain system. All the aspects mentioned above need to be evaluated on the toxic effects of various heavy metals in different types of plants in order to design future remediation strategies.
- To support more green remediation practice, the sustainable environmental remediation (SER) concept should be more considered and applied by minimizing cost of treatment and also environmental footprint (i.e. GHGs emissions, chemical and energy consumption) while maximizing societal benefits of a cleanup technology. A life cycle assessment (LCA) approach may assist in considering and selecting the best available technologies to mitigate the environmental burden of the remediation technologies throughout the life cycle. For instance, in terms of climate change resilience and sustainability assessment, a study conducted by

O'Connor et al. (2019) revealed that phytoremediation technique contributed only small life cycle ecological footprint and also showed substantial socioeconomic benefits of remediation. Moreover, compared to other hydroclimatic effects, the phytoremediation project was found to be resilient to moderate sea level rise scenario.

References

Angle JS, Baker AJM, Whiting SN, Chaney RL (2003) Soil moisture effects on uptake of metals by *Thlaspi, Alyssum, and Berkheya*. Plant Soil 256:325–332. https://doi.org/10.1023/A:1026137624250

Bhatia NP, Baker AJM, Walsh KB, Midmore DJ (2005) A role for nickel in osmotic adjustment in drought-stressed plants of the nickel hyperaccumulator *Stackhousia tryonii* Bailey. Planta 223: 134–139. https://doi.org/10.1007/s00425-005-0133-8

Biswas B, Qi F, Biswas J, Wijayawardena A, Khan M, Naidu R (2018) The fate of chemical pollutants with soil properties and processes in the climate change paradigm—a review. Soil Syst 2(3):51. https://doi.org/10.3390/soilsystems2030051

Booth S, Zeller D (2005) Mercury, food webs, and marine mammals: implications of diet and climate change for human health. Environ Health Perspect 113:521–526. https://doi.org/10.1289/ehp.7603

Bradford M (2013) Thermal adaptation of decomposer communities in warming soils. Front Microbiol 4:333. https://doi.org/10.3389/fmicb.2013.00333

Buch AC, Schmelz RM, Niva CC, Correia MEF, Silva-Filho EV (2017) Mercury critical concentrations to *Enchytraeus crypticus* (Annelida: Oligochaeta) under normal and extreme conditions of moisture in tropical soils—reproduction and survival. Environ Res 155:365–372. https://doi.org/10.1016/j.envres.2017.03.005

Castro HF, Classen AT, Austin EE, Norby RJ, Schadt CW (2010) Soil microbial community responses to multiple experimental climate change drivers. Appl Environ Microbiol 76(40):999–1007. https://doi.org/10.1128/AEM.02874-09

CCFAC (2002) Codex Alimentarius Commission. Report of the 34th session of the Codex Committee on Food Additives and Contaminants, Rotterdam, The Netherlands, March 11–15

Chen F, Yang Y, Ma Y, Hou H, Zhang SL, Ma J (2016) Effects of CO_2 leakage on soil bacterial communities from simulated CO_2-EOR areas. Environ Sci Process Impacts 18(5):547–554. https://doi.org/10.1039/c5em00571j

Classen AT, Sundqvist MK, Henning JA, Newman GS, Moore JAM, Cregger MA, Moorhead LC, Patterson CM (2015) Direct and indirect effects of climate change on soil microbial and soil microbial-plant interactions: what lies ahead? Ecosphere 6(8):130. https://doi.org/10.1890/ES15-00217.1

Crowther TW, Todd-Brown KEO, Rowe CW, Wieder WR, Carey JC, Machmuller MB, Snoek BL, Fang S, Zhou G, Allison SD (2016) Quantifying global soil carbon losses in response to warming. Nature 540(7631):104–108. https://doi.org/10.1038/nature20150

Disante KB, Fuentes D, Cortina J (2011) Response to drought of Zn-stressed Quercus suber L. seedlings. Environ Exp Bot 70:96–103. https://doi.org/10.1016/j.envexpbot.2010.08.008

European Union (2006) European Union regulation setting maximum levels for certain contaminants in foodstuffs. Commission Regulation (EC) No 1881/2006

Frey SD, Drijber R, Smith H et al (2008) Microbial biomass, functional capacity, and community structure after 12 years of soil warming. Soil Biol Biochem 40(11):2904–2907. https://doi.org/10.1016/j.soilbio.2008.07.020

Frohne T, Rinklebe J, Diaz-Bone RA (2014) Contamination of floodplain soils along the Wupper River, Germany, with As, Co, Cu, Ni, Sb, and Zn and the impact of pre-definite redox variations on the mobility of these elements. Soil Sediment Contam 23:779–799

FRTR (2001) Remediation technologies screening matrix and reference guide, version 3. http://www.frtr.gov/matrix2/top_page.html

Gelareh K, Hossein MH, Paul H, Babak MZ, Paul JM (2018) Effect of elevated atmospheric CO_2 concentration on growth and physiology of wheat and sorghum under cadmium stress. Commun Soil Sci Plant Anal 49(22):2867–2882. https://doi.org/10.1080/00103624.2018.1547388

Ghori Z, Iftikhar H, Bhatti MF, Nasar-um-Minullah, Sharma I, Kazi AG et al (2016) Phytoextraction: the use of plants to remove heavy metals from soil. In: Ahmad P (ed) Plant metal interaction: emerging remediation techniques. Elsevier, Amsterdam, pp 385–409

Goiun T, Armitage JM, Cousins IT et al (2013) Influence of global climate change on chemical fate and bioaccumulation: the role of multimedia models. Environ Toxicol Chem 32:20–31. https://doi.org/10.1002/etc.2044

Guo H, Zhu J, Zhou H, Sun Y, Yin Y, Pei D et al (2011) Elevated CO_2 levels affects the concentrations of copper and cadmium in crops grown in soil contaminated with heavy metals under fully open-air field conditions. Environ Sci Technol 45:6997–7003. https://doi.org/10.1021/es2001584

He ZL, Piceno Y, Deng Y, Xu MY, Lu ZM, DeSantis T, Andersen G, Hobbie SE, Reich PB, Zhou JZ (2012) The phylogenetic composition and structure of soil microbial communities shifts in response to elevated carbon dioxide. ISME J 6(2):259–272. https://doi.org/10.1038/ismej.2011.99

IPCC (2018) Summary for policymakers. In: Masson-Delmotte V, Zhai P, Pörtner H-O, Roberts D, Skea J, Shukla PR, Pirani A, Moufouma-Okia W, Péan C, Pidcock R, Connors S, Matthews JBR, Chen Y, Zhou X, Gomis MI, Lonnoy E, Maycock T, Tignor M, Waterfield T (eds) Global warming of 1.5°C. An IPCC special report on the impacts of global warming of 1.5°C above pre-industrial levels and related global greenhouse gas emission pathways, in the context of strengthening the global response to the threat of climate change, sustainable development, and efforts to eradicate poverty. World Meteorological Organization, Geneva, 32 pp

IPCC WG2 (2007) In: Parry ML, Canziani OF, Palutikof JP, van der Linden PJ, Hanson CE (eds) Climate change 2007: impacts, adaptation and vulnerability. Contribution of working group II to the fourth assessment report of the intergovernmental panel on climate change. Cambridge University Press, Cambridge

Jastrow JD, Miller RM, Matamala R, Norby RJ, Boutton TW, Rice CW, Owensby CE (2005) Elevated atmospheric carbon dioxide increases soil carbon. Glob Chang Biol 11(12):2057–2064. https://doi.org/10.1111/j.1365-2486.2005.01077.x

Jia Y, Tang SR, Wang RG, Ju XH, Ding YZ, Tu SX et al (2010) Effects of elevated CO_2 on growth, photosynthesis, elemental composition, antioxidant level, and phytochelatin concentration in Lolium mutiforum and Lolium perenne under Cd stress. J Hazard Mater 180:384–394. https://doi.org/10.1016/j.jhazmat.2010.04.043

Kim S, Kang H (2011) Effects of elevated CO_2 and Pb on phytoextraction and enzyme activity. Water Air Soil Pollut 219:365–375. https://doi.org/10.1007/s11270-010-0713-5

LeMonte JJ, Stuckey JW, Sanchez JZ, Tappero R, Rinklebe J, Sparks DL (2017) Sea level rise induced arsenic release from historically contaminated coastal soils. Environ Sci Technol 51: 5913–5922. https://doi.org/10.1021/acs.est.6b06152

Li ZY, Tang SR, Deng XF, Wang RG, Song ZG (2010) Contrasting effects of elevated CO_2 on Cu and Cd uptake by different rice varieties grown on contaminated soils with two levels of metals: implication for phytoextraction and food safety. J Hazard Mater 177:352–361. https://doi.org/10.1016/j.jhazmat.2009.12.039

Li D, Zhou D, Wang P, Li L (2011) Temperature affects cadmium-induced phytotoxicity involved in subcellular cadmium distribution and oxidative stress in wheat roots. Ecotoxicol Environ Saf 74:2029–2035. https://doi.org/10.1016/j.ecoenv.2011.06.004

Li Y, Zhang Q, Wang R, Gou X, Wang H, Wang S (2012) Temperature changes the dynamics of trace element accumulation in Solanum tuberosum L. Clim Chang 112:655–672. https://doi.org/10.1007/s10584-011-0251-1

Macdonald RW, Harner T, Fyfe J, Loeng H, Weingartner T (2003) AMAP assessment 2002: the influence of global change on hazardous substance pathways to, within, and from the Arctic. Arctic Monitoring and Assessment Programme (AMAP), Oslo
Margesin R, Schinner F (1997) Effect of temperature on oil degradation by a psychrotrophic yeast in liquid culture and in soil. FEMS Microbiol Ecol 24(3):243–249. https://doi.org/10.1111/j.1574-6941.1997.tb00441.x
McWatters RS, Rutter A, Rowe RK (2016) Geomembrane applications for controlling diffusive migration of petroleum hydrocarbons in cold region environments. J Environ Manag 181:80–94. https://doi.org/10.1016/j.jenvman.2016.05.065
Natali SM, Sanudo-Wilhelmy SA, Lerdau MT (2009) Plant and soil mediation of elevated CO_2 impacts on trace metals. Ecosystems 12:715–727. https://doi.org/10.1007/s10021-009-9251-7
Norby RJ, Iversen CM (2006) Nitrogen uptake, distribution, turnover, and efficiency of use in a CO_2-enriched sweetgum forest. Ecology 87:5–14. https://doi.org/10.1890/04-1950
Noyes PD, McElwee MK, Miller HD, Clark BW, Van Tiem LA, Walcott KC, Erwin KN, Levin ED (2009) The toxicology of climate change: environmental contaminants in a warming world. Environ Int 35:971–986. https://doi.org/10.1016/j.envint.2009.02.006
Nuccio EE, Anderson-Furgeson J, Estera KY, Pett-Ridge J, de Valpine P, Brodie EL, Firestone MK (2016) Climate and edaphic controllers influence rhizosphere community assembly for a wild annual grass. Ecology 97:1307–1318. https://doi.org/10.1890/15-0882.1
O'Connor D, Zheng X, Hou D, Shen Z, Li G, Miao G, O'Connell S, Guo M (2019) Phytoremediation: climate change resilience and sustainability assessment at a coastal brownfield redevelopment. Environ Int 130:104945. https://doi.org/10.1016/j.envint.2019.104945
Öborn I, Jansson G, Johnsson L (1995) A field study on the influence of soil pH on trace element levels in spring wheat (Triticum aestivum), potatoes (Solanum tuberosum) and carrots (Daucus carota). Water Air Soil Pollut 85:835–840
Ojuederie OB, Babalola OO (2017) Microbial and plant-assisted bioremediation of heavy metal polluted environments: a review. Int J Environ Res Public Health 14(12):1504. https://doi.org/10.3390/ijerph14121504
Pascual I, Antolín MC, García C, Polo A, Sánchez-Díaz M (2004) Plant availability of heavy metals in a soil amended with a high dose of sewage sludge under drought conditions. Biol Fertil Soils 40(5):291–299. https://doi.org/10.1007/s00374-004-0763-1
Polprasert C, Kittipongvises S (2010) Constructed wetlands and waste stabilization ponds. In: Treatise on water science, vol 4. Elsevier, pp 277–299. https://doi.org/10.1016/B978-0-444-53199-5.00090-7
Raich J, Schlesinger W (1992) The global carbon dioxide flux in soil respiration and its relationship to vegetation and climate. Tellus B 44(2):81–99. https://doi.org/10.1034/j.1600-0889.1992.t01-1-00001.x
Rajkumar M, Prasad MNV, Swaminathan S, Freitas H (2013) Climate change driven plant-metal-microbe interactions. Environ Int 53:74–86. https://doi.org/10.1016/j.envint.2012.12.009
Rodriguez JH, Klumpp A, Fangmeier A, Pignata ML (2011) Effects of elevated CO_2 concentrations and fly ash amended soils on trace element accumulation and translocation among roots, stems and seeds of Glycine max (L.) Merr. J Hazard Mater 187(1–3):58–66. https://doi.org/10.1016/j.jhazmat.2010.11.068
Rohr JR, Johnson P, Hickey CW, Helm RC, Fritz A, Brasfield S (2013) Implications of global climate change for natural resource damage assessment, restoration, and rehabilitation. Environ Toxicol Chem 32(1):93–101. https://doi.org/10.1002/etc.2036
Santala KR, Ryser P (2009) Influence of heavy-metal contamination on plant response to water availability in white birch, Betula papyrifera. Environ Exp Bot 66:334–340. https://doi.org/10.1016/j.envexpbot.2009.03.018
Sardans J, Peñuelas J, Estiarte M (2008a) Warming and drought change trace element bioaccumulation patterns in a Mediterranean shrubland. Chemosphere 70:874–885. https://doi.org/10.1016/j.chemosphere.2007.06.085

Sardans J, Penuelas J, Prieto P, Estiarte M (2008b) Changes in Ca, Fe, Mg, Mo, Na, and S content in a Mediterranean shrubland under warming and drought. J Geophys Res 13:1–11. https://doi.org/10.1029/2008JG000795

Sarma H, Forid N, Prasad R, Prasad MNV, Ma LQ, Rinklebe J (2021) Enhancing phytoremediation of hazardous metal(loid)s using genome engineering CRISPR–Cas9 technology. J Hazard Mater. https://doi.org/10.1016/j.jhazmat.2021.125493

Shaheen SM, Rinklebe J, Rupp H, Meissner R (2014) Lysimeter trials to assess the impact of different flood–dry-cycles on the dynamics of pore water concentrations of As, Cr, Mo and V in a contaminated floodplain soil. Geoderma 228–229:5–13. https://doi.org/10.1016/j.geoderma.2013.12.030

Silva CC, Guido ML, Ceballos JM et al (2008) Production of carbon dioxide and nitrous oxide in alkaline saline soil of Texcoco at different water contents amended with urea: a laboratory study. Soil Biol Biochem 40:1813–1822. https://doi.org/10.1016/j.soilbio.2008.03.004

Singh BK, Bardgett RD, Smith P, Reay DS (2010) Microorganisms and climate change: terrestrial feedbacks and mitigation options. Nat Rev Microbiol 8:779–790. https://doi.org/10.1038/nrmicro2439

Sonowal S, Nava AR, Joshi SJ, Borah SN, Islam NF, Pandit S, Prasad R, Sarma H (2022) Biosurfactants assisted heavy metals phytoremediation: green technology for the United Nations sustainable development goals. Pedosphere 2(1):198–210. https://doi.org/10.1016/S1002-0160(21)60067-X

Tan X, Machmuller MB, Wang Z, Li X, He W, Cotrufo MF, Shen W (2018) Temperature enhances the affinity of soil alkaline phosphatase to Cd. Chemosphere 196:214–222. https://doi.org/10.1016/j.chemosphere.2017.12.170

Tang X, Song Y, He X, Yi L (2019) Enhancing phytoremediation efficiency using regulated deficit irrigation pol. J Environ Stud 28(4):2399–2405. https://doi.org/10.15244/pjoes/91043

Thakare M, Sarma H, Datar S, Roy A, Pawar P, Gupta K, Pandit S, Prasad R (2021) Understanding the holistic approach to plant-microbe remediation technologies for removing heavy metals and radionuclides from soil. Curr Res Biotechnol. https://doi.org/10.1016/j.crbiot.2021.02.004

Thomas H (1997) Drought resistance in plants. In: Basra AS, Basra RK (eds) Mechanisms of environmental stress resistance in plants. Harwood, Netherlands, pp 1–42

UNEP/AMAP (2011) Climate change and pops: predicting the impacts. Report of the UNEP/AMAP expert group. Secretariat of the stockholm convention, Geneva, pp 62

United Nations (2019) Goal 13: take urgent action to combat climate change and its impacts. https://www.un.org/sustainabledevelopment/climate-change-2/

USEPA (1991) Understanding bioremediation a guidebook for citizens,. https://nepis.epa.gov/Exe/ZyPDF.cgi/10001JHQ.PDF?Dockey=10001JHQ.PDF

USEPA (2000a) Engineered approaches to in-situ bioremediation of chlorinated solvents: fundamentals and field applications. Office of Solid Waste and Emergency Response, Technology Innovation Office, Washington, D.C. EPA 542-R-00-008. http://www.clu-in.org

USEPA (2000b) Free water surface wetlands for wastewater treatment: a technology assessment. Office for Water Management, US Environmental Protection Agency, Washington, DC. https://nepis.epa.gov/Exe/ZyPURL.cgi?Dockey=20004923.TXT

USEPA (2005) EPA guideline—soil bioremediation. https://www.epa.sa.gov.au/files/8372_guide_soil.pdf

USEPA (2016) In-situ and ex-situ biodegradation technologies for remediation of contaminated sites. https://clu-in.org/download/contaminantfocus/dnapl/Treatment_Technologies/epa_2006_engin_issue_bio.pdf

Van der Putten WH (2012) Climate change, aboveground-belowground interactions and species range shifts. Annu Rev Ecol Evol Syst 43:365–383. https://doi.org/10.1146/annurev-ecolsys-110411-160423

Velki M, Ečimović S (2015) Changes in exposure temperature lead to changes in pesticide toxicity to earthworms: a preliminary study. Environ Toxicol Pharmacol 40(3):774–784. https://doi.org/10.1016/j.etap.2015.09.009

Venkatramanan V, Shah S, Prasad R (2020) Global climate change and environmental policy: resilient and smart agriculture. Springer, Singapore. (ISBN: 978-981-329-855-2). https://www.springer.com/gp/book/9789813298552

Venkatramanan V, Shah S, Prasad R (2021a) Sustainable bioeconomy: pathways to sustainable development goals. Springer, Singapore. (ISBN: 978-981-15-7320-0). https://www.springer.com/gp/book/9789811573200

Venkatramanan V, Shah S, Prasad R (2021b) Trade-offs between climate change and the sustainable development goals. Springer, Singapore. (ISBN: 978-981-15-7320-0). https://www.springer.com/gp/book/9789811573200

Veresoglou SD, Anderson IC, de Sousa NMF, Hempel S, Rillig MC (2016) Resilience of fungal communities to elevated CO_2. Microb Ecol 72(2):493–495. https://doi.org/10.1007/s00248-016-0795-8

Verta M, Salo S, Korhonen M, Porvari P, Paloheimo A, Munthe J (2010) Climate induced thermocline change has an effect on the methyl mercury cycle in small boreal lakes. Sci Tot Environ 408:3639–3647. https://doi.org/10.1016/j.scitotenv.2010.05.006

Wan R, Chen YG, Zheng X, Su YL, Li M (2016) Effect of CO_2 on microbial denitrification via inhibiting electron transport and consumption. Environ Sci Technol 50(18):9915–9922. https://doi.org/10.1021/acs.est.5b05850

Whiting SN, Neumann PM, Baker AJM (2003) Nickel and zinc hyperaccumulation by Alyssum murale and Thlaspi caerulescens (Brassicaceae) do not enhance survival and whole-plant growth under drought stress. Plant Cell Environ 26:351–360. https://doi.org/10.1046/j.1365-3040.2003.00959.x

Wu B, Shao HB, Wang ZP, Hu YD, Tang YJJ, Jun YS (2010) Viability and metal reduction of Shewanella oneidensis MR-1 under CO_2 stress: implications for ecological effects of CO_2 leakage from geologic CO_2 sequestration. Environ Sci Technol 44(23):9213–9218. https://doi.org/10.1021/es102299j

Wu Z, Dijkstra P, Koch GW et al (2011) Responses of terrestrial ecosystems to temperature and precipitation change: a metaanalysis of experimental manipulation. Glob Change Biol 17:927–942. https://doi.org/10.1111/j.1365-2486.2010.02302.x

Yang Z, Fang W, Lu X, Sheng G-P, Graham DE, Liang L, Wullschleger SD, Gu B (2016) Warming increases methylmercury production in an Arctic soil. Environ Pollut 214:504–509. https://doi.org/10.1016/j.envpol.2016.04.069

Yu T, Chen Y (2019) Effects of elevated carbon dioxide on environmental microbes and its mechanisms: a review. Sci Total Environ 655:865–879. https://doi.org/10.1016/j.scitotenv.2018.11.301

Zimmer C (2010) The microbe factor and its role in our climate future. http://e360.yale.edu/feature/the_microbe_factor_and_its_role_in_our_climate_future/2279/

Invasive Alien Plant Species: An Exploration of Social Aspect and Phytoremediation Acceptability

9

Prince K. Singh, Jay S. Yadav, Indrajeet Kumar, Umesh Kumar, and Rajesh K. Sharma

Abstract

Reduction and degradation of several types of soil contaminants such as heavy metals, pesticides, emerging contaminants, etc. are one of the major serious global challenges in the present scenario. For removal of heavy metals from the environment, various physical and chemical methods are used with some limitations like high cost, intensive labor, alteration of soil properties, etc. which make it problematic at application level. In the present time, science offers a green technique, i.e., phytoremediation; plants are capable of uptake, translocate, transform, and immobilize hazardous metals and reduced their toxicity and concentration in the environment. So, there is an increasing interest in utilization of "invasive alien species" among plant's kingdom in a productive way due to its easy availability, adaptability to changing climate, alternative mode of reproduction, phenotypic plasticity, and allelopathic in nature. These characteristics also make them more suitable for survival in high stress conditions. They can be used not only as a bioresource, but also as a method for the management of heavy metal polluted land. Invasive alien species are also source of medicines, fodders, and biofuel or bioenergy. The present chapter is focused on exploration of invasive alien plant species for various social aspects and phytoremediation of the heavy metal contaminated land.

Keywords

Invasive alien species · Phytoremediation · Heavy metals · Contaminated soil

P. K. Singh · J. S. Yadav · I. Kumar · U. Kumar · R. K. Sharma (✉)
Department of Botany, Institute of Science, Banaras Hindu University, Varanasi, India

R. Prasad (ed.), *Phytoremediation for Environmental Sustainability*,
https://doi.org/10.1007/978-981-16-5621-7_9

9.1 Introduction

An ecosystem is an integrated system of biotic (e.g., different flora, fauna, several types of microbes, etc.) and abiotic (e.g., soil, water, and different climatic gradients, etc.) components. Natural or anthropogenic activities can potentially disrupt the normal functioning of the ecosystem. Such types of disturbances help in the introduction of an alien species in a disturbed community which may replace or change different kinds of elements of the native flora, such alien organisms are known as invaders or invasive alien species (IAS). A single environmental variable and ecological attribute does not determine the success of invasive plant species (Rai and Singh 2020). Invasion ecology, intimately linked to the biology of climate change, land use pattern, conservation biology, health science, and the field of restoration is increasingly seen as transdisciplinary subject (Heshmati et al. 2019).

Pimentel et al. (2001) reported that serious threats posed by invasive alien species on native plant communities are degradation of biodiversity pool, ecosystem structure and function, availability of resources, population dynamics. Alien species are also known by several names such as foreign species, non-natives species, non-indigenous species, or introduced species. There are 1599 alien species in India belonging to 842 genera of 161 families (Khuroo et al. 2011). High-income countries or developed countries reported 30 times higher number of invasive plant species than low-income countries or developing countries (Seebens et al. 2018). Furthermore, hotspots of invasive alien plant species are mostly limited to high-income zone as European Union, North America, and Australian countries rather than Asian or African areas (Seebens et al. 2018).

Although alien species may be used in positive ways, but there are several demerits of alien species, e.g., threatening the biodiversity by abrading the gene pool, habitat loss or changing habitat conditions, degradation of endemic species, and reducing ecosystem functioning, economical loss, etc. In the present scenario, there is a compulsive need for management of alien species through its utilization, i.e., alien species have several characteristics which make its advance compared to another species like higher tolerating capacity to any kind of stress, upper level of adaptability, or survival capability to different environmental conditions, rapid growth rate, and short life span. Usance role of these alien species is meet by developing different new strategies and techniques for human welfare such as production of biofuels, animal feeds, or silage production, composts, fodders, fibers, and generation of bioactive compounds (Tessema 2012). The another major role of alien species is bioremediation.

Increased human population causes overexploitation of natural resources, disruption in biogeochemical cycle, and kind of anthropogenic activities such as urbanization, industrialization, use of different pesticides and fertilizers in agriculture, etc., poses severe threat to terrestrial, aquatic, and atmospheric environment (Behera and Prasad 2020). The emerging issue of present environmental problem is rapid and uncontrolled generation of various contaminants, e.g., toxic metals, organic compounds, metalloids, radionuclides, etc., through various anthropogenic sources that lead to deterioration of the quality of clean air, water, and fertile land (Thakare et al. 2021). Fifty-three elements in periodic table are listed as heavy metals which density >5 g/cm^3 and majority of which are considered as pollutants due to its non-degradable and persistent in nature (Prieto et al. 2018). On the basis of plant nutrition, heavy metals can be classified into two categories: essential and non-essential metals. Excess presence of essential as well as non-essential metals in plants leads to stress. Synthesis of reactive oxygen species under stress conditions causes several direct or indirect effects on morphological as well as physiological functions of plants. There are mainly two sources of heavy metals, i.e., natural (ores, volcanic eruption, etc.) and anthropogenic (transportation, industrialization, agriculture, domestic products, kinds of solid waste products, etc.). Environmental abundance of heavy metals is impacted by several factors, such as source of metals, rate of loading, pH, texture, organic matter content, redox potential, and composition of minerals in soil and different biological processes.

Sharma et al. (2008) reported that heavy metal imposes a health risk on the suburban's local resident of Varanasi, India due to consumption of wastewater irrigated vegetables. Heavy metals are significant environmental pollutants as they have hazardous effect on the ecosystem. To overcome this problem, several decontamination methods given as below are applied to manage the level of heavy metal contamination.

1. Physical methods: coagulation, adsorption, membrane filtration, biosorption, etc.
2. Chemical methods: chemical precipitation, electrochemical removal, ion exchange method, chemical oxidation or reduction, photocatalysis, etc.
3. Biological methods: In situ remediation, i.e., on-site treatments of contaminants, e.g., bioventing, biosparging, bioaugmentation, etc., ex situ remediation, i.e., off-site treatments of contaminants, e.g., bioreactor and phytoremediation, i.e., phytoextraction, phytovolatilization, phytofiltration, phytostabilization, phytodegradation, rhizo-remediation.

Bioremediation is a technique in which contaminants of soil and water are degraded or detoxified by biological methods that contain microbes as well as plants (Prasad et al. 2021). In phytoremediation, plants accumulate, translocate, transform or stabilize and degrade contaminants to reduce their toxicity in soil and water. Phytoremediation technique is novel, better clean-up, more cost-effective, ecofriendly, and esthetically acceptable techniques as compared to physical or chemical remediation techniques (Sarma et al. 2021; Sonowal et al. 2022). Phyto remediating plants contain several characteristics such as enhanced root system, high growth rate, growing capacity in nutrient poor conditions, short growth period, easily harvested, resistance to pathogens and heavy metals, etc., which improve its remediation capacity (Singh et al. 2017). All these characteristics are present in invasive alien species. There are several factors which affect the remediation capacity of invasive plants, e.g., soil composition, types and variety of plants, climatic factors, microbial pool in rhizospheric region, chemical forms and concentrations of heavy metals, etc. In the present time, the efficiency of phytoremediation techniques is improved through applications of different types of genetic engineering approaches such as transforming the plant genes which control the metal translocation and its homeostasis, defense mechanism against several oxidative stress and detoxification of xenobiotics. Phytoremediation efficiency can also be increased by use of nanoparticles known as "nanoremediation," e.g., use of Au nanoparticles in *Arabidopsis thaliana* improves the efficiency of remediation (Taylor et al. 2014; Prasad 2019a, b).

9.2 Social Aspects of Invasive Alien Species

Invasive alien species are a major cause for threat to natural ecosystems. For reducing its harmful effects, commercial utilization of such species is very essential. Therefore, the effective use for human well-being of invasive alien species is a good option for the future. There are a lot of aspects described below for applications of invasive alien plant species (Fig. 9.1).

9.2.1 In Herbal Medicines

Studies from India and several other countries reported the medicinal importance of invasive alien species. Kumar et al. (2019) reported the bioactive potential of *Ocimum americanum* and its biofabricated zinc oxide synthesize nanoparticle was found more potent with antimicrobial activity against Gram positive and negative bacteria. Saxena and Rao (2018) analyzed immense antioxidant activity in leaves of invasive weed *Malvastrum coromandelianum* (L.) Garcke. Application of GC-MS for aqua methanolic leaf extract of *M. coromandelianum* revealed that 29 bioactive compounds belong to several groups such as phenols, flavonoids, sterols, vitamins,

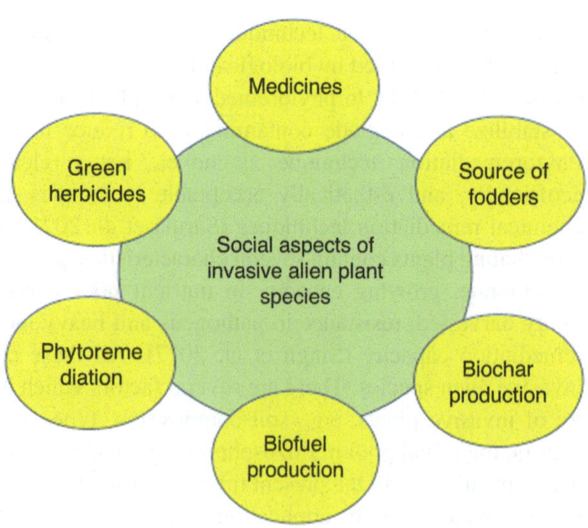

Fig. 9.1 Various social aspects of invasive alien plant species

fatty acids, etc. Some major phytochemicals detected by GC-MS were 9,12,15-octadecatrienoicacid (15.02%), methyl ester, (Z, Z, Z) (11.02%), 9,12,15-octadecadienoic acid, guanosine (17.05%), n-hexadecanoic acid (10.80%), trace amounts of 2,3-dihydro-3,5-dihydroxy-6-methyl, squalene, vitamin-E, neophytadiene, phytol, diosgenin, 4 hpyran-4-one, etc. Pappan and Thomas (2017) studied the medicinal use of 28 invasive alien species from 18 different families in the Mukkam municipality area of Kozhikode district, Kerala. Divakara and Prasad (2015) studied a total of 41 invasive alien species from Hazaribagh district of Jharkhand, India and reported 17 species with medicinal values. Saxena et al. (2020) studied different phytochemicals which occurred in *Malvastrum coromandelianum* L. like squalene, guanosine, diosgenin, malvastrone 1, β-sitosterol, stigmasterol, dotriacontanol, campesterol, *n*-hexadecanoic acid, etc. This plant may also be used as anti-inflammatory plants, anthelmintic plants, analgesics, larvicidal plants, antioxidants, antimicrobial and antidiarrheal drugs.

9.2.2 Livelihood Benefits: Utilized as Fodder

Ngorima and Shackleton (2019) showed that invasive alien species, *Acacia dealbata* growing naturally are used as fodder for livestock's in Eastern Cape Province. This plant is used as firewood, constructive timber, and also a part of traditional medicines. Rapid growth of *A. dealbata* exacerbates the growth of local plant varieties, so need greater effort to understand the growth of invasive plant and its regular removal from arable fields of household areas.

9.2.3 Chances and Opportunities in Biofuels Production

Due to increasing demand of fossil fuels, i.e., oils, its depletion rate becomes very higher. Fossil fuels are non-renewable energy resources, once degraded then impossible to its recovery. Biofuels are a good alternative of renewable energy resource. Biofuels are also called "Future fuel" which is blended with gasoline, i.e., gasohol (gasoline is blended with ethanol, which produced from plant biomass) and easily used in automobiles and others. *Parthenium hysterophorus* (congress grass), an invasive alien plant pretreated with *Trametes hirsute* (white rot fungus), is used for the production of bioethanol or biofuel energy (Rana et al. 2013). Fungal contamination altered its cellular structure and delignification of cell wall ascertained by scanning electron microscopy, x-ray diffraction, and Fourier transform infrared spectroscopy. Delignification causes higher availability of holocellulose (52.65 %) which make *P. hysterophorus* as a good source of biofuels.

9.2.4 Potential Use of Invasive Plant Species for Biochar Generation

Plant biomass becomes biochar after carbonization or pyrolysis, which is a high temperature treatment process in the absence of oxygen. Biochar is carbon-rich resource that is more porous and durable in nature. It is used as an immobilizer or stabilizer agent for reducing availability of heavy metals in contaminated soil and also in co-combustion process char blend with parent coal for improving its energy efficiency (Singh et al. 2020). Reza et al. (2019) studied that *Acacia holosericea* is an invasive alien species uses in production of biochar. In this experiment, analyses have done on both parameter in invasive plants, i.e., proximate analysis (moisture content, ash content, fixed carbon, and volatile matter) and ultimate analysis (carbon, hydrogen, and nitrogen content). Proximate analysis of *A. holosericea* revealed that the moisture content, ash contents, volatile matters, and fixed carbon were 9.56%, 3.91%, 65.12%, and 21.21%, respectively. Ultimate analysis of *A. holosericea* exhibits carbon content, hydrogen content, and nitrogen content as 44.03%, 5.67%, and 0.25%, respectively.

9.2.5 Biocontrol Agent: Green Herbicides

Weedy plants are one of the major threatening causes for loss of crop yields, biodiversity, different ecosystem services, etc. Climate change, the proliferation of resistance to synthetic herbicides, the loss of various ecological services, and several other factors may pose challenges and issues in area of weed ecology and management. Souza-Alonso et al. (2018) reported that residues of invasive alien species of *Acacia dealbata* and *Acacia longifolia* used in the place of synthetic herbicide reduce growth of dicotyledon weeds. Pot experiment performed under the green house conditions in which maize was taken as test crop. Residues of *A. dealbata* and *A. longifolia* applied to soil at different dose (1.5% and 3%) showed herbicidal effects of *Acacia* species on dicot weeds. As a result, *Acacia* species are a viable choice for weed control. The use of invasive alien species to combat weed plants is a long-term approach.

List of some invasive alien plant species, used as a natural resource in different areas is given in Table 9.1.

9.2.6 Phytoremediation Acceptability of Invasive Alien Species

Due to superior adaptive strategies of invasive alien species under any environmental stress conditions, they can be used as a phytoremediating plant to remediate heavy metals from contaminated soil. Mostly such plants are not used in food sector, so there is a less chance of food chain contamination by heavy metals. Abbas et al. (2019) found that in arid and semi-arid regions, the invasive tree *Prosopis glandulosa* can accumulate Cd, Pb, Cu, Zn, and Fe from sewage sludge treated

Table 9.1 Various uses of selected invasive alien species

S. No.	Alien plant species	Significance	References
1.	*Eichhornia crassipes*	Used in different nutrients and organic pollutants removal	Adelodun et al. (2021)
2.	*Anagallis arvensis* L.	Used as a pain killer from its leaf, fruit	Tripathi et al. (2020)
3.	*Tridax procumbens* L.	Used as a natural first aid in case of cuts, scratches, and injuries	Rex Immanuel (2020)
4.	*Opuntia ficus-indica*	Jam, syrup, chutney, and beer from fruits	Mdweshu and Maroyi (2020)
5.	*Prosopis juliflora*	For woody and herbaceous biomass, green building plaster	Linders et al. (2020), Sakthieswaran and Sophia 2020
6.	*Celosia argentea*	Contains many nutrients additives that protect dietary deficiencies and several chronic diseases	Adegbaju et al. (2019)
7.	*Eclipta prostrata*	Synthesis of bioactive compounds (flavonoid, terpenoids, and tannins)	Karuppaiah et al. (2019)
8.	*Euphorbia hirta*	Used in anticancerous purposes against neuroblastoma cell lines	Selvam et al. (2019)
9.	*Lantana camara*	Used in timber industry and production of biochar acts as a co-fuel with coal	Negi et al. (2019), Mundike et al. (2018)
10.	*Xanthium strumarium*	Medicinal resource in the primeval inhabitants of the Basin of Turpan	Sheng et al. (2018)
11.	*Ipomoea carnea*	Used in wound healing effect	Shukla et al. (2018)
12.	*Physalis minima*	Anti-inflammatory and cytotoxic constituent's inhibition characteristics	Wu et al. (2018)
13.	*Urena lobata*	Used as an antidiabetic effect	Wahyuningsih and Purnomo (2017)
14.	*Tribulus terrestris*	Used in inflammatory disease, pathogenic ailments in the digestive system	Chauhan et al. (2017)
15.	*Potamogeton illinoensis*	Help in mitigation or remediation of bisphenol A	Trueman and Erber (2013)
16.	*Salvinia molesta*	More feasible for remediating endosulfan pesticide from its aqueous solution	Harikumar et al. (2013)
17.	*Acacia dealbata*	Producing good quality of fire woods	Kull et al. (2011)

soil. Amendment of sewage sludge in soil significantly decreases soil pH as well as improves the soil organic matter content. Occurrence of Cd, Pb, Cu, Zn, and Fe in roots of *P. glandulosa* was more than their shoots.

9.3 Response of Invasive Alien Species Under Abiotic Stress

Invasive plant community structure is regulated by competitive ability of invasive plant and their abiotic stress tolerance potential as given in many ecological theories (Brooker and Callaghan 1998). There are several abiotic factors which affect the biological invasion, e.g., geographical locations, pH, temperature, salinity, and elevated level of CO_2, UV, O_3, and heavy metals, etc. due to presence of different abiotic stresses in environment affects several characteristics of invasive plant species.

9.3.1 Morphological Response

Kind of research studies have shown that the growth of invasive plant is influenced by abiotic stress. Silveira et al. (2018) found the effect of abiotic stress, i.e., warming trend on the freshwater hydrophyte, *Egeria densa* in its native range (Brazil) and introduced range (France). In both countries, these invasive plants form monospecific or pauci-specific and dense mats in water. With the increase in temperature in both conditions, i.e., native and introduced habitats, invasive plants showed morphological differences in terms of higher growth, i.e., length and biomass in native as compared to introduced habitats. Gentiili et al. (2018) reported the effect of different soil pH on growth of invasive plant, *Ambrosia artemisiifolia* (rageweed), is highly invasive and alien in Europe. In this experiment invasive plant grows in different soil pH value as 5 (acidic), 6 (sub-acidic), and 7 (neutral condition). Invasive plants show several changes as larger in length, number of inflorescences increases, and emitted pollen earlier mature at lower soil pH, i.e., 5 and 6. Changes in soil pH by different inorganic amendments may be suitable approach for controlling the growth of invasive weeds.

9.3.2 Physiological, Biochemical, and Molecular Responses

Invasive alien plants adopt several mechanisms to avoid different types of abiotic stress by altering the physiological, biochemical, and molecular processes. Naidoo and Naidoo (2018) showed the effect of drought conditions on invasive weed, *Chromolaena odorata* in terms of water relation and gas exchange. Uptake of CO_2 by plants under drought condition was decreased to 9.3 $\mu mol\ m^{-2}\ s^{-1}$ as compared to those of well-watered condition, i.e.,12.8 $\mu mol\ m^{-2}\ s^{-1}$. Leaf water potential and photosynthetic rate also reduced in invasive plants due to drought. Xiao et al. (2019) reported that a high latitude, invasive plants *Phytolacca americana* release triterpenoid saponin in reproductive tissues, which improve their resistance capacity against lower temperature. Xie et al. (2015) found that CBF pathway plays an important role in tolerance capacity of invasive plant to survive under cold areas. Lower temperature increases the demethylation rate of ICE1, which regulate the cold tolerance capacity of invasive plants.

Although heavy metals are one of the major components in enzymes and for survival of plants, without this important cofactor impossible to survive but when these heavy metals are present at higher concentration than the normal, they badly affect the plant growth. The elevated doses of Cr, Cd, Ni, Cu, Mn, Fe, etc. induced oxidative injury in plants by Haber–Weiss and Fenton reactions which result in the generation of reactive oxygen species (ROS) or free radicals that causes cell homeostasis disruption, DNA rupture, proteins or cell membrane degradation, and photosynthetic pigments damage and ultimately cell death (Kumar and Sharma 2018). To render the effects caused by oxidative stress which is induced by heavy metals, plants adopt either enzymatic or non-enzymatic strategies (Fig. 9.2). It has been well reported that under higher metal concentration, antioxidative enzymatic machinery is activated in plants which changes the gene expression for the generation of different enzymes such as catalase (CAT), superoxide dismutase (SOD), ascorbate peroxidase (APX), glutathione reductase (GR), and guaiacol peroxidase (GPX). The effects of free radicals formed during oxidative stress caused by heavy metals are eliminated by these enzymes. It has also been reported that non-enzymatic compounds such as several secondary metabolites, i.e., phenolics, flavonoids, terpenes, alkaloids, prolines, etc. are also increased under the metal stress and act as the scavengers of free radicals (Rastgoo et al. 2011; Sharma et al. 2012).

9.3.3 Bioaccumulation of Heavy Metals

Rapid urbanization, industrialization, and advancement in transport system, etc., threaten the quality of natural resources such as soil, water, and air. Phytoremediation is an ecofriendly, cost effective, esthetic, and green technology and became one of the best suitable strategies to remove the contaminants from the environment (Table 9.2). Chaney (1983) suggested the idea of phytoremediation strategy. India is considered under the top 12 megadiversity country in the world due to its geographical position and climatic conditions. India has 8% of the global biodiversity in only 2.4% land surface area of the world. Various approaches such as amendments or treatment of plants with different types of microbes, insertion of new gene of interest through different biotechnological approaches are used to increase the heavy metal accumulation capacity of plants.

Ahmed and Slima (2018) identified that fodder plant, *Corchorus olittorius* was used as heavy metals accumulator in wastewater irrigated soil such as Cd, Cu, Cr, Pb, Fe, Mn, Ni, and Zn. This plant accumulates Cd, Cr, Cu, Fe, and Zn in leaves (edible part) above the phytotoxic range and higher concentration of Cd, Cu, Cr, Pb, Fe, Mn, Ni, and Zn in roots as compared leaves. *Sonchus asper* act as a Cr hyperaccumulator when grown on tannery waste dump site soil (Nirola et al. 2018). Accumulation of Cr concentration was 212 mg/kg in shoot part from soil in which Cr concentration was only 41 mg/kg (Nirola et al. 2018). Water lettuce (*Pistia stratiotes* L.) become a good option for remediation of heavy metals contaminated sugar mill effluents as it accumulates Cd, Cu, Fe, Cr, Pb, Zn, and Mn with more than 1 bioaccumulation factor (Kumar et al. 2018). Sasidharan et al. (2013) reported that

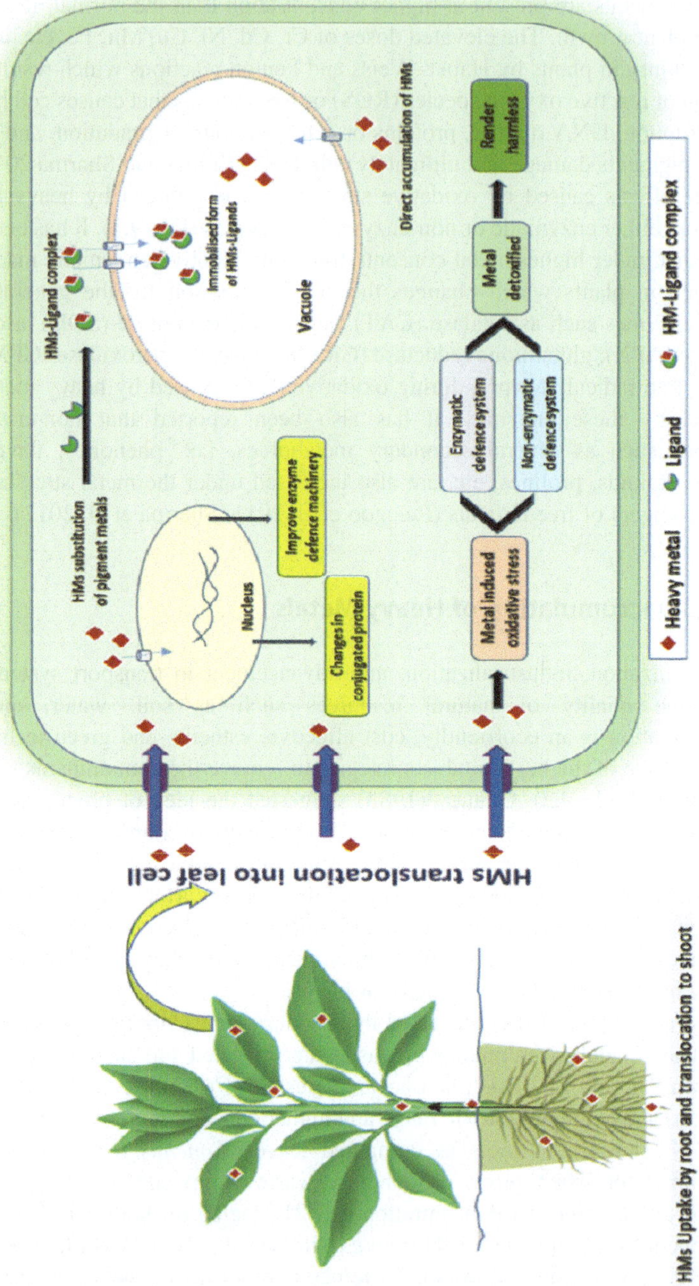

Fig. 9.2 Cellular mechanism for accumulation and detoxification of heavy metals in invasive plant. (Modified from Singh et al. 2017)

Table 9.2 Heavy metal accumulation in selected invasive alien species and their ecological significance

S. N.	Alien plant species	Heavy metals	Significance	References
1.	*Eichhornia crassipes*	Cd, Co, Cr, Cu, Fe, Mn, Ni, Pb, Zn	Used as an accumulator plant	Eid et al. (2021)
2.	*Ricinus communis* L.	Zn, Cu, Ni, Fe, Mn	Used as metal accumulator	Galal et al. (2021)
3.	*Ranunculus sceleratus* L.	Cd, Ni, Cu, Pb, Zn, Fe, Mn, Co	Have good potential of phytoremediation	Khalifa and Badr (2021)
4.	*Spartina densiflora*	As, Cd, Cu, Ni, Pb, Zn	Used as a metal accumulator	Mesa-Marín et al. (2020)
5.	*Ambrosia artemisiifolia*	Ba, Pb, Zn	Used as a phytoremediator plant	Ranđelović et al. (2020)
6.	*Amaranthus spinosus*	Cd, Pb	Good potential candidate for phytoremediation	Huang et al. (2019)
7.	*Eichhornia crassipes*	Cd, Co, Cr, Cu, Fe, Mn, Ni, Pb, Zn	Promising hydrophyte for remediation of heavy metals	Eid et al. (2019)
8.	*Pteris vittata*	Cd, Cr	Capability of decontamination of contaminated site	Prabhu et al. (2019)
9.	*Lantana camara*	Cd	Plant as a hyperaccumulator	Liu et al. (2019)
10.	*Chromolaena odorata, Bidens pilosa, Praxelis clematidea*	Cd	Plant as a hyperaccumulator	Wei et al. (2018)
11.	*Spartina alterniflora* (Smooth cordgrass)	Cr, Pb, Cu, Zn, and Mn	Used as heavy metal accumulator in a salt marsh	Chen et al. (2017)
12.	*Cassia tora* (L.) Roxb	Cr	Plant contains phytoremediation capability	Jena et al. (2016)
13.	Reynoutria × Bohemica	Fe, Ni, Cr, and Co	Used for phytoremediation of heavy metal polluted soils	Sirka et al. (2016)
14.	*Ceratophyllum demersum*	Pb	Beneficial in phytoremediation applications	Chen et al. (2015)
15.	*Catharanthus roseus*	Cr	Useful in the reclamation and remediation of chromium contaminated soil	Ahmad and Misra (2014)
16.	*Parthenium hysterophorus*	Pb, Ni, and Cd	Plants as a hyperaccumulator	Ahmad and Al-Othman (2013)
17.	*Ipomoea carnea*	Cd, Pb, Cu, Cr, Mn, and Ni	Potential applicability in remediation of contaminated sites	Pandey (2012)

(continued)

Table 9.2 (continued)

S. N.	Alien plant species	Heavy metals	Significance	References
18.	*Tamarix smyrnensis*	Pb, Cd	Used in remediation of metal contaminated sites	Kadukova et al. (2008)
19.	*Pteris vittata* (Ladder brake)	As	Extremely efficient in extracting arsenic from soils	Ma et al. (2001)

the concentration of heavy metals (Fe, Zn, Cu, Mn, Al, and Cr) in *Amaranthus tricolor* var. Arjun cultivated in Vembanad lake sediments, which were supplemented with compost and farmyard manure made from water hyacinth (*Eichhornia crassipes*). The water hyacinth compost amended sediments increased the accumulation of As, Cr, Ni and biomass of *A. tricolor*. But in case of As, Cr and Ni accumulation in *A. tricolor* were higher in *E. crassipes* compost amended sediments than those of farmyard compost.

9.4 Mechanism of Heavy Metal Uptake and Accumulation

The mechanism for removal of heavy metals or other contaminants by invasive plants from polluted soil is: (1) increase the mobilization of heavy metals in soil, (2) uptake of heavy metals by invasive plant, (3) translocation of accumulated heavy metals from roots to aerial parts, (4) heavy metal accumulation or sequestration in invasive plant tissues, (5) develop tolerance capacity to heavy metals by several physiological and biochemical adaptation. Heavy metals are taken up by invasive plants from soil matrix to its root part via two pathways. (1) Plants secret chemicals, e.g. siderophores into their rhizospheric region to solubilize and chelate the metals by complexation and (2) increase the bioavailability of heavy metals in soil by acidifying the rhizospheric region of plants by exudation of carboxylates or other chemicals which are readily taken up by plants (Wu et al. 2004; Hauser et al. 2005; Thakare et al. 2021). There are several types of molecules like chelating agents which control and regulate the overall translocation of heavy metals from soil solution to vacuolar region of cells (Fig. 9.2). For example, complexation of heavy metal ions by amino acids, organic acids containing atoms S, N, O acts as a ligand in their molecules (Shah and Nongkynrih 2007).

Heavy metals are taken up from the contaminated soil after mobilization by root cells, bound to cell wall, which acts as an ion exchanger (Blaylock 2000). After that heavy metals are transported by specialized transporter or hydrogen ion coupled carrier protein present in plasma membrane of root cell (Greipsson 2011). For example, zinc iron permease acts as a transporter for Zn and Fe transport (Clemens 2001). Due to stoichiometric similarity like atomic radius, oxidation states, etc., need for the same transmembrane transporter, non-essential heavy metals competes directly with essential metals (Thangavel and Subbhuraam 2004; Alford et al. 2010). Inside the plasma membrane, due to generation of negative potential

(−200 MV) in root epidermal cell, creating a strong driving force for uptake of cations through secondary transporter (Belouchi et al. 1997). Commonly insoluble metals within the plants move freely in the vascular system and disrupt the materials translocation. Plant immobilizes metals in the form of phosphates, carbonates, and sulfates and precipitate in certain extracellular and intracellular regions (Salt et al. 1995). Heavy metals are taken up from contaminated soil via root cell and transported to aerial parts through xylem vessels in a different way (Prasad 2004; Jabeen et al. 2009) and they are mostly stored in vacuolar region of cell. Heavy metals get accumulated in vacuolar region of the cells due to its lowest metabolic activities, i.e., avoiding or tolerating strategies for heavy metal by reducing their interaction with cellular metabolic mechanisms (Assuncao et al. 2003; Sheoran et al. 2011).

9.5 Potential Risks and Challenges in Applications of Invasive Alien Species

There are several limitations in remediation of contaminated soil using invasive alien species such as its growth period, unmanaged disposal, establishment of invasive alien species damages the native floral diversity, etc. Use of invasive alien species for human welfare may be a good approach for their utilization but due to some specific functional traits related to physiological function, biomass accumulation rate, high growth rate, short life span, good homeostatic capability to any stress conditions affects the survivalism of native or endemic species (Colautti et al. 2017). One of the most important reasons for spreading of invasive alien species is transportation from one country to another country during exporting and importing of materials. Screening of materials is an important step to overcome such problem. Maiti et al. (2008) studied that leachates and extracts of *Lantana camara* reduce germination rate, seed viability, and growing capacity of *Mimosa pudica* seed. *Eichhornia crassipes* have higher threatening capacity to degrade hydrophytic ecosystem, it can be doubled in amount just in 5 days and cover the water bodies with mat like covering and include up to two million plants per hectare. The growth of *E. crassipes* causes total cut off the sunlight penetration, increasing intraspecific competition cause eutrophication (Epstein 1998).

From the above, it is clear that invasive alien species causes degradation of native species in environment. Therefore, identification of invasive effects of alien species on locally growing native plants is necessary. Invasive alien varieties contain several characteristics such as wide spread distribution capability, ready for available natural resources, allelopathic in nature for avoiding predation, easily growing ability and survivor in unfavorable natural climatic and hydrological conditions, etc. The above mention characteristics of invasive alien species make them more suitable to changing climatic conditions and have capability to avoid all physiological, ecological, and climatic stresses, i.e., tolerance capacity to several disease, extra heat or cold conditions, drought or flood conditions, change in silviculture pattern and different types of human oriented chemicals, etc. Several steps may be taken for risk assessment of invasive alien species as decision making for its utilization, to understand its management risk, possible threats to ecosystem, and types of other derived challenges.

9.6 Conclusion

Recent research works on invasive alien plants improve our knowledge about its basic as well as applied aspects. There is a need to focus on application of invasive alien species as a tool for phytoremediation of wasteland or contaminated land. However, invasive alien plants can be utilized at large scale for the production of biofuels, fodders, drugs, biochar's, herbicides, etc. Several years ago, conservators tried to sustain the native plant species by different approaches such as retaining, restoring, and developing several strategies for rapid recovering ability in fire, flood, and drought prone areas. But in present time, scientists focus on invasive alien plant for its proper management and utilization in sustainable way due to its superior adaptive ability to biotic or abiotic stresses. Due to the negative aspect of invasive alien species such as threatening the agricultural landscape, forests, and wasteland it is necessary to establish appropriate strategies for economic consumption of invasive plant species in a positive manner without affected biodiversity. Further, management of invasive alien species should be done by using scientific approaches, societal interaction, and involvement of decision support frame workers. Invasive alien species, however, pose threats to ecosystems, but such species can be exploited for various societal uses as well as a potential tool for the remediation of heavy metal polluted sites.

References

Abbas AM, Elmabod SKA, Ashry SME, Soliman WS, Tayeh NE, Castillo JM (2019) Capability of the invasive tree *Prosopis glandulosa* Torr. to remediate soil treated with sewage sludge. Sustainability 11(9):2711. https://doi.org/10.3390/su11092711

Adegbaju OD, Otunola GA, Afolayan AJ (2019) Potential of *Celosia* species in alleviating micronutrient deficiencies and prevention of diet-related chronic diseases: a review. AIMS Agric Food 4(2):458–484. https://doi.org/10.3934/agrfood.2019.2.458

Adelodun AA, Olajire T, Afolabi NO, Akinwumiju AS, Akinbobola E, Hassan UO (2021) Phytoremediation potentials of Eichhornia crassipes for nutrients and organic pollutants from textile wastewater. Int J Phytoremediation:1–9. https://doi.org/10.1080/15226514.2021.1895719

Ahmad A, Al-Othman AAS (2013) Remediation rates and translocation of heavy metals from contaminated soil through *Parthenium hysterophorus*. Chem Ecol 30(4):317–327. https://doi.org/10.1080/02757540.2013.871269

Ahmad R, Misra N (2014) Evaluation of phytoremediation potential of *Catharanthus roseus* with respect to chromium contamination. Am J Plant Sci 5:2378–2388. https://doi.org/10.4236/ajps.2014.515251

Ahmed DA, Slima DF (2018) Heavy metal accumulation by *Corchorus olitorius* L. irrigated with wastewater. Environ Sci Pollut Res 25:14996–15005. https://doi.org/10.1007/s11356-018-1675-1

Alford ER, Pilon-Smits EAH, Paschke MW (2010) Metallophytes—a view from the rhizosphere. Plant Soil 337:33–50. https://doi.org/10.1007/s11104-010-0482-3

Assuncao AGL, Schat H, Aarts MGM (2003) *Thlaspi caerulescens*, an attractive model species to study heavy metal hyperaccumulation in plants. New Phytol 159:351–360. https://doi.org/10.1046/j.1469-8137.2003.00820.x

Behera BK, Prasad R (2020) Environmental technology and sustainability. Elsevier. (ISBN: 9780128191033). https://www.elsevier.com/books/environmental-technology-and-sustainability/behera/978-0-12-819103-3

Belouchi A, Kwan T, Gros P (1997) Cloning and characterization of the OsNramp family from Oryza sativa, a new family of membrane proteins possibly implicated in the transport of metal ions. Plant Mol Biol 33:1085–1092. https://doi.org/10.1023/A:1005723304911

Blaylock MJ (2000) Phytoextraction of metals. In: Phytoremediation of toxic metals: using plants to clean up the environment. Wiley, pp 53–70

Brooker RW, Callaghan TV (1998) The balance between positive and negative plant interactions and its relationship to environmental gradients: a model. Oikos 81:196–207. https://doi.org/10.2307/3546481

Chaney RL (1983) Plant uptake of inorganic waste. In: Land treatment of hazardous wastes

Chauhan S, Sharma D, Goel HC (2017) An in vitro evaluation of *Tribulus terrestris* L. fruit extract for exploring therapeutic potential against certain gut ailments. Indian J Exp Biol 56:430–436. http://nopr.niscair.res.in/handle/123456789/44535

Chen M, Zhang LL, Li J, He XJ, Cai JC (2015) Bioaccumulation and tolerance characteristics of a submerged plant (*Ceratophyllum demersum* L.) exposed to toxic metal lead. Ecotoxicol Environ Saf 122:313–321. https://doi.org/10.1016/j.ecoenv.2015.08.007

Chen L, Gao J, Zhu Q, Wang Y, Yang Y (2017) Accumulation and output of heavy metals in *Spartina alterniflora* in a salt marsh. Pedosphere. https://doi.org/10.1016/S1002-0160(17)60369-2

Clemens S (2001) Developing tools for phytoremediation: towards a molecular understanding of plant metal tolerance and accumulation. Int J Occup Med Environ Health 14:235–239

Colautti RI, Alexander JM, Dlugosch KM, Keller SR, Sultan SE (2017) Invasions and extinctions through the looking glass of evolutionary ecology. Philos Trans R Soc B Biol Sci 372(1712):20160031. https://doi.org/10.1098/rstb.2016.0031

Divakara BN, Prasad S (2015) Ethnomedicinal importance of invasive alien flora of Latehar and Hazaribagh districts: Jharkhand. Indian Forester 141(11):1172–1175

Eid EM, Shaltout KH, Moghanm FS, Youssef MS, El-Mohsnawy E, Haroun SA (2019) Bioaccumulation and translocation of nine heavy metals by *Eichhornia crassipes* in Nile Delta, Egypt: perspectives for phytoremediation. Int J Phytoremediation:1–10. https://doi.org/10.1080/15226514.2019.1566885

Eid EM, Shaltout KH, Almuqrin AH, Aloraini DA, Khedher KM, Taher MA, Alfarhan AH, Pico Y, Barcelo D (2021) Uptake prediction of nine heavy metals by Eichhornia crassipes grown in irrigation canals: a biomonitoring approach. Sci Total Environ 146887. https://doi.org/10.1016/j.scitotenv.2021.146887

Epstein P (1998) Nairobi weeds bring disease to the east African waterways. Lancet 351(9102):577. https://doi.org/10.1016/S0140-6736(05)78570-6

Galal TM, Essa B, Al-Yasi H (2021) Heavy metals uptake and its impact on the growth dynamics of the riparian shrub Ricinus communis L. along Egyptian heterogenic habitats. Environ Sci Pollut Res:1–14. https://doi.org/10.1007/s11356-021-13383-x

Gentiili R, Ambrosini R, Montagnani C, Caronni S, Citterio S (2018) Effect of soil pH on the growth, reproductive investment and pollen allergenicity of *Ambrosia artemisiifolia* L. Front Plant Sci 9:1335. https://doi.org/10.3389/fpls.2018.01335

Greipsson S (2011) Phytoremediation. Nat Educ Knowl:2–7

Harikumar PS, Jesitha K, Sreechithra M (2013) Remediation of endosulfan by biotic and abiotic methods. J Environ Prot 4(05):418. https://doi.org/10.4236/jep.2013.45050

Hauser L, Tandy S, Schulin R, Nowack B (2005) Column extraction of heavy metals from soils using the biodegradable chelating agent EDDS. Environ Sci Technol 39(17):6819–6824. https://doi.org/10.1021/es050143r

Heshmati I et al (2019) Forthcoming risk of *Prosopis juliflora* global invasion triggered by climate change: implications for environmental monitoring and risk assessment. Environ Monitoring Assess 191:72. https://doi.org/10.1007/s10661-018-7154-9

Huang Y, Xi Y, Gan L, Johnson D, Wu Y, Ren D, Liu H (2019) Effects of lead and cadmium on photosynthesis in *Amaranthus spinosus* and assessment of phytoremediation potential. Int J Phytoremediation:1–9. https://doi.org/10.1080/15226514.2019.1594686

Jabeen R, Ahmad A, Iqbal M (2009) Phytoremediation of heavy metals: physiological and molecular mechanisms. Bot Rev 75:339–364. https://doi.org/10.1007/s12229-009-9036-x

Jena P, Pradhan C, Patra HK (2016) Cr^{+6} induced growth, biochemical alterations and chromium bioaccumulation in *Cassia tora* (L.) Roxb. Ann Plant Sci 5(7):1368–1373. https://doi.org/10.21746/aps.2016.07.001

Kadukova J, Manousaki E, Kalogerakis N (2008) Pb and Cd accumulation and phyto-excretion by salt cedar (*Tamarix smyrnensis* Bunge). Int J Phytoremediation 10:31–46. https://doi.org/10.1080/15226510701827051

Karuppaiah B, Saraswathi P, Prabhu K, Rajendran M, Samuel S (2019) Preliminary phytochemical screening of *Eclipta prostrata*. World Journal of Pharm Res:783–789

Khalifa SM, Badr BM (2021) Habitats, phytochemical metabolites and phytoremediation potential of Ranunculus sceleratus L. J Pure Appl Sci 20(1):20–24. https://doi.org/10.51984/jopas.v20i1.963

Khuroo AA, Reshi ZA, Weber E, Reshid I (2011) Alien flora of India: taxonomic composition, invasion status and biogeographical affiliations. Biol Invasions 14:99–113. https://doi.org/10.1007/s10530-011-9981-2

Kull CA, Shackleton CM, Cunningham PJ, Ducatillon C, Dufour-Dror JM, Esler KJ, Friday JB, Gouveia AC, Griffin AR, Marchante E, Midgley SJ (2011) Adoption, use and perception of Australian acacias around the world. Divers Distrib 17(5):822–836. https://doi.org/10.1111/j.1472-4642.2011.00783.x

Kumar I, Sharma RK (2018) Production of secondary metabolites in plants under abiotic stress: an overview. Significances Bioeng Biosci 2(4):196–200. https://doi.org/10.31031/SBB.2018.02.000545

Kumar V, Singh J, Chopra AK (2018) Assessment of plant growth attributes, bioaccumulation, enrichment, and translocation of heavy metals in water lettuce (*Pistia stratiotes* L.) grown in sugar mill effluent. Int J Phytoremediation 20(5):507–521. https://doi.org/10.1080/15226514.2017.1393391

Kumar HN, Mohana NC, Nuthan BR, Ramesha KP, Rakshith D, Geetha N, Satish S (2019) Phyto-mediated synthesis of zinc oxide nanoparticles using aqueous plant extract of *Ocimum americanum* and evaluation of its bioactivity. SN Appl Sci 1(6):651. https://doi.org/10.1007/s42452-019-0671-5

Linders TE, Bekele K, Schaffner U, Allan E, Alamirew T, Choge SK, Eckert S, Hayi J, Muturi G, Mbaabu PR, Shiferaw H, Eschen R (2020) The impact of invasive species on social-ecological systems: relating supply and use of selected provisioning ecosystem services. Ecosyst Serv 41: 101055. https://doi.org/10.1016/j.ecoser.2019.101055

Liu S, Ali S, Yang R, Tao J, Ren B (2019) A newly discovered Cd-hyperaccumulator *Lantana camara* L. J Hazard Mater 371:233–242. https://doi.org/10.1016/j.jhazmat.2019.03.016

Ma LQ, Komar KM, Tu C, Zhang W, Cai Y, Kennelley ED (2001) A fern that hyperaccumulates arsenic. Nature 409(6820):579. https://doi.org/10.1038/35054664

Maiti PP, Bhakat RK, Bhattacharjee A (2008) Allelopathic effects of *Lantana camara* on physiobiochemical parameters of *Mimosa pudica* seeds. Allelopath J 22(1):59–67

Mdweshu L, Maroyi A (2020) Local perceptions about utilization of invasive alien species *Opuntia ficus-indica* in three local municipalities in the Eastern Cape Province, South Africa. Biodiversitas 21(4). https://doi.org/10.13057/biodiv/d210446

Mesa-Marín J, Redondo-Gómez S, Rodríguez-Llorente ID, Pajuelo E, Mateos-Naranjo E (2020) Microbial strategies in non-target invasive *Spartina densiflora* for heavy metal clean up in polluted saltmarshes. Estuar Coast Shelf Sci 238:106730. https://doi.org/10.1016/j.ecss.2020.106730

Mundike J, Collard FX, Gorgens JF (2018) Co-combustion characteristics of coal with invasive alien plant chars prepared by torrefaction or slow pyrolysis. Fuel 225:62–70. https://doi.org/10.1016/j.fuel.2018.03.024

Naidoo G, Naidoo KK (2018) Drought stress effects on gas exchange and water relations of the invasive weed *Chromolaena odorata*. Flora 248:1–9. https://doi.org/10.1016/j.flora.2018.08.008

Negi GCS, Sharma S, Vishvakarma SCR, Samant SS, Maikhuri RK, Prasad RC, Palni LMS (2019) Ecology and use *of Lantana camara* in India. Bot Rev 85:109–130. https://doi.org/10.1016/j.flora.2018.08.008

Ngorima A, Shackleton CM (2019) Livelihood benefits and costs from an invasive alien tree (*Acacia dealbata*) to rural communities in the Eastern Cape, South Africa. J Environ Manage 229:158–165. https://doi.org/10.1016/j.jenvman.2018.05.077

Nirola R, Biswas B, Megharaj M, Subramanian A, Thavamani P, Saint RAC (2018) Assessment of chromium hyper-accumulative behaviour using biochemical analytical techniques of greenhouse cultivated *Sonchus asper* on tannery waste dump site soils. Environ Sci Pollut Res 25: 26992–26999. https://doi.org/10.1007/s11356-018-2740-5

Pandey VC (2012) Invasive species based efficient green technology for phytoremediation of fly ash deposits. J Geochem Explor 123:13–18. https://doi.org/10.1016/j.gexplo.2012.05.008

Pappan A, Thomas B (2017) Contribution of invasive plants in herbal medicinal practices. Int J Herb Med 5(2):73–77

Pimentel D, McNair S, Janecka J, Wightman J, Simmonds C, Connell C, Wong E, Russel L, Zern J, Aquino T, Tsomondo T (2001) Economic and environmental threats of alien plant animal and microbe invasions. Agric Ecosyst Environ 84:1–20. https://doi.org/10.1016/S0167-8809(00)00178-X

Prabhu SG, Srinikethan G, Hegde S (2019) Spontaneous Cr (VI) and Cd (II) biosorption potential of native pinnae tissue of *Pteris vittata* L., a tropical invasive pteridophyte. Int J Phytoremediation 21(4):380–390. https://doi.org/10.1080/15226514.2018.1524845

Prasad MNV (2004) Phytoremediation of metals in the environment for sustainable development. Proc Indian Natl Sci Acad Part B 70:71–98

Prasad R (2019a) Plant nanobionics: advances in the understanding of nanomaterials research and applications. Springer International Publishing ISBN: 978-3-030-12495-3. https://www.springer.com/gp/book/9783030124953

Prasad R (2019b) Plant nanobionics: approaches in nanoparticles biosynthesis and toxicity. Springer International Publishing. ISBN: 978–3–030-16379-2 https://www.springer.com/gp/book/9783030163785

Prasad R, Nayak SC, Kharwar RN, Dubey NK (2021) Mycoremediation and environmental sustainability, vol 3. Springer. (ISBN: 978-3-030-54421-8). https://www.springer.com/gp/book/9783030544218

Prieto MJ, Acevedo SOA, Prieto GF (2018) Phytoremediation of soils contaminated with heavy metals. Biodivers Int J 2(4):362–376. https://doi.org/10.15406/bij.2018.02.00088

Rai PK, Singh JS (2020) Invasive alien plant species: their impact on environment, ecosystem services and human health. Ecol Indic 111:106020. https://doi.org/10.1016/j.ecolind.2019.106020

Rana S, Tiwari R, Arora A, Singh S, Kaushik R, Saxena AK, Dutta SC, Nain L (2013) Prospecting *Parthenium* sp. pretreated with *Trametes hirsuta*, as a potential bioethanol feed stock. Biocatal Agric Biotechnol 2:152–158. https://doi.org/10.1016/j.bcab.2013.02.002

Ranđelović D, Jakovljević K, Mišljenović T, Savović J, Kuzmanović M, Mihailović N, Jovanović S (2020) Accumulation of potentially toxic elements in invasive *Ambrosia artemisiifolia* on sites with different levels of anthropogenic pollution. Water Air Soil Pollut 231:1–14. https://doi.org/10.1007/s11270-020-04655-2

Rastgoo L, Alemzadeh A, Afsharifar A (2011) Isolation of two novel isoforms encoding zinc- and copper-transporting P1BATPase from Gouan (Aeluropus littoralis). Plant Omics J 4(7):377–383

Rex Immanuel R (2020) Phytotherapeutic utilization of *Tridax procumbens* L.: an Asteraceae weedy herb by farming communities in Kanyakumari District, Tamil Nadu, India. Plant Arch 20(1):819–830

Reza MS, Ahmed A, Caesarendra W, Bakar MSA, Shams S, Saidur R, Aslfattahi N, Azad AK (2019) *Acacia holosericea*: an invasive species for bio-char, bio-oil, and biogas production. Bioengineering 6(2):33. https://doi.org/10.3390/bioengineering6020033

Sakthieswaran N, Sophia M (2020) *Prosopis juliflora* fibre reinforced green building plaster materials—an eco-friendly weed control technique by effective utilization. Environ Technol Innov 20:101158. https://doi.org/10.1016/j.eti.2020.101158

Salt RCP, Pickering IJ, Raskin I (1995) Mechanisms of cadmium mobility and accumulation in Indian mustard. Plant Physiol 109:1427–1433. https://doi.org/10.1104/pp.109.4.1427

Sarma H, Forid N, Prasad R, Prasad MNV, Ma LQ, Rinklebe J (2021) Enhancing phytoremediation of hazardous metal(loid)s using genome engineering CRISPR–Cas9 technology. J Hazard Mater. https://doi.org/10.1016/j.jhazmat.2021.125493

Sasidharan NK, Azim T, Ambika D, Mathew S (2013) Water hyacinth for heavy metal scavenging and utilization as organic manure. Indian J Weed Sci 45:204–209

Saxena S, Rao PB (2018) GC-MS screening of bioactive constituents and antioxidant profiling in an invasive weed, *Malvastrum coromandelianum* (L.) Garcke. Pharma Innov J 7(4):738–746

Saxena S, Rawat DS, Rao PB (2020) *Malvastrum coromandelianum* (L.) Garcke: an invasive weed with multiple ethnopharmacological properties. https://doi.org/10.25258/phyto.12.1.2

Seebens H et al (2018) Global rise in emerging alien species results from increased accessibility of new source pools. Proc Natl Acad Sci U S A 115(10):E2264–E2273. https://doi.org/10.1073/pnas.1719429115

Selvam P, Vijayakumar T, Wadhwani A, Muthulakshmi L (2019) Bioreduction of silver nanoparticles from aerial parts of *Euphorbia hirta L.* (EH-ET) and its potent anticancerous activities against neuroblastoma cell lines. Indian J Biochem Biophys 56:132–136

Shah K, Nongkynrih JM (2007) Metal hyperaccumulation and bioremediation. Biol Plant 51:618–634. https://doi.org/10.1007/s10535-007-0134-5

Sharma RK, Agrawal M, Marshall FM (2008) Heavy metal (Cu, Zn, Cd and Pb) contamination of vegetables in urban India: a case study in Varanasi. Environ Pollut 154:254–263. https://doi.org/10.1016/j.envpol.2007.10.010

Sharma P, Jha AB, Dubey RS, Pessarakli M (2012) Reactive oxygen species, oxidative damage, and antioxidative defence mechanism in plants under stressful conditions. J Bot:1–26. https://doi.org/10.1155/2012/217037

Sheng P, Shang X, Jiang H (2018) Archaeobotanical evidence for early utilization of cockleburs (*Xanthium strumarium* L., Asteraceae) in the Xinjiang Uyghur Autonomous Region of China. Archaeol Anthropol Sci 11:2027–2038. https://doi.org/10.1007/s12520-018-0651-5

Sheoran V, Sheoran A, Poonia P (2011) Role of hyperaccumulators in phytoextraction of metals from contaminated mining sites: a review. Crit Rev Environ Sci Technol 41:168–214. https://doi.org/10.1080/10643380902718418

Shukla R, Gupta G, Kashaw SK, Jain AP, Lodhi S (2018) Wound healing effect of ethanolic extract from morning glory (*Ipomoea carnea Jacq.*) leaves by using different models in rats. Pak J Pharm Sci 31(4):1355–1361

Silveira MJ, Chollet S, Thiébaut G, Thomaz SM (2018) Abiotic factors, not herbivorous pressure, are primarily responsible for the performance of an invasive aquatic plant. Int J Limnol 54:12. https://doi.org/10.1051/limn/2018002

Singh H, Verma A, Kumar M, Sharma R, Gupta R, Kaur M, Negi M, Sharma SK (2017) Phytoremediation: a green technology to clean up the sites with low and moderate level of heavy metals. Austin Biochem 2(2):1012

Singh S, Kumar V, Dhanjal DS, Datta S, Bhatia D, Dhiman J, Samuela J, Prasad R, Singh J (2020) A sustainable paradigm of sewage sludge biochar: valorization, opportunities, challenges and future prospects. J Clean Prod. https://doi.org/10.1016/j.jclepro.2020.122259

Sirka VH, Jakovljevic K, Mihailovic N, Jovanovic S (2016) Heavy metal accumulation in invasive Reynoutria× bohemica Chrtek&Chrtkova in polluted areas. Environ Earth Sci 75(11):951. https://doi.org/10.1007/s12665-016-5763-6

Sonowal S, Nava AR, Joshi SJ, Borah SN, Islam NF, Pandit S, Prasad R, Sarma H (2022) Biosurfactants assisted heavy metals phytoremediation: green technology for the United Nations sustainable development goals. Pedosphere 2(1):198–210

Souza-Alonso P, Puig CG, Pedrol N, Freitas H, Rodríguez-Echeverría S, Lorenzo P (2018) Exploring the use of residues from the invasive Acacia sp. for weed control. Renew Agric Food Syst:1–12. https://doi.org/10.1017/S1742170518000170

Taylor AF, Rylott EL, Anderson CW, Bruce NC (2014) Investigating the toxicity, uptake, nano-particle formation and genetic response of plants to gold. PLoS One 9(4):e93793. https://doi.org/10.1371/journal.pone.0093793

Tessema YA (2012) Ecological and economic dimensions of the paradoxical invasive species—*Prosopis juliflora* and policy challenges in Ethiopia. J Econ Sustain Dev 3:62–70

Thakare M, Sarma H, Datar S, Roy A, Pawar P, Gupta K, Pandit S, Prasad R (2021) Understanding the holistic approach to plant-microbe remediation technologies for removing heavy metals and radionuclides from soil. Curr Res Biotechnol. https://doi.org/10.1016/j.crbiot.2021.02.004

Thangavel P, Subbhuraam C (2004) Phytoextraction: role of hyperaccumulators in metal contaminated soils. Proc Indian Natl Sci Acad Part B 70:109–130

Tripathi AN, Kumar P, Sati SC (2020) Uses of invasive alien plants in Kumaun Himalayan folk medicinal system. Int J Herb Med 8(1):27–31

Trueman RJ, Erber L (2013) Invasive species may offer advanced phytoremediation of endocrine disrupting chemicals in aquatic ecosystems. Emirates J Food Agric 25(9):648–656. https://doi.org/10.9755/ejfa.v25i9.16393

Wahyuningsih D, Purnomo Y (2017) Antidiabetic effect of *Urenalobata*: preliminary study on hexane, ethanolic, and aqueous leaf extracts. Med J Brawijaya 30:1–6. https://doi.org/10.21776/ub.jkb.2018.030.01.1

Wei H, Huang M, Quan G, Zhang J, Liu Z, Ma R (2018) Turn bane into a boon: application of invasive plant species to remedy soil cadmium contamination. Chemosphere 210:1013–1020. https://doi.org/10.1016/j.chemosphere.2018.07.129

Wu LH, Luo YM, Xing XR, Christie P (2004) EDTA-enhanced phytoremediation of heavy metal contaminated soil with Indian mustard and associated potential leaching risk. Agric Ecosyst Environ 102:307–318. https://doi.org/10.1016/j.agee.2003.09.002

Wu J, Lic X, Zhao J, Wang R, Xia Z, Li X, Liu Y, Xu Q, Khan IA, Yang S (2018) Anti-inflammatory and cytotoxic withanolides from *Physalis minima*. Phytochemistry 155:164–170. https://doi.org/10.1016/j.phytochem.2018.08.009

Xiao L, Herve MR, Carrillo J, Ding J, Huang W (2019) Latitudinal trends in growth, reproduction and defense of an invasive plant. Biol Invasions 21(1):189–201. https://doi.org/10.1007/s10530-018-1816-y

Xie HJ, Li H, Liu D, Dai WM, He JY, Lin S, Duan H, Liu LL, Chen SG, Song XL, Valverde BE, Qiang S (2015) ICE1 demethylation drives the range expansion of a plant invader through cold tolerance divergence. Mol Ecol 24(4):835–850. https://doi.org/10.1111/mec.13067

Phytotechnologies for Bioremediation of Textile Dye Wastewater

10

Dishant Patel, Kunal R. Jain, Datta Madamwar, and Chirayu Desai

Abstract

Phytotechnology is an economically viable as well as eco-friendly solution for wastewater treatment over conventional physical and chemical methods. Several studies have effectively used phytoremediation strategies to treat simple and complex inorganic/organic pollutants such as heavy metals, textile dyes, poly-aromatic hydrocarbon (PAHs), pesticides, crude oils, explosives, domestic wastes and wastewaters. Traditional phytoremediation strategies like plant soil bed, floating phyto-beds, and constructed wetland (CW) are well-established wastewater treatment technologies. These strategies are based on synergism between plant and its rhizospheric or endophytic microbial communities for enhanced wastewater treatment. Textile dye wastewater represents a serious environmental pollution problem and a public health concern. Removal of color and degradation of toxic dyes from textile wastewaters have become a huge challenge over the years. Phytotechnologies have been gradually developed to overcome the various limitations of traditional methods at large-scale operation. Large-scale CWs have been recently applied for on-site treatment of textile industry wastewaters. Bioaugmentation of specific microorganisms in rhizo-sphere of plants has assisted in the remediation process of textile wastewater in CW systems. Integration of microbial fuel cells with constructed wetland systems is achievable due to formation of various redox-gradients in phytoremediation

D. Patel · D. Madamwar · C. Desai (✉)
Department of Biological Sciences, P. D. Patel Institute of Applied Sciences, Charotar University of Science and Technology, CHARUSAT Campus, Changa, Anand, Gujarat, India
e-mail: chirayudesai.bt@charusat.ac.in

K. R. Jain
Post Graduate Department of Biosciences, Sardar Patel University, Satellite Campus, Bakrol, Gujarat, India

R. Prasad (ed.), *Phytoremediation for Environmental Sustainability*,
https://doi.org/10.1007/978-981-16-5621-7_10

systems. Applicability of integrated constructed wetland-microbial fuel cell (CW-MFC) systems has been recently demonstrated for the treatment of dyes and textile wastewater. Genetic engineering of transgenic plants capable of decolorization and degradation of specific dyes has shown promising results in phytoremediation systems. In this chapter, both traditional and advanced phytotechnologies have been discussed to understand and apply these strategies effectively for the remediation of textile dye contamination in soils and textile wastewaters.

Keywords

Phytoremediation · Textile dyes · Bioaugmentation · Constructed wetland · Wastewater

10.1 Introduction

Textile wastewater treatment and its management are still a major concern and a challenging task in developed as well as developing countries. Textile processing is a multi-step process, involving wet processes such as washing, dyeing, finishing, and other steps, as well as dry processes such as knitting, weaving, drying, fixing, etc. (Schonberger and Schafer 2003). These multi-step textile processing requires application of various simple and complex organic and inorganic compounds, including different dyes and enormous amount of freshwater. Astonishingly, color index enlists usage of nearly 8000 distinct chemicals and compound during textile processing (particularly in the dyeing process) (Banat et al. 1996). Therefore, textile industrial effluent is highly heterogeneous mixture of chemicals and unfixed toxic dyes in different concentrations. Thus, textile wastewater is always considered as one of the major sources of hazardous environmental pollution globally (Rawat et al. 2016).

The chemicals and dyes from textile effluents because of their xenobiotic origin are experimentally proven to be toxic exerting chronic and acute effects on living organisms (Platzek et al. 1999; Abbassi et al. 2013; Roy et al. 2022). The degree of toxicity increases mainly due to release of unused dyes. Upon entering into the environmental matrix, dyes are transformed into different intermediates. The transformations are dependent on microbial and various physico-chemical parameters of the environment receiving the effluents and dyes (Rawat et al. 2016). Theoretically, the transformation should yield less toxic or non-toxic intermediates. On contrary, in many occasions, this conversion led to the generation of more-toxic metabolites than the parent dye compounds (Gottlieb et al. 2003). Platzek et al. (1999) reported that in mammalian systems initially dyes are metabolized into carcinogenic intermediates by skin and/or gut microflora.

Realizing the toxic potential of textile effluents as one of the major environmental pollutants, different remediation technologies were developed in the past decades. Initially the focus of these technologies was to remove the color, as color in the open environment is the prime indicator of pollution. Various physico-chemical technologies were developed and used either in standalone or in integrated operations depending on the type of textile effluent and pollutant loads. These technologies were found to be efficient in color removal from the polluted environment. However, soon it was realized that these methods do not completely degrade the toxic dyes and simultaneously generate other toxic by-products requiring further treatment. Thus, in the search of new alternatives, biological treatment methods were developed. The alternative biological methods are not only simpler to apply as compared to physico-chemical technologies, they are also relatively economically viable and ecologically more acceptable.

Over the years, newer biological technologies have been developed such as bioaugmentation, biostimulation, natural attenuation, land-based treatments (solid-phase bioremediation), bacterial, fungal or algal based methods, development of various bioreactors, use of biocatalyst (enzymes), etc (Prasad 2017, 2018; Prasad et al. 2021a, b). These methods are routinely used in major textile wastewater treatment plants in combinations with physico-chemical methods either as secondary or tertiary treatments. Nevertheless, in the era of integrated biotreatment technologies the focus has been to remove color and COD of the effluent, however, less attention has been given for complete mineralization and detoxification of the textile effluents (Rawat et al. 2016).

Looking at the persistence of toxicity and incomplete degradation of metabolites, search for better technologies for treatment of textile effluent is still continued. For the treatment of sewage and municipal wastewater, tannery effluents and sites contaminated with heavy metals (such as dumping suites, mine field, etc.), phytotechnology or phytoremediation approaches utilizing specific plant species have proved to be highly successful. Similarly, lab scale and large-scale treatments using phytotechnologies have shown promising results in the remediation of textile dye wastewaters. Various experimental results now suggest that the removal of color and degradation of intermediates along with simultaneous reduction of toxicity from textile effluents can be effectively achieved in phytoremediation systems.

10.2 Traditional Phytotechnologies for Remediation of Textile Wastewater

10.2.1 Constructed Wetland Systems

Phytotechnologies utilize various plant-microbe combinations for efficient degradation of textile dyes during treatment of textile wastewaters in lab-scale or large-scale systems. Representative phytoremediation systems treating various textile dyes or textile wastewaters are described in Table 10.1. Constructed wetlands are infrastructures that are close to natural wetlands that efficiently remove total suspended solids (TSS), pollutants, and nutrients from wastewater without utilizing large amount of energy, which results in enhancement of the biodiversity and

Table 10.1 List of few phytoremediation technologies used for treatment of dye containing wastewater

S. No.	Dye or textile wastewater	Wastewater characteristics (COD, BOD)	System	Plants used	Treatment efficiency	References
1	Textile wastewater	COD range: 276–1379 mg/L BOD: 99–350 mg/L	Constructed wetland	*Phragmites australis*	Reduction in COD by 84% and BOD_5 by 66%	Bulc and Ojstr (2008)
2	Textile wastewater and mixture of five dyes	ADMI values 405 (textile wastewater) and 418 (mixture of dyes)	Hydroponic system	*Glandularia pulchella* (Sweet) Tronc	Reduction in ADMI by 94.7% and 94.8%, respectively	Kabra et al. (2012)
3	Textile wastewater and mixture of 10 dyes (3 mg/L each)	–	Plant tissue culture system	*Portulaca grandiflora*	Reduction in ADMI values by 54% and 74%, COD values by 25% and 70%, BOD values by 50% and 66% respectively in 40 h	Khandare et al. (2011)
4	Brilliant Blue G (50 mg/L)	–	Hydroponic system	*Petunia grandiflora*	86% decolorization in 36 h	Watharkar and Khandare (2012)
5	Textile wastewater	ADMI value: 4258 COD: 1600 mg/L, BOD: 282 mg/L	Hydroponic system	*Petunia grandiflora*	Reduction in ADMI by 83%, COD by 50%, and BOD by 57%	Watharkar and Khandare (2012)
6	Dye mixture 30 mg/L	ADMI value: 706 COD: 2420 mg/L, BOD: 163 mg/L	Hydroponic system	*Petunia grandiflora*	Reduction in ADMI 22%, COD by 33% and BOD by 50%	Watharkar and Khandare (2012)
7	Textile dye Basic Red 46 (BR46)	20 mg/L	Hydroponic system	*Hydrocotyle vulgaris*	Up to 95% dye degradation	Vafaei et al. (2013)

8	Textile wastewater	COD: 410 mg/L BOD: 207 mg/L	Phytoreactor system inoculated with two endophytic bacterial strains	*Typha domingensis*	Reduction in COD by 79% and BOD by 77% in 72 h	Shehzadi et al. (2014)
9	Simulated textile wastewater	–	Hydroponic system	Consortium of *Petunia grandiflora* and *Gaillardia grandiflora*	Reduction in ADMI by 94%, BOD by 69%, COD by 73% within 36 h	Watharkar and Jadhav (2014)
10	Diazo textile dye Brown 5R (200 mg/L)	–	Hydroponic system	*Ipomoea aquatica*	94% degradation of Brown 5R within 72 h	Rane et al. (2016)
11	Textile wastewater	ADMII value: 138 COD 990 mg/L BOD 740 mg/L	Soil bed with rhizofiltration reactor (510 L)	*Ipomoea aquatica* and *Ipomoea hederifolia*	Reduction in ADMI by 81%, COD by 77%, and BOD by 61% within 72 h	Rane et al. (2016)
12	Textile wastewater	ADMI: Value: 83 COD: 2149 mg/L BOD: 1389 mg/L	Large scale lagoon reactor system (60,000 L)	*Ipomoea aquatica*	Reduction in ADMI by 70%, COD by 87%, and BOD by 76% within 192 h	Rane et al. (2016)
13	Textile azo dye acid red 114	–	Plant tissue culture system	Hairy roots of *Ipomoea carnea Jacq.*	Decolorization of acid red 114 > 98%	Jha et al. (2016)
14	Textile wastewater	ADMI value 1035 ± 2.88, COD (mg/L) 1328 ± 2.30, BOD (mg/L) 1140 ± 5.19	Constructed wetland drenches	Consortium of *Typha angustifolia, Paspalum scrobiculatum*	Reduction in ADMI, COD, and BOD: 76, 70, and 75% respectively	Chandanshive et al. (2017)
15	Textile wastewater	ADMI value: 1285 ± 1.55 COD: 1438 ± 12.7 mg/L	Floating phyto-bed system	Consortium of *Fimbristylis dichotoma* and *Ammannia baccifera*	Reduction in ADMI by 79%, COD by72%, and BOD by 77%	Kadam et al. (2018)

(continued)

Table 10.1 (continued)

S. No.	Dye or textile wastewater	Wastewater characteristics (COD, BOD)	System	Plants used	Treatment efficiency	References
		BOD: 1230 ± 10.20 mg/L				
16	Artificial wastewater (ABRX3)	Reactive brilliant red X-3B (425 mg/L)	CW-MFC system	*Ipomoea aquatica*	Decolorization of 397.64 mg/L dye with 0.91 W/m^3 power density generation	Fang et al. (2018)
17	Synthetic wastewater	Acid Red 18 (AR 18) (0–500 mg/L)	CW-MFC system	*Typha latifolia*	91% decolorization with 0.9 mW/m^2 power density generation	Oon et al. (2018)
18	Textile wastewater	Color: 66 m^{-1} COD: 513 mg/L BOD: 283 mg/L	Floating treatment wetland macrocosms (1000 L)	*Phragmites australis* associated with three plant growth-promoting rhizobacteria (PGPR) strains	Reduction in color 86%, COD 92%, and BOD 91% 86	Tara et al. (2019)

restoration of ecosystems at the polluted sites (Zhao et al. 2013; Wu et al. 2014; Hernández-Crespo et al. 2017).

10.2.1.1 Different Types of Constructed Wetlands (CWs)

Based on wetland hydrology, CWs for wastewater treatment can be divided into two categories: subsurface flow (SSF) CWs and free water surface (FWS) CWs (Saeed and Sun 2013). FWS systems are pretty much same as the natural wetlands, in which shallow wastewater flows over any saturated substrate. Conversely, wastewater flows either horizontally or vertically in SSF CW systems, across the substrate supporting the plants for growth. SSF CWs depending upon its flow direction are divided into two types: horizontal flow (HF) and vertical flow (VF) CWs (Wu et al. 2015). Hybrid constructed wetlands are generally made up of more than two parallel CWs in series such as VF-HF CWs, HF-FWS CWs, HF-VF CWs, and FWS-HF have also been utilized for the efficient treatment of various wastewaters (Vymazal 2013).

10.2.1.2 Current Status of Application of CWs in Textile Wastewater Treatment

The application of CWs for azo dye containing wastewater treatment is an area which is still being explored (Nawab et al. 2018). The treatment method of constructed wetland system involves adsorption of the textile dyes present in the wastewater by the soil media, degradation by microbial activity, or phyto-accumulation by plant uptake (Saba et al. 2015). The major factors affecting the water quality in wetlands are the wetland plants. Being the major biological components of CWs, several reports have shown that the plants act as an intermedium for purification reactions and also as direct utilization sources of nitrogen and phosphorus (Liu et al. 2011). Textile dyes have been treated by a number of plant species like *Gaillardia grandiflora, Aster amellus, Portulaca grandiflora, Tagetes patula* (Khandare et al. 2011; Patil and Jadhav 2013; Chandanshive et al. 2018), *Petunia grandiflora* (Watharkar et al. 2013) as shown in Table 10.2. Many lab-scale CWs were reported for the treatment of simulated and real textile dye wastewater. Ong et al. (2010) reported lab scale up flow constructed wetland system for the treatment of Acid Orange 7 containing wastewater, using *Phragmites australis* plant species. In another study, lab scale sequential vertical and horizontal CW system is used for treating the textile wastewater. Three types of macrophytes such as *Phragmites australis*, *Dracaena sanderiana,* and *Asplenium platyneuron* were used in these systems (Saeed and Sun 2013).

10.2.1.3 Pilot Scale Studies

Several studies have demonstrated the potential of constructed wetland treatment technologies for on-site treatment of wastewater containing textile dyes. In one of the pilot studies by Chandanshive et al. (2017), a CW drench was planted with *Paspalum scrobiculatum, Typha angustifolia* and co-plantation (consortium-TP) were used for on-site treatment of wastewater containing textile dye. The in situ treatment of effluent by consortium-TP, *T. angustifolia,* and *P. scrobiculatum*

Table 10.2 List of plant species used for remediation of textile dye wastewater

S. No.	Plant species	System	Wastewater	References
1	*Salvinia molesta*	Constructed lagoon	Textile wastewater	Chandanshive et al. (2016)
2	*Ipomoea hederifolia*	Rhizofiltration coupled phytoreactor system	Textile wastewater	Rane et al. (2016)
3	*Ipomoea aquatica*	Rhizofiltration coupled phytoreactor system	Textile wastewater	Rane et al. (2016)
4	*Ipomoea carnea* Jacq.	Plant tissue culture system	Diazo dye acid red 114	Jha et al. (2016)
5	*Paspalum scrobiculatum*	Constructed drenches	Textile wastewater	Chandanshive et al. (2017)
6	*Typha angustifolia*	Constructed drenches	Textile wastewater	Chandanshive et al. (2017)
7	*Azolla pinnata*	Hydroponic system	Textile wastewater	Ugya et al. (2017)
8	*Lemna minor*	Pond microcosms	Synthetic textile wastewater	Yaseen and Scholz (2017)
9	*Eichhornia crassipes*	Hydroponic system	Textile wastewater	Wickramasinghe and Chandramali (2018)
10	*Pistia stratiotes*	Hydroponic system	Textile wastewater	Wickramasinghe and Chandramali (2018)
11	*Phragmites australis*	Floating treatment wetland	Textile wastewater	Tara et al. (2016)
12	*Typha domingensis*	Floating treatment wetland	Textile wastewater	Tara et al. (2016)
13	*Tagetes patula*	Constructed wetland	Textile wastewater	Chandanshive et al. (2018)
14	*Aster amellus*	Constructed wetland	Textile wastewater	Chandanshive et al. (2018)
15	*Portulaca grandiflora*	Constructed wetland	Textile wastewater	Chandanshive et al. (2018)
16	*Gaillardia grandiflora*	Constructed wetland	Textile wastewater	Chandanshive et al. (2018)
17	*Ammannia baccifera*	Floating phyto-beds in constructed tanks	Textile wastewater	Kadam et al. (2018)
18	*Asparagus densiflorus*	Vertical subsurface flow phytoreactor system	Textile wastewater	Watharkar et al. (2018)
19	*Brachiaria mutica*	Vertical flow constructed wetland	Textile wastewater	Hussein and Scholz (2017)
20	*Chara vulgaris*	Phytoreactor system	Textile wastewater	Mahajan et al. (2019)

(continued)

Table 10.2 (continued)

S. No.	Plant species	System	Wastewater	References
21	*Fimbristylis dichotoma*	Constructed wetland microbial fuel cell	Dyestuff wastewater	Rathour et al. (2019)
22	*Fimbristylis ferruginea* and *Elymus repens*	Constructed wetland microbial fuel cell	Textile dye wastewater	Patel et al. (2021)

showed a decrease in COD by 70%, 63%, and 65%, ADMI values by 76%, 59%, and 62%, BOD values by 75%, 63%, and 68%, TSS values by 47%, 31%, and 35%, TDS values by 57%, 39%, and 45% respectively, within 96 h. During this phytoremediation process heavy metals like arsenic, cadmium, lead, and chromium were also removed by 28–77%. This study clearly indicated the efficiency of the plant consortium over individual plant species in the treatment of textile wastewater. Recently, a comparative study on in situ textile wastewater treatment was carried out using garden ornamental plants such as *Portulaca grandiflora, Aster amellus, Gaillardia grandiflora,* and *Tagetes patula* which reduced ADMI values by 46%, 50%, 73%, and 59%, respectively, within 30 days in the constructed wetlands along with removal of heavy metals (Chandanshive et al. 2018). Tara et al. (2019) recently demonstrated in situ treatment of wastewater containing textile dyes by pilot-scale (1000 L) floating wetlands planted with *Phragmites australis*. This system reduced COD by 92%, BOD by 91%, and color by 86% in the real textile wastewater along with 87% heavy metal removal, respectively. In another study by Rane et al. (2016), two types of pilot-scale phytoremediation systems such as soil beds planted with *Ipomoea hederifolia* and rhizofiltration systems with *Ipomoea aquatica* were utilized in the treatment of textile dye wastewater. These pilot-scale phytoremediation systems efficiently treated about 510 L of textile wastewater in 72 h, with significant removal of ADMI, COD, BOD, and solids. Moreover, it was found in this study that macrophyte *I. aquatica* could also degrade a diazo-sulfonated textile dye Brown 5R and reduce its toxicity by three-folds on HepG2 cell lines. The *I. aquatica* based phytoremediation system was further replicated for an on-site treatment in constructed lagoon to achieve the treatment of 60,000 L of textile effluent. These studies have proved that phytoremediation approaches can be scaled-up for the on-site treatment of textile dye wastewaters.

10.2.2 Floating Treatment Wetland System

One of the reliable methods to treat various domestic, sewage, and industrial wastewaters are floating treatment wetlands (FTWs), which are a systematically designed and engineered for use in phytoremediation (Sasmaz et al. 2016; Arshad et al. 2017). Growing only a limited number of rooted plants to treat large amounts of effluents was one of the difficulties being faced previously in the phytoremediation systems. Floating treatment wetlands were developed to mitigate this challenge, designed in a unique way where plants can float freely on the wastewater. Moreover, this provided as a method for efficient in situ treatment of large amounts of wastewater because it has vegetations and roots of plants which are directly in contact with the wastewater. The large root surface area provided the maximum benefit in establishing a biofilm which was responsible for feeding on the additional nutrients from the wastewater and for the entrapment of suspended particles (Headley and Tanner 2011).

The floating treatment method has a very simple structural design: the phytoremediating plants are supported on a synthetic mat which is buoyant, and the roots of the plants hanging freely into the water. The FTW system provides increased bio-availability of pollutants to the surface area of the roots which are directly exposed to the water, which is not possible in the conventional CW system. Recently, floating phyto-beds (FPB) planted with *Ammannia baccifera, Fimbristylis dichotoma* along with their combination of "co-plantation consortium FA" were separately evaluated for the treatment of real textile wastewater by Kadam et al. (2018). Textile wastewater treated using phyto-bed with *A. baccifera* gave 64%, 68%, 67%, 48%, and 56% reductions in COD, BOD, ADMI, TSS, and TDS, respectively, after 9 days of treatment. Likewise, textile wastewater treated with *F. dichotoma* phyto-bed also showed the reduction in COD, BOD, ADMI, TSS, and TDS by 67%, 70%, 70%, 50%, and 62%, respectively. However, the phyto-beds with co-plantation were more effective in reduction of the parameters such as COD, BOD, ADMI, and TDS of textile effluent by 72%, 77%, 79%, and 66%, respectively. Additionally, plant consortium FA was more effective for removal of heavy metals (such as arsenic, lead, cadmium, and chromium) during the remediation of textile wastewater by FPB system (Kadam et al. 2018).

In a recent study, floating treatment wetland inoculated with a mixed culture of plant growth-promoting and dye degrading bacteria was used to treat synthetic textile dye wastewater (Nawaz et al. 2020). The FTWs planted with *P. australis* having a treatment capacity of 1000 L were used for the treatment of three different textile dye containing synthetic wastewater. Bacteria such as *Rhodococcus* sp. strain NT-39, *Acinetobacter junii* strain NT-15, and *Pseudomonas indoloxydans* strain NT-38 were used to make a bacterial consortium (10^8 CFU/mL of each strain). In this study, three different dyes such as Bemaplex Navy Blue DRD, Bemaplex Rubine DB, Bemaplex Black DRKP Bezma (500 mg/L each) were used to make three different synthetic wastewaters and three separate FTWs were used to treat these different dyes containing synthetic textile wastewaters. FTW system reduced 90% of COD, 85% of color, and 85% of TDS in the Bemaplex Navy Blue DRD containing synthetic textile dye wastewater. Likewise, in the treatment of Bemaplex Rubine DB containing wastewater 89.2% COD, 84.25% TDS, 83.25% color were reduced by FTW after the treatment. Similarly, 89.96% COD, 84.53% TDS, and 84% color were removed from Bemaplex Black DRKP Bezma containing wastewater after the treatment (Nawaz et al. 2020).

10.3 Emerging Hybrid Phytotechnologies

Hybrid phytotechnologies integrating microbial fuel cells (MFC) with phytoremediation for generation of bioelectricity in addition to wastewater treatment has been an emerging area of research. The MFC systems have been recently integrated with phytotechnologies for treatment of wastewaters consisting of textile dyes forming a constructed wetland-microbial fuel cell (CW-MFC) as described by Rathour et al. (2019). The main principal behind CW-MFC systems is that rhizodeposits of plants and organic pollutants in the wastewater are utilized by microorganisms as substrates, this process generates the electrons and protons which are required in the process of bioelectricity generation in the MFCs (Strik et al. 2008). The electrogenic bacteria breakdown the organic matter in wastewaters and transfer the electrons to the insoluble anode of the MFC which is often placed in an anaerobic environment, to accept these electrons. When the organic matter is oxidized at the anode, the protons released travel to the cathode via the wastewater, while the electrons travel via the external circuit. The protons and electrons are both utilized in a reduction reaction that takes place at the cathode, with oxygen serving as the terminal electron acceptor in most cases, due to its high redox potential and availability (Logan 2008). Similarly, if the anode of CW-MFC is kept in deeply submerged condition where atmospheric oxygen cannot reach, it will generate a redox gradient leading to bioelectricity generation (Doherty et al. 2015). The resultant CW-MFC technology formed by integration of constructed wetland with microbial fuel cells has been demonstrated to be efficient in textile dye wastewater biodegradation and generation of renewable energy in the form of bioelectricity (Yadav et al. 2012). Recently, Fang et al. (2016) observed 87.60% decolorization of azo dye methyl orange (MO) (450 mg/l) in a vertical flow CW-MFC system operated at 3 days HRT in the closed-circuit configuration. Moreover, this system successfully degraded the dimethyl-*p*-phenylenediamine (DMPD), a product formed after the bio-decolorization of MO in the CW-MFC system operated in a closed-circuit configuration. In this study, the maximum power generated by the CW-MFC system was 0.081 W/m^3 with 0.19% Coulombic efficiency. Similarly, Fang et al. (2018) employed CW-MFC systems planted with *Ipomoea aquatica* for the treatment of reactive brilliant red X-3B (ABRX3) (425 mg/L) containing artificial wastewater. Meanwhile, effect of cathode diameter on the performance of CW-MFC system was evaluated. Maximum decolonization of azo dye ABRX3 and COD reduction of 397.64 mg/L and 317.65 mg/L, respectively, were achieved with the cathode having diameter of 25 cm. Maximum power density produced by this CW-MFC system was 0.91 W/m^3 using a cathode with 25 cm diameter. Likewise, in a recent study by Oon et al. (2018), an up-flow constructed wetland-microbial fuel cell (UFCW-MFC) system planted with *Typha latifolia* was used for the treatment of Acid Red 18 (AR 18) containing synthetic wastewater. Different concentrations of AR 18 varied from 0 to 500 mg/L in the wastewater which was used to evaluate effect of dye concentration on the performance of the CW-MFC system. It was found in this study that the decolorization efficiency of the integrated UFCW-MFC was 91% at the highest concentration of AR 18 dye (500 mg/L) in the wastewater. This

CW-MFC system achieved a maximum power density of 0.9 mW/m^2 at 100 mg/L initial concentration of AR 18 dye in the wastewater. The success of these hybrid phytotechnologies demonstrates that in the future research these technologies can be replicated to engineer an eco-friendly large-scale integrated CW-MFC system for on-site treatment of textile dye wastewaters.

10.4 Enhancement of Phytoremediation Processes

10.4.1 Bacterial Bioaugmentation Strategy

As noted above various configurations of CW and CW-MFC systems were used by different researchers for the remediation of textile dyes/wastewater. However, to enhance the degradation of dye or dye wastewaters, many researchers have started using bioaugmentation in CW/CW-MFC systems. Bacteria were augmented either by mixing with influent wastewater or in the form of culture suspension (Hussain et al. 2018; Patel et al. 2021). Watharkar et al. (2013) observed the synergistic effect of *Bacillus pumilus* strain PgJ (rhizospheric bacterial isolate) and *Petunia grandiflora* Juss. (Plant tissue culture) in the decolorization of reactive azo dye Navy Blue RX (NBRX). Individually *Petunia grandiflora* and *B. pumilus* showed maximum decolorization of 80.01% and 76.80%, respectively. However, a consortium consisting of *B. pumilus* and *P. grandiflora* showed enhancement in dye decolorization by 96.86% within 36 h. In another study by Watharkar et al. (2015) a static hydroponic bioreactor was used for the treatment of textile wastewater using *Pogonatherum crinitum* plant and immobilized *Bacillus pumilus* cells. Three different systems were kept, i.e., system with only plant, bacterium reactor, and plant-bacterial consortium bioreactor (combined system). The system that consisted of only plant showed removal of BOD, COD, and ADMI, by 54%, 59%, and 74%, respectively. The *B. pumilus* (bacterium bioreactor) showed 31%, 42%, and 66% reductions of BOD, COD, ADMI, respectively. Whereas the combined system showed an increase in the removal of BOD, COD, and ADMI of the textile effluent by 70%, 78%, and 93%, respectively, within 12 days time period. This approach suggests that the use of bacteria in these phytoreactor systems increases efficiencies of textile wastewater bioremediation in the phytoreactors.

Hussain et al. (2018) used horizontal flow constructed wetlands (HFCWs) for treating the textile effluents. In this study, endophytic bacterial consortium was augmented in the system to enhance its efficiency. HFCWs vegetated with *Leptochloa fusca* showed 80% of COD, 29% of TSS, 76% of BOD, and 76% of color removal. However, bacterial consortium augmentation further improved the remediation ability of the system by showing the removal of COD, BOD, TDS and color by 86%, 78%, 35%, and 90%, respectively. In another study by Hussein and Scholz (2017), *Brachiaria mutica* was planted pilot-scale vertical flow CW system used for the treatment of textile wastewater. Among 20 different bacterial strains, five strains having higher color and COD reduction efficiency were selected to make a mixed bacterial culture. Overall treatment of wastewater was higher in wetland augmented with bacteria as compared to unaugmented wetland systems. The wetlands of *B. mutica* were augmented by endophytic bacteria showed increased dissolved oxygen by 7.31 mg/L as compared to the wetland without bacteria. The reduction rate of BOD and COD was similar as of DO: Wetlands of *B. mutica* decreased BOD and COD up to 71% and 79%, respectively, whereas a maximum reduction was observed in wetland containing *B. mutica* augmented with endophytic bacteria, i.e., 72% and 81%, respectively, after 48 h of treatment. Similarly, the average reduction in color and TDS was improved in the wetland augmented with bacteria. Bacterial augmentation increased the TDS and color removal by 4% and 18% after 48 h of treatment, respectively. Bacterial assisted phytoremediation of textile wastewater in CW system was also able to reduce the overall toxicity of textile dye wastewater which was confirmed by fish survival assays. These observation and results indicated that the application of bacterial-assisted CWs could be engineered in order to decrease the pollution load and wastewater toxicity.

Another recent pilot scale study by Tara et al. (2019) assessed the impact of bacterial augmentation on the efficacy of floating treatment wetlands (FTWs) of *Typha domingensis* and *Phragmites australis* for treatment of wastewater containing textile effluents. Among different isolated bacterial strains isolated, a consortium of most potent dye decolorizing bacterial strains such as *Pseudomonas indoloxydans, Acinetobacter junii,* and *Rhodococcus* sp. were used for bioaugmentation in this study. FTWs of both the plants *T. domingensis* and *P. australis* augmented with bacteria showed higher reduction in BOD from 249 to 31 and 48 mg/L and reduction in COD from 471 to 30 and 76 mg/L, respectively, than the unaugmented FTWs, along with removal of heavy metals like nickel, lead, iron, copper, and chromium after 8 days of experimentation. Treated wastewater also showed decrease in phytotoxicity. These results necessitates that further studies should be conducted on various types of CWs having different types of bacterial and plant species for development of an efficacious combinatorial approach to enhance phytoremediation of textile dye wastewaters.

In a recent study by Patel et al. (2021), a pilot-scale horizontal subsurface flow constructed wetland-microbial fuel cell (HSCW-MFC) was used for the treatment of wastewater containing textile dyes. Using the next generation sequencing (NGS) approach of 16S rRNA amplicon sequencing, it was found that the genera of *Exiguobacterium* and *Desulfovibrio* were dominant in the bacterial community DC5 with a relative abundance of 13.61% and 8.02%, respectively. These genera are well known for their textile dye degradation ability in bioelectrochemical systems. Upon augmenting the bacterial community DC5 in the HSCW-MFC system, treatment efficacy as well as bioelectricity generation of the system was enhanced. Results obtained in this study showed that before augmenting the bacterial community DC5 in the HSCW-MFC system, ADMI and COD removal obtained in the system were 90 $\pm$ 1.5% and 62 $\pm$ 2%, respectively. Whereas post bioaugmentation of bacterial community DC5 the ADMI and COD removal of the HSCW-MFC were significantly improved to 97.32% and 74.10%, respectively. Moreover, the maximum power generation in the CW-MFC-1 of the HSCW-MFC system was also found to increase from 177.3 to 197.94 mW/m^2 after bioaugmentation of bacterial community DC5. These results suggest that the performance of an integrated pilot-scale CW-MFC system can be improved by bioaugmentation with an electroactive bacterial community.

10.4.2 Application of Transgenic Plants

Though plants have the capacity to depollute xenobiotic compounds, they may lack certain catabolic pathways required for complete degradation of these compounds. However, these catabolic pathways are found in the plant-associated microorganisms. Development of transgenic plants having genes for effective dye remediating enzymes and stress-resistant phenotypes is another significant area of research (Abhilash et al. 2009). The phytoremediation of PCBs, dyes, herbicides, explosives, etc. can be enhanced by the transgenic plants expressing bacterial or mammalian genes involved in metabolism of xenobiotics. A transgenic *Arabidopsis* plant was established based on the overexpression of bacterial gene from *Citrobacter* sp. which encodes a triphenylmethane reductase (TMR) enzyme (Fu et al. 2013). This transgenic *Arabidopsis* was found to significantly degrade two triphenylmethane dyes which are widely used in textile industries, pharmaceutical products, biotic staining, and food industries, etc. these compounds are known to be resistant to biodegradation (Fu et al. 2013).

10.5 Factors Affecting the Phytotechnology

Like any other remediation technology, performance of phytotechnology is also dependent on several biotic and abiotic factors. Studying the factors affecting phytoremediation would aid in understanding and designing a successful strategy for effective bioremediation of textile effluents. The phytoremediation of textile wastewater can be influenced by biotic as well as abiotic factors such as suitable plants, temperature, water content, organic matter in the effluent, type of effluent to be treated, pH of effluent, dye concentration, bio-availability of dyes to plant roots, rhizospheric process, growth of microorganisms in rhizosphere, biomass of remediating plants etc. (Pilon-Smits 2005; Khandare and Govindwar 2015). Some relevant factors are further discussed in brief in the following sections.

10.5.1 Effluent Composition, Dye Concentration, and Hydraulics

One of the most significant parameters which affects the remediation of textile wastewater is its xenobiotic composition. The availability of textile dyes and other pollutants along with pH of the effluent are major contributing factors influencing wastewater treatment phytotechnologies. The presence of high amount of negatively charged organic matters (plants and microbial secretions, dead cells of microbes and plant parts, etc.) would increase the binding of positively charged dyes and become unavailable for uptake to plants (Pilon-Smits 2005; Khandare and Govindwar 2015). In one of the studies on degradation of organic pollutant-atrazine by poplar tress was affected by organic matter and dyes composites (Burken and Schnoor 1997).

Dye concentration in the textile effluent is another vital factor affecting phytoremediation. Effluents with lower dye concentration get easily decolorized, while higher concentrations show slow rate of dye degradation. Kagalkar et al. (2009) during their study on remediation of Direct Red 5B by *B. malcolmii* found that at lower concentration of 10–60 mg/L efficient dye removal was obtained as compared to higher concentration of the dye at 60–100 mg/L. In another study, during the treatment of Brilliant Blue R by *L. minor,* dye removal was found to decrease with an increase in the dye concentration from 2.5 to 10 mg/L (Kiliç et al. 2010). The possible reason for a decrease in phytoremediation at higher dye concentration might be due to the textile dye's toxicity on the remediating plants. The TSS and TDS content in the textile effluent also influences its treatment and hydraulic shock loads in the phytoremediation systems. Pophali et al. (2003) reported that high TDS disturbs oxygen transfer and interferes metabolism of pollutants. Phytotechnology was significantly affected by TSS and obstructs rhizofiltration. Therefore, higher TDS and TSS containing effluents may not be suitable for treatment in phytoremediation systems.

Hydraulic retention time (HRT) is one of the crucial factors affecting the treatment efficiency of the wetland phytoremediation systems. It defines the ratio of the wastewater volume (m^3) and the flow rate of the system (Stefanakis 2018). The HRT of the wetlands treating the textile dye wastewater varies from 4 to 20 days depending on the composition of the wastewater and weather conditions. Higher HRT increases the contact time between wastewater pollutants wetland component (plant roots, bacteria, packing material, and biofilm) and achieves better treatment efficiencies by the wetland phytoremediation systems (Akratos and Tsihrintzis 2007).

10.5.2 Plant Species

Plants species from various habitats are being used in phytoremediation of wastewater. Plants from arid lands, halophytes, garden ornamental plants, aquatic macrophytes are regularly being used for treatment of textile effluents. Table 10.2 describes different plants used in wastewater treatment especially in textile/dyestuff effluent. The plants listed in the table are non-medicinal as well as non-agricultural plants that have been reported for dye bioremediation. Plants such as *Typha angustifolia, Ipomoea aquatica, Ipomoea hederifolia, Typha domingensis,* and *Fimbristylis dichotoma* are very efficient for textile wastewater bioremediation in the different configurations of constructed wetland systems as demonstrated in many studies (Rane et al. 2016; Tara et al. 2016; Chandanshive et al. 2017; Rathour et al. 2019). In these studies majority of species of wild plants were in fact collected from wastewater contaminated sites from textile/dye industries and hence were acclimatized to long-term textile dye pollution. The main reason that these plants showed better treatment efficiencies is due to their ability to generate plant enzymes such as catalase, lignin peroxidase, veratryl alcohol oxidase (Chandanshive et al. 2016). In addition, the rhizospheric microbial community of these plants may have been well adapted to sustain high dye concentrations in the soils. Garden ornamental plants are also a lucrative choice for phytoremediation of textile wastewater as well as beautification of landscapes. Recently, Chandanshive et al. (2018) utilized ornamental plants such as *Portulaca grandiflora, Gaillardia grandiflora, Aster amellus,* and *Tagetes patula* in the bioremediation of textile wastewater. These plants were cultivated independently in the high rate transpiration system ridges with the dimensions of 91.4 m $\times$ 1.0 m. Among the plants used in this system, *G. grandiflora* was found to be most effective ornamental plant in terms of color removal as compared to other three plants used. *G. grandiflora* removed 73% of color from the textile wastewater within 30 days of treatment period.

10.5.3 Weathering

The natural procedures like volatilization, leaching, hydrolysis, evapotranspiration, biotransformation, etc. play a very important in phytoremediation (Herath and Vithanage 2015). The process of weathering directly affects availability of pollutants and their metabolism by plants as well as microbes (Cunningham and Ow 1996). The natural precipitations at many times dilute the pollutants which may increase their accessibility for remediation by plants, but simultaneously, possibilities of leaching and free release to neighboring water bodies can also arise. However, with better maintenance of sites under phytoremediation by irrigation, aeration, nutrient supply (i.e., providing manure), etc. can significantly enhance the phytoremediation processes (Khandare and Govindwar 2015).

10.6 Advantages and Limitations of Phytotechnologies

Phytotechnology is a solar energy driven technology, carbon neutral, and environmentally greener approach for treatment of wastewater (Ma et al. 2011; Khandare and Govindwar 2015). It is thought that plants provide roots, stems, leaves as ideal natural habitats for various microorganisms which enhance the bioremediation potential of plants (Khandare and Govindwar 2015). Plants with rapid growth, fibrous, and deep penetrating roots would be an ideal source for phytoremediation. Since plants are autotrophic and require low nutrient inputs, less maintenance requirement phytotechnologies can prove to be a really practical tool for restoration of sites polluted with dyes and dye intermediates (Chandanshive et al. 2018). It has been suggested that phytoremediation can also prevent erosion and leaching of pollutants from contaminated sites (Marques et al. 2009).

Phytoremediation using aquatic macrophytes and weeds like water hyacinths in a previous studies have showed the accumulation of various heavy metals. This shows that phytoremediation can be employed for effective treatment of several xenobiotic compounds (Sanmuga and Senthamil 2014; Khandare and Govindwar 2015). It has a direct on-site and in situ application, with cost effectiveness and it is an ecologically accepted greener technology as compared to conventional physico-chemical methods. The potency of phytoremediation can be further enhanced by supplementing additional and specific nutrients, microbial augmentation or by pre-treatment of wastewater by physico-chemical methods.

Though phytoremediation technology seems to be perfect solution it is an emerging technology and further detailed research is required for its effective utilization in the treatment of hazardous industrial wastewaters. As mentioned above, it is noteworthy to understand that numerous studies and results at laboratory scale have given encouraging results, the filed scale application especially for textile effluent is promising. One of the most important limitations of this technology is availability of plants and their growth times. Majority of plants mentioned above are seasonal and thus screening of various plants is required on large scale which can be made available throughout the year. Unlike other technologies, phytoremediation cannot be considered as either primary or secondary treatment. In many wastewater treatment plants (WTPs) phytotechnologies are being used as a tertiary or at polishing step before disposal of treated water into the environment.

Another significant disadvantage in phytotechnology is consideration of time required for treatment for one batch of effluent. As compared to physico-chemical methods or even for few other biological methods, phytoremediation is a slow process. To decrease the time span, either it should be integrated with other biological or chemical methods or must be augmented with acclimatized microorganisms.

10.7 Conclusion and Future Perspectives

Phytoremediation provides many advantages over conventional physicochemical methods and even over various other biological methods; however, still several unexplored areas in this technology warrant a thorough investigation. Phytotechnologies are well established for the tertiary treatment of sewage wastewaters; however, their potential for textile dye wastewater treatment has been realized only recently. As depicted in this chapter, encouraging results were observed in the treatment of textile dye wastewater in many laboratory-scale or pilot-scale studies. Further, it is highly imperative to execute and implement the obtained knowledge of phytotechnology into in situ or field-scale treatment of textile dye wastewater. Emerging hybrid phytotechnology such as CW-MFC provides a sustainable treatment solution for the textile dye wastewater by producing bio-electricity as a by-product. Recent studies successfully demonstrated that an efficient treatment of industrial effluents containing textile dyes in the pilot-scale CW-MFC systems can be achieved. Further studies are required for the field-scale or in situ bioremediation of textile dye containing wastewater using CW-MFC systems. The clear evidence of plant-microbe interactions has already increased the span of phytoremediation approaches in textile dye wastewater treatment. Overall the efficiencies of phytotechnologies for the bioremediation of wastewaters containing textile dyes can be further improved by the bioaugmentation of endophytic or rhizosphere bacterial strains with the ability of textile dye wastewater degradation. Pre-treatment of wastewaters containing textile dyes by physicochemical treatments such as Fenton and Photo-Fenton oxidation processes, UV-photolytic, and photocatalytic processes can increase the treatment efficiencies of hybrid phytotechnologies. Like other bioremediation technologies, in order to realize the actual potential of phytoremediation, a cross-disciplinary approach utilizing integrated or hybrid phytotechnologies should be the focus of future research in textile dye wastewater bioremediation.

Acknowledgement The authors are thankful to Department of Biotechnology, Ministry of Science and Technology, Government of India for funding the research project (Grant No. BT/PR18965/BCE/8/1401/2016).

References

Abbassi R, Yadav AK, Kumar N et al (2013) Modeling and optimization of dye removal using "green" clay supported iron nano-particles. Ecol Eng 61:366–370

Abhilash PC, Jamil S, Singh N (2009) Transgenic plants for enhanced biodegradation and phytoremediation of organic xenobiotics. Biotechnol Adv 27:474–488. https://doi.org/10.1016/j.biotechadv.2009.04.002

Akratos CS, Tsihrintzis VA (2007) Effect of temperature, HRT, vegetation and porous media on removal efficiency of pilot-scale horizontal subsurface flow constructed wetlands. Ecol Eng 29(2):173–191

Arshad A, Ali S, Khan SN et al (2017) Design of floating wetland for treatment of municipal wastewater and environmental assessment using emergy technique. Proc Int Acad Ecol Environ Sci 7:78–89

Banat IM, Nigam P, Singh D, Marchant R (1996) Microbial decolorization of textile-dyecontaining effluents: a review. Bioresour Technol 58(3):217–227

Bulc G, Ojstr A (2008) The use of constructed wetland for dye-rich textile wastewater treatment. J Hazard Mater 155:76–82. https://doi.org/10.1016/j.jhazmat.2007.11.068

Burken J, Schnoor J (1997) Uptake and metabolism of atrazine by poplar trees. Environ Sci Technol 31:1399–1406

Chandanshive VV, Rane NR, Gholave AR, Patil SM (2016) Efficient decolourization and detoxification of textile industry effluent by Salvinia molesta in lagoon treatment. Environ Res 150:88–96. https://doi.org/10.1016/j.envres.2016.05.047

Chandanshive VV, Rane NR, Tamboli AS, Gholave AR (2017) Co-plantation of aquatic macrophytes Typha angustifolia and Paspalum scrobiculatum for effective treatment of textile industry effluent. J Hazard Mater 338:47–56. https://doi.org/10.1016/j.jhazmat.2017.05.021

Chandanshive VV, Kadam SK, Khandare RV et al (2018) In situ phytoremediation of dyes from textile wastewater using garden ornamental plants, effect on soil quality and plant growth. Chemosphere 210:968–976. https://doi.org/10.1016/j.chemosphere.2018.07.064

Cunningham S, Ow D (1996) Promises and prospects of phytoremediation. Plant Physiol 110:715–719

Doherty L, Zhao Y, Zha X et al (2015) A review of a recently emerged technology: constructed wetland—microbial fuel cells. Water Res 85:38–45

Fang Z, Song H, Yu R, Li X (2016) A microbial fuel cell-coupled constructed wetland promotes degradation of azo dye decolorization products. Ecol Eng 94:455–463. https://doi.org/10.1016/j.ecoleng.2016.06.020

Fang Z, Cao X, Li X et al (2018) Biorefractory wastewater degradation in the cathode of constructed wetland-microbial fuel cell and the study of the electrode performance. Int Bioderer Biodegr 129:1–9. https://doi.org/10.1016/j.ibiod.2017.12.003

Fu X, Zhao W, Xiong A (2013) Phytoremediation of triphenylmethane dyes by overexpressing a Citrobacter sp. triphenylmethane reductase in transgenic. Arabidopsis 97(4):1799–1806. https://doi.org/10.1007/s00253-012-4106-0

Gottlieb A, Shaw C, Smith A et al (2003) The toxicity of textile reactive azo dyes after hydrolysis and decolourisation. J Biotechnol 101(1):49–56

Hernández-Crespo C, Gargallo S, Benedito-Durá V et al (2017) Performance of surface and subsurface flow constructed wetlands treating eutrophic waters. Sci Total Environ 595:584–593

Headley TR, Tanner CC (2011) Critical reviews in environmental science and technology constructed wetlands with floating emergent macrophytes: an innovative stormwater treatment technology. Crit Rev Env Sci Tec 42(21):2261–2310. https://doi.org/10.1080/10643389.2011.574108

Herath I, Vithanage M (2015) Phytoremediation in constructed wetlands. In: Phytoremediation. Springer, Cham, pp 243–263

Hussain Z, Arslan M, Hasan M et al (2018) Integrated perspectives on the use of bacterial endophytes in horizontal flow constructed wetlands for the treatment of liquid textile effluent: phytoremediation advances in the field. J Environ Manag 224:387–395. https://doi.org/10.1016/j.jenvman.2018.07.057

Hussein A, Scholz M (2017) Treatment of artificial wastewater containing two azo textile dyes by vertical-flow constructed wetlands. Environ Sci Pollut Res 25(7):6870–6889

Jha P, Jobby R, Desai NS (2016) Remediation of textile azo dye acid red 114 by hairy roots of Ipomoea carnea Jacq. and assessment of degraded dye toxicity with human keratinocyte cell line. J Hazard Mater 311:158–167. https://doi.org/10.1016/j.jhazmat.2016.02.058

Kabra AN, Khandare RV, Waghmode TR, Govindwar SP (2012) Phytoremediation of textile effluent and mixture of structurally different dyes by Glandularia pulchella (Sweet) Tronc. Chemosphere 87:265–272. https://doi.org/10.1016/j.chemosphere.2011.12.052

Kadam SK, Chandanshive VV, Rane NR, Patil SM (2018) Phytobeds with Fimbristylis dichotoma and Ammannia baccifera for treatment of real textile effluent: an in situ treatment , anatomical studies and toxicity evaluation. Environ Res 160:1–11. https://doi.org/10.1016/j.envres.2017.09.009

Kagalkar A, Jagtap U, Jadhav J et al (2009) Biotechnological strategies for phytoremediation of the sulfonated azo dye Direct Red 5B using Blumea malcolmii Hook. Bioresour Technol 100:4104–4110

Khandare RV, Govindwar SP (2015) Phytoremediation of textile dyes and effluents: current scenario and future prospects. Biotechnol Adv 33:1697–1714

Khandare RV, Kabra AN, Kurade MB, Govindwar SP (2011) Phytoremediation potential of Portulaca grandiflora Hook. (Moss-Rose) in degrading a sulfonated diazo reactive dye Navy Blue HE2R (Reactive Blue 172). Bioresour Technol 102:6774–6777. https://doi.org/10.1016/j.biortech.2011.03.094

Kiliç N, Duygu E, Dönmez G (2010) Triacontanol hormone stimulates population, growth and Brilliant Blue R dye removal by common duckweed from culture media. J Hazard Mater 182:525–530

Liu W, Zeng F, Jiang H, Yu H (2011) Total recovery of nitrogen and phosphorus from three wetland plants by fast pyrolysis technology. Bioresour Technol 102:3471–3479. https://doi.org/10.1016/j.biortech.2010.10.135

Logan BE (2008) Microbial fuel cells. Wiley

Ma Y, Prasad M, Rajkumar M, Freitas H (2011) Plant growth promoting rhizobacteria and endophytes accelerate phytoremediation of metalliferous soils. Biotechnol Adv 29:248–258

Mahajan P, Kaushal J, Upmanyu A, Bhatti J (2019) Assessment of phytoremediation potential of Chara vulgaris to treat toxic pollutants of textile effluent. J Toxicol 2019
Marques AP, Rangel AO, Castro PM (2009) Remediation of heavy metal contaminated soils: phytoremediation as a potentially promising clean-up technology. Crit Rev Environ SciTechnol 39(8):622–654
Nawab B, Esser KB, Jenssen PD et al (2018) Technical viability of constructed wetland for treatment of dye wastewater in Gadoon industrial estate, Khyber Pakhtunkhwa. Pakistan Wetlands 38(6):1097–1105
Nawaz N, Ali S, Shabir G et al (2020) Bacterial augmented floating treatment wetlands for efficient treatment of synthetic textile dye wastewater. Sustainability 12(9):3731. https://doi.org/10.3390/su12093731
Ong S, Uchiyama K, Inadama D et al (2010) Performance evaluation of laboratory scale up-flow constructed wetlands with different designs and emergent plants. Bioresour Technol 101:7239–7244. https://doi.org/10.1016/j.biortech.2010.04.032
Oon YL, Ong SA, Ho LN et al (2018) Up-flow constructed wetland-microbial fuel cell for azo dye, saline, nitrate remediation and bioelectricity generation: from waste to energy approach. Bioresour Technol 266:97–108. https://doi.org/10.1016/j.biortech.2018.06.035
Patel D, Bapodra SL, Madamwar D, Desai C (2021) Electroactive bacterial community augmentation enhances the performance of a pilot scale constructed wetland microbial fuel cell for treatment of textile dye wastewater. Bioresour Technol 125088. https://doi.org/10.1016/j.biortech.2021.125088
Patil AV, Jadhav JP (2013) Evaluation of phytoremediation potential of Tagetes patula L. for the degradation of textile dye reactive blue 160 and assessment of the toxicity of degraded metabolites by cytogenotoxicity. Chemosphere 92(2):225–232
Pilon-Smits E (2005) Phytoremediation. Annu Rev Plant Biol 56:15–39
Platzek T, Lang C, Grohmann G et al (1999) Formation of a carcinogenic aromatic amine from an azo dye by human skin bacteria in vitro. Hum Exp Toxicol 18(9):552–559
Pophali G, Kaula S, Mathur S (2003) Influence of hydraulic shock loads and TDS on the performance of large-scale CETPs treating textile effluents in India. Water Res 37:353–361
Prasad R (2017) Mycoremediation and Environmental Sustainability. 1. Springer International Publishing. ISBN: 978-3-319-68957-9. https://link.springer.com/book/10.1007/978-3-319-68957-9
Prasad R (2018) Mycoremediation and Environmental Sustainability. Springer International Publishing. vol 2. ISBN: 978-3-319-77386-5 https://www.springer.com/us/book/9783319773858
Prasad R, Kumar V, Singh J, Upadhyaya CP (2021a) Recent developments in microbial technologies. Springer ISBN: 978-981-15-4439-2. https://www.springer.com/gp/book/9789811544385
Prasad R, Nayak SC, Kharwar RN, Dubey NK (2021b) Mycoremediation and environmental sustainability, 3. Springer International Publishing. ISBN: 978-3-030-54421-8. https://www.springer.com/gp/book/9783030544218
Rane NR, Patil SM, Chandanshive VV et al (2016) Ipomoea hederifolia rooted soil bed and Ipomoea aquatica rhizofiltration coupled phytoreactors for efficient treatment of textile wastewater. Water Res 96:1–11. https://doi.org/10.1016/j.watres.2016.03.029
Rathour R, Patel D, Shaikh S, Desai C (2019) Eco-electrogenic treatment of dyestuff wastewater using constructed wetland-microbial fuel cell system with an evaluation of electrode-enriched microbial community structures. Bioresour Technol 285:121349. https://doi.org/10.1016/j.biortech.2019.121349
Rawat D, Mishra V, Sharma RS (2016) Detoxification of azo dyes in the context of environmental processes. Chemosphere 155:591–605
Roy A, Ananda Murthy HC, Ahmed H, Islam MN, Prasad R (2022) Phytogenic nanoparticles for degradation of dyes. J Renew Mater. https://doi.org/10.32604/jrm.2022.019410

Saba B, Jabeen M, Khalid A et al (2015) Effectiveness of rice agricultural waste, microbes and wetland plants in the removal of Reactive Black-5 Azo dye in microcosm constructed wetlands. Int J Phytoremediation 7(11):1060–1067. https://doi.org/10.1080/15226514.2014.1003787

Saeed T, Sun G (2013) A lab-scale study of constructed wetlands with sugarcane bagasse and sand media for the treatment of textile wastewater. Bioresour Technol 128:438–447

Sanmuga P, Senthamil P (2014) Water hyacinth (Eichhornia crassipes)—an efficient and economic adsorbent for textile effluent treatment—a review. Arab J Chem. https://doi.org/10.1016/j.arabjc.2014.03.002

Sasmaz A, Dogan IM, Sasmaz M (2016) Removal of Cr, Ni and Co in the water removal of Cr, Ni and Co in the water of chromium mining areas by using Lemna gibba L. and Lemna minor L. Water Environ J 30(3–4):235–242. https://doi.org/10.1111/wej.12185

Schönberger H, Schäfer T, (2003) Best available techniques in textile industry, Berlin Fed Environ Agency 47

Shehzadi M, Afzal M, Khan MU et al (2014) Enhanced degradation of textile effluent in constructed wetland system using Typha domingensis and textile effluent-degrading endophytic bacteria. Water Res 58:152–159. https://doi.org/10.1016/j.watres.2014.03.064

Stefanakis AI (2018) Introduction to constructed wetland technology. Wiley, Chichester, pp 1–21

Strik DPBTB, Bert HVMH, Snel JFH, Buisman CJN (2008) Green electricity production with living plants and bacteria in a fuel cell. Int J Energy Res 32(9):870–876. https://doi.org/10.1002/er.1397

Tara N, Iqbal M, Khan QM, Afzal M (2016) Bioaugmentation of floating treatment wetlands for the remediation of textile effluent. Water Environ Res 33(1):124–134. https://doi.org/10.1111/wej.12383

Tara N, Arslan M, Hussain Z et al (2019) On-site performance of floating treatment wetland macrocosms augmented with dye-degrading bacteria for the remediation of textile industry wastewater. J Clean Prod 217:541–548

Ugya AY, Imam TS, Hassan AS (2017) Phytoremediation of textile waste water using Azolla pinnata; a case study. World J Pharm Res 6(2)

Vafaei F, Movafeghi A, Khataee AR et al (2013) Potential of Hydrocotyle vulgaris for phytoremediation of a textile dye: inducing antioxidant response in roots and leaves. Ecotoxicol Environ Saf 93:128–134. https://doi.org/10.1016/j.ecoenv.2013.03.035

Vymazal J (2013) The use of hybrid constructed wetlands for wastewater treatment with special attention to nitrogen removal: a review of a recent development. Water Res 47:4795–4811. https://doi.org/10.1016/j.watres.2013.05.029

Watharkar AD, Jadhav JP (2014) Ecotoxicology and environmental safety detoxification and decolourization of a simulated textile dye mixture by phytoremediation using Petunia grandiflora and, Gailardia grandi flora: a plant–plant consortial strategy. Ecotoxicol Environ Saf 103: 1–8. https://doi.org/10.1016/j.ecoenv.2014.01.033

Watharkar AD, Khandare RV (2012) Phytoremediation potential of Petunia grandiflora Juss., an ornamental plant to degrade a disperse, disulfonated triphenylmethane textile dye Brilliant Blue G. Environ Sci Pol 20(2):939–949. https://doi.org/10.1007/s11356-012-0904-2

Watharkar AD, Rane NR, Patil SM et al (2013) Enhanced phytotransformation of Navy Blue RX dye by Petunia grandiflora Juss. with augmentation of rhizospheric Bacillus pumilus strain PgJ and subsequent toxicity analysis. Bioresour Technol 142:246–254. https://doi.org/10.1016/j.biortech.2013.05.044

Watharkar AD, Khandare RV, Waghmare PR et al (2015) Treatment of textile effluent in a developed phytoreactor with immobilized bacterial augmentation and subsequent toxicity studies on Etheostoma olmstedi fish. J Hazard Mater 283:698–704

Watharkar AD, Kadam SK, Khandare RV, Kolekar PD (2018) Asparagus densi florus in a vertical subsurface flow phytoreactor for treatment of real textile effluent: a lab to land approach for in situ soil remediation. Ecotoxicol Environ Saf 161:70–77. https://doi.org/10.1016/j.ecoenv.2018.05.078

Wickramasinghe S, Chandramali JKJ (2018) Potential of aquatic macrophytes Eichhornia crassipes, Pistia stratiotes and Salvinia molesta in phytoremediation of textile wastewater. Water Secur 4:1

Wu S, Kuschk P, Brix H et al (2014) Development of constructed wetlands in performance intensifications for wastewater treatment: a nitrogen and organic matter targeted review. Water Res 57:40–55

Wu H, Zhang J, Ngo HH et al (2015) A review on the sustainability of constructed wetlands for wastewater treatment: design and operation. Bioresour Technol 175:594–601

Yadav AK, Dash P, Mohanty A et al (2012) Performance assessment of innovative constructed wetland-microbial fuel cell for electricity production and dye removal. Ecol Eng 47:126–131

Yaseen DA, Scholz M (2017) Treatment of synthetic textile wastewater containing dye mixtures with microcosms. Environ Sci Pollut Res 25(2):1980–1997

Zhao Y, Collum S, Phelan M et al (2013) Preliminary investigation of constructed wetland incorporating microbial fuel cell: batch and continuous flow trials. Chem Eng J 229:364–370

11 Assessment of Pharmaceuticals in Water Systems: Sustainable Phytoremediation Strategies

Ana V. Dordio, A. J. Palace Carvalho, and A. P. Pinto

Abstract

The contamination of aquatic environments with pharmaceuticals has become over the latest years one of the top concerns in Environmental Science and in regard to Public Health and Safety policies. Although reported environmental concentrations of any single pharmaceutical compound are usually too low to induce acute ecotoxicological problems on its own, the prolonged exposure to these pseudo-persistent pollutants (which are originally designed to interfere with biochemical processes) is expected to potentially cause chronic effects in the long term. In addition, the wide variety of drugs already detected in the environment raises the possibility of cumulative effects of substances with similar modes of action or even of synergistic effects that may potentiate the harmful effects of some of the compounds. Therefore, the clarification of the current situation in terms of their removal from wastewaters under the currently used wastewater treatment processes, the impacts they may cause or are already causing in ecosystems and to human health, and the prospects for improvements of future

A. V. Dordio
Chemistry Department, School of Sciences and Technology, University of Évora, Évora, Portugal

MARE – Marine and Environmental Science Centre, University of Évora, Évora, Portugal

A. J. P. Carvalho
Chemistry Department, School of Sciences and Technology, University of Évora, Évora, Portugal

LAQV-REQUIMTE, University of Évora, Évora, Portugal

A. P. Pinto (✉)
Chemistry Department, School of Sciences and Technology, University of Évora, Évora, Portugal

Institute of Mediterranean Agricultural and Environmental Sciences (ICAAM), University of Évora, Évora, Portugal
e-mail: app@uevora.pt

R. Prasad (ed.), *Phytoremediation for Environmental Sustainability*,
https://doi.org/10.1007/978-981-16-5621-7_11

wastewater treatment plants design and operation are urgently needed. This chapter presents a review of the current knowledge on the sources, occurrence, and fate of a variety of classes of pharmaceuticals in the environment, WWTPs, sewage sludge and/or biosolids, and some crop plants and macrophytes. A summary of the most commonly detected pharmaceuticals and typical concentration levels at which they occur is presented, organized by therapeutical class. Wastewater treatment plants, which are the major source of pharmaceuticals in the aquatic environment, are analyzed in some detail, focusing on the efficiencies of pollutant removal that are typical of these conventional means. Ensuingly, alternative or complementary solutions provided by some advanced wastewater treatment technologies are briefly discussed. In this regard, a phytoremediation technology for wastewater treatment is gaining increasing acceptance and widespread use: the constructed wetlands systems, which are discussed in further detail in the final part of the text. The chapter concludes with an overall appreciation of this subject, pointing out some relevant topics that are still scarcely explored and, therefore, may lead to interesting new avenues of research in this field.

Keywords

Wastewater · Sewage sludge · Biosolids · Clean-up · Phytoremediation

11.1 Introduction

Pharmaceuticals are one of the cornerstones of the extraordinary improvements to human health care occurring over the last century. Accordingly, an enormous amount and variety of pharmaceuticals are released annually in the market and the trend is for a continuing growth in their consumption and of the release of new pharmaceutical substances (OECD 2018). However, notwithstanding the benefits they provide, the extensive use of pharmaceuticals also raises new problems, including environmental ones. In fact, in the last decades a broad diversity of pharmaceuticals started being detected in a variety of sample types, from treated wastewater, sludge, biosolids, manure, surface water, groundwater, plant and animal tissues to drinking water samples (Fent et al. 2006; Kummerer 2009; Fatta-Kassinos et al. 2011; Lapworth et al. 2012; Carvalho et al. 2014; Tasho and Cho 2016; Ebele et al. 2017; Al Farsi et al. 2017; Tran et al. 2018; Patel et al. 2019). Pharmaceutical's residues may already have been present as pollutants in aquatic environments for a longer period, but the problem has gone unnoticed until recently, due to the low concentrations at which they typically occur. It was only thanks to the improved ability of modern analytical chemistry to quantify pollutants down to trace levels (ng L^{-1} or μg L^{-1}), even in complex matrices such as environmental samples or wastewaters, which was made possible by the significant advances in new analytical methodologies and instrumentation over the last decades that awareness to this situation was raised.

Pharmaceuticals, as trace environmental pollutants, are relevant despite their usually low concentrations for several reasons (Fent et al. 2006; Petrovic and Barceló 2007; Enick and Moore 2007; Wang and Wang 2016; Yang et al. 2017; Tran et al. 2018; Patel et al. 2019):

- Pharmaceuticals are continuously introduced in the environment (for this reason they are referred to as "pseudo-persistent" pollutants), i.e. even if removal rates in wastewater treatment plants (WWTPs) are high (which is not usually the case) they are overcome by a continuous input due to their high consumption rates;
- Even if most pharmaceutical substances occur in the environment only at low concentrations, the huge variety of pharmaceutical substances currently in use potentially may produce large cumulative effects, or even worse, as several different substances may have similar modes of action or act on the same targets, they may present synergistic effects, thus intensifying their adverse action beyond what the low concentrations of any single substance would lead to predict;
- Pharmaceuticals are originally developed with the intention of performing a biological effect; their beneficial effects to treat diseases are, however, usually harmful to healthy individuals and, therefore, are potentially hazardous substances;
- Pharmaceuticals often have the same type of physicochemical behavior as other harmful xenobiotics (persistence in order to avoid the substance to be inactivated before having a curative effect, and lipophilicity in order to be able to traverse cell membranes); and.
- The low concentrations and high toxicity of pharmaceuticals make this type of pollutants, in general, very difficult to remove by conventional wastewater treatment processes.

Indeed, many of these compounds receive inefficient treatment in WWTPs because the latter were designed to deal with bulk pollutants and are not well suited to cope with the special characteristics of pharmaceuticals (as well as other micropollutants). Therefore, a substantial amount of pharmaceutical pollutants present in wastewater are eventually still present in the treated WWTP effluent and discharged with it in the receiving water bodies. This has been considered the main route for contamination of the aquatic environment by pharmaceuticals (Fent et al. 2006; Nikolaou et al. 2007; Aga 2008; Verlicchi et al. 2012b; Michael et al. 2013; Luo et al. 2014; Evgenidou et al. 2015; Petrie et al. 2015; Noguera-Oviedo and Aga 2016; Wang and Wang 2016; Yang et al. 2017; Tran et al. 2018). However, because pharmaceuticals are typically present in the environment at minute concentrations, their analysis and the monitoring of the environmental situation in that regard require sophisticated and laborious analytical tools for their separation and accurate quantification. Therefore, the presence of pharmaceuticals in the environment is emerging as a topic of major concern, but the full picture is still far from being clearly delineated. Nevertheless, as severe risks to the environment and human health resulting from an increased environmental exposure to pharmaceuticals are predictable (or even observed in some cases such as those of the antibiotics, whose presence in the environment has been the cause of the development of antibiotic resistance bacteria, among other harmful effects) there is an urgent need of finding ways to retain and remove these pollutants before they reach the receiving water bodies.

In the present chapter, pharmaceuticals are briefly described in terms of their chemical characteristics and behavior as pollutants, an overview is presented of the several sources of environment contamination with pharmaceuticals, of their possible environmental fates and of their ecotoxic effects. A summary of the most commonly detected pharmaceuticals, grouped by therapeutical class, as well as typical concentration levels in which they occur is also presented based on the available data collected from the literature. The major source of pharmaceutical contamination, the effluent discharge by WWTPs, is analyzed in more detail, with an assessment of the available data on typical pharmaceutical loads at WWTPs' input and discharge streams as well as typical efficiencies of removal of these pollutants at these conventional wastewater treatment facilities. The majority of published studies focus on the aqueous phase and, therefore, almost all available data on the pharmaceuticals that exit the WWTPs refer to pharmaceutical concentrations in WWTP effluents, whereas very scarce information is reported in regard to pharmaceuticals present in particulate phases, i.e. sludges and biosolids. However, the few data available on solid phases are also discussed briefly in this chapter, as this is an important topic given the common practice of biosolids application in soils as fertilizer and the consequent risks of soil contamination with pharmaceuticals. Finally, some of the advanced wastewater treatment technologies that have been considered as alternative or complementary to conventional wastewater treatment processes in an attempt to improve the removal of pharmaceutical from wastewaters are briefly discussed, analyzing some of the reasons why these have not yet been widely adopted. However, a phytoremediation technology for wastewater treatment is gaining increasing popularity, the constructed wetlands systems (CWS) are discussed in some more detail. Phytotechnologies have gained a good reputation as generally interesting low-cost and low-maintenance wastewater treatment technologies for non-conventional pollutants such as heavy metals and organic xenobiotics. Nowadays, CWS are becoming an alternative to conventional wastewater treatment processes or are being integrated in WWTPs as a secondary or tertiary treatment stage and may potentially become a cost-effective solution for the mitigation of much of the pharmaceutical (Dordio et al. 2010; Hijosa-Valsero et al. 2010, 2011, 2016, 2017; Ávila et al. 2010, 2014; Dordio and Carvalho 2011, 2013, 2017; Reyes-Contreras et al. 2012; Verlicchi and Zambello 2014; Zhang et al. 2014, 2018b; Li et al. 2014; Ávila and García 2015; Vymazal et al. 2017; Matamoros et al. 2017; Vo et al. 2018; Liu et al. 2019). The chapter concludes with an overall appreciation of this subject, pointing out some relevant topics that are still scarcely explored and, therefore, may lead to interesting new avenues of research in this field.

11.2 Pharmaceuticals in the Environment: Characteristics, Sources, and Fate

Pharmaceuticals, whether for human or veterinary use, are xenobiotic compounds that are designed to produce a biological effect on some part of the body of the individuals that ingest them (or use them via external application). Although it may be regarded as a class of chemical substances, the term "pharmaceuticals" is actually a general denomination that refers to the purpose with which they are used (i.e. for the diagnosis, prophylaxis, or therapy of a disease) and does not, in fact, imply a resemblance between pharmaceutical compounds in terms of their physical and chemical characteristics. Pharmaceuticals indeed comprise a wide variety of organic substances with very diverse properties. A few of these are common among many pharmaceuticals because they usually must perform their function in a same common medium, the cell. Therefore, they usually must be at least moderately soluble in aqueous media but still be able to traverse a hydrophobic medium (the lipidic cell membrane). In most other respects, pharmaceuticals span a large variety of families of chemical compounds and therefore present a wide diversity in most other physical and chemical properties.

Pharmaceuticals molecules can be large and chemically complex, varying widely in molecular weight (ranging typically from 200 to 1000 Da), structure, functionality, and shape, due to the diversity of processes in which pharmaceuticals must intervene. In general pharmaceuticals are polar amphiprotic molecules, frequently possessing more than one ionizable group, thus leading to the speciation of the compounds. Therefore, the degree of ionization and, consequently, many of their properties are pH dependent. As they are usually polar, pharmaceuticals are characterized by at least some moderate hydrophilicity that favors their solubility in water, which is the medium where they commonly must take effect. However, some of them also present some lipophilicity. It is important to note that the classification of pharmaceuticals according to their active substances, within subgroups of pharmaceuticals, also does not imply that they follow a definite chemical behavior. In fact, small changes in chemical structure may have significant effects on solubility, polarity, and other properties. This in turn may lead to pharmaceuticals from the same class (or even similar active substances) undergoing through a divergent environmental fate (i.e. the way, as they reach the environment, they will ultimately distribute among the different environmental compartments and subsequently be transformed/degraded, bioaccumulated, or stabilized). Polarity, water solubility, hydrophobicity, and volatility are some of the most important properties of pharmaceuticals that can contribute to their fate in aquatic environments. In addition, some other pharmaceuticals properties such as octanol–water partition coefficients (log K_{ow}), solid-water distribution coefficients (log K_d), organic carbon based sorption coefficients (log K_{oc}), and dissociation constants (pK_a) also have a shaping role in their environmental fate by influencing sorption, partitioning, hydrolysis, photodegradation, and biodegradation processes.

Only a fraction of the pharmaceuticals that are ingested by humans or animals are effectively absorbed by their bodies, being the remnant excreted via feces or urine or washed out (in the case of external application). The percentage of ingested drug that is absorbed by the body, as well as the portion that is subsequently metabolized, differs among different compounds. Pharmaceuticals can be released without suffering any kind of modifications (i.e. as the parent compound) or be excreted in modified forms of the original pharmaceutical (i.e. the metabolites) following its metabolic transformation in the organism (non-biological, human or microbial).

These modified compounds may differ only slightly from the parent substance, or they may exhibit more severe transformations, structural and chemical, leading altogether to whole new chemical compounds. Additionally, some metabolites may be easily reverted to the original parents or be further transformed in non-metabolic reactions, whereas some metabolites are rather stable compounds that resist most further transformations.

The excreted pharmaceuticals and metabolites that are introduced in domestic and hospital wastewaters or result from veterinary use (livestock and aquacultures) are, however, only one among other sources of pharmaceutical pollution. Pharmaceuticals, their metabolites and other transformation/degradation products can enter in the environment through a large and sometimes unexpected variety of routes (Fig. 11.1).

Improper disposal of unused or expired drugs, which therefore may escape adequate waste treatment (in landfills), may be a source of contamination of soil and, through leaching, of water bodies. In addition, at the manufacturing stage, some significant pharmaceutical pollution may also be produced, despite all the measures taken by the industry to limit and/or to mitigate it. In regard to the domestic wastewaters that are treated in municipal wastewater treatment plants (WWTPs), they too are an important (and probably the main) route of entry for pharmaceuticals (and metabolites or transformation products) in the environment despite the treatment they receive at the WWTPs (Fent et al. 2006; Nikolaou et al. 2007; Aga 2008; Verlicchi et al. 2012b; Michael et al. 2013; Luo et al. 2014; Evgenidou et al. 2015; Petrie et al. 2015; Noguera-Oviedo and Aga 2016; Tran et al. 2018). In fact, except for the most biodegradable ones, many pharmaceuticals are poorly removed by the conventional wastewater treatment processes used in most WWTPs, because these have been designed to deal with bulk pollutants as awareness for the problem of pharmaceutical pollution is a relatively recent issue (Fatta-Kassinos et al. 2011; Verlicchi et al. 2012b; Luo et al. 2014; Evgenidou et al. 2015; Gavrilescu et al. 2015; Tran et al. 2018). Conventional wastewater treatment processes, which are typically either of physical (screening, sedimentation, etc.) or biological (activated sludge, lagoon, etc.) nature, do not present sufficiently high efficiencies in the removal of this type of organic micropollutants due to the wide variety of their characteristics, to the low concentrations at which they are present in wastewaters, and to their generally low biodegradability (due to the chemical complexity of these molecules). Therefore, most of these contaminants are usually still present in the treated effluents from WWTPs, which represent the most important source point for aquatic exposure to pharmaceuticals (Fatta-Kassinos et al. 2011; Luo et al. 2014; Evgenidou et al. 2015; Barra Caracciolo et al. 2015; Noguera-Oviedo and Aga 2016; Tran et al. 2018; Patel et al. 2019). However, in addition to this major contribution, wastewaters from

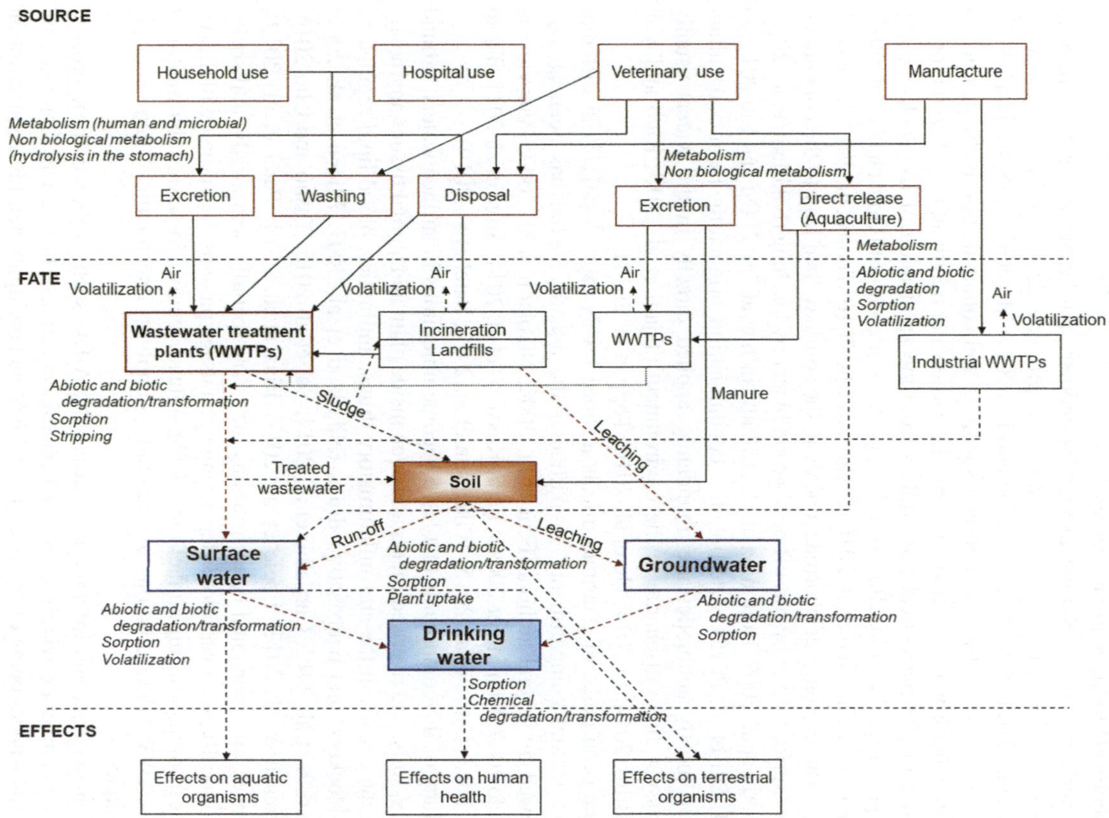

Fig. 11.1 Sources, pathways, and impacts of pharmaceuticals in the environment. (Sources: Halling-Sørensen et al. 1998; Heberer 2002; Farré et al. 2008; Lapworth et al. 2012)

hospitals (Sim et al. 2011; Lapworth et al. 2012; Verlicchi et al. 2012a; Frédéric and Yves 2014; Mendoza et al. 2015; Petrie et al. 2015), the pharmaceutical industry (Sim et al. 2011; Gadipelly et al. 2014; Tran et al. 2018), and landfill leachates (Eggen et al. 2010; Lapworth et al. 2012; Ramakrishnan et al. 2015; Masoner et al. 2016; Lu et al. 2016) are other minor (but significant) inputs of pharmaceutical contaminants to the water resources.

Veterinary use of pharmaceuticals is another significant, although sometimes overlooked, source of contamination of soil, groundwater as well as surface water (Kim et al. 2011; Sim et al. 2011; Du and Liu 2012; Mo et al. 2015; He et al. 2016; Tasho and Cho 2016). An important source of soil contamination is the practice of irrigating fields with reclaimed water if the wastewater treatment is inefficient for the removal of pharmaceuticals, as well as from the application of livestock manure, sewage sludge, and biosolids to soil as fertilizer or compost (Gottschall et al. 2012, 2013; García-Santiago et al. 2016; Tasho and Cho 2016; Topp et al. 2017; Tran et al. 2018). The presence of pharmaceuticals in the soil may then lead to the contamination of surface water by run-off or of groundwater by leaching (Nikolaou et al. 2007; Sabourin et al. 2009; Bottoni et al. 2010; Lapworth et al. 2012; Du and Liu 2012; Li 2014; Sui et al. 2015). Additionally, contamination originating from aquaculture (where mostly antibiotics are abundantly applied directly in the water) usually occurs via direct discharge in the environment (Tijani et al. 2013; Rico and Van den Brink 2014; Li 2014; He et al. 2016; Topp et al. 2017).

The occurrence of pharmaceuticals at trace levels (ngL^{-1}–μgL^{-1}) in different environmental compartments, in particular the aquatic media, has been already reviewed by several authors (Fent et al. 2006; Nikolaou et al. 2007; Petrovic and Barceló 2007; Kümmerer 2009b; Fatta-Kassinos et al. 2011; Silva et al. 2011; Tijani et al. 2013; Petrie et al. 2015; Wilkinson et al. 2017; Patel et al. 2019).

Many of the compounds that have become ubiquitous in surface waters, ground waters, soils, and river sediments and even inside plant and animal tissues are mostly from the classes of the anti-inflammatory drugs, antibiotics, blood lipid regulators, beta-blockers, and neuroactive drugs (Nikolaou et al. 2007; Miége et al. 2009; Lapworth et al. 2012; Carvalho et al. 2014; Sui et al. 2015; Tasho and Cho 2016; Wilkinson et al. 2017; Ebele et al. 2017; Riaz et al. 2018; Patel et al. 2019). However, environmental concentrations may vary spatially, temporally, and socio-economically, with variations depending upon usage patterns, locations (with heavy inputs from manufacturing facilities and hospitals), removal in wastewater treatment plants (WWTPs), dilution by rainfall, sampling uncertainties, and analysis techniques.

Non-steroidal anti-inflammatory drugs (NSAIDs) such as ibuprofen, naproxen, and diclofenac are widely consumed pharmaceuticals, a fact that is facilitated by these substances being over-the-counter drugs, and owing to their high consumption rates they are also very often detected in ground and surface waters (Fent et al. 2006; Petrovic and Barceló 2007; Sui et al. 2015; Wilkinson et al. 2017).

Among the blood lipid regulators, residues of fibrates (i.e. fibric acid derivatives such as clofibrate, bezafibrate, gemfibrozil, and fenofibrate; these have been highly prescribed drugs in the past but have now been largely substituted by other substances in many countries) also continue to be frequently detected in natural water bodies, particularly in the form of their bioactive metabolite clofibric acid. In fact, clofibric acid is one of the earliest pharmaceutical residues to be detected in aquatic environments and a notoriously persistent water contaminant of pharmaceutical origin, with a persistence in the environment that is estimated in 21 years (Khetan and Collins 2007). Currently, other pharmaceuticals from the blood lipid regulators class also commonly present in water bodies are the statins (e.g. atorvastatin, simvastatin, and lovastatin) and niacin (or nicotinic acid) (Sui et al. 2015; Tran et al. 2018; Patel et al. 2019).

Several beta-blockers are also frequently detected in surface waters. In many studies, the presence of atenolol, metoprolol, propranolol, and sotalol in environmental samples is typically reported, but among these, atenolol seems to be the most frequently found worldwide waters (Fent et al. 2006; Verlicchi et al. 2012b; Luo et al. 2014; Sui et al. 2015; Ebele et al. 2017; Tran et al. 2018).

Within the neuroactive drugs class, carbamazepine, fluoxetine, and some benzodiazepines such as diazepam are the most studied and frequently detected substances (Luo et al. 2014; Li 2014; Cunha et al. 2017; Ebele et al. 2017). In particular, carbamazepine has an especially recurrent presence in the aquatic environment due to a long history of clinical usage and a very recalcitrant behavior of this drug (Fent et al. 2006; Verlicchi et al. 2012b; Luo et al. 2014; Sui et al. 2015; Tran et al. 2018). Water contamination with antibiotics presents special significance among the wide variety of pharmaceutical residues detected in the environment not only due to their extensive use and the high frequency of their detection in environmentally relevant concentration levels, but also for the serious risks posed to the aquatic environments and to human health. In fact, the wide ubiquity of several classes of antibiotics such as sulfonamides (e.g. sulfamethoxazole), macrolides (e.g. roxithromycin, ciprofloxacin), tetracyclines (e.g. oxytetracycline), and fluoroquinolones (e.g. ofloxacin, ciprofloxacin) in aquatic environments as well as in soils and sediments has been confirmed repeatedly by numerous studies (Kummerer 2009; Kim et al. 2011; Du and Liu 2012; Michael et al. 2013; Larsson 2014; Gothwal and Shashidhar 2015; Goel 2015; Tasho and Cho 2016; Ebele et al. 2017; Tran et al. 2018; Xie et al. 2018).

In all countries with developed medical care systems, other types of compounds, such as X-ray contrast media or antimicrobial agents (e.g. triclosan, triclocarban), can also be expected to be present at appreciable concentrations in waters (Heberer 2002; Yang et al. 2017; Tran et al. 2018; Patel et al. 2019).

Several studies have also shown that the use of reclamation water from WWTP effluents for irrigation of crops or the use of biosolids and manure as fertilizer or compost could result in pharmaceutical contamination of the soils (Christou et al. 2017). Moreover, as some of these pharmaceuticals have the potential to be taken up by plants, there is a risk that crops grown on contaminated soil can also become contaminated and, thereby, becoming a threat to public health (Picó and Andreu 2007; Wu et al. 2010, 2015; Carvalho et al. 2014; Tasho and Cho 2016; Al Farsi et al. 2017; Xie et al. 2018; Riaz et al. 2018). Conversely, the uptake of pharmaceuticals by plants can be explored as an advantageous feature because it can be used to assist in the reduction of the pollutant load in contaminated water and soil in phytoremediation technologies (Dordio and Carvalho 2013; Zhang et al. 2014; Carvalho et al. 2014; Li et al. 2014; Dordio and Carvalho 2017). Table 11.1 presents some assessments of plant uptake of pharmaceuticals, grouped by therapeutical class, from soil and contaminated water, the latter usually obtained in hydroponic experiments.

As Table 11.1 illustrates, several plant species, both from crops and macrophyte species, have already being studied in an environment exposed to pharmaceuticals (either in soil or hydroponic conditions). Among crop plants, the most studied ones are carrots, while among macrophytes (the type of plants mostly used in phytoremediation) the most studied species include *Typha* and *Phragmites*. An important aspect that is assessed in some studies is the capability of the plants to uptake pharmaceuticals, because in regard to crop plants it may lead to the unwanted result of introducing pharmaceutical contaminants into the food chain whereas in regard to macrophytes it is a desirable property for the purpose of phytoremediation. Pharmaceuticals whose uptake by some plants has already been proven span most of the therapeutical classes, but the most frequently studied are the antibiotics (sulfonamides, tetracyclines, macrolides, and fluoroquinolones) (Carvalho et al. 2014; Azanu et al. 2016; Al Farsi et al. 2017; Madikizela et al. 2018).

Once reaching the environment, pharmaceuticals, their metabolites and transformation products may be submitted to a variety of biotic or abiotic processes that may responsible for their transport, transfer among the various environmental compartments, and transformation/degradation, ultimately determining their fate in the environment. These processes are potentially the same ones that also determine the environmental fate of other organic micropollutants, namely sorption, hydrolysis, biodegradation, redox reactions, photodegradation, volatilization, and precipitation/dissolution (Fig. 11.1) (Petrovic and Barceló 2007; Farré et al. 2008; Aga 2008; Caliman and Gavrilescu 2009; Kümmerer 2009b; Lapworth et al. 2012; Tijani et al. 2013; Li 2014; Wilkinson et al. 2017). Understanding pharmaceutical biodegradability, conjugation, and deconjugation, metabolic pathways, persistence, and sorption are essential to predict their environmental fate. Some of these pathways may contribute to reduce the concentration or availability (through their stabilization in inert forms) of some pharmaceuticals in the environment, or even to their full elimination, thereby lowering their potential to harm human health and aquatic life. However, some pharmaceutical metabolites or transformation products resulting from some of these processes may be more persistent and/or even more toxic than their parent compounds (Celiz et al. 2009; Fatta-Kassinos et al. 2011; Escher and Fenner 2011; Luo et al. 2014; Postigo and Richardson 2014; Barra Caracciolo et al. 2015; Bletsou et al. 2015; Noguera-Oviedo and Aga 2016; Yang et al. 2017; Patel et al. 2019).

Some of the major differences between pharmaceuticals and other common organic micropollutants (such as, for example, pesticides, PCBs, PAHs, or explosives) are that pharmaceutical molecules are in general designed to be sufficiently hydrophilic and water soluble (because they are usually supposed to function in that medium). This implies that aquatic environments are the most relevant ones in

Table 11.1 Removal and/or uptake of same pharmaceuticals by some crop plant and some macrophytes

Therapeutic class	Selected compounds	Crop plants—hydroponics		Crop plants—soil		Macrophytes—hydroponics or planted beds	
		Plants	References	Plants	References	Plants	References
Antibiotics	Amoxicillin			*Triticum aestivum* L. (wheat)	Franklin et al. (2016)		
				Daucus carota L. (carrot), *Lactuca sativa* L. (lettuce)	Azanu et al. (2016)		
	Azithromycin	*Lactuca sativa* (lettuce), *Spinacia oleracea* (spinach), *Daucus carota sativus* (carrots)	Jones-Lepp et al. (2010)				
	Chlortetracycline			*Oryza sativa* (rice), *Cichorium endivia* (sweet oat), *Cucumis sativus* (cucumber)	Liu et al. (2009)		
	Ciprofloxacin					*Juncus acutus* L. (spiny rush)	Christofilopoulos et al. (2016)
						Phragmites australis (common reed)	Liu et al. (2013)

(continued)

Table 11.1 (continued)

Therapeutic class	Selected compounds	Crop plants—hydroponics		Crop plants—soil		Macrophytes—hydroponics or planted beds	
		Plants	References	Plants	References	Plants	References
	Clindamycin	*Lactuca sativa* (lettuce), *Spinacia oleracea* (spinach), *Daucus carota sativus* (carrots)	Jones-Lepp et al. (2010)				
	Enrofloxacin					*Phragmites australis* (common reed)	Carvalho et al. (2014)
	Ofloxacin			*Triticum aestivum* L. (wheat)	Franklin et al. (2016)		
						Cyperus alternifolius (umbrella papyrus)	Yan et al. (2016)
	Oxytetracycline					*Phragmites australis* (common reed)	Liu et al. (2013)
	Roxithromycin					*Cyperus alternlifolius* (umbrella papyrus)	Yan et al. (2016)
		Lactuca sativa (lettuce), *Spinacia oleracea* (spinach), *Daucus carota sativus* (carrots)	Jones-Lepp et al. (2010)				

	Sulfadiazine	*Pisum sativum* (pea), *Cucumis sativus* (cucumber)	Tanoue et al. (2012)				
						Eichhornia crassipes (water hyacinth)	Lin and Li (2016)
	Sulfadimethoxine	*Hordeum vulgare* L. (barley)	Ferro et al. (2010)				
	Sulfamethazine			*Oryza sativa* (rice), *Cichaorium endivia* (sweet oat), *Cucumis sativus* (cucumber)	Liu et al. (2009)		
						Phragmites australis (common reed)	Liu et al. (2013)
	Sulfamethoxazole			*Daucus carota* L. (carrots), *Ipomoea batatas* (sweet potatoes)	Malchi et al. (2014)		
						Eichhornia crassipes (water hyacinth)	Lin and Li (2016)
						Cyperus alternlifolius (umbrella papyrus)	Yan et al. (2016)
						Juncus acutus L. (spiny rush)	Christofilopoulos et al. (2016)

(continued)

Table 11.1 (continued)

Therapeutic class	Selected compounds	Crop plants—hydroponics		Crop plants—soil		Macrophytes—hydroponics or planted beds	
		Plants	References	Plants	References	Plants	References
		Pisum sativum (pea), *Cucumis sativus* (cucumber)	Tanoue et al. (2012)				
				Solanum lycopersicum (tomato)	Christou et al. (2017)		
				Oryza sativa (rice), *Cichaorium endivia* (sweet oat), *Cucumis sativus* (cucumber)	Liu et al. (2009)		
				Cucumis sativus (cucumbers), *Solanum lycopersicum* (tomatoes)	Goldstein et al. (2014)		
	Sulfamonomethoxine	*Pisum sativum* (pea), *Cucumis sativus* (cucumber)	Tanoue et al. (2012)				
	Tetracycline					*Phragmites australis* (common reed)	Carvalho et al. (2014)
						Phragmites australis (common reed)	Arslan Topal (2015)
				Daucus carota L. (carrot), *Lactuca sativa* (lettuce)	Azanu et al. (2016)		

						Lemna gibba (gibbous duckweed)	Arslan Topal (2015)
						Chrysopogon zizanioides (vetiver grass)	Datta et al. (2013)
				Oryza sativa (rice), *Cichaorium endivia* (sweet oat), *Cucumis sativus* (cucumber)	Liu et al. (2009)		
	Trimethoprim			*Oryza sativa* (rice), *Cichaorium endivia* (sweet oat), *Cucumis sativus* (cucumber)	Liu et al. (2009)		
		Pisum sativum (pea), *Cucumis sativus* (cucumber)	Tanoue et al. (2012)				
				Solanum lycopersicum (tomato)	Christou et al. (2017)		

(continued)

Table 11.1 (continued)

Therapeutic class	Selected compounds	Crop plants—hydroponics		Crop plants—soil		Macrophytes—hydroponics or planted beds	
		Plants	References	Plants	References	Plants	References
Antifungal/ antimicrobials	Triclocarban					*Typha* spp. (cattails)	Zarate et al. (2012)
		Cucumis sativus (Cucumber), *Solanum lycopersicum* (tomato), *Brassica oleracea* (cabbage), *Abelmoschus esculentus* (okra), *Capsicum annuum* (pepper), *Solanum tuberosum* (potato), *Beta vulgaris* (beet), *Allium cepa* (onion), *Brassica oleracea* (broccoli), *Apium graveolens* (celery), *Asparagus officinalis* (asparagus)	Mathews et al. (2014)				
				Glycine max (L.) Merr. (soybean)	Wu et al. (2010)		
	Triclosan			*Hordeum vulgare* (barley), *Festuca pratense* (meadow fescue), *Daucus carota* ssp. (carrot)	Macherius et al. (2012)		

						Lemna minor (common duckweed)	Matamoros et al. (2012)
				Glycine max (L.) Merr. (soybean)	Wu et al. (2010)		
		Cucumis sativus (Cucumber), *Solanum lycopersicum* (tomato), *Brassica oleracea* (cabbage), *Abelmoschus esculentus* (okra), *Capsicum annuum* (pepper), *Solanum tuberosum* (potato), *Beta vulgaris* (beet), *Allium cepa* (onion), *Brassica oleracea* (broccoli), *Apium graveolens* (celery), *Asparagus officinalis* (asparagus)	Mathews et al. (2014)				
		Daucus carota (carrot), *Raphanus sativus* (radish), *Lactuca sativa*	Prosser et al. (2014)				

(continued)

Table 11.1 (continued)

Therapeutic class	Selected compounds	Crop plants—hydroponics		Crop plants—soil		Macrophytes—hydroponics or planted beds	
		Plants	References	Plants	References	Plants	References
		(lettuce), *Triticum aestivum* (spring wheat), *Glycine Max* (soybean)					
						Elodea canadensis (pondweed)	Matamoros et al. (2012)
Non-steroidal anti-inflammatory drugs (NSAIDs)	Acetaminophen	*Hordeum vulgare* L. (barley)	Ferro et al. (2010)				
						Lemna gibba (gibbous duckweed)	Allam et al. (2015)
						Phragmites australis (common reed)	Kotyza et al. (2010)
	Diclofenac			*Daucus carota* L. (carrots), *Ipomoea batatas* (sweet potatoes)	Malchi et al. (2014)		
		Hordeum vulgare L. (barley)	Ferro et al. (2010)				
				Solanum lycopersicum (tomato)	Christou et al. (2017)		

						Cyperus alternlifolius (umbrella papyrus)	Zhai et al. (2016)
						Elodea canadensis (pondweed)	Matamoros et al. (2012)
						Lemna minor (common duckweed)	Matamoros et al. (2017)
						Phragmites australis (common reed)	Kotyza et al. (2010)
						Lemna gibba (Gibbous duckweed)	Allam et al. (2015)
						Typha spp. (cattails)	Bartha et al. (2014)
		Pisum sativum (pea), *Cucumis sativus* (cucumber)	Tanoue et al. (2012)				
				Daucus carota L. (carrots), *Ipomoea batatas* (sweet potatoes)	Malchi et al. (2014)		
						Eichhornia crassipes (water hyacinth)	Amos Sibeko et al. (2019)

(continued)

Table 11.1 (continued)

Therapeutic class	Selected compounds	Crop plants—hydroponics		Crop plants—soil		Macrophytes—hydroponics or planted beds	
		Plants	References	Plants	References	Plants	References
	Ibuprofen			*Daucus carota* L. (carrots), *Ipomoea batatas* (sweet potatoes)	Malchi et al. (2014)		
		Hordeum vulgare L. (barley)	Ferro et al. (2010)				
				Cucumis sativus (cucumbers), *Solanum lycopersicum* (tomatoes)	Goldstein et al. (2014)		
						Elodea canadensis (pondweed), *Lemna minor* (common duckweed)	Matamoros et al. (2012)
						Eichhornia crassipes (water hyacinth)	Lin and Li (2016)
						Phragmites australis (common reed)	Kotyza et al. (2010)

						Typha spp. (cattails)	Dordio et al. (2010)
						Typha spp. (cattails)	Li et al. (2014)
						Eichhornia crassipes (water hyacinth)	Amos Sibeko et al. (2019)
	Indomethacin	*Pisum sativum* (pea), *Cucumis sativus* (cucumber	Tanoue et al. (2012)				
	Ketoprofen			*Daucus carota* L. (carrots), *Ipomoea batatas* (sweet potatoes)	Malchi et al. (2014)		
		Pisum sativum (pea), *Cucumis sativus* (cucumber)	Tanoue et al. (2012)				
				Cucumis sativus (cucumbers), *Solanum lycopersicum* (tomatoes)	Goldstein et al. (2014)		
	Naproxen			*Daucus carota* L. (carrots), *Ipomoea batatas* (sweet potatoes)	Malchi et al. (2014)		
				Cucumis sativus (cucumbers),			

(continued)

Table 11.1 (continued)

Therapeutic class	Selected compounds	Crop plants—hydroponics		Crop plants—soil		Macrophytes—hydroponics or planted beds	
		Plants	References	Plants	References	Plants	References
				Solanum lycopersicum (tomatoes)	Goldstein et al. (2014)		
						Elodea canadensis (pondweed), *Lemna minor* (common duckweed)	Matamoros et al. (2012)
						Scirpus validus (softstem bulrush)	Zhang et al. (2012)
						Eichhornia crassipes (water hyacinth)	Amos Sibeko et al. (2019)
β-Blockers	Atenolol					Typha spp. (cattails), *Phragmites australis* (common reed)	Dordio et al. (2009)
	Metoprolol			*Daucus carota* L. (carrots), *Ipomoea batatas* (sweet potatoes)	Malchi et al. (2014)		
Hormones	Progesterone					*Lemna gibba* (gibbous duckweed)	Allam et al. (2015)

Lipid regulators	Atorvastatin			*Lactuca sativa* (lettuce), *Dacus carota* subsp. *sativa* (carrot), and *Lycopersicon esculentum* (tomato)	D'Abrosca et al. (2008)		
	Bezafibrate			*Daucus carota* L. (carrots), *Ipomoea batatas* (sweet potatoes)	Malchi et al. (2014)		
				Cucumis sativus (cucumbers), *Solanum lycopersicum* (tomatoes)	Goldstein et al. (2014)		
	Clofibric acid			*Daucus carota* L. (carrots), *Ipomoea batatas* (sweet potatoes)	Malchi et al. (2014)		
				Cucumis sativus (cucumbers), *Solanum lycopersicum* (tomatoes)	Goldstein et al. (2014)		
						Elodea canadensis (pondweed), *Lemna minor*	Matamoros et al. (2012)

(continued)

Table 11.1 (continued)

Therapeutic class	Selected compounds	Crop plants—hydroponics		Crop plants—soil		Macrophytes—hydroponics or planted beds	
		Plants	References	Plants	References	Plants	References
						(common duckweed)	
						Scirpus validus (softstem bulrush)	Zhang et al. (2013a)
						Typha spp. (Cattails)	Dordio et al. (2009)
						Typha spp. (Cattails)	Dordio et al. (2010)
	Gemfibrozil			*Lactuca sativa* (lettuce), *Dacus carota* subsp. *sativa* (carrot), and *Lycopersicon esculentum* (tomato)	D'Abrosca et al. (2008)		
				Daucus carota L. (carrots), *Ipomoea batatas* (sweet potatoes)	Malchi et al. (2014)		
				Cucumis sativus (cucumbers), *Solanum lycopersicum* (tomatoes)	Goldstein et al. (2014)		

Psycho/ neuroactive drugs	Carbamazepine			*Daucus carota* L. (carrots), *Ipomoea batatas* (sweet potatoes)	Malchi et al. (2014)		
				Triticum aestivum L. (wheat)	Franklin et al. (2016)		
				Cucumis sativus (cucumbers), *Solanum lycopersicum* (tomatoes)	Goldstein et al. (2014)		
						Eichhornia crassipes (water hyacinth)	Lin and Li (2016)
						Cyperus alternlifolius (umbrella papyrus)	Yan et al. (2016)
						Typha spp. (cattails)	Dordio et al. (2010)
						Typha spp. (cattails)	Dordio and Carvalho (2011)
						Scirpus validus (softstem bulrush)	Zhang et al. (2013b)
				Glycine max (L.) Merr. (soybean)	Wu et al. (2010)		
		Pisum sativum (pea), *Cucumis sativus* (cucumber)	Tanoue et al. (2012)				
				Daucus carota L. (carrots), *Ipomoea batatas* (sweet potatoes)	Malchi et al. (2014)		

(continued)

Table 11.1 (continued)

Therapeutic class	Selected compounds	Crop plants—hydroponics		Crop plants—soil		Macrophytes—hydroponics or planted beds	
		Plants	References	Plants	References	Plants	References
	Chlordiazepoxide	*Raphanus sativus* (radish), *Beta vulgaris* (silverbeet)	Carter et al. (2018)				
	Clonazepam	*Raphanus sativus* (radish), *Beta vulgaris* (silverbeet)	Carter et al. (2018)				
	Diazepam	*Raphanus sativus* (radish), *Beta vulgaris* (silverbeet)	Carter et al. (2018)				
	Flurazepam	*Raphanus sativus* (radish), *Beta vulgaris* (silverbeet)	Carter et al. (2018)				
	Oxazepam	*Raphanus sativus* (radish), *Beta vulgaris* (silverbeet)	Carter et al. (2018)				
	Temazepam	*Raphanus sativus* (radish), *Beta vulgaris* (silverbeet)	Carter et al. (2018)				
	Triazolam	*Raphanus sativus* (radish), *Beta vulgaris* (silverbeet)	Carter et al. (2018)				
	Fluoxetine			*Glycine max* (L.) Merr. (soybean)	Wu et al. (2010)		

regard to the contamination with pharmaceuticals. Conversely, the spread of hydrophobic pharmaceuticals in aquatic environments is relatively limited and much slower. However, in that case they tend to accumulate in the fatty tissues of organisms (Ebele et al. 2017).

Pharmaceuticals are also designed to be chemically stable. In fact, some pharmaceuticals (e.g. clofibric acid, carbamazepine) can persist in the environment for many years and become biologically active through accumulation (Ebele et al. 2017).

The combination of high consumption and the properties of a significant water solubility and high resistance to degradation both by biotic and abiotic processes are the conditions that favor the introduction and persistence of pharmaceuticals in the aquatic environment. However, even when the susceptibility for (bio)degradation is moderate, some pharmaceuticals may reach steady-state levels in the environment (thus being known as pseudo-persistent pollutants) as result of their continuous introduction in sewage systems due to the continuingly high consumptions.

Before reaching the environment, pharmaceuticals have already passed through the digestive tracts of humans or animals and, in most cases, also through wastewater treatment processes. Two consequences of this pre-exposure to a special biotic environment and to biochemical metabolism can therefore be anticipated: (1) many pharmaceuticals will enter the aquatic environment in a modified form that is more stable in regard to biotic transformation or degradation and (2) those pharmaceuticals still remaining unaltered at the end of this path are probably highly resistant to biotic transformation or degradation. This understanding suggests certain inferences regarding the importance of abiotic processes for the fate of pharmaceutical compounds in the aquatic environment. Given the significant solubility in water of many pharmaceuticals, abiotic processes most likely to transform these water pollutants and more definitively remove them from the aquatic environment including hydrolysis and photodegradation. However, as most pharmaceuticals are, first of all, exposed to the digestive tract and, subsequently, remain for relatively long residence times in aqueous media within the WWTPs, hydrolysis reactions in such cases are less likely to play a relevant role in the fate of pharmaceuticals when they reach the aquatic environment. Conversely, direct photodegradation by sunlight may be an important elimination process for those pharmaceuticals that have significant absorbances in the spectrum region between 290 and 800 nm (Velagaleti 1997; Andreozzi et al. 2003; Boreen et al. 2003; Challis et al. 2014).

In addition to the physical and chemical properties of the pharmaceutical, environmental conditions (including temperature, sunlight, pH, content of organic matter in soils and sediments and redox conditions) can also influence the way abiotic and biotic processes affect its short-term behavior as well as its long-term environmental fate. Nevertheless, according to evidence accumulated over the years, many pharmaceuticals show, at least to some extent, a refractory behavior towards (bio)-degradation and transformation under ordinary conditions.

In summary, pharmaceuticals generally have the potential to reach and to persist in the aqueous environment for long periods. However, relatively little is known about the impending adverse effects to water organisms and to human health that can arise from the cumulative exposure to an extensively varied blend of pharmaceuticals and their metabolites which are becoming progressively disseminated throughout several environmental compartments (notwithstanding the usually diminutive concentrations at which they occur). The design of pharmaceutical molecules is targeted for interacting with specific biochemical pathways. As a side-effect, when introduced in the environment, it is plausible that pharmaceuticals interfere with analogous pathways of other organisms which possess similar target organs, tissues, cells, or biomolecules. Even in such cases where organisms lack matching receptors for a particular pharmaceutical molecule, it may still induce a disruptive effect caused by an alternative mode of action. In fact, it should be pointed out that the specific modes of action of many pharmaceuticals are not well characterized and in many cases there may be not only one but several different modes of action occurring simultaneously. Therefore, the ecotoxicity of most pharmaceuticals, as well as their metabolites and transformation products, is hard to assess or predict (Fent et al. 2006; Celiz et al. 2009; Evgenidou et al. 2015; Yang et al. 2017). Furthermore, there is a risk that the environmental contamination with pharmaceuticals may propagate to crops, as some of these substances, possessing favorable chemical properties, have the potential to be taken up by plants. In general, there is an even broader risk that vegetation may uptake and accumulate pharmaceuticals, which will then take part of the diet of herbivores and, subsequently, be passed along the food chain (although, for the most part, possibly in a transformed form).

11.3 Assessment of Pharmaceuticals in Wastewater Treatment Plants

The evaluation of the presence of pharmaceuticals in wastewaters, as well as its removal efficiency by WWTPs, has been the focus of several recent reviews, which clearly shows the importance conceded to this subject in the present days (Fent et al. 2006; Nikolaou et al. 2007; Miége et al. 2009; Kümmerer 2009b; Verlicchi et al. 2012b; Michael et al. 2013; Tijani et al. 2013; Luo et al. 2014; Evgenidou et al. 2015; Petrie et al. 2015; Wang and Wang 2016; Yang et al. 2017; Tran et al. 2018; Patel et al. 2019). A survey of the data collected in reviews on the occurrence of the most relevant and commonly detected pharmaceuticals is presented in Table 11.2. The data describe the typical concentration ranges of each pharmaceutical that are quantified in WWTPs' influent and effluent streams as well as the assessed removal efficiencies in WWTPs of each pharmaceutical.

A brief inspection of Table 11.2 reveals that pharmaceuticals typically occur in WWTP influents or effluents at concentration levels in the range of the ng L^{-1} to μg L^{-1}. However, a feature that also stands out in these data is a quite significant variability of the concentration levels that are reported by different studies. This variability may result from a poorer accuracy of the chemical analyses due to the difficulties posed by such low concentration levels and the complex compositions of wastewater matrices. However, concentration levels of pharmaceuticals in

Table 11.2 WWTP influent and effluent concentrations and respective efficiencies of removal in conventional WWTPs, for selected pharmaceuticals, grouped by therapeutic class (sources: Luo et al. 2014; Wang and Wang 2016; Yang et al. 2017; Tran et al. 2018; Patel et al. 2019)

Therapeutic class	Selected compounds	Influent ($\mu g\ L^{-1}$)	Effluent ($\mu g\ L^{-1}$)	Removal (%)
Antibiotics	Amoxicillin	ND-6.52	ND-1.67	69.9–100
	Chloramphenicol	<MQL-2.43	ND-1.05	<0–99
	Clarithromycin	<MQL-8.0	0.005–7	<0–99
	Erythromycin	ND-10	ND-2.84	<0–82.5
	Norfloxacin	ND-0.68	0.0139–0.36	31–93
	Ofloxacin	ND-1.27	<MQL-8.6	<0–99
	Oxytetracycline	<MQL-47	<MQL-4.2	29–96
	Roxithromycin	ND-0.13	ND-0.14	<0
	Sulfamethoxazole	<MQL-11.6	<MQL-1.8	<0–99
	Tetracycline	0.029–1.300	0.016–0.85	12–100
	Trimethoprim	0.06–6.80	<0.01–3.05	0–81.6
Antifungal/antimicrobials	Miconazole	>MQL-0.6	<MQL-0.036	<0–99
	Thiabendazole	<MQL-0.22	<MQL-0.14	<0–88
	Triclocarban	0.097–8.89	ND-5.86	<0–99
	Triclosan	<MQL-6.82	<MQL-0.43	<0–100
Non-steroidal anti-inflammatory drugs (NSAIDs)	Acetaminophen/paracetamol	1.57–292	ND-0.03	98.7–100
	Codeine	<MQL-32.3	<MQL-15.59	<0–98
	Diclofenac	<0.001–94.2	< MQL-5.2	<0–98
	Ibuprofen	<0.0004–603	ND-69	72–100
	Fenoprofen	<MQL-2.26	<MQL-0.41	98.6–100
	Ketoprofen	<0.004–8.56	<0.003–3.92	10.8–100
	Mefenamic acid	<0.017–3.20	<0.005–2.40	0–70.2
	Naproxen	<0.002–611	<0.002–33.9	43.3–98.6
	Salicylic acid	0.58–63.7	ND-0.50	89.6–100
β-Blockers	Atenolol	0.1–33.1	0.13–7.60	0–85.1
	Metoprolol	0.002–1.52	0.003–0.25	3–56.4
	Propranolol	0.05–0.64	0.01–0.615	< 0–44
Hormones	Estrone (E1)	<MQL-0.67	<MQL-0.100	<0–100
	Estriol (E3)	<MQL-0.8	ND-0.28	18–100
	17α-ethinylestradiol (EE2)	<MQL-0.67	<MQL-0.11	33–100
Blood lipid regulators	Bezafibrate	0.05–7.6	0.02–4.30	9.10–70.5

(continued)

Table 11.2 (continued)

Therapeutic class	Selected compounds	Influent (μg L^{-1})	Effluent (μg L^{-1})	Removal (%)
	Clofibric acid	0–0.74	0.042–0.33	0–93.6
	Gemfibrozil	0.10–17.1	<0.025–5.24	0–92.3
	Pravastatin	0.023–0.33	<MQL-0.4	n.r.
Psycho/neuroactive drugs	Alprazolam	0.019–0.049	0.011–0.034	n.r.
	Carbamazepine	<0.04–3.78	<0.05–4.60	0–62.3
	Diazepam	<MQL-0.2	<MQL-0.24	n.r.
	Fluoxetine	<MQL-0.03	<MQL-0.001	n.r.

MQL method quantification limit, *ND* not detected, *n.r* not reported

wastewaters are also strongly affected by many factors, some of which may present a significant spatial and temporal variability. This includes variations, over time, and throughout locations in the world, of the production/sales/consumption levels of the pharmaceuticals (depending on rates of production, sales volume and market strategies, local prescription and usage practices, spatial and seasonal distributions of disease prevalence, etc.), thus affecting the inputs of WWTPs downstream. In addition, variability may also be associated with some aspects that relate with WWTP design, operation, and environmental conditions that affect the characteristics of the final WWTP effluents (water consumption per person and per day, WWTP size, plant configuration especially the type of bioreactor, hydraulic retention time, solids retention time, temperature, rainfall, sunlight) and water catchment characteristics (e.g. land use, population size, and population density).

The short sample of studies presented in Table 11.2 illustrates the therapeutical classes of pharmaceuticals whose occurrence typically predominate in wastewaters: those pharmaceuticals that are most commonly detected in WWTPs are mainly analgesics and anti-inflammatory drugs, antibiotics, blood lipid regulators, beta-blockers, or psycho/neuroactive drugs (Fent et al. 2006; Nikolaou et al. 2007; Miége et al. 2009; Kümmerer 2009b; Verlicchi et al. 2012b; Michael et al. 2013; Tijani et al. 2013; Luo et al. 2014; Evgenidou et al. 2015; Petrie et al. 2015; Yang et al. 2017; Tran et al. 2018). Non-steroidal anti-inflammatory drugs (NSAIDs) usually arise as those with higher loads in WWTP influents, which may be attributed to the fact that these are over-the-counter drugs and, thus, are highly consumed pharmaceuticals. Within this therapeutical class, the active substances ibuprofen, naproxen, diclofenac, and ketoprofen are usually referred as the most frequently detected and at the highest concentrations in WWTP influents. Meanwhile, in the effluents leaving the WWTPs, and even though the concentrations of these compounds are notably lowered (because they are reasonably biodegradable and, thus, typically well removed in WWTPs), NSAIDs are frequently still present at levels quite far from negligible. Indeed, because they enter the WWTP at such high loads, the remnant at the exit of the WWTP, even after large removals, still remains a significant amount (Aga 2008; Caliman and Gavrilescu 2009; Kümmerer 2009b; Li 2014; Tran et al. 2018). Hence, the concentrations of some NSAIDs in effluents of

WWTPs are frequently higher than their predicted no-effect concentrations (PNECs) for the aquatic ecosystems and consequently, the discharges of WWTP effluents into the receiving water bodies may pose potential long-term risks.

Numerous reports also evidence an ubiquitous occurrence of many antibiotics in effluents of WWTPs, thus reaffirming the concern relative to this class of pharmaceuticals, in particular the recurrent worries associated with the development of antibiotic resistant bacteria (ARB) and antibiotic resistance genes (ARGs) (Michael et al. 2013; Bouki et al. 2013; Mo et al. 2015; He et al. 2016; Singer et al. 2016; Topp et al. 2017; Xie et al. 2018; Shao et al. 2018; Barancheshme and Munir 2018; Abidelfatah et al. 2019; Koch et al. 2021). Apart from resistance selection, antibiotics in the influents of WWTPs can also directly influence the activities of microorganisms and, consequently, of wastewater treatment performance (Gonzalez-Martinez et al. 2014; Tran et al. 2018). Furthermore, if after an unsuccessful treatment they leave the WWTP in their original form or as some toxic metabolites, the discharge of antibiotic-containing effluents into the receiving water bodies can also harm the aquatic organisms and environment (Kümmerer 2009a; Kim et al. 2011; Du and Liu 2012; Verlicchi et al. 2012b; Larsson 2014; Gothwal and Shashidhar 2015; Goel 2015; Bengtsson-Palme and Larsson 2016; Shao et al. 2018). It was assessed in some studies that concentrations of many antibiotics in WWTP effluents were over their predicted-no-effect concentrations (PNECs) for ecological toxicity to aquatic organisms (Bengtsson-Palme and Larsson 2016; Tran et al. 2018). Among the classes of antibiotics investigated, sulfonamides (e.g. sulfamethoxazole), fluoroquinolones (e.g. ciprofloxacin, norfloxacin, ofloxacin), macrolides (e.g. clarithromycin, erythromycin, roxithromycin), and trimethoprim were frequently detected in both WWTP influent and effluent samples worldwide. In contrast, the occurrence of β-lactams (e.g. amoxicillin), tetracyclines (e.g. tetracycline and oxytetracycline), and chloramphenicol in WWTP influents and effluents is less reported for North American and European countries, while they are still present in WWTP influents and effluents from some Asian countries (Tran et al. 2018). Although β-lactams are among the most widely used prescribed antibiotics, their frequent absence from wastewaters may be attributed to a high susceptibility to chemical or enzymatic hydrolysis (Watkinson et al. 2007; Le Minh et al. 2010; Tran et al. 2018). Chemical hydrolysis and/or chemical transformations of β-lactams antibiotics can seemingly take place under acidic or alkaline conditions or by reactions with weak nucleophiles, e.g. water or metal ions (Le Minh et al. 2010). In addition, β-lactam antibiotics can be enzymatically hydrolyzed by β-lactamases.

Antifungal and antimicrobial agents are increasingly being recognized as another class of pollutants of concern for aquatic environments, similarly to the case of antibiotics, as they too have potential for inducing the development of ARGs and ARB and, generally, for causing adverse effects on aquatic organisms (Bouki et al. 2013; Singer et al. 2016; Tran et al. 2018; Barancheshme and Munir 2018). This type of substances (including miconazole, thiabendazole, triclocarban, and triclosan) is widely used in some household products such as hair shampoos, dermal creams, soaps, toothpastes, and shower gels. Miconazole and thiabendazole are also commonly used in therapeutic products for the treatment of fungal infections in humans.

In WWTP influents, concentration levels of antimicrobial agents (i.e. triclocarban and triclosan) seem to be usually higher than those of antifungal compounds (e.g. miconazole, thiabendazole) by at least one order of magnitude (Table 11.2). Conversely, the levels of most antifungal and antimicrobial agents in effluents exiting the WWTPs typically vary from below MQL to a few hundreds of ng L^{-1}, being usually much lower than those in the influent, implying some extent of removal of this kind of pharmaceutical pollutants in WWTPs. Generally, the concentrations of triclosan and triclocarban in WWTP effluents are often higher than their PNECs for aquatic organisms (Tran et al. 2018).

A number of beta-blockers are also detected in WWTP influents and effluents, namely atenolol, metoprolol, propranolol, and sotalol, among which atenolol is the most frequently found worldwide and in highest concentrations, followed by metoprolol and propranolol (Maurer et al. 2007; Luo et al. 2014; Evgenidou et al. 2015; Godoy et al. 2015; Tran et al. 2018). The high levels of atenolol in wastewaters may be attributed to its high consumption and high excretion rates as an unchanged drug (50%) in comparison with other beta-blockers (e.g. the excretion as unchanged drug of metoprolol and propranolol is approximately of 15% and 0.5%, respectively) (Evgenidou et al. 2015).

Carbamazepine, fluoxetine, and diazepam are among the neuroactive pharmaceuticals, those substances that are more commonly detected in wastewaters (Luo et al. 2014; Cunha et al. 2017; Tran et al. 2018). The anti-epileptic carbamazepine is in particular one of the prominent cases among pharmaceutical pollutants, with an especially frequent presence in the aquatic environment, a fact that is the result of a long history of clinical usage and of its notoriously recalcitrant behavior.

Estrogenic hormones form another class of water contaminants causing serious concern because of the high potential of these substances for causing endocrine disruption and other severe ecotoxic effects such as negatively affecting the sexual and reproductive systems in wildlife, fish, and humans (Gabet-Giraud et al. 2010; Chang et al. 2011; Hamid and Eskicioglu 2012; Liu et al. 2015; Tran et al. 2018). Detection of estrogens in wastewaters and sludge has been reported (Bolong et al. 2009; Radjenovic et al. 2009; Hamid and Eskicioglu 2012; Liu et al. 2015; Tran et al. 2018) for both natural (i.e. estrone and 17b-estradiol) and synthetic (17a-ethinylestradiol) hormones. The concentrations of these natural and synthetic estrogens that have been found in WWTP effluents sometimes exceed their PNECs for ecological toxicity to aquatic organisms, implying possible risks to aquatic ecosystems (Tran et al. 2018).

Among pharmaceuticals with high consumption rates, those that are more recalcitrant to biodegradation in general show a frequent occurrence in treated WWTP effluents. Notwithstanding, among those that are more amenable to be biodegraded in WWTPs and, therefore, attain high removals during the wastewater treatment, may still be also present at non-negligible levels in the treated effluents, due to the high influent loads in which they arrive at WWTPs. Consequently, they still may be introduced as pollutants into the receiving water bodies upon the discharge of the contaminated treated effluent, even though their removal efficiencies in WWTPs may be considered reasonably high.

11.3.1 Occurrence of Pharmaceuticals in Sewage Sludge and Biosolids

Sewage sludge is the solid or semi-solid residue originated in the primary (physical and/or chemical), secondary (biological), and tertiary (often nutrient removal) treatment stages. Sludge materials that receive additional treatment in order to adequate it for application to soil as fertilizer are designated as biosolids (i.e. treated sewage sludge). In fact, sludge is rich in nutrients such as nitrogen and phosphorous and contains valuable organic matter that is useful when soils are depleted or subjected to erosion. Agricultural application of sewage sludge and biosolids to soils is the most economical outlet for sludge (in comparison with incineration) and has become a widespread method for its disposal. This re-use of sludge is generally regarded as a beneficial practice that should be encouraged as it can provide a long-term solution as long as the re-used sludge quality complies with the requirements of public health safety and of environmental protection. As a matter of fact, sludge tends to concentrate heavy metals and poorly biodegradable trace organics that are insoluble or adsorbed to particulate matter, as well as potentially pathogenic organisms (viruses, bacteria, etc.). Therefore, the occurrence and abundance of these contaminants in sludges and biosolids, their fate as well as the potential risks they ultimately pose to public health and to the environment need to be assessed.

Considering the hydrophobic/lipophilic nature and interactions with sludge particles (e.g. cation bridging, hydrogen bonding), it is believed that pharmaceuticals can sorb onto sludge during primary and secondary sedimentation. Moreover, for some pharmaceuticals, such as antibiotic fluoroquinolones, sorption to sewage sludge represents the main removal route during wastewater treatment (Giger et al. 2003; Picó and Andreu 2007; Lillenberg et al. 2009; Michael et al. 2013; Zhou et al. 2013; Frade et al. 2014; Tran et al. 2018; Riaz et al. 2018).

However, most studies on pharmaceuticals' occurrence and fate in WWTPs focus only the aqueous phase and, therefore, data describing the presence and behavior of pharmaceuticals in the sludge and biosolids are scarce. This may be due to the considerable complexity of the sludge matrix and, as consequence, to the difficulties of performing chemical analyses on that medium. Notwithstanding, the characterization of pharmaceuticals in particulate phases is essential for assessing their fate in the environment, although very few studies have been conducted on this matter to date.

In the few studies available in the literature, which have been compiled in a review by Tran et al. (2018), pharmaceuticals levels in sludge and biosolids were found to span a wide range of concentrations (from below the MQL to greater than mg/g dw). This high variability may be attributed to the complex dependence on many factors, some of which also having a large variability of its own, such as pharmaceuticals usage patterns over time and throughout world locations, pharmaceuticals physical and chemical properties (e.g. water solubility, log K_{ow}, pK_a, etc.) and molecular features, influent wastewater and sludge characteristics (pH, organic matter, and cation concentration), wastewater and sludge treatments, the operational conditions, and environmental conditions (Tran et al. 2018). For example, higher concentrations of several pharmaceuticals were observed in secondary sludge compared to those in primary sludge, which may be attributed to the occurrence of hydrolysis of pharmaceutical conjugates which regenerate their parent compounds or to a higher content of organic matter in secondary sludge (Urase and Kikuta 2005).

Among the pharmaceuticals assessed in sludges and/or biosolids in the few studies conducted so far, the most frequently targeted therapeutical groups are the antibiotics (often reported as typically the predominant class in sludges), in particular fluoroquinolones (e.g. ofloxacin), tetracyclines (e.g. oxytetracycline, minocycline, and tetracycline) and sulfonamides (e.g. sulfamethoxazole) (Picó and Andreu 2007; Lillenberg et al. 2009; Yang et al. 2012; Michael et al. 2013; Dorival-Garcia et al. 2013; Zhou et al. 2013; Frade et al. 2014; Tran et al. 2018; Riaz et al. 2018; Ezzariai et al. 2018), and the antimicrobial agents (e.g. triclocarban and triclosan) (Sabourin et al. 2009; Healy et al. 2017; Tran et al. 2018; Ezzariai et al. 2018). Reportedly, these pharmaceuticals can often be found in median concentrations in the upper 1000 ng/g dw (Tran et al. 2018), which is a cause of alert for the risk that it may provide selective pressure for the development of ARGs and ARB if those contaminated sludges and biosolids are used in agricultural activities and, thus, are a source of continuous exposure of the agricultural environment to these antibiotics and antimicrobial agents (Munir and Xagoraraki 2011; Topp et al. 2017; Xie et al. 2018; Shao et al. 2018; Abidelfatah et al. 2019; Pei et al. 2019). Conversely, β-lactams and chloramphenicol are usually rarely detected in sludges and/or biosolids, which may be due mainly to the fast degradation of these antibiotics during wastewater treatment as well as during the anaerobic digestion of the sludge.

Only few studies are available to date which report data on the occurrence of anti-inflammatories and even fewer reporting data on other therapeutic classes (namely neuroactive drugs, blood lipid regulators, hormones, β-blockers, etc.) (Maurer et al. 2007; Nieto et al. 2010; Jelic et al. 2011, 2012; Yu et al. 2011; Vieno and Sillanpää 2014; Tran et al. 2018).

Some less commonly prescribed pharmaceuticals are frequently detected in sludges and biosolids due to their recalcitrant behavior.

11.3.2 Why Are Pharmaceuticals Not Efficiently Removed in Conventional WWTPs?

As has been mentioned before, wastewater entering the municipal WWTPs typically contains a lot of different trace pollutant compounds (both of synthetic and natural origins). The degree to which such pollutants are removed after treatment in those WWTPs varies from near completion to almost none. Notwithstanding, most studies show that the removal of many pharmaceutical compounds in municipal WWTPs is clearly insufficient. Indeed, a significant fraction of the pharmaceuticals, their metabolites and transformation products entering the WWTPs are discharged with the final effluent into the aquatic environment or are present in sludges and biosolids (see Sect. 11.3.1).

The treatment processes in municipal WWTPs are designed to remove bulk constituents of wastewater, such as suspended solids, biodegradable organic matter, pathogens, and nutrients, by physical, chemical, and biological processes available along three or four consecutive stages of a conventional treatment (Fig. 11.2).

Conversely, conventional WWTPs were not designed to deal with pharmaceuticals or trace organic pollutants in general. Typically, there is very little elimination of most organic micropollutants at the preliminary and primary treatments of wastewaters, and it is also unlikely that many pharmaceuticals will be removed during screening or primary sedimentation (Jelic et al. 2011; Hamid and Eskicioglu 2012; Verlicchi et al. 2012b; Luo et al. 2014; Wang and Wang 2016; Yang et al. 2017; Tran et al. 2018). In fact, in some cases there may even be an increase of the concentrations of some pharmaceuticals during these stages, caused by the simultaneous presence of conjugated derivatives (metabolites) of these compounds in the raw influent that are reverted back into the parent compound during wastewater treatment (Carballa et al. 2004; Tran et al. 2018). Secondly, pharmaceuticals excreted via urine and feces may be enclosed in fecal particles and be gradually released during wastewater treatment, thus also resulting in an apparent increase in concentration inside the WWTP (Gobel et al. 2007; Tran et al. 2018).

Given the low biological activity at these initial stages, any pollutant removal in this phase of treatment will depend on the tendency of each pharmaceutical to sorb to the solids of the primary sludge as well as on the extent of the suspended solids removal in the primary sedimentation tanks (Zhang et al. 2008; Monteiro and Boxall 2010; Hamid and Eskicioglu 2012; Luo et al. 2014; Tran et al. 2018). Usually, at this point, only the more hydrophobic compounds are expected to transfer to the solid phase and little to no loss of polar drugs is expected. In general, elimination of any compound by sorption to sludge is considered relevant only when the log K_d for that compound is higher than ~2.5–2.7 (i.e. corresponding to $K_d > 300$–500 L kg^{-1}) (Ternes et al. 2004; Joss et al. 2005; Tran et al. 2018). The removal of pharmaceuticals may also be affected by some other factors such as pH, retention time, temperature, and amount and type of solids present in the wastewater (Ternes et al. 2004; Joss et al. 2005; Carballa et al. 2008; Hamid and Eskicioglu 2012; Verlicchi et al. 2012b; Luo et al. 2014; Wang and Wang 2016; Yang et al. 2017; Tran et al. 2018).

At secondary (biological) treatment by activated sludge, removal of pharmaceuticals may occur by the same mechanisms as do other organic micropollutants, including sorption to secondary sludge, chemical degradation, or transformation (such as hydrolysis or photolysis) and biotransformation/biodegradation (aerobic, anoxic and anaerobic) (Monteiro and Boxall 2010; Hamid and Eskicioglu 2012; Verlicchi et al. 2012b; Luo et al. 2014; Wang and Wang 2016; Yang et al. 2017; Tran et al. 2018). Biodegradation of pharmaceuticals in this stage may occur to various extents, from complete mineralization (although that is rarely the case) to incomplete degradation (i.e. yielding still somewhat complex and

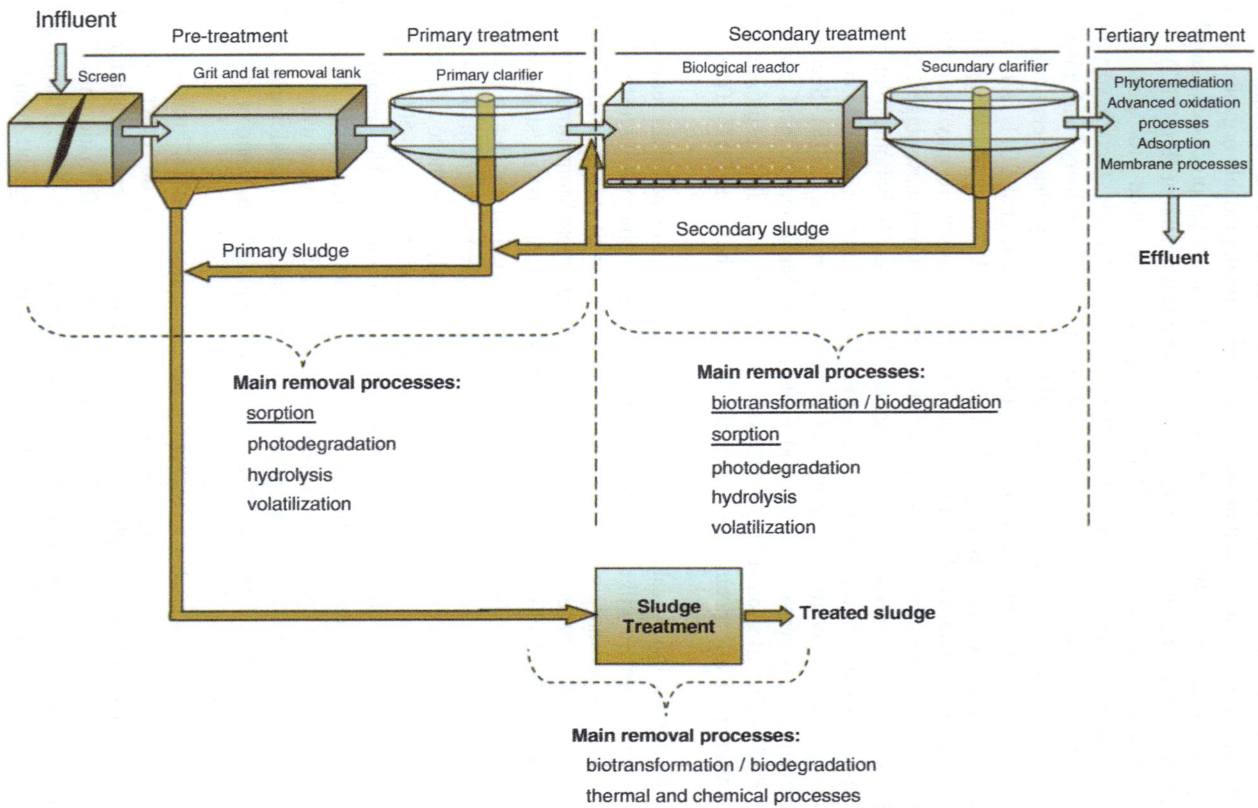

Fig. 11.2 Diagram of a conventional wastewater treatment plant and main removal processes of pharmaceuticals in each treatment stage

possibly still toxic transformation products instead of the simplest and mostly innocuous CO_2, etc.). In principle, two pathways for biodegradation may be possible, namely via metabolism or via co-metabolism. However, indications from numerous studies conducted so far suggest that biodegradation of pharmaceuticals in wastewater treatment processes takes place via co-metabolism rather than through metabolism. Indeed, the fact that many pharmaceuticals are toxic for microorganisms and are often present in wastewater at trace levels (ng/L – μg/L) makes them less suitable as an energy source and implies that most pharmaceuticals do not enter catabolic and anabolic pathways of microbial cells. In other words, the energy resulting from the biodegradation of pharmaceuticals is not sufficient to support microbial growth and induce the relevant enzymes/co-factors involved in the biodegradation. Therefore, the biodegradation of pharmaceuticals is highly dependent on the presence (and abundance) of primary substrates (e.g. ammonium, carbonate salts or organic carbon sources) as well as conditions for the development of microorganisms involved in co-metabolic biodegradation (Tran et al. 2018).

Additionally, considering the typically low volatility of most pharmaceuticals, compound loss through volatilization associated with the aeration (stripping) operation is expected to be negligible (Caliman and Gavrilescu 2009; Miége et al. 2009; Verlicchi et al. 2012b; Luo et al. 2014; Tran et al. 2018). In fact, Henry coefficients of at least $\sim 10^{-3}$ are generally regarded as the minimum requirement for significant stripping in a bioreactor with fine bubble aeration (Larsen et al. 2004). Pharmaceutical removals at this stage are also affected by environmental and operation conditions (Joss et al. 2005; Clara et al. 2005; Onesios et al. 2009; Verlicchi et al. 2012b; Luo et al. 2014; Tran et al. 2018).

Any pharmaceutical residues remaining in wastewaters after primary and secondary treatment may eventually be eliminated by tertiary or advanced treatments. However, in most countries only a reduced number of WWTPs include these stages of treatment. Advanced oxidation processes, membrane processes, and adsorption processes are some of the most common advanced treatment techniques that have been applied in wastewater treatment and demonstrated to be capable of removing pharmaceuticals to levels below detection limits (Fent et al. 2006; Snyder et al. 2007; Rosal et al. 2010; Dolar et al. 2012; Kit Chan et al. 2012; Feng et al. 2013; Ek et al. 2014; Rizzo et al. 2015; Rodriguez-Mozaz et al. 2015; Yang et al. 2017; Kanakaraju et al. 2018; de Andrade et al. 2018; Fonseca Couto et al. 2018; Pei et al. 2019). However, the effectiveness of some (or all) of these advanced techniques depends on the treatment conditions employed.

Notwithstanding, despite the sometimes high removal efficiencies that are attainable through these technologies, in most cases their implementation and operation are too expensive and complex for use on a large scale in wastewater treatment (Fent et al. 2006; Tahar et al. 2013). Moreover, the type of processes involved in some of these treatments may give origin to some transformation products that may in some cases be even more persistent or toxic than the parent compounds (Farré et al. 2008; Fatta-Kassinos et al. 2011; Postigo and Richardson 2014; Wang and Lin 2014; Evgenidou et al. 2015; Wang and Wang 2016; Yang et al. 2017).

Alternatively, the use of phytoremediation technologies such as constructed wetlands (CWs) for the removal of pharmaceuticals residues from wastewater is increasingly being seen as a more economic, while still very effective, option and has been increasingly studied and explored over the latest decades. In fact, these systems are becoming an option for secondary wastewater treatment systems or as treatment units for polishing secondary effluent from WWTPs. In addition to low cost, simple operation and maintenance (thereby not requiring highly skilled labor) and environmental friendliness are some of their most attractive characteristics.

11.4 Phytoremediation Strategies for Pharmaceuticals Clean-up: Constructed Wetlands

Constructed wetlands systems (CWS) are man-made structures that emulate natural wetlands for human use and benefits (Cooper et al. 1996; Vymazal et al. 1998; Kadlec and Wallace 2009; Dordio and Carvalho 2013). These systems consist of water saturated beds, containing (in addition to the water column) soil or other selected solid support matrix, emergent and/or submergent wetland vegetation, and microbial populations as the main components. For a long time, natural wetlands have been credited with the ability of depurating the water that inundated such areas. Based on this observation, the idea of constructing artificial wetlands was developed as an attempt to take advantage of many of the same processes that occur in natural wetlands, but within a more controlled environment, with systems designed for an enhanced water depurating action. In these engineered systems, it is sought to obtain an optimization of the operating conditions and selection of its components in order to achieve higher efficiencies, considering the roles played by each CWS component and drawing on some understanding of the mechanisms involved in the removal of pollutants in these systems. CWS optimization thus aims to potentiate the concerted action of all the components (support matrix, vegetation, and microbial population) through a variety of interdependent chemical, physical, and biological processes as illustrated in the scheme of Fig. 11.3.

In the past, CWS have been mainly used as wastewater treatment systems intended to serve as alternative or complementary systems to the conventional treatment for domestic wastewaters of small communities. Thus, initially CWS were mostly applied for the removal of bulk pollutants such as suspended solids, organic matter, excess of nutrients and pathogens. However, CWS are now also being used more often to provide a form of secondary or tertiary treatment for wastewaters. More recently, an increasing number of studies have been exploring the use of CWS to target more specific pollutants, especially those which are more recalcitrant to conventional wastewater treatment such as pharmaceuticals and other trace organic pollutants (Matamoros et al. 2008; Dordio et al. 2009, 2010; Hijosa-Valsero et al. 2010, 2016, 2017; Ávila et al. 2010, 2014; Reyes-Contreras et al. 2012; Dordio and Carvalho 2013; Verlicchi and Zambello 2014; Zhang et al. 2014, 2018b; Li et al. 2014; Ávila and García 2015; Vymazal et al. 2017; Matamoros et al. 2017; Liu et al. 2019). In fact, a wide variety of pharmaceuticals, spanning several different

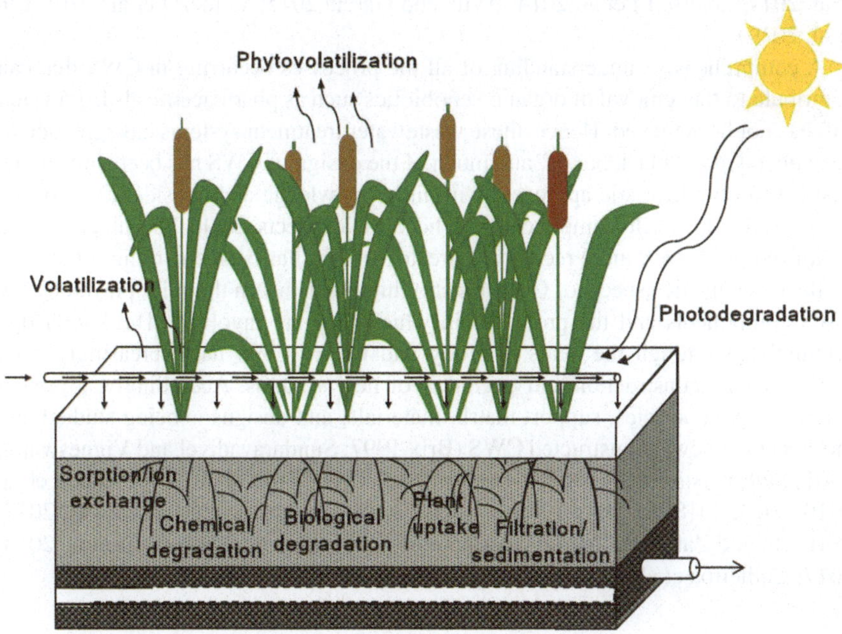

Fig. 11.3 Summary of the major physical, chemical, and biological processes controlling pollutant removal in a sub-surface flow CWS

therapeutic classes as well as various chemical structures and properties, have already been studied in respect to the capacities of CWS to remove them from water and wastewater (Matamoros et al. 2008, 2017; Dordio et al. 2009, 2010; Hijosa-Valsero et al. 2010, 2011, 2016, 2017; Ávila et al. 2010, 2014; Dordio and Carvalho 2011, 2013, 2017; Reyes-Contreras et al. 2012; Verlicchi and Zambello 2014; Zhang et al. 2014, 2018b; Li et al. 2014; Ávila and García 2015; Vymazal et al. 2017; Liu et al. 2019).

Studies have been performed on different types of CWS (namely Surface Flow Constructed Wetlands (SF), Horizontal Sub-surface Flow Constructed Wetlands (HSSF), Vertical Sub-surface Flow Constructed Wetlands (VSSF) and Hybrid Constructed Wetlands (hybrid CWS)), at different scales (microcosm scale, mesocosm (or pilot) scale, and full scale) as well as using different operating modes. The screening of different plant species and types of support matrix materials has also been a major study topic. So far, many of the studies conducted on this subject have demonstrated a noteworthy potential of CWS to remove a wide variety of pharmaceutical compounds and, thus, for providing an efficient and cost-effective solution for the decontamination of pharmaceutical-polluted wastewaters (Matamoros et al. 2008, 2017; Dordio et al. 2009, 2010; Hijosa-Valsero et al. 2010, 2011, 2016, 2017; Ávila et al. 2010, 2014; Dordio and Carvalho 2011, 2013, 2017; Reyes-Contreras et al. 2012; Verlicchi and Zambello 2014; Zhang et al. 2014, 2018b; Li et al. 2014; Ávila and García 2015; Vymazal et al. 2017; Liu et al. 2019).

A comprehensive understanding of all the processes occurring in CWS that can contribute to the removal of organic xenobiotics such as pharmaceuticals from water still has not be achieved. Hence, these wastewater treatment systems have frequently been operated as "black boxes" and much of the design of CWS has been done in the past based on a heuristic approach, with little knowledge and consideration for the roles played by each component and how their effects could be enhanced and optimized. However, more recently, a greater interest has been emerging for studies on the mechanistic aspects of CWS functioning, focusing on the roles played by the CWS components and the processes in which they are involved. The knowledge accumulated through the years of study and use of CWS has increasingly been applied in the construction and operation of new systems. Accordingly, a greater variety of plant species, support matrix materials, and designs is being studied and introduced in newly constructed CWS (Brix 1997; Sundaravadivel and Vigneswaran 2001; Stottmeister et al. 2003; Calheiros et al. 2009; Truu et al. 2009; Zhang et al. 2010, 2014, 2018a, b; Hijosa-Valsero et al. 2011; Dordio and Carvalho 2013, 2017; Verlicchi and Zambello 2014; Carvalho et al. 2014; Li et al. 2014; Avila et al. 2014, 2017; Calheiros et al. 2017; Liu et al. 2019).

11.5 Conclusion and Final Remarks

This chapter provides an overview of available data on the sources, occurrence, and fate of a variety of classes of pharmaceuticals in environment, WWTPs, sewage sludge and/or biosolids, and some crop plants and macrophytes.

Pharmaceuticals that are typically detected in environmental samples, and reported in a large number of studies, are mainly from the therapeutical classes of analgesics and anti-inflammatory drugs, blood lipid regulators, beta-blockers, psycho/neuroactive drugs, and antibiotics. Antibiotics in particular are one of the most studied classes, probably because of its well-known adverse impacts on the environment and public health. However, other therapeutical classes probably also have very negative environmental effects (e.g. hormones and regulators of the endocrine system) and, thus, a more extensive study of pharmaceuticals' ecotoxicity needs to be pursued in the future.

Since the main sources of environmental contamination with pharmaceuticals are the effluents of WWTPs, pharmaceuticals removal in WWTPs were also summarized. However, most studies aimed at detecting and quantifying pharmaceuticals in WWTPs almost always focus exclusively on the aqueous phase and very few studies have addressed also the particulate phase. Therefore, scarce data is available on the occurrence and fate of pharmaceuticals in sewage sludge and biosolids and that is a gap which is direly needed to be filled in future studies, especially considering that biosolids are frequently used as soil fertilizer, with an obvious potential to contaminate crops.

From the comparison from pharmaceuticals concentration levels in WWTP influents and effluents it can be concluded that WWTPs are frequently the source of water contamination with pharmaceuticals mainly because they often do not achieve sufficiently high removal efficiencies for this type of pollutants (although in some cases the main reason is the very high loads in the WWTPs influents of some frequently prescribed pharmaceuticals).

Another feature that stands out from an overview of the available data on pharmaceuticals occurrence in different WWTPs is the large variability (of some orders of magnitude) in the measured concentrations that are reported, both for WWTPs' influents and effluents. This could be attributed to various factors such as differences in population size/demographic density and in pharmaceuticals usage patterns in separate regions and different periods of the year (e.g. some epidemic surges of some diseases have a seasonal periodicity), or to differences in climatic conditions, etc. However, other factors that may also affect the precision of pharmaceutical concentration data are related with the chemical analysis itself, such as the adequacy of the analytical methods and instrumentation used (i.e. to address the challenge of quantifying trace levels within very complex matrices) and, in particular, the sampling strategies. The use of less suitable sampling schemes may represent one of the major weaknesses of the reported data on the occurrence of pharmaceuticals (and other types of pollutants). As such, an effort to improve on the sampling methods (e.g. by using a composite sampling strategy instead of the common grab sampling strategy) or removal calculation approaches (e.g. time-shifted mass balancing or fractionated approaches) should be considered on the analytical side to enhance the convergence of concentration data and removal performance of WWTPs.

Notwithstanding, the variability of WWTP performance data cannot, of course, be attributed solely to chemical analysis limitations. Efficiencies of pharmaceuticals removal in WWTPs differ substantially for different compounds because their chemical nature and associated physicochemical properties vary widely among them. In addition, WWTP performance is very dependent on details of their design, operation, and environmental conditions. Optimization of wastewater treatment processes remains a task with top priority. One of the main aspects to be addressed in this regard is the enhancement of biological treatment efficiency, which is frequently low for many pharmaceuticals. Improvements have been attempted under more favorable conditions, e.g. increasing contact times (i.e. hydraulic and solid retention times), optimizing temperature and fine-tuning redox conditions. Certainly, more effective optimizations may be achieved if backed up by a comprehensive knowledge of the fate of pharmaceuticals in WWTPs. Thus, the pursuit of a more profound understanding of the factors that affect the environmental fate of organic micropollutants such as pharmaceuticals is an essential line of research for achieving a successful mitigation of this type of pollution.

Removal of the most recalcitrant pharmaceuticals can be significantly improved by applying advanced treatment processes downstream to the conventional biological treatment, prior to effluent discharge. Adsorption processes, advanced oxidation processes, and membrane processes are some of these promising advanced technologies that have in many cases exhibited very high efficiencies in partially or fully (bio)degrading organic micropollutants (including pharmaceuticals) or removing them from the aqueous phase. However, two major issues preclude a full-scale application of some of these technologies: the relatively high costs generally involved in their implementation, operation and maintenance; and the fact that some of these processes yield final reaction products or lead to the formation of by-products whose ecotoxicity is not well known and are potentially hazardous. Indeed, the latter issue does not affect all advanced treatment processes, in particular it is not a problem that affects those processes that do not involve the occurrence of chemical reactions (e.g. adsorption processes). However, for those that do, the processes need to be studied in more detail in order to describe the optimal conditions that may favor mechanisms where the formation of such by-products is avoided or the conditions under which a complete decomposition may be achieved. In regard to economical considerations, attempts to lower the costs required to implement efficient wastewater treatment alternatives have been increasingly pursued by seeking and studying low cost reagents and materials (e.g. easily and widely available natural materials or agricultural wastes that may be used as efficient adsorbents) and by developing and/or optimizing cheaper technologies such as CWS.

In fact, CWS is becoming a relevant technology that is increasingly being introduced as an alternative (in the case of small communities) or complementary (as tertiary or polishing stages) treatment to the conventional wastewater processes. However, as living organisms (plants and microorganisms) are involved in the removal of pollutants in these systems, their responses to the various pollutants types and loads are more difficult to predict. Therefore, a prior study of the constructed wetlands' behavior with a given type of wastewater needs to be conducted (and possibly, a subsequent tweaking of its design—in terms of its plants and support matrix compositions—and/or operation) in order to assess its capacity to cope with the pollutants in question and, thus, assess its potential usefulness and reliability as a treatment option.

Some of the study topics with major relevance for enhancing the performance of constructed wetlands can be found by focusing on the roles played by each CW component and attempting to potentiate their action. For example, one can enunciate the following topics:

- the extent of pharmaceuticals uptake by plants and their subsequent metabolization within plant tissues, more data needs to be collected on this topic in order to better understand the role of plants in constructed wetlands, to characterize the fate of pharmaceuticals inside plants, and ultimately assess the risks posed by harvested and decaying plants and plant debris of constructed wetlands;
- the characterization of microorganism populations in constructed wetlands (including those endophytic to plants) as well as their processes of transformation of pharmaceuticals; the characteristics of these populations may eventually be modified and improved;

- finally, the evaluation of alternative (low cost) materials for the support matrix that may provide a fast-responding temporary retention of pharmaceuticals (by adsorption) while keeping them bioavailable for the slower biotic (i.e. provided by plants and microorganisms) removal processes that removed them more definitively; the role of this component may be useful to quickly respond to peak loads of pollutants, to moderate environmental conditions, and to mitigate the lower activities of the biotic components during the winter seasons (as the activity of adsorption processes is less sensitive to temperature variations).

References

Abidelfatah MN, Heitam F, Yeshayahu N (2019) The role of wastewater treatment plants in the environmental dissemination of antibiotic resistant bacteria (ARB) and resistance genes (ARG). J Water Resour Prot 11:14

Aga DS (2008) Fate of pharmaceuticals in the environment and in water treatment systems. CRC Press, Boca Raton

Al Farsi RS, Ahmed M, Al Busaidi A, Choudri BS (2017) Translocation of pharmaceuticals and personal care products (PPCPs) into plant tissues: a review. Emerg Contam 3:132–137

Allam A, Tawfik A, Negm A, Yoshimura C, Fleifle A (2015) Treatment of drainage water containing pharmaceuticals using duckweed (Lemna gibba). Energy Procedia 74:973–980

Amos Sibeko P, Naicker D, Mdluli PS, Madikizela LM (2019) Naproxen, ibuprofen, and diclofenac residues in river water, sediments and Eichhornia crassipes of Mbokodweni river in South Africa: an initial screening. Environ Forensic 20:129–138

Andreozzi R, Raffaele M, Nicklas P (2003) Pharmaceuticals in STP effluents and their solar photodegradation in aquatic environment. Chemosphere 50:1319–1330

Arslan Topal EI (2015) Uptake of tetracycline and metabolites in Phragmites australis exposed to treated poultry slaughterhouse wastewaters. Ecol Eng 83:233–238

Ávila C, García J (2015) Pharmaceuticals and personal care products (PPCPs) in the environment and their removal from wastewater through constructed wetlands. Compr Anal Chem 67:195–244

Ávila C, Pedescoll A, Matamoros V, Bayona JM, García J (2010) Capacity of a horizontal subsurface flow constructed wetland system for the removal of emerging pollutants: an injection experiment. Chemosphere 81:1137–1142

Avila C, Matamoros V, Reyes-Contreras C, Pina B, Casado M, Mita L, Rivetti C, Barata C, Garcia J, Bayona JM (2014) Attenuation of emerging organic contaminants in a hybrid constructed wetland system under different hydraulic loading rates and their associated toxicological effects in wastewater. Sci Total Environ 470:1272–1280

Ávila C, Pelissari C, Sezerino PH, Sgroi M, Roccaro P, García J (2017) Enhancement of total nitrogen removal through effluent recirculation and fate of PPCPs in a hybrid constructed wetland system treating urban wastewater. Sci Total Environ 584:414–425

Azanu D, Mortey C, Darko G, Weisser JJ, Styrishave B, Abaidoo RC (2016) Uptake of antibiotics from irrigation water by plants. Chemosphere 157:107–114

Barancheshme F, Munir M (2018) Strategies to combat antibiotic resistance in the wastewater treatment plants. Front Microbiol 8:2603

Barra Caracciolo A, Topp E, Grenni P (2015) Pharmaceuticals in the environment: Biodegradation and effects on natural microbial communities. A review. J Pharm Biomed Anal 106:25–36

Bartha B, Huber C, Schröder P (2014) Uptake and metabolism of diclofenac in Typha latifolia - how plants cope with human pharmaceutical pollution. Plant Sci 227:12–20

Bengtsson-Palme J, Larsson DGJ (2016) Concentrations of antibiotics predicted to select for resistant bacteria: proposed limits for environmental regulation. Environ Int 86:140–149

Bletsou AA, Jeon J, Hollender J, Archontaki E, Thomaidis NS (2015) Targeted and non-targeted liquid chromatography-mass spectrometric workflows for identification of transformation products of emerging pollutants in the aquatic environment. TrAC Trends Anal Chem 66:32–44

Bolong N, Ismail AF, Salim MR, Matsuura T (2009) A review of the effects of emerging contaminants in wastewater and options for their removal. Desalination 239:229–246

Boreen AL, Arnold WA, McNeill K (2003) Photodegradation of pharmaceuticals in the aquatic environment: a review. Aquat Sci 65:320–341

Bottoni P, Caroli S, Caracciolo AB (2010) Pharmaceuticals as priority water contaminants. Toxicol Environ Chem 92:549–565

Bouki C, Venieri D, Diamadopoulos E (2013) Detection and fate of antibiotic resistant bacteria in wastewater treatment plants: a review. Ecotoxicol Environ Saf 91:1–9

Brix H (1997) Do macrophytes play a role in constructed treatment wetlands? Water Sci Technol 35:11–17

Calheiros CSC, Duque AF, Moura A, Henriques IS, Correia A, Rangel AOSS, Castro PML (2009) Substrate effect on bacterial communities from constructed wetlands planted with *Typha latifolia* treating industrial wastewater. Ecol Eng 35:744–753

Calheiros CSC, Pereira SIA, Brix H, Rangel AOSS, Castro PML (2017) Assessment of culturable bacterial endophytic communities colonizing Canna flaccida inhabiting a wastewater treatment constructed wetland. Ecol Eng 98:418–426

Caliman FA, Gavrilescu M (2009) Pharmaceuticals, personal care products and endocrine disrupting agents in the environment—a review. Clean-Soil Air Water 37:277–303

Carballa M, Omil F, Lema JM, Llompart M, García-Jares C, Rodríguez I, Gómez M, Ternes T (2004) Behavior of pharmaceuticals, cosmetics and hormones in a sewage treatment plant. Water Res 38:2918–2926

Carballa M, Fink G, Omil F, Lema JM, Ternes T (2008) Determination of the solid-water distribution coefficient (Kd) for pharmaceuticals, estrogens and musk fragrances in digested sludge. Water Res 42:287–295

Carter LJ, Williams M, Martin S, Kamaludeen SPB, Kookana RS (2018) Sorption, plant uptake and metabolism of benzodiazepines. Sci Total Environ 628-629:18–25

Carvalho PN, Basto MCP, Almeida CMR, Brix H (2014) A review of plant-pharmaceutical interactions: from uptake and effects in crop plants to phytoremediation in constructed wetlands. Environ Sci Pollut Res 21:11729–11763

Celiz MD, Tso J, Aga DS (2009) Pharmaceutical metabolites in the environment: analytical challenges and ecological risks. Environ Toxicol Chem 28:2473–2484

Challis JK, Hanson ML, Friesen KJ, Wong CS (2014) A critical assessment of the photodegradation of pharmaceuticals in aquatic environments: defining our current understanding and identifying knowledge gaps. Environ Sci Process Impacts 16:672–696

Chang H, Wan Y, Wu S, Fan Z, Hu J (2011) Occurrence of androgens and progestogens in wastewater treatment plants and receiving river waters: comparison to estrogens. Water Res 45:732–740

Christofilopoulos S, Syranidou E, Gkavrou G, Manousaki E, Kalogerakis N (2016) The role of halophyte Juncus acutus L. in the remediation of mixed contamination in a hydroponic greenhouse experiment. J Chem Technol Biotechnol 91:1665–1674

Christou A, Karaolia P, Hapeshi E, Michael C, Fatta-Kassinos D (2017) Long-term wastewater irrigation of vegetables in real agricultural systems: concentration of pharmaceuticals in soil, uptake and bioaccumulation in tomato fruits and human health risk assessment. Water Res 109: 24–34

Clara M, Kreuzinger N, Strenn B, Gans O, Kroiss H (2005) The solids retention time-suitable design parameter to evaluate the capacity of wastewater treatment plants to remove micropollutants. Water Res 39:97–106

Cooper PF, Job GD, Green MB, Shutes RBE (1996) Reed beds and constructed wetlands for wastewater treatment. WRc Publications, Medmenham

Cunha DL, de Araujo FG, Marques M (2017) Psychoactive drugs: occurrence in aquatic environment, analytical methods, and ecotoxicity—review. Environ Sci Pollut Res 24:24076–24091

D'Abrosca B, Fiorentino A, Izzo A, Cefarelli G, Pascarella MT, Uzzo P, Monaco P (2008) Phytotoxicity evaluation of five pharmaceutical pollutants detected in surface water on germination and growth of cultivated and spontaneous plants. J Environ Sci Health Part A 43:285–294

Datta R, Das P, Smith S, Punamiya P, Ramanathan DM, Reddy R, Sarkar D (2013) Phytoremediation potential of vetiver grass [Chrysopogon Zizanioides (L.)] for tetracycline. Int J Phytoremediation 15:343–351

de Andrade JR, Oliveira MF, da Silva MGC, Vieira MGA (2018) Adsorption of pharmaceuticals from water and wastewater using nonconventional low-cost materials: a review. Ind Eng Chem Res 57:3103–3127

Dolar D, Gros M, Rodriguez-Mozaz S, Moreno J, Comas J, Rodriguez-Roda I, Barcelo D (2012) Removal of emerging contaminants from municipal wastewater with an integrated membrane system, MBR-RO. J Hazard Mater 239:64–69

Dordio A, Carvalho AJP (2011) Phytoremediation: an option for removal of organic xenobiotics from water. In: Golubev IA (ed) Handbook of phytoremediation. Nova Science Publishers, Hauppauge, pp 51–92

Dordio AV, Carvalho AJP (2013) Organic xenobiotics removal in constructed wetlands, with emphasis on the importance of the support matrix. J Hazard Mater 252–253:272–292

Dordio A, Carvalho AJP (2017) Removal processes of pharmaceuticals in constructed wetlands. In: Stefanakis A (ed) Constructed wetlands for industrial wastewater treatment. Wiley-Blackwell

Dordio A, Pinto J, Barrocas Dias C, Pinto AP, Palace Carvalho AJ, Teixeira DM (2009) Atenolol removal in microcosm constructed wetlands. Int J Environ Anal Chem 89:835–848

Dordio A, Carvalho AJP, Teixeira DM, Dias CB, Pinto AP (2010) Removal of pharmaceuticals in microcosm constructed wetlands using *Typha* spp. and LECA. Bioresour Technol 101:886–892

Dorival-Garcia N, Zafra-Gomez A, Camino-Sanchez FJ, Navalon A, Vilchez JL (2013) Analysis of quinolone antibiotic derivatives in sewage sludge samples by liquid chromatography-tandem mass spectrometry: comparison of the efficiency of three extraction techniques. Talanta 106: 104–118

Du LF, Liu WK (2012) Occurrence, fate, and ecotoxicity of antibiotics in agro-ecosystems. A review. Agron Sustain Dev 32:309–327

Ebele AJ, Abou-Elwafa Abdallah M, Harrad S (2017) Pharmaceuticals and personal care products (PPCPs) in the freshwater aquatic environment. Emerg Contam 3:1–16

Eggen T, Moeder M, Arukwe A (2010) Municipal landfill leachates: a significant source for new and emerging pollutants. Sci Total Environ 408:5147–5157

Ek M, Baresel C, Magner J, Bergstrom R, Harding M (2014) Activated carbon for the removal of pharmaceutical residues from treated wastewater. Water Sci Technol 69:2372–2380

Enick OV, Moore MM (2007) Assessing the assessments: pharmaceuticals in the environment. Environ Impact Assess Rev 27:707–729

Escher BI, Fenner K (2011) Recent advances in environmental risk assessment of transformation products. Environ Sci Technol 45:3835–3847

Evgenidou EN, Konstantinou IK, Lambropoulou DA (2015) Occurrence and removal of transformation products of PPCPs and illicit drugs in wastewaters: a review. Sci Total Environ 505: 905–926

Ezzariai A, Hafidi M, Khadra A, Aemig Q, El Fels L, Barret M, Merlina G, Patureau D, Pinelli E (2018) Human and veterinary antibiotics during composting of sludge or manure: global perspectives on persistence, degradation, and resistance genes. J Hazard Mater 359:465–481

Farré ML, Pérez S, Kantiani L, Barceló D (2008) Fate and toxicity of emerging pollutants, their metabolites and transformation products in the aquatic environment. TrAC Trends Anal Chem 27:991–1007

Fatta-Kassinos D, Meric S, Nikolaou A (2011) Pharmaceutical residues in environmental waters and wastewater: current state of knowledge and future research. Anal Bioanal Chem 399:251–275

Feng L, van Hullebusch ED, Rodrigo MA, Esposito G, Oturan MA (2013) Removal of residual anti-inflammatory and analgesic pharmaceuticals from aqueous systems by electrochemical advanced oxidation processes. A review. Chem Eng J 228:944–964

Fent K, Weston AA, Caminada D (2006) Ecotoxicology of human pharmaceuticals. Aquat Toxicol 76:122–159

Ferro S, Trentin AR, Caffieri S, Ghisi R (2010) Antibacterial Sulfonamides: accumulation and effects in barley plants. Fresenius Environ Bull 19:2094–2099

Fonseca Couto C, Lange LC, Santos Amaral MC (2018) A critical review on membrane separation processes applied to remove pharmaceutically active compounds from water and wastewater. J Water Process Eng 26:156–175

Frade VMF, Dias M, Teixeira ACSC, Palma MSA (2014) Environmental contamination by fluoroquinolones. Braz J Pharm Sci 50:41–54

Franklin AM, Williams CF, Andrews DM, Woodward EE, Watson JE (2016) Uptake of three antibiotics and an antiepileptic drug by wheat crops spray irrigated with wastewater treatment plant effluent. J Environ Qual 45:546–554

Frédéric O, Yves P (2014) Pharmaceuticals in hospital wastewater: their ecotoxicity and contribution to the environmental hazard of the effluent. Chemosphere 115:31–39

Gabet-Giraud V, Miège C, Choubert JM, Ruel SM, Coquery M (2010) Occurrence and removal of estrogens and beta blockers by various processes in wastewater treatment plants. Sci Total Environ 408:4257–4269

Gadipelly C, Pérez-González A, Yadav GD, Ortiz I, Ibáñez R, Rathod VK, Marathe KV (2014) Pharmaceutical industry wastewater: review of the technologies for water treatment and reuse. Ind Eng Chem Res 53:11571–11592

García-Santiago X, Franco-Uría A, Omil F, Lema JM (2016) Risk assessment of persistent pharmaceuticals in biosolids: dealing with uncertainty. J Hazard Mater 302:72–81

Gavrilescu M, Demnerová K, Aamand J, Agathos S, Fava F (2015) Emerging pollutants in the environment: present and future challenges in biomonitoring, ecological risks and bioremediation. New Biotech 32:147–156

Giger W, Alder AC, Golet EM, Kohler HPE, McArdell CS, Molnar E, Siegrist H, Suter MJF (2003) Occurrence and fate of antibiotics as trace contaminants in wastewaters, sewage sludges, and surface waters. Chimia 57:485–491

Gobel A, McArdell CS, Joss A, Siegrist H, Giger W (2007) Fate of sulfonamides, macrolides, and trimethoprim in different wastewater treatment technologies. Sci Total Environ 372:361–371

Godoy AA, Kummrow F, Pamplin PA (2015) Occurrence, ecotoxicological effects and risk assessment of antihypertensive pharmaceutical residues in the aquatic environment—a review. Chemosphere 138:281–291

Goel S (2015) Antibiotics in the environment: a review. Emerging micro-pollutants in the environment: occurrence, fate, and distribution. American Chemical Society, pp 19–42

Goldstein M, Shenker M, Chefetz B (2014) Insights into the uptake processes of wastewater-borne pharmaceuticals by vegetables. Environ Sci Technol 48:5593–5600

Gonzalez-Martinez A, Rodriguez-Sanchez A, Martinez-Toledo MV, Garcia-Ruiz MJ, Hontoria E, Osorio-Robles F, GonzalezGÇôLopez J (2014) Effect of ciprofloxacin antibiotic on the partial-nitritation process and bacterial community structure of a submerged biofilter. Sci Total Environ 476–477:276–287

Gothwal R, Shashidhar T (2015) Antibiotic pollution in the environment: a review. Clean-Soil Air Water 43:479–489

Gottschall N, Topp E, Metcalfe C, Edwards M, Payne M, Kleywegt S, Russell P, Lapen DR (2012) Pharmaceutical and personal care products in groundwater, subsurface drainage, soil, and wheat grain, following a high single application of municipal biosolids to a field. Chemosphere 87: 194–203

Gottschall N, Topp E, Edwards M, Payne M, Kleywegt S, Russell P, Lapen DR (2013) Hormones, sterols, and fecal indicator bacteria in groundwater, soil, and subsurface drainage following a high single application of municipal biosolids to a field. Chemosphere 91:275–286

Halling-Sørensen B, Nors Nielsen S, Lanzky PF, Ingerslev F, Holten Lützhøft HC, Jørgensen SE (1998) Occurrence, fate and effects of pharmaceutical substances in the environment—a review. Chemosphere 36:357–393

Hamid H, Eskicioglu C (2012) Fate of estrogenic hormones in wastewater and sludge treatment: a review of properties and analytical detection techniques in sludge matrix. Water Res 46:5813–5833

He Z, Cheng X, Kyzas GZ, Fu J (2016) Pharmaceuticals pollution of aquaculture and its management in China. JMol Liq 223:781–789

Healy MG, Fenton O, Cormican M, Peyton DP, Ordsmith N, Kimber K, Morrison L (2017) Antimicrobial compounds (triclosan and triclocarban) in sewage sludges, and their presence in runoff following land application. Ecotoxicol Environ Saf 142:448–453

Heberer T (2002) Occurrence, fate, and removal of pharmaceutical residues in the aquatic environment: a review of recent research data. Toxicol Lett 131:5–17

Hijosa-Valsero M, Matamoros V, Sidrach-Cardona R, Martín-Villacorta J, Bécares E, Bayona JM (2010) Comprehensive assessment of the design configuration of constructed wetlands for the removal of pharmaceuticals and personal care products from urban wastewaters. Water Res 44:3669–3678

Hijosa-Valsero M, Matamoros V, Sidrach-Cardona R, Pedescoll A, Martín-Villacorta J, García J, Bayona JM, Bécares E (2011) Influence of design, physico-chemical and environmental parameters on pharmaceuticals and fragrances removal by constructed wetlands. Water Sci Technol 63:2527

Hijosa-Valsero M, Reyes-Contreras C, Domínguez C, Bécares E, Bayona JM (2016) Behaviour of pharmaceuticals and personal care products in constructed wetland compartments: influent, effluent, pore water, substrate and plant roots. Chemosphere 145:508–517

Hijosa-Valsero M, Sidrach-Cardona R, Pedescoll A, Sanchez O, Becares E (2017) Role of bacterial diversity on PPCPs removal in constructed wetlands. In: Stefanakis A (ed) Constructed wetlands for industrial wastewater treatment. Wiley-Blackwell, Hoboken

Jelic A, Gros M, Ginebreda A, Cespedes-Sínchez R, Ventura F, Petrovic M, Barcelo D (2011) Occurrence, partition and removal of pharmaceuticals in sewage water and sludge during wastewater treatment. Water Res 45:1165–1176

Jelic A, Fatone F, Di Fabio S, Petrovic M, Cecchi F, Barcelo D (2012) Tracing pharmaceuticals in a municipal plant for integrated wastewater and organic solid waste treatment. Sci Total Environ 433:352–361

Jones-Lepp TL, Sanchez CA, Moy T, Kazemi R (2010) Method development and application to determine potential plant uptake of antibiotics and other drugs in irrigated crop production systems. J Agric Food Chem 58:11568–11573

Joss A, Keller E, Alder AC, Gobel A, McArdell CS, Ternes T, Siegrist H (2005) Removal of pharmaceuticals and fragrances in biological wastewater treatment. Water Res 39:3139–3152

Kadlec RH, Wallace SD (2009) Treatment wetlands. CRC Press, Boca Raton

Kanakaraju D, Glass BD, Oelgemöller M (2018) Advanced oxidation process-mediated removal of pharmaceuticals from water: a review. J Environ Manage 219:189–207

Khetan SK, Collins TJ (2007) Human pharmaceuticals in the aquatic environment: a challenge to green chemistry. Chem Rev 107:2319–2364

Kim KR, Owens G, Kwon SI, So KH, Lee DB, Ok YS (2011) Occurrence and environmental fate of veterinary antibiotics in the terrestrial environment. Water Air Soil Pollut 214:163–174

Kit Chan W, Jouët J, Heng S, Lun Yeung K, Schrotter JC (2012) Membrane contactor/separator for an advanced ozone membrane reactor for treatment of recalcitrant organic pollutants in water. J Solid State Chem 189:96–100

Koch N, Islam NF, Sonowal S, Prasad R, Sarma H (2021) Environmental antibiotics and resistance genes as emerging contaminants: methods of detection and bioremediation. Cur Res Microbial Sci. https://doi.org/10.1016/j.crmicr.2021.100027

Kotyza J, Soudek P, Kafka Z, Vanek T (2010) Phytoremediation of pharmaceuticals-preliminary study. Int J Phytoremediation 12:306–316

Kummerer K (2009) Antibiotics in the aquatic environment—a review—part I. Chemosphere 75: 417–434

Kümmerer K (2009a) Antibiotics in the aquatic environment—A review—Part II. Chemosphere 75: 435–441

Kümmerer K (2009b) The presence of pharmaceuticals in the environment due to human use—present knowledge and future challenges. J Environ Manage 90:2354–2366

Lapworth DJ, Baran N, Stuart ME, Ward RS (2012) Emerging organic contaminants in groundwater: a review of sources, fate and occurrence. Environ Pollut 163:287–303

Larsen TA, Lienert J, Joss A, Siegrist H (2004) How to avoid pharmaceuticals in the aquatic environment. J Biotechnol 113:295–304

Larsson DGJ (2014) Antibiotics in the environment. Upsala J Med Sci 119:108–112

Le Minh N, Khan SJ, Drewes JE, Stuetz RM (2010) Fate of antibiotics during municipal water recycling treatment processes. Water Res 44:4295–4323

Li WC (2014) Occurrence, sources, and fate of pharmaceuticals in aquatic environment and soil. Environ Pollut 187:193–201

Li YF, Zhu GB, Ng WJ, Tan SK (2014) A review on removing pharmaceutical contaminants from wastewater by constructed wetlands: design, performance and mechanism. Sci Total Environ 468:908–932

Lillenberg M, Yurchenko S, Kipper K, Herodes K, Pihl V, Sepp K, Lohmus R, Nei L (2009) Simultaneous determination of fluoroquinolones, sulfonamides and tetracyclines in sewage sludge by pressurized liquid extraction and liquid chromatography electrospray ionization-mass spectrometry. J Chromatogr A 1216:5949–5954

Lin YL, Li BK (2016) Removal of pharmaceuticals and personal care products by Eichhornia crassipe and Pistia stratiotes. J Taiwan Inst Chem Eng 58:318–323

Liu F, Ying GG, Tao R, Zhao JL, Yang JF, Zhao LF (2009) Effects of six selected antibiotics on plant growth and soil microbial and enzymatic activities. Environ Pollut 157:1636–1642

Liu L, Liu Yh, Liu Cx, Wang Z, Dong J, Zhu Gf, Huang X (2013) Potential effect and accumulation of veterinary antibiotics in Phragmites australis under hydroponic conditions. Ecol Eng 53:138–143

Liu Z, Lu G, Yin H, Dang Z, Rittmann B (2015) Removal of natural estrogens and their conjugates in municipal wastewater treatment plants: a critical review. Environ Sci Technol 49:5288–5300

Liu X, Guo X, Liu Y, Lu S, Xi B, Zhang J, Wang Z, Bi B (2019) A review on removing antibiotics and antibiotic resistance genes from wastewater by constructed wetlands: performance and microbial response. Environ Pollut 254(112):996

Lu MC, Chen YY, Chiou MR, Chen MY, Fan HJ (2016) Occurrence and treatment efficiency of pharmaceuticals in landfill leachates. Waste Manage

Luo Y, Guo W, Ngo HH, Nghiem LD, Hai FI, Zhang J, Liang S, Wang XC (2014) A review on the occurrence of micropollutants in the aquatic environment and their fate and removal during wastewater treatment. Sci Total Environ 473–474:619–641

Macherius A, Eggen T, Lorenz WG, Reemtsma T, Winkler U, Moeder M 2012 Uptake of Galaxolide, Tonalide, and Triclosan by Carrot, Barley, and Meadow Fescue

Madikizela LM, Ncube S, Chimuka L (2018) Uptake of pharmaceuticals by plants grown under hydroponic conditions and natural occurring plant species: a review. Sci Total Environ 636: 477–486

Malchi T, Maor Y, Tadmor G, Shenker M, Chefetz B (2014) Irrigation of root vegetables with treated wastewater: evaluating uptake of pharmaceuticals and the associated human health risks. Environ Sci Technol 48:9325–9333

Masoner JR, Kolpin DW, Furlong ET, Cozzarelli IM, Gray JL (2016) Landfill leachate as a mirror of today's disposable society: pharmaceuticals and other contaminants of emerging concern in final leachate from landfills in the conterminous United States. Environ Toxicol Chem 35:1–13
Matamoros V, Garcia J, Bayona JM (2008) Organic micropollutant removal in a full-scale surface flow constructed wetland fed with secondary effluent. Water Res 42:653–660
Matamoros V, Nguyen LX, Arias CA, Salvadó V, Brix H (2012) Evaluation of aquatic plants for removing polar microcontaminants: a microcosm experiment. Chemosphere 88:1257–1264
Matamoros V, Rodríguez Y, Bayona JM (2017) Mitigation of emerging contaminants by full-scale horizontal flow constructed wetlands fed with secondary treated wastewater. Ecol Eng 99:222–227
Mathews S, Henderson S, Reinhold D (2014) Uptake and accumulation of antimicrobials, triclocarban and triclosan, by food crops in a hydroponic system. Environ Sci Pollut Res 21:6025–6033
Maurer M, Escher BI, Richle P, Schaffner C, Alder AC (2007) Elimination of b-blockers in sewage treatment plants. Water Res 41:1614–1622
Mendoza A, Aceña J, Pérez S, López de Alda M, Barceló D, Gil A, Valcárcel Y (2015) Pharmaceuticals and iodinated contrast media in a hospital wastewater: a case study to analyse their presence and characterise their environmental risk and hazard. Environ Res 140:225–241
Michael I, Rizzo L, McArdell CS, Manaia CM, Merlin C, Schwartz T, Dagot C, Fatta-Kassinos D (2013) Urban wastewater treatment plants as hotspots for the release of antibiotics in the environment: a review. Water Res 47:957–995
Miége C, Choubert JM, Ribeiro L, Eusèbe M, Coquery M (2009) Fate of pharmaceuticals and personal care products in wastewater treatment plants—conception of a database and first results. Environ Pollut 157:1721–1726
Mo WY, Chen Z, Leung HM, Leung AOW (2015) Application of veterinary antibiotics in China's aquaculture industry and their potential human health risks. Environ Sci Pollut Res:1–12
Monteiro SC, Boxall ABA (2010) Occurrence and fate of human pharmaceuticals in the environment. Rev Environ Contam Toxicol 202(202):53–154
Munir M, Xagoraraki I (2011) Levels of antibiotic resistance genes in manure, biosolids, and fertilized soil. J Environ Qual 40:248–255
Nieto A, Borrull F, Pocurull E, Marce RM (2010) Occurrence of pharmaceuticals and hormones in sewage sludge. Environ Toxicol Chem 29:1484–1489
Nikolaou A, Meric S, Fatta D (2007) Occurrence patterns of pharmaceuticals in water and wastewater environments. Anal Bioanal Chem 387:1225–1234
Noguera-Oviedo K, Aga DS (2016) Lessons learned from more than two decades of research on emerging contaminants in the environment. J Hazard Mater
OECD (2018) OECD.Stat—pharmaceutical market. https://stats.oecd.org/Index.aspx?DataSetCode=HEALTH_PHMC. Accessed 16 Sep 2019
Onesios K, Yu J, Bouwer E (2009) Biodegradation and removal of pharmaceuticals and personal care products in treatment systems: a review. Biodegradation 20:441–466
Patel M, Kumar R, Kishor K, Mlsna T, Pittman CU, Mohan D (2019) Pharmaceuticals of emerging concern in aquatic systems: chemistry, occurrence, effects, and removal methods. Chem Rev 119:3510–3673
Pei M, Zhang B, He Y, Su J, Gin K, Lev O, Shen G, Hu S (2019) State of the art of tertiary treatment technologies for controlling antibiotic resistance in wastewater treatment plants. Environ Int 131:105026
Petrie B, Barden R, Kasprzyk-Hordern B (2015) A review on emerging contaminants in wastewaters and the environment: current knowledge, understudied areas and recommendations for future monitoring. Water Res 72:3–27
Petrovic M, Barceló D (2007) Analysis, fate and removal of pharmaceuticals in the water cycle. Elsevier, Amsterdam
Picó Y, Andreu V (2007) Fluoroquinolones in soil—risks and challenges. Anal Bioanal Chem 387: 1287–1299

Postigo C, Richardson SD (2014) Transformation of pharmaceuticals during oxidation/disinfection processes in drinking water treatment. J Hazard Mater 279:461–475
Prosser RS, Trapp S, Sibley PK (2014) Modeling uptake of selected pharmaceuticals and personal care products into food crops from biosolids-amended soil. Environ Sci Technol 48:11397–11404
Radjenovic J, Petrovic M, Barceló D (2009) Fate and distribution of pharmaceuticals in wastewater and sewage sludge of the conventional activated sludge (CAS) and advanced membrane bioreactor (MBR) treatment. Water Res 43:831–841
Ramakrishnan A, Blaney L, Kao J, Tyagi RD, Zhang TC, Surampalli RY (2015) Emerging contaminants in landfill leachate and their sustainable management. Environ Earth Sci 73: 1357–1368
Reyes-Contreras C, Hijosa-Valsero M, Sidrach-Cardona R, Bayona JM, Bécares E (2012) Temporal evolution in PPCP removal from urban wastewater by constructed wetlands of different configuration: a medium-term study. Chemosphere 88:161–167
Riaz L, Mahmood T, Khalid A, Rashid A, Ahmed Siddique MB, Kamal A, Coyne MS (2018) Fluoroquinolones (FQs) in the environment: a review on their abundance, sorption and toxicity in soil. Chemosphere 191:704–720
Rico A, Van den Brink PJ (2014) Probabilistic risk assessment of veterinary medicines applied to four major aquaculture species produced in Asia. Sci Total Environ 468–469:630–641
Rizzo L, Fiorentino A, Grassi M, Attanasio D, Guida M (2015) Advanced treatment of urban wastewater by sand filtration and graphene adsorption for wastewater reuse: effect on a mixture of pharmaceuticals and toxicity. J Environ Chem Eng 3:122–128
Rodriguez-Mozaz S, Ricart M, Kock-Schulmeyer M, Guasch H, Bonnineau C, Proia L, de Alda ML, Sabater S, Barcelo D (2015) Pharmaceuticals and pesticides in reclaimed water: efficiency assessment of a microfiltration-reverse osmosis (MF-RO) pilot plant. J Hazard Mater 282:165–173
Rosal R, Rodriguez A, Perdigon-Melon JA, Petre A, Garcia-Calvo E, Gomez MJ, Aguera A, Fernandez-Alba AR (2010) Occurrence of emerging pollutants in urban wastewater and their removal through biological treatment followed by ozonation. Water Res 44:578–588
Sabourin L, Beck A, Duenk PW, Kleywegt S, Lapen DR, Li H, Metcalfe CD, Payne M, Topp E (2009) Runoff of pharmaceuticals and personal care products following application of dewatered municipal biosolids to an agricultural field. Sci Total Environ 407:4596–4604
Shao S, Hu Y, Cheng J, Chen Y (2018) Research progress on distribution, migration, transformation of antibiotics and antibiotic resistance genes (ARGs) in aquatic environment. Crit Rev Biotechnol 38:1195–1208
Silva BF, Jelic A, López-Serna R, Mozeto AA, Petrovic M, Barceló D (2011) Occurrence and distribution of pharmaceuticals in surface water, suspended solids and sediments of the Ebro river basin, Spain. Chemosphere 85:1331–1339
Sim WJ, Lee JW, Lee ES, Shin SK, Hwang SR, Oh JE (2011) Occurrence and distribution of pharmaceuticals in wastewater from households, livestock farms, hospitals and pharmaceutical manufactures. Chemosphere 82:179–186
Singer AC, Shaw H, Rhodes V, Hart A (2016) Review of antimicrobial resistance in the environment and its relevance to environmental regulators. Front Microbiol 7:1728
Snyder SA, Adham S, Redding AM, Cannon FS, DeCarolis J, Oppenheimer J, Wert EC, Yoon Y (2007) Role of membranes and activated carbon in the removal of endocrine disruptors and pharmaceuticals. Desalination 202:156–181
Stottmeister U, Wiessner A, Kuschk P, Kappelmeyer U, Kastner M, Bederski O, Muller RA, Moormann H (2003) Effects of plants and microorganisms in constructed wetlands for wastewater treatment. Biotechnol Adv 22:93–117
Sui Q, Cao X, Lu S, Zhao W, Qiu Z, Yu G (2015) Occurrence, sources and fate of pharmaceuticals and personal care products in the groundwater: a review. Emerg Contam 1:14–24
Sundaravadivel M, Vigneswaran S (2001) Constructed wetlands for wastewater treatment. Crit Rev Environ Sci Technol 31:351–409

Tahar A, Choubert JM, Coquery M (2013) Xenobiotics removal by adsorption in the context of tertiary treatment: a mini review. Environ Sci Pollut Res 20:5085–5095

Tanoue R, Sato Y, Motoyama M, Nakagawa S, Shinohara R, Nomiyama K (2012) Plant uptake of pharmaceutical chemicals detected in recycled organic manure and reclaimed wastewater. J Agric Food Chem 60:10203–10211

Tasho RP, Cho JY (2016) Veterinary antibiotics in animal waste, its distribution in soil and uptake by plants: a review. Sci Total Environ 563–564:366–376

Ternes TA, Herrmann N, Bonerz M, Knacker T, Siegrist H, Joss A (2004) A rapid method to measure the solid-water distribution coefficient (Kd) for pharmaceuticals and musk fragrances in sewage sludge. Water Res 38:4075–4084

Tijani JO, Fatoba OO, Petrik LF (2013) A review of pharmaceuticals and endocrine-disrupting compounds: sources, effects, removal, and detections. Water Air Soil Pollut:224

Topp E, Larsson DGJ, Miller DN, Van den Eede C, Virta MPJ (2017) Antimicrobial resistance and the environment: assessment of advances, gaps and recommendations for agriculture, aquaculture and pharmaceutical manufacturing. FEMS Microbiol Ecol 94

Tran NH, Reinhard M, Gin KY-H (2018) Occurrence and fate of emerging contaminants in municipal wastewater treatment plants from different geographical regions—a review. Water Res 133:182–207

Truu M, Juhanson J, Truu J (2009) Microbial biomass, activity and community composition in constructed wetlands. Sci Total Environ 407:3958–3971

Urase T, Kikuta T (2005) Separate estimation of adsorption and degradation of pharmaceutical substances and estrogens in the activated sludge process. Water Res 39:1289–1300

Velagaleti R (1997) Behavior of pharmaceutical drugs (human and animal health) in the environment. Drug Inf J 31:715–722

Verlicchi P, Zambello E (2014) How efficient are constructed wetlands in removing pharmaceuticals from untreated and treated urban wastewaters? A review. Sci Total Environ 470–471:1281–1306

Verlicchi P, Al Aukidy M, Galletti A, Petrovic M, Barceló D (2012a) Hospital effluent: investigation of the concentrations and distribution of pharmaceuticals and environmental risk assessment. Sci Total Environ 430:109–118

Verlicchi P, Al Aukidy M, Zambello E (2012b) Occurrence of pharmaceutical compounds in urban wastewater: removal, mass load and environmental risk after a secondary treatment—a review. Sci Total Environ 429:123–155

Vieno N, Sillanpää M (2014) Fate of diclofenac in municipal wastewater treatment plant—a review. Environ Int 69:28–39

Vo HN-P, Bui XT, Nguyen TM-H, Koottatep T, Bandyopadhyay A (2018) Insights of the removal mechanisms of pharmaceutical and personal care products in constructed wetlands. Curr Pollut Rep 4:93–103

Vymazal J, Brix H, Cooper PF, Green MB, Haberl R (1998) Constructed wetlands for wastewater treatment in Europe. Backhuys Publishers, Leiden

Vymazal J, Dvoráková Brezinová T, Koželuh M, Kule L (2017) Occurrence and removal of pharmaceuticals in four full-scale constructed wetlands in the Czech Republic—the first year of monitoring. Ecol Eng 98:354–364

Wang XH, Lin AY-C (2014) Is the phototransformation of pharmaceuticals a natural purification process that decreases ecological and human health risks? Environ Pollut 186:203–215

Wang J, Wang S (2016) Removal of pharmaceuticals and personal care products (PPCPs) from wastewater: a review. J Environ Manage 182:620–640

Watkinson AJ, Murby EJ, Costanzo SD (2007) Removal of antibiotics in conventional and advanced wastewater treatment: implications for environmental discharge and wastewater recycling. Water Res 41:4164–4176

Wilkinson J, Hooda PS, Barker J, Barton S, Swinden J (2017) Occurrence, fate and transformation of emerging contaminants in water: an overarching review of the field. Environ Pollut 231:954–970

Wu C, Spongberg AL, Witter JD, Fang M, Czajkowski KP (2010) Uptake of pharmaceutical and personal care products by soybean plants from soils applied with biosolids and irrigated with contaminated water. Environ Sci Technol 44:6157–6161

Wu H, Zhang J, Ngo HH, Guo W, Hu Z, Liang S, Fan J, Liu H (2015) A review on the sustainability of constructed wetlands for wastewater treatment: design and operation. Bioresour Technol 175: 594–601

Xie WY, Shen Q, Zhao FJ (2018) Antibiotics and antibiotic resistance from animal manures to soil: a review. Eur J Soil Sci 69:181–195

Yan Q, Feng G, Gao X, Sun C, Guo Js, Zhu Z (2016) Removal of pharmaceutically active compounds (PhACs) and toxicological response of Cyperus alternifolius exposed to PhACs in microcosm constructed wetlands. J Hazard Mater 301:566–575

Yang SF, Lin CF, Wu CJ, Ng KK, Lin AYC, Hong PKA (2012) Fate of sulfonamide antibiotics in contact with activated sludge—sorption and biodegradation. Water Res 46:1301–1308

Yang Y, Ok YS, Kim KH, Kwon EE, Tsang YF (2017) Occurrences and removal of pharmaceuticals and personal care products (PPCPs) in drinking water and water/sewage treatment plants: a review. Sci Total Environ 596–597:303–320

Yu YY, Huang QX, Cui JL, Zhang K, Tang CM, Peng XZ (2011) Determination of pharmaceuticals, steroid hormones, and endocrine-disrupting personal care products in sewage sludge by ultra-high-performance liquid chromatography-tandem mass spectrometry. Anal Bioanal Chem 399:891–902

Zarate J, Schulwitz SE, Stevens KJ, Venables BJ (2012) Bioconcentration of triclosan, methyl-triclosan, and triclocarban in the plants and sediments of a constructed wetland. Chemosphere 88:323–329

Zhai J, Rahaman M, Ji J, Luo Z, Wang Q, Xiao H, Wang K (2016) Plant uptake of diclofenac in a mesocosm-scale free water surface constructed wetland by Cyperus alternlifolius. Water Sci, Technol

Zhang Y, Geißen SU, Gal C (2008) Carbamazepine and diclofenac: removal in wastewater treatment plants and occurrence in water bodies. Chemosphere 73:1151–1161

Zhang BY, Zheng JS, Sharp RG (2010) Phytoremediation in engineered wetlands: mechanisms and applications. Proc Environ Sci 2:1315–1325

Zhang DQ, Hua T, Gersberg RM, Zhu J, Ng WJ, Tan SK (2012) Fate of diclofenac in wetland mesocosms planted with Scirpus validus. Ecol Eng 49:59–64

Zhang DQ, Gersberg RM, Hua T, Zhu J, Ng WJ, Tan SK (2013a) Assessment of plant-driven uptake and translocation of clofibric acid by Scirpus validus. Environ Sci Pollut Res 20:4612–4620

Zhang DQ, Hua T, Gersberg RM, Zhu J, Ng WJ, Tan SK (2013b) Carbamazepine and naproxen: fate in wetland mesocosms planted with Scirpus validus. Chemosphere 91:14–21

Zhang D, Gersberg RM, Ng WJ, Tan SK (2014) Removal of pharmaceuticals and personal care products in aquatic plant-based systems: a review. Environ Pollut 184:620–639

Zhang L, Lyu T, Ramírez Vargas CA, Arias CA, Carvalho PN, Brix H (2018a) New insights into the effects of support matrix on the removal of organic micro-pollutants and the microbial community in constructed wetlands. Environ Pollut 240:699–708

Zhang X, Jing R, Feng X, Dai Y, Tao R, Vymazal J, Cai N, Yang Y (2018b) Removal of acidic pharmaceuticals by small-scale constructed wetlands using different design configurations. Sci Total Environ 639:640–647

Zhou XF, Zhang YL, Shi L, Chen JB, Qiang ZM, Zhang TC (2013) Partitioning of fluoroquinolones on wastewater sludge. Clean-Soil Air Water 41:820–827

Fluoride (F) Remediation Using Phytoremediation and Nanomaterials

12

Neha Singh and Suphiya Khan

Abstract

Fluoride (F) is the 13th commonly found compound in the earth's crust, which naturally occurs in soil, water, and air. For F remediation, several conventional technologies have been developed such as phytoremediation, electrokinetic systems, excavation, adsorption, reverse osmosis, and landfills. Phytoremediation is a good and highly accepted method for treating contaminated soil as it is cheap, eco-friendly, and effective technique utilizing green plants. This chapter reviews the advance in technique of phytoremediation of contaminated soil via nanomaterials. Nanomaterials can function in the phytoremediation system through directly removing pollutants, promoting plant growth, and increasing pollutant phytoavailability. Phytoextraction is the most effective and recognized phytoremediation strategy for remedying contaminated soil. Nanoscale zero-valent iron is the most studied nanomaterials for facilitating the phytoremediation due to its successful engineering applications in treating contaminated soil and groundwater.

Keywords

Fluoride · Phytoremediation · Phytoextraction · Nanotechnology · Nano-phytoremediation

N. Singh · S. Khan (✉)
Department of Bioscience and Biotechnology, Banasthali Vidyapith, Banasthali Tonk, Rajasthan, India

R. Prasad (ed.), *Phytoremediation for Environmental Sustainability*,
https://doi.org/10.1007/978-981-16-5621-7_12

12.1 Introduction

Fluoride (F) poses a foremost environmental issue these days due to its abundance as 13th in mostly found elements in earth crust. It is mostly found in ionic form in rocks and soils as fluorine contributes to the water and soil pollution. In India, millions of people belonging mostly to 17 provinces are influenced severely with skeletal, dental, and non-skeletal fluorosis problems. According to a report of WHO, maximum amount is 1.5 mg L^{-1} and its uncontrolled ingestion leads to skeletal and dental fluorosis, thyroid disorder, and infertility (Meenakshi and Maheshwari 2006; Bhatnagar et al. 2011). A natural source of F pollution includes volcanic gases, weathering of fluoride minerals (such as cryolite, mica, feldspar, etc.), marine aerosols, and rocks like fluorspar, fluorapatite, and hydroxylapatite. Additionally, individual activities like coal mining, semi-conductor production, and metal plating also contaminate our environment by releasing fluorides (Paudyal et al. 2013). Increasing consumption of contaminated water leads to various health issues like skeletal fluorosis, kidney damage, osteoporosis, thyroid related defects, and ultimately death (Kumari and Khan 2017). However, in the present scenario 24 countries lying in world F belt extending from Turkey to China and Japan via Iraq, Afghanistan, and India are largely affected by fluorosis.

In India, issues related to fluoride arise from subsurface water as fluoride leaches into these waters from rocks like apatite and fluorite due to geogenic reason (Brindha et al. 2011). Of the total 85 million tons of F, 12 million tons available only in India (Arlappa et al. 2013). Statistics recommend that among 19 most affected states of India, Rajasthan (19.34 mg L^{-1}) is majorly influenced by fluoride contamination (Vikas et al. 2013). Plants are harmed to fluoride through the air, water, and soil. In agricultural soils, F concentrations have been reported as 1000 and 5300 mg kg^{-1} (Singh et al. 2018). The more acidic or alkaline nature of soil is due to high F contamination and increases threat of F in shoots of plants. F also affects plants such as necrotic lesion, chlorosis, decrement in chlorophyll content, catalase activity, and germination rate (Baunthiyal and Ranghar 2015). F in soil is the result of dynamic balance between two geochemical processes, leaching and enrichment. Studies have reported high F contents in the soils of China (Guangdong province, 2860 mg kg^{-1}), Pakistan (Punjab, 16 mg kg^{-1}), India (135.5 mg kg^{-1}), and Sri Lanka (North-central, 411 mg kg^{-1}) (Saxena and Sewak 2015).

Classically known conventional technologies include phytoremediation, electro-kinetic systems, excavation, and landfills for the F remediation from the soil (Zhu et al. 2009). However, these methods possess limitations such as magnitude problem and its in situ remediation. Moreover, being cost-effective, environmental-friendly phytoremediation proved to be more promising as a long term suitable alternative. Phytoremediation is technique of utilizing plants for the eradication of various contaminants present in soil (Johnson et al. 2015). Moreover, traditional technique for contaminant remediation costs US \$ 10–1000/m^3 water; however, phytoremediation costs US \$ 0.05/m^3 water (Del Socorro and Zamora-Pedraza 2010). Several plants are considered suitable for phytoremediation which belong to families like Brassicaceae, Poaceae, Euphorbiaceae, and Caryophyllaceae

(Mahdavi et al. 2014). As far as F phytoremediation is concerned, *Acacia tortilis, Cassia fistula, Diapensia lapponica, Shortia galacifolia* (Kumari and Khan 2018), *Elodea nuttallii, Potamogeton malaianus, Ceratophyllum demersum, Myriophyllum verticillatum, Hydrilla verticillata* (Zhou et al. 2012), *Platanus* sp., *Taxodium distichum, Fraxinus pennsylvanica, Salix nigra, Liriodendron tulipifera, Salix willow,* and *Camellia sinensis* (Ruan et al. 2003), etc. are reported for soil and leachate decontamination (Kang et al. 2008). Plants like *Avena sativa* L., *Acer pseudoplanatus* L., and *Camellia sinensis* L. can accumulate more than the normal concentration of F (Katiyar et al. 2020). Another effective and appreciable solution is searching hyperaccumulator species for F remediation. Ideally, fibrous root systems and higher biomass of plants are best suitable for removal of F and thus, *Prosopis juliflora* is considered as most suitable hyperaccumulator plant as it can survive in broad range of soils like saline, sandy, alkaline, and rocky and its roots can reach great depths (Saini et al. 2012). *P. juliflora* is known for its high biomass content, capability to develop rapidly in nutrient deficit soil as different hyperaccumulators and long root system (Senthilkumar et al. 2005). Moreover, phytoremediation has some drawbacks: time taking, recycling of plant parts (Zhao et al. 2010). For increasing the efficiency of hyperaccumulator plants various amendments have been used such as chelating agents and microbes (PGPR) (Salt et al. 1998; Chaudhary and Khan 2014).

Different methods are developed for the eradication of F from water like coagulation, forward osmosis (FO), nanofiltration (NF), adsorption, reverse osmosis (RO), ion exchange electrocoagulation, and electrochemical oxidation (Gong et al. 2012). Among these techniques, adsorption is mostly favored as highly efficient, easy to use, and inexpensive (Jagtap et al. 2012). Different conventional adsorbents were identified but nanomaterials proved extremely competent for removal as they have high surface-to-volume ratio. Nanomaterials (NMs) possess significant surface areas, a high number of active surface sites, and high adsorption capacities, which make them a promising solution for the remediation of contaminated soils (Verma et al. 2021; Borah et al. 2022).

Methods being used for F removal from soil and water have various setbacks such as expensive, less effective, high required investment and cannot be properly disposed; so, there is a need of cheap and environment friendly methods. Nanoparticles (NPs) these days are recommended for environmental F remediation and resource management because of cheap cost, environmental-friendly nature, and effectiveness (Zare et al. 2013). NPs have large range of operations because of exclusive thermal, physical, chemical, and optical properties (Panigrahi et al. 2004). Environmental nanotechnology has provided an innovative frontier to combat the aforesaid issues of sustainable environment by reducing the non-requisite use of raw materials, electricity, excessive use of agrochemicals irrigation, and release of industrial effluents into water bodies after their treatment to minimize the discharge of pollutants into environment (Prasad et al. 2014, 2017; Yadav et al. 2021). According to these facts, it is intensely important to work on environment friendly and cost-effective removal strategies over F remediation.

12.2 F Sources in Environment

The prime cause of the F pollution in the atmosphere is the F-enriched minerals present such as fluorapatite, fluorspar, cryolite, and hydroxylapatite (Table 12.1). Both natural, i.e., F leaching from F rich minerals by geochemical factors and anthropogenic sources, i.e., plastics factories, brick and tile works, phosphate fertilizer plants, smelters cause F pollution in soil, water, and air (Bhattacharya and Samal 2018). Normally, F and OH^- ions are both of almost same ionic sizes and negatively charged. Therefore, when water having higher bicarbonate and carbonate content is passed through rocks which are rich in F, F ions are released from the rocks via several chemical reactions and thereby increase F concentration in ground water (Saxena and Ahmed 2001).

12.3 Effects of F Contamination on Life Forms

Various studies were conducted to detect F toxicity on soil and human health. F acts as a double sword as it is beneficial as well as dangerous to human health. According to WHO, the uppermost level of fluoride is >1.5 mg L^{-1}. F is beneficial to health up to certain limit (0.5 to 1.5 mg L^{-1}) as it checks tooth decaying and is thus necessary for forming dental enamels (Hussain et al. 2004). F above a certain limit (>1.5 mg L^{-1}) causes numerous problems like skeletal and dental fluorosis, osteoporosis, thyroid imbalance, non-functional pineal gland and impair kidney function and in some cases leads to death (Ozsvath 2006). The biological consequences of F concentration on human health are shown in Table 12.2. In plants various processes such as reduction in enzymatic activities, necrosis and chlorosis of leaf, and low biomass are severely affected by F (Oruc 2008). Several plants, e.g., *Cyamopsis*

Table 12.1 F bearing materials (Biswas et al. 2017)

Mineral	Chemical formula	% Fluorine
Sellaite	MgF_2	61
Fluorite	CaF_2	49
Cryolite	Na_3AlF_6	45
Bastnaesite	$(Ce,La)(CO_3)F$	9
Fluorapatite	$Ca_3(PO_4)_3F$	3–4

Table 12.2 Toxic .effects of fluoride on human health

Fluoride, ppm[a]	Medium	Effect
0.002	Air	Injury to vegetation
1	Water	Dental caries reduction
>2	Water	Mottled enamel
8	Water	10% osteosclerosis
50	Food and water	Thyroid changes
100	Food and water	Growth retardation
120	Food and water	Kidney damage

[a]In water-medium, ppm can be taken as equivalent to mg/L

tetragonoloba (cluster bean), *Oryza sativa* (paddy), *Cicer arietinum* (Bengal gram), *Populus deltoides* (poplar), and *Vigna radiata* (mung bean) were reported with high F toxicity (Baunthiyal et al. 2014).

12.4 Techniques Available for F Remediation

The various technologies are accounted for the alleviation of F from polluted soil and water. However, available techniques have certain limitations like an expensive, less effective, high required investments and absence of proper disposal; so, attempt should be made to manufacture economic materials. Therefore, these facts of limitation demand the techniques with certain approaches like cost-effective, eco-friendly, high biomass, high magnitude, etc. for remediation of F.

Several researches have been conducted in respect to overcome the limitations of existing of techniques for F removal from soil and water. Kumari and Khan 2018 confirmed the hypothesis that Fe_3O_4 NPs can enhance the F accumulation efficiency of plant *P. juliflora* and its biomass. NPs also efficiently increased the F accumulation efficiency of plant from 34.13 mg kg^{-1} to 63.07 mg kg^{-1} by reducing the Ca ions uptake (Kumari and Khan 2018).

12.4.1 Phytoremediation

Phytoremediation (phtyo refers to plant + latin suffix remedium refers restore or to clean) means multiple combination of plant-built technologies that can clean adulterated environment by employing normally emerging or genetically altered plants (Flathman and Lanza 1998) (Fig. 12.1).

Phytoremediation is a simple, clean, low cost, environmental-friendly green technology (Wei et al. 2004) and its by-product can be used in other wide ranges (Truong 1999, 2003). Phytoremediation is an eco-friendly technique that is utilized for the eradication of toxic soil by using plant and its components like root colonizing microbes which are responsible for converting toxic complexes into non-toxic derivatives (Sarma et al. 2021; Sonowal et al. 2022).

12.4.1.1 Mechanisms of Phytoremediation

Phytoremediation of metals generally happens through any of the following methods (Fig. 12.2): Phytoaccumulation, phytostabilization, phytodegradation, phytovolatilization, and rhizodegradation.

12.4.1.1.1 Phytoaccumulation

In phytoaccumulation, contaminants are taken up by plants from the soil and transferred to growing shoot through the root system, where these contaminants are then accumulated. In this process, the recovery of the extracted metals can be possibly done through harvesting the appropriate plant and hyperaccumulator plant are mostly used in this process. In this process, the absorbed contaminants get

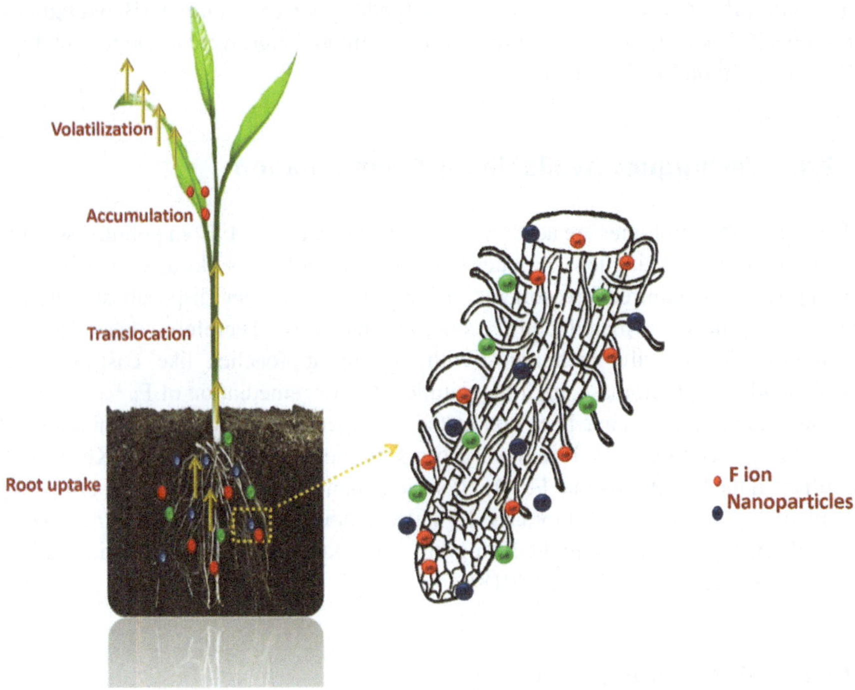

Fig. 12.1 Schematic diagram of phytoremediation process

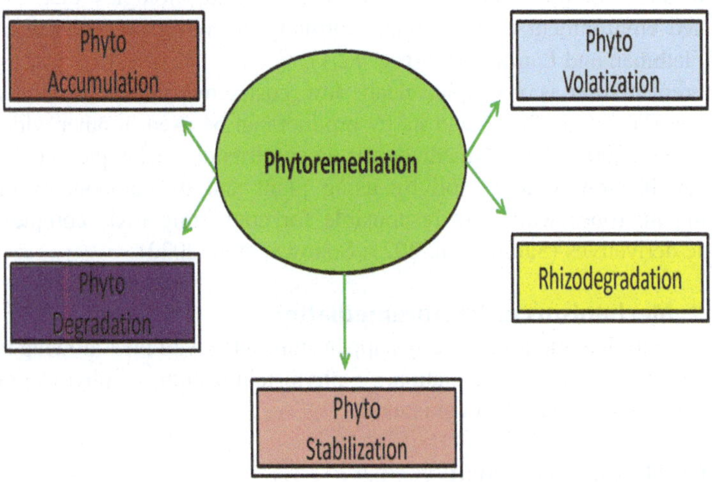

Fig. 12.2 Depiction of various methods in phytoremediation

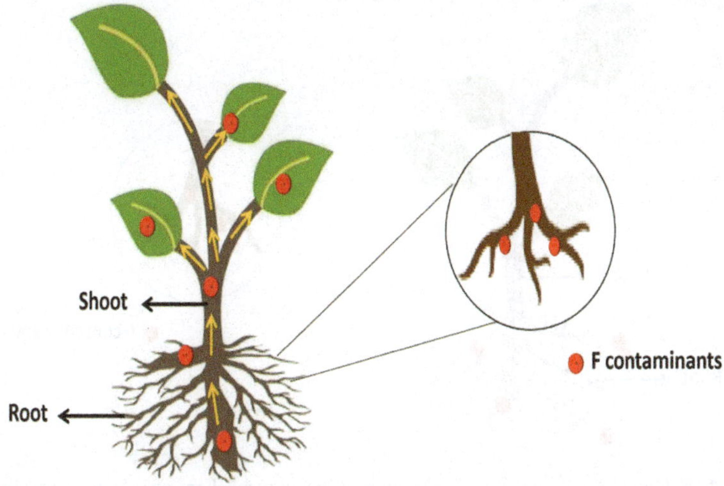

Fig. 12.3 Schematic representation of phytoaccumulation (also known as phytoextraction) mechanism

accumulated in the leaves, shoots, and various other parts of the plant instead of destroyed as illustrated in Fig. 12.3 (Rashid et al. 2014). This process is mostly utilized for the remediation of metallic and radionuclide wastes such as uranium, plutonium, and cesium.

The plants species like *Spirodela polyrhiza* (Zayed et al. 1998), *Myriophyllum aquaticum* (Harguinteguy et al. 2013), *Ludwigia palustris* (Anawar et al. 2008), *Pistia stratiotes* (Zayed et al. 1998), and *Mentha aquatic* (Zurayk et al. 2002) are reported for their high potential ability to collect heavy metals. Plant species used in this method of phytoremediation are generally grown into wetlands for its higher growth rate and increase taking up of contaminants.

12.4.1.1.2 Phytostabilization

It is this process, immobilization of organic and inorganic contaminants is done by microbial interactions of plant roots, by making them bind to soil particles resulting in reduced contaminants transfer to ground water. Different species of plants were used for stabilizing pollutants present at polluted sites by gathering through adsorption onto root surface, root hairs, or precipitation within the rhizosphere of plants by roots (Berti and Cunningham 2000). Phytostabilization is the process that exhibits the entry of contaminants in food chain by limiting the contaminants motion and eventually decreasing its bioavailability. Since report shows that phyto-stabilization prevents entry of contaminants into vegetative parts and arrests the contaminants within root zones as shown in Fig. 12.4.

The plant-associated microorganisms in this process not only reduce the metal uptake to upper vegetative parts by arresting the bioavailability of metal within the plant's rhizosphere but also promoting the metal tolerance capacity and plant

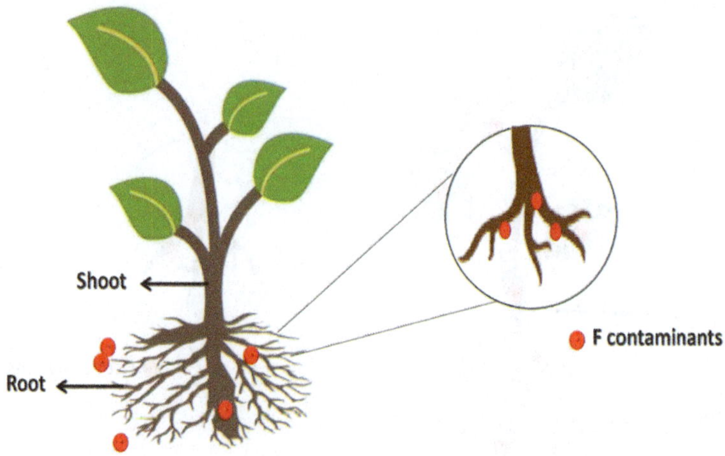

Fig. 12.4 Schematic diagram illustrates phytostabilization mechanism: plants immobilized the contaminants

growth. The organic acids were capable to reduce the bioavailability of metal and its toxicity, produced by soil living microorganisms. Soil microbes have a vital role in enhancement of phytostabilization.

12.4.1.1.3 Phytovolatilization

Phytovolatilization involves releasing of metals into the atmosphere in less toxic form after their extraction from soil (Moreno et al. 2004). Phytovolatilization can be a permanent solution for contaminated site because in this process, gaseous volatilized products cannot redeposit near the site. However, microorganisms have a dominant function in volatilization of contaminants from soil and ability of plant to be able to perform such functions was later proved. In other remediation techniques, the by-product of the remediation is utilized for different purposes. Though in cases of phytovolatilization, there is no sign of contaminants passing to other places. Phytovolatilization process should be avoided near highly populated cities having unusual patterns of weather as these sites could lead to uncontrollable release of volatile substances (Heaton et al. 1998). Several plant species from the aquatic environment are employed for eradication of selenium from polluted sites (Pilon-Smits et al. 1999) (Fig. 12.5).

12.4.1.1.4 Phytodegradation

In phytodegradation, contaminants are broken down into simpler less toxic products by utilizing plants. In phytodegradation, breakdown of contaminants occurs into two ways: In plant, metabolic process, and enzymes produced by the plant.

Products obtained from the broken down of contaminants are used for plant's faster growth. Several studies reported that certain (Class 4) plants show better efficiency in phytodegradation than some (Class 3) plant species (Khandare and Govindwar 2015). Various factors affect the phytodegradation including

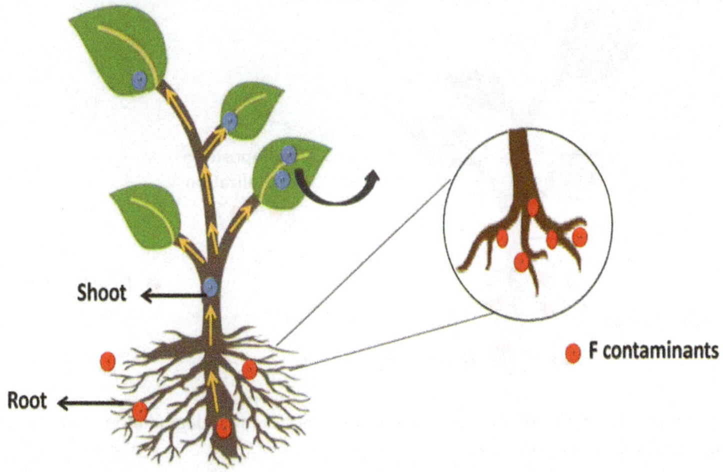

Fig. 12.5 Schematic representation of phytovolatilization mechanism

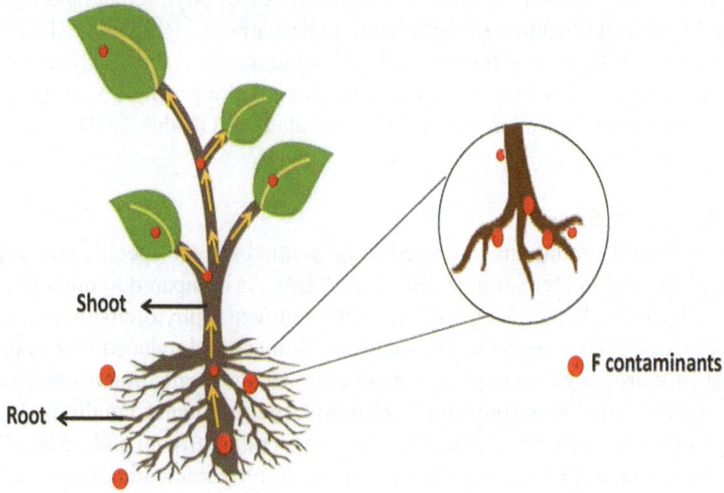

Fig. 12.6 Schematic illustration of phytodegradation mechanism: plant degrades the contaminants into less toxic compounds

(1) efficiency of pollutants uptake and (2) its concentration in the soil and ground water. Phytochemical properties of plants also determine the efficiency of contaminant uptake (Fig. 12.6).

12.4.1.1.5 Rhizodegradation

In this process, contaminants are breakdown within the plant zone, or rhizosphere as shown in Fig. 12.7. Rhizodegradation is performed by various microorganisms like

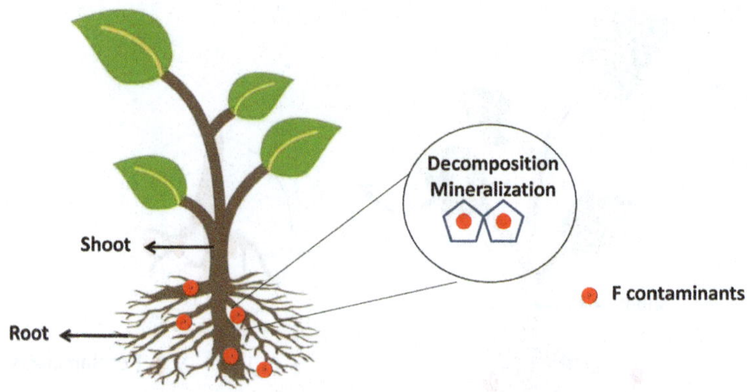

Fig. 12.7 Schematic illustration of rhizodegradation mechanism: degradation of contaminants by microbial activity in rhizosphere

bacteria flourished in the rhizosphere. In rhizosphere, microbed turns the contaminants into non-toxic products. This technique of phytoremediation generally utilized for the eradication of petroleum hydrocarbons. Rhizoremediation is an important part of phytoremediation and it can occur naturally or can also be triggered by the introduction of several microorganisms that degrade pollution or introducing plant growth promoting microbes (Gerhardt et al. 2009) (Table 12.3).

12.4.1.1.6 Limitations

There are several limitations that need to be studied on site-specific soil and plant conditions of phytoremediation (Danh et al. 2009). As compared to other treatments like microbiological, chemical, and physical treatments, phytoremediation is much slower process. It also results in low yields of biomass and reduced root system that does not support the efficient phytoremediation. Hence, contaminants leaching into aquatic system cannot be prevented. Extreme environmental conditions and soil toxicity in polluted land have a negative impact on plant growth and survival, hence environmental conditions play an important role in phytoremediation success (Danh et al. 2009). There are several other limitations of phytoremediation technologies that need focus and attentions in applying the strategy are as such:

- Metal should be present in bio-available form to plants because if metal is highly bound to the soil organic portions it becomes unavailable to plants.
- If water soluble metals are present, then, they move by root system without being accumulated.
- Specific metal hyperaccumulator species need for multiple metals contaminated water and soil that further require a broad range of research regard to its applications.

Table 12.3 The mechanism and significance of phytoremediation technology

Phytoremediation technique	Mode of action	Plant species	Pollutants	References
Phytoaccumulation	Hyperaccumulation in plants	*H. annuus* and *B. juncea*	Organic pollutants and metals (Cd, Co, Pb, Zn, Ag)	Subhashini and Swamy (2013) and Rafati et al. (2011)
Phytodegradation	Breakdown of contaminants	Algae, stonewort	Chlorinated solvents and petroleum products	Subhashini and Swamy (2013)
Phytovolatilization	Volatilization of contaminants through transpiration process	Poplars, *B. juncea*	Chlorinated solvents, metals (Hg, Se and As)	Subhashini and Swamy (2013) and Padmavathiamma and Li (2007)
Phytostabilization	Sorption of contaminants	Grasses, poplars and *B. juncea*	Inorganics	Subhashini and Swamy (2013) and Barcelo and Poschenrieder (2003)
Rhizodegradation	Decomposition of contaminants in rhizosphere	*A. smithii* and *B. gracilis*	Chlorinated solvents, petroleum products	Mukhopadhyay and Maiti (2010)

Table 12.4 Expenditure of various remediation techniques (Glass 1999)

Treatments	Expenditure (USS ton^{-1})
Vitrification	75–425
Land filling	100–500
Chemical treatments	100–500
Electrokinetics	20–200
Phytoextraction	5–40

The process of phytotechnology is cheap (Table 12.4), environmental user and hold the capacity to reclaim F contaminated site.

12.4.1.1.7 Selection Criteria of Plant for Phytoremediation

Several plants that are considered suitable for phytoremediation belong to families like Brassicaceae, Poaceae, Euphorbiaceae, and Caryophyllaceae (Mahdavi et al. 2014). Many hyperaccumulator plants are reported as *A. tortilis, Prosopis juliflora, Cassia fistula, Diapensia lapponica, Schizocodon soldanelloides, Shortia galacifolia, Tibouchina organensis, Psychotria jasminflora, Rudgea leucocephala, Camellia sinensis,* etc. (Bauntiyal and Ranghar 2015). For the phytoremediation, a variety of plant physical characteristics and biological processes are utilized. F enters in the plant systems mainly through two routes as direct absorption from the contaminated soil and airborne deposition through stomata that is more significant

(Davison 1983). The movement of F in plants takes place from the roots to shoots, leaves, and fruits. The F tolerance in different plant species varies due to the presence of calcium (Ca) in cell wall which behaves as a barrier against the accumulation of F. The mechanism of how F enters in cell is not reported. As reported in literature, the F uptake in plants enhances with the chloride deficiency so it is possible that chloride channels mediates the F cellular uptake (Miller et al. 1986). Cell membrane possess negative charge and low permeability that leads the F passes through the apoplast (intercellular spaces and cell walls) and symplast (cell membrane, plasmalemma) (Takmaz-Nisancioglu and Davison 1988). Various factors controlling the uptake of F by plants are (a) soil pH, (b) soil type, and (c) solubility of minerals like P and Ca. There are several approaches for enhancing the phytoremediation efficiency of the plants including plant symbiosis with bacteria and fungi as well as plant genetic engineering.

Selection of good F hyperaccumulator is a significant method for removal of F from endemic region of F. The translocation factor, bioconcentration factor, and enrichment value of the hyperaccumulator plant are >1. Bhargava and Bhardwaj (2011) reported that wheat when grown in 20 mg L^{-1} NaF treatment accumulates F more in roots (4.24 mg g^{-1}) than in leaves (1.45 mg g^{-1}). The study was performed on 17 plant species to determine the tolerant and highly tolerant species for F remediation (Del Socorro and Zamora-Pedraza 2010). Their results indicated that out of 17 plant species three species, i.e., *Camellia japonica, Saccharum officinarum,* and *Pittosporum tobira* have significant F uptake ability.

12.4.2 Nanomaterials

Nowadays nanomaterials have achieved a great interest. Nanomaterials literally mean a particle in the size range of 1–100 nm. "The first scientific report explaining the nature of NP was given by Michael Faraday in his pioneering work, Experimental relations of gold to light in 1857." They display exceptional larger surface area to the volume (Hasan 2015). NPs can be synthesized by two approaches as shown in Fig. 12.8.

Nanobioremediation is a method used for removing the pollutants from contaminated areas with the use of nanomaterials which are produced using "green synthesis" technique (Yadav et al. 2017). NPs consist of organic (proteins, viruses, polysaccharides, etc.) and inorganic (metals, aluminosilicates, iron, oxyhydroxides, etc.) compounds occurring naturally (Hough et al. 2011). Huge varieties of biological resources such as microorganisms (bacteria, yeast, fungi, viruses, and algae) and various other species of plants can be utilized for the synthesis of nanoparticles (Mohanpuria et al. 2008; Prasad 2014, Prasad et al. 2018a, b).

Nanobioremediation shows efficient potential to eradicate huge polluted sites in situ and ex situ, hence by reducing clean-up time and the contaminant concentration. Nanotechnology along with association of other technology can change the face of research to deal with key challenges and also seems to be promising approach for

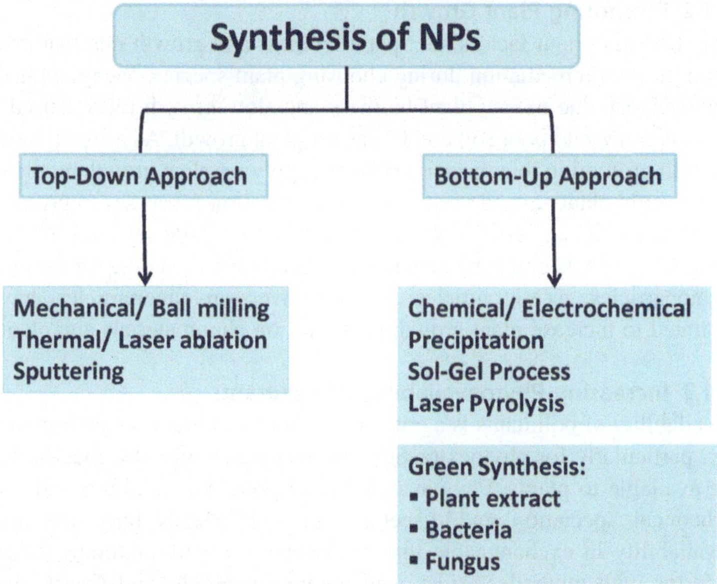

Fig. 12.8 Mechanism of NPs synthesis, i.e., Bottom-up and top-down

providing advances and innovative methods to clean up ecosystem (Singh and Khan 2018; Prasad and Aranda 2018).

12.4.2.1 Roles of Nanomaterials in Phytoremediation

The nanomaterials-assisted phytoremediation system includes three primary constitutes: plants, pollutants, and nanomaterials. On the one hand, nanomaterials can improve phytoremediation by directly acting on the pollutants and plants. On the other hand, the applied nanomaterials may be involved in the interactions between the pollutants and plants, indirectly affecting the final remediation efficiency. In conclusion, there are generally several aspects via which nanomaterials function in the process of phytoremediation, i.e., direct pollutant removal by nanomaterials, promoting plant growth, and increasing phytoavailability of pollutants (Song et al. 2019).

12.4.2.1.1 Direct Pollutant Removal by Nanomaterials

In phytoremediation process, nanomaterials are capable to remove pollutants directly from the soil, which reduces the burden of removing pollutants by plants. To directly remove the pollutants from soil, nanomaterials can function via adsorption or redox reactions (Mueller and Nowack 2010). For example, pollutants can be immobilized through adsorption by carbon nanotubes. This is similar to phytostabilization. As for removing pollutants through redox reactions, nano zero-valent iron (nZVI) is the most studied. Generally, nZVI can be used as an electron donor for reductive degradation or stabilization of pollutants.

12.4.2.1.2 Promoting Plant Growth

There are two important factors, i.e., plant biomass and growth rate that need to be addressed in phytoremediation during choosing plant species. Many applied plants are unsatisfactory due to low plant biomass and slow growth rates caused by low pollutant tolerance and poor soil conditions for plant growth. As a result, inoculating plant growth promoting rhizobacteria (PGPR), applying plant growth regulators, and using transgenic plants are all used in phytoremediation processes to promote plant growth (Prasad et al. 2015; Nahar et al. 2017; Yadu et al. 2018). Some nanomaterials, such as graphene quantum dots, carbon nanotubes, Ag nanoparticles, ZnO nanoparticles, nZVI particles, and upconversion nanoparticles, have been demonstrated to increase plant growth in studies on nanomaterials and plants.

12.4.2.1.3 Increasing Phytoavailability of Pollutants

Phytoavailability of pollutants is a critical element that influences phytoremediation efficacy, particularly for phytoextraction. Pollutants are only absorbed in the forms that are available to plants. Phytoavailability of pollutants is highly influenced by their chemical speciation and dispersion in soil. Metals have the maximum phytoavailability in exchangeable forms (dissolved in soil solution), followed by mixed forms with minerals, oxides, and organic materials, and finally crystalline forms (Liang et al. 2017). Additionally, soil physicochemical properties and plant physiological characteristics also affect the phytoavailability of pollutants (Ren et al. 2018). The phytoremediation process is frequently hampered by low phytoavailability.

12.4.2.2 Impact of Nanomaterials on Plants

During last decades, NPs have been utilized for domestic and industrialized goods in different applications. NPs are likely to be released in the atmosphere because of their great application in marketable products. NPs can pollute the atmosphere by different routes such as the inappropriate execution of manufacturing waste and inappropriate discarding of products by the consumers. The stability of NPs can be manipulated by the physical and chemicals parameters existence in diverse atmosphere. Thus, NPs might execute in a diverse way in dissimilar situation and thus their accessibility and reactivity in environment is affected (Levard et al. 2012).

The composition of NPs might also alter the properties and thereby altering their reactivity, translocation and penetration in the plant leading to unusual plant reactions to the identical NPs, for example, it is identified that NPs capping can manipulate the reactions of plant as compared to uncapped NPs (Barrios et al. 2016). Since plants are in regular contact with soil, water, and air, all of them may synthesize nanoparticles. Although the flora of a specific region is consumed by animals, so the NPs are relocated to them. There is a danger that NPs may possibly enter the food chain and turn out to be hazardous to human beings.

Uptake and aggregation of some NPs in flora can also increase shoot length and decrease the root length (Atha et al. 2012). The harmful reaction is based on the content, shape, and size of NPs (Siddiqi and Husen 2016). Many harmful effects of silver NPs have been reported (Gubbins et al. 2011).

NPs are reported as advantageous to plants up to a certain level in a number of studies (Table 12.5). Lately, Rico et al. (2015) revealed that CeO_2 NPs enhanced growth (*H. vulgare*) with no negative response. It is concluded that CeO_2 NPs are not toxic effect on cucumber (Jiang et al. 2010). The effectiveness of Fe_2O_3 NPs as fertilizer for *Arachis hypogaea* has been reported (Rui et al. 2016). It is reported that effect of Fe_2O_3 NPs and $Fe(NO_3)_3 \cdot 9H_2O$ salt is time reliant (Jeyasubramanian et al. 2016).

12.5 Nano-Phytoremediation

Nano-phytoremediation is a method which includes combination of nanotechnology and phytoremediation for exclusion of contaminants from polluted soil. Nano-phytoremediation can be utilized for the breakdown and remediation of TNT (2,4,6-trinitrotoluene)-contaminated soil. It includes combination of both approaches, i.e., nanotechnology and phytoremediation in order to uncontaminate the atmosphere or environment. Phytoremediation efficiency can be enhanced by nanotechnology and nanotechnology can also be utilized for eliminating pollutants from soil such as organic, inorganic pollutants, and heavy metals. Enzyme-based bioremediation in NPs can be utilized in combination with phytotechnology (Yadav et al. 2017). Nanomaterials are generally used for remediation process owing to their several distinctive properties such as large surface area and easy penetration into the contaminated sites (Borah et al. 2022). Schematic diagram illustrates the factors that affect the process of nano-phytoremediation as shown in Fig. 12.9.

12.6 Nano-Phytoremediation of Pollutants in Soil

Nano-phytoremediation technique is a combination of nanoparticles and phytoremediation used for the remediation of contaminates from soil. The integration of phytoremediation and nanotechnology plays a critical role in removing the contaminants from polluted soil. Study reports the several nanoparticles/ nanomaterials like heavy metal, organic, inorganic contaminant in soils. The magnetite nanoparticles (nFe_3O_4), nano zero-valent iron (nZVI), and bimetallic nanoparticles (Pd/Fe) can also damage organic pollutants such as lindane, atrazine, pentachlorophenol, chlorpyrifos, trichloroethylene (TCE), 2,4-dinitrotoluene, pyrene, ibuprofen, and polychlorinated biphenyls (PCBs) from polluted soil environment (Reddy et al. 2012; Singh et al. 2012). Nano-hydroxyapatite (NHAP), possessing high defluoridation capacity, has been widely used to remove fluoride (F) from polluted water, but little is known about how it affects the bioavailability and toxicity of soil F towards plants (Gan et al. 2021). Soil bacteria have been reported with positive response in some cases, as 1 or 10 mg kg^{-1} of NPs of iron oxide were used in soil, silver NPs (0.1–10 mg kg^{-1}) had a negative effect on soil (He et al. 2016).

Table 12.5 Positive and negative effect of different NPs uptake on plants

Plant	Concentration	Effect	References
Zea mays	Ag NPs: 40 g ha^{-1}	Improved quantitative yields of fodder maize	Berahmand et al. (2012)
Oryza sativa	Ag NPs: 0.30–60 mg L^{-1}	60 μg mL^{-1} penetrates the cells by destroying the cell structure, whereas 30 μg mL^{-1} was not able to destroy the root cells, up to 30 μg mL^{-1} accelerates root growth, whereas 60 μg mL^{-1} restricts the root ability to grow	Mirzajani et al. (2013)
Triticum aestivum, Vigna sinensis, Brassica juncea	Ag NPs: 50, 75 mg L^{-1}	Optimum growth promotion and increased root nodulation were observed at 50 ppm treatment (cowpea), improved shoot parameters were recorded at 75 ppm (*Brassica juncea*)	Pallavi et al. (2016)
Cucurbita pepo	Ag NPs: 250, 750 mg L^{-1}	Reduction in plant biomass and transpiration significantly reduced the pH	Hawthorne et al. (2012)
Elodea densa	Cu NPs: 0.025, 0.25, 0.5, 1, 5 mg L^{-1}	Catalase and superoxide dismutase activities increase by 1.5 to 2 times, stimulated photosynthesis up to 0.25 mg L^{-1} level whereas suppressed it above 1 mg L^{-1} concentration	Nekrasova et al. (2011)
Glycine max, Cicer arietinum	Cu NPs: 0, 5, 15, 30, 45, 60, 100, 200, 400, 600, 800, 1000, 1500, 2000 mg L^{-1}	A decline in root and shoot growth on above 100 mg L^{-1} concentration, a decline in root and shoot growth on above 45 mg L^{-1} concentration	Adhikari et al. (2012)
Lemna minor	Cu NPs: 10, 50, 100, 150, 200 mg L^{-1}	Increase in peroxidase, catalase, superoxide dismutase activity, increase in lipid peroxidation, inhibition of plant growth	Song et al. (2016)
Triticum aestivum	TiO_2 NPs: 100, 200, 300 mg L^{-1}	Titanium dioxide NPs at 0.02% increased different agronomic traits including gluten and starch content under water deficit condition	Jaberzadeh et al. (2013)
Triticum aestivum	TiO_2 NPs: 0, 20, 40, 60, 80, 100 mg kg^{-1}	Increase in root and shoot length with the treatment of 60 mg kg^{-1} or less. Decrease in root and shoot length above 60 mg kg^{-1} concentration	Rafique et al. (2014)
Solanum lycopersicum	TiO_2 NPs: 0, 100, 250, 500, 750, 1000 mg kg^{-1}	Up to 250 mg kg^{-1} promoted the plant height, root length, and biomass, lycopene content and fruit yield was maximum for 100 mg kg^{-1}, chlorophyll concentration increases up to 750 mg kg^{-1} of NPs	Raliya et al. (2015)

(continued)

Table 12.5 (continued)

Plant	Concentration	Effect	References
Triticum aestivum	CeO_2 NPs: 100, 400 mg kg^{-1}	400 mg kg^{-1} of NP decreased the chlorophyll content and increased catalase and superoxide dismutase activities, exposure to 200 mg kg^{-1} resulted in embryos with larger vacuoles, whereas 400 mg kg^{-1} resulted in reduced number of vacuoles	Du et al. (2015)
Zea mays	Fe_3O_4 NPs: 20, 50, 100 mg L^{-1}	Germination index was observed to be higher with 20 and 50 mg L^{-1} NP treatment whereas decreases with 100 mg L^{-1} treatment	Li et al. (2016)

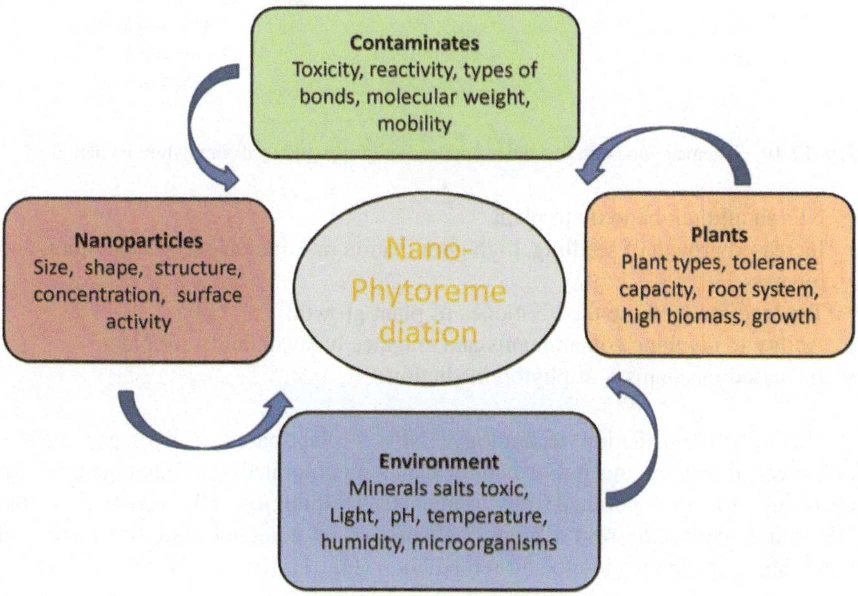

Fig. 12.9 Factors affecting the nano-phytoremediation technique during remediation of contamination

12.7 NPs Selection for Phytoremediation

In phytoremediation process, NPs utilized in the technique should have various properties as such:

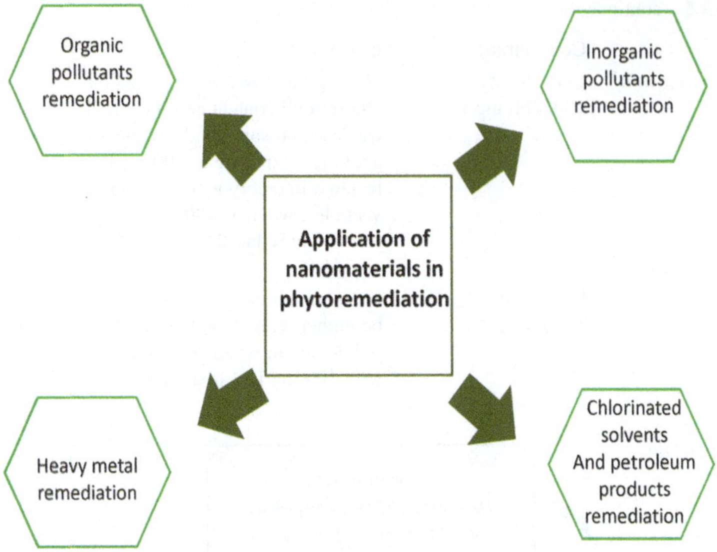

Fig. 12.10 Schematic diagram illustrates application of nanomaterials in phytoremediation

- NPs should not be toxic to plant.
- Increased growth of seedling, high germination rate, increment of root-shoot, and biomass.
- Tendency to increase the hormones of plant growth.
- Ability to combine contaminants and enhance bioavailability for plant.
- Increased mechanism of phytoremediation.

Nano-phytoremediation technology (NPs + plant-based technology) utilizes either genetically engineered or natural plants incorporated with nanomaterials for cleansing of the polluted environments. Despite of all advantages, the NPs/nanomaterials possess a huge challenge of its potential toxicity to soil and plant during its application to soil remediation (Fig. 12.10).

12.8 Conclusion

Nanotechnology provides green and economic alternatives for environmental management and cleanup. Many fungi, bacteria, and plants having the ability to accumulate enormous metal concentrations have been identified, they are termed as hyperaccumulators. Such types of fungi, bacteria, and plant species are of specific interest for removal of heavy metals from polluted areas. Nanomaterials can be used in various forms for the remediation of polluted environment. It is required to understand the mechanism for transport of nanomaterials into environment to check their toxicological effect on plants or the environment. Selecting the appropriate nanomaterials and species of plants for pollutants absorption is needed with optimization of agronomic management for highly efficient cleaning technique.

Acknowledgements We acknowledge DBT BIG-BIRAC Ref No. BIRAC/CCAMP0723/BIG-13/18 and DST major project entitled "Low Cost-Renewable Energy Driven (LC-RED) Water Treatment Solution Centre" Ref No. DST/TM/WTI/WIC/2 K17/124 for providing us the funding. We also acknowledge bioinformatics center for providing computational facilities.

References

Adhikari T, Kundu S, Biswas AK, Tarafdar JC, Rao AS (2012) Effect of copper oxide nano particle on seed germination of selected crops. J Agric Sci Technol A 2(6A):815–823

Anawar HM, Garcia-Sanchez A, Alam MTK, Rahman MM (2008) Phytofiltration of water polluted with arsenic and heavy metals. Int J Environ Pollut 33:292–312

Arlappa N, Aatif Qureshi I, Srinivas R (2013) Fluorosis in India: an overview. Int J Res Dev Health 1(2):97–102

Atha DH, Wang H, Petersen EJ, Cleveland D, Holbrook RD, Jaruga P, Dizdaroglu M, Nelson BC (2012) Copper oxide nanoparticle mediated DNA damage in terrestrial plant models. Environ Sci Technol 46(3):1819–1827

Barcelo J, Poschenrieder C (2003) Phytoremediation: principles and perspectives. Contrib Sci 2: 333–344

Barrios AC, Rico CM, Trujillo-Reyes J, Medina-Velo IA, Peralta-Videa JR, Gardea-Torresdey JL (2016) Effects of uncoated and citric acid coated cerium oxide nanoparticles, bulk cerium oxide, cerium acetate, and citric acid on tomato plants. Sci Total Environ 563:956–964

Baunthiyal M, Ranghar S (2015) Accumulation of fluoride by plants: potential for phytoremediation. Clean Soil Air Water 43(1):127–132

Baunthiyal M, Bhatt A, Ranghar S (2014) Fluorides and its effects on plant metabolism. Int J Agric Technol 10:1–27

Berahmand AA, Panahi AG, Sahabi H, Feizi H, Moghaddam PR, Shahtahmassebi N, Fotovat A, Karimpour H, Gallehgir O (2012) Effects silver nanoparticles and magnetic field on growth of fodder maize (*Zea mays* L.). Biol Trace Elem Res 149(3):419–424

Berti WR, Cunningham SD (2000) Phytostabilization of metals. In: Phytoremediation toxic met using plants to clean up environment, pp 71–88

Bhargava D, Bhardwaj N (2011) Phytotoxicity of fluoride on a wheat variety (*Triticum aestivum* var. raj. 4083) and its bioaccumulation at the reproductive phase. Asian J Exp Sci 25(1):37–40

Bhatnagar A, Kumar E, Sillanpää M (2011) Fluoride removal from water by adsorption—a review. Chem Eng J 171(3):811–840

Bhattacharya P, Samal AC (2018) Fluoride contamination in groundwater, soil and cultivated foodstuffs of India and its associated health risks: a review. Res J Recent Sci 7(4):36–47

Biswas G, Kumari M, Adhikari K, Dutta S (2017) A critical review on occurrence of fluoride and its removal through adsorption with an emphasis on natural minerals. Curr Pollut Rep 3(2):104–119

Borah SN, Koch N, Sen S, Prasad R, Sarma H (2022) Novel nanomaterials for nanobioremediation of polyaromatic hydrocarbons. In: Sharma H, Dominguez DC, Lee W-Y (eds) Emerging contaminants in the environment. Elsevier, pp 643–667

Brindha K, Rajesh R, Murugan R, Elango L (2011) Fluoride contamination in groundwater in parts of Nalgonda District, Andhra Pradesh, India. Environ Monit Assess 172:481–492

Chaudhary K, Khan S (2014) Effect of plant growth promoting rhizobacteria (PGPR) on plant growth and flouride (F) uptake by F hyperaccumulator plant *Prosopis juliflora*. Int J Recent Sci Res 5:0976–3031

Danh LT, Truong P, Mammucari R, Tran T, Foster N (2009) Vetiver grass, *Vetiveria zizanioides*: a choice plant for phytoremediation of heavy metals and organic wastes. Int J Phytoremediation 11:664–691

Davison AW (1983) Uptake, transport and accumulation of soil and airborne fluorides by vegetation. In: Fluorides: effects on vegetation, animals and humans, pp 61–82

Del Socorro Santos-Díaz M, Zamora-Pedraza C (2010) Fluoride removal from water by plant species that are tolerant and highly tolerant to hydrogen fluoride. Fluoride 43(2):150–156

Du W, Gardea-Torresdey JL, Ji R, Yin Y, Zhu J, Peralta-Videa JR, Guo H (2015) Physiological and biochemical changes imposed by CeO_2 nanoparticles on wheat: a life cycle field study. Environ Sci Technol 49(19):11884–11893

Flathman PE, Lanza GR (1998) Phytoremediation: current views on an emerging green technology. Soil Sedim Contam 71:415–432

Gan CD, Jia YB, Yang JY (2021) Remediation of fluoride contaminated soil with nano-hydroxyapatite amendment: response of soil fluoride bioavailability and microbial communities. J Hazard Mater 405:124694

Gerhardt KE, Huang XD, Glick BR, Greenberg BM (2009) Phytoremediation and rhizoremediation of organic soil contaminants: potential and challenges. Plant Sci 176:20–30

Glass DJ (1999) U.S. and International markets for phytoremediation, 1999–2000. D. Glass Associates, Needham, MA

Gong WX, Qu JH, Liu RP, Lan HC (2012) Effect of aluminum fluoride complexation on fluoride removal by coagulation. Colloids Surf A Physicochem Eng Asp 395:88–93

Gubbins EJ, Batty LC, Lead JR (2011) Phytotoxicity of silver nanoparticles to *Lemna minor* L. Environ Pollut 159(6):1551–1559

Harguinteguy CA, Schreiber R, Pignata ML (2013) *Myriophyllum aquaticum* as a biomonitor of water heavy metal input related to agricultural activities in the Xanaes River (Córdoba, Argentina). Ecol Indic 27:8–16

Hasan (2015) A review on nanoparticles: their synthesis and types. Res J Recent Sci 4:2277–2502

Hawthorne J, Musante C, Sinha SK, White JC (2012) Accumulation and phytotoxicity of engineered nanoparticles to *Cucurbita pepo*. Int J Phytoremediation 14(4):429–442

He S, Feng Y, Ni J, Sun Y, Xue L, Feng Y, Yang L (2016) Different responses of soil microbial metabolic activity to silver and iron oxide nanoparticles. Chemosphere 147:195–202

Heaton ACP, Rugh CL, Wang N, Meagher RB (1998) Phytoremediation of mercury- and methylmercury-polluted soils using genetically engineered plants. J Soil Contam 7:497–509

Hough RM, Noble RRP, Reich M (2011) Natural gold nanoparticles. Ore Geol Rev 42(1):55–61

Hussain J, Sharma KC, Hussain I (2004) Fluoride in drinking water in Rajasthan and its ill effects on human health. J Cell Tiss Res 4(2):263–273

Jaberzadeh A, Moaveni P, Moghadam HRT, Zahedi H (2013) Influence of bulk and nanoparticles titanium foliar application on some agronomic traits, seed gluten and starch contents of wheat subjected to water deficit stress. Notulae Botanicae Horti Agrobotanici Cluj Napoca 41(1):201–207

Jagtap S, Yenkie MK, Labhsetwar N, Rayalu S (2012) Fluoride in drinking water and defluoridation of water. Chem Rev 112(4):2454–2466

Jeyasubramanian K, Thoppey UUG, Hikku GS, Selvakumar N, Subramania A, Krishnamoorthy K (2016) Enhancement in growth rate and productivity of spinach grown in hydroponics with iron oxide nanoparticles. RSC Adv 6(19):15451–15459

Jiang M, Ohnuki T, Kozai N, Tanaka K, Suzuki Y, Sakamoto F, Kamiishi E, Utsunomiya S (2010) Biological nano-mineralization of Ce phosphate by *Saccharomyces cerevisiae*. Chem Geol 277(1–2):61–69

Johnson DM, Deocampo DM, El-Mayas H, Greipsson S (2015) Induced phytoextraction of lead through chemical manipulation of switchgrass and corn; role of iron supplement. Int J Phytoremediation 17(12):1192–1203

Kang D, Tsao D, Wang-Cahill F, Rock S, Schwab A, Banks M (2008) Assessment of landfill leachate volume and concentration of cyanide and fluoride during phytoremediation. Biorem J 12:32–45

Katiyar P, Pandey N, Sahu KK (2020) Biological approaches of fluoride remediation: potential for environmental clean-up. Environ Sci Pollut Res 27(12):13044–13055

Khandare RV, Govindwar SP (2015) Phytoremediation of textile dyes and effluents: current scenario and future prospects. Biotechnol Adv 33:1697–1714

Kumari S, Khan S (2017) Defluoridation technology for drinking water and tea by green synthesized Fe_3O_4/Al_2O_3 nanoparticles coated polyurethane foams for rural communities. Sci Rep 7(1):8070–8081

Kumari S, Khan S (2018) Effect of Fe_3O_4 NPs application on fluoride (F) accumulation efficiency of *Prosopis juliflora*. Ecotoxicol Environ Saf 166:419–426

Levard C, Hotze EM, Lowry GV, Brown GE Jr (2012) Environmental transformations of silver nanoparticles: impact on stability and toxicity. Environ Sci Technol 46(13):6900–6914

Li J, Hu J, Ma C, Wang Y, Wu C, Huang J, Xing B (2016) Uptake, translocation and physiological effects of magnetic iron oxide (γ-Fe_2O_3) nanoparticles in corn (*Zea mays* L.). Chemosphere 159: 326–334

Liang SX, Jin Y, Liu W, Li X, Shen SG, Ding L (2017) Feasibility of Pb phytoextraction using nano-materials assisted ryegrass: results of a one-year field-scale experiment. J Environ Manage 190:170–175

Mahdavi A, Khermandar K, Asbchin SA, Tabaraki R (2014) Lead accumulation potential in *Acacia victoria*. Int J Phytoremediation 16(6):582–592

Meenakshi, Maheshwari RC (2006) Fluoride in drinking water and its removal. J Hazard Mater 137:456–463

Miller GW, Pushnik JC, Giannini J (1986) A model for subcellular distribution of fluoride in higher plants: relation to in vitro concentrations required for enzyme inhibition. Stud Environ Sci 27: 241–251

Mirzajani F, Askari H, Hamzelou S, Farzaneh M, Ghassempour A (2013) Effect of silver nanoparticles on *Oryza sativa* L. and its rhizosphere bacteria. Ecotoxicol Environ Saf 88:48–54

Mohanpuria P, Rana NK, Yadav SK (2008) Biosynthesis of nanoparticles: technological concepts and future applications. J Nanopart Res 10(3):507–517

Moreno FN, Anderson CWN, Stewart RB et al (2004) Mercury phytoextraction and phytovolatilisation from hg-contaminated artisanal mine sites. In: Phytoremediation of mercury-contaminated mine wastes, pp 147–159

Mueller NC, Nowack B (2010) Nanoparticles for remediation: solving big problems with little particles. Elements 6:395–400

Mukhopadhyay S, Maiti SK (2010) Phytoremediation of metal enriched mine waste: a review. Am Eur J Agric Environ Sci 9(5):560–575

Nahar N, Rahman A, Nawani NN, Ghosh S, Mandal A (2017) Phytoremediation of arsenic from the contaminated soil using transgenic tobacco plants expressing ACR2 gene of *Arabidopsis thaliana*. J Plant Physiol 218:121–126

Nekrasova GF, Ushakova OS, Ermakov AE, Uimin MA, Byzov IV (2011) Effects of copper (II) ions and copper oxide nanoparticles on *Elodea densa planch*. Russ J Ecol 42(6):422–428

Oruc N (2008) Occurrence and problems of high fluoride waters in Turkey: an overview. Environ Geochem Health 30(4):315–323

Ozsvath DL (2006) Fluoride concentrations in a crystalline bedrock aquifer Marathon County, Wisconsin. Environ Geol 50:132–138

Padmavathiamma PK, Li LY (2007) Phytoremediation technology: hyper-accumulation metals in plants. Water Air Soil Pollut 184(1–4):105–126

Pallavi MCM, Srivastava R, Arora S, Sharma AK (2016) Impact assessment of silver nanoparticles on plant growth and soil bacterial diversity. 3 Biotech 6(2):254–263

Panigrahi S, Kundu S, Ghosh S, Nath S, Pal T (2004) General method of synthesis for metal nanoparticles. J Nanopart Res 6(4):411–414

Paudyal H, Pangeni B, Inoue K, Kawakita H, Ohto K, Ghimire K, Alam A (2013) Preparation of novel alginate based anion exchanger from *Ulva japonica* and its application for the removal of trace concentrations of fluoride from water. Bioresour Technol 148:221–227

Pilon-Smits EAH, de Souza MP, Hong G et al (1999) Selenium volatilization and accumulation by twenty aquatic plant species. J Environ Qual 28:1011

Prasad R (2014) Synthesis of silver nanoparticles in photosynthetic plants. J Nanoparticles 2014: 963961. https://doi.org/10.1155/2014/963961

Prasad R, Aranda E (2018) Approaches in bioremediation. Springer International, Berlin. https://www.springer.com/de/book/9783030023683

Prasad R, Kumar V, Prasad KS (2014) Nanotechnology in sustainable agriculture: present concerns and future aspects. Afr J Biotechnol 13(6):705–713

Prasad R, Kumar M, Varma A (2015) Role of PGPR in soil fertility and plant health. In: Egamberdieva D, Shrivastava S, Varma A (eds) Plant growth-promoting rhizobacteria (PGPR) and medicinal plants. Springer International, Cham, pp 247–260

Prasad R, Bhattacharyya A, Nguyen QD (2017) Nanotechnology in sustainable agriculture: recent developments, challenges, and perspectives. Front Microbiol 8:1014. https://doi.org/10.3389/fmicb.2017.01014

Prasad R, Jha A, Prasad K (2018a) Exploring the realms of nature for nanosynthesis. Springer International, Cham. (ISBN 978-3-319-99570-0) https://www.springer.com/978-3-319-99570-0

Prasad R, Kumar V, Kumar M, Wang S (2018b) Fungal nanobionics: principles and applications. Springer Nature Singapore, Singapore. (ISBN 978-981-10-8666-3) https://www.springer.com/gb/book/9789811086656

Rafati M, Khorasani N, Moattar F, Shirvany A, Moraghebi F, Hosseinzadeh S (2011) Phytoremediation potential of *Populus alba* and *Morus alba* for cadmium, chromium and nickel absorption from polluted soil. Int J Environ Res 5:961–970

Rafique R, Arshad M, Khokhar MF, Qazi IA, Hamza A, Virk N (2014) Growth response of wheat to titania nanoparticles application. NUST J Eng Sci 7(1):42–46

Raliya R, Nair R, Chavalmane S, Wang WN, Biswas P (2015) Mechanistic evaluation of translocation and physiological impact of titanium dioxide and zinc oxide nanoparticles on the tomato (*Solanum lycopersicum* L.) plant. Metallomics 7(12):1584–1594

Rashid A, Mahmood T, Mehmood F et al (2014) Phytoaccumulation, competitive adsorption and evaluation of chelators-metal interaction in lettuce plant. Environ Eng Manag J 13:2583–2592

Reddy A, Madhavi V, Reddy KG, Madhavi G (2012) Remediation of chlorpyrifos-contaminated soils by laboratory-synthesized zero-valent nano iron particles: effect of pH and aluminium salts. J Chem 2013:7

Ren X, Zeng G, Tang L, Wang J, Wan J, Liu Y, Deng R (2018) Sorption, transport and biodegradation—an insight into bioavailability of persistent organic pollutants in soil. Sci Total Environ 610:1154–1163

Rico CM, Barrios AC, Tan W, Rubenecia R, Lee SC, Varela-Ramirez A, Peralta-Videa JR, Gardea-Torresdey JL (2015) Physiological and biochemical response of soil-grown barley (*Hordeum vulgare* L.) to cerium oxide nanoparticles. Environ Sci Pollut Res 22(14):10551–10558

Ruan J, Ma L, Shi Y, Han W (2003) Uptake of fluoride by tea plants (*Camellia sinensis* L.) and the impact of aluminium. J Sci Food Agric 83:1342–1348

Rui M, Ma C, Hao Y, Guo J, Rui Y, Tang X, Zhao Q, Fan X, Zhang Z, Hou T, Zhu S (2016) Iron oxide nanoparticles as a potential iron fertilizer for peanut (*Arachis hypogaea*). Front Plant Sci 7:815–824

Saini P, Khan S, Baunthiyal M, Sharma V (2012) Organ-wise accumulation of fluoride in *Prosopis juliflora* and its potential for phytoremediation of fluoride contaminated soil. Chemosphere 89(5):633–635

Salt DE, Smith RD, Raskin I (1998) Phytoremediation. Annu Rev Plant Biol 49(1):643–668

Sarma H, Forid N, Prasad R, Prasad MNV, Ma LQ, Rinklebe J (2021) Enhancing phytoremediation of hazardous metal(loid)s using genome engineering CRISPR–Cas9 technology. J Hazard Mater. https://doi.org/10.1016/j.jhazmat.2021.125493

Saxena V, Ahmed S (2001) Dissolution of fluoride in groundwater: a water-rock interaction study. Environ Geol 40(9):1084–1087

Saxena KL, Sewak R (2015) Fluoride consumption in endemic villages of India and its remedial measures. Int J Eng Sci Invent 4:58–73

Senthilkumar P, Prince WS, Sivakumar S, Subbhuraam CV (2005) *Prosopis juliflora*—a green solution to decontaminate heavy metal (Cu and Cd) contaminated soils. Chemosphere 60(10):1493–1496

Siddiqi KS, Husen A (2016) Engineered gold nanoparticles and plant adaptation potential. Nanoscale Res Lett 11(1):400–409

Singh N, Khan S (2018) Nanobioremediation: an innovative approach to fluoride (F) contamination. In: Approaches in bioremediation. Springer, Cham, pp 343–353

Singh R, Misra V, Mudiam MKR, Chauhan LKS, Singh RP (2012) Degradation of γ-HCH spiked soil using stabilized Pd/Fe0 bimetallic nanoparticles: pathways, kinetics and effect of reaction conditions. J Hazard Mater 237–238:355–364

Singh G, Kumari B, Sinam G, Kumar N, Mallick S (2018) Fluoride distribution and contamination in the water, soil and plants continuum and its remedial technologies, an Indian perspective—a review. Environ Pollut 239:95–108

Song G, Hou W, Gao Y, Wang Y, Lin L, Zhang Z, Niu Q, Ma R, Mu L, Wang H (2016) Effects of CuO nanoparticles on *Lemna minor*. Bot Stud 57(1):30–10

Song B, Xu P, Chen M, Tang W, Zeng G, Gong J, Ye S (2019) Using nanomaterials to facilitate the phytoremediation of contaminated soil. Crit Rev Environ Sci Technol 49(9):791–824

Sonowal S, Nava AR, Joshi SJ, Borah SN, Islam NF, Pandit S, Prasad R, Sarma H (2022) Biosurfactants assisted heavy metals phytoremediation: Green technology for the United Nations sustainable development goals. Pedosphere 2(1):198–210. https://doi.org/10.1016/S1002-0160(21)60067-X

Subhashini V, Swamy AVVS (2013) Phytoremediation of Pb and Ni contaminated soils using *Catharanthus roseus* (L.). Univ J Environ Res Technol 3:465–472

Takmaz-Nisancioglu S, Davison AW (1988) Effects of aluminium on fluoride uptake by plants. New Phytol 109(2):149–155

Truong P (1999) Vetiver grass technology for mine rehabilitation. Bangkok 2:1–19

Truong P (2003) Vetiver system for water quality improvement. In: Proceedings of 3rd International Vetiver Conference, Oct. 6–9, Guangzhou, China, pp 61–74

Verma A, Roy A, Bharadvaja N (2021) Remediation of heavy metals using nanophytoremediation. In: Advanced oxidation processes for effluent treatment plants. Elsevier, Amsterdam, pp 273–296

Vikas C, Kushwaha R, Ahmad W, Prasannakumar V, Reghunath R (2013) Genesis and geochemistry of high fluoride bearing groundwater from a semi-arid terrain of NW India. Environ Earth Sci 68(1):289–305

Wei SH, Zhou QX, Wang X, Cao W, Ren LP, Song YF (2004) Potential of weed species applied to remediation of soils contaminated with heavy metals. J Environ Sci 16(5):868–873

Yadav KK, Gupta N, Kumar V, Singh JK (2017) Bioremediation of heavy metals from contaminated sites using potential species: a review. Indian J Environ Prot 37(1):65–84

Yadav N, Garg VK, Chhillar AK, Rana JS (2021) Detection and remediation of pollutants to maintain ecosustainability employing nanotechnology: a review. Chemosphere 280:130792

Yadu B, Chandrakar V, Korram J, Satnami ML, Kumar M, Keshavkant S (2018) Silver nanoparticle modulates gene expressions, glyoxalase system and oxidative stress markers in fluoride stressed *Cajanus cajan* L. J Hazard Mater 353:44–52

Zare K, Najafi F, Sadegh H (2013) Studies of ab initio and Monte Carlo simulation on interaction of fluorouracil anticancer drug with carbon nanotube. J Nanostruct Chem 3(1):71

Zayed A, Gowthaman S, Terry N (1998) Phytoaccumulation of trace elements by wetland plants: I. Duckweed. J Environ Qual 27:715–721

Zhao X, Wang J, Wu F, Wang T, Cai Y, Shi Y, Jiang G (2010) Removal of fluoride from aqueous media by Fe_3O_4@$Al(OH)_3$ magnetic nanoparticles. J Hazard Mater 173(1–3):102–109

Zhou J, Gao J, Liu Y, Ba K, Chen S, Zhang R (2012) Removal of fluoride from water by five submerged plants. Bull Environ Contam Toxicol 89:395–399

Zhu S, Zhang J, Dong T (2009) Removal of fluorine from contaminated field soil by anolyte enhanced electrokinetic remediation. Environ Earth Sci 59(2):379–384

Zurayk R, Sukkariyah B, Baalbaki R, Ghanem DA (2002) Ni phytoaccumulation in *Mentha aquatica* L. and *Mentha sylvestris* L. water. Air Soil Pollut 139:355–364

13 Sustainable Use of African Palm Shell Waste Applied to Paraben Adsorption from Aqueous Solutions

Astrid Roxanna Moreno-Marenco, Liliana Giraldo, and Juan Carlos Moreno-Piraján

Abstract

The utilization of agricultural waste residues has been explored as an alternative to costly conventional activated carbon production methods. This study addresses the potential use of agricultural waste from the processing of palm oil (African palm shell) to produce activated carbons modified with metal salts at 973 K and 1173 K, as adsorbents for methylparaben (MePB) and propylparaben (PrPB), a type of emerging contaminants present in personal care products (PCPs). The carbons obtained were able to retain parabens, but the highest adsorption was found for PrPB in carbon with the highest contribution of micropores to the total pore volume and with the highest content of acid surface groups, which were favored by the effect of the higher activation temperature.

Keywords

Adsorption · Methylparaben · Propylparaben · Agroindustrial waste · African palm shell · Activated carbon · Personal care products

A. R. Moreno-Marenco · L. Giraldo
Departamento de Química, Facultad de Ciencias, Universidad Nacional de Colombia, Bogotá, Colombia
e-mail: armorenom@unal.edu.co; lgiradogu@unal.edu.co

J. C. Moreno-Piraján (✉)
Departamento de Química, Facultad de Ciencias, Universidad de los Andes, Bogotá, Colombia
e-mail: jumoreno@uniandes.edu.co

R. Prasad (ed.), *Phytoremediation for Environmental Sustainability*, https://doi.org/10.1007/978-981-16-5621-7_13

13.1 Introduction: Water and Pollution

It is widely known that water is a limited natural resource, necessary for the development and sustainability of life, which constitutes an essential part of every ecosystem. However, the constant increase in demand for this resource due to its excessive use in various activities of our daily lives, results in the release of tons of biologically active substances through the water. Wastewater, whatever its origin, has not been considered beyond being a contaminated or altered supply with no other purpose than to dispose of. However, as mentioned in the 2017 edition of the United Nations World Water Development Report: "Wastewater: The Untapped resource" shows that this conception is beginning to change, not only because of the associated pollution problems, but because the scarcity of this resource has increased in many regions. Thus, the importance of the collection, treatment, and reuse of wastewater begins to be recognized, that is, in the management that is given to these, which generates essential social, environmental, and economic benefits for sustainable development (Ryder 2017).

The behaviors and adverse impacts of various chemical compounds such as heavy metals, colorants, pesticides, or polycyclic aromatic hydrocarbons, among others, have been known for a long time (*World Health Organization: Guidelines for Drinking-water Quality. Recomendations* 2008), which are mostly regulated. However, the development of more sensitive analytical methods has made it possible to warn of the presence of new or less known pollutants, whose accumulation of scientific evidence derived from new research is beginning to acquire greater relevance and concern about their impacts on public health or the environment (Recommendations Report: Contaminants of Emerging Concern Workgroup 2019). These pollutants are called emerging contaminant (EC) whose study appears among the priority research lines of the main organizations dedicated to the protection of public and environmental health, such as the World Health Organization (WHO), the Environmental Protection Agency (EPA) (OW/ORD Emerging Contaminants Workgroup 2008), or the European Commission (European Environment Agency 2012; Martin and Kortenkamp 2009). The risk associated with the presence of these pollutants in the environment is not due so much to their acute toxicity but to the development of resistance to pathogens and endocrine alterations due to continuous exposure, mainly of aquatic organisms to these pollutants. Many of these are classified as carcinogens, endocrine disruptors, or with other toxic effects for humans and the ecosystem (OW/ORD Emerging Contaminants Workgroup 2008; Pal et al. 2014; Schriks et al. 2010; WHO 2006).

The list of emerging contaminants includes a wide variety of compounds with different structures and uses, as well as their metabolites and transformation products that are part of pharmaceutical and personal care products (PCPs), among others (USEPA 2016). Global production of pharmaceutics and PCPs is estimated to increase 3% each year (Kwarciak-Kozłowska 2019). Reason why a continuous introduction of these pollutants into the environment is generated, since not everything is used or absorbed by the body, thus becoming part of the wastewater.

Table 13.1 Emerging contaminants most representative (Dey et al. 2019; Freyria et al. 2018)

Subgroup	Emerging contaminants
PCPs	
Parabens	Methyl-, ethyl-, propyl-, and butylparaben
Antiseptics/disinfectants	Triclosan, chlorophene, chloramines
Sunscreen agents	Benzophenones, benzylidene, homosalate
Fragrances	Nitro, polycyclic and macrocyclic fragrances, musk xylol
Insect repellents	*N*, *N*-diethyltoluamide
Pharmaceutics	
Antibiotics	Amoxicillin, ciprofloxacin, erythromycin, penicillin
Steroids and hormones	Estrogen, estrone, estriol, testosterone, 17β-estradiol
Nonsteroidal anti-inflammatory drugs (NSAIDs)	Aspirin, ibuprofen, naproxen, ketoprofen, paracetamol, diclofenac, morphine
Psychiatric	Benzodiazepines, barbiturates
Illicit drugs	Cocaine, codeine, heroin, methadone, amphetamines

Table 13.1 shows a list of some of the emerging contaminants found in PCPs and pharmaceutics.

The risk associated with the presence of these contaminants in the environment is due to development of resistance to pathogens and endocrine disfunction. Several studies have shown that the effects of endocrine disruptors chemicals (EDCs) on living organisms are multiple, the most alarming are related to the reproductive system, where the anomalous development of the reproductive organs has been evidenced in some fish, that leads to the acquisition of male and female genetic and phenotypic characteristics (Niemuth and Klaper 2015). In humans, it produces hormonal alterations, influences reproductive function, in addition to generating antimicrobial resistance (Sanderson et al. 2016). These compounds are found in low concentrations on the order of μg L^{-1}, ng L^{-1} even pg L^{-1} (Kalia 2019), where adverse estrogenic effects have been assigned at concentrations as low as 1 ng L^{-1} (Kuster et al. 2008). Among the compounds listed above, PCPs have acquired great notoriety due to their widespread use. And within these, parabens have specifically received attention since their use is widely distributed.

13.2 Parabens: What Are They?

Parabens (PBs) are esters of 4-hydroxybenzoic acid with alkyl or aril substituents ranging from methyl to butyl, benzyl, or phenyl. These PBs are widely used as preservatives in a wide variety of cosmetic products and pharmaceutics (Haman et al. 2015) although they are also often used in food and industrial products (Brand et al. 2017). Methylparaben (MePB) is widely used as a preservative in drugs, often combined with propylparaben (PrPB) to obtain a synergistic antimicrobial effect. Although PrPB is not approved in food if it is allowed in the manufacture of plastic materials and articles intended to come into contact with food (European

Table 13.2 Physical–chemical properties of parabens (Andersen and Larsen 2013; Błędzka et al. 2014; Yalkowsky et al. 2010)

Parameter	MePB	PrPB
N° CAS	99-76-3	94-13-3
Molecular formula	$C_8H_8O_3$	$C_{10}H_{12}O_3$
Molecular weight (g mol^{-1})	152.15	180.21
Water solubility (mg L^{-1}) at 25 °C	2.50×10^3	5.00×10^2
Log K_{ow}	1.66	2.71
pK_a	8.17	8.35
Area (nm^2)[a]	0.406	0.754
Structure	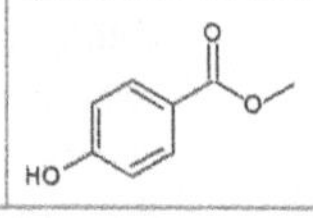	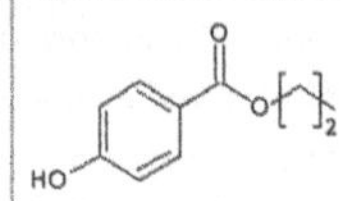

[a]Value determined using HyperChem software (version 8.0.7 for Windows)

Commission Regulation No. 10/2011) and can, through migration, also enter the food (Brand et al. 2017). In fact, PBs are present in 80% of PCPs (Błędzka et al. 2014), with MePB and PrPB being the main preservatives used in this type of products (Nowak et al. 2018), which is related to their greater presence in wastewater compared to other PBs. The maximum authorized concentration in ready-made preparations for the individual esters and their salts is 0.4% (w/w) for MePB and 0.14% for PrPB, while for paraben mixtures it is 0.8%, as established by the European Union (Brand et al. 2017; Hessel et al. 2019). Thus, the estimated exposure in children and adults is about 3 mg/kg/day and 0.2 mg/kg/day for these parabens, related to PCPs since the contribution of food is less than 1% (Brand et al. 2017).

PBs show chemical stability in a wide pH range (3.0–6.5) (Angelov et al. 2008) and broad spectrum of antimicrobial activity, which is proportional to the chain length of the ester group (Brand et al. 2017; Mackay et al. 2006). As the length of the alkyl chain increases, the hydrolysis resistance of aqueous paraben solutions increases (Masten 2005). However, the value of the octanol–water partition coefficient (log K_{ow}) increases, which results in a decrease in water solubility (Table 13.2), so that sodium salts of parabens are also frequently used in formulations. The values of the acid dissociation constant (pK_a) are around 8.3 and therefore, in aquatic environments they are in their free acid form (Andersen and Larsen 2013; Błędzka et al. 2014).

13.2.1 Environmental Impact of Parabens

Parabens are compounds that can mimic the effects of the main natural estrogen; therefore, they are listed in the EU list of possible endocrine disruptors (EDCs) within category 1, where substances for which endocrine disrupting activity has been documented a living organism is found in at least one study and is given the highest priority for further studies (Andersen and Larsen 2013). EDCs are known as a class of chemicals that have xenobiotic and exogenous origins while interfering by mimicking or inhibiting the normal activity of the hormonal system, causing alterations in the health of the reproductive system and metabolism in general (European Environment Agency 2012).

Among the alterations that have been related to the effects of EDCs, especially due to exposure to them during pregnancy, childhood, and puberty, are breast and prostate cancer, reproductive disorders such as infertility, metabolism disorders such as diabetes and obesity, autoimmune diseases, asthma, cardiovascular problems such as hypertension, mental disorders such as Alzheimer's and Parkinson's, and behaviors such as memory, motility, attention, among others. Some studies have shown the possible estrogenic effects of parabens, especially in fish (Diamanti-Kandarakis et al. 2009; European Environment Agency 2012). Some studies have shown the possible estrogenic effects of parabens, especially in fish (Andersen and Larsen 2013; Brausch and Rand 2015; Haman et al. 2015) and crustaceans (Lee et al. 2018) at high exposure levels. Other studies in young male rats have shown adverse effects on sperm production and testosterone levels after oral exposure to PrPB (Andersen and Larsen 2013). MePB is generally considered to have a much lower potential to cause endocrine disrupting effects compared to PrPB, in addition to several studies that have shown PrPB to have estrogenic and/or antiandrogenic effects in vivo and in vitro (Andersen and Larsen 2013; Andersen 2008).

EDC effects towards animals are well reported, although direct effects for humans are still debated and require further study. However, few studies suggest that the effect of exposure to EDCs on human health includes a decrease in male sperm count, an increase in testicular, prostate, ovarian, and breast cancer (Brand et al. 2017), as well as reproductive dysfunctions (Bolong et al. 2009). PBs can be degraded in the human body into various metabolites, being p-hydroxybenzoic acid the most common to all parabens, a product of its hydrolysis (Pugazhendhi et al. 2005), so it cannot be used directly to discriminate the metabolites of each paraben, added because they are not always completely metabolized, so a small fraction of the free parent substance can also be found. These parabens and their metabolites are mainly excreted in the urine, and as the length of the alkyl chain increases, the urinary excretion rate of *p*-hydroxybenzoic acid decreases (Brand et al. 2017; Hessel et al. 2019).

13.2.2 Remotion Treatments of PCPs from Water

Wastewater treatment technologies are classified according to their stage of operation and application in a large-scale system. In general terms they are classified as conventional and advanced treatment methods (Kaur et al. 2019). Conventional treatments including coagulation/flocculation and sedimentation are ineffective in removing PCPs (<30%) (Yang et al. 2017), so they can still be detected in the effluents after the conventional treatment process, although their concentrations are low (Wang and Wang 2016). Thus, PCPs residues have been found in the tissues of plants when bio-solids or manure-amended soils were used or when sewage was used for irrigation (Rajapaksha et al. 2014).

Due to the inefficiency of conventional treatment methods for the significant removal of PCPs, advanced methods have been developed employing oxidation, membrane filtration, phytoremediation, and adsorption processes. Oxidation treatments include ozonation, UV irradiation, photocatalysis with Titania, among others, which involve the generation of highly reactive radicals (especially • OH radicals). Membrane filtration techniques include nanofiltration (NF), ultrafiltration (UF), and reverse osmosis (RO), considering that the removal efficiency of NF/UF is affected by the physicochemical properties of PCPs, such as hydrophobicity, charge, and molecular weight. Phytoremediation for its part is a cost-efficient plant-based approach that takes advantage of the natural ability of certain plants to bioaccumulate, degrade, immobilize, degrade contaminants from the environment, and metabolize various molecules in their tissues. It is potentially the least harmful method because it uses living organisms and preserves the environment in a more natural state. However, it is limited to the surface and depth that the roots occupy, and plant survival is affected by the toxicity of the contaminated environment and the general state of the soil (Farraji et al. 2016; Hauptvogl et al. 2020). Within a sustainable bioeconomy, studies are also focusing on the development of adsorbents from phytoremediation residues for the adsorption of pollutants (Abu Hasan et al. 2020).

Finally, adsorption treatments mainly involve carbonaceous materials such as activated carbon, graphene, and carbon nanotubes; however, performance is also affected by the physicochemical properties of PCPs, as well as their solubility in water (Kaur et al. 2019). Within the methods of removal of PCPs in wastewater (Table 13.3) it is evident that each process has advantages, disadvantages, and limitations, so new methods are continually being improved or proposed.

Currently there are a series of processes that are used in the elimination of parabens. Oxidation methods such as ultraviolet photodegradation (Álvarez et al. 2020), persulfate oxidation and ozonization (Hernández-Leal et al. 2011; Kwarciak-Kozłowska 2019; Tay et al. 2010a) with a high percentage of elimination but the possible formation of toxic degradation by-products. The study by Tay et al. (2010b) discovered that the hydroxylation of parabens is the main reaction that occurred during ozonation with the detection of a number of compounds including hydroquinone and 4-Hydroxybenzoic acid. In another study by Canosa et al. (2006) it was reported the formation of chlorinated parabens during the chlorination process, as well as that the levels of chlorine generally contained in tap water are sufficient to produce significant amounts of its chlorinated by-products in a few minutes that can become more resistant to additional oxidation than the original parabens.

Because of this there is a growing need for more efficient, cost-effective, and safe methods for treating wastewater. Therefore, the adsorption method can offer a much safer way to remove parabens from water and wastewater, as it is a method free of harmful substances and it is more environmentally friendly. In general, a wide variety of adsorbent materials have been applied for the removal of parabens such as organic textile fibers (Ran et al. 2020), composites (Mashile et al. 2020), polymers

Table 13.3 Conventional and advanced methods to removal of PCPs in wastewater

Method	Advantages	Disadvantages	Ref.
Conventional treatments			
Coagulation/ flocculation/ sedimentation	• Simple (low technology) • Easy installation and operation • Low cost • Removal of lipophilic compounds	• Low elimination capacity (<20%) • Incomplete degradation resulting in toxic degradation products • High sludge production • Disposition problems	Bolong et al. (2009), Chang et al. (2009) and Kaur et al. (2019)
Tratamiento biológico	• Less amount of biomass per unit of substrate degraded • Economic	• The removal depends on the biodegradability of the compound • Variable elimination rate	Bolong et al. (2009) and Chang et al. (2009)
Advance treatments			
Oxidación avanzada (UV, O_3, TiO_2)	• High efficiency (>90%) • Complete mineralization of micropollutants	• Generation of subproducts with unknown effects • 50–80% removal with UV • Removal efficiency is generally proportional to dose • High cost due to longer contact time and higher dosage required • High energy requirement • UV lamp replacement frequency	Kaur et al. (2019), Kwarciak-Kozłowska (2019) and Morone et al. (2019)
Membrane filtration (RO, NF)	• High removal (almost complete with RO) • Removes wide range of contaminants (NF) • Small space requirement	• High energy consumption (RO) • Removal depends on the properties of membrane • Generation of huge volume of concentrates	Kaur et al. (2019), Kwarciak-Kozłowska (2019) and Morone et al. (2019)
Phytoremediation	• Suitable for various types of contaminants • Financial costs low (energy from solar radiation)	• Not applicable to all plants or contaminants • Process is slower than normal physicochemical methods • Applicable to moderately contaminated land • Contaminants cannot be completely removed	Farraji et al. (2016) and Hauptvogl et al. (2020)
Adsorption (AC)	• High efficiency for most (>90%) • No transformation by-products generated • Little environmental impact • Regeneration of adsorbent	• Cost of the adsorbent material • Dependence with pH • Adsorbent saturation • Less removal of polar compounds • Disposal of spent adsorbent • Nonselective adsorbent	Chang et al. (2009), Katsigiannis et al. (2015), Morone et al. (2019) and Rossner et al. (2009)

(Chin et al. 2010), fly ash (De Oliveira et al. 2020), magnetic nanoparticles (Chen et al. 2017), bioadsorbents (Mallek et al. 2018), among others. However, the adsorption of parabens using activated carbons has been little studied, compared to other emerging pollutants such as pharmaceutical compounds.

Mailler et al. (2014) studied the adsorption of a number of emerging pollutants including methyl butyl and benzyl paraben using powdered commercial activated carbon (PAC). They determined adsorption capacities greater than 70%, particularly for PrPB. They identified the adsorbent dose as the most influential operating parameter, which correlates with the performance of the process. Delgado et al. (2016) used a commercial granular activated carbon (GAC) which exhibited a high adsorption capacity for MePB (300 mg g^{-1}). The results presented by de los Ángeles Bernal-Romero et al. (2019) show that between 80% and 90% of MePB and PrPB can be removed from real water and noted that removal of both parabens was improved at higher doses of PAC. They concluded that the higher solubility in water and the lower log K_{ow} values could explain the lower adsorption capacity evidenced for MePB.

13.3 Activated Carbon: Properties and Production

13.3.1 Properties of Activated Carbon

Activated carbon (AC) is considered an amorphous solid consisting mainly of carbon atoms which join the other carbon atoms forming angles of 120° giving rise to flat sheets of hexagonal rings displaced from each other, forming a criss-cross structure of basal planes joined by forces. of Van der Waals. In fact, the folding of the hexagonal sheets takes place producing a rigid structure, with very little mobility, which avoids the ordering by creating interstices that give rise to the different types of porosity (Marsh and Rodríguez-Reinoso 2006). According to the IUPAC it is possible to make a classification of the pores according to their size as: micropores (≤2 nm), mesopores (between 2 and 50 nm), and macropores (>50 nm) (Thommes et al. 2015). Micropores contribute more to the high surface areas of activated carbon and provide high adsorption capacities for small molecules such as gases and most solvents. These in turn can be classified into two subcategories, for example, narrow micropores (<0.7 nm) and super micropores (0.7–2 nm) or primary micropores (<0.8 nm) and secondary micropores (0.8–2 nm) (Daud and Houshamnd 2010).

Graphene layers can present a large number of imperfections, impurities, non-aromatic rings, as well as edges that constitute highly energetic sites, associated with higher densities of unpaired electrons and therefore show a strong tendency to chemisorb other heteroatoms, such as oxygen, hydrogen, nitrogen, sulfur, etc., giving rise to stable surface compounds (Rouquerol et al. 2014b).

The presence of these surface groups determines the apparent chemical character of the activated carbon surface (Rodriguez-Reinoso and Molina-Sabio 1998), as well as its hydrophobic or hydrophilic character (Aburub and Wurster 2006). The exact nature of these surface groups is not fully established; however, it is known that there

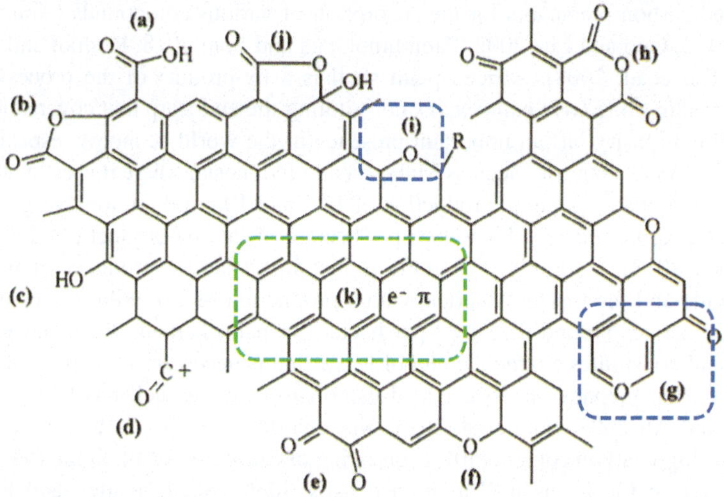

Fig. 13.1 Schematic representation of the oxygenated functional groups present on the surface of activated carbon: (**a**) carboxyl, (**b**) lactone, (**c**) hydroxyl, (**d**) carbonyl, (**e**) quinone, (**f**) ether, (**g**) pyrone, (**h**) carboxylic anhydride, (**i**) chromene, (**j**) lactol, and (**k**) π electron density in the basal planes of activated carbon. (Taken and modified from Bandosz et al. (Bandosz and Ania 2006))

are various types of surface groups. The acid character is associated with surface groups such as carboxylic acid, lactone, phenol, anhydride, and carbonyl in the form of quinone and hydroquinone, so much so that the pyrone and chromene type groups are related to the basic character. Conversely, the nonpolar surface of activated carbon has basic properties associated with regions rich in π electrons located in the basal graphene layers (Boehm 2002; Daud and Houshamnd 2010). Some studies have suggested that the basic property derived from the basal planes is weak compared to that derived from basic functional groups and that the increase in acid groups on the surface usually leads to a decrease in the basic groups (Daud and Houshamnd 2010).

In Fig. 13.1, the most common oxygenated surface groups present on the surface of activated carbon are shown, capable of forming specific interactions between the solute and the adsorbent. Highlighted in blue are the functional groups to which the basic character is attributed and in green are the apolar regions of the basal surface of activated carbon that intervene in nonspecific interactions with the apolar regions of the solute.

13.3.2 Agroindustrial Waste

Lignocellulosic residues from agriculture have been widely used for the preparation of activated carbons (Attia et al. 2008; da Silva Lacerda et al. 2015; Guo et al. 2003; Iwasaki et al. 2002; Torrellas et al. 2015; Tseng et al. 2006; Venkatramanan et al. 2021). Among these, African palm residues have demonstrated their potential as activated carbon precursors for the adsorption of various compounds (Arami-Niya et al. 2012; Guo and Lua 2002; Pamidimukkala and Soni 2018; Rashidi and Yusup 2017; Tan et al. 2008). African palm shell is a by-product of the processing of African palm oil (*Elaeis guineensis*), constituting the endocarp that covers the palm fruit. This industry has an important presence in the world economy, especially in Southeast Asian countries such as Malaysia and Indonesia, where it is estimated that in 2008, they had a palm oil production of 17.7 and 19.3 million tons, respectively, generating approximately 1.1 tonnes per hectare of this by-product (Abdullah and Sulaiman 2013). Currently, Colombia is the fourth largest producer of palm oil worldwide and the first in America, with a production of 1.6 million tons in 2017, which generates around 0.73 tons per hectare of palm kernels (Ruiz and Romero 2011) and constitutes a large amount of waste that is necessary to dispose of. In the literature, the preparation of activated carbon from palm shell has been reported (Daud and Ali 2004; Jung et al. 2014; Nizamuddin et al. 2015; Ruiz et al. 2015), given its high carbon content (50%) consisting of cellulose (17%), lignin (53%), and hemicellulose (30%), as well as high density, high volatile matter, and low ash content (Daud and Ali 2004; González-García 2018). Physicochemical characteristics that allow the production of activated carbon with high density and large pore volume.

The preparation of activated carbons can be carried out in two stages. The first stage of the production process is carbonization, where the precursor material is subjected to a heat treatment in an inert atmosphere (usually nitrogen) at temperatures below 1073 K. This treatment seeks to eliminate the volatile matter content of the precursor, decrease its density, and increase the carbon content, generating the initial porosity of the carbonized. In this stage there is a degradation of the lignocellulosic material due to the carbonization and aromatization of the carbon skeleton, giving rise to the initial porous structure. However, during the carbonization process, some of the pores in the resulting carbon are partially filled or blocked with tars, thus requiring an activation step to improve the textural characteristics. The second stage of the process is activation, whose objective is to transform the carbonized into a highly adsorbent material due to the increase and widening of its internal porosity.

13.3.3 Activation with Metallic Salts

Chemical activation using various agents is used to modify the characteristics of activated carbons because they act as dehydrating agents to inhibit tar formation during pyrolytic decomposition (Marsh and Rodríguez-Reinoso 2006; Molina-Sabio and Rodríguez-Reinoso 2004). Previous studies have shown the benefits of adding metal salts to catalyze the coal gasification reaction ($C - CO_2$ y $C - H_2O$), leading to materials with a broader pore size distribution (Gryglewicz and Lorenc-Grabowska 2004; Juárez-Galán et al. 2009; Molina-Sabio et al. 1994), which favor the adsorption process.

Liu et al. (2009) evaluated the catalytic effect on the degradation of cellulose and hemicellulose in corn stubble of several metallic salts among which are $MgCl_2$ and $CaCl_2$, showing a slight positive effect dissolving the hemicellulose fraction. During the carbonization of the impregnated precursor, the conversion of $CaCl_2$ is shown in the following equations (Mondal et al. 2007):

$$CaCl_{2(s)} + CO_{2(g)} + H_2O_{(l)} \leftrightharpoons CaCO_{3(s)} + 2HCl_{(g)} \quad (13.1)$$

$$CaCO_{3(s)} + C_{(s)} \leftrightharpoons CaO_{(s)} + 2CO_{(g)} \quad (13.2)$$

$$CaO_{(s)} + CO_{(g)} \rightarrow Ca_{(g)} + CO_{2(g)} \quad (13.3)$$

As the impregnated precursor is heated within the reactor in the absence of oxygen $CaCO_3$ is first produced, which is then converted to CaO by the carbon attached to the activated carbon. This CaO can form a layer on the surface of activated carbon. Rufford et al. (2010) performed the thermogravimetric analysis (TGA) of ground coffee treated with $MgCl_2$, the decomposition profile showed that the loss of water molecules ($MgCl_2.6H_2O$) that occurs in multiple steps dominates the TGA curve at temperatures below 200 °C. Between 210 °C and 510 °C there are several weight loss steps including $MgCl_2.6H_2O$ dehydration to $MgCl_2$, anhydrous (~ 300 °C) and gasification of the carbon precursor. At temperatures above 500 °C, $MgCl_2$ decomposes directly to MgO leaving MgO particles within the carbon matrix.

During the carbonization of the precursor impregnated with $MgCl_2$ the conversion occurs as shown in the following equations (Huang et al. 2011; Kirsh et al. 1987; Rongti et al. 2002):

$$MgCl_2 \cdot nH_2O_{(s)} \overset{203^\circ C}{\rightarrow} MgCl_2 \cdot (x-y)H_2O_{(s)} + yH_2O_{(l)} \quad (13.4)$$

$$MgCl_2 \cdot H_2O_{(s)} \overset{235^\circ C}{\rightarrow} Mg(OH)Cl + HCl_{(g)} \quad (13.5)$$

$$Mg(OH)Cl \overset{415^\circ C}{\rightarrow} MgO_{(s)} + HCl_{(g)} \quad (13.6)$$

$$MgO_{(s)} + C_{(s)} \rightarrow Mg_{(g)} + CO_{(g)} \quad (13.7)$$

$$MgO_{(s)} + CO_{(g)} \rightarrow Mg_{(g)} + CO_{2(g)} \quad (13.8)$$

The following indirect reaction can also occur with both salts:

$$CO_{2(g)} + C_{(s)} \rightarrow 2CO_{(g)} \quad (13.9)$$

Activation with dehydrating metals salts such as $CaCl_2$ and $MgCl_2$ have been less studied compared to $ZnCl_2$. However, some works have been done in which these

Table 13.4 Activated carbons from various biomass precursors using different activation conditions

Biomass precursor	Activating (concentration)/ relation (AA:P)	Activation temperature (time)	Surface texture properties	Ref.
Rice husk	$CaCl_2$ (0–2.5% w/w)/1 L: 100 g	873 K (4 h)	109–173[a]	Mondal et al. (2007)
Olive stone	$CaCl_2$ (7% w)/ NR	1097 K (4 h) 1023 K (12 h) 1073 K (6 h)	656[a]; 0.91[b]; 0.27[c]; 0.64[d] 669[a]; 1.22[b]; 0.28[c]; 0.94[d] 670[a]; 1.39[b]; 0.27[c]; 1.12[d]	Juárez-Galán et al. (2009)
Waste coffee grounds	$MgCl_2$ (NR)/1:1 mass ratio	1173 K (1 h)	123[a] 0.21[b] 0.01[c] 0.2[d]	Rufford et al. (2010)
Carnauba palm leaves/ macauba seeds endocarp/pine nut shell	$CaCl_2$ (1 M)/ 0.1 L: 8 g	773 K (1 h)	265–431[a] 0.12–0.25[b] 0.082–0.12[c]	da Silva Lacerda et al. (2015)
Palm shell	$MgCl_2$ $CaCl_2$ (3, 5 y 7% w/v)/ 2 mL:1 g	773–1073 K (6 h)	20–501[a]; 0.02–0.29[b] 19–453[a]; 0.03–0.25[b]	Acevedo et al. (2017)

P precursor, *AA* activating agent, *NR* no reported
[a]BET surface ($m^2\ g^{-1}$)
[b]Pore volume ($cm^3\ g^{-1}$)
[c]Micropore volume ($cm^3\ g^{-1}$)
[d]Mesopore volume ($cm^3\ g^{-1}$)

agents are used in the production of activated carbon showing micro- and mesoporosity development.

Table 13.4 summarizes some of the main results found in the literature on the use of calcium and magnesium salts as activating agents in obtaining activated carbons using lignocellulosic precursors and their effect on their textural properties. The results show an increase in the area with the increase in the concentration of salt, which favors the formation of porosity. Other results show that the decrease in salt concentration and the increase in temperature favor the development of porosity in solids, as well as with these activating agents it is possible to obtain micro-mesoporous carbons. Therefore, the properties of adsorbents depend not only on the nature of the activating agent and its concentration, but also on the activation temperature.

These two parameters are then evaluated on a series of activated carbons obtained from African palm shell (*Elaeis guineensis*) as precursor lignocellulosic material, chemically modified by impregnation with metallic salts solution of $MgCl_2$ (ACM1,

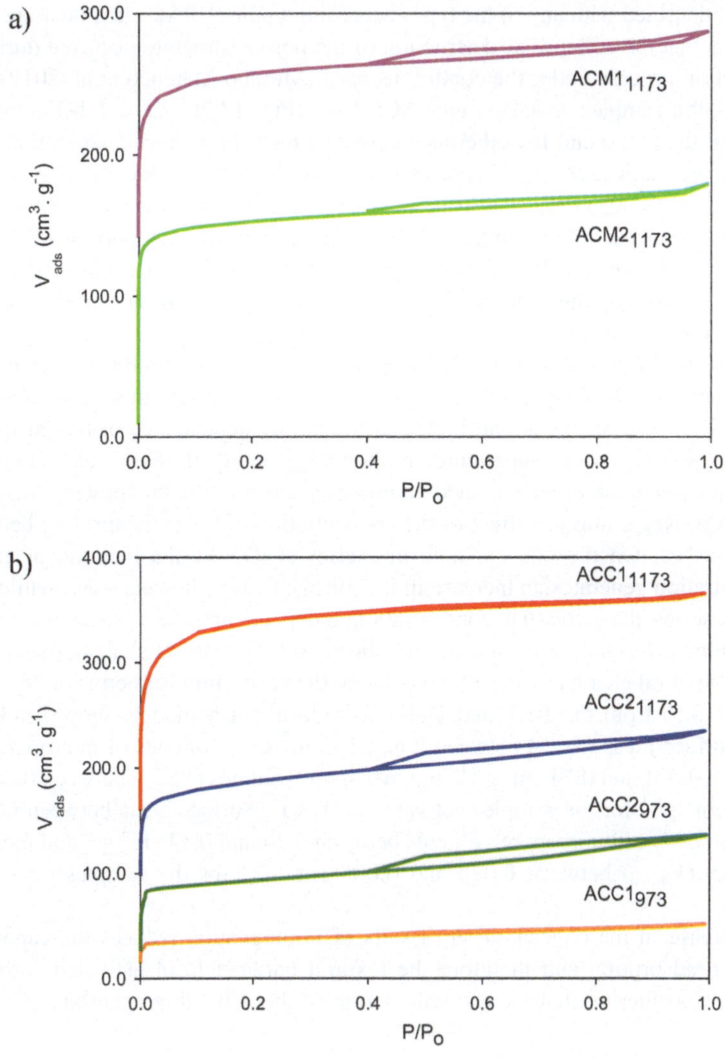

Fig. 13.2 Nitrogen adsorption isotherms of (**a**) ACM1 and ACM2; (**b**) ACC1, ACC2 at 77 K

ACM2) and $CaCl_2$ (ACC1, ACC2) at different concentrations (1% and 2% w/v) and carbonized in a carbon dioxide atmosphere at 973 K and 1173 K for 2 h.

Nitrogen adsorption isotherms for activated carbons are shown in Fig. 13.2. In Fig. 13.2a, the $ACM1_{1173}$ and $ACM2_{1173}$ carbons essentially exhibit type I adsorption isotherms, according to the IUPAC classification (Thommes et al. 2015), characteristic of microporous solids, due to the fact that the hysteresis loop formed in desorption is small, which is related to a low amount of mesopores, in this figure the isotherms of the $ACM1_{973}$ and $ACM2_{973}$, carbons were omitted, this because these samples presented a very low nitrogen adsorption and the isotherms were not clearly identified with any of the types according to the IUPAC classification, which suggests that the collapse or destruction of the porous structure occurred during the production process under the conditions used (Moreno-Marenco et al. 2019). Conversely, the samples $ACC2_{973}$ and $ACC2_{1173}$ (Fig. 13.2b) show a behavior composed of the type I and II isotherms, showing a more pronounced adsorption at low relative pressures ($P/P_o < 0.1$) corresponding to the filling of the micropores, while at higher pressures ($P/P_o > 0.2$) there is an increase in the slope corresponding to capillary condensation, accompanied by an H4 type hysteresis loop associated with solids whose pore size distribution is mainly in the range of micropores (Rouquerol et al. 2014a), although it is also characteristic of micro-mesoporous carbons (Thommes et al. 2015).

It should be noted that the coals obtained at 1173 K exhibit greater nitrogen adsorption, which suggests that a greater porosity develops at this temperature than at 973 K. These results are attributed to the strong dependence of the gasification reaction (C-CO_2) with temperature, as shown by Guo et al. (Guo and Lua 2002) in the characterization of an activated carbon prepared from palm kernel by activation with CO_2. Regarding the effect of the concentration of the activator, two behaviors are observed. On the one hand, in the series at 973 K, the increase in calcium concentration generates an increase in the nitrogen adsorption capacity, while in the 1173 K series, the increase in concentration, independent of salt, produces a decrease in the nitrogen adsorption capacity of solids and their textural characteristics.

The textural characteristics obtained from the adsorption isotherms of N_2 at 77 K (Table 13.5) from the BET and Dubinin-Radushkevich models show solids with BET surface areas (S_{BET}) between 2 and 392 $m^2\ g^{-1}$, volumes of micropores (V_o) between 0.001 and 0.14 $cm^3\ g^{-1}$, and mesopore volumes (V_{meso}) between 0.005 and 0.073 $cm^3\ g^{-1}$ for the samples activated at 973 K. Surface areas between 608 and 1370 $m^2\ g^{-1}$, and micropore volumes between 0.24 and 0.54 $cm^3\ g^{-1}$and mesopore volumes (V_{meso}) between 0.028 and 0.065 $cm^3\ g^{-1}$ for the samples activated at 1173 K.

Decrease in the concentration of salts of impregnation reduces the capacity of nitrogen adsorption and therefore the textural parameters of activated carbons at 973 K. Considering that the catalytic action of the activating agent increases with

Table 13.5 Textural parameters of activated carbons from N_2 isotherms at 77 K

	BET	DR		
Activated carbon	S_{BET} ($m^2\ g^{-1}$)	$V_{T\ 0.99}$ ($cm^3\ g^{-1}$)	V_o ($cm^3\ g^{-1}$)	V_{meso} ($cm^3\ g^{-1}$)
$ACM1_{973}$	2	0.006	0.001	0.005
$ACM2_{973}$	4	0.010	0.001	0.009
$ACC1_{973}$	156	0.082	0.056	0.026
$ACC2_{973}$	392	0.21	0.14	0.073
$ACM1_{1173}$	1087	0.44	0.41	0.030
$ACM2_{1173}$	608	0.28	0.24	0.038
$ACC1_{1173}$	1370	0.57	0.54	0.028
$ACC2_{1173}$	791	0.36	0.30	0.065

concentration, so there is a greater removal of carbon atoms from the precursor matrix, which favors the development of porosity in the material (Juárez-Galán et al. 2009; Silvestre-Albero et al. 2012; Vargas Delgadillo 2013). While at 1173 K there is an opposite effect with respect to the concentration of activating agent evidencing the decrease in the area and volume of micropore with the increase in the concentration of activating agent, although the volumes of mesoporous increase for $ACC2_{1173}$ and remain almost constant for $ACM2_{1173}$ (Fig. 13.3) suggesting that the increase in calcium concentration causes a greater removal of carbon atoms from the matrix which generates a wider porosity at the expense of microporosity and therefore the decrease of the surface area.

Also, it can be observed that the increase in the activation temperature brings an increase in the microporosity and therefore in the surface area, because gasification process is favored (Lua and Yang 2004); therefore, an increase in temperature is required to develop a highly porous structure. Regarding the nature of the impregnating salt (Fig. 13.3), it is evident that calcium activation develops mainly microporous carbons with a contribution of mesoporosity almost independently of the activation temperature, while activation with magnesium develops mainly mesoporous materials at 973 K.

In relation to the chemical characteristics of the activated carbons (Table 13.6) obtained at 973 K, a weakly basic character is presented, close to neutrality for all samples except the sample activated by magnesium $ACM1_{973}$, which has an acid

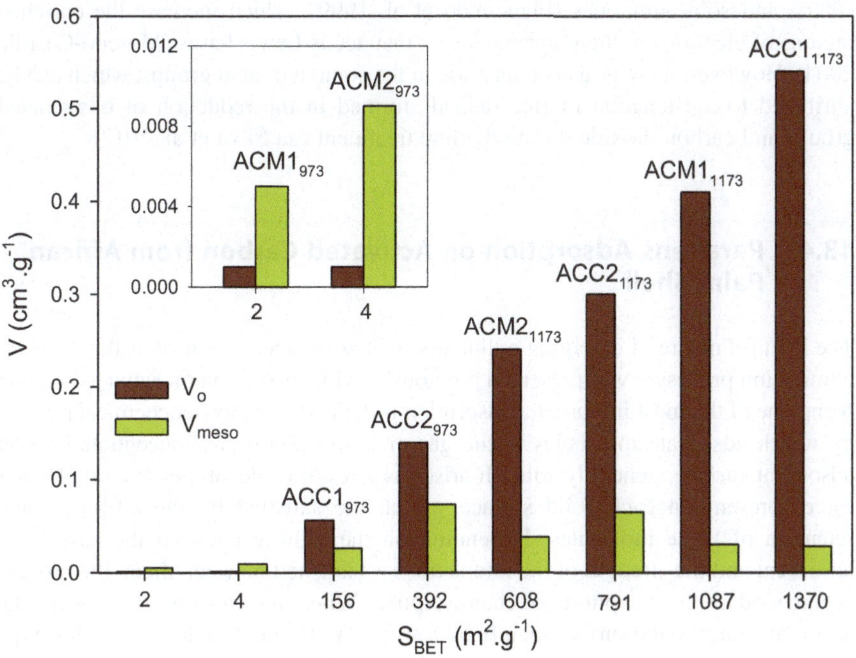

Fig. 13.3 Relation between superficial area and pore volume

Table 13.6 Chemical characterization of activated carbons

Activated carbon	Basic groups (mmol g^{-1})	Acid groups (mmol g^{-1})	Oxygen groups (mmol g^{-1})	pH_{PZC}
$ACM1_{973}$	0.014	0.46	0.48	5.6
$ACM2_{973}$	0.045	0.050	0.095	7.3
$ACC1_{973}$	0.12	0.18	0.30	7.4
$ACC2_{973}$	0.13	0.20	0.32	7.2
$ACM1_{1173}$	0.46	0.28	0.74	8.8
$ACM2_{1173}$	0.59	0.20	0.79	9.5
$ACC1_{1173}$	0.61	0.40	1.0	9.1
$ACC2_{1173}$	0.62	0.24	0.86	9.2

character, related to the presence of carboxylic acids, anhydrides, lactones, phenols, and carbonyl compounds (Boehm 2002). While at 1173 K the carbons have a greater basic character, which is related to the presence of oxygenated groups with structures such as pyrone and chromene (Boehm 2008), as well as enriched regions of delocalized π electrons in the graphene layers that act as bases of Lewis (Moreno-Castilla 2004), these being the major contributors to the basicity of activated carbon (Thommes et al. 2012). It should also take in mind that heat treatment eliminates surface functional groups that are susceptible to decomposition with temperature, such as carboxylic acids from 373 to 673 K, lactones and anhydrides from 463 to 900 K, phenols from 873 to 973 K, and at higher temperatures carbonyls, phenols, ethers, and some quinones (Figueiredo et al. 1999), which increase the enriched regions of electrons in the graphene layers that act as Lewis bases (Moreno-Castilla 2004). However, there is also an increase in the content of acid groups, which can be attributed to the reaction of free radicals formed in the reduction of oxygenated groups and carbon dioxide during thermal treatment (da Silva et al. 2017).

13.4 Parabens Adsorption on Activated Carbon from African Palm Shell

The harmful nature of emerging pollutants such as parabens, as well as the different elimination processes was presented previously, with adsorption on activated carbon being one of the most important. Adsorption is defined as a physicochemical process by which adsorbate molecules in the gas or liquid phase are concentrated on an adsorbent surface, generally solid. It arises as a result of decompensated molecular forces present on each solid surface, which are satisfied by the attraction and retention of those molecules. Depending on the affinity between the adsorbate-adsorbent and the strength of the interactions established between them, the process is classified as physisorption and chemisorption. When the interactions are weak, the adsorbate binds to the surface mainly by Van der Waals and London forces, this type of adsorption is nonspecific and occurs in any adsorbate-adsorbent system. Conversely, when the interactions are strong, the exchange of electrons is generated between the adsorbate molecules and the adsorbent surface, that is, it is carried out between the functional groups of the adsorbent and the pollutant (Moreno-Castilla 2004).

Adsorption isotherms are an experimental tool to diagnose the nature of adsorption processes and evaluate the adsorption capacity of activated carbons with a particular molecule. Below are the adsorption isotherms of the AC-PB systems studied (Figs. 13.4 and 13.5), which were measured by varying the initial concentration of each of the parabens in a range between 20 and 200 mg L^{-1} at 291 K. The adsorption isotherms obtained for the AC-MePB systems are shown in Fig. 13.4 and for the AC-PrPB systems they are shown in Fig. 13.5. In the curves obtained with the 1173 series carbons (Figs. 13.4b and 13.5b) it is observed that the adsorption of PBs increased abruptly, while with the carbons of the 973 series (Figs. 13.4a and 13.5a) the increases were moderate. This behavior is associated with the diffusion of PBs molecules to the micropores through the larger pores (meso- and macropores) and once the micro- and mesopores have been occupied, the possibility of the PB molecule to find an active site in which it can be retained (Sotelo et al. 2012). By increasing the concentration at equilibrium, progressive AC saturation occurs, which is evidenced as a plateau that is less pronounced in the 973 series.

The experimental data obtained from the adsorption isotherms were adjusted to the mathematical models of Langmuir and Freundlich. The Langmuir model assumes that the surface of the adsorbent is energetically homogeneous and that adsorption is a chemical process in which the coverage of the adsorbent surface occurs by formation of a monolayer, while the Freundlich model is appropriate to describe the adsorption of heterogeneous systems, with the possibility of intermolecular interactions between adsorbate molecules, regardless of the saturation of the adsorption surface and, therefore, indicates the appearance of physisorption (Shahbeig et al. 2013).

The resulting parameters of the models are summarized in Table 13.7. When comparing the fit to the models with all the systems (AC-PB) it is observed that they present a better fit to the Langmuir model, therefore considering the principles of this model suggests the formation of homogeneous energetic interactions with the basal planes of activated carbons, obtaining adsorption capacities (Q_{mL}) of 199.6 mg g^{-1} and 247.1 mg g^{-1} for MePB and PrPB, respectively, in both cases with carbon $ACC1_{1173}$. Conversely, the Langmuir adsorption coefficient (K_L) is related to the apparent adsorption energy, showing a lower favorability for the adsorption of MePB, probably due to the weak interaction between these molecules and the surface of the activated carbon. The Freundlich model was chosen to estimate the adsorption intensity (n_F) of the adsorbate on the adsorbent surface and the favorability of the process. The values obtained ($n_F > 2$) indicate that both, MePB and PrPB are adsorbed on the activated carbons, but it is particularly favorable for $ACC1_{1173}$ and $ACM1_{1173}$ with PrPB, which is related to the increase in affinity for the possible mechanisms heterogeneous which leads to a strong interaction between these carbons and PrPB. Similar results have been reported in the methylparaben and propylparaben adsorption (de los Ángeles Bernal-Romero et al. 2019; Mashile et al. 2020).

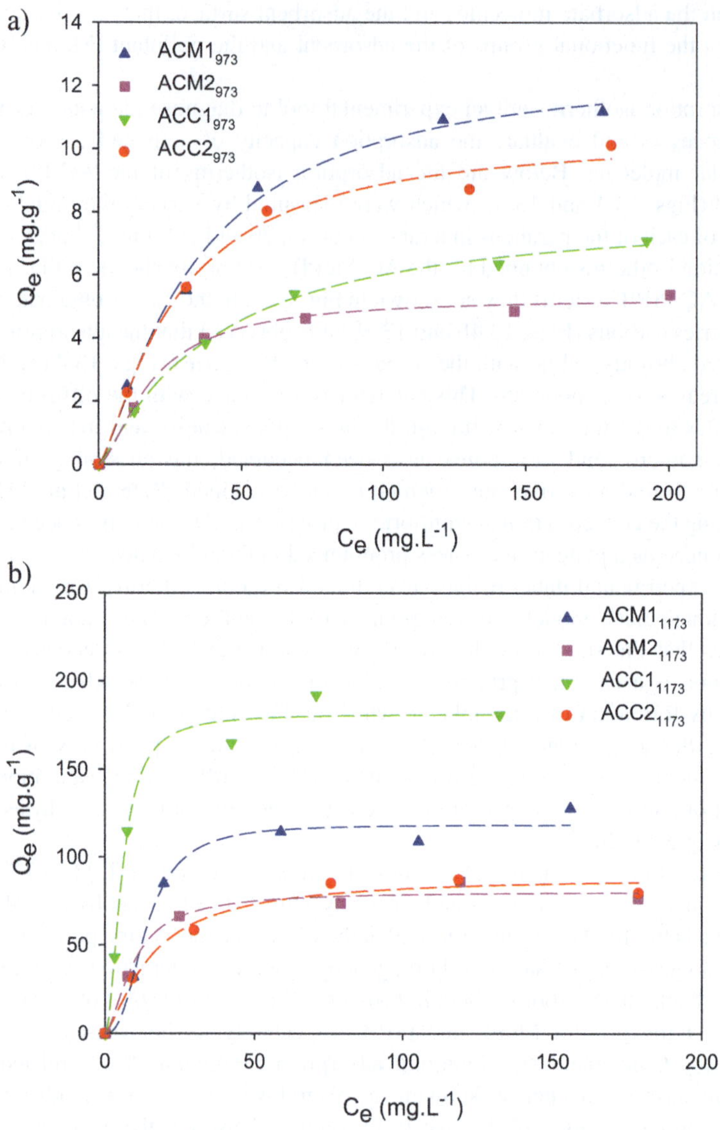

Fig. 13.4 Methylparaben adsorption isotherms of activated carbons obtained at (**a**) 973 K and (**b**) 1173 K

The adsorption process of organic contaminants is related to the porous structure of activated carbon because typically micropores are considered active sites of adsorption. In Fig. 13.6 the relationship between the adsorption capacity and the micropore volume of the activated carbons is shown, where the increase in the amount of paraben adsorbed is observed as the micropore volume increases. This

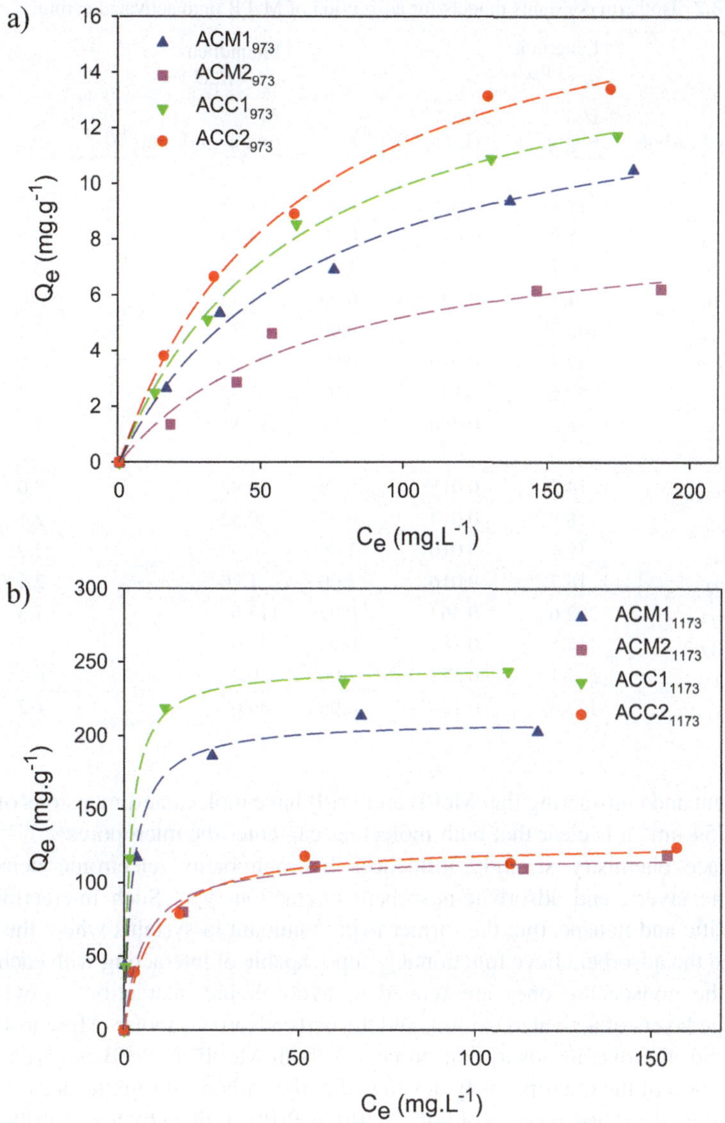

Fig. 13.5 Propylparaben adsorption isotherms of activated carbons obtained at (**a**) 973 K and (**b**) 1173 K

indicates that the microporous structure of activated carbons favors adsorption, due to the improved adsorption potential produced by the effect of the proximity of the adjacent pore walls (Hadi Madani et al. 2016) that enables dispersive π–π interactions with parabens. Likewise, it is important to mention that the adsorption capacity will depend on the accessibility of paraben to the internal surface of the

Table 13.7 Isotherm constants models for adsorption of MePB onto activated carbons

Activated carbon	Langmuir $q_e = \frac{Q_{mL} K_L C_e}{1+K_L C_e}$ Q_{mL} (mg g^{-1})	K_L (L mg^{-1})	r^2	Freundlich $q_e = K_F C_e^{1/nF}$ K_F (mg g^{-1}) (L mg^{-1})$^{n-1}$	n_F	r^2
MePB						
$ACM1_{973}$	14.4	0.023	0.99	1.37	2.4	0.96
$ACM2_{973}$	5.91	0.045	0.99	1.23	3.5	0.95
$ACC1_{973}$	8.7	0.021	1.00	0.81	2.4	0.98
$ACC2_{973}$	11.7	0.030	0.99	1.42	2.6	0.96
$ACM1_{1173}$	140.7	0.050	0.90	27.3	3.2	0.91
$ACM2_{1173}$	107.4	0.070	0.99	27.9	4.6	0.93
$ACC1_{1173}$	199.6	0.14	0.98	57.3	3.9	0.92
$ACC2_{1173}$	96.2	0.060	0.96	22.3	3.7	0.93
PrPB						
$ACM1_{973}$	14.2	0.015	0.99	0.82	2.0	0.99
$ACM2_{973}$	8.8	0.014	0.97	0.52	2.1	0.94
$ACC1_{973}$	16.1	0.016	1.00	0.98	2.0	0.98
$ACC2_{973}$	18.7	0.016	1.00	1.20	2.1	0.99
$ACM1_{1173}$	202.6	0.463	0.97	113.6	7.3	0.91
$ACM2_{1173}$	118.5	0.181	0.98	41.0	4.6	0.97
$ACC1_{1173}$	247.1	0.572	1.00	120.4	6.1	0.93
$ACC2_{1173}$	127.6	0.124	0.99	39.0	4.2	0.96

adsorbent and considering that MePB and PrPB have molecular dimensions of 0.406 and 0.754 nm^2 it is clear that both molecules can enter the micropores.

Surface chemistry strongly influences hydrophobicity, electronic density of graphene layers, and adsorbate-adsorbent interaction type. Such interactions can be specific and nonspecific, the former is predominant in systems where the adsorbate and the adsorbent have functional groups capable of interacting with each other, while the nonspecific ones are related to hydrophobic interactions between the graphene layers of activated carbon. and the parts adsorbate apolar (Moreno-Castilla 2004). So, the overall adsorption process of both MePB as PrPB is given by the contribution of the microporosity developed in the carbons during the activation. So the overall adsorption process of both MePB as PrPB is given by the contribution of the microporosity developed in the carbons during the activation treatments, as well as by the establishment of specific interactions. Depending on the characteristics of each system, both can contribute as observed for $ACC1_{1173}$ (Fig. 13.6a, b) or one of them prevails as in the case of $ACM1_{973}$ (Fig. 13.6a), where MePB adsorption is related to the higher content of oxygenated groups rather than microporosity which in this solid is very poorly developed, but there is a greater adsorption compared to the other carbons of the 973 series.

The oxygenated groups present on the surface of activated carbons can be acidic or basic in nature. As mentioned above the acid character is associated with the

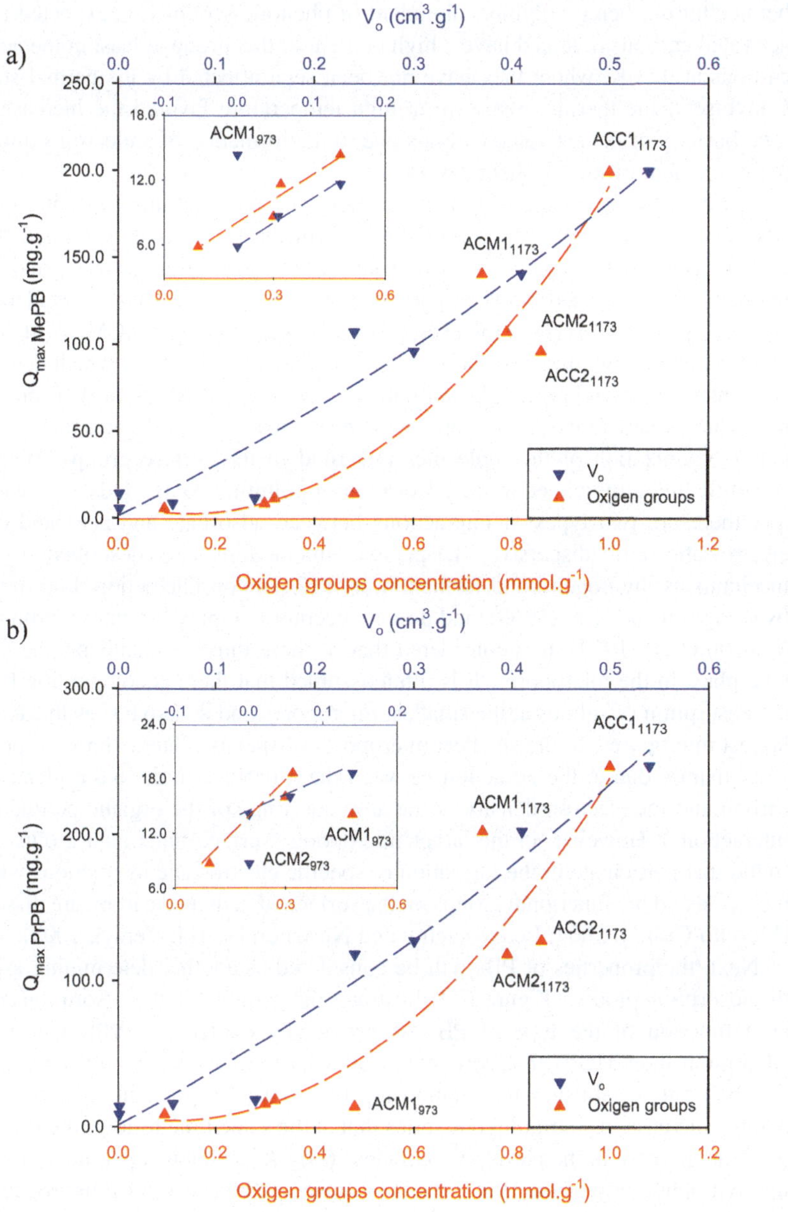

Fig. 13.6 Relation between oxygen groups concentration and micropore volume with adsorption capacity of (**a**) MePB and (**b**) PrPB

presence of carboxylic acids, lactones, phenols, and anhydrides, while the basic character is associated with the pyrone, chromene, ether, quinone, and carbonyl groups. Given that African palm shell is constituted by lignin, cellulose, and hemicellulose, being polymers abundant in phenolic groups it is expected that the activated carbons obtained have a high content in this group at least in the activated carbons at 973 K, where they have not been seen affected by its thermal stability. Conversely, the thermal treatment at high temperature favored the increase in the total basicity of the activated carbons due to the formation of π electrons during the rearrangement of the graphene layers.

In Fig. 13.7 the relationship between acid and basic groups concentration with adsorption capacity of PBs is presented. From these results it is evident that the surface chemistry plays a relevant role in the adsorption process and in the interactions established between parabens and activated carbons, as evidenced by $ACC1_{1173}$ and $ACM1_{1173}$. However, it is noteworthy that with $ACM1_{973}$ it does not manifest itself with high adsorption when compared to $ACC1_{1173}$ which has a lower concentration of acid groups, because the process with $ACM1_{973}$ is also conditioned by textural characteristics associated with the decrease in surface area that prevents a better organization of the molecules that bind to the surface groups. Within the scientific literature related to the adsorption of polluting compounds, it is suggested that there are two types of interactions between adsorbate and activated carbon: electrostatic and dispersive. Dispersive interactions are described by three mechanisms: hydrogen bond formation, π–π dispersion interaction, both proposed by Coughlin and Ezra (1968), and donor–acceptor complex formation, proposed by Mattson et al. (1969). It is considered that of these three mechanisms, the last two take place in the micropores. It is often assumed that there is competition between the adsorption of solutes at the smallest micropores and at active sites located at the largest micropores. In the smallest micropores, dispersive interactions are predominant, mainly due to the attraction between the π orbitals in the basal planes of the carbon and the electron density in the aromatic rings of the organic pollutant (π–π interactions). However, in the larger micropores, surface functional groups can be found that participate in the formation of specific electrostatic interactions when the molecule and the functional groups on the surface of activated carbon are dissociated (Moreno-Castilla 2004; Podkościelny and Nieszporek 2011; Terzyk 2004).

Next, the properties of PBs will be considered as another determining aspect in the adsorption process. Figure 13.8 illustrates the variation in the adsorption capacity as a function of the type of PB. As previously evidenced, PrPB has a higher adsorption than MePB that correlates well with the increase in molecular weight. Traube's rule establishes the relationship between adsorption in aqueous solution and the increase in a homologous series that at the same time follows the same trend with the increase in the partition coefficient (Log K_{ow}), being a parameter related to the hydrophobicity of the PBs. This is because parabens can form hydrogen bonds with water molecules in solution, but as the size of the paraben molecule increases, its hydrophobicity increases, which generates the repellency of water, but favors interactions of dispersive type between the π electrons of the aromatic ring of paraben and those of the aromatic structure of activated carbon (Hamdaoui and Naffrechoux 2007). These results are in agreement with other studies of paraben adsorption on various adsorbents (Chen et al. 2017; Chin 2013; Chin et al. 2010). Conversely, the Lundelius rule establishes an inverse dependence between the

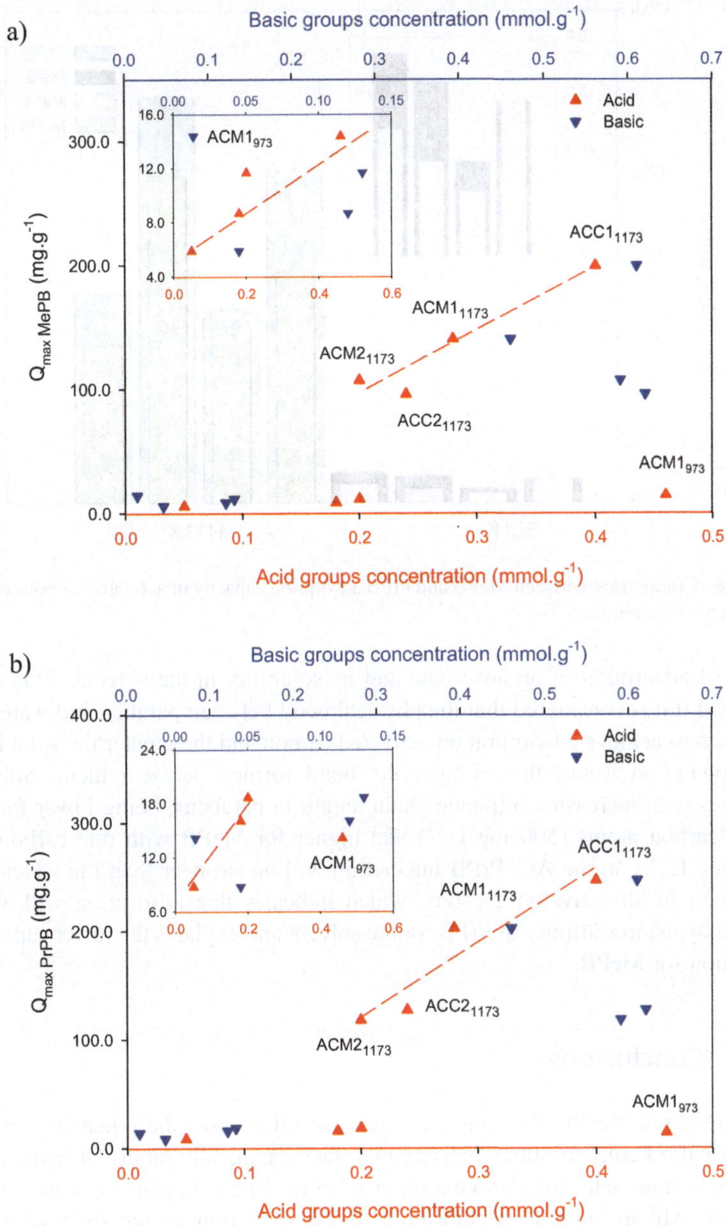

Fig. 13.7 Relation between acid and basic groups concentration with adsorption capacity of (**a**) MePB and (**b**) PrPB

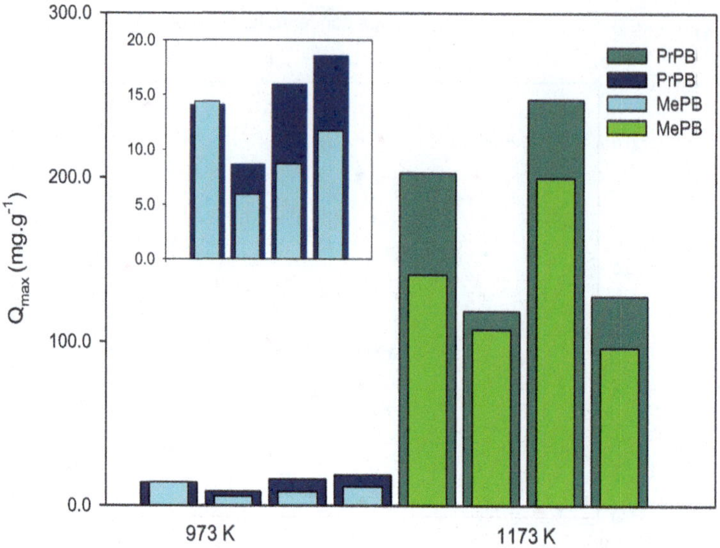

Fig. 13.8 Comparison between MePB and PrPB adsorption capacity of activated carbons obtained at different temperatures

degree of adsorption of an adsorbate and its solubility in the solvent. This can be explained if it is considered that the physical bond between paraben and water must be broken to achieve adsorption on activated carbon and the greater the solubility of a compound in water, the stronger the bond formed between them. Solubility decreases with increasing aliphatic chain length in parabens, being lower for PrPB with 3 carbon atoms (500 mg L^{-1}) and higher for MePB with one carbon atom (2500 mg L^{-1}), so the AC–PrPB interaction will be stronger than the water–PrPB interaction in all activated carbons, which indicates that adsorption will also be induced by the low affinity of PrPB for the solvent and explains the lower capacity of adsorption for MePB.

13.5 Conclusions

In this research, the main results were presented that show the potential use of the African palm kernel to obtain activated carbons chemically modified with calcium and magnesium salts for the adsorption of emerging pollutants in water such as parabens, which could also contribute to the reduction in the disposal of this by-product generated in the obtaining of palm oil.

The differences in the textural characteristics found among the activated carbons show that the concentration of activating agent and the activation temperature are determining factor in the preparation of the activated carbons, because the modifications on the surface in the samples were greater with the increase in temperature and decrease in the concentration of impregnating agent, particularly in the series activated at 1173 K. This means that for a low salt concentration, a high temperature favors the gasification process, which is more catalyzed by the presence of calcium. While at high salt concentration, the gasification process is favored when the temperature decrease represents a factor that contributes to the formation of a porous structure. Similarly, it was evidenced that the greatest effect on adsorption is related to the concentration of the impregnating salt during preparation, rather than to the nature of the metal salt. So, the selection of the synthesis conditions will continue to be the most important factor to tailoring the final microstructure, textural properties, adsorption capacity, and surface chemistry of the produced carbons.

Although it is noteworthy that this greater interaction and therefore greater adsorption is given by the contribution of surface chemistry and microporosity developed in the carbons due to the effect of metal salts and thermal treatments, the greater adsorption of PrPB respect to MePB is correlated with the increase of hydrophobicity (log K_{ow}), molecular weight, and decrease in water solubility of PrPB, where dispersive-type interactions are favored between the π electrons of the aromatic ring of paraben and those of the aromatic structure of activated carbon.

Acknowledgments The authors thank the Framework Agreement between the Universidad Nacional de Colombia and the Universidad de Los Andes and the act of agreement established between the Chemistry Departments of the two universities. They also wish to thank the Colciencias Scholarship "Doctorados Nacionales 2015" Convocation 727 and the funding according to the project "Thermodynamic study of the parabens adsorption from aqueous solution on activated carbons modified with metallic salts" FP44842-135-2017. Prof. Dr. Juan Carlos Moreno-Piraján also appreciates the support provided by the Vice-rectory of research of the Universidad de los Andes (Bogotá, Colombia) with the program "Convocatoria de Proyectos de investigación y Creación – Nuevo como Regionalización" number INV-2019-91-1905.

References

Abdullah N, Sulaiman F (2013) The oil palm wastes in Malaysia. In: Matovic MD (ed) Biomass now—sustainable growth and use. Queen's University, Kingston, ON, pp 75–100. https://doi.org/10.5772/55302

Abu Hasan H, Farooq Alshekhli A, Muhamad MH, Sheikh Abdullah SR (2020) Development of adsorbent from phytoremediation plant waste for methylene blue removal. J Ecol Eng 21(8):207–215. https://doi.org/10.12911/22998993/126

Aburub A, Wurster DE (2006) Phenobarbital interactions with derivatized activated carbon surfaces. J Colloid Interface Sci 296:79–85. https://doi.org/10.1016/j.jcis.2005.08.035

Acevedo S, Giraldo L, Moreno-Piraján JC (2017) Adsorption of CO2 onto activated carbons prepared by chemical activation with metallic salts. Int J Chem React Eng 15:1–11. https://doi.org/10.1515/ijcre-2017-0029

Álvarez MA, Ruidíaz-Martínez M, Cruz-Quesada G, López-Ramón MV, Rivera-Utrilla J, Sánchez-Polo M, Mota AJ (2020) Removal of parabens from water by UV-driven advanced oxidation processes. Chem Eng J 379:122334. https://doi.org/10.1016/j.cej.2019.122334

Andersen FA (2008) Final amended report on the safety assessment of methylparaben, ethylparaben, propylparaben, isopropylparaben, butylparaben, isobutylparaben and benzylparaben as used in cosmetic products. Int J Toxicol 27(4):1–82. https://doi.org/10.1080/10915810802548359

Andersen DN, Larsen PB (2013) Survey of parabens (Part of the LOUS-review Environmental Project No. 1474 No. Environmental Project No. 1474, 2013). Copenhagen. http://www2.mst.dk/Udgiv/publications/2013/04/978-87-93026-02-5.pdf

Angelov T, Vlasenko A, Tashkov W (2008) HPLC determination of pKa of parabens and investigation on their lipophilicity parameters. J Liq Chromatogr Relat Technol 31:188–197. https://doi.org/10.1080/10826070701738787

Arami-Niya A, Wan Daud WMA, Mjalli S, Abnisa F, Shafeeyan MS (2012) Production of microporous palm shell based activated carbon for methane adsorption: modeling and optimization using response surface methodology. Chem Eng Res Des 90(6):776–784. https://doi.org/10.1016/j.cherd.2011.10.001

Attia A, Girgis B, Fathy N (2008) Removal of methylene blue by carbons derived from peach stones by H3PO4 activation: batch and column studies. Dyes Pigments 76:282–289. https://doi.org/10.1016/j.dyepig.2006.08.039

Bandosz TJ, Ania CO (2006) Surface chemistry of activated carbons and its characterization. In: Bandosz TJ (ed) Activated carbon surfaces in environmental remediation, vol 7, 1st edn. Elsevier, Amsterdam, pp 159–229. https://doi.org/10.1016/S1573-4285(06)80013-X

Błędzka D, Gromadzińska J, Wąsowicz W (2014) Parabens. From environmental studies to human health. Environ Int 67:27–42. https://doi.org/10.1016/j.envint.2014.02.007

Boehm H (2002) Surface oxides on carbon and their analysis: a critical assessment. Carbon 40:145–149

Boehm H-P (2008) Surface chemical characterization of carbons from adsorption studies. In: Adsorption by carbons. Elsevier, Amsterdam, pp 301–323

Bolong N, Ismail AF, Salim MR, Matsuura T (2009) A review of the effects of emerging contaminants in wastewater and options for their removal. Desalination 238(1–3):229–246. https://doi.org/10.1016/j.desal.2008.03.020

Brand W, Boon PE, Hessel EVS, Meesters JAJ, Weda M, Schuur AG (2017) Exposure to and toxicity of methyl-, ethyl-and propylparaben, Netherlands. https://doi.org/10.21945/RIVM-2017-0028

Brausch JM, Rand GM (2015) A review of personal care products in the aquatic environment: environmental concentrations and toxicity. Chemosphere 82:1518–1532. https://doi.org/10.1016/j.chemosphere.2010.11.018

Canosa P, Rodríguez I, Rubí E, Negreira N, Cela R (2006) Formation of halogenated by-products of parabens in chlorinated water. Anal Chim Acta 575:106–113. https://doi.org/10.1016/j.aca.2006.05.068

Chang HS, Choo KH, Lee B, Choi SJ (2009) The methods of identification, analysis, and removal of endocrine disrupting compounds (EDCs) in water. J Hazard Mater 172(1):1–12. https://doi.org/10.1016/j.jhazmat.2009.06.135

Chen HW, Chiou CS, Chang SH (2017) Comparison of methylparaben, ethylparaben and propylparaben adsorption onto magnetic nanoparticles with phenyl group. Powder Technol 311:426–431. https://doi.org/10.1016/j.powtec.2017.01.060

Chin YP (2013) Adsorption of parabens in aqueous solution onto β-cyclodextrin cross-linked polymer. University of Malaya, Kuala Lumpur. http://studentsrepo.um.edu.my/id/eprint/4247

Chin YP, Mohamad S, Abas M, Bin R (2010) Removal of parabens from aqueous solution using B-cyclodextrin cross-linked polymer. Int J Mol Sci 11:3459–3471. https://doi.org/10.3390/ijms11092459

Coughlin RW, Ezra FS (1968) Role of surface acidity in the adsorption of organic pollutants on the surface of carbon. Environ Sci Technol 2(4):291–297

da Silva Lacerda V, López-Sotelo JB, Correa-Guimarães A, Hernández-Navarro S, Sánchez-Báscones M, Navas-Gracia LM et al (2015) Rhodamine B removal with activated carbons obtained from lignocellulosic waste. J Environ Manage 155:67–76. https://doi.org/10.1016/j.jenvman.2015.03.007

da Silva WL, Salomão AA, Vila MM, Tubino M (2017) Influence of water and ultraviolet irradiation on the induction period of the oxidation of biodiesel. J Braz Chem Soc 28(4):676–680. https://doi.org/10.5935/0103-5053.20160202

Daud WMAW, Ali WSW (2004) Comparison on pore development of activated carbon produced from palm shell and coconut shell. Bioresour Technol 93(1):63–69. https://doi.org/10.1016/j.biortech.2003.09.015

Daud WMAW, Houshamnd AH (2010) Textural characteristics, surface chemistry and oxidation of activated carbon. J Nat Gas Chem 19(3):267–279. https://doi.org/10.1016/S1003-9953(09)60066-9

de los Ángeles Bernal-Romero M, Boluda-Botella N, Prats Rico D (2019) Removal of emerging pollutants in water treatment plants: adsorption of methyl and propylparaben onto powdered activated carbon. Adsorption 25(5):983–999. https://doi.org/10.1007/s10450-019-00120-7

De Oliveira FF, Moura KO, Costa LS, Vidal CB, Loiola AR, Nascimento RF (2020) Reactive adsorption of parabens on synthesized micro- and mesoporous silica from coal fly ash: pH effect on the modification process. ACS Omega 5(7):3346–3357. https://doi.org/10.1021/acsomega.9b03537

Delgado NY, Capparelli AL, Marino DJ, Navarro AF, Peñuela GA, Ronco AE (2016) Adsorption of pharmaceuticals and personal care products on granular activated carbon. J Surf Eng Mater Adv Technol 6:183–200. https://doi.org/10.4236/jsemat.2016.64017

Dey S, Bano F, Malik A (2019) Pharmaceuticals and personal care product (PPCP) contamination—a global discharge inventory. In: Prasad MNV, Vithanage M, Kapley A (eds) Pharmaceuticals and personal care products: waste management and treatment technology, pp 1–26. https://doi.org/10.1016/b978-0-12-816189-0.00001-9

Diamanti-Kandarakis E, Bourguignon J, Guidice L, Hauser R, Prins G, Soto A, Gore A et al (2009) Endocrine-disrupting chemicals: an Endocrine Society scientific statement. Endocr Rev 30(4):293–342. https://www.nicnas.gov.au/news-and-events/Topics-of-interest/subjects/endocrine-disrupting-chemicals

European Environment Agency (2012) The impacts of endocrine disrupters on wildlife, people and their environments. The Weybridge+15 (1996–2011) report. EEA Technical report (Vol. 2/2012). ISSN 1725-2237 ISBN 978-92-9213-307-8. https://doi.org/10.2800/41462

Farraji H, Zaman N, Tajuddin R, Faraji H (2016) Advantages and disadvantages of phytoremediation: a concise review. Int J Environ Technol Sci 2:69–75

Figueiredo J, Pereira MF, Freitas MM, Órfão JJ (1999) Modification of the surface chemistry of activated carbons. Carbon 37:1379–1389. https://doi.org/10.1016/S0008-6223(98)00333-9

Freyria FS, Geobaldo F, Bonelli B (2018) Nanomaterials for the abatement of pharmaceuticals and personal care products from wastewater. Appl Sci 8:170. https://doi.org/10.3390/app8020170

González-García P (2018) Activated carbon from lignocellulosics precursors: a review of the synthesis methods, characterization techniques and applications. Renew Sustain Energy Rev 82:1393–1414. https://doi.org/10.1016/j.rser.2017.04.117

Gryglewicz G, Lorenc-Grabowska E (2004) Mesoporous activated carbons from ca and Fe exchanged sub-bituminous and bituminous coals. Carbon 42(3):688–691. https://doi.org/10.1016/j.carbon.2004.01.001

Guo J, Lua AC (2002) Characterization of adsorbent prepared from oil-palm shell by CO2 activation for removal of gaseous pollutants. Mater Lett 55(5):334–339. https://doi.org/10.1016/S0167-577X(02)00388-9

Guo Y, Yang S, Fu W, Qi J, Li R, Wang Z, Xu H (2003) Adsorption of malachite green on micro- and mesoporous rice husk-based active carbon. Dyes Pigments 56:219–229. https://doi.org/10.1016/S0143-7208(02)00160-2

Hadi Madani S, Hu C, Silvestre-Albero A, Biggs MJ, Rodríguez-Reinoso F, Pendleton P (2016) Pore size distributions derived from adsorption isotherms, immersion calorimetry, and isosteric heats: a comparative study. Carbon 96:1106–1113. https://doi.org/10.1016/j.carbon.2015.10.072

Haman C, Dauchy X, Rosin C, Munoz JF (2015) Occurrence, fate and behavior of parabens in aquatic environments: a review. Water Res 68:1–11. https://doi.org/10.1016/j.watres.2014.09.030

Hamdaoui O, Naffrechoux E (2007) Modeling of adsorption isotherms of phenol and chlorophenols onto granular activated carbon. Part I Two-parameter models and equations allowing determination of thermodynamic parameters. J Hazard Mater 147(1–2):381–394. https://doi.org/10.1016/j.jhazmat.2007.01.021

Hauptvogl M, Kotrla M, Prčík M, Pauková Ž, Kováčik M, Lošák T (2020) Phytoremediation potential of fast-growing energy plants: challenges and perspectives—a review. Pol J Environ Stud 29(1):505–516

Hernández-Leal L, Temmink H, Zeeman G, Buisman CJN (2011) Removal of micropollutants from aerobically treated grey water via ozone and activated carbon. Water Res 45:2887–2896. https://doi.org/10.1016/j.watres.2011.03.009

Hessel EVS, Boon PE, den Braver-Sewradj SP, Meesters JAJ, Weda M, Brand W (2019) Review on butylparaben: exposure, toxicity and risk assessment. Netherlands. https://doi.org/10.21945/RIVM-2018-0161

Huang Q, Lu G, Wang J, Yu J (2011) Thermal decomposition mechanisms of MgCl2·6H2O and MgCl2·H2O. J Anal Appl Pyrolysis 91:159–164. https://doi.org/10.1016/j.jaap.2011.02.005

Iwasaki S, Fukuhara T, Abe I, Yanagi J, Mouri M, Iwashima Y, Shinohara O et al (2002) Adsorption of alkylphenols onto microporous carbons prepared from coconut shell. Synth Met 125(2):207–211. https://doi.org/10.1016/S0379-6779(01)00530-6

Juárez-Galán JM, Silvestre-Albero A, Silvestre-Albero J, Rodríguez-Reinoso F (2009) Synthesis of activated carbon with highly developed "mesoporosity.". Micropor Mesopor Mater 117:519–521. https://doi.org/10.1016/j.micromeso.2008.06.011

Jung SH, Oh SJ, Choi GG, Kim JS (2014) Production and characterization of microporous activated carbons and metallurgical bio-coke from waste shell biomass. J Anal Appl Pyrolysis 109:123–131. https://doi.org/10.1016/j.jaap.2014.07.003

Kalia VC (2019) Pharmaceutical and personal care product contamination: a global scenario. In: Vara Prasad MN, Vithanage M, Kapley A (eds) Pharmaceuticals and personal care products: waste management and treatment technology. Elsevier, Amsterdam, pp 27–61. https://doi.org/10.1016/b978-0-12-816189-0.00002-0

Katsigiannis A, Noutsopoulos C, Mantziaras J, Gioldasi M (2015) Removal of emerging pollutants through granular activated carbon. Chem Eng J 280:49–57. https://doi.org/10.1016/j.cej.2015.05.109

Kaur H, Hippargi G, Pophali GR, Bansiwal AK (2019) Treatment methods for removal of pharmaceuticals and personal care products from domestic wastewater. In: Vara Prasad MN, Vithanage M, Kapley A (eds) Pharmaceuticals and personal care products: waste management and treatment technology. Elsevier, Amsterdam, pp 129–150. https://doi.org/10.1016/b978-0-12-816189-0.00006-8

Kirsh Y, Yariv S, Shoval S (1987) Kinetic analysis of thermal dehydration and hydrolysis of MgCl2.6H2O by DTA and TG. J Thermal Anal 32:393–408. https://doi.org/10.1007/bf01912692

Kuster M, López de Alda M, Hernando M, Petrovic M, Martin-Alonso J, Barceló D (2008) Analysis and occurrence of pharmaceuticals, estrogens, progestogens and polar pesticides in sewage treatment plant effluents, river water and drinking water in the Llobregat river basin (Barcelona, Spain). J Hydrol 358:112–123. https://doi.org/10.1016/j.jhydrol.2008.05.030

Kwarciak-Kozłowska A (2019) Removal of pharmaceuticals and personal care products by ozonation, advance oxidation processes, and membrane separation. In: Vara Prasad MN, Vithanage M, Kapley A (eds) Pharmaceuticals and personal care products: waste management and treatment technology. Elsevier, Amsterdam, pp 151–171. https://doi.org/10.1016/b978-0-12-816189-0.00007-x

Lee J, Bang SH, Kim YH, Min J (2018) Toxicities of four parabens and their mixtures to *Daphnia magna* and *Aliivibrio fischeri*. Environ Health Toxicol 33:e2018018. https://doi.org/10.5620/eht.e2018018

Liu L, Sun J, Cai C, Wang S, Pei H, Zhang J (2009) Corn Stover pretreatment by inorganic salts and its effects on hemicellulose and cellulose degradation. Bioresour Technol 100(23):5865–5871. https://doi.org/10.1016/j.biortech.2009.06.048

Lua AC, Yang T (2004) Effect of activation temperature on the textural and chemical properties of potassium hydroxide activated carbon prepared from pistachio-nut shell. J Colloid Interface Sci 274:594–601. https://doi.org/10.1016/j.jcis.2003.10.001

Mackay D, Shiu WY, Ma K, Lee SC (2006) Handbook of physical-chemical properties and environmental fate for organic chemicals, 2nd edn. CRC, Taylor & Francis Group, Boca Raton, FL. https://doi.org/10.1201/9781420044393

Mailler R, Gasperi J, Coquet Y, Deshayes S, Zedek S, Cren-Olivé C et al (2014) Study of a large scale powdered activated carbon pilot: removals of a wide range of emerging and priority micropollutants from wastewater treatment plant effluents. Water Res 72:315–330. https://doi.org/10.1016/j.watres.2014.10.047

Mallek M, Chtourou M, Portillo M, Monclús H, Walha K, Ben Salah A, Salvadó V (2018) Granulated cork as biosorbent for the removal of phenol derivatives and emerging contaminants. J Environ Manage 223(June):576–585. https://doi.org/10.1016/j.jenvman.2018.06.069

Marsh H, Rodríguez-Reinoso F (2006) Activated carbon. Elsevier, Amsterdam. https://doi.org/10.1016/B978-008044463-5/50018-2

Martin O, Kortenkamp A (2009) State of the art assessment of endocrine disrupters. Comparative analysis of endpoints and assays by human health and wildlife endpoint. European Commission. http://ec.europa.eu/environment/chemicals/endocrine/pdf/annex3_comparative_analysis.pdf

Mashile GP, Mpupa A, Nqombolo A, Dimpe KM, Nomngongo PN (2020) Recyclable magnetic waste Tyre activated carbon-chitosan composite as an effective adsorbent rapid and simultaneous removal of methylparaben and propylparaben from aqueous solution and wastewater. J Water Process Eng 33(July 2019):101011. https://doi.org/10.1016/j.jwpe.2019.101011

Masten SA (2005) Butylparaben. Review of toxicological literature butylparaben. http://ntp.niehs.nih.gov/ntp/htdocs/chem_background/exsumpdf/butylparaben_508.pdf

Mattson JS, Mark HB, Malbin MD, Weber WJ, Crittenden JC (1969) Surface chemistry of active carbon: specific adsorption of phenols. J Colloid Interface Sci 31:116–130. https://doi.org/10.1016/0021-9797(69)90089-7

Molina-Sabio M, Rodríguez-Reinoso F (2004) Role of chemical activation in the development of carbon porosity. Colloids Surf A Physicochem Eng Asp 241:15–25. https://doi.org/10.1016/j.colsurfa.2004.04.007

Molina-Sabio M, Pérez V, Rodríguez-Reinoso F (1994) Impregnation of activated carbon with chromium and copper salts: effect of porosity and metal content. Carbon 32(7):1259–1265. https://doi.org/10.1016/0008-6223(94)90111-2

Mondal P, Majumder CB, Mohanty B (2007) Removal of trivalent arsenic (as(III)) from contaminated water by calcium chloride (CaCl2)-impregnated rice husk carbon. Ind Eng Chem Res 46(8):2550–2557. https://doi.org/10.1021/ie060702i

Moreno-Castilla C (2004) Adsorption of organic molecules from aqueous solutions on carbon materials. Carbon 42(1):83–94. https://doi.org/10.1016/j.carbon.2003.09.022

Moreno-Marenco AR, Giraldo L, Moreno-Piraján JC (2019) Relación entre la capacidad de adsorción y la entalpía de inmersión de carbones activados modificados químicamente en soluciones acuosas de metilparabeno. Afinidad 587(September):213–220. https://www.raco.cat/index.php/afinidad/article/view/361481

Morone A, Mulay P, Kamble SP (2019) Removal of pharmaceutical and personal care products from wastewater using advanced materials. In: Vara Prasad MN, Vithanage M, Kapley A (eds) Pharmaceuticals and personal care products: waste management and treatment technology. Elsevier, Amsterdam, pp 173–212. https://doi.org/10.1016/b978-0-12-816189-0.00008-1

Niemuth N, Klaper R (2015) Emerging wastewater contaminant metformin causes intersex and reduced fecundity in fish. Chemosphere 135:38–45. https://doi.org/10.1016/j.chemosphere.2015.03.060

Nizamuddin S, Jayakumar NS, Sahu JN, Ganesan P, Bhutto AW, Mubarak NM (2015) Hydrothermal carbonization of oil palm shell. Korean J Chem Eng 32(9):1789–1797. https://doi.org/10.1007/s11814-014-0376-9

Nowak K, Ratajczak-Wrona W, Górska M, Jabłońska E (2018) Parabens and their effects on the endocrine system. Mol Cell Endocrinol 474:238–251. https://doi.org/10.1016/j.mce.2018.03.014

OW/ORD Emerging Contaminants Workgroup (2008) Aquatic life criteria for contaminants of emerging concern. Part I General challenges and recommendations. US-EPA. https://www.epa.gov/sites/production/files/2015-08/documents/white_paper_aquatic_life_criteria_for_contaminants_of_emerging_concern_part_i_general_challenges_and_recommendations_1.pdf

Pal A, He Y, Jekel M, Reinhard M, Gin KYH (2014) Emerging contaminants of public health significance as water quality indicator compounds in the urban water cycle. Environ Int 71:46–62. https://doi.org/10.1016/j.envint.2014.05.025

Pamidimukkala PS, Soni H (2018) Efficient removal of organic pollutants with activated carbon derived from palm shell: spectroscopic characterisation and experimental optimisation. J Environ Chem Eng 6(2):3135–3149. https://doi.org/10.1016/j.jece.2018.04.013

Podkościelny P, Nieszporek K (2011) Adsorption of phenols from aqueous solutions: equilibria, calorimetry and kinetics of adsorption. J Colloid Interface Sci 354(1):282–291. https://doi.org/10.1016/j.jcis.2010.10.034

Pugazhendhi D, Pope GS, Darbre PD (2005) Oestrogenic activity of p-hydroxybenzoic acid (common metabolite of paraben esters) and methylparaben in human breast cancer cell lines. J Appl Toxicol 25(4):301–309. https://doi.org/10.1002/jat.1066

Rajapaksha A, Vithanage M, Lim J, Ahmed M, Zhang M, Lee S, Ok YS (2014) Invasive plant-derived biochar inhibits sulfamethazine uptake by lettuce in soil. Chemosphere 111:500–504. https://doi.org/10.1016/j.chemosphere.2014.04.040

Ran J, Li M, Zhang C, Xue F, Tao M, Zhang W (2020) Synergistic adsorption for parabens by an amphiphilic functionalized polypropylene fiber with tunable surface microenvironment. ACS Omega 5:2920–2930. https://doi.org/10.1021/acsomega.9b03765

Rashidi NA, Yusup S (2017) Potential of palm kernel shell as activated carbon precursors through single stage activation technique for carbon dioxide adsorption. J Clean Prod 168:474–486. https://doi.org/10.1016/j.jclepro.2017.09.045

Recommendations Report: Contaminants of Emerging Concern Workgroup (2019) Arlington, Virginia. https://www.asdwa.org/wp-content/uploads/2019/05/ASDWA-ACWA-Report-on-Contaminants-of-Emerging-Concern-2019.pdf

Rodriguez-Reinoso F, Molina-Sabio M (1998) Textural and chemical characterization of microporous carbons. Adv Colloid Interface Sci 76–77:271–294. https://doi.org/10.1016/S0001-8686(98)00049-9

Rongti L, Wei P, Sano M, Li J (2002) Kinetics of reduction of magnesia with carbon. Thermochim Acta 390:145–151. https://doi.org/10.1016/S0040-6031(02)00324-6

Rossner A, Snyder SA, Knappe DRU (2009) Removal of emerging contaminants of concern by alternative adsorbents. Water Res 43(15):3787–3796. https://doi.org/10.1016/j.watres.2009.06.009

Rouquerol F, Rouquerol J, Sing KSW (2014a) Assessment of mesoporosity. In: Adsorption by powders and porous solids, 2nd edn. Elsevier, Amsterdam, pp 191–213. https://doi.org/10.1016/B978-0-12-598920-6.50008-7

Rouquerol F, Rouquerol J, Sing KSW, Llewellyn P, Maurin G (2014b) Adsorption by powders and porous solids, 2nd edn. Elsevier, Amsterdam. https://doi.org/10.1017/CBO9781107415324.004

Rufford TE, Hulicova-Jurcakova D, Zhu Z, Lu GQ (2010) A comparative study of chemical treatment by FeCl3, MgCl2, and ZnCl2 on microstructure, surface chemistry, and double-

layer capacitance of carbons from waste biomass. J Mater Res 25(8):1451–1459. https://doi.org/10.1557/jmr.2010.0186
Ruiz R, Romero HM (2011) The growth of the oil palm industry in Colombia (April 2016)
Ruiz H, Zambrano M, Giraldo L (2015) Production and characterisation of activated carbon from oil-palm shell for carboxylic acid. Orient J Chem 31(2):753–762. https://doi.org/10.13005/ojc/310217
Ryder G (2017) The United Nations world water development report, 2017: wastewater: the untapped resource. France. https://unesdoc.unesco.org/ark:/48223/pf0000247153
Sanderson H, Fricker C, Brown S, Majury A, Liss S (2016) Antibiotic resistance genes as an emerging environmental contaminant. Environ Rev 24:205–218. https://doi.org/10.1139/er-2015-0069
Schriks M, Heringa MB, van der Kooi MME, de Voogt P, van Wezel AP (2010) Toxicological relevance of emerging contaminants for drinking water quality. Water Res 44(2):461–476. https://doi.org/10.1016/j.watres.2009.08.023
Shahbeig H, Bagheri N, Ghorbanian SA, Hallajisani A, Poorkarimi S (2013) A new adsorption isotherm model of aqueous solutions on granular activated carbon. World J Model Simul 9(4):243–254
Silvestre-Albero A, Gonçalves M, Itoh T, Kaneko K, Endo M, Thommes M et al (2012) Well-defined mesoporosity on lignocellulosic-derived activated carbons. Carbon 50:66–72. https://doi.org/10.1016/j.carbon.2011.08.007
Sotelo JL, Rodríguez AR, Mateos MM, Hernández SD, Torrellas SÁ, Rodríguez JG (2012) Adsorption of pharmaceutical compounds and an endocrine disruptor from aqueous solutions by carbon materials. J Environ Sci Health Pt B Pest Food Contam Agric Wastes 47(7):640–652. https://doi.org/10.1080/03601234.2012.668462
Tan I, Ahmad A, Hameed B (2008) Adsorption of basic dye using activated carbon prepared from oil palm shell: batch and fixed bed studies. Desalination 225:13–28. https://doi.org/10.1016/j.desal.2007.07.005
Tay KS, Rahman NA, Abas M, Bin R (2010a) Kinetic studies of the degradation of parabens in aqueous solution by ozone oxidation. Environ Chem Lett 8:331–337. https://doi.org/10.1007/s10311-009-0229-7
Tay KS, Rahman NA, Abas M, Bin R (2010b) Ozonation of parabens in aqueous solution: kinetics and mechanism of degradation. Chemosphere 81(11):1446–1453. https://doi.org/10.1016/j.chemosphere.2010.09.004
Terzyk AP (2004) Molecular properties and intermolecular forces-factors balancing the effect of carbon surface chemistry in adsorption of organics from dilute aqueous solutions. J Colloid Interface Sci 275:9–29. https://doi.org/10.1016/j.jcis.2004.02.011
Thommes M, Cychosz KA, Neimark AV (2012) Advanced physical adsorption characterization of nanoporous carbons. In: Novel carbon adsorbents. Elsevier, Amsterdam, pp 107–139. https://doi.org/10.1016/B978-0-08-097744-7.00004-1
Thommes M, Kaneko K, Neimark AV, Olivier JP, Rodriguez-Reinoso F, Rouquerol J, Sing KSW (2015) Physisorption of gases, with special reference to the evaluation of surface area and pore size distribution (IUPAC technical report). Pure Appl Chem 87(9–10):1051–1069. https://doi.org/10.1515/pac-2014-1117
Torrellas SÁ, García Lovera R, Escalona N, Sepúlveda C, Sotelo JL, García J (2015) Chemical-activated carbons from peach stones for the adsorption of emerging contaminants in aqueous solutions. Chem Eng J 279:788–798. https://doi.org/10.1016/j.cej.2015.05.104
Tseng R, Tseng S, Wu F (2006) Preparation of high surface area carbons from corncob with KOH etching plus CO2 gasification for the adsorption of dyes and phenols from water. Colloids Surf A Physicochem Eng Asp 279:69–78. https://doi.org/10.1016/j.colsurfa.2005.12.042
USEPA (2016) Summary of Nominations for the Fourth Contaminant Candidate List (CCL 4)
Vargas Delgadillo DP (2013) Preparación, caracterización y funcionalización de materiales carbonosos para la adsorción de CO2. Universidad Nacional de Colombia, Bogotá

Venkatramanan V, Shah S, Prasad R, Shah M (2021) Life cycle assessment of lignocellulosic waste biorefinery. In: Shah S, Venkatramanan V, Prasad R (eds) Biovalorization of waste: trends and perspectives. Springer Nature, Singapore, pp 327–347

Wang J, Wang S (2016) Removal of pharmaceuticals and personal care products (PPCPs) from wastewater: a review. J Environ Manage 182:620–640. https://doi.org/10.1016/j.jenvman.2016.07.049

WHO (2006) Guidelines for drinking water quality. Recommendations, vol 1. WHO, Geneva. http://www.who.int/water_sanitation_health/dwq/gdwqvol32ed.pdf

World Health Organization (2008) Guidelines for Drinking-water Quality. Recomendations, vol 1, 3rd edn. WHO Library Cataloguing-in-Publication Data, Geneva. https://doi.org/10.1248/jhs1956.35.307

Yalkowsky SH, He Y, Jain P (2010) Handbook of aqueous solubility data, 2nd edn. CRC Press, Taylor & Francis Group, Boca Raton. https://doi.org/10.1017/CBO9781107415324.004

Yang Y, Ok YS, Kim KH, Kwon EE, Tsang YF (2017) Occurrences and removal of pharmaceuticals and personal care products (PPCPs) in drinking water and water/sewage treatment plants: A review. Sci Total Environ 596–597:303–320. https://doi.org/10.1016/j.scitotenv.2017.04.102

Removal of Indoor Pollutants (VOCs): Phytoremediation Applications and Adsorption Studies Using Immersion Calorimetry

14

Diana Hernández-Monje, Liliana Giraldo, and Juan Carlos Moreno-Piraján

Abstract

Phytoremediation and adsorption on activated carbons are used for removing some environmental indoor pollutants (VOCs), this compounds as organic solvents help for producing products such as cleaning products, paints, adhesives, waxes, varnishes, detergents, coatings, and inks. This chapter is going to mention some phytoremediation applications, particularly for benzene and toluene BTEX, mentioning some plants that have presented considerable removal percentages, also how these pollutants enter to the plant and their transformations once they have penetrated the vegetal organism. Then, volatile organic compounds adsorption on modified activated carbons and their energetic interaction characterization by immersion calorimetry (describing this interesting and unconventional technique). Finally, this manuscript describes the influence of the physicochemical properties of the porous solids and the characteristics of contaminants in the immersion enthalpies.

Keywords

VOCs · Phytoremediation · Adsorption · Activated carbon · Immersion enthalpy

D. Hernández-Monje · L. Giraldo
Departamento de Química, Facultad de Ciencias, Universidad Nacional de Colombia—Sede Bogotá, Bogotá, Colombia

J. C. Moreno-Piraján (✉)
Departamento de Química, Facultad de Ciencias, Universidad de los Andes, Bogotá, Colombia
e-mail: jumoreno@uniandes.edu.co

R. Prasad (ed.), *Phytoremediation for Environmental Sustainability*,
https://doi.org/10.1007/978-981-16-5621-7_14

14.1 Introduction

Volatile organic compounds (VOCs) constitute an environmental problem that also affects the health of the people who are exposed to them. The increase of the VOCs emissions and the regulations that have been generated make necessary the implementation of means and technologies for their removal; until now, several processes have emerged (Zhang et al. 2017).

One of them is phytoremediation in which plants are used to remove contaminants from the indoors even if the concentration is low. Some authors have found that ornamental plants can take away VOCs from indoor air. For BTEX VOCs removal plants as *Zamioculcas zamiifolia, Sansevieria trifasciata, Schefflera arborícola, Monster acuminate, Spathiphyllum wallisii, Scindapsus aureus, Epipremnum aureum, Chamaedorea seifrizii, Ixora ebarbata, Philodendron domesticum,* and *Dracaena sanderiana* have been used (Parseh et al. 2018; Sriprapat and Thiravetyan 2013; Teiri et al. 2018). The mechanisms that plants use to take pollutants out are phytostabilization, phytoextraction, rhizodegradation, phytovolatilization, phytofiltration, and phytodegradation where this last one is the most used (Parseh et al. 2018; Thakare et al. 2021; Sarma et al. 2021; Sonowal et al. 2022). Conversely, the adsorption has been recognized as an efficient and economic control strategy for VOCs that allows to recover the adsorbent and the adsorbate. Among the porous solids, activated carbon is widely used due to its high adsorption capacity, large surface area and porosity, versatility, stability, and selectivity (Zhou et al. 2017; Yang et al. 2018; Gallego et al. 2013; Li et al. 2011).

Immersion calorimetry is an interesting technique for determining energy transfer that occurs when a solid and a liquid come into contact. This technique allows calculating the immersion enthalpy and the thermal effects resulting from immersing a solid in a solvent (generally of non-polar type) that does not interact chemically with the solid, the enthalpy of immersion can be related to the surface area of the solid according to the models developed by Dubinin and Stoeckli. Also, the effect of surface chemistry on solid–liquid interactions for adsorbent materials as activated carbon can be studied in their modifications by immersion enthalpy to know the heat involved in the interactions between the activated carbon and the wetting liquid. Conversely, immersion calorimetry is used in different areas since it provides complementary information to the studies of gas and liquid phase adsorption isotherms (Moreno-Piraján et al. 2012; Giraldo et al. 2018).

This manuscript refers to two methods for removing environmental indoor pollutants (VOCs). First, describing the phytoremediation and some applications for particular volatile organic compounds like BTEX. Then, this work addresses the adsorption of VOCs (particularly hexane, cyclohexane, benzene, and toluene) on modified activated carbons and their energetic interaction through the immersion calorimetry. It also mentions the influence of both the chemical and textural properties of the porous solids and the characteristics of adsorbates in the enthalpies of immersion that were obtained. These VOCs were chosen due to their use as organic solvents used in the production of products such as cleaning products, adhesives, waxes, paints, detergents, varnishes, coatings, inks, detergents, among others and because they are molecules that differ in its arrangement and molecular size.

14.2 Volatile Organic Compounds (VOCs): Use and Harmful Effects

VOCs are organic compounds that have a low boiling point, high vapor pressure, and reactivity with respect to photochemical reactions. These come from biogenic or anthropogenic emissions: those of natural origin are produced mainly from wetlands, forests, oceans, and volcanoes, while anthropogenic emissions are generated in manufacturing, petrochemical, vehicle emissions, and even in everyday activities such as building buildings, painting, smoking, among others; in turn, they can be found in a large number of household products such as detergents, waxes, varnishes, solvents, detergents, cleaning products, or paints. It has also been shown that they are emitted during the use of electronic devices such as photocopiers or printers and in the use of fossil fuels (Zhang et al. 2017; Salar-García et al. 2017; Mirzaei et al. 2016; Sarigiannis et al. 2011; Berenjian et al. 2012; Huang et al. 2016; Nurmatov et al. 2015; Kamal et al. 2016; Cheng et al. 2016). Adverse effects of exposure to these substances include neurotoxicity, myelotoxicity, conjunctivitis, dermatitis, irritation of the respiratory tract, and central nervous system (CNS) disease (Lee et al. 2013; Betancur-Sánchez et al. 2017; Lacerda et al. 2012; Costa et al. 2012).

14.2.1 Benzene

It is used as a solvent for inks, paints, lacquers, varnishes, waxes, resins, plastics, rubbers, fats, and oils, in the extraction of seed oils, among others. In turn, it is used as a gasoline additive; however, due to its high toxicity, at present, it is only added when there is no adequate substitute (Sarigiannis et al. 2011; UNAM 2016; US EPA, OAR 2016).

As for its effects on health, it is a toxic substance since it is carcinogenic, mutagenic, and neurotoxic. In fact, the International Agency for Research on Cancer (IARC) has classified it as a group 1 carcinogen (confirmed as a human carcinogen) for all exposure routes (Tsai 2016).

14.2.2 Toluene

The main use of toluene is its addition to gasoline to improve the degree of octane. It is also used as a solvent in paints, synthetic fragrances, coatings, adhesives, inks, and cleaning products, in the production of polymers for the manufacture of nylon, plastic bottles of soda, polyurethanes, dyes, pharmaceuticals, and cosmetics for nails (Moro et al. 2012; ATSDR 2017).

With regard to the health-related conditions that this compound can cause, it is carcinogenic, mutagenic, and neurotoxic. It has been found that the central nervous system (CNS) is the main target for the toxicity of toluene, both in humans and in animals in acute (short-term) and chronic (long-term) exposures (Moro et al. 2012; Lee et al. 2013; Betancur-Sánchez et al. 2017; Park et al. 2016; Agency for Toxic Substances and Disease Registry 2015; Hong-li et al. 2017).

14.2.3 Cyclohexane

It is mainly used as a solvent for substances such as lacquers, resins, fats, waxes, oils, bitumen, and rubber. It is also used in the leather industry, the manufacture of perfumes, adhesives, nylon production, paint, and varnish remover.

As for the adverse effects on health, few studies have been made in this regard, but it has been found that this solvent can be a central nervous system depressant. Its exposure can irritate the skin, mucous membranes, and eyes; when inhaled, it can generate headaches, dizziness, nausea, lightheadedness, and even fainting, respiratory, and throat conditions. Extreme acute exposure can cause nausea, vomiting, lack of coordination, coma, and even death (Luttrell and Lyiza 2010; INSHT 2009).

14.2.4 Hexane

It is used in the extraction of edible seeds and vegetable crops (for example, soybeans, peanuts, corn); as a cleaning agent (degreaser) in the printing industry; as a solvent and in the formulation of some adhesive products, lacquers, varnishes, inks, cements, and paints. It is also found in consumer products such as gasoline, drying adhesives, and cement (ATSDR 2010). Acute and short-term exposure (by inhalation) causes effects on the central nervous system (CNS) (ATSDR 2010; Betancur-Sánchez et al. 2017; Park et al. 2016; Bates et al. 2016).

14.3 Phytoremediation for Decreasing the VOCs Concentration from Indoors

Phytoremediation is a process where biological agents as plants are used to detoxify environments with volatile organic compounds and then improve the quality of the air, especially indoors (Sarma et al. 2021; Sonowal et al. 2022). For this, some ornamental plants have been used for removal of VOCs from indoors (Sriprapat and Thiravetyan 2013; Teiri et al. 2018).

Volatile organic compounds degradation could be performed with the help of rhizosphere and another type of microorganisms. Plants take VOCs during the gas exchange process through stomata and could transform them into amino acids. This process depends on the physicochemical properties and the total surface area of the plant, as well as on the presence of trichomes on leaves, besides on the matrix of soil

(Gawrońska and Bakera 2015; Torpy et al. 2018; Brilli et al. 2018; Irga et al. 2013; Li et al. 2018; Khaksar et al. 2016; Wei et al. 2021; Teiri et al. 2018).

One of the most studied volatile organic compounds are benzene and toluene (Parseh et al. 2018; Limmer et al. 2018; Sriprapat and Thiravetyan 2013; Wei et al. 2021; Brilli et al. 2018; Treesubsuntorn and Thiravetyan 2012; Kim et al. 2012; Hörmann et al. 2017; Sriprapat et al. 2014).

In the work of Parseh et al. (2018) the removal of benzene was performed with *Schefflera arboricola* and *Spathiphyllum wallisii* during 3 days (inlet benzene concentration between 10.5 and 29.5 μg m^{-3}). It was found that the removal efficiency (RE) was similar for both plants (94% and 93%, respectively), part of this RE was due to abiotic agents and possible soil absorption with presence of microorganisms. It is also mentioned that leaves, cuticles, and stomata take the pollutants and then VOC is diffused into the spaces of the leaf cells; later they are transformed by the plant tissues or absorbed by films of water. According to results, efficiency decrease with the pollutant concentration, maybe because when there are higher concentrations of benzene, this compound can damage cell membranes and block stomata and intercellular space. So, the phytoremediation with these two plants can be an effective method to remove benzene at low concentrations; however, the authors mention that due to the lack of studies in this field, more research is needed to evaluate other plants and the VOCs mechanisms of removal.

Sriprapat and Thiravetyan (2013) studied the benzene and toluene adsorption using *Zamioculcas zamiifolia*, where the uptake of benzene and toluene was 0.96 and 0.93 mmol m^{-2} per leaf unit area at 72 h of exposure, respectively. The benzene was removed faster than toluene because of its size; also, 80% of C_6H_6 and 76% of C_7H_8 were removed through the stomata and the least part was removed through cuticles (C_6H_6: 20%; C_7H_8: 24%).

Also, in Brilli et al. (2018) it is mentioned that the lipid composition of the cell epidermis membrane and the cuticular wax are important for the removal of hydrophobic VOCs as benzene, and also the plant enzymes can generate the hydroxylation and cleavage of the aromatic rings of toluene and benzene.

These processes are better described by Ugrekhelidze et al. (1997) where they have mentioned that benzene and toluene enter through both sides of the hypostomatous leaf in *A. campestre, Malus domestica,* and *Vitis vinifera* plants. Then, there could be cleavage for the aromatic ring and the C atoms are mostly added to nonvolatile organic acids and a lower portion to amino acids. *Spinacia oleracea* generates benzene oxidation through its chloroplasts mainly under light conditions. Also, hydroxylation takes place as first step of aromatic hydrocarbon conversion for higher plants. They suggest that phenol and pyrocatechol could be the first intermediate molecules in C_6H_6 oxidative cleavage in plants. On the other hand, they mention that muconic acid might be the primary product for the cleavage of the aromatic ring and this muconic acid could generate fumaric acid, causing the involving of benzene into organic acids metabolism. For toluene, the oxidative cleavage could occur for oxidation of methyl group to carboxyl one and a ring hydroxylation, or for ring hydroxylation without the methyl group oxidation, generating α-carboxymuconic acid or α-methylmuconic acid.

Other species studied for benzene removal are *Ixoraebarbata craib, Chamaedorea seifrizii, Sansevieria trifasciata, Scindapsus aureus, Philodendron domesticum, Epipremnum aureum, Monster acuminate, and Dracaena sanderiana* and it was found that they showed between 43% and 77% of benzene removal where the last plant was the one with greater removal capacity (from 60% to 77%) at 72 h. The uptake of benzene was 46% by crude wax and 54% by stomata, showing take out efficiency at day and night.

Twelve plants were studied by Sriprapat et al. (2014) for the removal of toluene (removal values between 1.25 and 2.68 μmol). They found that *Sansevieria trifasciata* was the most efficient plant for this phytoremediation process and this molecule could enter through the plant's cuticle. Conversely, the wax of *S. trifasciata* and *Sansevieria hyacinthoides* showed higher take out of toluene probably due to the help of the hexadecanoic acid contained in it. Also, for this species the removal quantities seem not to be not correlated to the stoma quantity.

According to above, phytoremediation can be a useful tool for the removal of environmental indoor pollutants like VOCs, where several species are accurate to take out benzene or toluene from indoor spaces being taken through leaves, cuticles, and stomata. Later transformed by plant enzymes by means of hydroxylation and cleavage processes of the aromatic rings to be added to nonvolatile organic acids or to amino acids.

Also, although phytoremediation can remove VOCs, there are some species of plants that show drawbacks at higher concentrations of the pollutant, since it can affect the plant cells and decrease its effectiveness. In turn, several authors agree that a more exhaustive study of the removal and transformation mechanisms of VOCs is required, as well as a broadening of the range of plant species that could be used as decontaminating agents for indoor environments.

14.4 Adsorption on Activated Carbon and Its Application in the Removal of VOCs

For the removal of VOCs, control mechanisms have emerged, they can be divided into recovery methods and methods of destruction; the latter mainly converts VOCs into CO_2 and H_2O; however, the recovery methods are more economical, require less energy, and are less polluting. Adsorption is a method of recovery that is considered favorable due to its low cost and high efficiency, for which activated carbon has been widely used due to its versatility, selectivity, surface area, variety of porous structure, high capacity, and fast adsorption kinetics (Moreno-Piraján et al. 2011; Zhou et al. 2017; Bradley 2011; Wang et al. 2014b). The adsorption capacity depends on its physicochemical properties: surface area, the pore size distribution, as well as the chemical composition of the surface. For the adsorption of benzene, toluene, cyclohexane, and hexane, activated carbons have been found with adsorption capacities between 0.4 and 314.84 mg g^{-1} (Yao et al. 2013; Pak et al. 2016; Mazlan et al. 2016; Liu et al. 2016; Lopes et al. 2015; Yang et al. 2018; Martínez De Yuso et al. 2013; Pei and Zhang 2012; Tham et al. 2011; Tazibet et al. 2013).

The importance of modifying activated carbons in order to obtain differences in their physicochemical properties to evaluate the intensity of the interaction with the contaminants of interest is relevant since it allows obtaining information regarding the affinity that may exist between the adsorbent and the adsorbate so that it can be determined which type of porous solid is most suitable for the removal of the study VOCs.

14.4.1 Isothermal Immersion Calorimetry

As mentioned above, calorimetry is a technique that allows to evaluate the intensity of the interaction between the pollutant studied and the porous solid that adsorbs it; since it is not a conventional technique it can show interesting and complementary results to the characterization of the adsorbent material and processes such as the adsorption isotherms of the molecules of interest. This is why some important aspects of how it is used to calculate the thermodynamic parameter of the immersion enthalpy will be described.

In the isothermal immersion calorimeter, there is a considerable exchange of energy between the cell and the surroundings (the cell and the surroundings are at the same constant temperature). It contains a thermal resistance R_T, very small, with a heat capacity of the surroundings is infinitely high, so the temperature of the cell (T_C) and the temperature of the surroundings (T_S) can remain constant over time, but without manifesting heat flow. In real determinations there is a flow of energy between the cell and the surroundings that is detected by means of thermal sensors located between them. This flow is due to the small temperature difference between T_A and T_C during the calorimetric experience, this quantity depends on the geometry of the cell, the type of insulation of the thermal sensors, the amount of heat released per unit of time, the thermal conductivity. Although this small temperature difference exists, it is considered an isothermal process if each of them remains constant throughout the process that causes the flow of energy.

The surroundings and the cell are connected by means of the thermal resistance R_T, relating the heat flow with the temperature difference dQ/dt. This difference is given by.

$$\frac{dQ}{dt} = \frac{\Delta T}{R_T} \tag{14.1}$$

Integrating:

$$Q = \frac{1}{R_T} \int \Delta T(t) dt \tag{14.2}$$

For the same amount of heat:

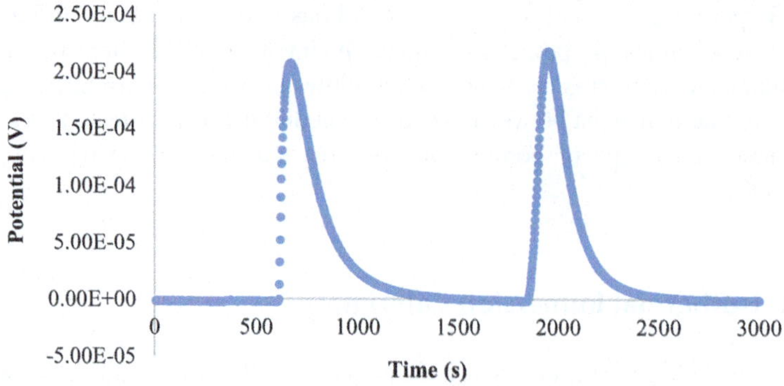

Fig. 14.1 Calorimetric curve determined by means of an isothermal immersion calorimeter

$$\int \frac{\Delta T(t)\mathrm{d}t}{R_{\mathrm{T}}} = \text{constant} \tag{14.3}$$

The conduction of heat inside a real instrument is complex, then it makes very difficult to calculate R_{T}, which quantitatively connects the difference of the measured temperature with the corresponding heat flow, so that the resistance is determined by means of calibration. The reciprocal value of the thermal resistance is the calibration factor $K_{(t)}$.

$$Q = K \int \Delta T(t)\mathrm{d}t \tag{14.4}$$

Generally, the calibration factor can be recorded as constant within the temperature range in which the calorimetric experiment is carried out (Giraldo and Moreno-Piraján 2007).

A calorimetric curve is obtained when the measurement is made, this captures the variation of the electric potential as a function of time (Fig. 14.1).

The calorimetric curves (as shown in Fig. 14.1) contain two peaks: the first corresponds to the immersion of the solid (activated carbon, in this case) into the liquid (pollutant), cell rupture, and wetting of the sample and the second to the electric calibration process of the calorimeter; this calibration is carried out by heating the system with an electrical resistance. The diagram of a heat conduction microcalorimeter is presented in Fig. 14.2.

Once the calorimetric curves are determined, they are used to calculate the immersion enthalpy which is proportional to the area under the curve of the immersion peak.

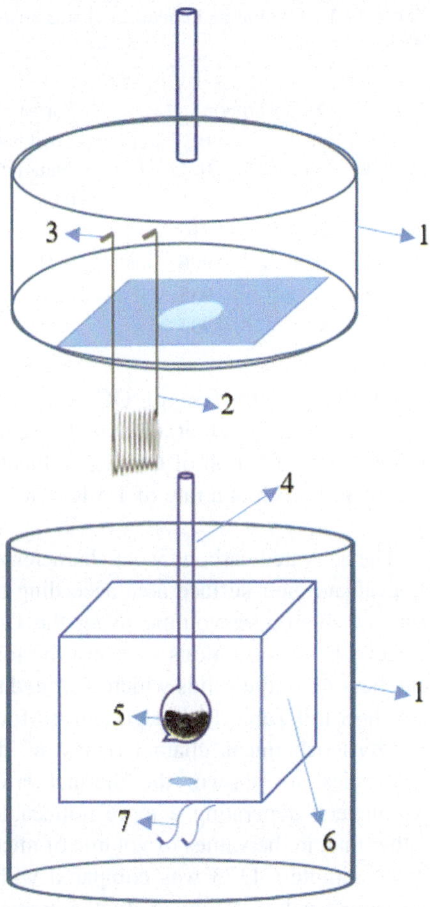

Fig. 14.2 Heat conduction microcalorimeter diagram: 1. Insulation cover; 2. Resistance; 3. Connection to the power source; 4. Glass ampoule with fragile peak; 5. Sample; 6. Sensors; 7. System connection to the multimeter interface. (Adapted from (Moreno-Piraján et al. 2011), p. 171)

14.4.2 Influence of the Adsorbents and Adsorbates Properties in Their Interaction: Enthalpic Determination

To evaluate the influence of adsorbent and adsorbates in the adsorption process a study of calorimetric results for five samples of activated carbon that differ in their textural properties and surface chemistry will be shown. The modification of these adsorbent materials is described below (Hernández-Monje et al. 2019a):

- GC: Prepared from coconut shell, it was sieved to 1 mm as particle size, washed with distilled water, and dried for 24 h at 363 K, later stored in containers under a nitrogen atmosphere.
- C1173: A fraction of GC was subjected to thermal treatment in nitrogen atmosphere for 10 h at a rate of 1.5 K min-1 and then for 1 h at 1173 K.
- OC: A portion of GC was subjected to an oxidation process with a solution of HNO_3 6 M.

Table 14.1 Textural and chemical characteristics of activated carbons (Hernández-Monje et al. 2019a)

Sample	N_2 adsorption		Surface chemical groups		Hydrophobic factor (H_f)
	W_o (cm^3g^{-1})	BET area (m^2g^{-1})	Total basicity (mmol g^{-1})	Total acidity (mmol g^{-1})	$\frac{\Delta H_{im} C_6H_6}{\Delta H_{im} H_2O}$
GC	0.34	841	0.08	0.20	2.14
OC	0.32	810	0.05	0.39	1.43
C1173	0.36	996	0.31	0.05	4.48
C1023	0.35	935	0.26	0.06	3.44
C723	0.35	903	0.11	0.28	2.02

- OC1023: A fraction of OC was subjected to thermal treatment in nitrogen atmosphere for 8 h at a rate of 1.5 K min^{-1} and then for 1 h at 1023 K.
- OC723: A fraction of OC was subjected to thermal treatment in nitrogen atmosphere for 5 h at a rate of 1.5 K min^{-1} and then for 1 h at 723 K.

The activated carbons were characterized using N_2 adsorption isotherms at 77 K to evaluate their surface area according to BET (Brunauer–Emmett–Teller) model and the micropore volume using the Dubinin–Radushkevich (D-R) model; conversely, Boehm titrations were also performed in order to calculate the total acidity and basicity of the solids which is related to the surface chemical groups. The results are shown in Table 14.1 (Hernández-Monje et al. 2019a).

Physicochemical characteristics of the five solids are in Table 14.1, two tendencies are shown: the thermal effect increases the values of the textural parameters, generating a more noticeable increase in the values of surface area, rather than in the values of volume of micropore. An increase of 23% was presented if the sample C1173 was compared with OC for the BET area and 13% for the micropore volume; conversely, the increase in the temperature of modification also decreases the acid character of the porous solids and increases the total basicity thereof and increases their hydrophobicity (according to the hydrophobic factor). However, the chemical modification with HNO_3, responsible for the addition of oxygenated groups to the structure of the adsorbing materials, generates the opposite effect, so that it decreases the surface area and the pore volume in terms of physical characteristics and increasing the content of these surface groups increases the total acidity of the coals and decreases the basic character of the coals, making this sample present the lowest values of hydrophobic factor. These tendencies may be due to the thermal stability of the surface oxygenated groups, so at higher temperatures greater heteroatom removal in the adsorbent (Rodríguez-Estupiñán et al. 2013; Belhachemi and Addoun 2011; Wang et al. 2014a; Yin et al. 2007; Mangun et al. 1999).

According to the results described above, it would be expected that the porous solids with which they will have greater interaction with the VOCs of interest are those subjected to a higher temperature of thermal treatment, this hypothesis will be corroborated with the results that will be shown below. In turn, they will show how the interactions are modified to the extent that the adsorbate studied is modified.

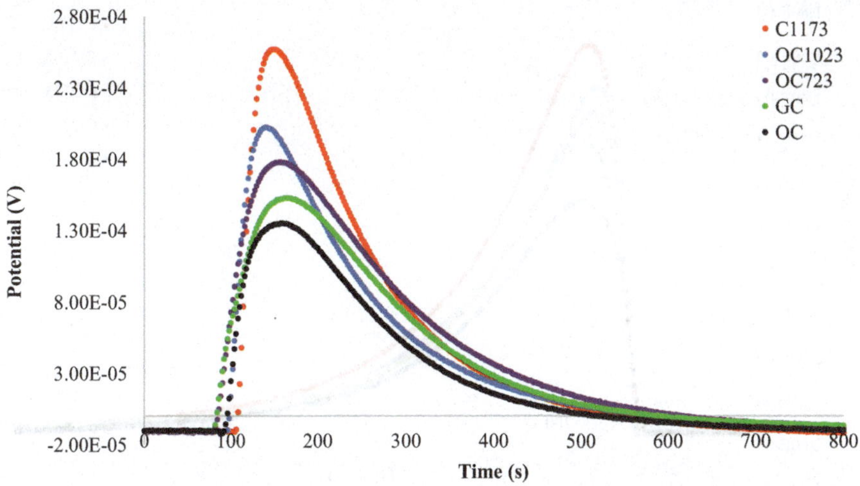

Fig. 14.3 Calorimetric curves to the immersion of the five activated carbons into benzene

Figure 14.3 shows the calorimetric curves product of the immersion of the activated carbons in benzene (Hernández-Monje et al. 2019b). As mentioned above, the area under the curve is directly proportional to the enthalpy value of the immersion and therefore to the intensity of the adsorbent–adsorbate interaction. For all the samples the interaction is exothermic, since the immersion process generates an increase in the evaluated potential where the area under the curve becomes larger for the samples that have higher temperatures of thermal treatment and decreases with the chemical modification with the acid, as mentioned in the hypothesis described above. In the case of benzene, it occurs because when this molecule (aromatic compound) is put into contact with the activated carbon there is an interaction between the regions with high electronic density located in the graphene layers and the π electrons of the molecule, particularly when activated carbon is treated at higher temperatures, since the removal of oxygenated groups favors specific interactions between such graphenic layers with the aromatic ring of benzene, when the amount of benzene molecules increases on the surface, they tend to stack together, generating a structural rearrangement, giving rise to a greater intermolecular π–π interaction (García et al. 2004; Wang et al. 2015).

The calorimetric curves product of the immersion of the activated carbons into toluene are in Fig. 14.4 (Hernández-Monje et al. 2019b). They show the same tendency, higher area under the curve then higher interactions for the samples with thermal treatment, where C1173 shows the highest value, but with less intensity. In the case of toluene the interaction is similar to benzene, because it is also an aromatic compound, so there is attraction between the electronic density of the aromatic ring of the molecule and the π orbitals on the basal planes of the carbon, besides, it has a dipole moment that allowed it to interact also with polar sites present on the surface of the activated carbon, increasing the interaction energy between the adsorbent and

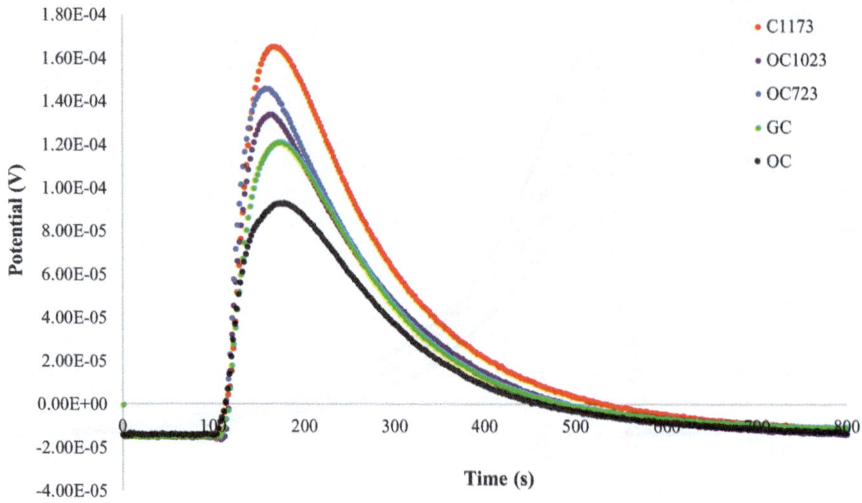

Fig. 14.4 Calorimetric curves to the immersion of the five activated carbons into toluene

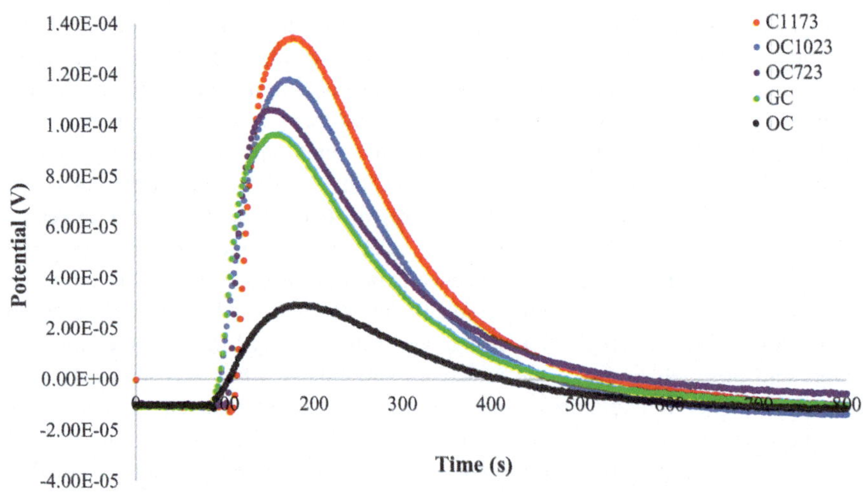

Fig. 14.5 Calorimetric curves to the immersion of the five activated carbons into cyclohexane

the adsorbate, but this interaction could be lower than the interaction with benzene because the presence of the methyl generated restrictions for its entry and subsequent interaction with the microporous structure (Wibowo et al. 2007; Lopes et al. 2015; Villacañas et al. 2006; Anuradha et al. 2014).

For the interaction between the activated carbons and cyclohexane, the calorimetric curves of the immersion of the solids into the liquid are presented in Fig. 14.5. Again, the same tendency is shown: the highest interaction occurs with C1173, then with OC1023, OC723, later with the starting carbon GC, and finally, the lowest

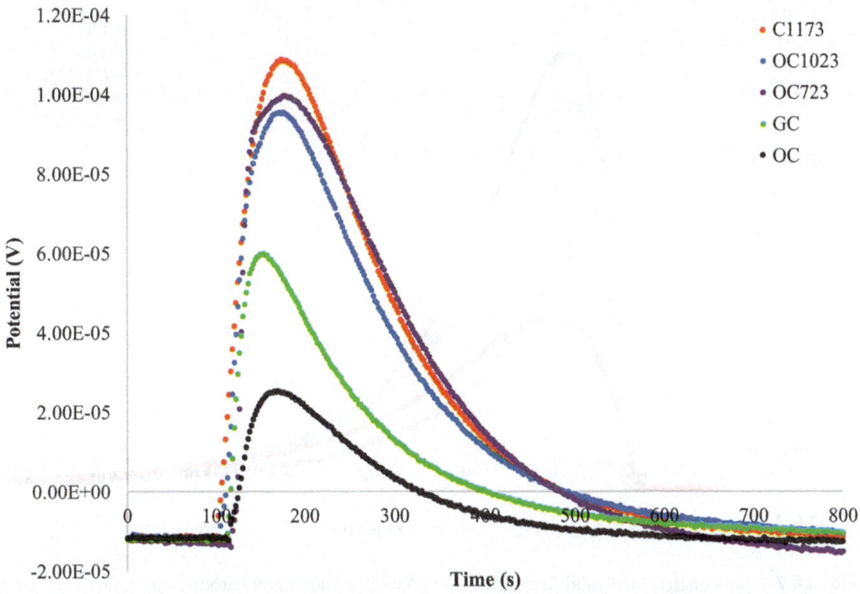

Fig. 14.6 Calorimetric curves to the immersion of the five activated carbons into hexane

interaction is with OC; however, the areas under the curve are lower than benzene and toluene, this could be because cyclohexane is a neutral hydrophobic compound, so it began to cover the basal plane system of the activated carbon but with less intensity because it does not have delocalized π electrons and the shape of this molecule is different from benzene, because C_6H_6 is planar and cyclohexane is warped, where the dispersive interactions are responsible for the cyclohexane–carbon interaction. The lowest value corresponds to OC, since the attractive forces between molecules and the activated carbon decreased because of the increase of the oxygenated surface functional groups, then, the dispersive interactions had lower intensity (Arafat et al. 2004; Fomin et al. 2015; Wang et al. 2015).

Fig. 14.6 shows the calorimetric curves for the immersion of the samples (activated carbons) into hexane. For this hydrocarbon, the areas under the curve present, one more time, the same tendency: low values to the sample with chemical modification with HNO_3 and high values for the solids subjected to the thermal modifications, where the highest value was for C1173.

The interaction hexane-activated carbon was also of dispersive type but with less intensity than if the liquid of immersion was benzene, toluene, or cyclohexane because this is a compound that behaves as an elongated and plane cylinder, this made its entry into the porous structure somewhat restricted, making that the number of molecules interacting with the porous network of the solid was smaller. In turn, it was the VOC that has less affinity with the activated carbon structure since it is an open chain aliphatic compound. On the other hand, the adsorbent–adsorbate interaction includes three types of interactions: attractive dispersion forces (van der

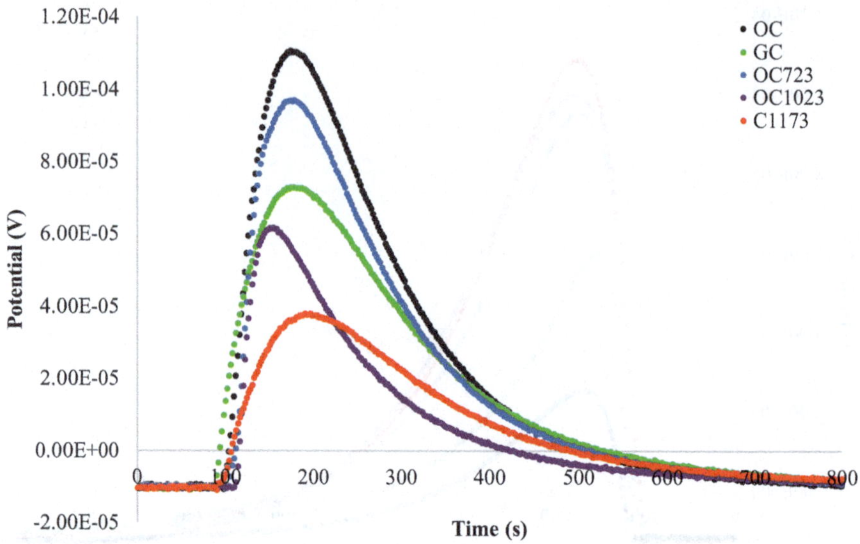

Fig. 14.7 Calorimetric curves to the immersion of the five activated carbons into water

Waals), short range repulsive forces and electrostatic forces; dispersion forces can be caused for fluctuations in electron density of the atoms which induces an electrical dipole moment in neighboring atoms, generating an attraction between the atoms. All VOCs are non-polar, but the dipole moment, for example, of toluene is four times higher than for n-hexane, this made that the contribution of the interactions between the samples and this hydrocarbon was lower than for the other compounds (Wang et al. 2015; Martínez De Yuso et al. 2013).

Figure 14.7 shows the calorimetric curves for the immersion of C1173, OC1023, OC723, GC, and OC into a polar solvent: water. These determinations were made despite the fact that water is not a volatile organic compound, for two reasons: the first one is that it is interesting to show how the adsorbent–adsorbate interaction is modified when the process is carried out with a polar molecule, and the second is that the relationship between the immersion enthalpy of the activated carbons into a reference non-polar solvent (benzene) and a reference polar solvent (water) allows to determine an interesting factor called the hydrophobic factor whose values are shown in Table 14.1.

According to the areas under the curve, as expected, the behavior is completely opposite: the greatest interaction is generated with the sample without thermal treatment exposed to nitric acid and there is an inversely proportional relationship between the thermal modification temperature and the area under the water immersion curve in the porous solids; this occurs because the liquid is a polar molecule, then the adsorbate–adsorbent interaction can occur by direct electrostatic interactions between the water molecules and the carbon surface or inductive-type interactions between the water molecules and the oxygenated surface groups, in turn, surface quadrupoles located in the basal plane of carbon can be generated. Then, as

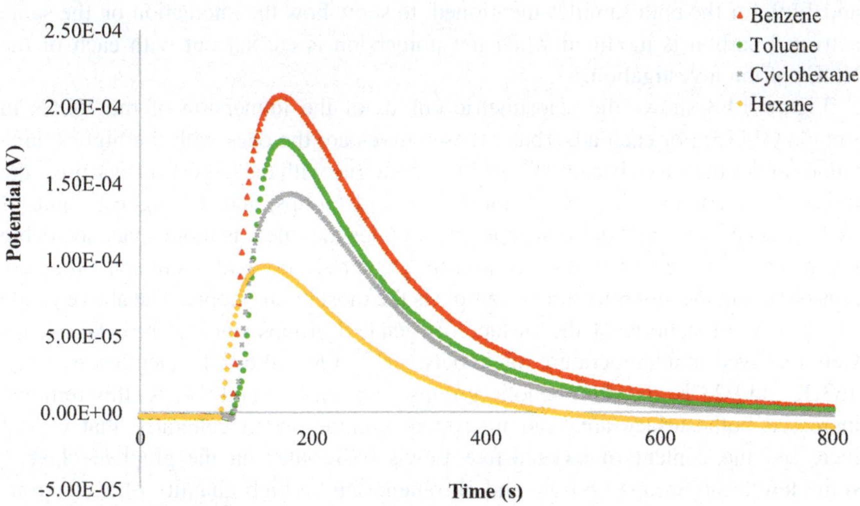

Fig. 14.8 Calorimetric curves to the immersion of activated carbon C1173 in benzene, toluene, cyclohexane, and hexane

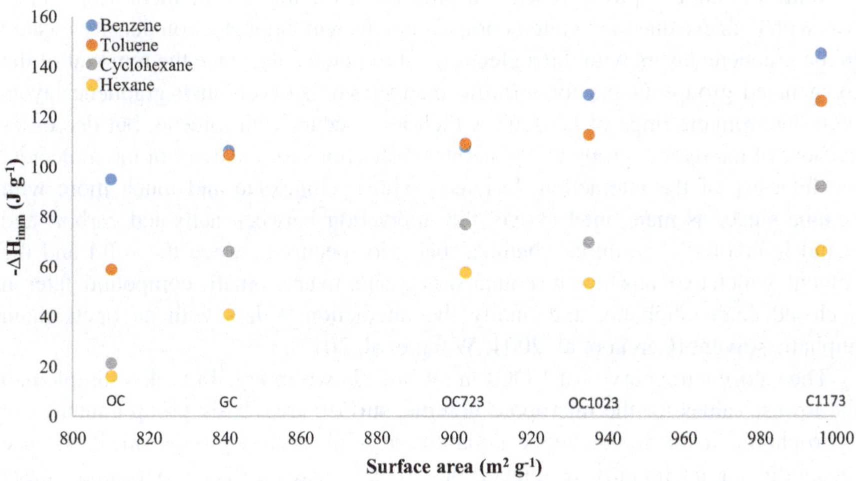

Fig. 14.9 Immersion enthalpy of activated carbons in solvents in function of surface area solids

the number of oxygenated groups increases, this type of interactions will increase, while the increase in temperature generates removal of heteroatoms, so that the energy involved in the process decreases (Nwaka et al. 2016; Brennan et al. 2002; Hernández-Monje et al. 2016).

Since the trend with respect to the pollutants evaluated is that the greatest interaction occurs with the sample C1173 and the lowest with OC, the calorimetric curves of benzene, toluene, cyclohexane, and hexane will be presented in Figs. 14.8

and 14.9, for the both samples mentioned, to show how the interaction on the same activated carbon is modified when the immersion is carried out with each of the VOCs of the investigation.

Figure 14.8 shows the calorimetric curves of the immersion of the VOCs in sample C1173. For each adsorbate, these curves are the ones with the highest area values under the curve because C1173 is the sample with the largest surface area, the largest pore volume, the lowest content of acid groups, and the highest value of hydrophobic factor and basic groups; this implies that there is more space available for the entry of the molecules and, in turn, greater affinity with them since they are non-polar and the aforementioned sample is the most hydrophobic. The above is due to the thermal stability of the surface oxygenated groups since carboxylic groups were removed in a temperature range between 373 K and 673 K; lactones between 463 K and 923 K and the phenolic groups between 873 and 973 K, this removal increased both surface area and micropore volume and it generated that C1173 increased the content of oxygen-free Lewis basic sites on the graphene layers, some few basic groups (pyrone and chromene) and a high quantity of π electrons on the basal plans of the carbon (Shafeeyan et al. 2010; Montes-Morán et al. 2012; Bhatnagar et al. 2013; Abdulrasheed et al. 2018; Lo et al. 2014; Daud and Houshamnd 2010; Goncharuk 2015).

With this kind of physicochemical properties, the highest interaction of C1173 was with benzene due to the interaction of regions with high electron density located in the graphene layers with the π electrons of the molecule, since the removal of the oxygenated groups favors the specific interactions between such graphene layers with the aromatic rings of benzene, which also occurs with toluene, but decreases because of the methyl group that generates restrictions for the entry of the molecule; the intensity of the interaction decreases with cyclohexane and much more with hexane since, as mentioned above, the interaction between activated carbon and liquid is favored if a similar chemical behavior occurs between the solid and the solvent, which explains that the enthalpy is greater in an aromatic compound, later in a closed chain aliphatic, and finally the interaction is less with an open chain aliphatic solvent (García et al. 2004; Wang et al. 2015).

The calorimetric curves of VOCs in OC are shown in Fig. 14.9, this sample had the lowest values for the micropore volume, surface area, basic group content and hydrophobic factor, and the highest quantity of acidic surface groups, this is because the modification with nitric acid incorporated oxygenated surface groups in the basal carbon planes by means of free radicals, according to the proposed mechanism, besides, another sequence could generate carbonyl groups, also, an α-ketone substituent could break the C–C bond of the aliphatic part and subsequent oxidation would lead to the formation of carboxylic acids; in turn, the formation of hydroxyl groups is reported, however, the carboxylic group is the functional group most tended to be formed, this explained why the textural properties decreased and the acidity of the samples increased, these two factors decreased the intensity of the interactions with VOCs due to the restriction to the entrance because of the heteroatoms and the low affinity of the sample with these compounds since it has a greater hydrophilic character that makes it more affine to polar substances than to non-polar molecules,

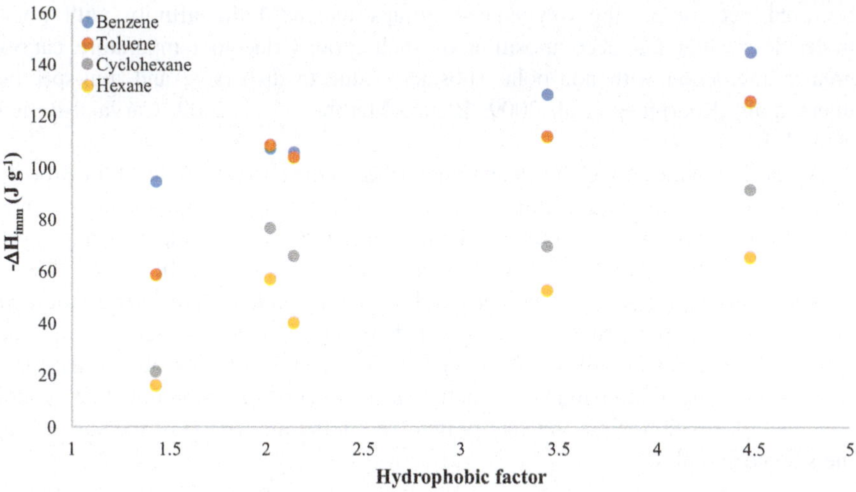

Fig. 14.10 Immersion enthalpy of activated carbons in solvents in function of the hydrophobic factor solids

which was corroborated by the calorimetric curves shown in Fig. 14.7. The tendency of the interaction between the VOCs and OC is the same that for C1173: benzene>toluene>cyclohexane>hexane, where the explanation of the behavior is the same that is mentioned above but with less intensity due to the physicochemical characteristics of the sample (Vinke et al. 1994; Ternero-Hidalgo et al. 2016; Shen et al. 2008; Shafeeyan et al. 2010).

The immersion enthalpies of the solvents as a function of the surface area of the solids are in Fig. 14.9 where there is a directly proportional relationship between both variables. With respect to adsorbates, the greatest interaction occurred with the aromatic compounds, followed by the closed chain aliphatic, and finally the open chain aliphatic, while with respect to the samples, what was mentioned above occurs, as far as the modification temperature increased the interaction increased too, but this interaction decreased in the sample subjected to oxidation with nitric acid. This indicated that the oxygenated groups added in the chemical modification were later removed with the action of temperature, allowing a greater accessibility of the molecules to the porous structure of the solid, generating higher values of surface area and also the possibility that more molecules interacted with the surface and thus increased the values of the immersion enthalpies (Bansal and Goyal 2005).

Finally, Fig. 14.10 shows the relationship between $-\Delta H_{imm}$ of VOCs into the solids and the hydrophobic factor of the activated carbons. As the hydrophobic character increased, the interaction with the volatile organic compounds studied also increased since the hydrophobic factor increased proportionally with the thermal activation and decreased with the chemical activation as indicated in the figure, since for the sample OC723 increased by 41%, for OC1023 it multiplied its value 2.4 times and for C1173 it increased its value three times with respect to OC. This occurred because as the oxygenated groups increased the affinity with polar molecules, while the decomposition of such groups due to temperature causes greater interaction with non-polar substances due to dispersive and non-specific interactions (Rodríguez et al. 2009; Blanco-Martínez et al. 2009; Carvajal-Bernal et al. 2018).

According to the above, this chapter described some characteristics of the process of adsorption, the use and health consequences of four volatile organic compounds of non-polar type with difference in their structure and molecular arrangement (benzene, toluene, cyclohexane, and hexane) and how they can be removed with activated carbon, for this, five samples of this type of solids with differences in their physicochemical properties were presented as an example, so that it could be evidenced by means of different parameters, changes in the adsorbent–adsorbate interaction using immersion calorimetry as an interesting technique that would generate information about the energy involved in the immersion of the solvents in the studied samples.

14.5 Conclusions

- Phytoremediation is useful for the removal of VOCs from indoors, especially at low concentrations, where the plants can take the pollutants through leaves, cuticles, and stomata. For the case of benzene and toluene that are one of the most common molecules studied, the plant enzymes generate hydroxylation and cleavage processes for the aromatic rings that allow them to be added to amino acids or nonvolatile organic acids. Besides, more exhaustive study of the removal and transformation mechanisms of VOCs is required, as well as a broadening of the range of plant species that could be used.
- Conversely, for the adsorption process, activated carbons are useful adsorbent materials for the removal of volatile organic compounds; if these pollutants are non-polar, it was evidenced that the adsorbent–adsorbate interaction increases when:
  - The solids show high values of surface area and micropore volume in their textural properties since this increasesthe space available for the entrance of the molecules.
  - The adsorbent materials contain high values of hydrophobicity and basicity in their chemical characteristics, also low content of acidic groups, since the low content of heteroatoms favors non-specific interactions (van der Waals type).
  - The adsorbates are of the aromatic type due to the interaction between the regions with high electronic density located in the graphene layers and the π electrons of the molecule.
  - The adsorbates have planar arrangement (benzene and toluene) since they can be stacked in the activated carbon structure. The adsorbent–adsorbate interaction decreases when the arrangement is warped (cyclohexane) and finally the lower interaction takes place when the adsorbate has an elongated cylinder order (hexane), since this configuration restricts a little the entrance to the pores and the subsequent interaction with them.

Acknowledgements The authors thank the Framework Agreement between the Universidad de los Andes and the Universidad Nacional de Colombia and the act of agreement established between the Chemistry Departments of the two universities. Besides, the authors thank the Colciencias Scholarship "Doctorados Nacionales 2016" Convocation 757. Prof. Dr. Juan Carlos Moreno-Piraján also appreciates the support provided by the Vice-rectory of research of the Universidad de los Andes (Bogotá, Colombia) with the program "Convocatoria de Proyectos de investigación y Creación – Nuevo como Regionalización" number INV-2019-91-1905.

References

Abdulrasheed AA, Jalil AA, Triwahyono S, Zaini MAA, Gambo Y, Ibrahim M (2018) Surface modification of activated carbon for adsorption of SO2 and NOX: a review of existing and emerging technologies. Renew Sustain Energy Rev 94:1067–1085. https://doi.org/10.1016/j.rser.2018.07.011

Agency for Toxic Substances and Disease Registry (2015) Public health statement: toluene. Publ Health Serv 2015. https://www.atsdr.cdc.gov/phs/phs.asp?id=159&tid=29

Anuradha S, Raj KJA, Elangovan T, Viswanathan B (2014) Adsorption of VOC on steam activated carbon derived from coconut Shell charcoal. Indian J Chem Technol 21:345–349

Arafat HA, Ahnert F, Pinto NG (2004) On the adsorption of aromatics on oxygenated activated carbon in nonaqueous adsorption media. Sep Sci Technol 39:43–62. https://doi.org/10.1081/ss-120027400

ATSDR (2010) Toxicological profile for N-hexane. ATSDR Toxicol Profiles 2010. http://www.atsdr.cdc.gov/toxprofiles/tp113.pdf

ATSDR (2017) Toxicological profile for toluene. Agency for Toxic Substances and Disease Registry (ATSDR), Atlanta, GA. https://www.atsdr.cdc.gov/toxprofiles/tp56.pdf

Bansal RC, Goyal M (2005) In: Bansal RC, Goyal M (eds) Activated carbon adsorption. CRC, Boca Raton

Bates MN, Reed BR, Liu S, Eisen EA, Katharine Hammond S (2016) Solvent exposure and cognitive function in automotive technicians. Neurotoxicology 57:22–30. https://doi.org/10.1016/j.neuro.2016.08.009

Belhachemi M, Addoun F (2011) Effect of heat treatment on the surface properties of activated carbons. J Chem 8:992–999. https://doi.org/10.1155/2011/649254

Berenjian A, Chan N, Malmiri HJ (2012) Volatile organic compounds removal methods: a review. Am J Biochem Biotechnol 8:220–229. https://doi.org/10.3844/ajbbsp.2012.220.229

Betancur-Sánchez AM, Vásquez-Trespalacios EM, Sardi-Correa C (2017) Impaired colour vision in workers exposed to organic solvents: a systematic review. Archivos de La Sociedad Española de Oftalmología (English Edition) 92(1):12–18. https://doi.org/10.1016/j.oftale.2016.09.003

Bhatnagar A, Hogland W, Marques M, Sillanpää M (2013) An overview of the modification methods of activated carbon for its water treatment applications. Chem Eng J 219:499–511. https://doi.org/10.1016/j.cej.2012.12.038

Blanco-Martínez DA, Giraldo L, Moreno-Piraján JC (2009) Immersion enthalpy of carbonaceous samples in aqueous solutions of monohydroxilated phenols. J Therm Anal Calorim 96(3):853–857. https://doi.org/10.1007/s10973-009-0057-9

Bradley RH (2011) Recent developments in the physical adsorption of toxic organic vapours by activated carbons. Adsorp Sci Technol 29:1–28. https://doi.org/10.1260/0263-6174.29.1.1

Brennan JK, Thomson KT, Gubbins KE (2002) Adsorption of water in activated carbons: effects of pore blocking and connectivity. Langmuir 18:5438–5447. https://doi.org/10.1021/la0118560

Brilli F, Fares S, Ghirardo A, de Visser P, Calatayud V, Muñoz A, Annesi-Maesano I et al (2018) Plants for sustainable improvement of indoor air quality. Trends Plant Sci 23:507–512. https://doi.org/10.1016/j.tplants.2018.03.004

Carvajal-Bernal A, Gómez-Granados F, Giraldo L, Moreno-Piraján J (2018) A study of the interactions of activated carbon-phenol in aqueous solution using the determination of immersion enthalpy. Appl Sci 8:843–859. https://doi.org/10.3390/app8060843

Cheng Y, He H, Yang C, Zeng G, Li X, Chen H, Guanlong Y (2016) Challenges and solutions for biofiltration of hydrophobic volatile organic compounds. Biotechnol Adv 34:1091–1102. https://doi.org/10.1016/j.biotechadv.2016.06.007

Costa TL, Barboni MTS, de Araújo Moura AL, Bonci DMO, Gualtieri M, de Lima Silveira LC, Ventura DF (2012) Long-term occupational exposure to organic solvents affects color vision, contrast sensitivity and visual fields. PLoS One 7:1–9. https://doi.org/10.1371/journal.pone.0042961

Daud WMAW, Houshamnd AH (2010) Textural characteristics, surface chemistry and oxidation of activated carbon. J Nat Gas Chem 19(3):267–279. https://doi.org/10.1016/S1003-9953(09)60066-9

Fomin YD, Ryzhov VN, Tsiok EN (2015) The behavior of cyclohexane confined in slit carbon nanopore. J Chem Phys 143(18):184702. https://doi.org/10.1063/1.4935197

Gallego E, Roca FJ, Perales JF, Guardino X (2013) Experimental evaluation of VOC removal efficiency of a coconut shell activated carbon filter for indoor air quality enhancement. Build Environ 67:14–25. https://doi.org/10.1016/j.buildenv.2013.05.003

García T, Murillo R, Cazorla-Amorós D, Mastral AM, Linares-Solano A (2004) Role of the activated carbon surface chemistry in the adsorption of phenanthrene. Carbon 42(8–9):1683–1689. https://doi.org/10.1016/j.carbon.2004.02.029

Gawrońska H, Bakera B (2015) Phytoremediation of particulate matter from indoor air by *Chlorophytum Comosum* L. plants. Air Qual Atmos Health 8(3):265–272. https://doi.org/10.1007/s11869-014-0285-4

Giraldo L, Moreno-Piraján JC (2007) Instrumentación Calorimétrica Aplicada a La Determinación de Entalpias de Inmersión de Sólidos Porosos. In: Moreno-Piraján JC (ed) *Preparación, Caracterización y Aplicaciones*. Ed. Uniandes, Bogotá, pp 281–297

Giraldo L, Rodríguez-Estupiñán P, Moreno-Piraján JC (2018) Calorimetry of immersion in the energetic characterization of porous solids. In: Moreno-Piraján JC (ed) Calorimetry—design, theory and applications in porous solids. Intech Open, London, pp 35–53. https://doi.org/10.5772/intechopen.71051

Goncharuk OV (2015) The heat of immersion of modified silica in polar and nonpolar liquids. J Therm Anal Calorim 120(2):1365–1373. https://doi.org/10.1007/s10973-015-4438-y

Hernández-Monje D, Giraldo L, Moreno-Piraján JC (2016) Estudio Entálpico de La Inmersión de Carbones Activados Granulares Modificados En Benceno, Hexano y Ciclohexano. Afinidad 74: 1–8

Hernández-Monje D, Giraldo L, Moreno-Piraján J (2019a) Immersion enthalpy of activated carbon–cyclohexane and activated carbon–hexane. difference in the solid–liquid interaction enthalpy due to the structure of the solvent. *Processes* 7(4):180. https://doi.org/10.3390/pr7040180

Hernández-Monje D, Giraldo L, Moreno-Piraján JC (2019b) Immersion enthalpy of benzene/cyclohexane and toluene/cyclohexane binary mixtures into modified activated carbons. J Therm Anal Calorim 138:2565–2575. https://doi.org/10.1007/s10973-019-08478-0

Hong-li W, Sheng-ao J, Sheng-rong L, Hu Q-y, Li L, Shi-kang T, Cheng H, Li-ping Q, Chang-hong C (2017) Volatile organic compounds (VOCs) source profiles of on-road vehicle emissions in China. Sci Total Environ 607–608:253–261. https://doi.org/10.1016/j.scitotenv.2017.07.001

Hörmann V, Brenske KR, Ulrichs C (2017) Suitability of test chambers for analyzing air pollutant removal by plants and assessing potential indoor air purification. Water Air Soil Pollut 228(10):402–414. https://doi.org/10.1007/s11270-017-3586-z

Huang Y, Steven H, Lu Y, Niu R, Xu L, Cao J (2016) Removal of indoor volatile organic compounds via photocatalytic oxidation: a short review and Prospect. Molecules 21:56–76. https://doi.org/10.3390/molecules21010056

INSHT (2009) Documentación Toxicológica Para El Establecimiento Del Límite de Exposición Profesional Del Ciclohexano. http://www.insht.es/InshtWeb/Contenidos/Documentacion/TextosOnline/Valores_Limite/Doc_Toxicologica/Ficheros/DLEP10.pdf

Irga PJ, Torpy FR, Burchett MD (2013) Can hydroculture be used to enhance the performance of indoor plants for the removal of air pollutants? Atmos Environ 77(October):267–271. https://doi.org/10.1016/j.atmosenv.2013.04.078

Kamal MS, Razzak SA, Hossain MM (2016) Catalytic oxidation of volatile organic compounds (VOCs)—a review. Atmos Environ 140:117–134. https://doi.org/10.1016/j.atmosenv.2016.05.031

Khaksar G, Treesubsuntorn C, Thiravetyan P (2016) Effect of endophytic *Bacillus Cereus* ERBP inoculation into non-native host: potentials and challenges for airborne formaldehyde removal. Plant Physiol Biochem 107(October):326–336. https://doi.org/10.1016/j.plaphy.2016.06.020

Kim KJ, Yoo EH, Kays SJ (2012) Decay kinetics of toluene phytoremediation stimulation. Hort Sci 47(8):1195–1198. https://doi.org/10.21273/hortsci.47.8.1195

Lacerda EMDCB, Lima MG, Rodrigues AR, Teixeira CEC, De Lima LJB, Ventura DF, Silveira LCDL (2012) Psychophysical evaluation of achromatic and chromatic vision of workers chronically exposed to organic solvents. J Environ Public Health 2012:1–7. https://doi.org/10.1155/2012/784390

Lee EH, Paek D, Kho YL, Choi K, Chae HJ (2013) Color vision impairments among shipyard workers exposed to mixed organic solvents, especially xylene. Neurotoxicol Teratol 37:39–43. https://doi.org/10.1016/j.ntt.2013.02.005

Li L, Liu S, Liu J (2011) Surface modification of coconut shell based activated carbon for the improvement of hydrophobic VOC removal. J Hazard Mater 192:683–690. https://doi.org/10.1016/j.jhazmat.2011.05.069

Li S, Tosens T, Harley PC, Jiang Y, Kanagendran A, Grosberg M, Jaamets K, Niinemets Ü (2018) Glandular trichomes as a barrier against atmospheric oxidative stress: relationships with ozone uptake, leaf damage, and emission of LOX products across a diverse set of species. Plant Cell Environ 41(6):1263–1277. https://doi.org/10.1111/pce.13128

Limmer MA, Wilson J, Westenberg D, Lee A, Siegman M, Burken JG (2018) Phytoremediation removal rates of benzene, toluene, and chlorobenzene. Int J Phytoremediation 20(7):1413330–1413339. https://doi.org/10.1080/15226514.2017.1413330

Liu HB, Yang B, Xue ND (2016) Enhanced adsorption of benzene vapor on granular activated carbon under humid conditions due to shifts in hydrophobicity and Total micropore volume. J Hazard Mater 318:425–432. https://doi.org/10.1016/j.jhazmat.2016.07.026

Lo J, Velo-gala I, Sa M (2014) Surface modifications of activated carbon by gamma irradiation. Carbon 67:236–249. https://doi.org/10.1016/j.carbon.2013.09.087

Lopes C, da Anna S, de Carvalho SML, Brasil DSB, de Alcântara Mendes R, Lima MO (2015) Surface modification of commercial activated carbon (CAG) for the adsorption of benzene and toluene. Am J Anal Chem 6(6):528–538. https://doi.org/10.4236/ajac.2015.66051

Luttrell WE, Lyiza C (2010) Cyclohexane. J Chem Health Saf 17(3):33–34. https://doi.org/10.1016/j.jchas.2010.03.009

Mangun CL, Benak KR, Daley MA, Economy J (1999) Oxidation of activated carbon fibers: effect on pore size, surface chemistry, and adsorption properties. Chem Mater 11:3476–3483. https://doi.org/10.1021/cm990123m

Martínez de Yuso A, Izquierdo MT, Valenciano R, Rubio B (2013) Toluene and N-hexane adsorption and recovery behavior on activated carbons derived from almond shell wastes. Fuel Process Technol 110:1–7. https://doi.org/10.1016/j.fuproc.2013.01.001

Mazlan MAF, Uemura Y, Yusup S, Elhassan F, Uddin A, Hiwada A, Demiya M (2016) Activated carbon from rubber wood sawdust by carbon dioxide activation. Proc Eng 148:530–537. https://doi.org/10.1016/j.proeng.2016.06.549

Mirzaei A, Leonardi SG, Neri G (2016) Detection of hazardous volatile organic compounds (VOCs) by metal oxide nanostructures-based gas sensors: a review. Ceram Int 42:15119–15141. https://doi.org/10.1016/j.ceramint.2016.06.145

Montes-Morán MA, Dimas S, Angel Menéndez J, Fuente E (2012) The basicity of carbons. In: Tascón JMD (ed) Novel carbon adsorbents, 1st edn. Elsevier, Oxford, pp 173–203. https://doi.org/10.1016/b978-0-08-097744-7.00006-5

Moreno-Piraján JC, Giraldo L, García-Cuello V, Delgadillo DPV, Rodríguez-Estupiñán P, Murillo-Acevedo Y, Cantillo M (2011) Thermodynamic of the interactions between gas-solid and solid-liquid on carbonaceous materials. In: Thermodynamics—interaction studies: solids, liquids and gases, pp 164–195. https://doi.org/10.5772/31875

Moreno-Piraján JC, Blanco D, Giraldo L (2012) Relation between the adsorbed quantity and the immersion enthalpy in catechol aqueous solutions on activated carbons. Int J Mol Sci 13:44–55. https://doi.org/10.3390/ijms13010044

Moro AM, Brucker N, Charão M, Bulcão R, Freitas F, Baierle M, Nascimento S et al (2012) Evaluation of genotoxicity and oxidative damage in painters exposed to low levels of toluene. Mutat Res 746:42–48. https://doi.org/10.1016/j.mrgentox.2012.02.007

Nurmatov UB, Tagiyeva N, Semple S, Devereux G, Sheikh A (2015) Volatile organic compounds and risk of asthma and allergy: a systematic review. Eur Respir Rev 24:92–101. https://doi.org/10.1183/09059180.00000714

Nwaka D, Tahmasebi A, Lu T, Jianglong Y (2016) The effects of pore structure on the behavior of water in lignite coal and activated carbon. J Colloid Interface Sci 477:138–147. https://doi.org/10.1016/j.jcis.2016.05.048

Pak SH, Jeon MJ, Jeon YW (2016) Study of sulfuric acid treatment of activated carbon used to enhance mixed VOC removal. Int Biodeter Biodegr 113:195–200. https://doi.org/10.1016/j.ibiod.2016.04.019

Park H, Park HD, Jang JK (2016) Exposure characteristics of construction painters to organic solvents. Saf Health Work 7(1):63–71. https://doi.org/10.1016/j.shaw.2015.10.004

Parseh I, Teiri H, Hajizadeh Y, Ebrahimpour K (2018) Phytoremediation of benzene vapors from indoor air by Schefflera Arboricola and Spathiphyllum Wallisii plants. Atmos Pollut Res 9(6):1083–1087. https://doi.org/10.1016/j.apr.2018.04.005

Pei J, Zhang JS (2012) Determination of adsorption isotherm and diffusion coefficient of toluene on activated carbon at low concentrations. Build Environ 48(1):66–76. https://doi.org/10.1016/j.buildenv.2011.08.005

Rodríguez G, Giraldo L, Moreno JC (2009) Entalpías de Inmersión de Telas de Carbon Activado Como Parámetro de Caracterización Fisicoquímica. Rev Colombiana de Química 38(3):435–447

Rodríguez-Estupiñán P, Giraldo L, Moreno-Piraján JC (2013) Modified surface chemistry of activated carbons. Correlation with immersion enthalpy. J Therm Anal Calorim 114(1):245–251. https://doi.org/10.1007/s10973-012-2932-z

Salar-García MJ, Ortiz-Martínez VM, Hernández-Fernández FJ, de los Ríos AP, Quesada-Medina J (2017) Ionic liquid technology to recover volatile organic compounds (VOCs). J Hazard Mater 321:484–499. https://doi.org/10.1016/j.jhazmat.2016.09.040

Sarigiannis DA, Karakitsios SP, Gotti A, Liakos IL, Katsoyiannis A (2011) Exposure to major volatile organic compounds and carbonyls in European indoor environments and associated health risk. Environ Int 37(4):743–765. https://doi.org/10.1016/j.envint.2011.01.005

Sarma H, Forid N, Prasad R, Prasad MNV, Ma LQ, Rinklebe J (2021) Enhancing phytoremediation of hazardous metal(loid)s using genome engineering CRISPR–Cas9 technology. J Hazard Mater. https://doi.org/10.1016/j.jhazmat.2021.125493

Shafeeyan MS, Wan M, Wan A, Houshmand A, Shamiri A (2010) A review on surface modification of activated carbon for carbon dioxide adsorption. J Anal Appl Pyrolysis 89(2):143–151. https://doi.org/10.1016/j.jaap.2010.07.006

Shen W, Li Z, Liu Y (2008) Surface chemical functional groups modification of porous carbon. Recent Patents Chem Eng 1(1):27–40. https://doi.org/10.2174/2211334710801010027

Sonowal S, Nava AR, Joshi SJ, Borah SN, Islam NF, Pandit S, Prasad R, Sarma H (2022) Biosurfactants assisted heavy metals phytoremediation: green technology for the United Nations sustainable development goals. Pedosphere 2(1):198–210. https://doi.org/10.1016/S1002-0160(21)60067-X

Sriprapat W, Thiravetyan P (2013) Phytoremediation of BTEX from indoor air by Zamioculcas Zamiifolia. Water Air Soil Pollut 224(3):1482–1490. https://doi.org/10.1007/s11270-013-1482-8

Sriprapat W, Suksabye P, Areephak S, Klantup P, Waraha A, Sawattan A, Thiravetyan P (2014) Uptake of toluene and ethylbenzene by plants: removal of volatile indoor air contaminants. Ecotoxicol Environ Saf 102(1):147–151. https://doi.org/10.1016/j.ecoenv.2014.01.032

Tazibet S, Boucheffa Y, Lodewyckx P (2013) Heat treatment effect on the textural, hydrophobic and adsorptive properties of activated carbons obtained from olive waste. Micropor Mesopor Mater 170:293–298. https://doi.org/10.1016/j.micromeso.2012.12.008

Teiri H, Pourzamani H, Hajizadeh Y (2018) Phytoremediation of VOCs from indoor air by ornamental potted plants: a pilot study using a palm species under the controlled environment. Chemosphere 197(April):375–381. https://doi.org/10.1016/j.chemosphere.2018.01.078

Ternero-Hidalgo JJ, Rosas JM, Palomo J, Valero-Romero MJ, Rodríguez-Mirasol J, Cordero T (2016) Functionalization of activated carbons by HNO3 treatment: influence of phosphorus surface groups. Carbon 101:409–419. https://doi.org/10.1016/j.carbon.2016.02.015

Thakare M, Sarma H, Datar S, Roy A, Pawar P, Gupta K, Pandit S, Prasad R (2021) Understanding the holistic approach to plant-microbe remediation technologies for removing heavy metals and radionuclides from soil. Curr Res Biotechnol. https://doi.org/10.1016/j.crbiot.2021.02.004

Tham YJ, Latif PA, Abdullah AM, Shamala-Devi A, Taufiq-Yap YH (2011) Performances of toluene removal by activated carbon derived from durian shell. Bioresour Technol 102:724–728. https://doi.org/10.1016/j.biortech.2010.08.068

Torpy F, Clements N, Pollinger M, Dengel A, Mulvihill I, He C, Irga P (2018) Testing the single-pass VOC removal efficiency of an active Green Wall using methyl ethyl ketone (MEK). Air Qual Atmos Health 11(2):163–170. https://doi.org/10.1007/s11869-017-0518-4

Treesubsuntorn C, Thiravetyan P (2012) Removal of benzene from indoor air by Dracaena Sanderiana: effect of wax and stomata. Atmos Environ 57(September):317–321. https://doi.org/10.1016/j.atmosenv.2012.04.016

Tsai W-T (2016) Toxic volatile organic compounds (VOCs) in the atmospheric environment: regulatory aspects and monitoring in Japan and Korea. Environments 3:23–30. https://doi.org/10.3390/environments3030023

Ugrekhelidze D, Korte F, Kvesitadze G (1997) Uptake and transformation of benzene and toluene by plant leaves. Ecotoxicol Environ Saf 37(1):24–29. https://doi.org/10.1006/eesa.1996.1512

UNAM (2016) Hoja De Seguridad Benceno. http://www.quimica.unam.mx/IMG/pdf/5benceno.pdf

US EPA, OAR, Office of Air Quality Planning and Standards (2016) Benzene | technology transfer network air toxics web site | US EPA. EPA website. https://www.epa.gov/sites/production/files/2016-9/documents/benzene.pdf

Villacañas F, Pereira MFR, Órfão JJM, Figueiredo JL (2006) Adsorption of simple aromatic compounds on activated carbons. J Colloid Interface Sci 293(1):128–136. https://doi.org/10.1016/j.jcis.2005.06.032

Vinke P, van der Eijk M, Verbree M, Voskamp AF, van Bekkum H (1994) Modification of the surfaces of a gas activated carbon and a chemically activated carbon with nitric acid, hypochlorite, and ammonia. Carbon 32(4):675–686

Wang R, Lang J, Yan X (2014a) Effect of surface area and heteroatom of porous carbon materials on electrochemical capacitance in aqueous and organic electrolytes. Sci China Chem 57:1570–1578. https://doi.org/10.1007/s11426-014-5123-x

Wang S, Liang Z, Long C, Li A (2014b) Enhanced adsorption and desorption of VOCs vapor on novel Micro-mesoporous polymeric adsorbents. J Colloid Interface Sci 428:185–190. https://doi.org/10.1016/j.jcis.2014.04.055

Wang G, Dou B, Zhang Z, Wang J, Liu H, Hao Z (2015) Adsorption of benzene, cyclohexane and hexane on ordered mesoporous carbon. J Environ Sci (China) 30:65–73. https://doi.org/10.1016/j.jes.2014.10.015

Wei Z, Van Le Q, Peng W, Yang Y, Yang H, Haiping G, Lam SS, Sonne C (2021) A review on phytoremediation of contaminants in air, water and soil. J Hazard Mater 403 (February):123658–123665. https://doi.org/10.1016/j.jhazmat.2020.123658

Wibowo N, Setyadhi L, Wibowo D, Setiawan J, Ismadji S (2007) Adsorption of benzene and toluene from aqueous solutions onto activated carbon and its acid and heat treated forms: influence of surface chemistry on adsorption. J Hazard Mater 146:237–242. https://doi.org/10.1016/j.jhazmat.2006.12.011

Yang X, Yi H, Tang X, Zhao S, Yang Z, Ma Y, Feng T, Cui X (2018) Behaviors and kinetics of toluene adsorption -desorption on activated carbons with varying pore structure. J Environ Sci 67(July):104–114. https://doi.org/10.1016/J.JES.2017.06.032

Yao X, Liu J, Gong G, Yu J, Xie Q (2013) Preparation and modification of activated carbon for benzene adsorption by steam activation in the presence of KOH. Int J Min Sci Technol 23(3):395–401. https://doi.org/10.1016/j.ijmst.2013.05.015

Yin CY, Aroua MK, Daud WMAW (2007) Review of modifications of activated carbon for enhancing contaminant uptakes from aqueous solutions. Sep Purif Technol 52:403–411. https://doi.org/10.1016/j.seppur.2006.06.009

Zhang X, Gao B, Creamer AE, Cao C, Li Y (2017) Adsorption of VOCs onto engineered carbon materials: a review. J Hazard Mater 338(September):102–123. https://doi.org/10.1016/j.jhazmat.2017.05.013

Zhou L, Yu Q, Cui Y, Xie F, Li W, Li Y, Chen M (2017) Adsorption properties of activated carbon from Reed with a high adsorption capacity. Ecol Eng 102(102):443–450. https://doi.org/10.1016/j.ecoleng.2017.02.036

Phytoremediation: A Tool for Environmental Sustainability

15

Neerja Srivastava

Abstract

Environment is a very significant and essential part for the survival of both man and other biotic organisms. The existence as well as security of the entire components is primarily based on the conservation of the physical environment. Due to industrial revolution, pollution in the environment has amplified enormously. Rise in population also causes strain on the environment with many commercial activities such as logging and mining. In fact, the elimination of harmful pollutants with any known method is just not sufficient. Therefore, the best practice for maintaining ecological balance is to use all the wastes in a recyclable manner which will assist the biotic and abiotic components to maintain visually attractive as well as healthy and perfect environment.

A novel holistic approach for "sustainable phytoremediation" or "phytomanagement," is nowadays being recommended where economically as well as ecologically precious, natural colonizer species are being utilized for the remediation of contaminated sites, instead of introduced species. There is a broad variety of naturally colonizing vegetation on contaminated as well as waste dump sites which have phytoremediation potential. Of these, certain plants are suitable for sustainable phytoremediation in terms of creating a multifunctional ecosystem. Natural vegetation on contaminated sites as well as waste dump sites is the right choice for selecting a suitable candidate for phytoremediation plans. If scientific experts can choose ecologically and socioeconomically significant plants, like aromatic and energy plants among the natural vegetation, then sustainability in phytoremediation can be achieved.

N. Srivastava (✉)
Department of Biochemistry, IBSBT, CSJM University, Kanpur, Uttar Pradesh, India
e-mail: neerjasrivastava@csjmu.ac.in

R. Prasad (ed.), *Phytoremediation for Environmental Sustainability*,
https://doi.org/10.1007/978-981-16-5621-7_15

Keywords

Phytoremediation · Pollutants · Environmental sustainability · Sustainable phytoremediation

15.1 Introduction

Earth is gradually being polluted with inorganic and organic compounds mainly due to human activities. While inorganic contaminants are present as natural elements in the Earth's crust and atmosphere, human activities like industry, mining, motorized traffic, agriculture, logging, as well as military operations promote their discharge and accumulation in the environment, which leads to toxicity (Nriagu 1979). Organic contaminants in the environment are generally man-made and xenobiotic means not usually formed or expected to be found in organisms, of which several are toxic and/or carcinogenic. Organic pollutants are produced in the environment through accidental releases like fuels or solvents, industrial activities like chemical, petrochemical, agriculture like pesticides, herbicides, and military actions like explosives, chemical weapons besides others. In fact, contaminated sites mostly possess combination of both organic and inorganic pollutants (Ensley 2000). Around 6–8 billion dollars a year is spent on environmental remediation in the US, and 25–50 billion dollars per year worldwide (Glass 1999; Tsao 2003). Most of the remediation is still being done through traditional procedures like excavation and reburial, capping, and soil washing and burning. But, new developing biological remediation procedures, like phytoremediation, are generally easy to perform and economical. Phytoremediation includes variety of technologies which utilize plants to eliminate, lessen, degrade, or immobilize environmental contaminants from soil as well as water, therefore converting polluted sites in a relatively clean, nontoxic environment. Phytoremediation is based upon natural processes in which plants detoxify inorganic as well as organic pollutants, through degradation, sequestration, or transformation (Pilon-Smits and Freeman 2006; Thakare et al. 2021; Sarma et al. 2021; Sonowal et al. 2022). The uses of plants in contamination remediation have been tested since the 1970s, and in the 1980s, the governmental and commercial sectors started recognizing the concept of phytoremediation (Lu et al. 2018). Gradually this technology has been widely explored over the years, and there are over 100 soil heavy metal remediation pilot/field projects using the phytoremediation technology that have been reported (USEPA 2016). Based on the applicability, phytoremediation techniques are subdivided into different classes (Ali et al. 2013; Khalid et al. 2017; Mahar et al. 2016; Rezania et al. 2016; Yadav et al. 2018, Prabakaran et al. 2019).

Besides their conventional role for production of food, feed, fuel, and fibers, green plants can be employed to store toxic metals as well as organic contaminants from polluted soils and water for cleanup reasons, to stop further deterioration of our surroundings and to ameliorate the damage caused through increasingly industrialized society. The utilization of plants particularly selected or produced for the restoration of contaminated land and brownfields, water purification, and even elimination of indoor or outdoor air pollutants is becoming indispensable to achieve sustainable development (Conesa et al. 2008). Plants signify better environmentally suitable and cheaper technique for site renewal in comparison to physicochemical strategies, even if the time period needed to achieve the target is mostly a restrictive factor. Plants are already remediating our environment continuously, universally, working as "green livers," even if we do not identify or see it. Trace element storing species can accumulate arsenic, cadmium, cobalt, manganese, nickel, lead, selenium, thallium, or zinc up to 100 or 1000 times more than normally stored through plants (Al-Najar et al. 2005; Behmer et al. 2005; Caille et al. 2005; Comino et al. 2005; Jiang et al. 2005; McGrath et al. 2006; Zhao et al. 2006). People have started employing plants which hyperaccumulates particular metals in remediation processes in past few decades. Contrary to it, crops with a decreased ability to store toxic metals as well as organic contaminants in edible portion should be valued to increase food security. While crop plants with increased ability to store essential minerals in simple assimilated form can assist in giving nutritious food to the fast-growing global population and enhance human welfare via well-adjusted mineral nutrition. The concept of enriching food crops with the essential minerals needed for a balanced diet is comparatively new. Like in the case of iron and zinc deficiencies which are currently the prime nutritional ailments all over the world and most of the people get it through eating plants, enhancing the iron and/or zinc concentration in crop plants could improve their health significantly. Most metals which can be hyperaccumulated are also essential nutrients, and food fortification as well as phytoremediation are therefore two sides of the same coin (White and Broadley 2005). United Nations Environment Programme proposed "phytotechnologies as ecotechnologies related with the utilization of plants to settle environmental difficulties in a crisis management through prevention of site degradation, remediation and regeneration of damaged ecosystems, regulation of environmental processes, observation and valuation of the environmental quality." Phytotechnologies utilize natural methods and can be employed for remediating damaged lands like quarries and road sides, exclusion of unnecessary nutrient loads, i.e., phytoamelioration and the cleaning of wastewater such as road runoff, municipal as well as industrial wastes, landfill leachates, stormwater, surface, and seepage water. Phytotechnologies provide effective tools and environment-friendly solutions for remediating polluted sites and water, enhancement in food chain security, as well as development of renewable energy sources, which contributes towards sustainable utilization of water and land management (Domínguez et al. 2008; Schwitzguebel et al. 2009).

15.2 Phytoremediation

The term "phytoremediation" was derived from the Greek phyto, meaning "plant," and the Latin suffix remedium, "able to cure" or "restore," by Ilya Raskin in 1994, and is employed to mention those plants which can remediate polluted medium (Vamerali et al. 2010). Phytoremediation is also known as green remediation, botanoremediation, agroremediation, or vegetative remediation and can be described as an in situ remediation approach that utilizes plants and accompanying microbiota, soil amendments, and agronomical practices to eliminate, restrict, or make environmental pollutants harmless (Cunningham and Ow 1996; Helmisaari et al. 2007; Srivastava 2016).

Phytoremediation is a novel emerging field of science and technology (Salt et al. 1998) which utilizes plants to remediate contaminated soil, groundwater as well as wastewater. Phytoremediation is described as the utilization of green plants with grasses and woody species, to eliminate, restrict, or transform environmental pollutants like heavy metals, metalloids, trace elements, organic compounds, and radioactive compounds risk-free in soil or water. This definition comprises all plant-influenced biological (Zouboulis and Katsoyiannis 2005), chemical as well as physical methods that help in the intake, compartmentalization, decomposition, and metabolism of pollutants, through plants, soil microbes, or plant and microbial interactions. Phytoremediation takes advantage of the exclusive as well as selective uptake capacity of plant root systems, along with the translocation, bioaccumulation, and pollutant accumulation/decomposition capacity of the whole plant body. Plant-dependent soil remediation schemes can be seen as biological treatment schemes with a widespread, self-expanding uptake network, the root system which increases the underground ecosystem for successive fruitful application. Phytoremediation averts excavation as well as transportation of contaminated media which decreases the danger of dispersing the pollution and has the capacity to remediate sites contaminated with several varieties of contaminants. Certain disadvantages related with phytoremediation are dependance on the growing environment needed by the plant like climate, geology, altitude, temperature, extensive operations need accessibility of agricultural tools and information; plant resistance to the contaminant influences the remediation success; pollutants accumulated in senescing tissues may be discharged back into the surroundings in particular seasons; period taken to treat sites is more than other techniques; and pollutant solubility may be enhanced which leads to more environmental degradation and the probability of leakage (Mudhoo et al. 2010).

15.3 Mechanisms of Phytoremediation

There are various methods through which plants clean or remediate polluted sites. The plants uptake pollutants via the root system which possess the key mechanisms for averting toxicity. The root system offers large surface area which absorbs and stores water and nutrients vital for growth besides other nonessential pollutants (Raskin and Ensley 2000). There are several processes through which plants can influence pollutant quantity in soil, sediments, as well as water. While, there are several similarities in some of these processes, but the categorization varies (Table 15.1). Every process affects the amount, movement, or toxicity of pollutants, as the use of phytoremediation is projected to do (USEPA 2000).

Table 15.1 Types of phytoremediation (Susarla et al. 2002)

S. no	Phytoremediation type	Pollutants treated
1.	Phytoextraction/ phytoaccumulation	Cd, Cr, Pb, Ni, Zn, and other heavy metals, Se, radionuclides; BTEX (benzene, ethyl benzene, toluene, and xylenes), pentachlorophenol, short-chained aliphatic compounds, and other organic compounds
2.	Rhizofiltration	Heavy metals, organic chemicals, and radionuclides
3.	Phytovolatilization	Chlorinated solvents (tetrachloroethane, trichloromethane and tetrachloromethane); Hg and Se
4.	Phytostabilization	Heavy metals in mine tailings ponds, phenols and chlorinated solvents (tetrachloromethane and trichloromethane)
5.	Phytodegradation/ Phyto-transformation	Munitions (DNT, HMX, nitrobenzene, nitroethane, nitromethane, nitrotoluene, picric acid, RDX, TNT), atrazine; chlorinated solvents (chloroform, carbon tetrachloride, hexachloroethane, tetrachloroethene, trichloroethene, dichloroethene, vinyl chloride, trichloroethanol, dichloroethanol, trichloroacetic acid, dichloroacetic acid, monochloroacetic acid, tetrachloromethane, trichloromethane), DDT; dichloroethene; methyl bromide; tetrabromoethene; tetrachloroethane; other chlorine and phosphorus-based pesticides; polychlorinated biphenols, other phenols, and nitriles
6.	Rhizodegradation/ Phytostimulation	Polycyclicaromatic hydrocarbons; BTEX (benzene, ethylbenzene, toluene, and xylenes); other petroleum hydrocarbons; atrazine; alachlor; polychlorinated biphenyl (PCB); tetrachloroethane, trichloroethane; and other organic compounds

15.3.1 Phytoextraction

This process is also known as phytoaccumulation, where metal pollutants in the soil are taken up through plant roots in the aerial parts of the plants. Phytoextraction is mainly utilized for the remediating polluted soils (Zhang et al. 2010). This strategy employs plants to take up, collect, as well as precipitate harmful metals from polluted soils into the aerial parts like shoots, leaves. Detection of metal hyperaccumulator species shows that plants have the capacity for eliminating metals from polluted soils (Wuana et al. 2010). A hyperaccumulator is a plant species with capacity of storing 100 times more metal in comparison to a general non-accumulating plant. Metals like Ni, Zn, and Cu are the ideal elements for elimination through phytoextraction as they are preferred by most of plants (about 400) which uptake and absorb huge quantities of metals. There are various benefits of phytoextraction. The expenses in phytoextraction are quite less in comparison to traditional processes. One more advantage is that pollutant is permanently eliminated from the soil. Besides this, the quantity of waste material which has to be discarded is significantly reduced (EPA 2000) which is almost up to 95%, and in certain cases, the pollutant can be recycled from the pollutant plant biomass. The application of hyperaccumulator species is restricted because of slow growth, shallow root system as well as insignificant biomass yield. Besides this, the plant biomass should be collected and discarded appropriately. There are various reasons which restrict the range of metal phytoextraction like bioavailability of metals inside the rhizosphere, rate of metal uptake through roots, percentage of metal "fixed" inside the roots, rate of xylem loading/translocation into shoots, and cellular resistance to harmful metals. The process is also generally restricted to metals as well as other inorganic material in soil or sediment. For making remediation process possible, the plants should (1) extract heavy metals in big amount in the roots, (2) transfer the heavy metal in the surface biomass, as well as (3) produce a huge amount of plant biomass. Besides this, treated plants should have processes for detoxification and/or resisting greater level of metals stored in their shoots (Brennan and Shelley 1999).

15.3.2 Rhizofiltration

This is employed for treating extracted groundwater, surface water as well as wastewater with less pollutants. In this process, there is adsorption or precipitation in plant roots or absorption of pollutants around the root zone. Rhizofiltration is generally utilized in either in situ or extracted groundwater, surface water, or wastewater for eliminating metals or other inorganic materials. Rhizofiltration can be employed for lead (Pb), cadmium (Cd), copper (Cu), nickel (Ni), zinc (Zn), and chromium (Cr), which are mainly held within the roots. Rhizofiltration is just like phytoextraction, but the plants are utilizing polluted groundwater in place of soil. To adjust the plants, when a huge root system is produced, polluted water is collected from a waste site and transported to the plants where it is replaced for their water source. The plants are then grown in the polluted region where the roots extract the water as well as pollutants. When the roots become saturated with pollutants, they are collected. Sunflower, Indian mustard, tobacco, rye, spinach, and corn have proven their potential to eliminate lead from water, of which sunflower has the highest ability. The benefit linked with rhizofiltration is its capacity to employ both terrestrial and aquatic plants for either in situ or ex situ utilizations. Additional benefit is that pollutants are not being translocated into the shoots. Therefore, species other than hyperaccumulators should be employed. Terrestrial plants are favored as they possess fibrous as well as much bigger root system, enhancing the root area. Drawbacks of rhizofiltration are: the requirement of continuous adjustment of pH, requirement of plants to be grown first in a greenhouse or nursery, regular harvesting as well as plant disposal, and need of decent knowledge of the chemical speciation/ interactions.

15.3.3 Phytovolatilization

This process utilizes plants to take up pollutants from the soil, converting them into volatile forms and transpiring them into the surroundings. Phytovolatilization may also diffuse pollutants from the stems or other plant organs so that the pollutant moves through before reaching the leaves. Phytovolatilization may occur with pollutants found in soil, sediment, or water. Hg is the main metal pollutant where this process is employed. It occurs with volatile organic compounds also like trichloroethene, as well as inorganic chemicals which have volatile forms, like Se and As. The benefit of this process is that the pollutant, like mercuric ion, may be converted into a less harmful compound. The drawback is that the Hg discharged into the environment is expected to be recycled through precipitation and then redeposited back into lakes and oceans, repeating the formation of methylmercury through anaerobic bacteria.

15.3.4 Phytostabilization

This is in situ inactivation and is employed for treating soil, sediment, as well as sludge. In this process, some plant species are utilized to immobilize pollutants in the soil and groundwater by absorption and storage through roots, adsorption on roots, or precipitation inside rhizosphere. This method reduces the movement of the pollutant and inhibits transport in the groundwater, and it also decreases bioavailability of metal in the food chain. This process can also be employed for restoring vegetation cover at places where natural vegetation is unable to survive because of large amounts of metals in surface soils or physical disruptions to surface materials. Metal-resistant species is utilized to reestablish vegetation at pollutant locations, which reduces the possible transfer of contaminants by wind erosion and transfer of exposed surface soils as well as leakage of soil pollutants in the groundwater. Phytostabilization can take place via sorption, precipitation, or reduction in metal valence. It is valuable for the remediation of Pb, As, Cd, Cr, Cu as well as Zn. Benefit of this process is the differences in soil chemistry and environment caused by presence of plant. These alterations in soil chemistry may encourage adsorption of pollutants in the plant roots or soil or precipitate metals in the plant root. Phytostabilization is successful in attending metals as well as other inorganic pollutants in soil and sediments. Certain benefits linked with this technique are that removal of dangerous material/biomass is not needed and it is very effectual when quick immobilization is required to conserve ground as well as surface waters (Zhang et al. 2009). The plant's presence also restricts soil erosion and reduces the quantity of water present in the system. But this remediation technique has various key drawbacks like: pollutant residual in soil, massive use of fertilizers, or soil modifications, which need compulsory observation as well as the stabilization of the pollutants basically because of soil amendments.

15.3.5 Phytodegradation

This is also called as phytotransformation. It decomposes complex organic molecules into simple molecules or integrates these molecules into plant tissues

(Trap et al. 2005). In the phytodegradation process, pollutants are decomposed after their uptake by the plant. Like phytoextraction and phytovolatilization, in this process also plant uptake usually takes place only when the solubility and hydrophobicity of pollutant drop into a definite suitable range. Phytodegradation remediates certain organic pollutants, like chlorinated solvents, herbicides as well as munitions, and it can attend pollutants in soil, sediment, or groundwater.

15.3.6 Rhizodegradation

This is also known as phytostimulation. It decomposes pollutants inside the plant root zone, or rhizosphere. It is thought to be performed through bacteria or other microbes. There are almost 100 times more microorganisms in rhizosphere soil in comparison to the soil outside the rhizosphere. Microbes are present more in the rhizosphere as plant secretes sugars, amino acids, enzymes as well as other substances which can induce growth of bacteria. The roots also have extra surface area for development of microbes and a route for transfer of oxygen from the surroundings. The restricted nature of rhizodegradation implies that it is mainly valuable in polluted soil and found to be little bit successful in remediating a large variety of usual organic chemicals like petroleum hydrocarbons, polycyclic aromatic hydrocarbons (PAHs), chlorinated solvents, pesticides, polychlorinated biphenyls (PCBs), benzene, toluene, ethylbenzene, as well as xylenes. It can be viewed as plant-supported bioremediation, the activation of microbial as well as fungal decomposition through discharge of exudates/enzymes in the rhizosphere (Zhuang et al. 2005; Sharma and Pandey 2014).

15.4 Environmental Sustainability

Sustainability is the ability to tolerate. The term "sustainability" is originated from the Latin sustinere (tenere, to hold; sus, up). In ecology, this means in what way biological systems stay diverse as well as fruitful all times. For human beings, it is the capacity for long-period maintenance of welfare, which is ultimately based upon the welfare of the nature and the responsible utilization of natural resources (http://en.wikipedia.org/wiki/Environmental Sustainability Index). Environmental sustainability is a method which ensures that existing methods of dealings with the surroundings are followed with the concept of keeping the surroundings as pure as naturally possible on the basis of perfect actions. An "unsustainable condition" arises when the entire resources of nature are utilized up earlier than it can be restored. Sustainability needs that humans just utilize natural resources at a speed at which they can be restored naturally. Hypothetically, the long-term consequence of environmental decomposition is the failure to nurture human life. Globally such decomposition could indicate loss of humanity (http://www.IndependentlySustainableRegion 2010). A healthy environment is that which gives essential commodities as well as facilities to humans and other creatures in its ecosystem. This can be attained by two routes and comprises finding ways of decreasing adverse human influence and increasing the welfare and life of all living creatures including plants as well as animals in the environment. Daly (1990) proposed three distinctive conditions for environmental sustainability: renewable sources should give a sustainable yield, i.e., the rate of yield should not be more than the rate of renewal; for nonrenewable sources, there should be equal growth of renewable replacements; waste production should not be more than the acclimatizing potential of the surroundings. It is essential to also distinctly describe what is the significance of the environment for humans who are in the center of it and are influenced positively as well as negatively according to their actions in the environment. Therefore, Bankole (2008) stated that "Environment" denotes the physical settings of man, where he is component as well as dependent for his functions such as physiological activities, production, as well as utilization. His physical surrounding ranges from air, water, and land to natural sources such as metals, energy carriers, soil, plants, animals, and ecosystems. For urban human being, major portion of his environment is man-made. But still, the nonnatural surroundings like buildings and roads as well as tools like clothes and automobiles are the outcome of both efforts and natural resources (Ezeonu et al. 2012).

15.5 The Sustainable Phytoremediation

For sustainable ecological as well as agricultural progress, it is essential to remediate polluted regions, and the entry of contaminants into the food chain should be reduced. Because of this, the plant-dependent remediation processes designated as phytoremediation got much recognition in the last few decades. It is an easy, dynamic, cheaper, requires less hard work, commonly accepted, well-suited, environmental-friendly, sustainable, dependable, as well as promising technique which can be applied in huge areas, especially when local, environmentally, and socioeconomically useful plants are utilized for the treatment of contaminated areas with which revenue is also generated through production of phytoproducts of polluted regions (Pandey et al. 2015, 2016). Phytoremediation is helpful in treating large number of contaminants and is about ten times less costly in comparison to traditional methods (do Nascimento and Xing 2006). Plants have the intrinsic potential to nullify both organic and inorganic contaminants through various methods like bioaccumulation, translocation, and degradation, therefore working as a crucial sink for biologically harmful contaminants (Pandey and Bajpai 2019).

It is well recognized that heavy metals are unable to be decomposed and demolished. They bioaccumulate via food chain and bring huge possibility of human health dangers. Among all the existing methods, phytoremediation is a cheap process for treating the polluted regions. Plants can treat contaminants by various methods such as adsorption, transport with translocation, hyperaccumulation or transformation as well as mineralization (Meagher 2000). A variety of naturally growing plant species have developed on heavy metals polluted areas. However, just a few of them are helpful in phytoremediation and makes a multifunctional ecosystem. The properties which make any species valuable in phytoremediation have rapid growth with capacity to store greater biomass, simple and quick proliferation, abundant root system, more metal storing ability, resistance to severe local soil conditions, and unacceptable by cattle (Pandey et al. 2012a). It is also required that they should be perennial, as well as should be capable of starting ecological succession. Additionally, the species chosen for remediation should also be beneficial in yield of commodities and facilities to the society. Extra advantages are carbon sequestration, increase in substrate quality, pleasant scenery, and biodiversity protection (Pandey 2002, 2013). Overlooking the problems of cost of inputs as well as maintenance, most of the existing research to date endorse introduced plant species for phytoremediation. For example, Vamerali et al. (2010) reported that worldwide introduced crop species are manly involved in phytoremediation. It obviously demonstrates that not naturally growing plants have not got much attention for the phytoremediation of polluted areas. But, employing introduced crop species to phytoremediate has several environmental, financial, and public challenges. The introduced crops need inputs as well as maintenances of their establishment on the severe environments that exist in heavy metals polluted regions. Moreover, if the introduced crops are edible, then there will be severe risk of heavy metals going in the food chain and ultimately affecting human health. These difficulties can be overcome by employing naturally developing species which can be inedible but financially as well as socially valuable for the public (Pandey and Singh 2011). Through our scientific work in creating appropriate information in this area and connecting this information to our action can assist us in decreasing the human health dangers and gaining further from the phytoremediation endeavors (Fig. 15.1).

15.5.1 Ecologically and Economically Useful Species

Naturally growing species are best and perfect choice for phytoremediation of polluted regions. If workers, by interdisciplinary work, are capable to select environmentally as well socioeconomically significant plant species or profitable crops like aromatic plants and energy crops among naturally growing species, then we can attain sustainable phytoremediation. Certain environmentally and socioeconomically significant plant species are munj (*Saccharum munja*), Kans (*S. spontaneum*), etc., which are excluders that restrict heavy metals toxicity. In the same way, certain aromatic plants like vetiver (*Vetiveria zizanioides*), lemon grass (*Cymbopogon flexuosus*), tulsi (*Ocimum basilicum*) are stress resistant in nature. The major product of aromatic crops is essential oil which is free from heavy metal dangers (Khajanchi et al. 2013). The potential energy crops such as *Ricinus communis* (Pandey 2013), *Jatropha curcas* (Pandey et al. 2012b), and *Miscanthus giganteus* (Nsanganwimana et al. 2014) have the capacity for phytoremediation of polluted areas with a variety of ecological as well as ecosystem services. All of these species are perennial as well as inedible by cattle. They are also environmentally suitable for phytoremediating heavy metals in contaminated areas,

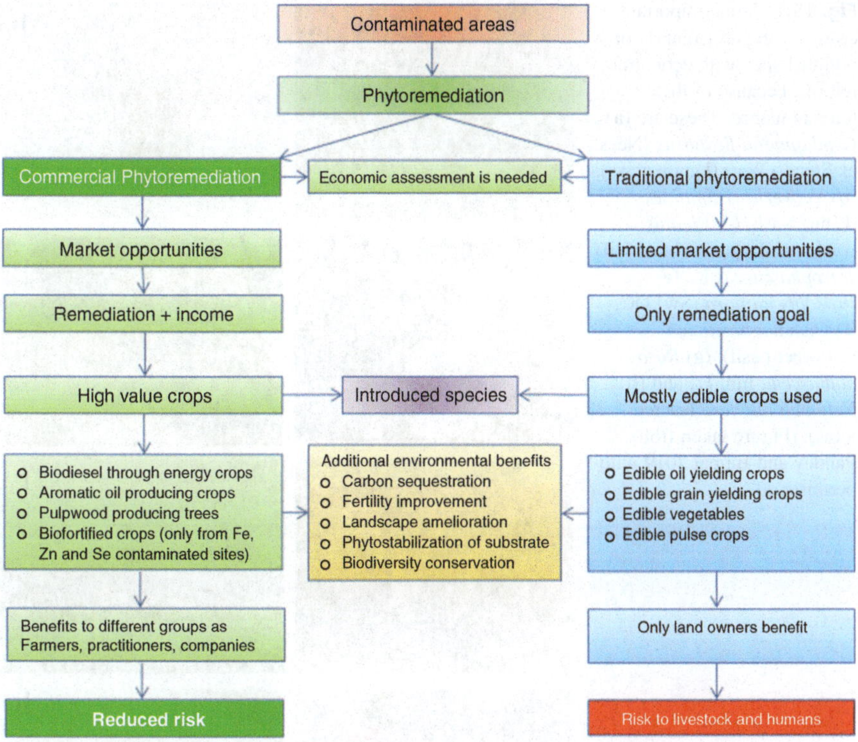

Fig. 15.1 A conceptual diagram showing comparison between the traditional phytoremediation and novel approach for sustainable phytoremediation. (Figure taken from Pandey and Souza-Alonso 2019 with permission)

and therefore present a new chance for their application in phytoremediation. More precisely, there is not much danger in utilization of major product of these species, e.g., essential oils and biodiesel (Pandey et al. 2015) (Fig. 15.2).

15.5.2 Plant Species Involved in Phytoremediation

Many plants had been identified and tested for phytoremediation work. Highly useful terrestrial as well as aquatic plant species have been recognized after challenging lab as well as field studies which are listed below (Table 15.2).

Certain other species are *Elodea canadensis*, *Ceratophyllum demersum*, *Potamogeton* spp., *Myriophyllum* spp., *Spartina alterniflora, Pinus sylvestris, Poa alpine,* and *Bouteloua gracilis* (Rice et al. 1997; Watanabe 1997). Lot of them are still wild, but others are domesticated because of their food value. They are high salt as well as toxicity resistant, have large root binding system, and were tested for restoration process. A variety of them quickly absorb, volatilize, and/or metabolize substances like tetrachloroethane, trichloroethylene, metachlor, atrazine,

Fig. 15.2 Some important crops for the remediation of polluted sites with economic returns, because of their tolerant nature. These are (**a**) *Cymbopogon flexuosus* (Nees ex Steud.) Wats (lemon grass), (**b**) *Vetiveria zizanioides* (Linn) Nash, (**c**) *Ricinus communis* L. (castor bean), (**d**) *Jatropha curcas* L., (**e**) *Prosopis juliflora* (Sw) DC, (**f**) *Ocimum basilicum* L. (sweet basil), (**g**) *Rosa damascena* mill L., and (**h**) *Nelumbo nucifera* (sacred lotus). (Figure taken from Pandey and Bajpai 2019 with permission)

nitrotoluenes, anilines, dioxins as well as several petroleum hydrocarbons. Perfect plants for this work are members of grass family Gramineae with Cyperaceae as well as the members of Brassicaceae specially the Brassica, Alyssum and Thlaspi, and Salicaceae particularly willow and poplar trees. Grasses like the vetiver, clover and rye grass, Bermuda grass, tall fescue are specifically very effectual in treating soils polluted through heavy metals as well as crude oil (Kim 1996). Sunflower plants (*Helianthus annus*) were planted on large scale around Chernobyl (erstwhile USSR), where nuclear tragedy in 1985 discharged huge quantity of radioactive substances into the surroundings. The land as well as soil of the region area was severely polluted. Sunflower is observed to take up radionuclides from soil to clean it up. This phytoremediation method has a cost of about 2 dollar per hectare for remediating the soil which might be costing millions of dollars through other methods. Duckweeds can "absorb" and "adsorb" total of dissolved gases as well as other substances, with heavy metals, from the wastewater. Just in 2–3 weeks, the condition of wastewater enhances considerably in terms of biological oxygen demand as well as dissolved

Table 15.2 Highly useful plant species for phytoremediation (Adapted from Sinha et al. 2007)

S. no.	Plant name	S. no.	Plant name
1.	Vetiver grass (*Vetiveria zizanioides*)	16.	White radish (*Raphanus sativus*)
2.	Bermuda grass (*Cynodon dactylon*)	17.	Catnip (*Nepeta cataria*)
3.	Bahia grass (*Paspalum notatum*)	18.	Big bluestem (*Andropogan gerardii*)
4.	Sunflower oil plant (*Helianthus annus*)	19.	Indian grass (*Sorghastrum nutans*)
5.	Poplar tree (*Populus* spp.)	20.	Canada wild rye (*Elymus canadensis*)
6.	Mustard oil plant (*Brassica juncea*)	21.	Nightshade (*Solanum nigrum*)
7.	Periwinkle (*Catheranthus roseus*)	22.	Wheat grass (*Agropyron cristatum*)
8.	Cumbungi (*Typha angustifolia*)	23.	Alfalfa (*Medicago sativa*)
9.	Water hyacinth (*Eichhornia crassipes*)	24.	Tall fescue (*Festuca arundinacea*)
10.	Duck weed (*Lemna minor*)	25.	Lambsquarters (*Chenopodium berlandieri*)
11.	Red mulberry (*Morus rubra*)	26.	Reed grass (*Phragmites australis*)
12.	Kochia (*Kochia scoparia*)	27.	Tall wheatgrass (*Thynopyron elongatum*)
13.	Foxtail barley (*Hordeum jubatum*)	28.	Rhodes grass (*Chloris guyana*)
14.	Switch grass (*Panicum variegatum*)	29.	Flatpea (*Lathyrus sylvestris*)
15,	Musk thistle (*Carduus nutans*)	30.	Carrot (*Daucus carota*)

oxygen values, heavy metals, and suspended solids and can be utilized for irrigation, industrial uses, and aquaculture. It decontaminates the wastewater having high concentration of P, NO_3^- as well as K till the water is clean with P and N contents falling close to 0.5 mg/L in just 20 days. Many microbes reside in the roots of water hyacinths in symbiotic relationships which flourish on minerals as well as organic pollutants present in the effluents. Water hyacinth can eradicate heavy metals by 20–100%. Within 24 h, the weed can remove more than 75% of Pb from polluted water. It also absorbs Cd, Ni, Cr, Zn, Cu, Fe as well as pesticides and various harmful substances from the sewage. Within 7 days of exposure, it can reduce about 97% BOD and eliminate over 90% of NO_3^- and PO^{4-}. It can also eliminate radioactive pollutants (Sinha et al. 2007).

15.6 Conclusions

In spite of the variety of possible choices, phytoremediation is still in its initial stages. Most of the studies have been done in labs in comparatively ideal situation for brief periods of time. There is need of better exhaustive studies in fields for more time periods to clearly know about the possible function of phytoremediation. There is a limitation in phytoremediation method that a particular phytoremediation treatment cannot be applied in all situations with a specific chemical pollutant due to diverse site-specific circumstances of soil and climate which may not be appropriate for the target plant. Plants also have interaction with and are influenced by other living beings like insects, pests, and pathogens, and plants exposure to pollutants and linked stresses can make the phytoremediation more vulnerable to these other agents, which subsequently affects the result of phytoremediation efforts. In addition to it, phytoremediation usually is limited to those areas where the quantity of pollutants is not dangerous to the plants planned for remediation. Lastly, the pollutants must be available to the tissue accountable for uptake like root system in plants. Consequently, in situ phytoremediation utilizing living plants is limited to areas favorable to development of the particular plant with the pollutant present within the potential root area of the specific plant.

References

Ali H, Khan E, Sajad MA (2013) Phytoremediation of heavy metals-concepts and applications. Chemosphere 91(7):869–881

Al-Najar H, Kaschl A, Schulz R, Roemheld V (2005) Effect of thallium fractions in the soil and pollution origins on Tl uptake by hyperaccumulator plants: a key factor for the assessment of phytoextraction. Int J Phytoremediation 7:55–67

Bankole OP (2008) Major environmental issues and the need for environmental statistics and indicators in Nigeria. In: Proceedings of the ECOWAS workshop on Environmental Statistics, Abuja, Nigeria

Behmer ST, Lloyd CM, Raubenheimer D, Stewart-Clark J, Knight J, Leighton RS, Harper FA, Smith JAC (2005) Metal hyperaccumulation in plants: mechanisms of defence against insect herbivores. Funct Ecol 19:55–66

Brennan MA, Shelley ML (1999) A model of the uptake, translocation, and accumulation of lead (Pb) by maize for the purpose of phytoextraction. Ecol Eng 12(3–4):271–297

Caille N, Zhao FJ, McGrath SP (2005) Comparison of root absorption, translocation and tolerance of arsenic in the hyperaccumulator *Pteris vittata* and the nonhyperaccumulator *Pteris tremula*. New Phytol 165(3):755–761

Comino E, Whiting SN, Neumann PM, Baker AJM (2005) Salt (NaCl) tolerance in the Ni hyperaccumulator Alyssum murale and the Zn hyperaccumulator *Thlaspi caerulescens*. Plant and Soil 270(1):91–99. https://doi.org/10.1007/s11104-004-1233-0

Conesa HM, Schulin R, Nowack B (2008) Mining landscape: a cultural tourist opportunity or an environmental problem? Ecol Econ 64(4):690–700

Cunningham SD, Ow DW (1996) Promises and prospects of phytoremediation. Plant Physiol 110(3):715–719

Daly HE (1990) Toward some operational principles of sustainable development. Ecol Econ 2(1):1–6

do Nascimento CWA, Xing B (2006) Fitoextração: uma revisão sobre disponibilidade induzida e acumulação de metais em plantas. Sci Agric 63(3):299–311

Domínguez MT, Marañón T, Murillo JM, Schulin R, Robinson BH (2008) Trace element accumulation in woody plants of the Guadiamar Valley, SW Spain: a large-scale phytomanagement case study. Environ Pollut 152(1):50–59

Ensley BD (2000) Rationale for use of phytoremediation. In: Raskin I, Ensley BD (eds) Phytoremediation of toxic metals: using plants to clean up the environment. Wiley, New York, NY

EPA (2000) A citizen's guide to phytoremediation. United States Environmental Protection Agency, Washington, DC, p 6

Ezeonu CS, Tagbo R, Anike EN, Oje OA, Onwurah IN (2012) Biotechnological tools for environmental sustainability: prospects and challenges for environments in Nigeria-a standard review. Biotechnol Res Int 2012:1–27

Glass DJ (1999) US and international markets for phytoremediation, 1999–2000. D Glass Associates, Needham, MA

Helmisaari HS, Salemaa M, Derome J, Kiikkilä O, Uhlig C, Nieminen TM (2007) Remediation of heavy metal contaminated forest soil using recycled organic matter and native woody plants. J Environ Qual 36(4):1145–1153

Jiang RF, Ma DY, Zhao FJ, McGrath SP (2005) Cadmium hyperaccumulation protects Thlaspi caerulescens from leaf feeding damage by thrips (*Frankliniella occidentalis*). New Phytol 167(3):805–814

Khajanchi L, Yadava RK, Kaurb R, Bundelaa DS, Khana MI, Chaudharya M, Meenaa RL, Dara SR, Singha G (2013) Productivity, essential oil yield, and heavy metal accumulation in lemon grass (*Cymbopogon flexuosus*) under varied wastewater-groundwater irrigation regimes. Ind Crop Prod 45:270–278

Khalid S, Shahid M, Niazi NK, Murtaza B, Bibi B, Dumat C (2017) A comparison of technologies for remediation of heavy metal contaminated soils. J Geochem Explor 182:247–268

Kim I (1996) Harnessing the green clean. Chem Eng J 103:39–41

Lu B, Xu Z, Li J, Chai X (2018) Removal of water nutrients by different aquatic plant species: an alternative way to remediate polluted rural rivers. Ecol Eng 110:18–26

Mahar A, Wang P, Ali A, Awasthi MK, Lahori AH, Wang Q, Li R, Zhang Z (2016) Challenges and opportunities in the phytoremediation of heavy metals contaminated soils: a review. Ecotoxicol Environ Saf 126:111–121

McGrath SP, Lombi E, Gray CW, Caille N, Dunham SJ, Zhao FJ (2006) Field evaluation of cd and Zn phytoextraction potential by the hyperaccumulators *Thlaspi caerulescens* and Arabidopsis halleri. Environ Pollut 141(1):115–125

Meagher RBP (2000) Phytoremediation of toxic elemental and organic pollutants. Curr Opin Plant Biol 3(2):153–162

Mudhoo A, Sharma SK, Lin ZQ, Dhankher OP (2010) Phytoremediation of arsenic-contaminated environment an overview. In: Sharma SK, Mudhoo A (eds) Green chemistry for environmental sustainability. Taylor and Francis Group, Boca Raton, London, New York, p 127

Nriagu JO (1979) Global inventory of natural and anthropogenic emissions of trace metals to the atmosphere. Nature 279(5712):409–411

Nsanganwimana F, Pourrut B, Mench M, Francis Douay F (2014) Suitability of Miscanthus species for managing inorganic and organic contaminated land and restoring ecosystem services. A review. J Environ Manage 143:123–134

Pandey DN (2002) Sustainability science for mine-spoil restoration. Curr Sci 83:792–793

Pandey VC (2013) Suitability of *Ricinus communis* L. cultivation for phytoremediation of fly ash disposal sites. Ecol Eng 57:336–341

Pandey VC, Bajpai O (2019) Phytoremediation: from theory toward practice. In: Phytomanagement of polluted sites. Elsevier, Amsterdam, pp 1–49

Pandey VC, Singh K (2011) Is *Vigna radiata* suitable for the revegetation of fly ash landfills? Ecol Eng 37(12):2105–2106

Pandey VC, Souza-Alonso P (2019) Market opportunities: in sustainable phytoremediation. In: Phytomanagement of polluted sites. Elsevier, Amsterdam, pp 51–82

Pandey VC, Singh K, Singh JS, Kumar A, Singh B, Singh RP (2012a) *Jatropha curcas*: a potential biofuel plant for sustainable environmental development. Renew Sustain Energy Rev 16(5):2870–2883

Pandey VC, Singh K, Singh RP, Singh B (2012b) Naturally growing *Saccharum munja* on the fly ash lagoons: a potential ecological engineer for the revegetation and stabilization. Ecol Eng 40: 95–99

Pandey VC, Pandey DN, Singh N (2015) Sustainable phytoremediation based on naturally colonizing and economically valuable plants. J Clean Prod 86:37–39

Pandey VC, Bajpai O, Singh N (2016) Energy crops in sustainable phytoremediation. Renew Sustain Energy Rev 54:58–73
Pilon-Smits EA, Freeman JL (2006) Environmental cleanup using plants: biotechnological advances and ecological considerations. Front Ecol Environ 4(4):203–210
Prabakaran K, Li J, Anandkumar A, Leng Z, Zou CB, Du D (2019) Managing environmental contamination through phytoremediation by invasive plants: a review. Ecol Eng 138:28–37
Raskin I, Ensley BD (2000) Recent developments for in situ treatment of metal contaminated soils. In: Phytoremediation of toxic metals: using plants to clean up the environment. Wiley, New York
Rezania S, Taib SM, Md Din MFM, Dahalan FA, Kamyab H (2016) Comprehensive review on phytotechnology: heavy metals removal by diverse aquatic plants species from wastewater. J Hazard Mater 318:587–599
Rice PJ, Anderson TA, Anhalt JC, Coats JR (1997) Phytoremediation of atrazine and metachlor contaminated water with submerged and floating aquatic plants. In: Proceedings of the of the 12th Annual Conference on Hazardous Waste Research, Kansas City, MO, May 19–22, p 52
Salt DE, Smith RD, Raskin I (1998) Phytoremediation. Annu Rev Plant Physiol Plant Mol Biol 49(1):643–668
Sarma H, Forid N, Prasad R, Prasad MNV, Ma LQ, Rinklebe J (2021) Enhancing phytoremediation of hazardous metal(loid)s using genome engineering CRISPR–Cas9 technology. J Hazard Mater. https://doi.org/10.1016/j.jhazmat.2021.125493
Schwitzguebel JP, Kumpiene J, Comino E, Vanek T (2009) From green to clean: a promising and sustainable approach towards environmental remediation and human health for the 21st century. Agrochimica 53(4):1–29
Sharma P, Pandey S (2014) Status of phytoremediation in world scenario. Int J Environl Bioremediat Biodegrad 2(4):178–191
Sinha RK, Herat S, Tandon PK (2007) Phytoremediation: role of plants in contaminated site management. In: Environmental bioremediation technologies. Springer, Berlin, Heidelberg, pp 315–330
Sonowal S, Nava AR, Joshi SJ, Borah SN, Islam NF, Pandit S, Prasad R, Sarma H (2022) Biosurfactants assisted heavy metals phytoremediation: green technology for the United Nations sustainable development goals. Pedosphere 2(1):198–210. https://doi.org/10.1016/S1002-0160(21)60067-X
Srivastava N (2016) Phytoremediation of heavy metals contaminated soils through transgenic plants. In: Phytoremediation: management of environmental contaminants, vol 3, pp 345–391
Susarla S, Medina VF, McCutcheon SC (2002) Phytoremediation: an ecological solution to organic chemical contamination. Ecol Eng 18(5):647–658
Thakare M, Sarma H, Datar S, Roy A, Pawar P, Gupta K, Pandit S, Prasad R (2021) Understanding the holistic approach to plant-microbe remediation technologies for removing heavy metals and radionuclides from soil. Curr Res Biotechnol 3:84–98. https://doi.org/10.1016/j.crbiot.2021.02.004
Trap S, Kohler A, Larsen LC, Zambrano KC, Karlson U (2005) Phytotoxicity of fresh and weathered diesel and gasoline to willow and poplar tree's. J Soil Sediment 1:71–76
Tsao DT (2003) Overview of phytotechnologies. Adv Biochem Eng Biotechnol 78:1–50
United States Environmental Protection Agency (2000) Introduction to phytoremediation. EPA, United States Environmental Protection Agency, Office of Research and Development, Cincinnati, OH
USEPA (2016) Phytotechnology project profiles. United States Environmental Protection Agency, Washington, DC
Vamerali T, Marianna B, Giuliano M (2010) Field crops for phytoremediation of metal-contaminated land: a review. Environ Chem Lett 8:1–17
Watanabe ME (1997) Phytoremediation on the brink of commercialization. Environ Sci Technol 31(4):182A–186A

White PJ, Broadley MR (2005) Biofortifying crops with essential mineral elements. Trends Plant Sci 10(12):586–593

Wikipedia and the Free (2010) Encyclopedia—"Sustainability". http://en.wikipedia.org/wiki/Sustainability

Wuana RA, Okieimen FE, Imborvungu JA (2010) Removal of heavy metals from a contaminated soil using organic chelating acids. Int J Environ Sci Technol 7(3):485–496

Yadav KK, Gupta N, Kumar A, Reece LM, Singh N, Rezania S, Khan SA (2018) Mechanistic understanding and holistic approach of phytoremediation: a review on application and future prospects. Ecol Eng 120:274–298

Zhang H, Dang Z, Zheng LC, Yi XY (2009) Remediation of soil cocontaminated with pyrene and cadmium by growing maize (*Zea mays* L.). Int J Environ Sci Technol 6(2):249–258

Zhang X, Xia H, Li Z, Zhuang P, Gao B (2010) Potential of four forage grasses in remediation of Cd and Zn contaminated soils. Bioresour Technol 101(6):2063–2066

Zhao FJ, Jiang RF, Dunham SJ, McGrath SP (2006) Cadmium uptake, translocation and tolerance in the hyperaccumulator *Arabidopsis halleri*. New Phytol 172(4):646–654

Zhuang P, Ye ZH, Lan CY, Xie ZW, Shu WS (2005) Chemically assisted phytoextraction of heavy metal contaminated soils using three plant species. Plant and Soil 76(1–2):153–162

Zouboulis AI, Katsoyiannis IA (2005) Recent advances in the bioremediation of arsenic-contaminated groundwaters. Environ Int 31(2):213–219

16 Role of Phytoremediation as a Promising Technology to Combat Environmental Pollution

Mehmooda Munazir, Rahmatullah Qureshi, Mubashrah Munir, and Hamid Mukhtar

Abstract

A wide variety of pollutants such as heavy metals, organic, and inorganic wastes are continuously being added to the environmental components globally. These pollutants are stressing our environment and badly eroding the biotic components of our ecosystems. Besides this, these are hazardous to human health. Phytoremediation is a promising environment friendly technology that has gained attention of researchers across the globe from the past few decades. Phytoremediation (also known as "green remediation" and "botanical bioremediation") utilizes plants to reduce, remove, degrade, or immobilize environmental toxins, primarily those of anthropogenic origin aiming at restoring polluted sites to a condition useable for private or public applications. Some of the heavy metals and pollutants such as lead, chromium, cadmium, copper, nickel, mercury, zinc, strontium, boron, selenium, arsenic, thallium, uranium, calcium, cobalt, manganese, nitrates, herbicides, and chlorinated compounds are highly toxic and lethal even in trace amounts which may be teratogenic, mutagenic, endocrine disruptive as well as behavioral and neurotoxic in nature. With ever-increasing urbanization

M. Munazir
Department of Botany, Government College Women University, Sialkot, Punjab, Pakistan

R. Qureshi (✉)
Department of Botany, Pir Mehr Ali Shah Arid Agriculture University, Rawalpindi, Punjab, Pakistan
e-mail: rahmatullahq@uaar.edu.pk

M. Munir
Department of Biological Sciences, University of Veterinary and Animal Sciences, Lahore, Punjab, Pakistan

H. Mukhtar
Institute of Industrial Biotechnology, Government College University (GCU), Lahore, Pakistan

R. Prasad (ed.), *Phytoremediation for Environmental Sustainability*,
https://doi.org/10.1007/978-981-16-5621-7_16

and advancement in technology, the addition of pollutants is expected to continue by many folds. Phytoremediation has been found effective in remedying the high concentration of these pollutants from the soil and groundwater. Some plant species have interestingly been found effective in absorbing radioactive and toxic elements from air as well. The concept of phytoremediation was well-known, and various plants are being used by the Neanderthal man for wastewater treatment from thousand years ago. Some of the species such as *Avena sativa, Brassica juncea, B. napus, Hordeum vulgare, Panicum virgatum, Thlaspi caerulescens,* and *Viola calaminaria* have successfully been used to absorb environmental pollutants. From the past decade, several methods of phytoremediation like phytoextraction/phytoaccumulation, phytotransformation, phytostabilization, phytostimulation, phytorhizodegradation, phytodegradation, and phytovolatilization have been under investigation. Besides, the role of different factors that affect phytoremediation such as EDTA, CDTA, DTPA, EDDS, NTA, HEDTA, EGTA, and citric acid have also been studied by various researchers globally. This chapter is an endeavor to provide a comprehensive overview on all aforementioned aspects of phytoremediation along with future prospects of this technology. In addition, limitations and advantages of the said technique are also discussed in detail that would help the readers to find answers to various questions pertaining to this potential technique.

Keywords

Phytoremediation · Phytotransformation · Phytostabilization · Phytostimulation · Chelating agents

16.1 Introduction

Phytoremediation is a promising environment-friendly technology that has gained attention of researchers across the globe from the past few decades. This is plant-based technology used either naturally or genetically engineered plants for cleaning up the polluted environments (Cunningham et al. 1997; Flathman and Lanza 1998; Sarma et al. 2021; Sonowal et al. 2022). This is supposed to be a low-cost technology that utilizes plants to reduce, remove, degrade, or immobilize environmental toxins, primarily those of anthropogenic origin for restoring polluted sites to a condition useable for private or public applications (Ensley 2000). Though the term, phytoremediation is a quite new discovery; however, it is practiced since ages (Cunningham et al. 1997; Brooks 1998). The use of semiaquatic plants for recycling the radionuclide-polluted water was found in practice in Russia at the initiating time of nuclear period (Timofeev-Resovsky et al. 1962). A number of plants have capability to accumulate significant amount of metals in their tissues while growing on metal deposited soils without showing toxicity (Baker et al. 1991; Entry et al. 1999). The effectiveness of phytoremediation depends on the type of pollutant, bioavailability, and soil properties (Cunningham and Ow 1996). Some of the heavy metals and pollutants such as lead, chromium, cadmium, copper, nickel, mercury, zinc, strontium, boron, selenium, arsenic, thallium, uranium, calcium, cobalt, manganese, nitrates, herbicides, and chlorinated compounds are highly toxic and lethal even in trace amounts which may be teratogenic, mutagenic, endocrine disruptive as well as behavioral and neurotoxic in nature (Duffus 2002). With ever-increasing urbanization and advancement in technology, the addition of pollutants is expected to continue by many folds. Phytoremediation has been found effective in remedying the high concentration of these pollutants from the soil and groundwater (Lone et al. 2008).

16.2 Environmental Pollution and the Need of Remediation

Healthy, prosperous, and successful life on earth is dependent on healthy environment. But the quality of environment has been deteriorated by means of environmental pollution. The environmental pollution can be defined as "addition of unwanted and undesirable elements to the biotic and abiotic components of an environment by means of anthropogenic activities which ultimately decrease the quality of life." The scarcity of drinking water and loss of soil fertility are the initial results of pollution. The situation becomes worse when it enters at the food chain level. It is pathetic to note that drinking water is not healthy in most of the parts of the world owing to the contamination by various environmental pollutants (Daud et al. 2017).

The environment is comprised of two types of components, i.e., abiotic and biotic. The three major abiotic components include air, water, and soil. The biotic components, on the other side, include human beings, flora, fauna, and the microbes. The abiotic components are affected first by environmental pollution in which they directly affect the biotic components. Addition of contaminants to the environment has been taking place since human existence on the planet. There are two types of heavy metal contamination, i.e., natural or anthropogenic caused by human beings (Fig. 16.1). However, majority of heavy metals are mainly added by human beings themselves, thus making the environment unfit for leading good quality of life. For example, industrial wastes badly pollute our environment. Distillery industries are one of the examples of such industries that add polluted water to the soil. This water contains a mixture of organic and inorganic pollutants which may gain entry to food chain and directly affect the quality of life (Chowdhary et al. 2019; Thakare et al. 2021; Prasad 2021).

The type and quantity of contaminants vary in different countries. Intensity of severity is found higher in the developing and poorly developed nations, since they are careless about their environment. Industrial effluents are usually present in the surrounding areas without any treatment, thus become major health hazard for the people dwelling in such areas. The water, soil, and air are badly polluted, and the contaminants can easily gain entry to the food chain. Regretfully, a huge number of deaths occur every year due to diseases and illnesses caused by environmental

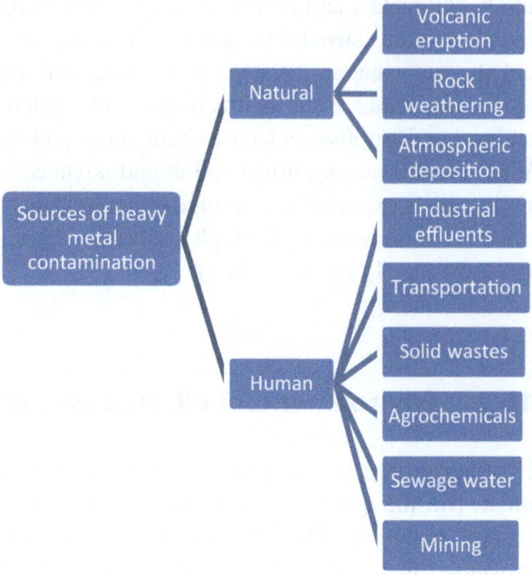

Fig. 16.1 Sources of heavy metal contamination

pollution. According to a report, nine million deaths occurred during 2015–2016 as result of environmental pollution (Gangamma 2018).

The severity of problem is increased where the people are usually illiterate, less educated, and totally unaware of the consequences of pollution. They manage to work in small industries and factories without bothering the extent of pollution that they are exposed to. Hence the problems of poor are aggravated by poor standards of life and health issues. The governments and administrative units, in such countries, are usually less concerned about the issues of environment. As a result, the environment gets more and more polluted without any check and control. Among different contaminants, heavy metals like lead, chromium, cadmium, copper, nickel, mercury, zinc, strontium, boron, selenium, arsenic, thallium, uranium, calcium, cobalt, manganese, nitrates, herbicides, and chlorinated compounds are highly toxic (Santos et al. 2018). Followings are sources of these toxic materials released in the environmental systems (Kanwar et al. 2020):

- Mining and smelters may cause the addition of As, Cd, Pb, and Hg metals.
- Various industries may add As, Cd, Cr, Co, Cu, Hg, Ni, and Zn metals.
- Atmospheric deposition may result into addition of As, Cd, Cr, Cu, Pb, Hg, and U.
- Agrochemicals may deposit As, Cd, Cu, Pb, Se, U, and Zn.
- Solid/liquid waste may cause addition of As, Cd, Cr, Cu, Pb, Hg, and Zn.

Besides, few bacteria may also add toxic mercury (mono- and/or dimethylmercury) to the environment that eventually polluted drinking water and food materials (Kumar et al. 2017). According to the United Kingdom Environment Agency (UKEA), there are some 1300 plus mining places that polluted soil and water reservoirs by adding different kinds of heavy metals like copper, cadmium, lead, and zinc (Foulds et al. 2014). Besides, the cosmetics and chemical fertilizers are also accountable for heavy metal pollution (Callender 2004).

Consequences of environmental pollution range from minor and negligible to serious problems for human. Among different problems that arise as a result of environmental pollution, the deterioration of human health is most eye-catching and alarming. A large number of diseases like renal dysfunction, alimentary canal problems strike human race every year, causing serious and irreversible health damage and even to death at times (Briggs 2003). Contamination of food is one of the major hazards that affect humanity worldwide. Environmental pollution has one more serious role and a potential threat to cause change at genetic level in any biotic component that resulted in life-threatening diseases and irreversible damages. Cancer is one of such devastating diseases, which owes large number of casualties every year in almost every part of the world (Boffetta 2006).

16.3 Types of Environmental Contaminants

It has been estimated that the pollution caused by heavy metals may surpass the other contaminants if it goes unchecked. A wide variety of contaminants exist that affect the quality of life to a great deal. The solid wastes and nuclear discharges are usually ranked as the worst pollutants, followed by heavy metals (Chen et al. 2003a). Following are the major types of pollutants that are predominantly found:

16.3.1 Inorganic Contaminants

An element which is found in periodic table cannot be further broken down into simpler parts. It is an entity in itself that has the potential to react with other elements to form compounds of various natures. The heavy metals are one of the major contaminants of soil that adversely affect the quality of soil and cause serious pollution, mostly affect the street and road-side soils (Christoforidis and Stamatis 2009; Li et al. 2001). Due to pure form, they cannot be broken down so they remain as such in the medium causing serious damages to the environment. Heavy metals, in minor quantities, work with enzymatic system of the plants to regulate physiological processes of plants, but at higher concentrations, they have negative impacts on plant growth and development. Arsenic, cadmium, zinc, copper, lead, iron, helium, neon, and solvents acetone, ethyl acetate, butanol, ethanol, methanol, deuterated water, hexane, chloroform, quercetin, and lots of chemicals are used in the laboratories of research institutes and hospitals (Charlesworth et al. 2017). Brief details about some of the heavy metals that pollute environment are given below:

16.3.1.1 Chromium

Chromium is abundantly found as part of rocks. In addition, it is found in the form of complexes with metals like lead (Pb), calcium (Ca), potassium (K), phosphorus (P), copper (Cu), aluminum (Al), sulfur (S), and others. It is found in different valent forms, the most reactive being Cr (VI) and Cr (III). Its natural forms are not serious environmental hazard as they are complexes of varying natures. Chromium may be released from rocks by natural weathering of rocks but this process is usually slow. On the other hand, anthropogenic activities add chromium to the environment as a reactive entity leading to serious health issues. It is mainly used in industries like alloying, tanning of animal hides, textile industry, lumber, and pigments. It has been observed that chromium in the form of chromate ions is most toxic due to its high solubility and ability to penetrate living membranes but its other forms like hydroxides, oxides, and sulphates are less toxic due to less solubility (Oliveira 2012). So, solubility of its chemical forms plays main role in its extent of toxicity. The contribution of leather industry as a source of chromium is now ranked first. Industrial cities such as Sialkot and Faisalabad (famous for their leather and textile industry) add the highest level of chromium to the environment compared to other cities of Pakistan. It is also worthwhile to mention that the incidences of cancer in such cities have increased many folds in the last decade.

16.3.1.2 Lead

Lead is another heavy metal which is toxic to living beings in many ways. The highest amount of lead is added by various industries to the environment leading to soil, water, and air pollution. Lead poisoning is a serious concern, and children are more severely affected by lead toxicity as compared with adults especially because they are not aware of the potential damages that it may bring. Lead toxicity may lead to a variety of health disorders ranging from abdominal pain, irritation, or lethargy to comma, anemia, and even neurological disorders (Hai et al. 2018).

16.3.1.3 Arsenic

Arsenic is one of the toxic contaminants of soil which is usually found in industrial wastes. It has the potential to damage human, plant, and animal health if it gains entry to the food chain (Prasad et al. 2013). Its bioavailability is predominantly affected by soil pH and can be increased by addition of organic chelating agents preferably citric acid (González et al. 2019).

16.3.1.4 Cadmium

Cadmium is added to the environment by natural processes from the earth's crust; however, anthropogenic activities lead to increase at higher levels. Processing of different metals like zinc, iron, and aluminum is the major way of its addition to environment. Besides, it is added through cigarette smoke, coal, and oil combustion from power plants and phosphate-based fertilizer applications. Higher cadmium levels not only affect plant physiology (especially respiration and transpiration), but they cause damages to microbial world and organisms dwelling in water (i.e., fish). The health damages from cadmium toxicity range from minor to major issues. Its higher levels may lead to cancer, birth defects, anemia, kidney damage, etc. Cadmium is mentioned as a "red list" metal as it may have serious impacts on health.

16.3.2 Organic Contaminants

An organic contaminant is the one that can be metabolized by plants and converted into inorganic constituents. Almost 30% of photosynthates of a plant are released into the rhizosphere by roots. A variety of phytochemicals and sugars are released by plant roots. The microbes that are associated with roots utilize such metabolites to gain energy. Such contaminants sometimes become serious hazard for the surrounding environment. They may gain entry to the water table and affect quality of biotic life.

16.4 In-Practice Strategies to Combat Environmental Pollution

Human beings have now realized the role of pollution in deteriorating and damaging health and overall life quality, and efforts are being made worldwide to protect environment of further damage. Following four main strategies are being used to control environmental pollution:

- To control the addition of contaminants to the environment. In this way, environmental pollution can be reduced by adopting procedures and strategies that add minimum pollutants. This strategy can play vital role in reducing environmental deterioration.
- To render such pollutants harmless or less toxic in an effective way by using different physical, chemical, and biological means. Extensive research may prove effective if all these means are studies in-depth.
- Increased awareness campaigns among masses thus educating them about environmental pollution and its disastrous results.
- Strict legislation and effective monitoring at the government level.

Some of the countries such as China, India, and Pakistan use the sewage water for growing vegetables and fruits due to lack of knowledge and awareness leading to transfer of contaminants to the food chain.

The abovementioned second strategy has been adopted by researchers worldwide to combat environmental pollution.

16.4.1 Bioremediation

American Environmental Protection Agency (EPA) defined bioremediation agents as "microbial cultures, enzymes and nutrient additives that significantly increase the rate of biodegradation to mitigate the effects of various pollutants" (Nichols 2001). Bioremediation can be performed both in situ and ex situ. In situ bioremediation can be performed at the contaminated site, while ex situ contamination involves removal of contaminated materials from one site and transfer to the other site after treatment.

Both approaches are successfully used but usually in situ approach is preferred being cost effective (April et al. 2000).

Some of the microorganisms have always been playing role as decomposers which decompose organic matter of all kinds and convert it to simpler inorganic substances. Similarly, some of the species of fungi and bacteria degrade hydrocarbons in the environment (Nilanjana and Chandran 2011). This is a natural decontamination process that takes place on its own (Venosa and Zhu 2003; Vinothini et al. 2015). The first reports on successful use of bioremediation report back to 1974 when a bacterial strain *Pseudomonas putida* was used to remove petroleum from contaminated sites. In those years, efforts were focused on identification of various microbial species in cleaning petroleum spills. The research work on microbial species got more attention and progressed further till 2000 A.D. Since this process takes place nicely at a fast rate under controlled conditions in vitro but it is comparatively quite slow in the field conditions (Venosa et al. 1996; Mearns 1997). So, the research concentrated more on phytoremediation, and the importance of plants as natural remediated captivated thoughts of researchers.

16.4.2 Phytoremediation

Phyto is derived from "*phyton*" which is a Greek word that means "plant," and "remediation" is derived from "*remedium*" a Latin word that means "to restore balance" (Cunningham et al. 1996). So, the term "phytoremediation" can roughly be defined as "a technique to restore balance by using plants." It can be properly defined as "a set of methods/technologies that employ living plants to clean up soil, air, and water contaminated with hazardous contaminants." Or it can be defined as "the use of plants along with other mechanical techniques to scavenge, remove or detoxify environmental contaminants" (Prasad 2017, 2018). The concept of phytoremediation was well known, and various plants are being used by the Neanderthal man for wastewater treatment for thousands of years ago (Rastogi and Nandal 2020).

The term "phytoremediation" was coined in 1991 by Ilya Ruskin. This technique has proved effective in removal of organic and inorganic pollutants from environment (Etim 2012). This technology has been widely accepted as an eco-friendly technology by researchers, academicians, and general public of different continents (Ghazaryan et al. 2019). The use of plants to deal with environmental pollution is not new. This concept has always been with human but it is practically caught more attention gradually. About 300 years ago, plants were used to clean water of contaminants. This technique has gained popularity across globe to such an extent that plenty of research activities, publications, conferences, and symposia are devoted to it every year for the past 50 years.

This technique employs natural physiological processes of plants (Etim 2012). Plants are unique organisms as they contain unique set of metabolic processes that can be used to tackle with environmental pollution. Growing plants in a contaminated matrix (soil, water, organic/inorganic debris) can fix problems of environmental pollution (USEPA 2000). They work in unique ways by fixing the contaminants in their bodies (immobilizing/binding the contaminants), degrading them by using their cellular machinery, or converting them in to less harmful or totally harmless forms (conversion). Once the plants have performed their role, they can either be removed or disposed off in appropriate ways. Over the years, plants have thankfully evolved in a way that they can deal with environmental pollutants by using their metabolic or physiological processes. Human beings can benefit from such processes. Plants can be used to clean up metals, herbicides, pesticides, solvents/toxic chemicals, explosives, crude oils, polyaromatic hydrocarbons, and landfill leachates from soil and/or water. They can be used to clean up river basins and even ocean bottoms that are accidentally polluted by oil spills, etc.

16.5 Types of Phytoremediation

Plants can work in different ways to deal with environmental contaminants. Phytoremediation can either be used alone or in combination with different chemical and/or mechanical procedures to clean up the environment (Etim 2012). The root system of plants plays vital role in absorbing contaminants from soil. The roots contain systems to protect themselves from harmful concentrations of contaminants. The roots provide larger surface area for absorption of such substances hence trees are considered better for this purpose as they have larger rhizosphere. Trees can play strong role in areas where contaminants are found in deeper layers of soil by pulling water up from their deep and wider rhizosphere. In some cases, the roots release few substances into their rhizosphere that play role in aggregation of soil particles, hence affecting the rate of absorption of contaminants from soil.

16.5.1 Phytoextraction

It is the type of remediation in which plants absorb contaminants from the environment in a harvestable form. It involves absorption of contaminants from soil and accumulation in above ground parts (preferably crown parts/foliage) of plant (Fig. 16.2). The soil can be used for growth of other plant species after proper remediation. The roots absorb the substances from soil and concentrate them in the above ground parts. Those plant species which can concentrate higher levels of contaminants in their bodies are called "hyperaccumulators." While those plant species which accumulate less contaminant may be cropped repeatedly to remove the medium of a contaminant. Such plants which hyper-accumulate toxic metals are generally regarded as metallophytes. Examples of such species include *Salix* and *Populus*. Phytoextraction has gained in popularity for the past few decades. It has been found effective in removal of inorganic substances or heavy metals. The contaminants usually accumulate in different plant parts. In case the plants have accumulated the contaminants up to a certain level, it becomes impossible for them to accumulate beyond that amount, and hence the remaining contaminants may leach down to the deeper soil layers.

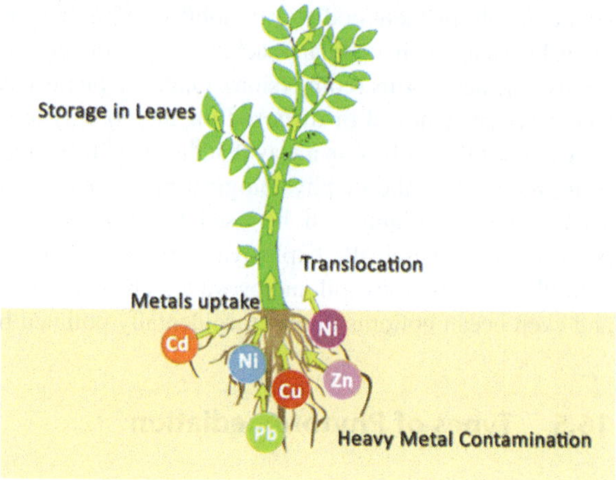

Fig. 16.2 Generalized figure to describe the process of phytoextraction in a plant. (Anoopkumar et al. 2020)

This strategy is helpful in remediation of soils contaminated with heavy metals, and a huge number of studies exist that reported use of different plant species to remediate soils affected with such metals. For example, Kaviani et al. (2019) used phytoremediation strategy to remediate Ni-contaminated soil. Their study is eye-catching as they not only measured the phytoextraction potential of a plant species (*Salicornia iranica*) but also measured the detoxification capacity of the said plant. This task was accomplished by measuring glutathione-S-transferase (GST) expression in addition to the other physiological parameters. They reported that the total chlorophyll and carotenoids were reduced after exposure to high dose of Ni (i.e., 500 mg/kg) recorded at different time intervals. They also reported a higher GST expression. The plant could accumulate higher levels of Ni in roots and aerial parts. The root and shoot lengths were reduced. The results showed that this plant species can be used for remediation of Ni-contaminated soils.

It is even more interesting to note that phytoremediation can be successfully applied where there is just single metal present in the medium. In case more than one metal are present, then only those metals can be absorbed which do not compete with each other for absorption by the roots. Not all metals in a medium can be absorbed simultaneously by the plant roots. This is because few metals antagonize the other metals absorption. This fact has been reported by many researchers. For example, in a latest research by Singha et al. (2019), iron plaque formation was observed on the roots of an aquatic macrophyte, *Pistia stratiotes* L. This iron plaque was formed by ferrous ions in the industrial wastewater. This plaque favored the extraction of iron and potassium and reduced the absorption of calcium from water. The absorption of cadmium was also suppressed but when the concentration of cadmium was raised to 500 μmol then its absorption by plaque containing roots increased. Cadmium detoxification was also observed in plants with iron plaque formation on roots. Kanwar et al. (2020) has reported various plant species which are used for eliminating different kind of heavy metals across the world (Table 16.1).

Helianthus annuus is known to absorb arsenic (Raab et al. 2005). *Pteris vittata* is another plant species that can accumulate arsenic (Fayiga et al. 2004). Willow (Salix smithiana) is good extractor of copper, zinc, and cadmium (Kacalkova et al. 2009). It has the ability to quickly transport the metal from points of absorption to upper parts of the plant. In addition, it produces high biomass which can be utilized for energy production. Alpine penny cress (*Thlaspi caerulescens*) has the ability to accumulate cadmium and zinc at higher levels (Cosio et al. 2004). But this species does not accumulate copper. *Salix viminalis* has been found effective in accumulating cadmium, another toxic metal (Mleczek et al. 2009). Different metal chelators can be used to increase efficiency of absorption of metals from medium. Among them, EDTA is one of the most famous and highly experimented chelators.

16.5.2 Phytovolatilization

Phytovolatilization involves the removal (by volatilization) of contaminants from a medium (from soil or water) and release in to the air (Limmer and Burken 2016). The substances are not released as such in to the air, rather they are converted in to less toxic and less harmful substances before their release. The contaminants are usually volatilized at the surfaces of leaves or stems but they may get evaporated from roots. Selenium and mercury are the metals that can be phytovolatilized by such plants. Plants with higher rates of transpiration can be effective in this regard, e.g., Poplar trees (Fig. 16.3).

16.5.3 Phytotransformation/Phytodegradation

Phytotransformation is also known as phytodegradation. In this process, the toxic elements are decomposed by the plants and rendered nontoxic or less harmful (Fig. 16.4). This method is advantageous as it can scavenge toxic substances from soil, water, and air also. In this method, the substances are not completely broken down into their simplest components instead they are transformed by the machinery of plants from one form to the other (Bock et al. 2002). The compounds that come in contact with plant are broken down either inside the body of plant or in the rhizosphere. It has been observed that special enzymes are released by such plants into the rhizosphere that decompose the organic matter in the matrix. It is effective in removal of various types of solvents from any matrix. We all know very well that plants are generally regarded as "lungs of nature" as they add oxygen to the environment. But the role of plants in this method resembles that of 'human liver' where liver has the role to detoxify the human body of harmful or toxic substances. So, the plants are usually regarded as "green liver" due to their metabolic capabilities that render different compounds nontoxic hence cleaning the environment of such contaminants. This method is under investigation by some of the research groups

Table 16.1 Uptake of various heavy metals by the higher plants (after Kanwar et al. 2020)

Toxic metal	Plant	Medium	Uptake of heavy metal (mg/kg)	References
As	*Pteris vittata* L.	Soil and water	8331	Kalve et al. (2011)
	Pteris ryukyuensis Tagawa	Soil	3647	Srivastava et al. (2006)
	Pteris quadriaurita Retz.		2900	
	Pteris biaurita L.		2000	
	Pteris cretica L.		1800	
	Eleocharis acicularis (L.) Roem. & Schult.	Water	1470	Sakakibara et al. (2011)
	Sedum alfredii Hance	–	9000	Xiong et al. (2004)
	Prosopis laevigata (Humb. & Bonpl. ex Willd.) M.C.Johnst.	–	8176	Buendía-González et al. (2010)
	Arabis gemmifera (Matsum.) Makino	–	5600	Kubota and Takenaka (2003)
	Salsola kali L.	Water	2075	de la Rosa et al. (2004)
	Azolla pinnata R.Br.	Water	740	Rai (2008)
	Deschampsia cespitosa (L.) P. Beauv.	Water	236.2	Kucharski et al. (2005)
	Corrigiola telephiifolia Pourr.	Soil	2110	García-Salgado et al. (2012)
Ni	*Alyssum bertolonii* Desv. [Syn. *Odontarrhena bertolonii* (Desv.) Jord. & Fourr.]	Soil	10,900	Li et al. (2003)
	Alyssum caricum T.R.Dudley & Hub.-Mor. [Syn.*Odontarrhena carica* (T.R. Dudley & Hub.-Mor.) Španiel, Al-Shehbaz, D.A.German & Marhold]		12,500	
	Alyssum corsicum Rob. ex Gren. & Godr. [Syn. *Odontarrhena robertiana* (Bernard ex Gren. & Godr.) Španiel, Al-Shehbaz, D.A.German & Marhold]		18,100	
	Alyssum pterocarpum T.R.Dudley [Syn. *Odontarrhena pterocarpa* (T.R.Dudley) Španiel, Al-Shehbaz, D.A.German & Marhold]		13,500	
	Alyssum heldreichii Hausskn. Syn. *Odontarrhena heldreichii* (Hausskn.) Španiel, Al-Shehbaz, D.A.German & Marhold	Soil	11,800	Bani and Pavlova (2010)
	Alyssum markgrafii O.E. Schulz [synonym *of Odontarrhena chalcidica*	Soil	19,100	

(continued)

Table 16.1 (continued)

Toxic metal	Plant	Medium	Uptake of heavy metal (mg/kg)	References
	(Janka) Španiel, Al-Shehbaz, D.A. German & Marhold]			
	Alyssum murale M.Bieb. [synonym of *Odontarrhena alpestris* (L.) Ledeb.]	Soil	4730–20,100	
	Alyssum serpyllifolium Desf.	Soil	10,000	Prasad (2005)
	Isatis pinnatiloba P.H. Davis	Soil	1441	Altinozlu et al. (2012)
Cd	*Phytolacca americana* L.	Soil	10,700	Peng et al. (2008)
	Sedum alfredi Hance		9000	Xiong et al. (2004)
	Prosopis laevigata (Humb. & Bonpl. ex Willd.) M.C.Johnst.	Soil	8176	Buendía-González et al. (2010)
	Arabis gemmifera (Matsum.) Makino [Syn. *Arabidopsis halleri* subsp. *gemmifera* (Matsum.) O'Kane & Al-Shehbaz]	–	5600	Kubota and Takenaka (2003)
	Salsola kali L.	Water	2075	de la Rosa et al. (2004)
	Azolla pinnata R.Br.	Water	740	Rai (2008)
	Deschampsia cespitosa (L.) P.Beauv.	Soil	236.2	Kucharski et al. (2005)
	Rorippa globosa (Turcz. ex Fisch. & C. A.Mey.)	Soil	>100	Wei et al. (2008)
	Thlaspi caerulescens J. Presl & C.Presl [Syn. *Noccaea caerulescens* (J.Presl & C.Presl) F.K.Mey.]	Soil	263	Lombi et al. (2001)
	Azolla pinnata R.Br.	Water	740	Rai (2008)
	Pteris vittata L.	Water and soil	20,675	Kalve et al. (2011)
	Eleocharis acicularis (L.) Roem. & Schult.	Water	11,200	Sakakibara et al. (2011)
	Thlaspi calaminare (Lej.) Lej. & Courtois [Syn. *Noccaea caerulescens* subsp. *calaminaris* (Lej.) Holub]	Soil	10,000	Sheoran et al. (2009)
	Deschampsia cespitosa (L.) P.Beauv.	Soil	966.5–3614	Kucharski et al. (2005)
Hg	*Achillea millefolium* L.	Soil	18.275	Wang et al. (2012)
	Marrubium vulgare L.	Soil	13.8	Rodriguez et al. (2003)
	Rumex induratus Boiss. & Reut.	Soil	6.45	Rodriguez et al. (2003)

(continued)

Table 16.1 (continued)

Toxic metal	Plant	Medium	Uptake of heavy metal (mg/kg)	References
	Silene vulgaris (Moench) Garcke	Soil	4.25	Pérez-Sanz et al. (2012)
	Festuca rubra L.	Soil	3.17	Rodriguez et al. (2003)
	Poa pratensis L.	Soil	2.74	Sas-Nowosielska et al. (2008)
	Helianthus tuberosus L.		1.89	
	Armoracia rusticana G. Gaertn., B.Mey. &		0.97	
	Juncus maritimus Lam.	–	0.315	Zheng et al. (2016)
	Cicer arietinum L.	Soil	0.2	Wang et al. (2012)
	Eleocharis acicularis (L.) Roem. & Schult.	Water and soil	20,200	Sakakibara et al. (2011)
	Aeollanthus biformifolius De Wild. [Syn. *Aeollanthus subacaulis* var. *linearis* (Burkill) Ryding]	Soil	13,700	Chaney et al. (2010)
	Ipomoea alpina Rendle [Syn. *Ipomoea linosepala* subsp. *alpina* (Rendle) Lejoly & Lisowski]	–	12,300	Mitch (2002)
	Haumaniastrum katangense (S.Moore) P.A.Duvign. & Plancke	Soil	8356	Sheoran et al. (2009)
	Pteris vittata L.	Soil	91.975	Wang et al. (2012)
Cr	*Pteris vittata* L.	Soil and water	20,675	Kalve et al. (2011)
Pb	*Medicago sativa* L.	Soil	43,300	Koptsik (2014)
	Brassica juncea (L.) Czern.		10,300	
	Brassica nigra (L.) W.D.J.Koch		9400	
	Helianthus annuus L.		5600	
	Betula occidentalis Hook.		1000	
	Euphorbia cheiradenia Boiss. & Hohen.	Soil	1138	Chehregani and Malayeri (2007)
	Deschampsia cespitosa (L.) P.Beauv.	Soil	966.5	Kucharski et al. (2005)
	Euphorbia cheiradenia Boiss. & Hohen	Soil	1138	Chehregani and Malayeri (2007)

across the globe, with special emphasis on phytodegradation of organic compounds, e.g., methyl-tert-butyl ether, herbicides, tri-chloroethylene, industrial substances, xenobiotics (Newman and Reynolds 2004).

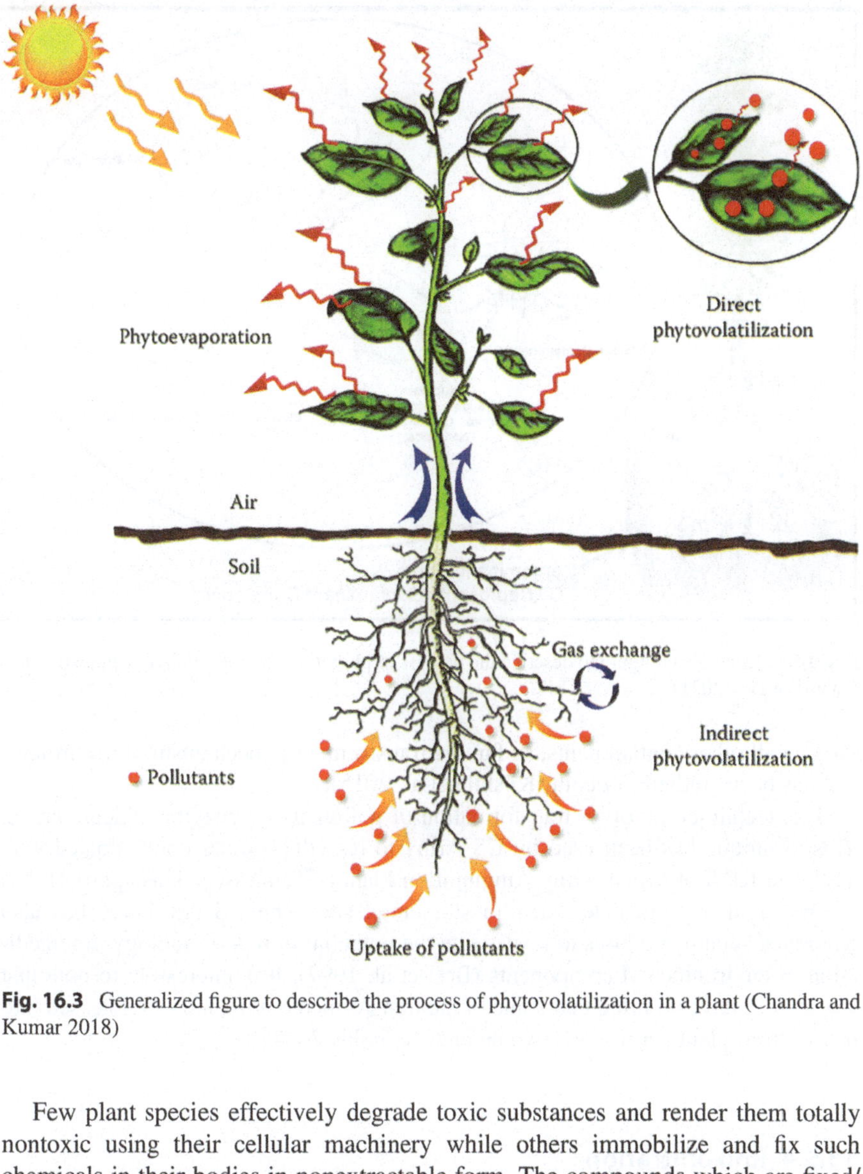

Fig. 16.3 Generalized figure to describe the process of phytovolatilization in a plant (Chandra and Kumar 2018)

Few plant species effectively degrade toxic substances and render them totally nontoxic using their cellular machinery while others immobilize and fix such chemicals in their bodies in nonextractable form. The compounds which are fixed/ stored in the plant body are dealt in a way that they do not affect the health of the plant itself. It is also interesting to note that, in some cases, the microbes in association with few plant species have the ability to metabolize and decompose such compounds in the rhizosphere. The recent studies are focused on finding the mechanisms of transformation in different plant species that are good at such transformation. The studies done so far reflect that there are three stages/phases of this transformation that usually start with adding polarity to the contaminants.

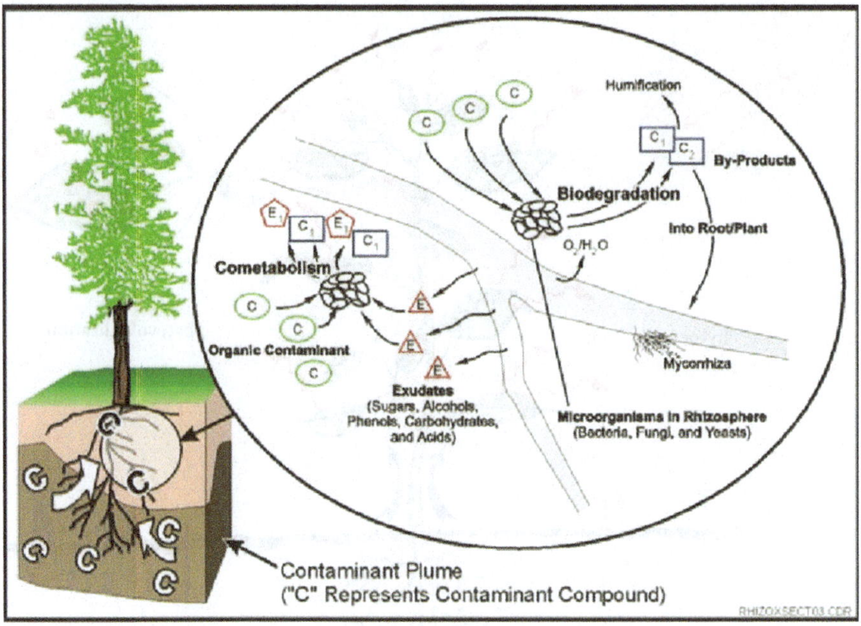

Fig. 16.4 Generalized figure to describe the process of phytotransformation/phytodegradation in a plant (Longley 2021)

Among all other contaminants, so far, the transformation mechanism of trinitrotoluene has been studied in-depth (Kiiskila et al. 2015).

This technique involves transformation of pollutants by enzymatic degradation. This technique has been used by US Army to remediate water contaminated with TNT and RDX at Milan Army Ammunition Plant at Tennessee. This approach has the potential to remediate water in situ or ex situ. The US Air Force has also employed such procedures to investigate the potential of this technology in remediation of environmental components (Best et al. 1997). It is interesting to note that few studies have reported that some of the transformed compounds are released into the air from plant surfaces (Newman and Reynolds 2004).

16.5.4 Rhizofiltration

It may be defined as "a process in which plant roots are employed to filter water/soil of contaminants." The pollutants are either absorbed (concentrated in the plant body) or adsorbed to the roots (Fig. 16.5). It actually involves transfer of pollutants from soil to the plant roots. The plants are generally grown first in a greenhouse (in pots) or in a hydroponic system. Then, this process can be performed either directly at the contaminated site to free water of contaminants or contaminated water can be collected and taken to the site where plants are growing (i.e., off-site area). Few

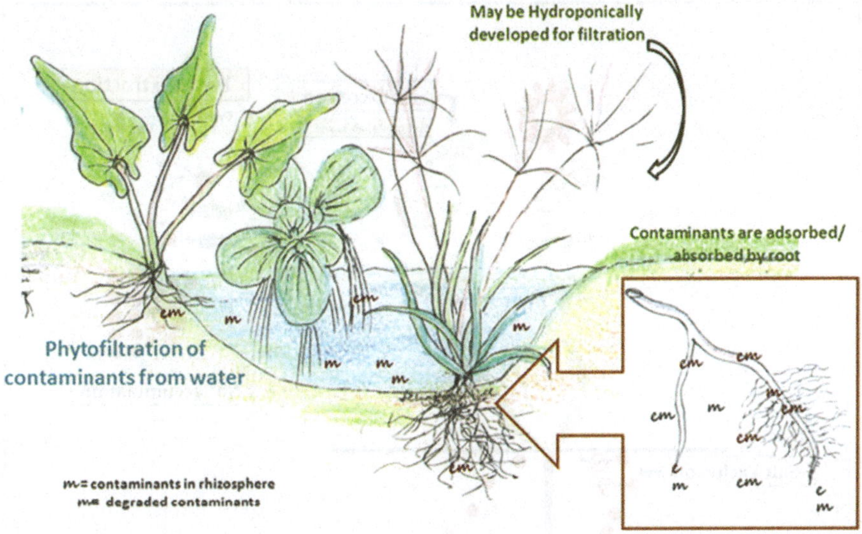

Fig. 16.5 Generalized figure to describe the process of rhizofiltration in a plant (Datta et al. 2013)

plant species that have gained popularity in this regard include *Helianthus annuus* L., *Brassica juncea* (L.) Czern., *Phaseolus vulgaris* L. var. *vulgaris,* and a number of members from Poaceae family. The metals that have been effectively removed so far include copper, zinc, chromium, cadmium, lead, and uranium. Among different radioactive metals, uranium (238 U) has become a serious concern for all nations as its mining and other activities keep adding it to the atmosphere. Some plant species are good at absorbing this radioactive contaminant. Such plants usually absorb it through their roots (Gupta et al. 2019). In a study, sunflower efficiently absorbed almost 80% of uranium from water contaminated with the said metal. Interestingly, the heavy metal was absorbed within just 24 h. It reflects the strength of this technique in removal of heavy metals from contaminated water.

16.5.5 Phytodesalination

This method may be described as "a method that employs the plants tolerant to higher concentrations of salts to clean a medium of excess salts" (Fig. 16.6). Such plants are generally regarded as "halophytes." The soil, after removal of salts, can be used for agricultural purpose. This technique has been studied extensively especially in countries with saline areas in order to free soil of excess salts and improve its qualities for growing food crops. Efforts have been made to identify salt-tolerant genes from different organisms including microbes and introduction of such genes into selected plant species to be used for phytodesalination (Walid et al. 2012).

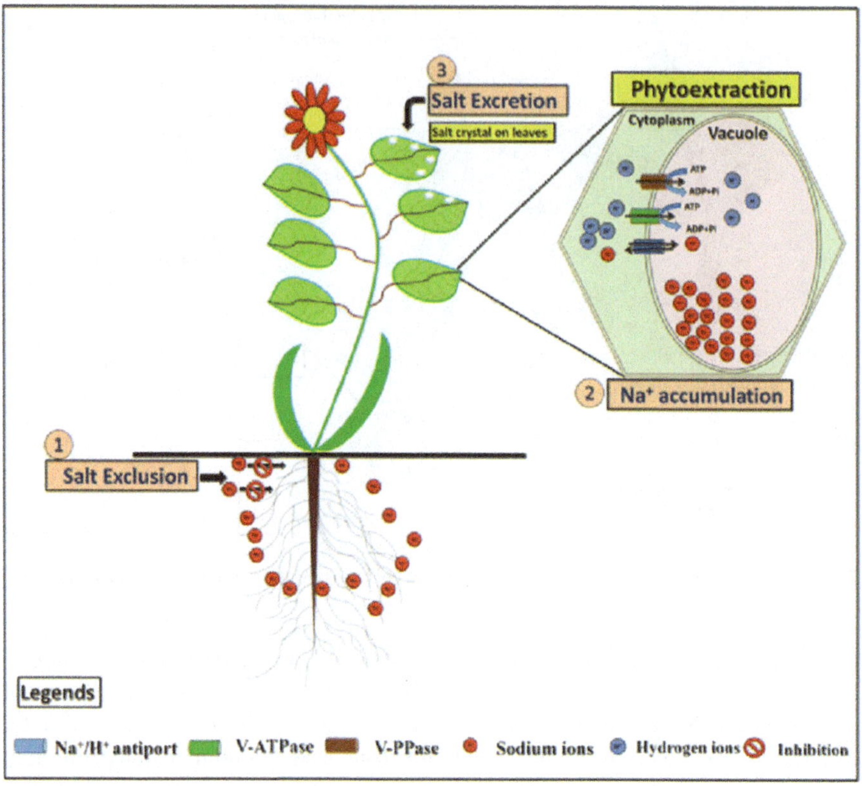

Fig. 16.6 Generalized figure to describe the process of phytodesalination in a plant. (Saddhe et al. 2020)

16.5.6 Phytostimulation

In this method, the plant roots enhance and support microbial growth in the near surface zone of earth crust, and the contaminants are degraded by those microbes. This approach is named as "plant-assisted remediation" by some researchers. The concept of this technique revolves around an increase in microbial growth and activity in rhizosphere and resultant degradation of contaminants in soil (Fig. 16.7). Since this process is bound to happen around any plant species provided that the soil structure favors microbial growth and degradation in rhizosphere.

It has been observed that the plant roots increase the growth of microbes in three major ways; (1) by adding organic matter to the rhizosphere (by death and decay of roots), (2) by respiration (thus adding oxygen), and (3) by secretion of roots exudates. Since the growth of microbes is enhanced by activity of plant roots so this method is dependent on plant growth in that contaminated area (Hussain and Hasnain 2011).

Fig. 16.7 Generalized figure to describe the process of phytostimulation (Datta et al. 2013)

16.5.7 Phytostabilization

In this method, heavy metals are stabilized and reduced at the plant–root surface and their mobility is reduced. It is just like captivating something and not letting it go anywhere else (Fig. 16.8). Soil erosion and leeching of contaminants are higher in areas with little vegetative cover. So, growth of contaminant-tolerant plants in such area helps fix the contaminants in the root zone. In this way, the spread and exposure of such contaminants to surrounding environment is reduced. The soil is gradually reclaimed by vegetative cover. Once the soil properties are restored to normal, different crops can be cultivated in routine (Bolan et al. 2011).

16.5.8 Hydraulic Control

In this method, tree roots control/limit water movement by strong pumping action. The water cannot move to deeper soil layers, rather it moves up by hydraulic pressure of roots. Greater volumes of water can be pulled by trees on daily basis. For examples, 30 gallons of water are pulled up by pumping action of Poplar tree

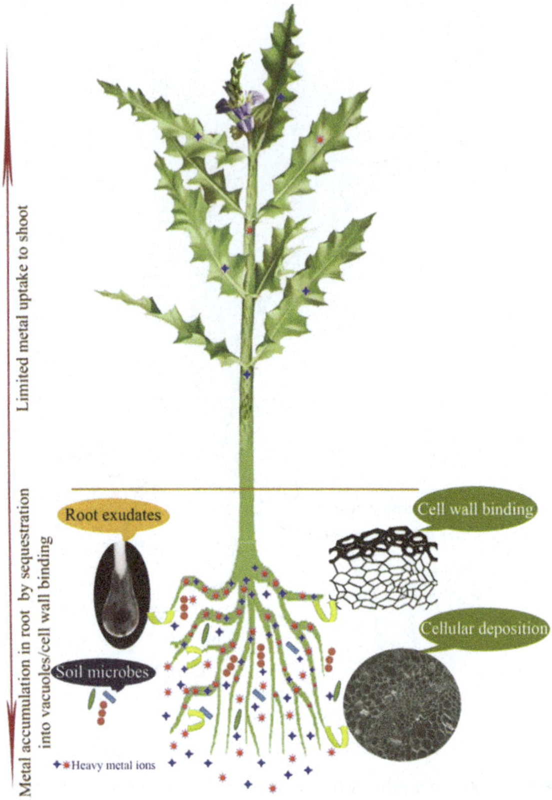

Fig. 16.8 Generalized figure to describe the process of phytostabilization (Shackira and Puthur 2019)

roots per day. Similarly, cotton wood tree can pull up to 350 gallons of water per day. It has been documented by Environmental Protection Agency of USA that Poplar trees have great potential to limit the leaching down of various contaminants by pulling water up. Poplar trees have great potential to remediate soil of toxic elements (Castro-Rodríguez et al. 2016). These trees have been shown to remediate soil of high levels of nitrates (O'Neill and Gordon 1994).

16.6 Factors Affecting Phytoremediation

Following factors have been studied so far that can affect the rate of phytoremediation. These factors, in simple words, play role in enhancement of efficiency of phytoremediation. These factors are discussed below:

16.6.1 Plant Species

The selection of plant species and proper planning of phytoremediation involves the study of the following:

1. The type of contaminants (organic/inorganic/metallic/mixtures of metallic, organic, and/or inorganic).
2. The nature of contaminated medium (soil/water/air/organic or inorganic debris/combination of soil, water and/or organic, and inorganic debris/sediments).
   (a) In case of soil, the depth of the soil that is contaminated helps determine the right type of plant species. For example, if the top layer of the soil is contaminated, then the use of trees is not favorable, rather shallow rooted plant species can be useful in this regard. While, in case of contamination of deeper soil layers, trees can play effective role in remediation of soil.
   (b) If contaminants are found in water, then it can be a water reservoir like a pond, lake, river, stream, ocean, or waterfall. For example, if contaminants have leeched down to the level of groundwater table, then tree species can play proper role in pumping the water up using their extensive root system.
   (c) Waste effluents of factories/industries can either be organic or inorganic debris or mixtures of one or more types of contaminants.
   (d) Wastes from residential areas (solids/liquids).
3. The climatic and environmental conditions of the contaminated site.
4. The nature of solid wastes, metallic, or nonmetallic.
5. The nature and type of soil particles along with its physical and chemical characteristics. Since the soil is mostly the medium of growth for most of the plants.
6. In case of air, the study of air quality index is also important.

It has been observed that the efficiency/rate of phytoremediation differs among different plant species. It has also been shown that different plant parts have varying potentials/tendencies to accumulate contaminants in them. For example, it has been reported by Firdaus-e-Bareen et al. (2019) that heavy metal phytoextraction capacity was different in two selected plant species, i.e., sorghum and pearl millet. They found that pearl millet had higher efficiency of extraction of heavy metals. And at the same time, different plant parts showed different specificity for accumulation of metals with roots ranked the first followed by stem and leaves.

Different plant parts accumulate different levels of heavy metals. It was observed by many researchers. For example, Khan et al. (2019) reported higher metal accumulation in roots than in foliar parts of *Petunia hybrida* L. copper got accumulated in roots and the other metals (Cd, Cr, Cu, Ni, and Pb) in above ground parts of the same plant. They reported that the plant underwent through heavy stress in contaminated water. Its physiological processes were disturbed to a great deal under heavy metal stress. The quality of plant aesthetics decreased when it was exposed to higher concentrations of heavy metals. Yet the said plant can be used for phytoextraction of selected metals. Similarly, some other species including *Helianthus annuus*, mustard plants (especially *Brassica juncea*), *Apocynum canabinum*, *Pteris vittata*, *Salix* spp., *Beta vulgaris*, Ragweed (*Ambrosia artimisifolia*), *Populus* spp. have also been found effective in this regard.

A comprehensive study of mechanisms/physiological processes of plant species should also be conducted prior to phytoremediation. It is important to mention here that few contaminants, if continue to stay in a medium/matrix (for some time at least), they support the growth of various types of microbial species. Hence, the selection of plant species depends on those microbial species also. In such a case, tolerance and response level of plant species to such factors must be studied prior to planning a phytoremediation project. It is also noteworthy that plant species may be selective in remedying a medium. In addition, a variety of plant species may be needed to grow simultaneously in a medium based on the type of contaminants found in it. Rotation of different plant species at the same site/area may be needed, depending upon the type of contaminants. Among different plant species halophytes are well known for their role as phytoremediators. The role of halophytes is acknowledged worldwide. They are known to remediate coastal and other areas due to their hyperaccumulation potential.

16.6.2 Soil Amendments

The research has accelerated on finding the mechanisms, strategies, and methods to enhance the phytoremediation efficiency/rate. Amendment of soils is considered as one of the major ways to do so. Many researchers across the world have conducted such researches which involve addition of different chemical, microbial, and other agents to the rhizosphere. It is noteworthy that researchers have proved that soil amendments have remarkable role in improvement of phytoremediation rate. Soil amendments that have been experimented so far include the use of chelating agents of various natures (ranging from organic to inorganic, natural to synthetic), microbial species and even varying combinations of such agents. All these agents are discussed below:

16.6.3 Chelating Agents

Chelating agents are those substances that chelate with the contaminants in soil/water and enhance their accumulation and bioavailability in the medium. Chelating agents mostly function by formation of coordination complexes with metals in the medium. Hence, we can say that the phytoremediation efficiency is enhanced by such agents. There are two major types of chelating agents, i.e., organic and inorganic. The inorganic ones are usually synthetic agents. The inorganic ones have been considered more effective in increasing bioavailability of contaminants in the medium (Dineshkumar et al. 2019). Prominent examples of inorganic chelating agents include EDTA, DOTA, EDDS, DTPA, EDDHA. On the other hand, organic ones are those that are derived from living organisms. Examples of such agents include proteins, carbohydrates, nucleic acids, and various types of organic acids. The organic acids that have been experimented so far to enhance phytoremediation rate include 2,3 Dihydroxy benzoic acid, citric acid, homo citric acid, and gluconic acid.

16.6.3.1 Organic Chelating Agents

It is the interaction of roots, soil particles, dissolved, and un-dissolved materials in the soil liquids that decides the overall nature of rhizosphere and absorption of contaminants from soil. Organic acids play crucial role in deciding the soil pH. The acidic environment has effects on solubility and subsequent bioavailability of contaminants in the soil. For example, lead oxides, carbonates, and sulphates are readily soluble in acidic medium (Traina and Laperche 1999).

Rhizosphere is modified not only by proton contributing properties of acids but by their action as ligands that chelate with metals. Redox reactions also occur in the rhizosphere as soon as the acids are added to it. Such reactions can affect metal mobility in the rhizosphere (Violante et al. 2010). These acids are usually weak and their pK_a values range from 3 to 9. Their molecular weight varies greatly. The lowest molecular weight of organic acids is possessed by oxalic and citric acids (Wei et al. 2009). The acids with lower pK_a values generally have carboxylic functional group while those with higher values have phenolic group. Irrespective of their functional groups, they generally increase metal solubility in the soil water and affect weathering process. Various researchers from different countries have been working on this aspect, e.g., Wu et al. (2012) tried to assess the effect of organic amendments on phytoremediation using *Sedum* spp. Among organic chelating agents, following three acids have been researched the most:

16.6.3.1.1 Citric Acid

Citric acid is a well-known organic agent that is famous among masses for its health promoting effects. It is a metabolite of almost all living aerobic organisms. It has high antioxidant activity and is usually found in high concentrations in fruits especially those with bitter/sour taste (like lemon and orange). It is a weak acid and is used by general public to add sour taste to foods. Its molecular formula is $C_6H_8O_7$ (Kaushik 2015). It is a tricarboxylic acid. It has great potential to chelate heavy metals in the medium and get absorbed by plant roots at a greater pace due to its small-sized molecules. The effect of citric acid supplementation has been found effective by few researchers, e.g., Turgut et al. (2004) and Chen et al. (2003b).

16.6.3.1.2 Oxalic Acid

Oxalic acid is yet another organic acid that is being studied for its chelating tendencies for different metals in the soil solution. It is a metabolite of human, algae, and plants as well. It is a dicarboxylic acid, produced in the living cells by metabolism of ascorbic or glyoxylic acid. Interestingly, the bodies cannot metabolize it and it is excreted out of the body as a waste product. It is a good reducing agent and chelates with metals thus increasing their phytoavailability in the soil solution (Wang et al. 2019). Oxalic acid has been shown to increase bioavailability of cadmium in the soil (Hou et al. 2019). Oxalates have been reported to impart tolerance to plants growing in soils contaminated with aluminum, lead, cadmium, and zinc (Rajendra and Yashbir 2017).

16.6.3.1.3 Gluconic Acid

Gluconic acid is also a naturally produced mild organic acid found mainly in honey, fruits, teas, and wine. Its molecular formula is $C_6H_{12}O_7$ and pK_a value is 3.7 (Kaushik 2015). It is generally produced in the bodies of microorganisms (e.g., *Aspergillus niger* and Gluconobacter) by degradation of glucose (Ramachandran et al. 2006). Gluconic acid and its derivative sodium gluconate have huge applications in food and pharmaceutical fields. It effectively chelates with metals including iron, aluminum, and calcium. It works best in alkaline environment.

In an interesting study by Hu et al. (2019), citric acid, oxalic acid, and EDDS (ethylenediamine disuccinic acid) were applied to the soil contaminated with higher uranium levels. They grew Macleaya cordata in those affected soils and observed the effect of aforementioned chelating agents. Citric acid was found to be the most effective chelating agent while oxalic acid was not much efficient in increasing the bioavailability of uranium in soil. The antioxidant system of the plant performed well against the oxidative stress caused by both chemical entities applied. Similarly, cadmium availability is improved by addition of oxalic acid in contaminated soils. Bioavailability of other metals has also been studied in organic acid supplemented soils.

In addition to the abovementioned organic acids, malic acid, tartaric acid, homocitric acid, and 2,3-Dihydroxy benzoic acid have also been investigated for their possible effects on phytoremediation. But only few studies can be observed in this regard. It is also interesting to note that organic acids, not always, increase the phytoremediation rate. Sometimes, their role has been either negligible or negative. This situation reflects that more studies are needed to properly investigate their role in remediation of contaminated sites/media.

16.6.3.2 Synthetic Chelating Agents

The effect of synthetic chelating agents on efficiency of phytoremediation of different plant species growing in soils contaminated with different metals has been studied extensively. Among different synthetic agents, EDTA has been experimented the most for its role in phytoremediation. It has been well-established that EDTA has strongly positive role in this regard as it does accelerate remediation of contaminated sites. Jiang et al. (2019) provided the first successful evidence in 2019 that lead absorption was enhanced by addition of EDTA to soil. The bamboo plants efficiently absorbed higher lead concentrations quickly when EDTA was employed as a chelating agent. They used such higher concentrations of lead up to 0–1500 mg/kg, and EDTA was used from 250 to 1000 mg/kg.

Recent research by Gul et al. (2019) also supported the view that EDTA has strongly positive effect on phytoremediation. They studied the effect of EDTA supplementation on phytoremediation efficiency of two selected plant species (*Pelargonium hortorum* and *P. zonale*) from Pb and Cd contaminated soil was evaluated.

Different concentrations of lead (0 to 1500 mg/kg), cadmium (0 to 150 mg/kg), and EDTA (0 to 5 mmol/kg) were employed. They reported marked difference in phytoextraction efficiency of both plant species where *P. hortorum* showed higher phytoremediation potential. In addition, the EDTA supplemented soils lead to higher phytoextraction of contaminants from soil.

In another recent study by Dou et al. (2019) reported an enhanced uptake of cadmium by plant species *Bidens pilosa* when EDTA was added as a chelating agent. They also demonstrated that cadmium could be efficiently absorbed at equal rate irrespective of its type (sulfate/phosphate/chloride of cadmium). The only factor that affected the rate of uptake was the chelating agent itself. In another valuable study by Chaturvedi et al. (2019), phytoremediation potential of *Brassica oleracea* L. and *Raphanus sativus* L. plant species was investigated. These plants were grown in soils contaminated with different heavy metals including zinc and lead. They documented that the metal uptake was enhanced when chelating agent was added. This study also supported the idea of soil amendment to improve heavy metal uptake by plants.

In yet another study, the metal uptake, tolerance to extracted metal, and biomass was increased when soil was amended with addition of EDTA as chelating agent. This study was conducted by Wasino et al. (2019). They employed EDTA as chelating agent for three heavy metals cadmium, zinc, and lead. They used plant species *Chrysopogon zizanioides* and *C. nemoralis* for phytoremediation. They recorded an increase in absorption efficiency of the said plant species when EDTA was employed. Both species showed significant remediation potential in case of soil contaminated with all three metals.

It must be kept in mind while experimenting with such agents that applied concentration does carry weightage in affecting solubility, extractability, translocation, and other parameters of phytoremediation. For example, EDTA and other chelating agents do increase phytoremediation efficiency but their higher concentrations have opposite effect. The study of Yu et al. (2019) is worth mentioning here. They reported that higher levels of EDTA had negative impacts on remediation of soil contaminated with manganese using *Polygonum pubescens*.

It is also noteworthy that addition of EDTA is not always fruitful. It may miraculously increase uptake of a metal by a plant species but at the same time it may not increase uptake of another metal. For example, Ghazaryan et al. (2019) found that the uptake of copper was enhanced in EDTA supplemented medium while there was no such increase in uptake of molybdenum by the same plant species under the same conditions. It shows that prior to planning a project, a pre-hand study must be done to investigate the effect of different factors that may influence phytoremediation efficiency of a plant species.

Interestingly, studies do exist which compare the effects of synthetic and natural chelating agents on phytoremediation (Wu et al. 2004). For example, Wu et al. (2004) reported that EDTA enhanced the uptake of two of the four experimented heavy metals, i.e., Cu and Pb by *Brassica juncea* while organic acids including oxalic, citric, or malic acid had no remarkable effect. All chelating agents were used at equal concentration, i.e., 3 mmol kg^{-1}. These results show superiority of the use of EDTA over organic acids in influencing the rate of phytoremediation. DOTA (1,4,7-Tetraazacyclododecane 1,4,7,10 Tetra acetic acid) is yet another popular synthetic agent that has shown its effects on phytoremediation. It is also known as Tetraxetan. Its IUPAC name is 2-[4,7,10-tris (carboxymethyl)-1,4,7,10-tetrazacyclododec-1-yl] acetic acid, and its molecular formula is $C_{16}H_{28}N_4O_8$. It is mostly used to chelate with lanthridine (Kaushik 2015). Other synthetic agents have also been investigated for their potential effects on phytoremediation but the highest impacts have been produced so far by EDTA that has already been discussed in this chapter.

16.6.3.3 Combined Effect of Organic and Inorganic Chemical Agents

Some other researchers have documented the idea of soil amendments to increase bioavailability of contaminants in soil. Majority of them agreed upon the use of fertilizers to support plant growth and subsequent increased absorption of contaminants. In a study, Zhang et al. (2019) has experimented the use of EDTA and silicon-based fertilizers to see their effect on phytoextraction by rice plants. They observed an increase in phytoextraction of cadmium by rice plants when EDTA and fertilizers were applied simultaneously. Shahid et al. (2019) had experimented the effect of EDTA and citric acid was evaluated on physio-biochemical traits of young and old bean leaves under cadmium stress. They reported that EDTA enhanced Cd uptake and accumulation and decreased its toxicity by controlling different physio-biochemical traits. But citric acid surprisingly reduced uptake of heavy metal. It shows that EDTA enhances metal uptake and protects plants against its damages while citric acid which is an organic acid reduces metal uptake.

Some researchers are now looking for some suitable additives that reduce the toxic effects of metals and EDTA on plants. In a latest and interesting study conducted by Revathi and Subhashree (2019), sodium nitroprusside has been used to see its effect on phytoremediation efficiency and physiological processes of the plant. It was observed by them that in absence of sodium nitroprusside, the antioxidant activity in plant increased upon addition of EDTA and heavy metal which means that the plant needed to get rid of free radicals. But when sodium nitroprusside was added, the antioxidant enzyme activities (catalase, superoxide dismutase, ascorbate peroxidase, and glutathione reductase) reduced. It clearly shows that the additives like sodium nitroprusside can reduce the stress levels in plants and indirectly enhance phytoremediation potential of the plant.

16.6.4 Microorganisms

Plant growth promoting rhizobacteria, several fungal species, mycorrhiza, endophytes, and algae have been shown to increase the rate of phytoremediation (Umesh et al. 2016). Among these, plant growth promoting rhizobacteria (PGPRs) play role in strengthening the plants against abiotic stresses by improving the efficacy of soil and plant growth promotion (Prasad et al. 2015). They also have role in influencing the crop sustainability by increasing production of various enzymes, nitrogen fixation, and solubilization of phosphorus and potassium in the soil. In a study, they have been proved to improve cadmium bioavailability in leguminous plants by bioaccumulation and by formation of complexes and chelates (Jebara et al. 2019).

Plant–microbe interactions have been extensively studied for the past two to three decades. But the research on this aspect has accelerated enormously for the past few years. At the same time, plant microbe and metal interaction has also seen extensive interest and research. Plant microbes usually include different bacterial species and mycorrhizae. Such microbes make symbiotic associations with plant roots, thus getting mutual benefits. This shows that plant-associated microbes can play vital role in remediation of contaminated environment. There are various studies available that comply with this statement and prove it. For example, it has been shown by Jan et al. (2019) that soil under high cadmium stress can be remediated by rice seedlings with the help of *Bacillus cereus,* a bacterial species.

Since this view has been well-established that biogeochemical interactions play vital role in bioavailability and uptake of environmental contaminants, researchers have been looking for appropriate microbial species to investigate their effect on phytoremediation potential of plants. Since microbial species are plant-species-specific and the properties of soil are also crucial to their growth and sustenance and overall performance, optimization of conditions for microbial-phytoremediation is the point of focus for researchers. Most of the researchers agree that different types of microbial metabolites have role in adjustment of rhizosphere and hence subsequent phytoremediation. Such metabolites include indole-3-acetic acid, organic acids, siderophores, and 1-amino-cyclo-propane-1-carboxylic acid deaminase (Rajkumar et al. 2012). Recently, microorganisms are being genetically engineered with two main objectives. The first one is to increase their efficiency of pollution control (which is by modification in their innate metabolic characteristics), and the other one is to regulate plant growth. Both these strategies ultimately strengthen phytoremediation efficiency (Mishra et al. 2019).

16.6.5 Combination of Chelating Agents and Microorganisms

Some researchers have reported that use of combinations of different factors improved phytoremediation efficiency. For example, Asilian et al. (2019) reported that combination of chemical and microbial approaches enhanced phytoremediation efficiency of maize plants. They used a bacterial (*Pseudomonads fluoresce*) and a fungal species (*Piriformospora indica*) along with Tween-80 surfactant for this purpose. They grew maize seeds in cadmium polluted soil and observed the status of plant growth and cadmium levels in plant tissues. Their study provided evidence for higher phytoremediation efficiency of maize plants after combined application of microbial and chemical factors.

In yet another latest study, Yasin et al. (2019) tested the effect of EDTA and a bacterium *Enterobacter* sp. CS2 on phytoextraction efficiency of a plant species *Impatiens balsamina* L. from soil. The researchers used soil contaminated with industrial effluents carrying different concentrations of nickel (Ni). The seeds of the said species were soaked in this contaminated soil for 50 days and Ni-tolerance index, bioconcentration, and translocation factor were observed. They expertly found out that Ni reduced plant growth and development in absence of both factors, i.e., bacterium and chelating agent. But the plants had higher tolerance level for the metal when the soil was supplemented with bacterial species *Enterobacter* sp. CS2. In addition, EDTA supplementation enhanced metal uptake by plant. Hence their study clearly indicated that combination of microbial and chelating agent supplementation has direct effect on efficiency of phytoremediation.

16.6.6 Combined Effect of Organic and Synthetic Chelating Agents

Few studies have focused on evaluation of combined effect of organic and inorganic chelating agents. Guo et al. (2018) reported application of chelating agents with the potherb *Brassica juncea* while growing in soil contaminated with a smelter. Two heavy metals viz., zinc and cadmium, were found in the soil which was efficiently absorbed by the plants after application of EDTA in combination with citric acid and oxalic acid. The chelating agents were added to soil 3–4 weeks after sowing. The accumulation of both metals was enhanced almost 1.5–3 folds in different experiments. While the plant physiology went through heavy stress and the antioxidant enzymes were produced at a higher concentration. Their study provided interesting results. For example, the highest phytoremediation efficiency was observed with single chelating agent, i.e., EDTA alone which was followed by combination of EDTA and organic acid. It is interesting to note that only organic acids were added to the medium before phytoremediation, and there was no significant increase in phytoremediation efficiency. McBride et al. (2019) noticed higher solubility and phytoavailability of cadmium and zinc prior the application of organic acids only (i.e., without EDTA) but there was no enhancement in uptake of any metal by *Phytolacca americana*. They elaborated that this reduced uptake might be due to presence of competent metals (copper and manganese) in the medium or less bioavailability of resultant metal complexes.

16.6.7 Plant Growth Regulators

Phytohormones, generally known as plant growth regulators (PGRs), are amazing compounds that have crucial role in the life of plants and without them, plants cannot exist. The reason being, they influence every cellular process from its formation, sustenance, growth, development, division, and so on (Rostami and Azhdarpoor 2019). It has been well-established and well-understood that plant hormones protect plants against all sort of biotic and abiotic stresses that hit their lives. They are involved in signaling and absorption of metals from soil too. Hence, their role in absorption of contaminants especially that of heavy metals is also under investigation. It has been proved that exogenous application of PGRs has positive impacts on plant growth and development and to alleviate heavy metals stress and management of their toxicity by great many researchers including Zhu et al. (2012, 2013), Agami and Mohamed (2013), and Masood et al. (2016). Among PGRs, auxins, cytokinins, gibberellins, and salicylic acids have shown potential in increasing the rate/efficiency of phytoremediation in plants; while brassinosteroids have been documented to play the same role in microalgae. The extent of their effectiveness depends on the type of plant species and its physiological conditions, their concentrations used, and the environmental conditions. Their direct effect is on the growth of plant which is enhanced thus adding to biomass of plants. The efficient plant growth increases the efficiency of absorption of contaminants from their environment (Bajguz 2019).

16.6.8 Intercropping Different Plant Species

The concept of intercropping is not new. Growing different plant species in a shared place, under the same field conditions has unique consequences. This technique is different from crop rotation in which more than one different plant species are grown in the same field area one after the other. Crop rotation is said to retain soil fertility. While, intercropping involves growth of different species simultaneously, this may have role in affecting the rhizosphere. This concept of intercropping has been widely used in case of phytoremediation. Recently, in the early months of 2019, few studies have reported that intercropping hyperaccumulator plant species which can have positive influence on phytoremediation potential of such plants. Shuzhen et al. (2019) reported a hyperaccumulator plant species *Sedum plumbizincicola* which was intercropped with *Oxalis corniculata* and *Buxus sinica* in soils affected with higher Cd levels. They used organic acids as chelating substances to increase efficiency of remediation, and they obtained positive results in this regard when oxalic acid was used at a higher concentration of 11 mmol kg^{-1}. Oxalic acid increased bioavailability of Cd in the soil, which was then absorbed by these aforementioned plant species.

16.6.9 Alterations in Plant Genome

It is quite evident that few plant species are more tolerant to metal toxicity. This tolerance may rightly be attributed to the genetic characteristics. Since the genome of a living being is the backbone of life. The genes are responsible for all morphological and biochemical characteristics in a living being. So, keeping this information in view, it can be concluded that any alteration in the genome may enhance the tolerance of a plant species for specific contaminants. Two different strategies such as: introduction of mutant in plant genome and introduction of tolerant genes to plant genome can be used to improve genetic traits of metal tolerant plants.

16.6.9.1 Introduction of Mutations in Plant Genome

Recently, the role of mutations in plant genome has also been investigated. Since mutations may lead to unique results. Experimentation carried out by Navarro-León et al. (2019) is valuable in this regard as they studied the role of a gene in affecting the efficiency of phytoextraction of cadmium from contaminated soil. They used *Brassica rapa* for phytoremediation considering greater efficiency of said plant species in tolerating heavy metals. They reported that the TILLING mutants of *B. rapa* (BraA.hma4a-3) had greater tolerance for Cd as they reflected lower reduction in biomass and higher quantities of the said metal in their foliar parts. This study clearly demonstrated that mutations in selected genes can be beneficial as they enhance phytoremediation efficiency of plants.

16.6.9.2 Introduction of Tolerant Genes to Plant Genome

Plant species that are more tolerant to heavy metal toxicity are being screened for their genes. Such genes may be introduced to other plant species which are better suited for phytoremediation projects. For example, *Prosopis juliflora* has strong tolerance for heavy metals. Keeran et al. (2019) have reported that this plant can be taken as ideal for gene mining for phytoremediation. Its genes can be transferred to other species for better performance. The research efforts are continuously being made to produce metal tolerant plants yet such plants have not reached field level. Some researchers have been trying to search such tolerant or heavy metal responsive genes from different plant species. For example, Abou-Elwafa et al. (2019) performed such an analysis on 107 accessions of sorghum using a set of 181 microsatellite markers. They reported 14 phytoremediation and heavy metals tolerance QTLs in the said plant species and 19 heavy metals stress tolerance genes.

16.6.10 Electrokinetics

The use of electrokinetics and phytoremediation approaches one after the other at different time intervals can influence the rate of phytoremediation. It has been observed by some researchers that alternate application of these two methods can substantially remediate the contaminated site in shorter span of time. For example, in a study by Chang et al. (2019), circulation-enhanced electrokinetics-phytoremediation (using corn plants)-circulation-enhanced electrokinetics was applied alternately to lead-contaminated area. They reported a reduction of 63% in initial lead levels.

16.7 Advantages of Phytoremediation

Following are the advantages of using phytoremediation to combat environmental pollution (Bock et al. 2002):

- Can not only be performed in situ but ex situ.
- Low cost.
- Solar energy-driven technique (natural process).
- In situ remediation of contaminated areas and subsequent restoration of a site.
- The plants can be easily monitored.
- It is much effective in areas with lower levels of contamination (less polluted areas).
- A huge variety of contaminants can be dealt by using this single technique, just right selection of plant species is crucial to this process.
- The fertility of soil is either maintained or improved.
- Effective where most of the other methods especially mechanical ones fail to work.
- Accepted and understood easily by public.
- The possibility of recovery and reuse of metals.
- Strengthens natural ecosystem.
- Eco-friendly technology.
- It may play its role in alteration of environmental or climatic conditions of an area depending on the type of plant species and the dimensions of area under cover. For example, if a huge area is affected by radiation, the phytoremediation strategies may involve the growth of big bare land area under cultivation of one or more plant species to free that zone of radiations, thus contributing to the overall weather patterns of the region.
- Phytoremediation has proved successful in reducing the levels of explosive compounds in soil and water. Few submerged plant species have effectively reduced the levels of RDX up to 40% which was simply doubled, i.e., up to 80% after addition of microbial species in the matrix. In the same way, 5% reduction in TNT concentration was found by submerged and floating plant species.
- Contaminants of various types have been successfully controlled from damaging the environment using phytoremediation. Yet further studies are required on this aspect to establish it as a promising approach in combating environmental pollution.
- Large areas of soil/water bodies can be remediated by using plants.
- Plants can be easily monitored.
- Possibility of recovery and reuse of contaminants.
- Expensive and complicated equipment are not required.
- Social acceptance.

16.8 Disadvantages/Limitations of Phytoremediation

Following are the disadvantages or limitations of using phytoremediation:

- Phytoremediation can work effectively in shallow groundwater, soils, or sediments. The area where roots cannot penetrate will remain affected of contaminants as such.
- Higher concentrations of contaminants can be fatal to the plant as well as the consumers of those plants. For example, the phytotransformers can be fatal to small animals like snails, so special care must be taken to keep animals from such plants prior to and during phytoremediation projects.
- Slow technique as compared with other conventional methods of waste management.
- The plant species with slow growth and higher biomass can be problematic.
- Only suitable for fully or at least partially hydrophilic contaminants.
- Contaminants may reach groundwater since the plant roots may fail at some points in absorbing the contaminants.
- The characteristics of soil and climatic and/or environmental conditions of the area can influence the rate and quality of remediation.
- If algae are used for this purpose, then excessive growth of algae in the top-most water layer may block the entry of light in deeper layers of water body hence suffocating the life underneath.
- Need of special attention on safe disposal of plants after remediation of environmental components, and special care is needed to prevent food chain from contamination since plants are the primary producers.
- Input of human resource that frequently and attentively keeps an eye on growth of plants and associated procedures in process.
- High input of labor and energy.
- Needs long-term commitment, care, and continuous study until the medium is restored to its normal.
- Long time period is required for complete recovery of contaminated environmental components (at least one growing season of planted species is required).
- The deeper layers of soil cannot be remediated as plant roots cannot penetrate to such depths (generally limited to top 3 ft of soil and top 10 ft of water).
- Addition of chelating agents is needed in some cases (in order to weaken/break the bonding between soil particles and the contaminant thus enhancing their availability for absorption).
- Large land areas are needed for phytoremediation.
- In some cases, contamination may shift from one environmental component to the other.
- The contaminants that get fixed in plant body may gain entry to food chain by herbivores, pollinators, and other consumers of plants.
- Climatic or any other factors may affect the growth of different plant species.
- It has been reported by almost every study on the said aspect that chlorophyll content decreased in every plant species that was exposed to heavy metal stress.

This reduction ultimately affects total photosynthates and hence plant life. This is a severe limitation of this technique as it depends mainly on plant.

16.9 Recent Research Trends in Phytoremediation

The recent research trends involve use of nanotechnology to enhance efficiency of phytoremediation (Zhu et al. 2019). The nanomaterials may remove contaminants from environment, promote plant growth and development, and increase availability of contaminants for absorption by plant roots. But it has been observed that still this technology is under investigation for its potential miraculous uses in phytoremediation. Most commonly used nanoparticles in this regard are that of iron. This nanoparticle has been under investigation to enhance environmental remediation. It is hypothesized that the said particle can increase phytoavailability of contaminants to the plant roots. For example, in a recent research, Mokarram-Kashtiban et al. (2019) have shown that zero valent nano particles of iron had positive effects on phytoremediation while higher concentrations had the reverse effect. Research on the use of nanoparticles is scarce and needs validation through in-depth study.

The use of phytoremediation is no doubt an effective method to remediate soil. But at the same time, slow speed of the said process has been taken as a serious limitation of it. Few researchers have come up with a solution by experimenting different methods in combination with phytoremediation to see if the efficiency of remediation increases. Interestingly, the efficiency increased when electrokinetic bioremediation method was used (Kim et al. 2005). They reported that use of electrokinetic bioremediation enhanced the efficiency of remediation. This approach was appreciated in the circle of researchers and they started different experimentations that employed phytoremediation in combination with other techniques to get better results in a short time span. In one of such latest studies, a group of researchers from Taiwan used circulation-enhanced electrokinetic-phytoremediation-circulation-enhanced electrokinetic approach to speed up the remediation process (Chang et al. 2019). Their research proved successful as they reported that the contaminated soil was remediated at a faster pace. They reported that the soils in Taiwan were contaminated with lead which could be removed efficiently by using this approach.

Since the soils are usually contaminated by more than one factor, research on phytoremediation strategies that hit more than one contaminant are being conducted. For example, in a recently published research work by Huang et al. (2019), oxalic acid-activated phosphate rock and bone meal were applied to copper- and lead-contaminated soil to investigate their effect on immobilization of both these contaminants.

Following are the main research topics under investigation nowadays:

- Selection of appropriate plant species for phytoremediation.
- Optimization of conditions for phytoremediation.

- Studies on the role of chelating agents in increasing bioavailability of contaminants (with special emphasis on heavy metals).
- Studies to enhance the rate of phytoremediation with special emphasis on different factors that can influence this process.
- Studies to find out metal-specific chelating agents to increase bioavailability of metals.
- Experimentation on the use of combinations of different factors for phytoremediation, e.g., use of combinations of organic/synthetic/natural agents.
- Research on hyperaccumulator plant genes with emphasis on mechanism of hyperaccumulation.
- Introduction of hyperaccumulator genes to other plant species, i.e., formation of transgenic plants.
- Studies on combined effect of mechanical, physical, chemical, and microbial agents in addition to phytoremediation to accelerate remediation of contaminated soils.
- Intercropping of different plant species to see its effect on phytoremediation.
- Crop rotation to enhance phytoremediation rate.

16.10 Future Prospects and Recommendations

Phytoremediation is being extensively studied at the level of universities and research institutes, yet its application is still limited to few areas on the globe. Phytoremediation being a natural process can be thought of as a miraculous method of remedying different environmental components and to find out solution to chronic environmental pollution in various world regions or planet as a whole. Efforts are being made in remedying environment with special emphasis on metals (especially uranium, arsenic, lead, chromium, and cadmium), pesticides, solvents, explosives, and oils of various kinds. Further research may lead to its full acceptance and wider applications, hence dealing environmental pollution in an effective manner. This technique can be foreseen as a strong weapon in combating environmental pollution worldwide. It is recommended that a researcher must gain proper knowledge of physiological processes and molecular mechanisms along with biological and engineering strategies as it can polish the quality of phytoremediation. Moreover, field trails should be performed to find out solutions to various problems and to find answers to various questions. Interestingly, on one side, a number of plant species can play vital roles in phytoremediation; while others cannot tolerate contaminants at all, an in-depth study of different plant species must be undertaken to prepare a list (preferably a database of plant species) that can be used for specific phytoremediation projects. Such a database should contain all relevant information with information on case studies to help select a plant species for phytoremediation in an area in case it is needed urgently. This may save time, energy, and resources to a great deal. Though this technique seems promising in dealing environmental pollution, care must be taken in order to gain maximum benefits and to eliminate its negative impacts on the biotic components of environment especially human beings. So, it is suggested that the use of such plant species which may lead to allergies of various kinds especially pollen allergy should not be brought into cultivation close to residential areas as they may solve one problem and affect human population in other ways. Proper study of the contaminated area must be conducted prior to planning a phytoremediation project. Such study may include investigation of climate, environment, weather conditions, and soil and/water characteristics. Type and properties of flora and fauna belonging to contaminated area must be focused. Initial testing of soil and/or water and a full-length experiment on a selected sample out of the infected site should be carried out to see the potential or chances of success in remediating the contaminated site. Most importantly, proper study must be conducted on proper disposal or dispositioning of the resultant plants. In case, the resultant contaminated plants are burnt, the properties of air and ash must be studied to see the effect of burning on metabolic products of contaminants. Extreme care must be taken in dealing plant samples so that the human and animal population is not harmed by them. The biodegraded compounds or by-products may reach groundwater or the food chain. So in-depth research is needed not only on proper disposal of such contaminants/contaminated plants but also the metabolic processes (including all biochemical reactions) must be studied so that the ecosystem can be protected from the possible damages that may be caused by them. Preferably those plants which grow by vegetative means must be preferred for phytoremediation. This is because the sex structures including pollens may get infected by contaminants. In case plants with sexual mode of reproduction are used, the palynological studies of such plants must be carried out to see if the contaminants got accumulated in pollens or other sex structures. The human beings and animals may come in contact with such pollens thus getting hurt in one way or the other. The hydrology and soil profile of contaminated site carry great weightage in planning a project so extreme care must be taken to study these parameters first.

16.11 Conclusion

An alarming, nonstop, and uncontrolled increase in release of wastes of diverse nature in the environment has been observed due to anthropogenic activities including increased urbanization, extreme lack of awareness of hazards of environmental pollution, deforestation, industrialization, and an increase in chemical and atomic warfare. This ever-increasing pollution is becoming the cause of concern for every nation on this planet as continuing healthy and comfortable life on this planet is becoming challenging. The consequences could be harsh if the situation goes unchecked. So, this matter must be taken seriously, and quick fix to the problem must be sorted out. On one side, strategies must be adopted to minimize the entry of pollutants to the environment, and on the other, such means must be investigated that could control the situation in an environment-friendly manner. Phytoremediation seems a method of choice as it has the potential to remediate the environment in most environment-friendly way. Though the researchers across globe have made efforts in this regard, yet in-depth research is needed on various aspects with special emphasis on understanding the mechanisms of phytoremediation, physiological processes of plants, mechanisms of uptake, translocation, accumulation, and tolerance. This would be made possible if interdisciplinary research is performed involving experts from physical, chemical, and biological sciences.

References

Abou-Elwafa SF, Amin AEAZ, Shehzad T (2019) Genetic mapping and transcriptional profiling of phytoremediation and heavy metals responsive genes in sorghum. Ecotoxicol Environ Saf 173: 366–372. https://doi.org/10.1016/j.ecoenv.2019.02.022

Agami RA, Mohamed GF (2013) Exogenous treatment with indole-3-acetic acid and salicylic acid alleviates cadmium toxicity in wheat seedlings. Ecotoxicol Environ Saf 94:164–171

Altinozlu H, Karagoz A, Polat T, Unver I (2012) Nickel hyperaccumulation by natural plants in Turkish serpentine soils. Turk J Bot 36:269–280. https://doi.org/10.3906/bot-1101-10

Anoopkumar AN et al (2020) Phytoextraction of heavy metals. In: Inamuddin, Ahamed MI, Lichtfouse E, Asiri AM (eds) Methods for bioremediation of water and wastewater pollution. Environmental chemistry for a sustainable world, vol 51. Springer, Cham. https://doi.org/10.1007/978-3-030-48985-4_12

April TM, Foght JM, Currah RS (2000) Hydrocarbon-degrading filamentous fungi isolated from flare pit soils in northern and western Canada. Can J Microbiol 46(1):38–49

Asilian E, Ghasemi-Fasaei R, Ronaghi A (2019) Chemical- and microbial-enhanced phytoremediation of cadmium-contaminated calcareous soil by maize. Toxicol Ind Health 35(5):378–386. https://doi.org/10.1177/0748233719842752

Bajguz A (2019) Brassinosteroids in microalgae: application for growth improvement and protection against abiotic stresses. In: Hayat S, Yusuf M, Bhardwaj R, Bajguz A (eds) brassinosteroids: plant growth and development. Springer, Singapore. https://doi.org/10.1007/978-981-13-6058-9_2

Baker AJM, Reeves RD, McGrath SP (1991) In situ decontamination of heavy metal polluted soils using crops of metal-accumulating plants: a feasibility study. In: Hinchee RL, Olfenbuttel RF (eds) In situ bioreclamation. Butterworth–Heinemann, Boston, pp 600–605

Bani A, Pavlova D (2010) Nickel hyperaccumulation by the species of Alyssum and Thlaspi (Brassicaceae) from the ultramafic soils of the Balkans. Bot Serbica 34(1):3–14

Best EP, Sprecher SL, Fredrickson HL (1997) Screening submersed plant species for phytoremediation of explosives-contaminated groundwater from the Milan Army ammunition plant, Milan, Tennessee. Final report. United States: N. p., web. https://www.osti.gov/biblio/574991screening-submersed-plant-species-phytoremediation-explosives-contaminated-groundwater-from-milan-army-ammunition-plant-milan-tennessee-final-report

Bock C et al (2002) Advances in phytoremediation: phytotransformation. In: Reible D, Demnerova K (eds) Innovative approaches to the on-site assessment and remediation of contaminated sites. NATO science series (series IV: earth and environmental sciences), vol 15. Springer, Dordrecht

Boffetta P (2006) Human cancer from environmental pollutants: the epidemiological evidence. 608(2):157–162. https://doi.org/10.1016/j.mrgentox.2006.02.015

Bolan NS, Park JH, Robinson B et al (2011) Chapter four. In: Sparks DL (ed) Phytostabilization: a green approach to contaminant containment, vol 112, pp 2–231. https://doi.org/10.1016/B978-0-12-385538-1.00004-4

Briggs D (2003) Environmental pollution and the global burden of disease. Br Med Bull 68(1): 1–24. https://doi.org/10.1093/bmb/ldg019

Brooks RR (1998) General introduction. In: Brooks RR (ed) Plants that hyperaccumulate heavy metals:their role in phytoreediation, microbiology, archaeology, mineral exploration and phytomining. CAB International, New York, pp 1–14

Buendía-González L, Orozco-Villafuerte J, Cruz-Sosa F, Barrera-Díaz C, Vernon-Carter E (2010) Prosopis laevigata a potential chromium (VI) and cadmium (II) hyperaccumulator desert plant. Bioresour Technol 101:5862–5867. https://doi.org/10.1016/j.biortech.2010.03.027

Callender E (2004) Heavy metals in environment-historical trends. Treatise Geochem 9:67–105. https://doi.org/10.1016/B978-0-08-095975-7.00903-7

Castro-Rodríguez V, García-Gutiérrez A, Canales J et al (2016) Poplar trees for phytoremediation of high levels of nitrate and applications in bioenergy. Plant Biotechnol J:299–312. https://doi.org/10.1111/pbi.12384

Chandra R, Kumar V (2018) Phytoremediation: a green sustainable technology for industrial waste management. Taylor & Francis, Boca Raton, FL

Chaney RL, Broadhurst C, Centofanti T (2010) Phytoremediation of soil trace elements. In: Hooda PS (ed) Trace elements in soils. Wiley, Chichester, pp 311–352

Chang J-H, Dong C-D, Shen S-Y (2019) The lead contaminated land treated by the circulation-enhanced electrokinetics and phytoremediation in field scale. J Hazard Mater 368:894–898. https://doi.org/10.1016/j.jhazmat.2018.08.085

Charlesworth S, Ayangbenro AS, Babalola OO (2017) A new strategy for heavy metal polluted environments: a review of microbial biosorbents. Int J Environ Res Public Health 14(1):94. https://doi.org/10.3390/ijerph14010094

Chaturvedi R, Favas P, Pratas J et al (2019) EDTA-assisted metal uptake in *Raphanus sativus* L. and *Brassica oleracea* L. assessment of toxicity and food safety. Bull Environ Contam Toxicol 103:490–495. https://doi.org/10.1007/s00128-019-02651-9

Chehregani A, Malayeri BE (2007) Removal of heavy metals by native accumulator plants. Int J Agric Biol 9:462–465

Chen C, Shi P, Mao Q (2003a) Application of remote sensing techniques for monitoring the thermal pollution of cooling-water discharge from nuclear power plant. J Environ Sci Health Pt A Toxic Hazard Subst Environ Eng 38(8):1659–1668. https://doi.org/10.1081/ESE-120021487

Chen YX, Lin Q, Luo YM et al (2003b) The role of citric acid on the phytoremediation of heavy metal contaminated soil. Chemosphere 50:807–811. https://doi.org/10.1016/S0045-6535(02)00223-0

Chowdhary P, Raj A, Bharagava RN (2019) Environmental pollution and health hazards from distillery wastewater and treatment approaches to combat the environmental threats: a review. Chemosphere 194:229–246. https://doi.org/10.1016/j.chemosphere.2017.11.163

Christoforidis A, Stamatis N (2009) Heavy metal contamination in street dust and roadside soil along the major national road in Kavala's region, Greece. Geoderma 151(3–4):257–263. https://doi.org/10.1016/j.geoderma.2009.04.016

Cosio C, Martinoia E, Keller C (2004) Hyperaccumulation of cadmium and zinc in *Thlaspi caerulescens* and *Arabidopsis halleri* at the leaf cellular level. Plant Physiol 134:716–725. https://doi.org/10.1104/pp.103.031948

Cunningham SD, Ow DW (1996) Promises and prospect of phytoremediation. Plant Physiol 110:715–719. https://doi.org/10.1104/pp.110.3.715

Cunningham SD, Anderson TA, Schwab PA et al (1996) Phytoremediation of soils contaminated with organic pollutants. Adv Agron 56:55–114

Cunningham SD, Shann JR, Crowley DE, Anderson TA (1997) Phytoremediation of contaminated water and soil. In: Kruger EL, Anderson TA, Coats JR (eds) Phytoremediation of soil and water contaminants. ACS symposium series 664. American Cfhemical Society, Washington, DC, pp 2–19

Datta S, Chatterjee S, Mitra A, Veer V (2013) Phytoremediation protocols: an overview. https://doi.org/10.1007/978-3-642-35564-6_1

Daud MK, Nafees M, Ali S et al (2017) Drinking water quality status and contamination in Pakistan. Biomed Res Int 2017:7908183. https://doi.org/10.1155/2017/7908183

de la Rosa G, Peralta-Videa JR, Montes M, Parsons JG, Cano-Aguilera I, Gardea-Torresdey JL (2004) Cadmium uptake and translocation in tumbleweed (*Salsola kali*), a potential cd-hyperaccumulator desert plant species: ICP/OES and XAS studies. Chemosphere 55:1159–1168. https://doi.org/10.1016/j.chemosphere.2004.01.028

Dineshkumar M, Seenuvasan M, Sarojini G (2019) Phytoremediation strategies on heavy metal removal. In: Bui XT, Chiemchaisri C, Fujioka T, Varjani S (eds) Water and wastewater treatment technologies. Energy, environment, and Sustainability. Springer, Singapore, pp 81–101. https://doi.org/10.1007/978-981-13-3259-3_5

Dou X, Dai H, Skuza L et al (2019) *Bidens pilosa L.* hyperaccumulating Cd with different species in soil and the role of EDTA on the hyperaccumulation. Environ Sci Pollut Res Int 26: 25668. https://doi.org/10.1007/s11356-019-05831-6

Duffus JH (2002) Heavy metals—a meaningless term? Pure Appl Chem 74(5):793–807

Ensley BD (2000) Rational for use of phytoremediation. In: Raskin I, Ensley BD (eds) Phytoremediation of toxic metals: using plants to clean- up the environment. Wiley, New York, pp 3–12

Entry JA, Watrud LS, Reeves M (1999) Accumulation of 137Cs and 90Sr from contaminated soil by three grass species inoculated with mycorrhizal fungi. Environ Pollut 104:449–457. https://doi.org/10.1016/S0269-7491(98)00163-8

Etim EE (2012) Phytoremediation and its mechanisms: a review. Int J Environ Bioenergy 2(3):120–136

Fayiga AO, Ma LQ, Cao X et al (2004) Effects of heavy metals on growth and arsenic accumulation in the arsenic hyperaccumulator *Pteris vittata* L. Environ Pollut 132(2):289–296. https://doi.org/10.1016/j.envpol.2004.04.020

Firdaus-e-Bareen, Rafiq K, Shafiq M et al (2019) Uptake and leaching of Cu, Cd, and Cr after EDTA application in sand columns using sorghum and pearl millet. Pol J Environ Stud 28(4): 2065–2077. https://doi.org/10.15244/pjoes/84834

Flathman PE, Lanza GR (1998) Phytoremediation: current views on an emerging green technology. J Soil Contam 7(4):415–432. https://doi.org/10.1080/10588339891334438

Foulds SA, Brewer PA, Macklin MG, Haresign W, Betson RE, Rassner SME (2014) Flood-related contamination in catchments affected by historical metal mining: an unexpected and emerging hazard of climate change. Sci Total Environ 476:165–180. https://doi.org/10.1016/j.scitotenv.2013.12.079

Gangamma S (2018) Commission on pollution: action plans and human resource development in India. Lancet 391(10138):2414–2415

García-Salgado S, García-Casillas D, Quijano M, Bonilla-Simón M (2012) Arsenic and heavy metal uptake and accumulation in native plant species from soils polluted by mining activities. Water Air Soil Pollut 223:559. https://doi.org/10.1007/s11270-011-0882-x

Ghazaryan KA, Movsesyan HS, Minkina TM et al (2019) The identification of phytoextraction potential of *Melilotus officinalis* and *Amaranthus retroflexus* growing on copper- and molybdenum-polluted soils. Environ Geochem Health:1–9. https://doi.org/10.1007/s10653-019-00338-y

González H, Fernández D, Fernández-Fuego D et al (2019) Effect of pH and citric acid on the growth, arsenic accumulation, and phytochelatin synthesis in *Eupatorium cannabinum* L., a promising plant for phytostabilization. Environ Sci Pollut Res 26:26242. https://doi.org/10.1007/s11356-019-05657-2

Gul I, Manzoor M, Silvestre J, Rizwan M et al (2019) EDTA-assisted phytoextraction of lead and cadmium by Pelargonium cultivars grown on spiked soil. Int J Phytoremediation 21(2): 101–110. https://doi.org/10.1080/15226514.2018.1474441

Guo D, Ali A, Ren C et al (2018) EDTA and organic acids assisted phytoextraction of Cd and Zn from a smelter contaminated soil by potherb mustard (*Brassica juncea*, Coss) and evaluation of its bioindicators. Ecotoxicol Environ Saf 167:396–403. https://doi.org/10.1016/j.ecoenv.2018.10.038

Gupta DK, Chatterjee S, Mitra A et al (2019) Uranium and plants: elemental translocation and phytoremediation approaches. In: Gupta D, Walther C (eds) Uranium in plants and the environment. Radionuclides and heavy metals in the environment. Springer, Cham, pp 149–161

Hai DN, Tung LV, Van DK et al (2018) Lead environmental pollution and childhood lead poisoning at ban Thi commune, bac Kan Province, Vietnam. Biomed Res Int 2018:5156812. https://doi.org/10.1155/2018/5156812

Hou S, Wang X, Shafi M et al (2019) Remediation efficacy of *Sedum plumbizincicola* as affected by intercropping of landscape plants and oxalic acid in urban cadmium contaminated soil. J Soil Sediment 19:3512–3520. https://doi.org/10.1007/s11368-019-02309-3

Hu N, Lang T, Ding D et al (2019) Enhancement of repeated applications of chelates on phytoremediation of uranium contaminated soil by *Macleaya cordata*. J Environ Radioact 199–200:58–65. https://doi.org/10.1016/j.jenvrad.2018.12.023

Huang G, Gao R, You J et al (2019) Oxalic acid activated phosphate rock and bone meal to immobilize Cu and Pb in mine soils. Ecotoxicol Environ Saf 174:401–407. https://doi.org/10.1016/j.ecoenv.2019.02.076

Hussain A, Hasnain S (2011) Phytostimulation and biofertilization in wheat by cyanobacteria. J Ind Microbiol Biotechnol 38(1):85–92. https://doi.org/10.1007/s10295-010-0833-3

Jan M, Shah G, Masood S et al (2019) Enhanced phytoremediation ability of rice seedlings under cadmium toxicity. Biomed Res Int 2019:8134651. https://doi.org/10.1155/2019/8134651

Jebara SH, Ayed SA, Chiboub M et al (2019) Chapter 19. Phytoremediation of cadmium-contaminated soils by using legumes inoculated by efficient and cadmium-resistant plant growth-promoting bacteria. In: Hasanuzzaman M, Prasad MNV, Fujita M (eds) Cadmium toxicity and tolerance in plants. Academic Press, New York, pp 479–493

Jiang M, Liu S, Li Y et al (2019) EDTA-facilitated toxic tolerance, absorption and translocation and phytoremediation of lead by dwarf bamboos. Ecotoxicol Environ Saf 170:502–512. https://doi.org/10.1016/j.ecoenv.2018.12.020

Kacalkova L, Tlustos P, Szakova J (2009) Phytoextraction of cadmium, copper, zinc and mercury by selected plants. Plant Soil Environ 55(7):295–304. https://www.agriculturejournals.cz/publicFiles/09193.pdf

Kalve S, Sarangi BK, Pandey RA, Chakrabarti T (2011) Arsenic and chromium hyperaccumulation by an ecotype of *Pteris vittata*–prospective for phytoextraction from contaminated water and soil. Curr Sci India 100:888–894

Kanwar VS, Sharma A, Srivastav AL, Rani L (2020) Phytoremediation of toxic metals present in soil and water environment: a critical review. Environ Sci Pollut Res 27:44835–44860

Kaushik P (2015) Use of natural organic and synthetic chelating agents for efficient phytoremediation. Int J Enhanced Res Sci Technol Eng 4(6):99–101

Kaviani E, Niazi A, Moghadam A et al (2019) Phytoremediation of Ni-contaminated soil by *Salicornia iranica*. Environ Technol 40(3):270–281. https://doi.org/10.1080/09593330.2017.1387179

Keeran NS, Balasundaram U, Govindan G, Parida AK (2019) Prosopis juliflora: a potential Plant for Mining of genes for genetic engineering to enhance phytoremediation of metals. In: Prasad MNV (ed) Transgenic plant technology for remediation of toxic metals and metalloids. India. Academic Press, Telangana, pp 381–393

Khan AHA, Butt TA, Mirza CR et al (2019) Combined application of selected heavy metals and EDTA reduced the growth of *Petunia hybrida* L. Sci Rep 9:4138. https://www.nature.com/articles/s41598-019-40540-7

Kiiskila JD, Das P, Sarkar D et al (2015) Phytoremediation of explosive-contaminated soils. Curr Pollut Rep 1:23–34. https://doi.org/10.1007/s40726-015-0003-3

Kim SJ, Park JY, Lee YJ et al (2005) Application of a new electrolytecirculation method for the ex situ electrokinetic bioremediation of a laboratory-prepared pentadecane contaminated kaolinite. J Hazard Mater 118(1–3):171–176. https://doi.org/10.1016/j.jhazmat.2004.10.016

Koptsik G (2014) Problems and prospects concerning the phytoremediation of heavy metal polluted soils: a review. Eurasian Soil Sci 47:923–939. https://doi.org/10.1134/S1064229314090075

Kubota H, Takenaka C (2003) Field note: Arabis gemmifera is a hyperaccumulator of Cd and Zn. Int J Phytoremediation 5:197–201. https://doi.org/10.1080/713779219

Kucharski R, Sas-Nowosielska A, Małkowski E, Japenga J, Kuperberg J, Pogrzeba M, Krzyak J (2005) The use of indigenous plant species and calcium phosphate for the stabilization of highly metal-polluted sites in southern Poland. Plant and Soil 273:291–305. https://doi.org/10.1007/s11104-004-8068-6

Kumar P, Kim KH, Bansal V, Lazarides T, Kumar N (2017) Progress in the sensing techniques for heavy metal ions using nanomaterials. J Ind Eng 54:30–34. https://doi.org/10.1016/j.jiec.2017.06.010

Li X, Poon C-S, Liu PS (2001) Heavy metal contamination of urban soils and street dusts in Hong Kong. Appl Geochem 16(11–12):1361–1368. https://doi.org/10.1016/S0883-2927(01)00045-2

Li YM, Chaney R, Brewer E et al (2003) Development of a technology for commercial phytoextraction of nickel: economic and technical considerations. Plant Soil 249:107–115. https://doi.org/10.1023/A:1022527330401

Limmer M, Burken J (2016) Phytovolatilization of organic contaminants. Environ Sci Technol 50(13):6632–6643. https://doi.org/10.1021/acs.est.5b04113

Lombi E, Zhao FJ, Dunham SJ, McGrath SP (2001) Phytoremediation of heavy metal-contaminated soils: natural hyperaccumulation versus chemical enhanced phytoextraction. J Environ Qual 30: 1919–1926

Lone MI, He Z-l, Stoffella PJ et al (2008) Phytoremediation of heavy metal polluted soils and water: progresses and perspectives. J Zhejiang Univ Sci B 9(3):210–220. https://doi.org/10.1631/jzus.B0710633

Longley K (2021) The feasibility of poplars for phytoremediation of TCE contaminated groundwater: a cost-effective and natural alternative means of groundwater treatment

Masood A, Khan MI, Fatma M et al (2016) Involvement of ethylene in gibberellic acid-induced sulfur assimilation, photosynthetic responses, and alleviation of cadmium stress in mustard. Plant Physiol Biochem 104:1–10. https://doi.org/10.1016/j.plaphy.2016.03.017

McBride MB, Zhou Y et al (2019) Cadmium and zinc bioaccumulation by *Phytolacca americana* from hydroponic media and contaminated soils. Int J Phytoremediation:1215–1224. https://doi.org/10.1080/15226514.2019.1612849

Mearns AJ (1997) Cleaning oiled shores: putting bioremediation to the test. Spill Sci Technol Bull 4(4):209–217. https://doi.org/10.1016/S1353-2561(98)00026-7

Mishra A, Mishra SP, Arshi A et al (2019) Plant-microbe interactions for bioremediation and phytoremediation of environmental pollutants and agro-ecosystem development. In: Bharagava RN, Saxena G (eds) Bioremediation of industrial waste for environmental safety, vol 2. Springer, Singapore, pp 415–436. https://doi.org/10.1007/978-981-13-3426-9_17

Mitch ML (2002) Phytoextration of toxic metals: a review of biological mechanism. J Environ Qual 31:109–120. https://doi.org/10.2134/jeq2002.0109

Mleczek M, Łukaszewski M, Kaczmarek Z et al (2009) Efficiency of selected heavy metals accumulation by *Salix viminalis* roots. Environ Exp Bot 65(1):48–53. https://doi.org/10.1016/j.envexpbot.2008.03.003

Mokarram-Kashtiban S, Hosseini SM, Kouchaksaraei MT et al (2019) The impact of nanoparticles zero-valent iron (nZVI) and rhizosphere microorganisms on the phytoremediation ability of white willow and its response. Environ Sci Pollut Res Int 26(11):10776–10789. https://doi.org/10.1007/s11356-019-04411-y

Navarro-León E, Oviedo-Silva J, Ruiz JM et al (2019) Possible role of HMA4a TILLING mutants of Brassica rapa in cadmium phytoremediation programs. Ecotoxicol Environ Saf 180:88–94. https://doi.org/10.1016/j.ecoenv.2019.04.081

Newman LA, Reynolds CM (2004) Phytodegradation of organic compounds. Curr Opin Biotechnol 15:225–230. https://doi.org/10.1016/j.copbio.2004.04.006

Nichols WJ (2001) The U.S. environmental protect agency: National oil and hazardous substances pollution contingency plan, subpart J product schedule (40 CFR 300.900). In: Proceedings of the International oil spill conference. American Petroleum Institute, Washington, DC, pp 1479–1483

Nilanjana D, Chandran P (2011) Microbial degradation of petroleum hydrocarbon contaminants: an overview. Biotechnol Res Int 2011:941810,13p. https://doi.org/10.4061/2011/941810

O'Neill GJ, Gordon AM (1994) The nitrogen filtering capability of *Carolina poplar* in an artificial riparian zone. J Environ Qual 23:1218–1223. https://doi.org/10.2134/jeq1994.00472425002300060014x

Oliveira H (2012) Chromium as an environmental pollutant: insights on induced plant toxicity. J Bot 2012:1–8. https://doi.org/10.1155/2012/375843

Peng K, Luo C, Lou L, Li X, Shen Z (2008) Bioaccumulation of heavy metals by the aquatic plants *Potamogeton pectinatus* L. and *Potamogeton malaianus* Miq. And their potential use for contamination indicators and in wastewater treatment. Sci Total Environ 392(1):22–29. https://doi.org/10.1016/j.scitotenv.2007.11.032

Pérez-Sanz A, Millán R, Sierra MJ, Alarcón R, García P, Gil-Díaz M (2012) Mercury uptake by *Silene vulgaris* grown on contaminated spiked soils. J Environ Manage 95:S233–S237. https://doi.org/10.1016/j.jenvman.2010.07.018

Prasad MNV (2005) Nickelophilous plants and their significance in phytotechnologies. Brazilian J Plant Physiol 17:113–128. https://doi.org/10.1590/S1677-04202005000100010

Prasad R (2017) Mycoremediation and environmental sustainability. 1. Springer International Publishing. ISBN: 978-3-319-68957-9. https://link.springer.com/book/10.1007/978-3-319-68957-9

Prasad R (2018) Mycoremediation and environmental sustainability. 2. Springer International Publishing. ISBN: 978-3-319-77386-5. https://www.springer.com/us/book/9783319773858

Prasad R (2021) Environmental pollution and remediation. Springer, Berlin. (ISBN: 978-981-15-5499-5). https://www.springer.com/gp/book/9789811554988

Prasad KS, Ramanathan AL, Paul J, Subramanian V, Prasad R (2013) Biosorption of arsenite (as^{+3}) and arsenate (as^{+5}) from aqueous solution by Arthrobacter sp. biomass. Environ Technol 34: 2701–2708

Prasad R, Kumar M, Varma A (2015) Role of PGPR in soil fertility and plant health. In: Egamberdieva D, Shrivastava S, Varma A (eds) Plant growth-promoting Rhizobacteria (PGPR) and medicinal plants. Springer International, Cham, pp 247–260

Raab A, Schat H, Meharg AA et al (2005) Uptake, translocation and transformation of arsenate and arsenite in sunflower (*Helianthus annuus*): formation of arsenic-phytochelatin complexes during exposure to high arsenic concentrations. New Phytol 168(3):551–558. https://doi.org/10.1111/j.1469-8137.2005.01519.x

Rai PK (2008) Phytoremediation of Hg and Cd from industrial effluents using an aquatic free floating macrophyte *Azolla pinnata*. Int J Phytoremediation 10:430–439

Rajendra P, Yashbir S (2017) Oxalic acid/oxalates in plants: from self-defence to phytoremediation. Curr Sci 112:1665–1667. https://doi.org/10.18520/cs/v112/i08/1665-1667

Rajkumar M, Sandhya S, Prasad MNV (2012) Perspectives of plant-associated microbes in heavy metal phytoremediation. Biotechnol Adv 30(6):1562–1574. https://doi.org/10.1016/j.biotechadv.2012.04.011

Ramachandran S, Fontanille P, Pandey A et al (2006) Gluconic acid: properties, applications and microbial production. Food Technol Biotechnol 44(2):185–195

Rastogi M, Nandal M (2020) Toxic metals in industrial wastewaters and phytoremediation using aquatic macrophytes for environmental pollution control: an eco-remedial approach. In: Bharagava R, Saxena G (eds) Bioremediation of industrial waste for environmental safety. Springer, Singapore. https://doi.org/10.1007/978-981-13-3426-9_11

Revathi S, Subhashree V (2019) Comparative study on the effect of nitric oxide, chromium and EDTA on oxidative stress responses in Agathi (*Sesbanic grandiflora* L. Pers.). Appl Ecol Environ Res 17(2):3269–3291. https://doi.org/10.15666/aeer/1702_32693291

Rodriguez L, Lopez-Bellido F, Carnicer A, Alcalde-Morano V (2003) Phytoremediation of mercury-polluted soils using crop plants. Fresen Environ Bull 9:328–332. https://doi.org/10.1007/s11356-019-06563-3

Rostami S, Azhdarpoor A (2019) The application of plant growth regulators to improve phytoremediation of contaminated soils: a review. Chemosphere 220:818–827. https://doi.org/10.1016/j.chemosphere.2018.12.203

Saddhe A, Manuka R, Nikalje G, Penna S (2020). Halophytes as a potential resource for phytodesalination. https://doi.org/10.1007/978-3-030-17854-3-92-1

Sakakibara M, Ohmori Y, Ha NTH, Sano S, Sera K (2011) Phytoremediation of heavy metal-contaminated water and sediment by *Eleocharis acicularis*. Clean Soil Air Water 39:735–741. https://doi.org/10.1002/clen.201000488

Santos J, Herrmann B, Mieske B, Krag LA, Haase S, Stepputtis D (2018) The efficiency of sieve-panels for bycatch separation in Nephrops trawls. Fish Manag Ecol 25(6):464–473. https://doi.org/10.1111/fme.12323

Sarma H, Forid N, Prasad R, Prasad MNV, Ma LQ, Rinklebe J (2021) Enhancing phytoremediation of hazardous metal(loid)s using genome engineering CRISPR–Cas9 technology. J Hazard Mater. https://doi.org/10.1016/j.jhazmat.2021.125493

Sas-Nowosielska A, Galimska-Stypa R, Kucharski R, Zielonka U, Małkowski E, Gray L (2008) Remediation aspect of microbial changes of plant rhizosphere in mercury contaminated soil. Environ Monit Assess 137:101–109. https://doi.org/10.1007/s10661-007-9732-0

Shackira AM, Puthur JT (2019) Phytostabilization of heavy metals: understanding of principles and practices. In: Srivastava S, Srivastava A, Suprasanna P (eds) Plant-metal interactions. Springer, Cham. https://doi.org/10.1007/978-3-030-20732-8_13

Shahid M, Shamshad S, Bakar A et al (2019) Comparative effect of organic amendments on physio-biochemical traits of young and old bean leaves grown under cadmium stress: a multivariate analysis. Environ Sci Pollut Res 26(12):11579–11590. https://doi.org/10.1007/s11356-018-2689-4

Sheoran V, Sheoran AS, Poonia P (2009) Phytomining: a review. Miner Eng 22:1007–1019. https://doi.org/10.1016/j.mineng.2009.04.00

Shuzhen H, Xin W, Mohammad S et al (2019) Remediation efficacy of *Sedum plumbizincicola* as affected by intercropping of landscape plants and oxalic acid in urban cadmium contaminated soil. J Soil Sediment:1–9. https://doi.org/10.1007/s11368-019-02309-3

Singha KT, Sebastian A, Prasad MNV (2019) Iron plaque formation in the roots of *Pistia stratiotes* L.: importance in phytoremediation of cadmium. Int J Phytoremediation 21(2):120–128. https://doi.org/10.1080/15226514.2018.1474442

Sonowal S, Nava AR, Joshi SJ, Borah SN, Islam NF, Pandit S, Prasad R, Sarma H (2022) Biosurfactants assisted heavy metals phytoremediation: green technology for the United Nations sustainable development goals. Pedosphere 2(1):198–210. https://doi.org/10.1016/S1002-0160(21)60067-X

Srivastava M, MaJorge LQ, Santos G, Santos JG (2006) Three new arsenic hyperaccumulating ferns. Sci Total Environ 364(1-3):24–31

Thakare M, Sarma H, Datar S, Roy A, Pawar P, Gupta K, Pandit S, Prasad R (2021) Understanding the holistic approach to plant-microbe remediation technologies for removing heavy metals and radionuclides from soil. Curr Res Biotechnol 3:84–98. https://doi.org/10.1016/j.crbiot.2021.02.004

Timofeev-Resovsky EA, Agafonov BM, Timofeev-Resovsky NV (1962) Fate of radioisotopes in aquatic environments (in Russian). Proc Biol Inst USSR Acad Sci 22:49–67

Traina SJ, Laperche V (1999) Contaminant bioavailability in soils, sediments, and aquatic environments. PNAS 96(7):3365–3371. https://doi.org/10.1073/pnas.96.7.3365

Turgut C, Katie Pepe M, Cutright TJ (2004) The effect of EDTA and citric acid on phytoremediation of cd, Cr, and Ni from soil using *Helianthus annuus*. Environ Pollut 131(1):147–154. https://doi.org/10.1016/j.envpol.2004.01.017

Umesh J, Vishwas B, Gaëlle S et al (2016) Role of microbes and plants in phytoremediation: potential of genetic engineering. In: Squires VR (ed) Ecological restoration. Nova Science, Hauppauge, NY, pp 89–109

United States Environmental Protection Agency (USEPA) (2000) Introduction to phytoremediation. EPA 600/R-99/107. U.S. Environmental Protection Agency, Office of Research and Development, Cincinnati, OH. https://cfpub.epa.gov/si/si_public_record_report.cfm?Lab=NRMRL&dirEntryId=63433

Venosa AD, Zhu X (2003) Biodegradation of crude oil contaminating marine shorelines and freshwater wetlands. Spill Sci Technol Bull 8(2):163–178. https://doi.org/10.1016/S1353-2561(03)00019-7

Venosa AD, Suidan MT, Suidan MT (1996) Bioremediation of an experimental oil spill on the shoreline of Delaware Bay. Environ Sci Technol 30(5):1764–1775. https://doi.org/10.1021/es950754r

Vinothini C, Sudhakar S, Ravikumar R (2015) Biodegradation of petroleum and crude oil by pseudomonas putida and *Bacillus cereus*. Int J Curr Microbiol App Sci 4(1):318–329. https://www.semanticscholar.org/paper/Biodegradation-of-petroleum-and-crude-oil-by-putida-Vinothini-Sudhakar/e3143da5502f2b7b230fbe14a087e44a600c9c77

Violante A, Cozzolino V, Perelomov L et al (2010) Mobility and bioavailability of heavy metals and metalloids in soil environments. J Soil Sci Plant Nutr 10(3):268–292. https://doi.org/10.4067/S0718-95162010000100005

Walid Z, Rabhi M, Ferchichi S et al (2012) Phytodesalination: a solution for salt-affected soils in arid and semi-arid regions. J Arid Land 22:299–302

Wang K, Liu Y, Song Z, Khan ZH, Qiu W (2019) Effects of biodegradable chelator combination on potentially toxic metals leaching efficiency in agricultural soils. Ecotoxicol Environ Safe 182:109399. https://doi.org/10.1016/j.ecoenv.2019.109399

Wang Y, Qiao M, Liu Y, Zhu Y (2012) Health risk assessment of heavy metals in soils and vegetables fromwastewater irrigated area, Beijing-Tianjin city cluster, China. J Environ Sci 24(4):690–698. https://doi.org/10.1016/s1001-0742(11)60833-4

Wasino R, Likitlersuang S, Janjaroen D (2019) The performance of vetivers (Chrysopogon zizaniodes and Chrysopogon nemoralis) on heavy metals phytoremediation. Lab Invest 21(7):624–633. https://doi.org/10.1080/15226514.2018.1546275

Wei S, Zhou Q, Saha UK (2008) Hyperaccumulative characteristics of weed species to heavy metals. Water Air Soil Pollut 192:173–181

Wei L, Chen C, Xu Z (2009) The effect of low-molecular-weight organic acids and inorganic phosphorus concentration on the determination of soil phosphorus by the molybdenum blue reaction. Biol Fertil Soils 45(7):775–779. https://doi.org/10.1007/s00374-009-0381-z

Wu L, Luo YM, Xing XR, Christie P (2004) EDTA-enhanced phytoremediation of heavy metal contaminated soil with Indian mustard and associated potential leaching risk. Agr Ecosyst Environ 102(3):307–318. https://doi.org/10.1016/j.agee.2003.09.002

Wu L, Li Z, Akahane I et al (2012) Effects of organic amendments on Cd, Zn and cu bioavailability in soil with repeated phytoremediation by *Sedum plumbizincicola*. Int J Phytoremediation 14(10):1024–1038. https://doi.org/10.1080/15226514.2011.649436

Xiong YH, Yang XE, Ye ZQ, He ZL (2004) Characteristics of cadmium uptake and accumulation by two contrasting ecotypes of *Sedum alfredii* Hance. J Environ Sci Health Pt A Toxic Hazard Subst Environ Eng A39(11–12):2925–2940

Yasin NA, Khan WU, Ahmad SR (2019) Effect of Enterobacter sp. CS2 and EDTA on the phytoremediation of Ni-contaminated soil by *Impatiens balsamina*. Pol J Environ Stud 28(1):425–433. https://doi.org/10.15244/pjoes/76179

Yu F, Li Y, Li F et al (2019) The effects of EDTA on plant growth and manganese (Mn) accumulation in *Polygonum pubescens* Blume cultured in unexplored soil, mining soil and tailing soil from the Pingle Mn mine, China. Ecotoxicol Environ Saf 173:235–242. https://doi.org/10.1016/j.ecoenv.2019.01.086

Zhang P, Zhao D, Liu Y et al (2019) Cadmium phytoextraction from contaminated paddy soil as influenced by EDTA and Si fertilizer. Environ Sci Pollut Res 26(23):23638–23644. https://doi.org/10.1007/s11356-019-05654-5

Zheng W, Obrist D, Weis D, Bergquist BA (2016) Mercury isotope compositions across north American forests. Global Biogeochem Cycles 30:1475–1492

Zhu XF, Jiang T, Wang ZW et al (2012) Gibberellic acid alleviates cadmium toxicity by reducing nitric oxide accumulation and expression of IRT1 in *Arabidopsis thaliana*. J Hazard Mater 239–240:302–307

Zhu XF, Wang ZW, Dong F et al (2013) Exogenous auxin alleviates cadmium toxicity in *Arabidopsis thaliana* by stimulating synthesis of hemicellulose 1 and increasing the cadmium fixation capacity of root cell walls. J Hazard Mater 263:398–403. https://doi.org/10.1016/j.jhazmat.2013.09.018

Zhu Y, Xu F, Liu Q et al (2019) Nanomaterials and plants: positive effects, toxicity and the remediation of metal and metalloid pollution in soil. Sci Total Environ 662:414–421. https://doi.org/10.1016/j.scitotenv.2019.01.234

Exploring the Potential of Plant Growth-Promoting Rhizobacteria (PGPR) in Phytoremediation

17

Ketaki V. Ramani, Kruti G. Dangar, and Disha B. Changela

Abstract

Evidently, PGPR hold enormous prospects in improved and sustainable crop production including reduced use of chemical inputs. Phytoremediation process uses plants to take up pollutants such as heavy metals, organic pollutants and mixed waste pollutants present in its vicinity and are further processed and reduced by the plants. This process is a greener alternative and causes least harm to the environment as compared to the chemical methods available for removal of pollutants from the contaminated sites. Phytoremediation can be enhanced further by employing the rhizobacteria. Several species of rhizobacteria play an important role in plant growth enabling higher pollutant uptake. The PGPR promote plant growth and development by using their own metabolism through a variety of direct or indirect techniques such as biological nitrogen fixation, enhanced nutrient supply in the rhizosphere, siderophore and phytohormones production. Owing to its beneficial aspects, several PGPR-based products are available in the market. In this chapter, we explore the ability of PGPR to eliminate pollutants such as heavy metals and organic pollutants along with the strategies used for phytoremediation and metagenomic approach to identify rhizosphere microbial population.

Keywords

PGPR · Phytoremediation · Heavy metals · Organic pollutants · Metagenomics

K. V. Ramani
Department of Microbiology, School of Science, RK University, Rajkot, Gujarat, India

K. G. Dangar (✉) · D. B. Changela
Department of Biochemistry, Saurashtra University, Rajkot, Gujarat, India

R. Prasad (ed.), *Phytoremediation for Environmental Sustainability*,
https://doi.org/10.1007/978-981-16-5621-7_17

17.1 Introduction

The advent of technology has made human life easier and access to the needs faster. As anthropological activities increase making human life better, they inadvertently anguish nature leading to drastic increase in levels of pollutants. Environment pollution affects the quality of air, water and soil inadvertently affecting the health of mankind. One way to sustainably curb the soil pollution is by phytoremediation. Phytoremediation employs plants to reduce pollutants which are taken up by immobilization or extraction. Several factors determine the efficacy of phytoremediation such as contaminant uptake by plant root and its further translocation to the other plant parts and its degradation in the plant. According to Ji et al. (2007), a cost saving of 50–80% can be achieved when phytoremediation is applied compared to conventional technologies. Plant growth-promoting rhizobacteria (PGPR) residing in the rhizosphere interact with plants and can be explored to enhance phytoremediation. Plants' roots secret an array of chemical substances known as root exudates which act as source of food for the microorganism in the rhizosphere. PGPR in turn affect the plant growth directly or indirectly, by releasing phytohormones, increasing the nutrient uptake through fixation or immobilization and decreasing the deleterious effects of pathogenic microorganisms on plants (Shrivastava et al. 2014). Epiphytes, nitrogen-fixing bacteria, mycorrhizal fungi are some of the well-known examples of plant root–microbe interaction (Prasad et al. 2015). In this chapter, we review the impact of PGPR on phytoremediation of soil polluted by various toxic pollutants, strategies employed by PGPRs for phytoremediation and discuss future directions.

Estimates of soil microbial diversity range from thousands to a million microbial "species" in a few grams of soil (Gans et al. 2005). Rhizosphere is the region in the soil that surrounds the plant roots and serves as the habitat to several microorganisms. The plant species, plant developmental stage and soil type are the key factors governing the composition of rhizosphere microbial communities prevailing in a particular ecological niche. The bacterial diversity in the rhizosphere depends on physicochemical composition of soil, its pH, partial pressure of oxygen (pO_2) and water potential. (Singh et al. 2020). There exists a symbiotic relationship in the rhizosphere region between the microbial community and the plant roots. The plant roots in turn secret a variety of chemicals such as ions, enzymes, metabolites, terpenes, flavonoids and water that attract the microbial community.

Soil harbours a plethora of microorganisms of which some affect the plant growth favourably and are known as plant growth-promoting rhizobacteria (PGPR). *Rhizobium, Azotobacter, Azospirillum, Beijerinkia, Enterobacter, Derxia, Pseudomonas* and *Bacillus* are some of the reported microorganisms. The PGPR benefit the plants by supporting seedling emergence, seed germination, nutrient uptake, nitrogen fixation and disease suppression. They are known to synthesize enzymes such as peroxidases, phosphatases, monooxygenases, dehalogenases and nitroreductases which degrade hazardous compounds present in soil and transform them into simpler compounds such as CO_2 and water to be released in the environment. Water holding capacity, partial pressure of oxygen (pO_2) and soil health in terms of micronutrients present and the pH are enhanced by species like *Pantoea agglomerans, Rhizobium* spp. and *Pseudomonas* spp. (Tewari and Arora 2014; Naseem and Bano 2014). PGPR accelerate plant growth under stress conditions by increasing plant tolerance to elevated salt, PHC and/or trace metal levels, as well as other environmental stressors such as saturated soil or drought conditions. This leads to rapid growth of plants, including their roots. The vigorous plant growth that ensues leads to greater proliferation of naturally existing microbes in the soil, resulting in a very active rhizosphere that is typical of soils with normal plant growth. The substantial root biomass that accumulates in the soil provides a sink which allows for rapid partitioning of salt ions out of the soil, and their subsequent accumulation in the foliar tissues of some plants.

PGPR such as *Chromo bacterium, Pseudomonas, Serratia, Erwinia, Agrobacterium, Arthrobacter, Caulobacter, Flavobacterium, Azotobacter, Azospirillum, Bacillus* and *Burkholderia* populate the region around the plant roots in the rhizosphere and are classified as extracellular plant growth-promoting rhizobacteria (ePGPR). *Bradyrhizobium, Mesorhizobium, Allorhizobium, Azorhizobium* and *Rhizobium* enter the root interior to establish endophytic populations in specialized nodular structures thereby benefiting the host plants and are known as intracellular plant growth-promoting rhizobacteria (iPGPR). (Compant et al. 2005; Bhattacharyya and Jha 2012). Besides bacterial species, fungal species such as *Trichoderma, Aspergillus, Alternaria* and *Penicillium* can also be used as PGPR. These are broadly known as mycorrhizae. Mycorrhiza aids plants in acquiring micronutrients (e.g., Zn and Cu) and water from the soil. They also provide tolerance to environmental stresses and in return receive carbon from the plant (Wipf et al. 2019).

Rhizobium inoculants have been commercially produced worldwide, mainly in the developed countries. Currently, the PGPR are used as inoculants along with charcoal. A variety of PGPR-based products are marketed to improve phytoremediation. Chemicals such as silicon dioxide, salicylic acid, titanium oxide nanoparticles are used to enhance the effect of PGPR. Pollutants including heavy metals such as Cd, Pb, Fe, Al and Ni, and organic soil pollutants including many insecticides and herbicides are known to be treated by the use of PGPR.

In the last three decades, the study of inter- and intra-relationships between plant and diverse microbes such as viruses, fungi, archaea, bacteria and oomycetes has not been fully determined (Singh et al. 2019). Still, the struggles with cultivable approaches are that only an elfin portion of the microbes is cultivated and the rest of communities needs to be identified with the help of various omics and culture-independent approaches. Metagenomics is one of the best approaches to explore plant–microbes' relation, and their active role in sustainable agriculture. Figure 17.1 depicts the rhizosphere microbial diversity identified using the metagenomics approach.

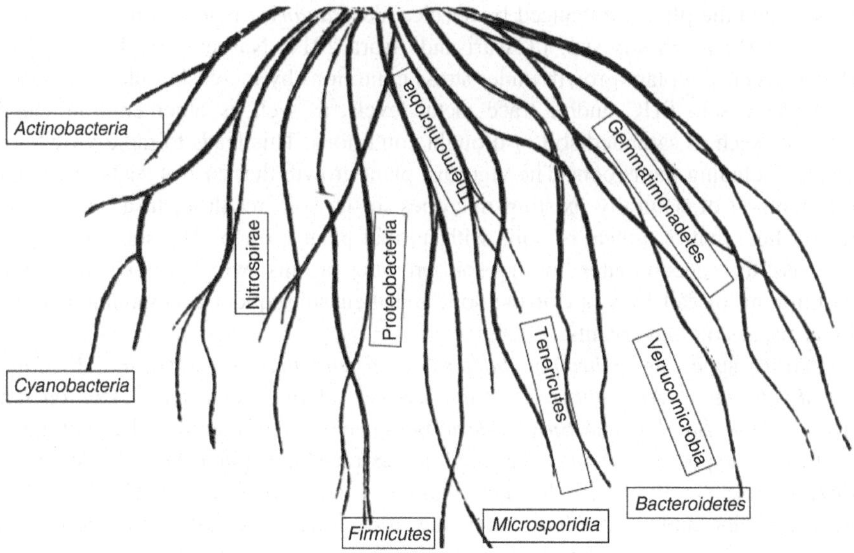

Fig. 17.1 Major microbial diversity identified in the rhizosphere region by metagenomic studies

17.2 Phytoremediation of Heavy Metals by PGPR

Utilization of mineral resources is imperative to technological development which has deleterious effects on ecosystems and communities therein. The majority of metals arise from mining, smelting, fertilizers and combustion to name a few. Toxic metals in elemental forms cannot be degraded further. Plants have the ability to extract these toxic metals from the soil. The metals being present in the water and in the soil move through the plant and end up in the leaves where they are left as they do not volatilize out. Some of the heavy metals are similar in size and charge to the nutrients taken up by plants which are unable to differentiate between them easily. These are thus taken by channels in plants. Some metals such as chromium 6 when taken up by plants are metabolized and converted to chromium 3 state. Metals such as lead are not metabolized by the plants and are shunted to the shoots and leaves of the plants where they accumulate. Reeves et al. (2018) published a global database of hyperaccumulator plants which includes 721 plant species. Hyperaccumulators are plants known for their unique properties to accumulate about hundreds or thousands of times greater heavy metals than normal for most plants. Plants can take up to 40% of the dry weight in metals allowing for easier disposal or where possible recycling. The key to phytoremediation is to concentrate the metals from a large quantity into a small quantity. Hence phytoremediation can be used in a massive area with very low concentration of heavy metal.

Sunflower (*Helianthus annuus* L.) belongs to the family *Asteraceae*. and is one of the most widely studied plants with respect to phytoremediation owing to its potential to accumulate the heavy metals in roots, stem and leaves. Khan et al. (2018) demonstrated *H. annuus'* potential to accumulate heavy metals (Zn, Cu and Pb) in sunflower shoots. Their study showed the combined treatment of PGPR and/or salicylic acid resulted in accumulation of Cd and Ni in plant shoot along with increased chlorophyll (67%), carotenoid (70%), leaf protein (64%), sugar (64%) and phenolic (62%) contents and lower leaf proline (62%) content, malondialdehyde (MDA) (64%) and antioxidant enzymes (67%). Inoculation of *Pseudomonas putida (P. putida)* have been shown to enhance Ni uptake up to 46% by *Eruca sativa* and in turn increasing the root (34%) and shoot (41%) length of *E. sativa* as demonstrated by Kamran et al. (2016). Consortium of rhizobacteria has often used to increase the accumulation of heavy metals. The application of rhizobacterial consortium to *Scirpus grossus* by Ismail et al. (2020) showed a plant growth at 26% and 29% for plant height and dry weight, respectively. *Scirpus grossus* also accumulated Fe (48%) and Al (19%) in the constructed wetlands indicating the phytoremediation ability of PGPR in accumulating heavy metals and promoting plant growth.

To aid in the process of phytoremediation, chelants are often used to increase the bioavailability of the heavy metal. Ju et al. (2020) studied the effect of S, S-ethylenediaminedisuccinic acid (EDDS), a chelant for phytoremediation of Cu-contaminated soil by co-inoculation of *Paenibacillus mucilaginosus* and *Sinorhizobium meliloti* using alfalfa. Their work showed 1.2 times higher Cu uptake by alfalfa roots along with a reduction in oxidative damage. The microbial biomass carbon and nitrogen also increased.

17.3 Phytoremediation of Organic Pollutants Through PGPR

The organic pollutants present in the soil are unique in terms of physicochemical properties and toxicological mode of action. Some of these pollutants are toxic, persistent, bioaccumulative and prone to long-range transport. These include: petroleum hydrocarbons like alkanes, alkenes, polychlorinated biphenyls (PCBs), polychlorinated dibenzo-p-dioxins and polychlorinated dibenzofurans, benzene, toluene, ethylbenzene, and xylene, collectively known as BTEX, polycyclic aromatic hydrocarbons (PAHs) and several herbicides such as alachlor, acetochlor; fungicides such as lindane, procymidone, and penconazole and insecticides such as endosulfan, heptachlor and endrin which can negatively impact terrestrial ecosystems. These usually arise as a consequence of human activities like domestic sewage (raw or treated), industrial effluent dumping and poor agricultural practices. A large variety of consumer items consists of these pollutants which eventually leach into surrounding materials and now can be detected in several living organisms causing deleterious health effects. Microbes form an important part of consortiums that assist in degrading contaminants. These include *Acinetobacter, Alcaligenes, Arthrobacter, Bacillus, Beijerinckia, Flavobacterium, Methylosinus, Mycobacterium, Myxococcus, Nitrosomonas, Nocardia, Penicillium, Phanerochaete, Pseudomonas, Rhizoctonia, Serratia, Trametes and Xanthobacter* (Dhir 2017).

Rani et al. (2021) demonstrated the degradation potential of PGPR strain *Bacillus* sp. PRB101 (89% at 5 mg kg^{-1} of soil) of endosulfan at 120 days after sowing *Solanum lycopersicum* along with an increase in plant biomass. The consortium of *Microbacterium resistens* strain ND2 and *Bacillus pumilus* strain ND3 showed capability to degrade polycyclic aromatic hydrocarbons (PAHs) up to 89.1%. Phytoremediation of PAHs was significantly better with *Lepironia articulata.* They also assisted the plant in enhancing the phytoremediation process of PAH-contaminated wastewater as reported by Sbani et al. (2021). Diuron is an herbicide or algicide which is an active pollutant present in the water, soil and other sediments. This herbicide was mineralized by the three-member bacterial consortium; *Arthrobacter sulfonivorans, Variovorax soli* and *Advenella* sp. and achieved the mineralization from 22.9 to 69.0% (Villaverde et al. 2017). Table 17.1 shows some of the current cases wherein the polluted sites have been remediated with the aid of PGPRs.

17.4 Strategies Employed by PGPR in Phytoremediation

17.4.1 Plant Growth Promotion

Rhizobacteria play a major role in the phytoremediation process by increasing the phytoremediation efficiency. Rhizobacteria augment phytoremediation by enhancing plant growth. In a study by He et al. (2020), it was found that inoculation of PGPR namely *Bacillus* sp. QX8 and QX13 significantly promoted the growth of plant *Solanum nigrum* which assisted in enhancement of phytoremediation efficiency. Pot experiments demonstrated that inoculation with PGPR enhanced the dry weight of shoots and roots. Khan et al. (2017) found four nickel-tolerant bacteria that demonstrated the synthesis of indole acetic acid (IAA), siderophore production and phosphate solubilization. The bacteria also significantly enhanced the root and shoot biomass along with assisting phytoremediation. Enhanced phytoremediation efficiency along with plant growth promotion was studied by Liu et al. (2014). In this study, the isolate *Klebsiella* sp. D5A illustrated profound plant growth-promoting activity along with phytoremediation of petroleum-contaminated saline-alkali soil. Abdelkrim et al. (2020) studied the beneficial effects of PGPR on growth and phytoremediation potential of *Lathyrus sativus* plants. The results indicated significant increase in shoot and root dry weights along with increase in nodule numbers when compared with uninoculated plants. Soil total nitrogen and available phosphorus were also significantly enhanced post PGPR inoculation.

17.4.2 Improved Phytoextraction

Phytoextraction is defined as uptake of hazardous elements through roots from the soil and followed by its translocation to the biomass of a plant (Ali et al. 2013). Plant-associated bacteria are known to augment phytoextraction by altering the

Table 17.1 Recent cases of PGPR aided phytoremediation of diverse pollutants

S. no.	PGPR	Plant	Strategies/mechanisms	Pollutants	Study area	References
1.	*Planomicrobium chinense* strain P1, *Bacillus cereus* strain P2	*Helianthus annus*	Enhanced heavy metal accumulation, plant growth, chlorophyll content and antioxidant enzymes	Cd, Pb and Ni	Karak, Pakistan	Khan et al. (2018)
2.	*Pseudomonas fluorescens*	*Trifolium repens*	Enhanced plant growth, Cd uptake and accumulation and chlorophyll content	Cd	Tehran Province, Iran	Zand et al. (2020)
3.	*R. leguminosarum*, *Pseudomonas fluorescens*, *Luteibacter* sp., *Variovorax* sp.	*Lathyrus sativus*	Improved growth parameters and shoot nitrogen content; enhanced phytoextraction, uptake and accumulation	Pb and Cd	Tunisia	Abdelkrim et al. (2020)
4.	*Klebsiella pneumoniae*	*Scirpus triqueter*	Increased dissipation rates of pyrene and removal rates of Ni	Pyrene-Ni	Shanghai, China	Zhang et al. (2020)
5.	*Bacillus velezensis*, *Bacillus proteolyticus*, *Lysinibacillus* sp.	*Scirpus grossus*	Enhanced plant growth promotion	Pb	Selangor, Malaysia	Kamaruzzaman et al. (2020)
6.	*Stenotrophomonas maltophilia* sp., *Agrobacterium* sp.	*Arundo donax* L.	Phytovolatilization and epigenetic modifications	Arsenic	Fisciano, Italy	Guarino et al. (2020)
7.	*Kocuria* sp. CRB15	*Brassica nigra*	High IAA production, P solubilization, ammonia production, hydrogen cyanide production; increase in root and shoot elongation	Copper	Rakha copper mine, Jharkhand, India	Hansda and Kumar (2017)
8.	*Bacillus methylotrophicus* SMT38, *Bacillus aryabhattai* SMT48, *Bacillus aryabhattai* SMT50, *Bacillus licheniformis* SMT51	*Spartina maritima*	Enhanced root growth and heavy metal accumulation	Arsenic, cadmium, copper, nickel, lead and zinc	Tinto river estuary, SW Spain	Mesa-Marín et al. (2018)

(continued)

Table 17.1 (continued)

S. no.	PGPR	Plant	Strategies/mechanisms	Pollutants	Study area	References
9.	*Enterobacter* sp. strain EG16	*Hibiscus cannabinus*	Phytostabilization, good plant growth, low plant accumulation of metals and reduced metal bioavailability in soil	Cadmium and iron	Dabao Mountain, Guangdong, China	Chen et al. (2017)
10.	*Pseudomonas putida* PTCC1694, *Bacillus megaterium* PTCC1656, *Azotobacter chroococcum, Proteus vulgaris* PTCC1079, *Bacillus subtilis* PTCC1715	Cabbage varieties	Increase in amounts of plant biomass, Pb and Cd concentrations in the root and shoot, Pb and Cd uptake performances, mobility, phytoavailability and translocation factor	Cadmium and lead	Zanzan, Iran	Abdollahi et al. (2020)
11.	*Klebsiella* sp. D5A and *Pseudomonas* sp. SB	*Testuca arundinacea* L.	Increase in biomass and hydrocarbon removal	Petroleum	Shengli oilfield, Shandong province, China	Hou et al. (2015)
12.	*Pseudomonas libanensis* TR1 and *Pseudomonas reactans* Ph3R3	*Brassica oxyrrhina*	Improved plant growth and metal accumulation under drought stress	Copper and zinc	Braganca, Portugal	Ma et al. (2016)
13.	*Stenotrophomonas* species and *Sphingobium* species	*Phragmites australis*	Increased the efficiency of phyto-based process	4-n-nonylphenol, mono-ethoxylated nonylphenols and di-ethoxylated nonylphenols	Italy	Di Gregorio et al. (2015)
14	*Staphylococcus, Bacillus, Pantoea* and *Salmonella*	*Spartina maritima*	Higher tolerance toward metal (loid)s and greater metal biosorption	Lead, zinc and copper	Odiel estuary, Spain	Paredes-Páliz et al. (2016)

15.	*Trichoderma asperellum* LZ1 and *Serratia sp.* LX2	*Lolium perenne* L. and *Lactuca versicolor*	Enhanced growth and P accumulation	Phosphate mining wasteland soil	Yichang, China	Guo et al. (2021)
16.	*Brevundimonas diminuta* MYS6	*Helianthus annuus* L.	Enhanced Cu uptake in roots and shoots along with increased root and shoot length, fresh and dry plant biomass and leaf chlorophyll	Copper	Punjab, India	Rathi and Yogalakshmi (2021)
17.	*Pseudomonas putida* MU02	*Ocimum gratissimum*	Phytostabilization and tolerance of plant from contaminant toxicity	Zinc and crude oil	Phichit, Thailand	Choden et al. (2021)

solubility, availability and translocation of noxious metals and by discharging chelating agents (Ma et al. 2016). In a study by Gullap et al. (2014), the impact of PGPR along with phosphorus fertilizer in phytoextraction of heavy metals like lead (Pb), nickel (Ni), boron (B), manganese (Mn) and zinc (Zn) from polluted soil was checked. The study revealed that the time required for removal of these heavy metals was decreased by application of PGPR and phosphorus fertilizer. Liu et al. 2015 reported the effect of *Phyllobacterium myrsinacearum* on the phytoextraction of heavy metals by *Sedum alfredii* and *Medicago sativa* L. The inoculation increased the phytoextraction of Pb, Cd and Zn by shoots.

17.4.3 Metal Accumulation in Plants (Bioaccumulation)

There is an increase in accumulation of metals in plant parts post rhizobacteria inoculation. This occurs because of change in soil chemistry and increase in the metal solubility and bioavailability (Burd et al. 2000; Lasat 2002; Thakare et al. 2021). Generally, metals are found in insoluble forms in soil. Rhizobacteria have been reported to reduce the soil pH which favours nutrient and metal uptake by plant through transforming them into accessible forms (Whiting et al. 2001; Carlot et al. 2002; Zhuang et al. 2007). PGPR have been reported to produce enzymes, siderophores and organic acids which augments metal uptake and prevents phytotoxic effects of pollutants (Yousaf et al. 2010). In a study by Kamran et al. 2016, the effect of PGPR *P. putida* was investigated with respect to growth of *E. sativa* along with nickel uptake. It was found that Ni uptake was enhanced in plants inoculated with *P. putida* as compared to noninoculated plants. This enhanced Ni uptake was attributed to IAA, siderophore and ACC deaminase activity in growing media. In another study by Mousavi et al. 2018, the role of siderophore producing *B. safensis* and *Pseudomonas fluorescens* inoculation on *H. annuus* growth and metal accumulation was investigated. Here it was found that microbial inoculations solubilized and enhanced the accumulation of Pb and Zn.

17.4.4 Biodegradation

PGPR along with plants has been well documented to be effective in removal/degradation of pollutants. In a study by Rani et al. (2021), the impact of *Solanum lycopersicum* and PGPR strains on endosulfan degradation in soil was evaluated. The study concluded that inoculation of PGPR demonstrated a favourable influence on the degradation of endosulfan along with enhanced plant biomass. Hou et al. (2015) have reported enhancement in phytoremediation of petroleum contaminated soil by PGPR. When a tall fescue plant was inoculated by two PGPR strains, the plant biomass was enhanced and petroleum hydrocarbons (especially aliphatic hydrocarbons and polycyclic aromatic hydrocarbons) were removed. In another study by Xun et al. (2015), combination of PGPR and arbuscular mycorrhizal fungi (AMF) on phytoremediation of saline-alkali soil contaminated by petroleum was checked. The inoculation with PGPR and AMF enhanced the degradation rate of petroleum hydrocarbons. Also, the soil quality was enhanced by increasing the activities of soil enzymes like urease, sucrase and dehydrogenase. There was also an increase in dry weight and stem height of inoculated plants as compared to uninoculated plants.

17.5 Exploring Rhizosphere Bionetwork by Metagenomics Approach

Several researchers have tried to explore the rhizospheric microorganism using the metagenomics approach. Figure 17.2 depicts a general mechanism of identification of PGPR based on metagenomic approaches. Wu et al. (2018) studied the

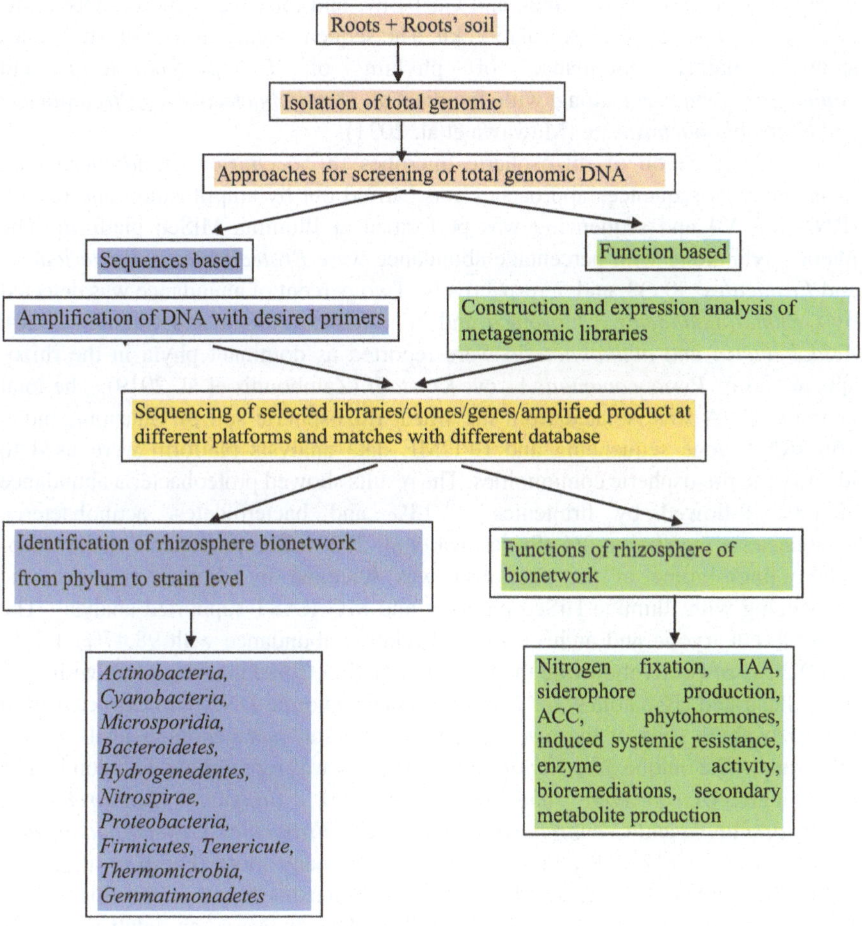

Fig. 17.2 A general mechanism of identification of PGPR based on metagenomic approaches

rhizosphere microbiome from the soil sample using the BioFast soil genomic DNA extraction kit. The result indicated the presence of *Pseudomonadaceae, Burkholderiaceae, Sphingomonadaceae and Streptomycetaceae.* The sequence reads showed the various defence mechanisms, acid survival, metabolism, transport and catabolism and presented significant insights into diverse community and functional attributes of the rhizosphere microbiome. Ramadan et al. (2020) explored the r*hizosphere* bacteria in *Calotropis procera* by the sequenced-based metagenomic 16S rRNA gene analysis of V3-V4 regions. Total 2173 operational taxonomic units were assigned out of which 24 OUT were selected with abundance. The most dominant phyla were *Acidobacteria* and *Actinobacteria.* While most abundant species were *Nocardia cyriacigeorgica*, *Rhodospirillales bacterium* WX 36 and *Arthrobacter crystallopoietes* along with 20 unknown species. This indicates that still so many rhizosphere species remain non-cultivable. In Kenya, the rhizospheres of mangrove species were identified by metagenomic analysis. The soil samples were collected from Mida Creek and Gazi Bay, and total metagenomic DNA was isolated by Power Soil DNA isolation kit. The sequence analysis of both study sites showed mainly dominance of phylum of *Deltaproteobacteria* and *Gammaproteobacteria* along with families of *Desulfobacteraceae*, *Pirellulaceae* and *Syntrophobacterace*ae (Muwawa et al. 2021).

The identification of rhizospheric microbes of *Saccharum arundinaceum* by metagenomics sequences approaches was carried out by amplification and of 16S rRNA V3–V4 and sequencing was performed at Illumina MiSeq platform. The major phyla along with percentage abundance were *Proteobacteria, Bacteriodetes* and *Firmicutes,* 50, 33, and 5, respectively. Two percent of abundance was detected by *Gemmatimonadetes*, *Chloroflexi* and *Tenericutes* (Kumar and Chandra 2020). *Acidobacteria* and *actinobacteria* were reported as dominant phyla in the rhizosphere soil of *Pyrus communis* L. cv. *Krystalli* (Zambounis et al. 2019). The total genomic DNA was isolated from the wheat rhizospheric soil of Ghazipur, India. 16S rRNA gene sequencing and EPI2ME data analysis platform were used to identify the rhizospheric communities. The results showed proteobacteria abundance at 68% followed by firmicutes at 13% and bacteroidetes, actinobacteria, acidobacteria together at 3% (Srivastava et al. 2020). The phylogeny of the rhizosphere microbiome of the sunflower was screened by shotgun metagenomic sequencing with Illumina HiSeq platform and MG-RAST supported analysis. The bacteria, eukaryotic and archaea showed relative abundance with 98.47%, 1.23% and 0.20%, respectively, while the most dominating Conexibacter genera with 17% were illustrated (Babalola et al. 2020). The metagenomic DNA was extracted from the rhizosphere soils of the maize plant and shotgun metagenomics analysis was performed. The unique diversity of rhizosphere soil was represented with dominance of fungi (*Ascomycota* and *Basidiomycota), archaea (Euryarchaeota, Thaumarchaeota* and *Crenarchaeota)* and bacteria *Firmicutes*, *Gemmatimonadetes*, *Acidobacteria*, *Chloroflexi*, *Planctomycetes*) and many more (Fadiji et al. 2021). Table 17.2 shows the diverse Phyla of microorganisms found in the rhizosphere region with the help of metagenomic studies. The metagenomic analysis clearly

Table 17.2 List of rhizospheric microbial diversity identified with the aid of metagenomic approach

Sr no.	Phylum	Plant	Site	References
1	*Actinobacteria*	*Arabidopsis thaliana*	Cologne and Eifel, Germany	Schlaeppi et al. (2014)
2.	*Bacteroidetes*	*Arachis hypogaea*	Bhavnagar, India	Yousuf et al. (2012)
3.	*Cyanobacteria*	*Paspalum scrobiculatum* L. (Kodo millet)	Jagdalpur, India	Prabha et al. (2019)
4.	*Firmicutes*	*Tabebuia billbergii*	Cerros de Amotape-Tumbes National Park, northern Peru	Llacsa et al. (2019)
5.	*Gemmatimonadetes*	*Cotton*	Alwar, India	Singh et al. (2020)
6.	*Hydrogenedentes*	*Amilaceous, Zea mays* L.	Acobamba (Huancavelica, Peru)	Correa-Galeote et al. (2016)
7.	*Microsporidia*	*Avicennia marina*	Thuwal, Saudi Arabia	Simões et al. (2015)
8.	*Nitrospirae*	*Zea mays*	Ventersdorp in the north West Province, South Africa	Molefe et al. (2021)
9.	*Proteobacteria*	*Saccharum arundinaceum*	Unnao, Uttar Pradesh, India	Kumar and Chandra (2020)
10.	*Tenericutes*	*Saccharum arundinaceum*	Unnao, Uttar Pradesh, India	Kumar and Chandra (2020)
11.	*Thermomicrobia*	*Paspalum scrobiculatum* L. *(Kodo millet)*	Jagdalpur, India	Prabha et al. (2019)
12.	*Verrucomicrobia*	*Calotropis procera*	Jeddah, Saudi Arabia	Ramadan et al. (2021)

showed that roots and root soils of plants act as a hotspot and are reflected to be most favourable for plant growth.

17.6 Future Prospects

Applying a PGPR consortium can have an enhanced effect on the phytoremediation process for a variety of pollutants. Studies can be focused on the soil ecology along with the phytoremediation by plant and rhizosphere. Improving the efficacy of PGPR inoculants and chelants used can have a major impact on the pollutant uptake by the plant. Metagenomics approach can be employed for understanding the mechanism of PGPRs and plant interaction for more sustainable agricultural practices. PGPR-based formulations for phytoremediation purposes need to be developed along with biocontrol and plant growth-promoting products that are available on the market. Advanced genetic engineering tools may be employed to develop a single PGPR strain with multiple traits to promote plant growth and enhance phytoremediation.

17.7 Conclusion

PGPR show promising results in enhancing phytoremediation of several pollutants. The application of PGPR for phytoremediation of toxic pollutants is sustainable and cheaper as compared to the conventional remediation alternatives available. Several plant species can be employed along with a consortium of PGPR depending upon soil composition, environmental conditions and pollutants. PGPR strains employ several mechanisms to promote plant growth, although studies should be focused on the relative contribution of each mechanism responsible for effective plant growth promotion. Metagenomic studies can reveal non-cultivable microorganisms in the rhizosphere region which plays an important role in plant growth promotion. More research on PGPR needs to be diverted towards the effect of the PGPR consortium, about their ecology and symbiotic relationship with plants in the rhizosphere. Besides this, reproducibility of the effects of microbial inoculants needs to be tested across a wide range of soil types and environmental conditions.

References

Abdelkrim S, Jebara SH, Saadani O et al (2020) In situ effects of *Lathyrus sativus*- PGPR to remediate and restore quality and fertility of Pb and Cd polluted soils. Ecotoxicol Environ Saf 192:110260. https://doi.org/10.1016/j.ecoenv.2020.110260

Abdollahi S, Golchin A, Shahryari F, Alamdari P (2020) PGPR inoculation of a contaminated soil affects plant growth and phytoavailability of Cd and Pb. Arch Agron Soil Sci:1–8. https://doi.org/10.1080/03650340.2020.1843157

Ali H, Khan E, Sajad MA (2013) Phytoremediation of heavy metals—concepts and applications. Chemosphere 91(7):869–881. https://doi.org/10.1016/j.chemosphere.2013.01.075

Babalola OO, Alawiye TT, Rodriguez Lopez CM, Ayangbenro AS (2020) Shotgun metagenomic sequencing data of sunflower rhizosphere microbial community in South Africa. Data Brief 31: 105831. https://doi.org/10.1016/j.dib.2020.105831

Bhattacharyya PN, Jha DK (2012) Plant growth-promoting rhizobacteria (PGPR): emergence in agriculture. World J Microbiol Biotechnol 4:1327–1350. https://doi.org/10.1007/s11274-011-0979-9

Burd GI, Dixon DG, Glick BR (2000) Plant growth-promoting bacteria that decrease heavy metal toxicity in plants. Can J Microbiol 3:237–245. https://doi.org/10.1139/w99-143

Carlot M, Giacomini A, Casella S (2002) Aspects of plant-microbe interactions in heavy metal polluted soil. Acta Biotechnol 22(1–2):13–20. https://doi.org/10.1002/1521-3846(200205)22:1/2%3C13::AID-ABIO13%3E3.0.CO;2-9

Chen Y, Yang W, Chao Y et al (2017) Metal-tolerant *Enterobacter sp.* strain EG16 enhanced phytoremediation using *Hibiscus cannabinus* via siderophore-mediated plant growth promotion under metal contamination. Plant and Soil 413(1–2):203–216. https://doi.org/10.1007/s11104-016-3091-y

Choden D, Pokethitiyook P, Poolpak T, Kruatrachue M (2021) Phytoremediation of soil co-contaminated with zinc and crude oil using *Ocimum gratissimum* (L.) in association with *Pseudomonas putida* MU02. Int J Phytoremediation 23(2):181–189. https://doi.org/10.1080/15226514.2020.1803205

Compant S, Reiter B, Sessitsch A et al (2005) Endophytic colonization of *Vitis vinifera* L. by plant growth-promoting bacterium Burkholderia sp. strain PsJN. Appl Environ Microbiol 71(4):1685–1693. https://doi.org/10.1128/AEM.71.4.1685-1693.2005

Correa-Galeote D, Bedmar EJ, Fernández-González AJ, Fernández-López M (2016) Bacterial communities in the rhizosphere of *Amilaceous maize (Zea mays L.)* as assessed by pyrosequencing. Front Plant Sci 7:1016. https://doi.org/10.3389/fpls.2016.01016

Dhir B (2017) Bioremediation technologies for the removal of pollutants. In: Kumar R, Sharma A, Ahluwalia S (eds) Advances in environmental biology. Springer, Singapore. https://doi.org/10.1007/978-981-10-4041-2_5

Di Gregorio S, Giorgetti L, Castiglione MR et al (2015) Phytoremediation for improving the quality of effluents from a conventional tannery wastewater treatment plant. Int J Environ Sci Technol 12(4):1387–1400. https://doi.org/10.1007/s13762-014-0522-2

Fadiji AE, Kanu JO, Babalola OO (2021) Metagenomic profiling of rhizosphere microbial community structure and diversity associated with maize plant as affected by cropping systems. Int Microbiol 5:1–1. https://doi.org/10.1007/s10123-021-00169-x

Gans J, Wolinsky M, Dunbar J (2005) Computational improvements reveal great bacterial diversity and high metal toxicity in soil. Science 309(5739):1387–1390. https://doi.org/10.1126/science.1112665

Guarino F, Miranda A, Castiglione S, Cicatelli A (2020) Arsenic phytovolatilization and epigenetic modifications in *Arundo donax* L assisted by a PGPR consortium. Chemosphere 251:126310. https://doi.org/10.1016/j.chemosphere.2020.126310

Gullap MK, Dasci M, Erkovan Hİ, Koc A et al (2014) Plant growth-promoting rhizobacteria (PGPR) and phosphorus fertilizer-assisted phytoextraction of toxic heavy metals from contaminated soils. Commun Soil Sci Plant Anal 45(19):2593–2606. https://doi.org/10.1080/00103624.2014.929702

Guo S, Feng B, Xiao C et al (2021) Phosphate-solubilizing microorganisms to enhance phytoremediation of excess phosphorus pollution in phosphate mining wasteland soil. Biorem J 25:271–281. https://doi.org/10.1080/10889868.2021.1884528

Hansda A, Kumar V (2017) Cu-resistant Kocuria sp. CRB15: a potential PGPR isolated from the dry tailing of Rakha copper mine. 3 Biotech 7(2):1–11. https://doi.org/10.1007/s13205-017-0757-y

He X, Xu M, Wei Q et al (2020) Promotion of growth and phytoextraction of cadmium and lead in *Solanum nigrum* L. mediated by plant-growth-promoting rhizobacteria. Ecotoxicol Environ Saf 205:111333. https://doi.org/10.1016/j.ecoenv.2020.11133

Hou J, Liu W, Wang B et al (2015) PGPR enhanced phytoremediation of petroleum contaminated soil and rhizosphere microbial community response. Chemosphere 138:592–598. https://doi.org/10.1016/j.chemosphere.2015.07.025

Ismail N, Abdullah SRS, Idris M et al (2020) Applying rhizobacteria consortium for the enhancement of Scirpus grossus growth and phytoaccumulation of Fe and Al in pilot constructed wetlands. J Environ Manage 267:110643. https://doi.org/10.1016/j.jenvman.2020.110643

Ji G, Sun T, Ni J (2007) Surface flow constructed wetland for heavy oil-produced water treatment. Bioresour Technol 98:436–441. https://doi.org/10.1016/j.biortech.2006.01.017

Ju W, Liu L, Jin X et al (2020) Co-inoculation effect of plant-growth-promoting rhizobacteria and rhizobium on EDDS assisted phytoremediation of cu contaminated soils. Chemosphere 254:126724. https://doi.org/10.1016/j.chemosphere.2020.126724

Kamaruzzaman MA, Abdullah SRS, Hasan HA et al (2020) Characterisation of Pb-resistant plant growth-promoting rhizobacteria (PGPR) from *Scirpus grossus*. Biocatal Agric Biotechnol 23: 101456. https://doi.org/10.1016/j.bcab.2019.101456

Kamran MA, Eqani SAMAS, Bibi S et al (2016) Bioaccumulation of nickel by *E. sativa* and role of plant growth promoting rhizobacteria (PGPRs) under nickel stress. Ecotoxicol Environ Saf 126: 256–263. https://doi.org/10.1016/j.ecoenv.2016.01.002

Khan WU, Yasin NA, Ahmad SR et al (2017) Role of Ni-tolerant *Bacillus* spp. and *Althea rosea* L. in the phytoremediation of Ni-contaminated soils. Int J Phytoremediation 19(5):470–477. https://doi.org/10.1080/15226514.2016.1244167

Khan N, Zandi P, Ali S et al (2018) Impact of salicylic acid and PGPR on the drought tolerance and phytoremediation potential of *Helianthus annus* [published correction appears in front Microbiol. 2019 Sep 27;10:2222]. Front Microbiol 9:2507. https://doi.org/10.3389/fmicb.2018.02507

Kumar V, Chandra R (2020) Metagenomics analysis of rhizospheric bacterial communities of *Saccharumarundinaceum* growing on organometallic sludge of sugarcane molasses-based distillery. 3 Biotech 10:316. https://doi.org/10.1007/s13205-020-02310-5

Lasat MM (2002) Phytoextraction of toxic metals: a review of biological mechanisms. J Environ Qual 31(1):109–120. PMID: 11837415

Liu W, Hou J, Wang Q et al (2014) Isolation and characterization of plant growth-promoting rhizobacteria and their effects on phytoremediation of petroleum-contaminated saline-alkali soil. Chemosphere 117:303–308. https://doi.org/10.1016/j.chemosphere.2014.07.026

Liu ZF, Ge HG, Li C et al (2015) Enhanced phytoextraction of heavy metals from contaminated soil by plant co-cropping associated with PGPR. Water Air Soil Pollut 226:29. https://doi.org/10.1007/s11270-015-2304-y

Llacsa LX, Solis-Castro RL, Mialhe E et al (2019) Metagenomic analysis of the bacterial and fungal community associated to the rhizosphere of *Tabebuia chrysantha* and *T. billbergii*. Curr Microbiol 76:1073–1080. https://doi.org/10.1007/s00284-019-01725-5

Ma Y, Rajkumar M, Zhang C, Freitas H (2016) Inoculation of *Brassica oxyrrhina* with plant growth promoting bacteria for the improvement of heavy metal phytoremediation under drought conditions. J Hazard Mater 320:36–44. https://doi.org/10.1016/j.jhazmat.2016.08.009

Mesa-Marín J, Del-Saz NF, Rodríguez-Llorente ID et al (2018) PGPR reduce root respiration and oxidative stress enhancing *Spartina maritima* root growth and heavy metal Rhizoaccumulation. Front Plant Sci 9:1500. https://doi.org/10.3389/fpls.2018.01500

Molefe RR, Amoo AE, Babalola OO (2021) Metagenomic insights into the bacterial community structure and functional potentials in the rhizosphere soil of maize plants. J Plant Interact 16(1):258–269. https://doi.org/10.1080/17429145.2021.1936228

Mousavi SM, Motesharezadeh B, Hosseini HM, Alikhani H, Zolfaghari AA (2018) Root-induced changes of Zn and Pb dynamics in the rhizosphere of sunflower with different plant growth promoting treatments in a heavily contaminated soil. Ecotoxicol Environ Saf 147:206–216. https://doi.org/10.1016/j.ecoenv.2017.08.045

Muwawa EM, Obieze CC, Makonde HM, Jefwa JM et al (2021) 16S rRNA gene amplicon-based metagenomic analysis of bacterial communities in the rhizospheres of selected mangrove species from Mida Creek and Gazi Bay, Kenya. PLoS One 16(3):e0248485. https://doi.org/10.1371/journal.pone.0248485

Naseem H, Bano A (2014) Role of plant growth-promoting rhizobacteria and their exopolysaccharide in drought tolerance of maize. J Plant Interact 2a9(1):689–701. https://doi.org/10.1080/17429145.2014.902125

Paredes-Páliz KI, Caviedes MA, Doukkali B et al (2016) Screening beneficial rhizobacteria from *Spartina maritima* for phytoremediation of metal polluted salt marshes: comparison of gram-positive and gram-negative strains. Environ Sci Pollut Res Int 23(19):19825–19837. https://doi.org/10.1007/s11356-016-7184-1

Prabha R, Singh DP, Gupta S, Gupta VK et al (2019) Rhizosphere metagenomics of *Paspalum scrobiculatum* L. *(Kodo millet)* reveals rhizobiome multifunctionalities. Microorganisms 7(12):608. https://doi.org/10.3390/microorganisms7120608

Prasad R, Kumar M, Varma A (2015) Role of PGPR in soil fertility and plant health. In: Egamberdieva D, Shrivastava S, Varma A (eds) Plant growth-promoting rhizobacteria (PGPR) and medicinal plants. Springer International, Cham, pp 247–260

Ramadan AM, Nazar MA, Gadallah NO (2020) Metagenomic analysis of rhizosphere bacteria in desert plant *Calotropisprocera*. Geomicrobiol J 7:1–9. https://doi.org/10.1080/01490451.2020.1860166

Ramadan AM, Nazar MA, Gadallah NO (2021) Metagenomic analysis of rhizosphere bacteria in desert plant Calotropisprocera. Geomicrobiol J 38(5):375–383. https://doi.org/10.1080/01490451.2020.1860166

Rani R, Kumar V, Gupta P, Chandra A (2021) Potential use of *Solanum lycopersicum* and plant growth promoting rhizobacterial (PGPR) strains for the phytoremediation of endosulfan stressed. Chemosphere 279:130589. https://doi.org/10.1016/j.chemosphere.2021.130589

Rathi M, Yogalakshmi KN (2021) *Brevundimonas diminuta* MYS6 associated *Helianthus annuus* L. for enhanced copper phytoremediation. Chemosphere 263:128195. https://doi.org/10.1016/j.chemosphere.2020.128195

Reeves RD, Baker AJM, Jaffré T et al (2018) A global database for plants that hyperaccumulate metal and metalloid trace elements. New Phytol 218(2):407–411. https://doi.org/10.1111/nph.14907

Sbani NH, Abdullah SR, Idris M, Hasan HA et al (2021) PAH-degrading rhizobacteria of *Lepironia articulata* for phytoremediation enhancement. J Water Process Eng 39:101688. https://doi.org/10.1016/j.jwpe.2020.101688

Schlaeppi K, Dombrowski N, Oter RG et al (2014) Quantitative divergence of the bacterial root microbiota in *Arabidopsis thaliana* relatives. Proc Natl Acad Sci U S A 111(2):585–592. https://doi.org/10.1073/pnas.1321597111

Shrivastava S, Prasad R, Varma A (2014) Anatomy of root from eyes of a microbiologist. In: Morte A, Varma A (eds) Root Engineering, vol 40. Springer-Verlag, Berlin Heidelberg, pp 3–22

Simões MF, Antunes A, Ottoni CA, Amini MS et al (2015) Soil and rhizosphere associated fungi in gray mangroves (*Avicennia marina*) from the Red Sea—a metagenomic approach. GPB 13(5):310–320. https://doi.org/10.1016/j.gpb.2015.07.002

Singh D, Raina TK, Kumar A, Singh J (2019) Plant microbiome: a reservoir of novel genes and metabolites. Plant Gene 18:100177. https://doi.org/10.1016/j.plgene.2019.100177

Singh RP, Johri AK, Dua M (2020) Metagenomic analysis of microbial diversity in cotton rhizosphere soil in Alwar, India. Microbiol Resour Announc 9(49):e00987–e00920. https://doi.org/10.1128/MRA.00987-20

Srivastava R, Srivastava AK, Ramteke PW, Gupta VK et al (2020) Metagenome dataset of wheat rhizosphere from Ghazipur region of eastern Uttar Pradesh. Data Brief 28:105094. https://doi.org/10.1016/j.dib.2019.105094

Tewari S, Arora NK (2014) Multifunctional exopolysaccharides from *Pseudomonas aeruginosa* PF23 involved in plant growth stimulation, biocontrol and stress amelioration in sunflower under saline conditions. Curr Microbiol 69(4):484–494. https://doi.org/10.1007/s00284-014-0612-x

Thakare M, Sarma H, Datar S, Roy A, Pawar P, Gupta K, Pandit S, Prasad R (2021) Understanding the holistic approach to plant-microbe remediation technologies for removing heavy metals and radionuclides from soil. Curr Res Biotechnol 3:84–98. https://doi.org/10.1016/j.crbiot.2021.02.004

Villaverde J, Rubio-Bellido M, Merchán F, Morillo E (2017) Bioremediation of diuron contaminated soils by a novel degrading microbial consortium. J Environ Manage 188:379–386. https://doi.org/10.1016/j.jenvman.2016.12.020

Whiting SN, de Souza MP, Terry N (2001) Rhizosphere bacteria mobilize Zn for hyperaccumulation by *Thlaspi caerulescens*. Environ Sci Technol 35(15):3144–3150. https://doi.org/10.1021/es001938v

Wipf D, Krajinski F, van Tuinen D et al (2019) Trading on the arbuscular mycorrhiza market: from arbuscules to common mycorrhizal networks. New Phytol 223(3):1127–1142. https://doi.org/10.1111/nph.15775

Wu L, Wang J, Wu H, Chen J et al (2018) Comparative metagenomic analysis of rhizosphere microbial community composition and functional potentials under *Rehmanniaglutinosa* consecutive monoculture. Int J Mol Sci 19(8):2394. https://doi.org/10.3390/ijms19082394

Xun F, Xie B, Liu S, Guo C (2015) Effect of plant growth-promoting bacteria (PGPR) and arbuscular mycorrhizal fungi (AMF) inoculation on oats in saline-alkali soil contaminated by petroleum to enhance phytoremediation. Environ Sci Pollut Res Int 22(1):598–608. https://doi.org/10.1007/s11356-014-3396-4

Yousaf S, Andria V, Reichenauer TG, Smalla K, Sessitsch A (2010) Phylogenetic and functional diversity of alkane degrading bacteria associated with Italian ryegrass (*Lolium multiflorum*) and Birdsfoot trefoil (*Lotus corniculatus*) in a petroleum oil-contaminated environment. J Hazard Mater 184(1–3):523–532. https://doi.org/10.1016/j.jhazmat.2010.08.067

Yousuf B, Keshri J, Mishra A, Jha B (2012) Application of targeted metagenomics to explore abundance and diversity of CO2-fixing bacterial community using cbbL gene from the rhizosphere of *Arachis hypogaea*. Gene 506(1):18–24. https://doi.org/10.1016/j.gene.2012.06.083

Zambounis A, Osathanunkul M, Madesis P (2019) Metagenome data of bacterial diversity in pear (*Pyruscommunis L.*) rhizospheres associated with *Phytophthora* infection and amino acid treatment. Data Brief 26:104396. https://doi.org/10.1016/j.dib.2019.104396

Zand AD, Mikaeili Tabrizi A, Vaezi Heir A (2020) Application of titanium dioxide nanoparticles to promote phytoremediation of cd-polluted soil: contribution of PGPR inoculation. Biorem J 24(2–3):171–189. https://doi.org/10.1080/10889868.2020.1799929

Zhang X, Su C, Liu X, Liu Z, Liang X, Zhang Y, Feng Y (2020) Effect of plant-growth-promoting rhizobacteria on phytoremediation efficiency of *Scirpus triqueter* in pyrene-Ni co-contaminated soils. Chemosphere 241:125027. https://doi.org/10.1016/j.chemosphere.2019.125027

Zhuang X, Chen J, Shim H, Bai Z (2007) New advances in plant growth-promoting rhizobacteria for bioremediation. Environ Int 33(3):406–413. https://doi.org/10.1016/j.envint.2006.12.005

Phycoremediation: Treatment of Pollutants and an Initiative Towards Sustainable Environment

18

Ankita Bhatt, Komal Agrawal, and Pradeep Verma

Abstract

The expanding population with simultaneous rise in industrialisation and urbanisation has deleterious and hazardous impact on both the ecosystem and mankind. It results in severe environmental pollution including generation of excessive wastes, heavy metal pollution and increased discharge of industrial effluents. Various factors like high cost, energy and the lower treatment efficiency have limited the use of traditional physical and chemical methods of waste treatment. This has raised an environmental concern among the researchers worldwide, and the scientific community is now shifting its focus towards biological and eco-friendly alternatives for waste treatment. The concept of phycoremediation has been gaining impetuous and involves the use of microalgae or macroalgae for bioremediation with simultaneous biomass production. The algae are used widely owing to their ease of cultivation. They are the primary producers of the aquatic ecosystem, grow fast and exhibit a high photosynthetic efficiency. In addition to their role as a potential bioremediating agent, algae are used for feed and food, for medicinal purposes, as a source of various pharmaceutical and cosmeceutical products and as a renewable feedstock for biofuel production. Thus, the present chapter would focus on the exploration of algae for bioremediation of various pollutants, its limitations and future prospects.

Ankita Bhatt and Komal Agrawal contributed equally with all other contributors.

A. Bhatt · K. Agrawal · P. Verma (✉)
Bioprocess and Bioenergy Laboratory, Department of Microbiology, Central University of Rajasthan, Bandarsindri, Kishangarh, Ajmer, Rajasthan, India
e-mail: pradeepverma@curaj.ac.in

R. Prasad (ed.), *Phytoremediation for Environmental Sustainability*,
https://doi.org/10.1007/978-981-16-5621-7_18

Keywords

Phycoremediation · Pollutants · Algae · Wastewater · Biomass · Greywater ·
Ankita Bhatt and Komal Agrawal contributed equally to this work.

18.1 Introduction

Due to rapid urbanisation and industrialisation, there has been a rise in the pressure on natural resources. This has resulted in an increase in generation and discharge of untreated wastewater from various industries and other non-industrial sectors, thereby leading to pollution of water bodies, deterioration of aquatic life and hazardous impacts on mankind. The issue of adequate wastewater treatment has thus emerged as a global concern, and efficient treatment techniques are required to combat this slow degradation of water resources and ecosystem. The conventional physical and chemical treatment techniques are expensive, involve high energy and provide low treatment efficiency. Hence, biological methods employing microalgae are now being studied for the treatment of various wastewaters. The employment of microalgae as bioremediating agents is called phycoremediation. Various benefits associated with microalgae-assisted remediation, namely passive cultivation process, faster growth rates and high photosynthetic efficiency facilitate the wide employment of this technique. Further, the treatment process can be coupled with biofuel production, carbon dioxide mitigation and generation of other value-added products like polyunsaturated fatty acids, vitamins and antioxidants (Aziz et al. 2017; Agrawal et al. 2020; Bhardwaj et al. 2020; Goswami et al. 2020a, b; Mehariya et al. 2021a, b; Rawat et al. 2021). Thus, the present chapter would focus on the exploration of algae for bioremediation of different kinds of wastewater (Fig. 18.1) followed by the limitations and future prospects of phycoremediation.

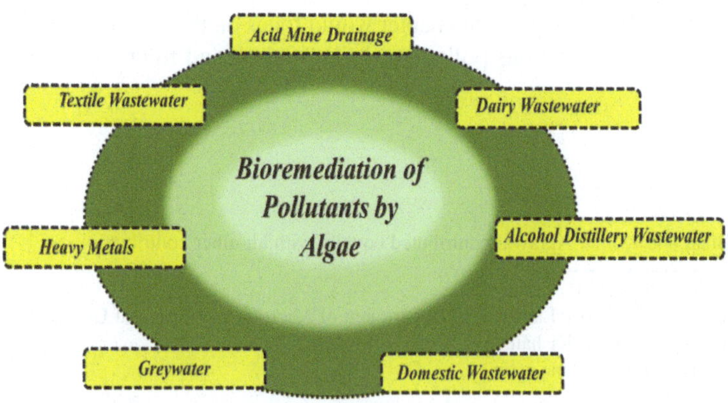

Fig. 18.1 Schematic representation of various pollutants treated by algae

18.2 Acid Mine Drainage (AMD)

The mining sector serves a prime role in the gross domestic product of many countries. Various rock interactions lead to the production of a hazardous and acidic effluent which gets stockpiled in tailing dams. This water which consists of chemicals of sulphate oxidation produced during mining activity is termed as acid rock drainage (ARD) or acid mine drainage (AMD) (Simate and Ndlovu 2014). The saline drainage and neutral or base mine drainage are also produced from the mine waters. The former is result of intrusion of seawater in mining while the latter has higher pH values owing to bicarbonate exchange and lower rock permeability. The high concentration of heavy metals in the acid mine drainage is the prime concern among the environmentalist because of the presence of active bacteria and high leaching capacity (Bwapwa et al. 2017). Hence, effective methods for AMD treatment and sequestration of heavy metals are required to prevent its adverse effects in the environment.

18.2.1 Sources and Characteristics of Acid Mine Drainage

Many ore stockpiles, mine pits, waste rock dumps, tailing deposits, slags and heap leach pads (Fig. 18.2) comprise major sources of AMD (Johnson and Hallberg 2005). Acid mine drainage is usually produced as a result of various anthropogenic activities particularly the oxidative decomposition of exposed pyrite which results in the production of aqueous sulphuric acid (2 units) and ferrous iron ions from the

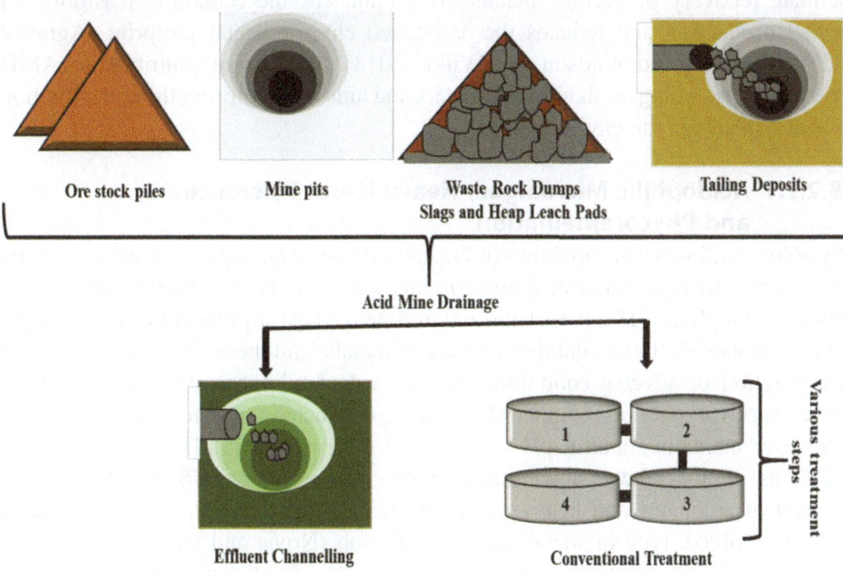

Fig. 18.2 Various sources of acid mine drainage and methods available for treatment

solid pyrite. The ferrous ions in the presence of oxygen further get oxidised to produce ferric ions which interact with pyrite, thereby enhancing the acidity of water. The acidification of the effluent allows for establishment of various acidophilic microbial populations (Costello 2003). The principle heavy metal contaminants present in AMD are copper (Cu), nickel (Ni), lead (Pb), arsenic (As), iron (Fe), manganese (Mn), aluminium (Al), zinc (Zn) and cadmium (Cd) (Chekroun and Baghour 2013). Cadmium was found as the most bioavailable and mobile heavy metal present in the AMD. Also, dissolved oxygen in water is reduced due to the presence of various microbes which promotes geochemical processes with simultaneous mineral oxidation (Bhattacharya et al. 2006).

18.2.2 Methods for Treatment of Acid Mine Drainage

Basically, two strategies are present for remediation of AMD, namely effluent channelling and the conventional treatment process. The former involves movement of effluent through wetlands (natural or constructed) which are comprised of microbial populations that have the ability to remediate the wastewater in a passive way while the latter includes the collection of effluent followed by it biological and chemical treatment in a centralised treatment plant (Fig. 18.2). Several benefits offered by the employment of algae as bioremediating agent include easier manipulation, lower costs, simple recovery methods and no production of secondary waste (Kalin et al. 2006). Remediation of AMD is described as cultivation of algae in the polluted effluent followed by separation of water and algal biomass. The recovered biomass is further used for biofuel production whereas the separated water is dried to facilitate recovery of various metals. This enhances the economic feasibility of biofuel production and reduces the associated environmental footprint (Agrawal and Verma 2022; Edmundson and Wilkie 2013). Microalgae contribute to AMD remediation by acting as alkalinity boosters and aids in monitoring the concentration of heavy metals in the environment.

18.2.2.1 Acidophilic Microalgae, Heavy Metal Tolerance and Phycoremediation

Physiology and survival capability of the natural microbial population are limited by the extreme environment of acid mine drainage, thereby facilitating the evolution of various acidophiles. The pH of the environment plays a prime role as lower pH values enhance the bioavailability of various metals, and these low values also lead to generation of adverse conditions for survival of microbes. Microalgae exhibit higher survival rates as compared to the cyanobacteria because the microalgal cytoplasm maintains neutral pH whereas the external acidic pH in cyanobacteria makes its photosynthetic apparatus prone to severe damage (Brock 1973). The removal of sulphates and heavy metals by microalgal is determined by the algal species involved, type of metal, its concentration (Novis and Harding 2007) and conditions of light intensity and temperature (Elbaz-Poulichet et al. 2000; Brake et al. 2004). The two main mechanisms employed by algae for heavy metal remediation are absorption and adsorption. The microalgae remove metals from wastewater with simultaneous utilisation of various nutrients to support algal growth. They are known to bioaccumulate the metals within the intercellular spaces or the vacuoles in their cells, and aqueous solutions of microalgal cultures are considered as most effective for removal of metal ions present in lower concentrations (Afkar et al. 2010; Kumar and Gaur 2011; Chen et al. 2012; Tripathy et al. 2021). Various metals like nickel, cobalt, cadmium, iron, chromium and copper are removed from the polluted environment by employing species like *Oedogonium rivulare* and *Cladophora glomerata* (Vymazal 1984). Chlorophyta is defined as the order which was the highest ability for evolution of heavy metals followed by the Phaeophyta and Rhodophyta (Al-Shwafi and Rushdi 2008). Mehta and Gaur (2005) demonstrated that lifeless algal biomass exhibits higher metal adsorption capacities as compared to the live algal cultures. A freshwater algae species *Stigeoclonium* sp. can effectively remove zinc and can tolerate zinc concentrations of 10 mmol (Pawlik-Skowronska 2001). Microalgae serve as cheap source for adsorption of heavy metals due to the presence of multilayer algal cell wall structure (Bilal et al. 2013; Gupta et al. 2015) and their ability to survive in both marine and freshwater bodies (Anastopoulos and Kyzas 2015).

18.2.2.2 Microalgae–Bacteria Biofilms and MFCs

The synergistic interaction between microalgae and bacteria serves as an innovative approach for efficient bioremediation of acid mine drainage. The microbial consortia are associated with lower energy requirements as it involves mutual exchange of metabolic intermediates like carbohydrates, proteins, oxygen production and carbon dioxide uptake. The use of this technique is limited by the survival possibility of individual microbe in large-scale treatment processes (Abinandan et al. 2018). Microbial fuel cells having the ability to generate power density of 290 MW/m^2 had been generated by Cheng et al. 2007. The pH is most important factor which determines remediation of acid mine drainage by the use of microbial fuel cells. The lower pH of AMD facilitates the process of oxidation, and insoluble Fe (III) is recovered at the anode. The development of microalgal bacterial biofilms provides benefit in MFC-based AMD remediation as carbon dioxide is supplied by the heterotrophic bacteria whereas the microalgae provide the organic matter during their interaction in the MFC. During the process of Fe (II) oxidation, electrons are transferred directly to the anode by iron oxidizing bacterial members of the biofilm. This leads to production of larger currents (Nevin and Lovley 2000). Due to the acidic surroundings, there is high proton transfer to the cathode, thereby resulting in high pH at the cathode end of MFC (Lefebvre et al. 2011). The native algal population from AMD can be utilised at the cathode cell as it serves as a good acceptor of electrons. This would also result in reduction of substrate cost as sufficient algal biomass generated at the cathodic end could be exploited as substrate for the anodic compartment in MFC (Gajda et al. 2015).

18.3 Dairy Wastewater (DWW)

The dairy industry is one of the most economically significant industries in the agricultural sector. It experienced a drastic growth owing to a steady increase in demand of milk and milk products all around the globe (Chokshi et al. 2016). India shares about 13.1% of total milk produced in the world owing to the presence of various small- and large-scale dairy industries (Kothari et al. 2012). Water is used as a prime processing medium in these industries and is utilised for sanitation, cooling, cleaning, heating and floor washing, hence a large amount of wastewater is generated from these industries.

18.3.1 Sources and Characteristics of Dairy Wastewater

The dairy industry produces wastewater from two main activities, namely milk production and dairy farming (Kamarudin et al. 2015). The wastewater generated from dairy farming activity consists of both solid and liquid waste having the capacity to destroy the ecological balance of aquatic systems. This is due to the presence of livestock manure containing high concentrations of phosphorus and nitrogen (Staples et al. 1981). Various organic materials are also present in the dairy wastewater that are hazardous for environment and lead to a significant rise in chemical oxygen demand (COD) and biochemical oxygen demand (BOD) of water bodies (Ramasamy and Abbasi 2000). The wastewater is characterized by a pH of 4.7–11 (Passeggi et al. 2009) and presence of various metals like iron (Fe), nickel (Ni), manganese (Mn), sodium (Na), chlorine (Cl), potassium (K), calcium (Ca) and magnesium (Mg). The total nitrogen and phosphorus content vary as 14–830 mg/L (Rico et al. 1991) and 9–280 mg/L (Gavala et al. 1999), respectively, with the BOD and COD defined as 40–48,000 mg/L and 80–95,000 mg/L, respectively. Other nutrients, detergents, milk solids, lactose, fats and sanitizing agents (USDA-SCS 1992) are also present in DWW. The microalgal cultivation in these nutrient rich wastewater leads to an economic feasible production of algal biomass (Fig. 18.3) with simultaneous nutrient recovery through microalgal accumulation (Hena et al. 2015; Lu et al. 2016).

18.3.2 Methods for Treatment of Dairy Wastewater

Many physicochemical, mechanical and biological methods are present to treat various types of agro-industrial wastewaters (Liu et al. 2016). The widely employed treatment techniques are successful in removal of only organic pollutants without having any effect on inorganic pollutants (Markou and Georgakakis 2011). Also, in some cases, the amount of inorganic materials is increased by employment of certain kinds of treatment methods (Hansen et al. 1998). The utilisation of effluents from these treatment plants for the purpose of irrigation over longer periods has further demonstrated serious adverse effects on the crops (Hamilton et al. 2007). Hence,

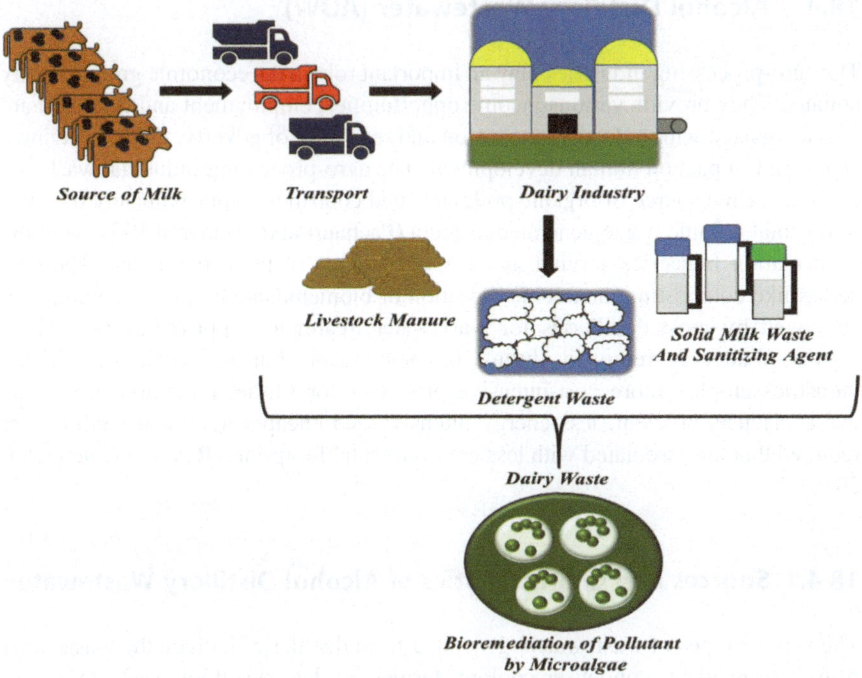

Fig. 18.3 The production of various dairy waste and its effective treatment by microalgae

environmentalists are now shifting to alternative treatment methods for dairy wastewater.

18.3.2.1 Microalgae-Assisted Remediation of DWW

Recently, phycoremediation of wastewater has gained importance due to the simultaneous biofuel production and wastewater remediation (Ogbonna et al. 2000; Lowrey et al. 2015). Since, secondary pollutants are not generated in the microalgal remediation, phycoremediation serves as an eco-friendly process with the benefit of reuse of generated algal biomass (Rawat et al. 2011; Posadas et al. 2014; Chen et al. 2015). The high-rate algal ponds (HRAPs) are known to possess great potential for dairy wastewater treatment (Craggs et al. 2004). The biomass productivity and the lipid yields are enhanced from microalgae cultivation in dairy industry effluents (Woertz et al. 2009; Kothari et al. 2012). Various freshwater chlorophytes employed for treatment of dairy wastewater are *Scenedesmus* sp. (Gentili 2014), *Chroococcus* sp. (Prajapati et al. 2014) and *Desmodesmus* sp. (Samorì et al. 2013).

18.4 Alcohol Distillery Wastewater (ADW)

The agro-processing industries play an important role in the economic growth of any country. They provide various income opportunities, employment and trade and are thus associated with industrial expansion and reduction of poverty, thereby having a significant impact on human development. The agro-processing industrial wastewater is the prime source of organic pollutants and contributes approximately 65–70% of the total organic waste generated in India (Pachauri and Sridharan 1998) with the fermentation industries serving as the major source of these pollutants. The processes like composting, anaerobic digestion or biomethanation and incineration are employed by most distilleries for wastewater treatment. Approximately 70% of organic matter is removed during biomethanation but the ethanol-producing industries employ more cost-intensive processes for further reduction of organic matter. Hence, efficient, less energy intensive and cheaper treatment methods are required that are associated with less environmental footprints (Ray and Ghangrekar 2019).

18.4.1 Sources and Characteristics of Alcohol Distillery Wastewater

The various types of wastewaters generated from distilleries include the wastewater from bottom plant, condenser cooling, fermenter cleaning, floor wash, fermenter cooling and spent wash (Pant and Adholeya 2007; Mohana et al. 2009; Chowdhary et al. 2017). These are collectively known as the alcohol distillery wastewater (ADW). For every litre of alcohol produced, approximately 8–15 L of ADW is generated (Saha et al. 2005). The factors like nature of substrate for alcohol production, wastewater handling and efficiency of process determine the concentration of pollutants and characteristics of wastewater generated from distilleries (Solovchenko 2019). The characteristics of the feedstock are attributed to presence of naturally occurring sugars, cellulose and starch. Stillage is a high strength wastewater that is produced from fermentation distilleries and possesses high percentage of organic or inorganic matters, contains unconvertible organic fractions and is characterised by a low pH value. The molasses stillage is characterised by dark-brown colour, inorganic impurities, BOD and COD as 40–65 g/L and 80–140 g/L, respectively. The grain stillage is described by a pH of 3.4–4.1 and COD as 40–60 g/L. It can lead to severe water and land pollution and therefore requires efficient treatment before being discharged into the environment (Ray and Ghangrekar 2019).

18.4.2 Methods for Treatment of ADW

18.4.2.1 Microalgae: An Evolving Technique for ADW Treatment

The phycoremediation of ADW emerged over the last two decades and involves utilisation of microalgae along with higher plants (Mata et al. 2012) and/or cyanobacteria and heterotrophic bacteria (Satyawali and Balakrishnan 2008) to achieve efficient bioremediation. The use of microalgae provides various benefits such as need of lower aeration in aerobic treatment. This is due to the oxidation of organic molecules by oxygen generated during microalgal photosynthesis which further supports the survival of heterotrophic bacteria (Muñoz and Guieysse 2006). The lower aeration further results in lower energy investment in this process. The simultaneous removal of nutrients (nitrogen and phosphorus) (Aslan and Kapdan 2006) and sequestration of carbon dioxide (Van Den Hende et al. 2012) by algal biomass serves as another crucial advantage offered by the process of phycoremediation. The chlorophytes, especially members of genus *Chlorella*, are mostly employed for ADW remediation owing to their mixotrophic nature and higher stress-tolerance abilities (Perez-Garcia et al. 2011; Alcántara et al. 2014). Besides the green microalgae, some species of cyanobacteria, particularly *Oscillatoria boryana*, is also employed for decolourisation of distillery spent wash which involves utilisation of melanoidin (a recalcitrant biopolymer) as a source of carbon and nitrogen by the microorganism (Kalavathi et al. 2001). As demonstrated by Patel et al. 2001, *Oscillatoria* sp., *Lyngbya* sp. and *Synechocystis* sp. exhibit 96%, 81% and 26% decolorization efficiency, respectively, in ADW treatment. The treatment efficiency of a microalgae and an aquatic plant was combined by Valderrama et al. (2002). Most of the organic matter was removed by *Chlorella vulgaris* whereas the aquatic plant *Lemna minuscule* was employed for polishing treatment of the effluent. Further, the microalgae have the ability to acquire large amount of nutrients from their surroundings by process called "luxury uptake" (De Mazancourt and Schwartz 2012). This feature acts as a beneficial aspect because it helps in removal of chemically bound phosphorus and nitrogen (Olguín 2003) from the wastewater, thereby negating the chances of eutrophication, algal blooms, hypoxia and biodiversity loss (Smil 2000; Anderson et al. 2002). Thus, algal cultivation in ADW serves as an efficient approach for nutrient removal from wastewater (Goswami et al. 2021a; Coppens et al. 2014; Solovchenko et al. 2016). Also, the generated biomass exhibits a great potential to serve as a renewable feedstock for production of various biofuels (Hu et al. 2008; Chisti 2010) as the pelleted dry biomass can be utilised for production of solid form of fuel, biogas can be obtained from anaerobic digestion whereas biodiesel and other liquid biofuels can be produced from carbohydrate-enriched algal biomass (Georgianna and Mayfield 2012). Other value-added products like astaxanthin (Dhankhar et al. 2012), essential polyunsaturated fatty acids, e.g., arachidonic (Crawford et al. 2003), eicosapentaenoic acid, linolenic (Wang et al. 2012), and other long-chain fatty acids (Cohen and Khozin-Goldberg 2010) and β-carotene (Goswami et al. 2021b; Takaichi 2011) can also be produced from ADW-grown microalgae.

18.5 Domestic Wastewater

The domestic wastewater serves as a cost-effective and easily available nutrient medium for microalgal cultivation for phycoremediation and biofuel production. Also, the nutrients can be recycled by utilisation of de-oiled algal biomass as fertilizer (Xin et al. 2010; Arora et al. 2016). The phosphorus content, nitrogen content and COD of various wastewaters have been reduced by phycoremediation and coupled biodiesel production employing members of genera *Chlorella*, *Chlamydomonas* and *Scenedesmus* (Feng et al. 2011; Pittman et al. 2011; Kothari et al. 2012).

18.5.1 Sources and Characteristics of Domestic Wastewater

The wastewater which is generated from the human activities in households is called as the domestic wastewater. It consists of greywater (from bathing, washing, kitchen) and blackwater (from toilets). The physical characteristics of domestic wastewater are described in terms of its temperature, pH, odour and colour (Rawat et al. 2011). The age of wastewater is represented by its colour. Owing to the presence of different kinds of suspended, dissolved material and various hydrogen solids, the domestic wastewater possesses a peculiar odour (Metcalf and Eddy 1987). Various factors like pH, saturation level of gases, alkalinity and conductivity are determined by the temperature of wastewater. The biological reactions of the aquatic organisms and other chemical characteristics are altered by the wastewater temperature. Also, the population of undesirable fungi and planktonic organisms can be increased by a rise in the temperature. The organic material consisting of carbon, oxygen, hydrogen and other metals like phosphorus, ammonia, (Jorgensen and Weatherley 2003) iron or sulphur constitutes a large proportion of the domestic wastewater with lipids, proteins, oils, carbohydrates and urea being the principal components. Chloride, iron, hydrogen, phosphorus, sulphur, nitrogen and heavy metals form the inorganic fraction of the wastewater (Muttamara 1996) while the biological characteristics include the various aquatic animals and other species of micro- and macro-organisms. This water serves as an appropriate growth medium for microbes in both aerobic and anaerobic treatment setups (Abeliovich 1986).

18.5.2 Methods for Treatment of Domestic Wastewater

18.5.2.1 Phycoremediation for Nutrient Removal from Domestic Wastewater

The efficient and promising bioremediation of domestic wastewater by employment of various algal species like *Scenedesmus*, *Chlorella*, *Phormidium*, *Chlamydomonas*, *Botryococcus* and *Spirulina* have been reported by Kong et al. (2010), and Rawat et al. (2011). A consortium of 15 native microalgal species was demonstrated to effectively remediate the wastewater as shown by Chinnasamy et al. (2010). These species achieved a 96% nutrient removal rate with 9.2–17.8 tons/ha/year biomass production and 6.82% lipid content in treated wastewater. Biodiesel could be formed from approximately 63.9% of the obtained algal oil. When the phycoremediation was done for short cultivation periods, a rapid decline in the amounts of nitrates, metals, and phosphate was observed which signifies that efficient nutrient removal from domestic wastewater could be attained with microalgae (Wang et al. 2010). They play a prime role during the tertiary treatment stages in maturation ponds or in facultative and aerobic ponds that are responsible for treatment of small- to middle-scale domestic wastewater. It is also observed that the starved microalgae were more efficient at nitrogen uptake as compared to the well-fed ones (Oswald 1988). The space requirements for phycoremediation of wastewater were shown to be reduced with the employment of these hyper-concentrated algal cultures, known as the 'activated sludge' and appropriate nutrient removal was attained in a very short time; within an hour (Lavoie and De la Noue 1985). The success of phycoremediation depends upon the species involved and the prevalent local environmental conditions. The large amounts of phosphorus and nitrogen present in the wastewater are used by the microalgal strains for synthesis of proteins, nucleic acids and phospholipids. Ammonia precipitation or ammonia stripping can increase the pH associated with photosynthesis, thereby further enhancing the rate of nutrient removal (Oswald 2003).

18.6 Greywater

The untreated wastewater discharge from baths, laundry, wash basin, showers, dishwasher, washing machines, kitchen sinks, school excluding toilet waste and office buildings (Mohamed et al. 2014) is collectively called as the greywater which contributes up to 70% of municipal wastewater by volume (Friedler 2004). The issue of freshwater scarcity can be dealt with utilisation of greywater as the alternative water source. The consumption of potable water can be reduced by 29–47% by utilisation of greywater for irrigation purposes, toilet flushing and floor washing in various households (Al-Jayyousi 2003; Hourlier et al. 2010; Galvis et al. 2014). Due to high amount of organic and inorganic contaminants present, the discharge of greywater is a major environmental issue with severe negative impacts on water bodies and the environment (Mohamed et al. 2017; Jais et al. 2017).

18.6.1 Sources and Characteristics of Greywater

The physicochemical properties of greywater are defined in terms of temperature, colour, odour, organic and inorganic constituents (Atiku et al. 2016). The colour of greywater is dependent upon its age with greyish appearance present during the storage period. Due to various biological reactions between different constituents of the wastewater, the peculiar odours are produced (Rawat et al. 2011). The alkalinity, saturation level of gases, conductivity and pH are determined by the temperature of greywater. Its chemical characteristics are dependent upon the source of production of wastewater. Proteins, carbohydrates, fats and volatile acids comprise approximately 70% of the organic carbon present in wastewater (Abdel-Raouf et al. 2012). Higher contents of proteins, lipids, cooking oils, phosphorus (50–70 mg/L) and nitrogen (20–40 mg/L) are found in the kitchen waste. The blood in meats which is washed in the sinks of kitchen serves as the prime source of nitrogen in wastewater while the nappies washed in the bathrooms discharge nitrate and phosphorus is released into the wastewater from soap and detergent contents used in the houses (Maimon et al. 2010; Donner et al. 2010). The obnoxious odours and turbidity are a result of grease and oil from the cooking oils used in the kitchen (Abid-Baig et al. 2003). A large number of chemical reactions between the pollutants result in a higher COD and BOD. The dissolved oxygen of the receiving water is reduced by high BOD concentrations which further leads to lowering of pH values, thereby inhibiting microbial growth and causing death of aquatic animals. High amounts of suspended solids are also present in the greywater due to various activities like cleaning dirty floors and clothes washing. Calcium, sodium, magnesium, sulphur, potassium, phosphate, chlorine, ammonium salts, heavy metals (zinc, cadmium, cobalt, nickel, iron, silver, mercury, copper, arsenic) and bicarbonate comprise the inorganic portion of greywater (Lim et al. 2010). The detergents are primarily responsible for release of the aforementioned heavy metals in greywater (Ledin et al. 2001; Leal et al. 2007).

18.6.2 Methods for Treatment of Greywater

18.6.2.1 Recycle of Greywater for Production of Microalgal Biomass

The physical, chemical and biological characteristics of greywater determine the potential of recycling it as a cultivation medium for the microalgal biomass. The nutrients namely nitrogen, phosphorus and other trace elements present in greywater are the prime factors for the microalgal growth and biomass production (Pahazri et al. 2016). Other vital parameters like temperature, light intensity and pH can be adjusted. Al-Gheethi et al. (2017), have studied the interaction between the bacterial and microalgal populations present in the greywater. The microalgal growth was enhanced by the carbon dioxide released from the bacterial cells whereas some microalgal strains have antibacterial abilities and thus inhibit bacterial growth. Similarly, some bacterial species also possess algicidal properties. Thus, selection of appropriate microalgal species is a key process in the use of greywater as growth medium for algal cultivation. Further, the utilisation of algal biomass can be limited by the presence of pathogenic bacteria that get harvested along with the algal biomass. This can be solved by sterilising the greywater before recycling (Al-Gheethi et al. 2019). The environmental conditions also determine the growth of microalgae in wastewaters (Wurochekke et al. 2019). Thus, most efficient treatment is obtained by the employment of native algal strains since they are already acclimatised to the harsh environmental conditions and thus have high survival capabilities relative to the other microbial populations present in greywater. The commonly employed microalgal species include *Botryococcus braunii*, *Scenedesmus dimorphic*, *Chlorella vulgaris*, *Phormidium* sp. and *Spirulina* sp. The nutrient removal can also be increased by starvation of the algal cells as nutrient-deficient conditions would induce the cells to go into a dormant state. Thus, phycoremediation can be enhanced under conditions of nutrient deficiency (Mohamed et al. 2017).

18.7 Heavy Metals (HMs)

There has been a drastic rise in the concentration of heavy metals in the environment due to rapid industrialisation and urbanisation, thus their effective remediation has now become a matter of global concern. HMs are described as elements with high molecular weight (Jais et al. 2017) possessing potential toxicity at even low concentrations (Ahmad et al. 2020; Kamal et al. 2010). Some HMs like Fe^{2+}, Mn^{2+}, Ni^{2+}, Cu^{2+}, $Mo^{2+}$and Zn^{2+} are crucial for the microalgae in low concentrations and are therefore called as trace elements while others like Al^{3+}, Sn^{2+}, Au^{3+}, Cd^{2+}, Pb^{2+} and Sr^{2+} do not possess any biological function and are thus toxic in nature (Nies 1999; Rathnayake et al. 1999). Their atomic density is greater than 4 g/cm^3, belong to the category of metalloids or metals (Kumar et al. 2015) and most are present as natural constituents of earth crust (Ahmad et al. 2020).

18.7.1 Sources and Characteristics of Heavy Metals

The common heavy metal contaminants present in the wastewater include cadmium, mercury, copper, chromium and zinc (Pathak et al. 2019; Thakare et al. 2021). The anthropogenic activities are mostly responsible for the release of these conservative pollutants in the water bodies. They demonstrate a Lewis acid behaviour and cannot be degraded by the metabolic pathways; hence, they have a tendency to bioaccumulate inside the living organisms. They adversely affect the fauna and flora of the environment. The various negative impacts on human health are represented in Table 18.1. They have various negative impacts like they lead to a reduction in the enzymatic activity, decrease production of chlorophyll and inhibit seed germination and photosynthesis (Ahmad et al. 2020). This demands for adequate treatment of wastewater laden with various heavy metals.

Table 18.1 Effects of common heavy metals on human health

Heavy metal	Effects	References
Arsenic (As)	Effects various organ systems like respiratory, nervous, cardiovascular, immune, hepatic, endocrine, renal and reproductive system	Abdul et al. (2015)
Cadmium (Cd)	Kidney and bone damage, carcinogen	Godt et al. (2006)
Copper (Cu)	Liver disease, Alzheimer's disease, neurological defects and oxidative stress	Uriu-Adams and Keen (2005)
Lead (Pb)	Encephalopathy, renal disease, anaemia, gastrointestinal diseases, cognitive and behavioural effects and reproductive system damage	Goyer (1990)
Mercury (Hg)	Kidney diseases, rheumatoid arthritis, effects on circulatory and nervous system	Clarkson (1993)
Zinc (Zn)	Hunger loss, neurological problems	Plum et al. (2010)

18.7.2 Methods for Treatment of Heavy Metal Wastewater

18.7.2.1 Phycoremediation and the Underlying Mechanisms for Removal of HMs

The process of phycoremediation has emerged as an eco-friendly treatment method and relies on the utilisation of microalgae for nutrient removal to treat wastewater. This nutrient removal is achieved by assimilation process whereas microalgal strains employ biosorption and bioaccumulation for the removal of heavy metals (Jais et al. 2017). Various microalgal species that represent high affinities for polyvalent metals are utilized in the treatment processes (de Bashan and Bashan 2010). Magnesium, potassium, calcium, cobalt, strontium, zinc, arsenic, lead, aluminium, molybdenum, vanadium, nickel, iron, copper and manganese are the HMs which are commonly removed by microalgae of the genus *Chlamydomonas* sp., *Chlorella* sp., *Chlorococcum* sp., *Scenedesmus* sp., *Cyclotella* sp., *Spirogyra* sp., *Lyngbya* sp., *Spirulina platensis* and *Stigeoclonium* sp. (Brinza et al. 2007).

Two main mechanisms are utilised by microalgae for removal of HMs namely bioaccumulation living cells and biosorption by the dead biomass. Different methods are utilised by the microalgal cells for uptake of HMs which are then metabolised by various pathways (Ajayan et al. 2011). Biosorption is the most commonly employed method which further consists of two steps as demonstrated by Monteiro et al. (2012). The first stage is characterised by a passive and rapid removal of heavy metals taking place at the microalgal cell surface while the second stage occurs inside the cell and is relatively slow as compared to the first one. In the first process, electrostatic interactions are employed for adsorption of HMs to the functional groups present on the surface of microalgal cells. Various processes like ion exchange, physical adsorption, coordination, chelation, complexation, chemisorption, entrapment, diffusion and microprecipitation are involved in this non-metabolic, reversible process that can occur inside both living or non-living cells. The second metabolism dependent and irreversible process is confined to the living cells. In this process, metal ions are transported across the cell membrane barrier followed by their accumulation inside the cells (Monteiro et al. 2011). The HMs can also be eliminated by non-viable microalgae in a relatively shorter duration as compared to the living cells. Further, toxicity problems are not generated by the dead biomass, and their absorption capability is slightly lower than the living cells. Owing to the lower costs associated with use of dead biomass, this process is widely employed in the bioremediation processes (Sandau et al. 1996). The kind of heavy metal ion, the algal species, the prevalent conditions and nature of biological system involved determine the metal biosorption efficiency. This method involves intracellular and extracellular metal binding (Aksu 1998). The non-living algal biomass can thus serve as a potential bioremediation agent for removal of toxic heavy metals from wastewaters.

18.8 Textile Wastewater (TWW)

The wastewater generated from various textile industries acts as a major source of water pollution (Baban et al. 2010). To impart appropriate quality to the fabrics, various dyes like basic or acidic dyes, reactive dyes, diazo dyes, azo dyes, metal-complex dyes and anthraquinone-based dyes are employed in these industries. These dyes are hazardous and toxic in nature and have severe adverse impact on the biological activity of aquatic organisms. Thus, efficient treatment of TWW is required before it is being discharged into the environment (Pathak et al. 2014).

18.8.1 Sources and Characteristics of Textile Wastewater

The most hazardous components of TWW that demand appropriate treatment are the dyes that are utilized during various stages in these industries. In a textile industry, a cloth passes through various stages, namely sizing, scouring, bleaching, dying, printing followed by the last stage, i.e., finishing and large amounts of wastewaters are generated at each step (Pathak et al. 2014). The physiochemical characteristics of TWW are described in terms of its colour, odour, temperature, salinity, pH, BOD, COD, total dissolved solids (TDS), total nitrogen (TN) and phosphorus (TP) content (Fazal et al. 2018). The TWW has a strong odour, dark colour and consists various organic compounds that are non-biodegradable in nature (Mantzavinos and Psillakis 2004). Some heavy metals like copper, chromium, arsenic and zinc are also present in TWW (Nicolo et al. 2016; Rajasimman et al. 2017). The COD and BOD values vary according to the dyes used because each dye and its metabolite possess a different structure (Zollinger 2003). Also, TN and TP are found in the range as 21–57 mg/L and 1.0–9.7 mg/L, respectively, in TWW (Cai et al. 2013).

18.8.2 Methods for Treatment of Textile Wastewater

18.8.2.1 Physical and Chemical Treatment Methods for TWW

Various physical and chemical methods employed for treatment of TWW are electro flocculation, flotation, membrane filtration, electrochemical destruction, electro kinetic coagulation, ion-exchange, precipitation, irradiation and ozonation (Pathak et al. 2014). The use of these methods is limited by their lower treatment efficiency, incurred higher costs, not environment friendly and inappropriateness to treat different kinds of dye wastewater (Robinson et al. 2001). This has led to a shift towards the biological approaches and use of microalgae for remediation purposes.

18.8.2.2 Microalgae-Assisted TWW Remediation

Various microalgal strains like *Chlorella vulgaris*, *Scenedesmus* sp., *Chlorella pyrenoidosa*, *Oscillatoria tenuisin* and *Spirogyra* sp. have been employed for phycoremediation of TWW (Andrade et al. 2018; Andrade and Andrade 2018). Some common dyes, and the microalgal strains employed for their bioremediation

Table 18.2 Microalgal strains employed for remediation of textile dyes

Dyes	Microalgal strains	References
Basic green 4	*Chlamydomonas* sp.	Khataee et al. (2009)
Congo red	*Chlorella vulgaris*	Hernandez-Zamora et al. (2015)
Malachite green	*Chlorella* sp., *Cosmarium* sp.	Khataee et al. (2010)
Methylene blue	*Dunaliella salina*	Abd-El-Kareema and Tahab (2012)
Orange G	*Acutodesmus obliques*	Sarwa and Verma (2013)
Remazol blue	*Phormidium* sp.	Sadettin and Dönmez (2007)
Rhodamine B	*Coelastrella* sp.	Baldev et al. (2013)

are represented in Table 18.2. Microalgae have the potential for simultaneous wastewater treatment, biofuel production, carbon dioxide mitigation and production of high-added value products like vitamins, pigments, polyunsaturated fatty acids and antioxidants (Garcia Segura et al. 2018). They utilize the nutrients and dyes present in TWW for its growth.

Two main mechanisms used by microalgae for TWW remediation are bioaccumulation (or bioconversion) and biosorption process (Fazal et al. 2018). The dyes are used as carbon source and further converted into metabolites by algal species in the bioconversion method whereas the dyes get absorbed to surface of algal cells in biosorption. These mechanisms can be employed simultaneously for TWW remediation (Chu et al. 2009). Both living cells and dead biomass can be utilized for treatment and colour removal (Forgacs et al. 2004; David Noel et al. 2014) from wastewater. However, dead biomass can remove dyes only by the process of adsorption. The living and non-viable biomass of *Spirogyra* sp. was shown to effectively remove reactive dye, Synazol, from TWW in a study done by Khalaf 2008. The basic dyes are removed effectively by living cells of *Caulerpa scalpelliformis* and *Caulerpa lentillifera* by the mechanism of biosorption (Marungrueng and Pavasant 2006; Aravindhan et al. 2007). The azo dye tectilon yellow 2G was shown to be removed by *Chlorella vulgaris* which converts it into aniline with a removal efficiency of 63–69% (Acuner and Dilek 2004). Similarly, biosorption can also be used for TWW remediation. For example, *Spirulina platensis* served as a biosorbent for the removal of dye reactive red (RR-120) with the maximum biosorption capacity of 482.2 mg/g and attained 97% removal efficiency (Cardoso et al. 2012). Malachite green was shown to be removed from biomass of *Cosmarium* sp. (Daneshvar et al. 2007) while *Scenedesmus quadricauda* had ability to eliminate remazol brilliant blue R (RBBR) (Ergene et al. 2009).

18.9 Limitations and Future Prospects of Phycoremediation of Wastewater

The use of phycoremediation for different kinds of wastewater is limited by various drawbacks like the cost-effectiveness of biomass harvestation, presence of other micropollutants in wastewater and efficiency of algae-mediated wastewater treatment in cold climatic conditions (Lavrinovics and Juhna 2017). The microscopic size of algal cells, very low amount of dry weight in total suspension and the negatively charged cell surface are the prime factors that complicate the harvest of microalgal cells, thereby increasing the cost associated with biomass harvest and wastewater treatment (Grima et al. 2003; Milledge and Heaven 2013). This can be overcome by utilisation of artificial aquatic food-web for harvesting biomass as this technique consumes lower energy and is thus a cost-effective approach. Besides various organic and inorganic materials, other micropollutants like pharmaceutical compounds, pathogenic bacteria, household chemistry are also present in the wastewaters. They pass through the conventional treatment systems and thus get released into the water bodies, thereby harming the aquatic ecosystem (Schwarzenbach et al. 2006). The application of microalgae for remediation of such micropollutants is still under study and needs further research (Ansa et al. 2012; Mani and Kumar 2014). Further, the low temperature conditions and shorter daylight hours limit the effectiveness of phycoremediation in colder environments. Hence, the use of algae for removal of pollutants from wastewater in temperate and cold climatic conditions demands more research (Lavrinovics and Juhna 2017). Thus, more research is required to explore the techniques for obtaining high biomass to achieve cost-effective commercialisation of treatment processes and other industrial applications.

Competing Interests All the authors declare that they have no competing interests.

References

Abd-El-Kareema MS, Tahab HM (2012) Decolorization of malachite green and methylene blue by two microalgal species. Int J Chem Environ Eng 3:297–302

Abdel-Raouf N, Al-Homaidan AA, Ibraheem IBM (2012) Microalgae and wastewater treatment. Saudi J Biol Sci 19:257–275. https://doi.org/10.1016/j.sjbs.2012.04.005

Abdul KSM, Jayasinghe SS, Chandana EP, Jayasumana C, De Silva PMC (2015) Arsenic and human health effects: a review. Environ Toxicol Pharmacol 40:828–846. https://doi.org/10.1016/j.etap.2015.09.016

Abeliovich A (1986) Algae in wastewater oxidation ponds. In: Richmond A (ed) Handbook of microbial mass culture. CRC, Boca Raton, pp 331–338

Abid-Baig M, Mir M, Bahatti ZL, Baig MA (2003) Removal of oil and grease from industrial effluents. EJEAF Che 2:577–585

Abinandan S, Subashchandrabose SR, Venkateswarlu K, Megharaj M (2018) Microalgae-bacteria biofilms: a sustainable synergistic approach in remediation of acid mine drainage. Appl Microbiol Biotechnol 102:1131–1144. https://doi.org/10.1007/s00253-017-8693-7

Acuner E, Dilek FB (2004) Treatment of tectilon yellow 2G by *Chlorella vulgaris*. Process Biochem 39:623–663. https://doi.org/10.1016/S0032-9592(03)00138-9

Afkar A, Ababna H, Fathi AA (2010) Toxicological response of the green alga *Chlorella vulgaris* to some heavy metals. Am J Environ Sci 6:230–237

Agrawal K, Verma P (2022) An overview of various algal biomolecules and its applications. In: An integration of phycoremediation processes in wastewater treatment. Elsevier, pp 249–270. https://doi.org/10.1016/B978-0-12-823499-0.00006-7

Agrawal K, Bhatt A, Bhardwaj N, Kumar B, Verma P (2020) Algal biomass: potential renewable feedstock for biofuels production–part I. In: Biofuel production technologies: critical analysis for sustainability. Springer, Singapore, pp 203–237. https://doi.org/10.1007/978-981-13-8637-4_8

Ahmad S, Pandey A, Pathak VV, Tyagi VV, Kothari R (2020) Phycoremediation: algae as eco-friendly tools for the removal of heavy metals from wastewaters. In: Bioremediation of industrial waste for environmental safety. Springer, Singapore, pp 53–76. https://doi.org/10.1007/978-981-13-3426-9_3

Ajayan KV, Selvaraju M, Thirugnanamoorthy K (2011) Growth and heavy metals accumulation potential of microalgae grown in sewage wastewater and petrochemical effluents. Pak J Biol Sci 14:805–811. https://doi.org/10.3923/pjbs.2011805.811

Aksu Z (1998) Biosorption of heavy metals by microalgae in batch and continuous systems. In: Wong YS, Tam NFY (eds) Algae for waste water treatment. Springer, Berlin, pp 37–53. https://doi.org/10.1007/978-3-662-10863-5_3

Alcántara C, Fernández C, García-Encina P, Muñoz R (2014) Mixotrophic metabolism of *Chlorella sorokiniana* and algal-bacterial consortia under extended dark-light periods and nutrient starvation. Appl Microbiol Biotechnol 1–12:6125–6125. https://doi.org/10.1007/s00253-014-6125-5

Al-Gheethi AA, Mohamed RM, Jais NM, Efaq AN, Abd Halid A, Wurochekke AA, Amir-Hashim MK (2017) Influence of pathogenic bacterial activity on growth of *Scenedesmus* sp. and removal of nutrients from public market wastewater. J Water Health 15:741–756. https://doi.org/10.2166/wh.2017.080

Al-Gheethi AAS, Noman EA, Mohamed RMSR, Apandi NM, Yaakob MA, Pahazri F, Kassim AHM (2019) Recycle of greywater for microalgae biomass production. In: Management of greywater in developing countries. Springer, Cham, pp 205–226. https://doi.org/10.1007/978-3-319-90269-2_11

Al-Jayyousi OR (2003) Greywater reuse: towards sustainable water management. Desalination 56:181–192. https://doi.org/10.1016/S0011-9164(03)00340-0

Al-Shwafi NA, Rushdi AI (2008) Heavy metal concentrations in marine green brown, and red seaweeds from coastal waters of Yemen, the Gulf of Aden. Environ Geol 55:653–660. https://doi.org/10.1007/s00254-007-1015-0

Anastopoulos I, Kyzas GZ (2015) Progress in batch biosorption of heavy metals onto algae. J Mol Liq 209:77–86. https://doi.org/10.1016/j.molliq.2015.05.023

Anderson DM, Glibert PM, Burkholder JM (2002) Harmful algal blooms and eutrophication: nutrient sources, composition, and consequences. Estuaries 25:704–726. https://doi.org/10.1007/BF02804901

Andrade CJ, Andrade LM (2018) Microalgae for bioremediation of textile wastewater. An overview. MOJ Food Process Technol 6:432–433

Andrade LM, Andrade CJ, Dias M, Nascimento CAO, Mendes MA (2018) *Chlorella* and *Spirulina* microalgae as sources of functional foods, nutraceuticals, and food supplements; an overview. MOJ Food Process Technol 6:1–14

Ansa EDO, Lubberding HJ, Gijzen HJ (2012) The effect of algal biomass on the removal of faecal coliform from domestic wastewater. Appl Water Sci 2:87–94. https://doi.org/10.1007/s13201-011-0025-y

Aravindhan R, Rao JR, Nair BU (2007) Removal of basic yellow dye from aqueous solution by sorption on green alga *Caulerpa scalpelliformis*. J Hazard Mater 142:68–76. https://doi.org/10.1016/j.jhazmat.2006.07.058

Arora N, Patel A, Sartaj K, Pruthi PA, Pruthi V (2016) Bioremediation of domestic and industrial wastewaters integrated with enhanced biodiesel production using novel oleaginous microalgae. Environ Sci Pollut Res 23:20997–21007. https://doi.org/10.1007/s11356-016-7320-y

Aslan S, Kapdan IK (2006) Batch kinetics of nitrogen and phosphorus removal from synthetic wastewater by algae. Ecol Eng 28:64–70. https://doi.org/10.1016/j.ecoleng.2006.04.003

Atiku H, Mohamed RMSR, Al-Gheethi AA, Wurochekke AA, Kassim AHM (2016) Harvesting of microalgae biomass from the phycoremediation process of greywater. Environ Sci Pollut Res 23:24624–24641. https://doi.org/10.1007/s11356-016-7456-9

Aziz N, Prasad R, Ibrahim AIM, Ahmed AIS (2017) Promising applications for the production of biofuels through algae. In: Patra JK, Vishnuprasad CN, Das G (eds) Microbial biotechnology. Springer Nature Singapore, Singapore, pp 81–103

Baban A, Yediler A, Ciliz NK (2010) Integrated water management and CP implementation for wool and textile blend processes. Clean 38(1):84–90. https://doi.org/10.1002/clen.200900102

Baldev E, Mubarak Ali D, Ilavarasi A, Pandiaraj D, Ishack KSS, Thajuddin N (2013) Degradation of synthetic dye, Rhodamine B to environmentally non-toxic products using microalgae. Colloids Surf B Biointerfaces 105:207–214. https://doi.org/10.1016/j.colsurfb.2013.01.008

Bhardwaj N, Agrawal K, Verma P (2020) Algal biofuels: an economic and effective alternative of fossil fuels. In: Microbial strategies for techno-economic biofuel production. Springer, Singapore, pp 207–227. https://doi.org/10.1007/978-981-15-7190-9_7

Bhattacharya J, Islam M, Cheong YW (2006) Microbial growth and action: implications for passive bioremediation of acid mine drainage. Mine Water Environ 25:233–240

Bilal M, Shah JA, Ashfaq T, Gardazi SMH, Tahir AA, Pervez A, Haroon H, Mahmood Q (2013) Waste biomass adsorbents for copper removal from industrial wastewater—a review. J Hazard Mater 263:322–333. https://doi.org/10.1016/j.jhazmat.2013.07.071

Brake SS, Hasiotis ST, Dannely HK (2004) Diatoms in acid mine drainage and their role in the formation of iron-rich stromatolites. Geomicrobiol J 21:331–340. https://doi.org/10.1080/01490450490454074

Brinza L, Dring MJ, Gavrilescu M (2007) Marine micro and macroalgal species as biosorbents for heavy metals. Environ Eng Manag J 6(3):237–251

Brock TD (1973) Lower pH limit for the existence of blue-green algae: evolutionary and ecological implications. Science 179:480–483. https://doi.org/10.1126/science.179.4072.480

Bwapwa JK, Jaiyeola AT, Chetty R (2017) Bioremediation of acid mine drainage using algae strains: a review. S Afr J Chem Eng 24:62–70. https://doi.org/10.1016/j.sajce.2017.06.005

Cai T, Park SY, Li Y (2013) Nutrient recovery from wastewater streams by microalgae: status and prospects. Renew Sustain Energy Rev 19:360–369. https://doi.org/10.1016/j.rser.2012.11.030

Cardoso NF, Lima EC, Royer B, Bach MV, Dotto GL, Pinto LA, Calvete T (2012) Comparison of *Spirulina platensis* microalgae and commercial activated carbon as adsorbents for the removal of reactive red 120 dye from aqueous effluents. J Hazard Mater 241:146–153. https://doi.org/10.1016/j.jhazmat.2012.09.026

Chekroun KB, Baghour M (2013) The role of algae in phytoremediation of heavy metals: a review. J Mater Environ Sci 4:873–880

Chen C, Chang H, Kao P, Pan J, Chang J (2012) Biosorption of cadmium by CO_2-fixing microalga *Scenedesmus obliquus* CNWN. Bioresour Technol 105:74–80. https://doi.org/10.1016/j.biortech.2011.11.124

Chen G, Zhao L, Qi Y (2015) Enhancing the productivity of microalgae cultivated in wastewater toward biofuel production: a critical review. Appl Energy 137:282–291. https://doi.org/10.1016/j.apenergy.2014.10.032

Cheng S, Dempsey BA, Logan BE (2007) Electricity generation from synthetic acid-mine drainage (AMD) water using fuel cell technologies. Environ Sci Technol 41(23):8149–8153. https://doi.org/10.1021/es0712221

Chinnasamy S, Bhatnagar A, Hunt RW, Das KC (2010) Microalgae cultivation in a wastewater dominated by carpet mill effluents for biofuel applications. Bioresour Technol 101:3097–3105. https://doi.org/10.1016/j.biortech.2009.12.026

Chisti Y (2010) Fuels from microalgae. Biofuels 1:233–235. https://doi.org/10.4155/bfs.10.9

Chokshi K, Pancha I, Ghosh A, Mishra S (2016) Microalgal biomass generation by phycoremediation of dairy industry wastewater: an integrated approach towards sustainable biofuel production. Bioresour Technol 221:455–460. https://doi.org/10.1016/j.biortech.2016.09.070

Chowdhary P, Yadav A, Kaithwas G, Bharagava RN (2017) Distillery wastewater: a major source of environmental pollution and its biological treatment for environmental safety. In: Singh R, Kumar S (eds) Green technologies and environmental sustainability. Springer International, Cham, pp 409–435. https://doi.org/10.1007/978-3-319-50654-8_18

Chu WL, See YC, Phang SM (2009) Use of immobilised *Chlorella vulgaris* for the removal of colour from textile dyes. J Appl Phycol 21:641–648. https://doi.org/10.1007/s10811-008-9396-3

Clarkson TW (1993) Mercury: major issues in environmental health. Environ Health Perspect 100: 31–38. https://doi.org/10.1289/ehp.9310031

Cohen Z, Khozin-Goldberg I (2010) Searching for PUFA-rich microalgae. In: Cohen Z, Ratledge C (eds) Single cell oils, 2nd edn. American Oil Chemists' Society, Champaign, pp 201–224. https://doi.org/10.1016/B978-1-893997-73-8.50014-1

Coppens J, Decostere B, Van Hulle S et al (2014) Kinetic exploration of nitrate-accumulating microalgae for nutrient recovery. Appl Microbiol Biotechnol 98:8377–8387. https://doi.org/10.1007/s00253-014-5854-9

Costello C (2003) Acid mine drainage: innovative treatment technologies. US Environmental Protection Agency Office of Solid Waste and Emergency Response, Washington, DC

Craggs RJ, Sukias JP, Tanner CT, Davies-Colley RJ (2004) Advanced pond system for dairy-farm effluent treatment. N Z J Agric Res 47:449–460. https://doi.org/10.1080/00288233.2004.9513613

Crawford MA, Golfetto I, Ghebremeskel K, Min Y, Moodley T, Poston L, Phylactos A, Cunnane S, Schmidt W (2003) The potential role for arachidonic and docosahexaenoic acids in protection against some central nervous system injuries in preterm infants. Lipids 38:303–315. https://doi.org/10.1007/s11745-003-1065-1

Daneshvar N, Ayazloo M, Khataee AR, Pourhassan M (2007) Biological decolorization of dye solution containing malachite green by microalgae *Cosmarium* sp. Bioresour Technol 98:1176–1182. https://doi.org/10.1016/j.biortech.2006.05.025

David Noel S, Rajan MR, Sivakumar P (2014) Cyanobacteria as a potential source of phycoremediation from textile industry effluent. J Microbiol Biotechnol Res 4:30–35. https://doi.org/10.4172/2155-6199.1000260

De Bashan LE, Bashan Y (2010) Immobilized microalgae for removing pollutants: review of practical aspects. Bioresour Technol 101:1611–1627. https://doi.org/10.1016/j.biortech.2009.09.043

De Mazancourt C, Schwartz MW (2012) Starve a competitor: evolution of luxury consumption as a competitive strategy. Theor Ecol 5:37–49. https://doi.org/10.1007/s12080-010-0094-9

Dhankhar J, Kadian SS, Sharma A (2012) Astaxanthin: a potential carotenoid. Int J Pharm Sci Res 3:1246–1125

Donner E, Eriksson E, Revitt DM, Scholes L, Holten LHC, Ledin A (2010) Presence and fate of priority substances in domestic greywater treatment and reuse systems. Sci Total Environ 408: 2444–2451. https://doi.org/10.1016/j.scitotenv.2010.02.033

Edmundson SJ, Wilkie AC (2013) Landfill leachate—a water and nutrient resource for algae-based biofuels. Environ Technol 34:1849–1857. https://doi.org/10.1080/09593330.2013.826256

Elbaz-Poulichet F, Dupuy C, Cruzado A, Velasqaez Z, Achterberg EP, Braungardt CB (2000) Influence of sorption process by iron oxides and algae fixation on arsenic and phosphate cycle in an acidic estuary (Tinto River, Spain). Water Res 34:3222–3230. https://doi.org/10.1016/S0043-1354(00)00073-7

Ergene A, Ada K, Tan S, Katircioğlu H (2009) Removal of Remazol brilliant blue R dye from aqueous solutions by adsorption onto immobilized *Scenedesmus quadricauda*: equilibrium and kinetic modeling studies. Desalination 249:1308–1314. https://doi.org/10.1016/j.desal.2009.06.027

Fazal T, Mushtaq A, Rehman F, Khan AU, Rashid N, Farooq W, Rehman MSU, Xu J (2018) Bioremediation of textile wastewater and successive biodiesel production using microalgae. Renew Sustain Energy Rev 82:3107–3126. https://doi.org/10.1016/j.rser.2017.10.029

Feng Y, Li C, Zhang D (2011) Lipid production of *Chlorella vulgaris* cultured in artificial wastewater medium. Bioresour Technol 102:101–105. https://doi.org/10.1016/j.biortech.2010.06.016

Forgacs E, Cserháti T, Oros G (2004) Removal of synthetic dyes from wastewaters: a review. Environ Int 30:953–971. https://doi.org/10.1016/j.envint.2004.02.001

Friedler E (2004) Quality of individual domestic greywater streams and its implication for onsite treatment and reuse possibilities. Environ Technol 25:997–108. https://doi.org/10.1080/09593330.2004.9619393

Gajda I, Greenman J, Melhuish C, Ieropoulos I (2015) Self-sustainable electricity production from algae grown in a microbial fuel cell system. Biomass Bioenergy 82:87–93. https://doi.org/10.1016/j.biombioe.2015.05.017

Galvis A, Zambrano DA, van der Steen NP, Gijzen HJ (2014) Evaluation of pollution prevention options in the municipal water cycle. J Clean Prod 66:599–609. https://doi.org/10.1016/j.jclepro.2013.10.057

Garcia Segura S, Ocon JD, Chong MN (2018) Electrochemical oxidation remediation of real wastewater effluents–a review. Process Saf Environ Prot 113:48–67. https://doi.org/10.1016/j.psep.2017.09.014

Gavala HN, Kopsinis H, Skiadas IV, Stamatelatou K, Lyberatos G (1999) Treatment of dairy wastewater using an Upflow anaerobic sludge blanket reactor. J Agric Eng Res 73:59–63. https://doi.org/10.1006/jaer.1998.0391

Gentili FG (2014) Microalgal biomass and lipid production in mixed municipal, dairy, pulp and paper wastewater together with added flue gases. Bioresour Technol 169:27–32. https://doi.org/10.1016/j.biortech.2014.06.061

Georgianna DR, Mayfield SP (2012) Exploiting diversity and synthetic biology for the production of algal biofuels. Nature 488:329–335. https://doi.org/10.1038/nature11479

Godt J, Scheidig F, Grosse-Siestrup C, Esche V, Brandenburg P, Reich A, Groneberg DA (2006) The toxicity of cadmium and resulting hazards for human health. J Occup Med Toxicol 1:22. https://doi.org/10.1186/1745-6673-1-22

Goswami RK, Mehariya S, Verma P, Lavecchia R, Zuorro A (2020a) Microalgae-based biorefineries for sustainable resource recovery from wastewater. J Water Process Eng 40:101747. https://doi.org/10.1016/j.jwpe.2020.101747

Goswami RK, Agrawal K, Mehariya S, Molino A, Musmarra D, Verma P (2020b) Microalgae-based biorefinery for utilization of carbon dioxide for production of valuable bioproducts. In: Chemo-biological systems for CO_2 utilization. CRC Press, pp 203–228

Goswami RK, Agrawal K, Verma P (2021a) Phycoremediation of nitrogen and phosphate from wastewater using Picochlorum sp.: a tenable approach. J Basic Microbiol:1–17. https://doi.org/10.1002/jobm.202100277

Goswami RK, Agrawal K, Verma P (2021b) An overview of microalgal carotenoids: advances in the production and its impact on sustainable development. In: Bioenergy research: evaluating strategies for commercialization and sustainability. Wiley Online Library, pp 105–128. https://doi.org/10.1002/9781119772125.ch6

Goyer RA (1990) Lead toxicity: from overt to subclinical to subtle health effects. Environ Health Perspect 86:177–181. https://doi.org/10.1289/ehp.9086177

Grima EM, Belarbi EH, Fernández FA, Medina AR, Chisti Y (2003) Recovery of microalgal biomass and metabolites: process options and economics. Biotechnol Adv 20(7–8):491–515. https://doi.org/10.1016/S0734-9750(02)00050-2

Gupta VK, Nayak A, Agarwal S (2015) Bioadsorbents for remediation of heavy metals: current status and their future prospects. Environ Eng Res 20:1–18

Hamilton AJ, Stagnitti F, Xiong X, Kreidl SL, Benke KK, Maher P (2007) Wastewater irrigation: the state of play. Vadose Zone J 6:823–840. https://doi.org/10.2136/vzj2007.0026

Hansen KH, Angelidaki I, Ahring BK (1998) Anaerobic digestion of swine manure: inhibition by ammonia. Water Res 32:5–12. https://doi.org/10.1016/S0043-1354(97)00201-7

Hena S, Fatihah N, Tabassum S, Ismail N (2015) Three stage cultivation process of facultative strain of *Chlorella sorokiniana* for treating dairy farm effluent and lipid enhancement. Water Res 80:346–356. https://doi.org/10.1016/j.watres.2015.05.001

Hernandez-Zamora M, Cristiani-Urbina E, Martínez-Jerónimo F, Perales-Vela HV, Ponce-Noyola-T, del Carmen M-HM, Cañizares-Villanueva RO (2015) Bioremoval of the azo dye Congo red by the microalga *Chlorella vulgaris*. Environ Sci Pollut Res 22:10811–10823. https://doi.org/10.1007/s11356-015-4277-1

Hourlier F, Masse A, Jaouen P, Lakel A, Gerente C, Faur C, Le Cloirec P (2010) Formulation of synthetic greywater as an evaluation tool for wastewater recycling technologies. Environ Technol 31:215–223. https://doi.org/10.1080/09593330903431547

Hu Q, Sommerfeld M, Jarvis E, Ghirardi M, Posewitz M, Seibert M, Darzins A (2008) Microalgal triacylglycerols as feedstocks for biofuel production: perspectives and advances. Plant J 54: 621–639. https://doi.org/10.1111/j.1365-313X.2008.03492.x

Jais NM, Mohamed RMSR, Al-Gheethi AA, Hashim MA (2017) The dual roles of phycoremediation of wet market wastewater for nutrients and heavy metals removal and microalgae biomass production. Clean Techn Environ Policy 19:37–52. https://doi.org/10.1007/s10098-016-1235-7

Johnson DB, Hallberg KB (2005) Acid mine drainage remediation options: a review. Sci Total Environ 338:3–14. https://doi.org/10.1016/j.scitotenv.2004.09.002

Jorgensen TC, Weatherley LR (2003) Ammonia removal from wastewater by ion exchange in the presence of organic contaminants. Water Res 37:1723–1728. https://doi.org/10.1016/S0043-1354(02)00571-7

Kalavathi DF, Uma L, Subramanian G (2001) Degradation and metabolization of the pigment—melanoidin in distillery effluent by the marine cyanobacterium *Oscillatoria boryana* BDU 92181. Enzyme Microb Technol 29:246–251. https://doi.org/10.1016/S0141-0229(01)00383-0

Kalin M, Fyson A, Wheeler WN (2006) The chemistry of conventional and alternative treatment systems for the neutralization of acid mine drainage. Sci Total Environ 366:395–408. https://doi.org/10.1016/j.scitotenv.2005.11.015

Kamal S, Prasad R, Varma A (2010) Soil microbial diversity in relation to heavy metals. In: Sherameti I, Varma A (eds) Soil heavy metals, vol 19. Springer-Verlag, Berlin, Heidelberg, pp 31–64

Kamarudin KF, Tao DG, Yaakob Z, Takriff MS, Rahaman MSA, Salihon J (2015) A review on wastewater treatment and microalgal by-product production with a prospect of palm oil mill effluent (POME) utilization for algae. Der Pharma Chem 7:73–89

Khalaf MA (2008) Biosorption of reactive dye from textile wastewater by nonviable biomass of *Aspergillus Niger* and *spirogyra* sp. Bioresour Technol 99:6631–6634. https://doi.org/10.1016/j.biortech.2007.12.010

Khataee AR, Zarei M, Pourhassan M (2009) Application of microalga *Chlamydomonas* sp. for biosorptive removal of a textile dye from contaminated water: modelling by a neural network. Environ Technol 30:1615–1623. https://doi.org/10.1080/09593330903370018

Khataee AR, Zarei M, Pourhassan M (2010) Bioremediation of malachite green from contaminated water by three microalgae: neural network modeling. CLEAN Soil Air Water 38:96–103. https://doi.org/10.1002/clen.200900233

Kong Q-x, Li L, Martinez B, Chen P, Ruan R (2010) Culture of microalgae *Chlamydomonas reinhardtii* in wastewater for biomass feedstock production. Appl Biochem Biotechnol 160:9–18. https://doi.org/10.1007/s12010-009-8670-4

Kothari R, Pathak VV, Kumar V, Singh DP (2012) Experimental study for growth potential of unicellular alga *Chlorella pyrenoidosa* on dairy waste water: an integrated approach for treatment and biofuel production. Bioresour Technol 116:466–470. https://doi.org/10.1016/j.biortech.2012.03.121

Kumar D, Gaur JP (2011) Metal biosorption by two cyanobacterial mats in relation to pH, biomass concentration, pre-treatment and reuse. Bioresour Technol 102:2529–2535. https://doi.org/10.1016/j.biortech.2010.11.061

Kumar KS, Dahms HU, Won EJ, Lee JS, Shin KH (2015) Microalgae—a promising tool for heavy metal remediation. Ecotoxicol Environ Saf 113:329–352. https://doi.org/10.1016/j.ecoenv.2014.12.019

Lavoie A, De la Noue J (1985) Hyper-concentrated cultures of *Scenedesmus obliquus*: a new approach for wastewater biological tertiary treatment? Water Res 19:1437–1442. https://doi.org/10.1016/0043-1354(85)90311-2

Lavrinovics A, Juhna T (2017) Review on challenges and limitations for algae-based wastewater treatment. Construct Sci 20:17–25

Leal L, Zeeman G, Temmink H, Buisman C (2007) Characterisation and biological treatment of greywater. Water Sci Technol 56(5):193–200. https://doi.org/10.2166/wst.2007.572

Ledin A, Eriksson E, Henze M (2001) Aspects of groundwater recharge using grey wastewater. In: Lettinga G (ed) Decentralised sanitation and reuse. IWA, London, p 650

Lefebvre O, Shen Y, Tan Z, Uzabiaga A, Chang IS, Ng HY (2011) Full-loop operation and cathodic acidification of a microbial fuel cell operated on domestic wastewater. Bioresour Technol 102: 5841–5848. https://doi.org/10.1016/j.biortech.2011.02.098

Lim S, Chu W, Phang S (2010) Use of *Chlorella vulgaris* for bioremediation of textile wastewater. J Bioresour Technol 101:7314–7322. https://doi.org/10.1016/j.biortech.2010.04.092

Liu C, Subashchandrabose S, Ming H, Xiao B, Naidu R, Megharaj M (2016) Phycoremediation of dairy and winery wastewater using *Diplosphaera* sp. MM1. J Appl Phycol 28:3331–3341. https://doi.org/10.1007/s10811-016-0894-4

Lowrey J, Brooks MS, Mc Ginn PJ (2015) Heterotrophic and mixotrophic cultivation of microalgae for biodiesel production in agricultural wastewaters and associated challenges—a critical review. J Appl Phycol 27:1485–1498. https://doi.org/10.1007/s10811-014-0459-3

Lu Q, Zhou W, Min M, Mat X, Ma Y, Chen P, Zheng H, Doan YTT, Liu H, Chen C, Urriola PE, Shurson GC, Ruan R (2016) Mitigating ammonia nitrogen deficiency in dairy wastewaters for algae cultivation. Bioresour Technol 201:33–40. https://doi.org/10.1016/j.biortech.2015.11.029

Maimon A, Tal A, Friedler E, Gross A (2010) Safe on-site reuse of greywater for irrigation a critical review of current guidelines. Environ Sci Technol 44:3213–3220. https://doi.org/10.1021/es902646g

Mani D, Kumar C (2014) Biotechnological advances in bioremediation of heavy metals contaminated ecosystems: an overview with special reference to phytoremediation. Int J Environ Sci Technol 11:843–872. https://doi.org/10.1007/s13762-013-0299-8

Mantzavinos D, Psillakis E (2004) Biodegradation of textile dyeing industry wastewater using modified anaerobic sequential batch reactor–start-up, parameter optimization and performance analysis. J Taiwan Inst Chem Eng 72:171-181.Hancement of biodegradability of industrial wastewaters by chemical oxidation pre-treatment. J Chem Technol Biotechnol 79:431–435. https://doi.org/10.1016/j.jtice.2017.01.027

Markou G, Georgakakis D (2011) Cultivation of filamentous cyanobacteria (blue-green algae) in agro-industrial wastes and wastewaters: a review. Appl Energy 88:3389–3401. https://doi.org/10.1016/j.apenergy.2010.12.042

Marungrueng K, Pavasant P (2006) Removal of basic dye (Astrazon blue FRGL) using macroalga *Caulerpa lentillifera*. J Environ Manage 78:268–274. https://doi.org/10.1016/j.jenvman.2005.04.022

Mata TM, Melo AC, Simões M et al (2012) Parametric study of a brewery effluent treatment by microalgae *Scenedesmus obliquus*. Bioresour Technol 107:151–158. https://doi.org/10.1016/j.biortech.2011.12.109

Mehariya S, Goswami RK, Verma P, Lavecchia R, Zuorro A (2021a) Integrated approach for wastewater treatment and biofuel production in microalgae biorefineries. Energies 14:2282. https://doi.org/10.3390/en14082282

Mehariya S, Goswami RK, Karthikeysan OP, Verma P (2021b) Microalgae for high-value products: a way towards green nutraceutical and pharmaceutical compounds. Chemosphere:280, 130553. https://doi.org/10.1016/j.chemosphere.2021.130553

Mehta SK, Gaur JP (2005) Use of algae for removing heavy metal ions from wastewater: progress and prospects. Crit Rev Biotechnol *25*:113–152. https://doi.org/10.1080/07388550500248571

Metcalf E, Eddy H (1987) Wastewater engineering: treatment, disposal, reuse. Tata McGraw-Hill, New York

Milledge JJ, Heaven S (2013) A review of the harvesting of micro-algae for biofuel production. Rev Environ Sci Biotechnol 12:165–178. https://doi.org/10.1007/s11157-012-9301-z

Mohamed RMSR, Wurochekke AA, Chan CM, Kassim AH (2014) The use of natural filter media added with peat soil for household greywater treatment. Int J Eng Technol 2:33–38. https://doi.org/10.7603/s40707-013-0011-x

Mohamed RM, Al-Gheethi AA, Aznin SS, Hasila AH, Wurochekke AA, Kassim AH (2017) Removal of nutrients and organic pollutants from household greywater by phycoremediation for safe disposal. Int J Energy Environ Eng 8:259–272. https://doi.org/10.1007/s40095-017-0236-6

Mohana S, Acharya BK, Madamwar D (2009) Distillery spent wash: treatment technologies and potential applications. J Hazard Mater 163:12–25. https://doi.org/10.1016/j.jhazmat.2008.06.079

Monteiro CM, Castro PML, Malcata FX (2011) Microalga-mediated bioremediation of heavy metal–contaminated surfacewaters. In: Khan MS, Zaidi A, Goel R, Musarrat J (eds) Biomanagement of metal-contaminated soils: environmental pollution. Springer, Dordrecht, pp 365–385. https://doi.org/10.1007/978-94-007-1914-9_16

Monteiro CM, Castro PML, Malcata FX (2012) Metal uptake by microalgae: underlying mechanisms and practical applications. Biotechnol Prog 28:299–311. https://doi.org/10.1002/btpr.1504

Muñoz R, Guieysse B (2006) Algal-bacterial processes for the treatment of hazardous contaminants: a review. Water Res 40:2799–2815. https://doi.org/10.1016/j.watres.2006.06.011

Muttamara S (1996) Wastewater characteristics. Resour Conserv Recycl 16:145–159

Nevin KP, Lovley DR (2000) Lack of production of electron-shuttling compounds or solubilization of Fe (III) during reduction of insoluble Fe (III) oxide by *Geobacter metallireducens*. Appl Environ Microbiol 66:2248–2251. https://doi.org/10.1128/AEM.66.5.2248-2251.2000

Nicolo MS, Guglielmino SPP, Solinas V, Salis A (2016) Consequences of microbial interactions with hydrocarbons, oils, and lipids; production of fuels and chemicals. Springer, Berlin, pp 1–20

Nies DH (1999) Microbial heavy metal resistance. Appl Microbiol Biotechnol 51:730–750. https://doi.org/10.1007/s002530051457

Novis PM, Harding JS (2007) Extreme Acidophiles: freshwater algae associated with acid mine drainage. In: Seckbach J (ed) Algae and cyanobacteria in extreme environment. Springer, Dordrecht, pp 443–463. https://doi.org/10.1007/978-1-4020-6112-7_24

Ogbonna JC, Yoshizawa H, Tanaka H (2000) Treatment of high strength organic wastewater by a mixed culture of photosynthetic microorganisms. J Appl Phycol 12:277–284. https://doi.org/10.1023/A:1008188311681

Olguín EJ (2003) Phycoremediation: key issues for cost-effective nutrient removal processes. Biotechnol Adv 22:81–91. https://doi.org/10.1016/S0734-9750(03)00130-7

Oswald WJ (1988) The role of microalgae in liquid waste treatment and reclamation series. Algae Hum Affairs:255–281

Oswald WJ (2003) My sixty years in applied algology. J Appl Phycol 15(2):99–106

Pachauri RK, Sridharan PV (eds) (1998) Looking back to think ahead: GREEN India 2047 (growth with resource enhancement of environment and nature). TERI, New Delhi

Pahazri N, Mohamed RMS, Al-Gheethi AA, Hashim A (2016) Production and harvesting of microalgae biomass from wastewater, a critical review. Environ Technol Rev 5:39–56. https://doi.org/10.1080/21622515.2016.1207713

Pant D, Adholeya A (2007) Biological approaches for treatment of distillery wastewater: a review. Bioresour Technol 98:2321–2334

Passeggi M, López I, Borzacconi L (2009) Integrated anaerobic treatment of dairy industrial wastewater and sludge. Water Sci Technol 59:501–506. https://doi.org/10.2166/wst.2009.010

Patel A, Pawar R, Mishra S, Tewari A (2001) Exploitation of marine cyanobacterial for removal of colour from distillery effluent. Indian J Environ Prot 21:1118–1121

Pathak VV, Singh DP, Kothari R, Chopra AK (2014) Phycoremediation of textile wastewater by unicellular microalga *Chlorella pyrenoidosa*. Cell Mol Biol 60:35–40

Pathak VV, Ahmad S, Kothari R (2019) Implication of algal microbiology for wastewater treatment and bioenergy production. In: Environmental biotechnology: for sustainable future. Springer, Singapore, pp 263–286. https://doi.org/10.1007/978-981-10-7284-0_10

Pawlik-Skowronska B (2001) Phytochelatin production in freshwater algae *Stigeoclonium* in response to heavy metals contained in mining water, effects of some environmental factors. Aquat Toxicol 52:241–249. https://doi.org/10.1016/S0166-445X(00)00144-2

Perez-Garcia O, Escalante FM, de Bashan LE, Bashan Y (2011) Heterotrophic cultures of microalgae: metabolism and potential products. Water Res 45:11–36. https://doi.org/10.1016/j.watres.2010.08.037

Pittman JK, Dean AP, Osundeko O (2011) The potential of sustainable algal biofuel production using wastewater resources. Bioresour Technol 102:17–25. https://doi.org/10.1016/j.biortech.2010.06.035

Plum LM, Rink L, Haase H (2010) The essential toxin: impact of zinc on human health. Int J Env Res Publ Health 7:1342–1365. https://doi.org/10.3390/ijerph7041342

Posadas E, Bochon S, Coca M, García-González MC, García-Encina PA, Muñoz R (2014) Microalgae-basedagro-industrial wastewater treatment: a preliminary screening of biodegradability. J Appl Phycol 26:2335–2345. https://doi.org/10.1007/s10811-014-0263-0

Prajapati SK, Kumar P, Malik A, Vijay VK (2014) Bioconversion of algae to methane and subsequent utilization of digestate for algae cultivation: a closed loop bioenergy generation process. Bioresour Technol 158:174–180. https://doi.org/10.1016/j.biortech.2014.02.023

Rajasimman M, Babu SV, Rajamohan N (2017) Biodegradation of textile dyeing industry wastewater using modified anaerobic sequential batch reactor–start-up, parameter optimization and performance analysis. J Taiwan Inst Chem Eng 72:171–181. https://doi.org/10.1016/j.jtice.2017.01.027

Ramasamy EV, Abbasi SA (2000) Energy recovery from dairy waste-waters: impacts of biofilm support systems on anaerobic CST reactors. Appl Energy 65:91–98. https://doi.org/10.1016/S0306-2619(99)00079-3

Rathnayake VN, Mallavarapu M, Bolan N, Naidu R (1999) Tolerance of heavy metals by gram-positive soil bacteria. World Acad Sci Eng Technol 53:1185–1189

Rawat I, Kumar RR, Mutanda T, Bux F (2011) Dual role of microalgae: phycoremediation of domestic wastewater and biomass production for sustainable biofuels production. Appl Energy 88:3411–3424. https://doi.org/10.1016/j.apenergy.2010.11.025

Rawat J, Gupta PK, Pandit S, Prasad R, Pande V (2021) Current perspectives on integrated approaches to enhance lipid accumulation in microalgae. 3 Biotech 11:303. https://doi.org/10.1007/s13205-021-02851-3

Ray SG, Ghangrekar MM (2019) Comprehensive review on treatment of high-strength distillery wastewater in advanced physico-chemical and biological degradation pathways. Int J Environ Sci Technol 16:527–546. https://doi.org/10.1007/s13762-018-1786-8

Rico JL, García PA, Fdz-Polanco F (1991) Anaerobic treatment of cheese production wastewater using a UASB reactor. Bioresour Technol 37:271–276. https://doi.org/10.1016/0960-8524(91)90194-O

Robinson T, McMullan G, Marchant R, Nigam P (2001) Remediation of dyes in textile effluent: a critical review on current treatment technologies with a proposed alternative. Bioresour Technol 77:247–255. https://doi.org/10.1016/S0960-8524(00)00080-8

Sadettin S, Dönmez G (2007) Simultaneous bioaccumulation of reactive dye and chromium (VI) by using thermophile *Phormidium* sp. Enzyme Microb Technol 41:175–180. https://doi.org/10.1016/j.enzmictec.2006.12.015

Saha N, Balakrishnan M, Batra V (2005) Improving industrial water use: case study for an Indian distillery resources. Conserv Recycl 43:163–174. https://doi.org/10.1016/j.resconrec.2004.04.016

Samorì G, Samorì C, Guerrini F, Pistocchi R (2013) Growth and nitrogen removal capacity of *Desmodesmus communis* and of a natural microalgae consortium in a batch culture system in view of urban wastewater treatment: part I. Water Res 47:791–801. https://doi.org/10.1016/j.watres.2012.11.006

Sandau EE, Sandau P, Pulz (1996) Heavy metal sorption by microalgae. Acta Biotechnol 16:227–235. https://doi.org/10.1002/abio.370160402

Sarwa P, Verma SK (2013) Decolourization of Orange G dye by microalgae *Acutodesmus obliquues* strain PSV2 isolated from textile industrial site. Int J Appl Sci Biotechnol 1:247–252. https://doi.org/10.3126/ijasbt.v1i4.9141

Satyawali Y, Balakrishnan M (2008) Wastewater treatment in molasses-based alcohol distilleries for COD and color removal: a review. J Environ Manage 86:481–497. https://doi.org/10.1016/j.jenvman.2006.12.024

Schwarzenbach RP, Escher BI, Fenner K, Hofstetter TB, Johnson CA, Von Gunten U, Wehrli B (2006) The challenge of micropollutants in aquatic systems. Science 313:1072–1077. https://doi.org/10.1126/science.1127291

Simate GS, Ndlovu S (2014) Acid mine drainage: challenges and opportunities. J Environ Chem Eng 2:1785–1803. https://doi.org/10.1016/j.jece.2014.07.021

Smil V (2000) Phosphorus in the environment: natural flows and human interferences. Annu Rev Energy Environ 25:53–88

Solovchenko A (2019) Microalgal treatment of alcohol distillery wastewater. In: Emerging and eco-friendly approaches for waste management. Springer, Singapore, pp 171–188. https://doi.org/10.1007/978-981-10-8669-4_9

Solovchenko A, Verschoor AM, Jablonowski ND, Nedbal L (2016) Phosphorus from wastewater to crops: an alternative path involving microalgae. Biotechnol Adv 34:550–564. https://doi.org/10.1016/j.biotechadv.2016.01.002

Staples CR, Fahey GC Jr, Rindsig RB, Berger LL (1981) Evaluation of dairy waste fiber as a roughage source for ruminants. J Dairy Sci 64:662–671. https://doi.org/10.3168/jds.S0022-0302(81)82627-6

Takaichi S (2011) Carotenoids in algae: distributions. Biosynth Funct Mar Drugs 9:1101–1118. https://doi.org/10.3390/md9061101

Thakare M, Sarma H, Datar S, Roy A, Pawar P, Gupta K, Pandit S, Prasad R (2021) Understanding the holistic approach to plant-microbe remediation technologies for removing heavy metals and radionuclides from soil. Curr Res Biotechnol 3:84–98. https://doi.org/10.1016/j.crbiot.2021.02.004

Tripathy A, More RD, Gupta S, Samuel J, Singh J, Prasad R (2021) Present and future prospect of algae: a potential candidate for sustainable pollution mitigation. Open Biotechnol J. https://benthamopen.com/EPUB/BMS-TOBIOTJ-2020-HT2-1761-7

Uriu-Adams JY, Keen CL (2005) Copper, oxidative stress, and human health. Mol Aspects Med 26:268–298. https://doi.org/10.1016/j.mam.2005.07.015

USDA-SCS (1992) Agricultural waste management field handbook. U.S. Department of Agriculture-Soil Conservation Service (USDA-SCS), Washington, DC

Valderrama LT, Del Campo CM, Rodriguez CM, de Bashan LE, Bashan Y (2002) Treatment of recalcitrant wastewater from ethanol and citric acid production using the microalga *Chlorella vulgaris* and the macrophyte *Lemna minuscula*. Water Res 36:4185–4192. https://doi.org/10.1016/S0043-1354(02)00143-4

Van Den Hende S, Vervaeren H, Boon N (2012) Flue gas compounds and microalgae: biochemical interactions leading to biotechnological opportunities. Biotechnol Adv 30:1405–1424. https://doi.org/10.1016/j.biotechadv.2012.02.015

Vymazal J (1984) Short term uptake of heavy metals by periphyton algae. Hydrobiologia 119:171–179. https://doi.org/10.1007/BF00015208

Wang L, Min M, Li Y, Chen P, Chen Y, Liu Y, Wang Y, Ruan R (2010) Cultivation of green algae *chlorella* sp. in different wastewaters from municipal wastewater treatment plant. Appl Biochem Biotechnol 162:1174–1186. https://doi.org/10.1007/s12010-009-8866-7

Wang X, Lin H, Gu Y (2012) Multiple roles of dihomo-gamma-linolenic acid against proliferation diseases. Lipids Health Dis 11:25. https://doi.org/10.1186/1476-511X-11-25

Woertz I, Feffer A, Lundquist T, Nelson Y (2009) Algae grown on dairy and municipal wastewater for simultaneous nutrient removal and lipid production for biofuel feedstock. J Environ Eng 135:1115–1122

Wurochekke AA, Mohamed RMSR, Al-Gheethi AAS, Noman EA, Kassim AHM (2019) Phycoremediation: a green technology for nutrient removal from greywater. In: Management of greywater in developing countries. Springer, Cham, pp 149–162. https://doi.org/10.1007/978-3-319-90269-2_8

Xin L, Hong-ying H, Ke G, Ying-xue S (2010) Effects of different nitrogen and phosphorus concentrations on the growth, nutrient uptake, and lipid accumulation of a freshwater microalga *Scenedesmus* sp. Bioresour Technol 101:5494–5500. https://doi.org/10.1016/j.biortech.2010.02.016

Zollinger H (2003) Color chemistry: syntheses, properties, and applications of organic dyes and pigments. Wiley, Hoboken, NJ

Phytoremediation: Mechanistic Approach for Eliminating Heavy Metal Toxicity from Environment

19

Sujoy Sarkar, Sahana Basu, Ram Prasad, and Gautam Kumar

Abstract

Heavy metals (HMs) are environmental and food chain contaminants having chronic and epidemic effects on human health. Introduction of HMs in the food chain takes place by their excessive uptake from soil through the crop plants, making it a global issue of concern to take necessary steps to counteract the problem. The HMs also cause toxicities to plants by affecting their growth and productivity. With the continuously changing global climatic conditions, the HM contamination in the soil is exaggerating, thereby resulting in the considerable yield reduction of major crop species. Furthermore, HM-induced soil pollution associated with the improper fertilization practices appears as a serious threat to the sustainable agriculture. It is therefore, a serious worldwide concern to minimize the HM toxicity in crop plants. Phytoremediation is a promising plant-based, cost-effective, and eco-friendly approach for the effective removal of the HMs from the environment. Several plants known as metallophytes accumulate higher level of HMs without having any toxic effects and, therefore, can be used to remove large amounts of HMs from the soil. The present chapter summarizes the mechanisms of HM uptake, translocation, and detoxification in plants. The mechanism adopted by the metallophytes in HM hyperaccumulation and their role in ameliorating the HM toxicity has also been discussed.

S. Sarkar · G. Kumar (✉)
Department of Life Science, Central University of South Bihar, Panchanpur, Gaya, Bihar, India
e-mail: gautam@cub.ac.in

S. Basu
Department of Biotechnology, Assam University, Silchar, Assam, India

R. Prasad
Department of Botany, Mahatma Gandhi Central University, Motihari, Bihar, India

R. Prasad (ed.), *Phytoremediation for Environmental Sustainability*,
https://doi.org/10.1007/978-981-16-5621-7_19

Keywords

Heavy metal (HM) · Detoxification · Metallophytes · Phytoremediation · Detoxification

19.1 Introduction

Progressive industrialization, modern agricultural practices, and increased anthropogenic activities due to urbanization are emerging as potential causes for heavy metal (HM) contamination in the environment (Singh et al. 2016). The HM contamination leading to toxicity in animals and humans has become a major concern in the last few decades. Unrestricted usage of pesticides and chemical fertilizers in agriculture, compost wastes, smelting industries, and metal mining are increasing the HM contamination throughout the world. HMs cause toxicity to the plants with significant negative influence on their growth and productivity (Arif et al. 2016). HMs are metallic chemical elements with high densities, atomic weights and numbers (Nagajyoti et al. 2010; Kamal et al. 2010). They are the natural components of the Earth's crust. HMs are nondegradable and create toxic effect even at very low concentrations. Examples of some common HMs are arsenic (As), cadmium (Cd), chromium (Cr), lead (Pb), mercury (Hg), and thallium (Tl). Alternatively, some HMs, such as copper (Cu), selenium (Se), and zinc (Zn), are considered as the trace elements playing a crucial role in the metabolic processes of plants and animals. However, these trace elements can lead to poisoning at higher concentrations. The HM contamination affects the soil microbial communities, influences the biogeochemical cycles, and directly impacts the ecological niche and diversity of soil bacterial communities (Thakare et al. 2021). It is therefore, a serious worldwide concern to take necessary steps to counteract the problem of HM toxicity in the environment.

Phytoremediation is a cost-effective and mutualistic eco-friendly approach with direct associations of the HM-tolerant plants and the HM-polluted soils (Muthusaravanan et al. 2018; Sarma et al. 2021; Sonowal et al. 2022). It is a plant-based technique, which involves the use of plants to extract and remove the HM pollutants or lower their bioavailability in the soil (Marques et al. 2009). Plants have the ability to absorb ionic compounds from the soil even at low concentrations through their root systems. They extend their root system into the soil matrix and establish the rhizosphere ecosystem to accumulate the HMs and modulate their bioavailability, through which they reclaim the polluted soil and stabilize the soil fertility. Phytoremediation is an autotrophic system powered by solar energy, therefore, simple to manage, and the cost of installation and maintenance is low. Being environment-friendly, it can reduce the HM pollutants from the ecosystem. Additionally, it can be applied over a large-scale field and can also easily be disposed. It prevents the erosion and metal leaching by stabilizing the HMs and reduces the risk of spreading the HM contaminants. It also improves the soil fertility by releasing various organic matters to the soil (Yan et al. 2020). Several studies have unraveled the molecular mechanisms underlying the HM tolerance in plants and have developed techniques to improve the phytoremediation efficiency of plants. The present chapter highlights the mechanisms of HM uptake, translocation, and detoxification in plants. The strategies adopted by the plants to improve the HM bioavailability, accumulation, and tolerance have also been discussed in this chapter, which may contribute in developing phytoremediation techniques to eliminate HM toxicity.

19.2 Plants' Responses to Heavy Metal Toxicity

Mineral nutrients are one of the key regulators of plant growth and productivity. A number of metals are important for the growth of plants. However, their essentiality depends on their concentrations in plant (different stages) and environment. Furthermore, some trace elements (mainly Fe, Zn, Cu, Ni, Co, and Mo) are essential for plant and cellular biochemistry being involved in cell protection, gene regulation, and signal transduction. However, excess concentrations of these elements than their optimum levels may cause toxicities to plants by retarding plant growth and yield. Other heavy metals (As, Cd, Hg, Pb, and Cr) are biologically nonessential and show toxicity even at low concentrations (DalCorso et al. 2019).

Heavy metals interfere with metabolic reactions in plant systems. HM toxicity reduces the plant growth, photosynthetic activities, mineral nutrition, and activity of essential enzymes. They are cytotoxic and carcinogenic to humans at low concentrations. HMs induce the production of reactive oxygen species (ROS) causing oxidative stress in plants. The ROS causes oxidation of DNA, proteins, and lipids leading to the cell death (Ojuederie and Babalola 2017). HM tolerance in plants is mediated by the vacuolar compartmentalization and sequestration of the HMs within the plant cells. Plants also develop antioxidant defense system which protects cells through effective scavenging of ROS. Understanding the mechanisms of HM detoxification is essential for searching the potential HM-tolerant plant species which can be used for the removal of HM from the contaminated sites.

19.3 Mechanism of Uptake, Translocation, and Detoxification of Heavy Metals in Plants

The HMs are usually present as insoluble forms in the soil. The uptake of HM in plants is governed by different factors, including the water content, pH, and organic substances. Higher water content increases the solubility of the HMs consequently, increasing their bioavailability. Wang et al. (2015) have reported the flooded conditions to intensify the bioavailability of arsenic (As) resulting in the efficient As uptake in rice. High soil temperature also increases the solubility and bioavailability of the HMs, thereby increasing its uptake in plants (Arao et al. 2018). The pH also enhances the dissolution of HMs by acidifying the rhizosphere by increasing the proton secretion from the roots (Peng et al. 2005). The organic substances' exude from the roots also increases the bioavailability of HMs to the plants. Cieslinski et al.

(1998) have revealed the presence of organic acids in the rhizosphere to increase the solubility and availability of cadmium (Cd). Additionally, root proliferation enhances the HM uptake in plants (Whiting et al. 2000).

19.3.1 Heavy Metal Uptake and Translocation

The bioavailable HMs are absorbed by the root hairs and driven across the plasma membrane of the root epidermal cells. Different HMs employ various ways to enter the plant roots. The HM uptake in roots generally occurs through apoplastic or symplastic route (Yan et al. 2020). The apoplastic pathway mediates the movement of HMs through the cell wall and intercellular spaces. Kidwai et al. (2019) have revealed that increased lignification in the roots act as an apoplastic barrier for the As entry in root cells resulting in the reduced As accumulation. On the other hand, the symplastic pathway includes plasma membrane-localized nonspecific ion channels or transporters. Arsenate As(V), the major form of As enters the plant root tissue via the phosphate (Pi) transporters (Shi et al. 2019). The Cd^{2+} uptake in plants takes place through the transporters involved in the Mg^{2+}, Ca^{2+}, Fe^{2+}, Zn^{2+}, and Cu^{2+} uptake (Ismael et al. 2019). The lead (Pb) uptake occurs via the Ca^{2+} permeable channels on roots (Pourrut et al. 2011). After entering the root cells, HMs are translocated to the aerial parts of the plants through xylem vessels.

19.3.2 Heavy Metal Detoxification

Detoxification of HMs is an important requirement for employing the phytoremediation approach (Viehweger 2014). The improved HM detoxification process in the hyperaccumulators allows them to persist under the HM-contaminated sites without having any toxic effect. Plants adopt avoidance or tolerance strategies to cope with the HM toxicity through HM homeostasis. Avoidance is the simplest strategy, which acts as the first line of defense at extracellular level to restrict HM uptake from the soil and prevent their translocation into aerial tissues (Dalvi and Bhalerao 2013). It comprises of different strategies—root sorption, ion precipitation, and exclusion of the HMs (Yan et al. 2020). Root sorption is the first step of HM avoidance, where HMs are immobilized in the rhizosphere by forming HM complex with different ligands (e.g., amino acid, organic acid). The precipitation of HM ions occurs by the alteration of the rhizosphere pH due to root exudates. Exclusion of the HMs is mediated by the barrier between the root and the shoot systems that restrict the aerial translocation of HMs. Arbuscular mycorrhizal (AM) fungi immobilize the HMs by binding with insoluble glycoprotein (glomalin) produced by AM hyphae, thereby inhibiting HM entry in plants (Basu and Kumar 2020a, 2021a). Presence of cell wall polysaccharide-derived functional groups (e.g., carboxyl, hydroxyl) favors ion-exchange with the wall counter-ions leading to increased HM binding capacity, which reduces the HM entry in the protoplast (Parrotta et al. 2015). HMs are also

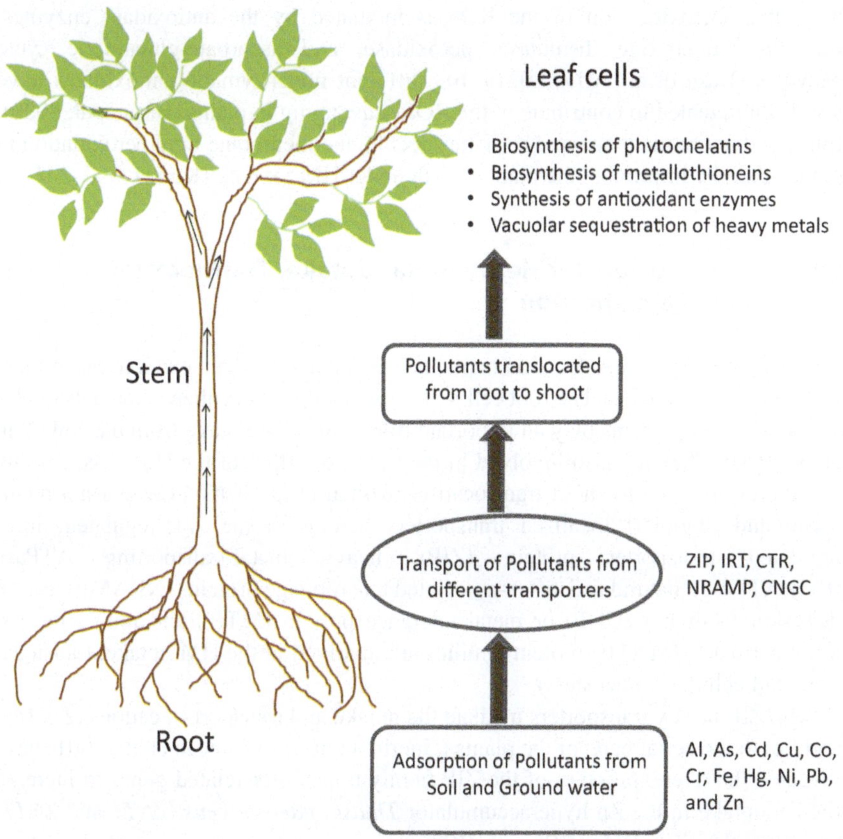

Fig. 19.1 Mechanism of heavy metal (HM) detoxification in metallophytes (hyperaccumulators). Detoxification of HMs within the cytosol occurs by their chelation with inorganic or organic ligands (amino acids, organic acids, phytochelatins, metallothioneins). Detoxification of HM-induced reactive oxygen species is mediated by the antioxidant enzymes (superoxide dismutase, peroxidase, ascorbate peroxidase). Metal homeostasis and tolerance in the hyperaccumulators is mediated by the vacuolar sequestration of the HMs within the plant cells

immobilized by reacting with the polygalacturonic acid present in the pectin of cell wall, consequently preventing their entry into the root cells.

The tolerance strategy acts as the second line of defense at intracellular level through inactivation, chelation, and vacuolar partitioning of the HMs (Fig. 19.1). Detoxification of HMs within the cytosol occurs by their chelation with inorganic or organic ligands. The organic ligands include amino acids, organic acids, phytochelatins (PCs), metallothioneins (MTs), and polyphenols, proteins, or pectins of cell wall. Arsenic detoxification in plants is associated with the binding of arsenite with the phytochelatins (PC) or glutathione followed by their vacuolar sequestration (Aborode et al. 2016). The MTs also play an important role in the HM detoxification (Kumar et al. 2012). HMs also induce the ROS production causing oxidative stress in plants. Detoxification of the ROS is mediated by the antioxidant enzymes, including superoxide dismutase, peroxidase, and ascorbate-glutathione cycle enzymes (Basu et al. 2017, 2021a, b). Different nonenzymatic antioxidants have also been revealed to contribute in the ROS scavenging in plants (Basu et al. 2020). Enhanced antioxidant enzyme activities decline the membrane lipid peroxidation in plants, thereby improving the plant growth under HM toxicity (Kumar et al. 2021a).

19.3.3 Transporters for Heavy Metal Uptake, Translocation, and Detoxification

The HM uptake and translocation in plants is facilitated by the metal ion transporters and complexing agents. The root cell's plasma membrane-localized channels or H^+-coupled carrier proteins play an important role in the HM uptake from the soil (Yan et al. 2020). They are also involved in the influx or efflux of the HM ions, thereby facilitating the root-to-shoot translocation (Komal et al. 2015). The plasma membrane and tonoplast localized transporters belong to the zinc-regulated, iron-regulated transporter protein (ZIP), heavy metal-transporting ATPase (P1B-ATPase), natural resistance-associated macrophage proteins (NRAMP), cation diffusion facilitator (CDF) or metal tolerance protein (MTP), and multidrug and toxin extrusion (MATE) protein families are involved in the HM uptake, translocation, and cellular homeostasis.

The ZIP family transporters mediate the uptake and transport of cations (Zn, Mn, and Fe) to the aerial parts of the plants (Guerinot 2000). Assuncao et al. (2001) have revealed the overexpression of the ZIP family transporter-related genes to increase the Zn uptake in the Zn hyperaccumulator *Thlaspi caerulescens* (*ZNT1* and *ZNT2*) and *Arabidopsis halleri* (*ZIP6* and *ZIP9*).

The P1B-type ATPases belong to the heavy metal transporting ATPases (HMAs) transporter family, which are involved in the transport of HMs (Cd, Co, Pb, and Zn) to the plasma membrane or the vacuolar sequestration of the HMs and play a vital role in metal homeostasis and tolerance (Hanikenne and Baurain 2014). The HMA3 (P1B-ATPase) localized on the tonoplast is responsible for the vacuolar compartmentation of HMs (Liu et al. 2017); whereas, the HMA4 carries out the aerial translocation of Cd and Zn (Wang et al. 2019). The overexpression of the *BjHMA4* (from *Brassica juncea*) has been shown to promote the HM tolerance in rice and wheat by inducing the efflux of Cd and Zn from the root cytoplasm into the xylem vessels (Wang et al. 2019).

The NRAMPs are a ubiquitous family of metal transporters responsible for the uptake and transport of various HMs (As, Cd, Co, Cu, Fe, and Mn) in different plant species (Nevo and Nelson 2006). Cailliatte et al. (2009) reported the NRAMP6 to contribute in the Cd transport. Later, Cailliatte et al. (2010) also revealed the plasma membrane-localized AtNRAMP1 to mediate the transport of Fe and Mn in Arabidopsis. Tiwari et al. (2014) reported the plasma membrane-localized OsNRAMP1 to facilitate the arsenite (AsIII) mobilization to the aerial parts of rice through xylem loading. Bastow et al. (2018) showed the tonoplast-localized

NRAMP3 and NRAMP4 to mediate the mobilization of vacuolar Fe in germinating seed.

The CDF or MTP transporter family is involved in the regulation of HM (Cd, Co, Mn, Ni, and Zn) homeostasis through the vacuolar sequestration or transport to the extracellular space (Ricachenevsky et al. 2013). The tonoplast-localized MTP1 and MTP4 have been reported to be the Zn^{2+}/H^{+} and Cd^{2+}/H^{+} antiporters involved in the vacuolar Zn and Cd sequestration in cucumber (Migocka et al. 2014). Comparative analyses of *A. thaliana* and Zn hyperaccumulators *A. halleri* and *T. caerulescens* have revealed higher expressions of MTP1, MTP8, and MTP11 in the hyperaccumulator species with enhanced Zn homeostasis (van de Mortel et al. 2006).

The MATE transporters also play crucial role in translocation of HMs (Al, Mn, and Zn). Dong et al. (2019) revealed the *CcMATE4* and *CcMATE34* (from *Cajanus cajan*) to be upregulated in the roots of pigeon pea under the Al, Mn, and Zn stresses. Ma et al. (2018) showed the *GsMATE* (from *Glycine soja*) overexpression to cause increased Al tolerance in *A. thaliana*.

19.4 About Phytoremediation

Phytoremediation includes several strategies for the detoxification of the HM-contaminated soils (Fig. 19.2). Various plant species used for different phytoremediation strategies for removal of the HM contaminants are presented in Table 19.1.

19.4.1 Phytoextraction

Phytoextraction (or phytoaccumulation) is the method where plants are used to remove the HM from the polluted soil and water through their uptake and accumulation into the harvestable plant parts (Suman et al. 2018). In this process, plants absorb the contaminants from soil or water together with other necessary nutrients required for plants' growth. The absorbed contaminants are translocated and accumulated in the aboveground plant tissues but are not destroyed (Rashid et al. 2014). Phytoextraction of HMs includes different steps: (1) HM mobilization in rhizosphere, (2) HM uptake by plant roots, (3) root-to-shoot translocation of HMs, and (4) vacuolar sequestration of HMs (Ali et al. 2013).

Phytoextraction is the most important phytoremediation procedure for HM removal from the contaminated soil. Selection of suitable plant species is crucial for the efficacious phytoextraction. The plant species should possess (1) extraordinary tolerance to the HM toxicity, (2) enhanced extractability and HM accumulation capacity in aboveground tissues, (3) high growth rate and high biomass, (4) extensive root and sufficient shoot system, (5) easily cultivable and environmental stress resistant, and (6) pathogen and pest resistance, herbivore repulsive to avoid HM flow into the food chain (Ali et al. 2013). Therefore, hyperaccumulator plants having

Fig. 19.2 Schematic diagram illustrating different mechanisms of phytoremediation for removal of heavy metals (HMs) from the contaminated sites. Phytoextraction mediates the extraction and removal of HMs from contaminated soil and water, phytostabilization mediates the reduction of HM bioavailability through belowground immobilization, phytovolatilization mediates the conversion of toxic HMs into less-toxic forms and releases them as volatile compounds into atmosphere, and phytodegradation mediates breakdown of toxic HMs into less-toxic forms

higher HM accumulation ability in the aboveground tissues and higher biomass production are appropriate for the phytoremediation of HM-contaminated sites. A number of edible crops accumulate high quantity of HM. However, edible crops are not recommended for phytoremediation as the HMs accumulated in their edible parts may contaminate the food chain. Therefore, nonedible hyperaccumulators should be selected for safe and effective phytoremediation of HMs.

Table 19.1 List of different plant species used for different phytoremediation strategies for removal of the HM contaminants

Phytoremediation process	Pollutants	Substrate	Detoxification mechanism	Phytoremediating plants
Phytoextraction (extraction and removal of HMs from contaminated soil and water)	Inorganics: Ag, As, Au, Cd, Co, Cr, Hg, Mo, Ni, Pb, Zn Radionuclides: Cs, Sr, U	Soil	Hyperaccumulation, uptake, and concentration of metals via soils, direct uptake into the plant tissue with subsequent removal of the plants. Plants are used to accumulate contaminants in the above ground, harvestable biomass	*Alyssum heldreichii, Astragalus racemosus, Berkheya coddii, Eleocharis acicularis, Hordeum vulgare, Pteris vittata, Thlaspi caerulescens, Zea mays*
Phytovolatilization (conversion of toxic HMs into less-toxic forms and release them as volatile compounds into atmosphere)	Inorganics: As, Hg, Se Organics: Chlorinated solvents	Soil and groundwater	Pollutants are converted inside plants to a gaseous state and released into the atmosphere via the evapotranspiration process	*Alternanthera philoxeroides, Arabidopsis thaliana, Artemisia princeps, Bidens frondosa, Bidens pilosa, Brassica juncea, Cynodon dactylon, Digitaria sanguinalis, Erigeron canadensis, Liriodendron tulipifera, Medicago sativa, Phragmites australis, Populus* sp., *Typha latifolia*
Phytostabilization (reduction of HM bioavailability through belowground immobilization)	Inorganics	Soil, ground water, mine tailing	Complexation; root exudates cause metal to precipitate soils, groundwater, and mine tailing and become less available. Pollutants are retained in the soil	*Amaranthus spinosus, Ludwigia palustris, Mentha aquatic, Myriophyllum aquaticum, Solanum nigrum, Spinacia oleracea, Populus cathayana, Populus przewalskii, Populus yunnanensis*
Phytofiltration (use of plant roots, shoots, or seedlings to remove HMs from contaminated groundwater and aqueous waste)	Organics/inorganics	Surface water and water pumped	Rhizosphere accumulation, uptake of metals into plant roots	*Bolboschoenus robustus, Helianthus annuus, Helianthus tuberosus, Nicotiana tabacum, Juncus xiphioides, Myriophyllum aquaticum, Spinacia oleracea, Typha latifolia*

(continued)

Table 19.1 (continued)

Phytoremediation process	Pollutants	Substrate	Detoxification mechanism	Phytoremediating plants
Phytodegradation (breakdown of toxic HMs to simpler less-toxic forms)	Organics: Chlorinated solvents, herbicides, phenols	Soil, ground water within rhizosphere	Degradation in plants enhances microbial degradation in rhizosphere. Pollutants are converted to less harmful substances	*Canna glauca*, *Colocasia esculenta*, *Cyperus papyrus*, *Pteris vittata*, *Typha angustifolia*

Phytoextraction technique is extensively employed to remove radioactive (Shahandeh and Hossner 2002) and metallic wastes (Kamal et al. 2004). Aquatic macrophytes like *Centella asiatica* and *Eichhornia crassipes* have been found to remove copper 99.6 and 97.3%, respectively (Mokhtar et al. 2011). Accumulation of HMs by the hyperaccumulators depends on the HM bioavailability within the rhizosphere, HM uptake rate by roots, proportion of fixed HM within the roots, rate of xylem loading/translocation to shoots, and cellular HM tolerance (Etim 2012).

Phytoextraction is performed with or without addition of chelate complexant for removal of HMs that remain sorbed to the solid soil components (Yan et al. 2020). Addition of chelating agents induces the formation of HM–chelate complexes preventing their sorption and precipitation. Thus the chelating agents maintain the bioavailability of HM for uptake by the mettalophytes. Chelating agents having strong affinity for the targeted HM enhance the phytoaccumulation ability of hyperaccumulators. However, the used chelate must be biodegradable for its rapid removal from the polluted site.

19.4.2 Phytostabilization

Phytostabilization is the method of phytoremediation where HM-tolerant plants are used to immobilize HMs belowground through their accumulation into plant roots (Mendez and Maier 2008). This process can also occur through precipitation of the HMs within the rhizosphere, adsorption onto the root surface, absorption, and vacuolar sequestration inside the root cells (Gerhardt et al. 2017). Phytostabilization decreases the bioavailability of the HMs by inhibiting their migration into the ecosystem and preventing their entry into the food chain. This process also serves as a filtration barrier against the root-to-shoot translocation of HM and is advantageous over phytoextraction as it does not require the disposal of the hazardous biomass (Lorestani et al. 2013).

Phytostabilization requires the selection of excessive HM-tolerant plant species (Yan et al. 2020). Plants should be easily maintainable under field conditions with fast growth rate and production of profuse biomass to cover the HM-contaminated site. Phytostabilization also requires dense rooting systems with increased root surface and depth for the stabilization of soil structure, and prevention of soil erosion through the HM immobilization. Improvement of the phytostabilization efficiency also requires the addition of the inorganic or organic amendments, which can improve the contaminated soil quality by enhancing the organic matter and essential nutrient contents consequently promoting plant colonization and water-holding capacity. This also alters the soil pH and redox state, thereby reducing the solubility and bioavailability of HM and also changing the HM speciation (Burges et al. 2018). For instance, application of *Gliricidia sepium* biomass as soil amendment elevated the phytostabilizing nature of *Zea mays* thereby remediating the Pb-contaminated soil (Muthusaravanan et al. 2018). Phytostabilization can also be promoted by the soil microbes including plant growth-promoting rhizobacteria (PGPR) and AM fungi. Dual application of silicon along with AM fungi has also been found to mitigate HM stress in crop plants (Basu and Kumar 2021b). These improve HM immobilization efficiency through production of chelators and adsorption of HM on cell walls, thereby stimulating the processes of precipitation (Ma et al. 2011). Madhaiyan et al. (2007) revealed HM-tolerant methylotrophic bacteria *Burkholderia* sp. and *Magnaporthe oryzae* to reduce Cd and Ni toxicity in tomato plants. Tamburini et al. (2017) examined the phytostabilization potential of strains belonging to *Amycolatopsis*, *Novosphingobium*, *Pseudomonas*, *Streptomyces*, and *Variovorax*, among which the *Variovorax* strain was found to be useful in the process of bioaugmentation in the mine areas, thereby promoting germination and plant growth.

19.4.3 Phytovolatilization

Phytovolatilization is the approach of phytoremediation where plants are used for uptake and conversion of toxic soil HMs into relatively less-toxic form subsequently releasing them into the atmosphere through transpiration as volatile compounds (Moreno et al. 2004). Detoxification of organic pollutants and HMs (As, Hg, and Se) is accomplished with this process (Mahar et al. 2016). Several studies revealed *Chara canescens*, *Brassica juncea* (Banuelos and Meek 1990), and aquatic plant *Typha latifolia* (LeDuc and Terry 2005) to be potential volatilizers of selenium (Se). Through the process of phytovolatilization, inorganic Se is converted into less-toxic volatile dimethyl selenide (DMSe) that can be dispersed into the air. Similarly, toxic Hg is converted to less-toxic volatile mercuric oxide and evaporated into the atmosphere (Bizily et al. 2000). Plant species *T. latifolia* has also been revealed to volatilize As, Cd, Co, Cr, Mn, Ni, and Zn (Varun et al. 2011). Phytovolatilization is advantageous over the other phytoremediation techniques as there is no need for the harvesting or disposal of the HM hyperaccumulating plants. Therefore, it is considered as a permanent solution for the HM removal as the volatilized products usually do not redeposit at the contaminated site.

19.4.4 Phytofiltration

Phytofiltration is the approach of phytoremediation where plants are used to remove HMs from contaminated groundwater or waste waters. This process can be of different types based on the plant parts used for the remediation practice—rhizofiltration (root), caulofiltration (shoot), and blastofiltration (seedling). Rhizofiltration includes the exudation from the roots that alters the rhizosphere pH leading to the HM precipitation on plant roots, thereby restricting the HMs to contaminate the underground water (Yan et al. 2020). During this process, HMs are adsorbed onto the root surface or absorbed by the roots. Therefore, the metallophytes used for this process contain dense root systems and huge biomass. They are initially grown hydroponically in clear water for developing the huge root system. Following the initial development, the plants are acclimatized to the HM-polluted environment by substituting the clear water with polluted water and subsequently transferred to the contaminated water for the HM removal. The hyperaccumulators are harvested and disposed after their roots become saturated with HMs.

Aquatic macrophytes like azolla, cattail, water hyacinth, poplar, and duckweed are usually used for remediation of the wetlands (Rezania et al. 2016). Arsenic-hyperaccumulating ferns *Pteris vittata* and *P. cretica* have been found to remove As from drinking water through phytofiltration. These plants have higher HM tolerance and HM accumulating capacity, rapid growth rate, and high biomass production. Several terrestrial plant species including *B. juncea* and *Helianthus annuus* have also been found to be used for rhizofiltration due to their longer and hairy root systems (Tome et al. 2008).

19.4.5 Phytodegradation

Phytodegradation is the approach of phytoremediation where plants are used to breakdown the toxic HMs to simpler less-toxic forms either through the plants' metabolic process inside or the enzymes produced by plants (Muthusaravanan et al. 2018). This process facilitates the degradation of pesticides, chlorinated solvents, and several inorganic/organic compounds. Moderately hydrophobic organic pollutants including short-chain aliphatic hydrocarbons, chlorinated solvents, benzene, ethyl benzene, toluene, and xylene at shallow depths are efficiently removed by this process. Phytodegradation is affected by few factors, such as concentration of HMs present in the soil, HM uptake efficiency, and the amount of water present in the ground. Phytodegradation is mediated by a number of the enzymes like nitroreductase, nitrilase, dehalogenase, laccase, and peroxidase. Rajakaruna et al. (2006) have revealed the aquatic plant species *Myriophyllum aquaticum* to produce nitroreductase enzyme that facilitates the reduction of trinitrotoluene (TNT).

19.5 Phytoremediation of Different Heavy Metals

Metallophytes are the HM hyperaccumulators having the natural ability to accumulate large amounts of toxic HMs from the contaminated soil, making them exclusive to be exploited in phytoremediation to clean up the environment. The hyperaccumulators possess unique HM tolerance strategies than the non-hyperaccumulators, which make them suitable for the phytoremediation. The metallophytes have higher proficiency of HM uptake, root-to-shoot translocation, and detoxification. Abundant studies have been performed on metallophytes for understanding their strategies of HM tolerance. Approximately 450 plant species across 45 angiosperm families (e.g., *Asterraceae, Brassicaceae, Euphorbiaceae, Fabaceae, Lamiaceae, and Scrophulariaceae*) have been recognized as hyperaccumulators (Suman et al. 2018). Some metallophytes can accumulate more than two elements. For instance, *Sedum alfredii* can accumulate Cd, Pb, and Zn (Yan et al. 2020). Details of different heavy metal hyperaccumulating plants are summarized in Table 19.2.

19.5.1 Aluminum

Aluminum (Al) being one of the most abundant elements is very toxic for plants and animals. Chronic Al intoxication causes osteomalacia fractures, encephalopathy, chronic renal failure, Parkinsonism dementia, and Alzheimer's disease in human (Exley 2016). It is also carcinogenic. Elevated mobile Al concentration is the main reason for phytotoxicity of acid soils resulting in the inhibition of plant growth, nutrient uptake, and productivity. The mechanisms of Al tolerance have been studied in barley, wheat, soybean, maize, and Arabidopsis, which include exudation of organic acids and H^+ ions from roots and secretion of mucilage to immobilize Al in the rhizosphere (Kochian et al. 2015; Belimov et al. 2020). Internal Al detoxification in plants involves induction of antioxidant activities, efflux of Al from the root tissues, and vacuolar sequestration of Al.

Symbiotic microorganisms, including PGPR, play an important role in counteracting Al toxicity on plants. Inoculation of maize plants grown in acid soil with P-solubilizing *Burkholderia* sp. has been revealed to decrease the Al accumulation in roots, promoting root elongation, and thereby combating the Al toxicity (Arora et al. 2017). Negative effects of Al toxicity on nodule initiation and inhibition of nitrogen fixation have been reported in pea and soybean (Jaiswal et al. 2018; Basu and Kumar 2020b). *Rhizobium* sp. isolated from nodule of chick pea has been found to be able to bind Al^{3+} due to production of siderophores, suggesting capability of this bacterium to protect the plant against Al toxicity (Sujkowska-Rybkowska and Borucki 2015). On the other hand, Al-tolerant symbiotic AM fungi present in the acid soils have also been found to alleviate Al toxicity in plants (Seguel et al. 2013).

19.5.2 Arsenic

Arsenic (As), a major environmental and food chain contaminant, is a major concern since the last few decades (Zhao et al. 2010). Introduction of As in the food chain takes place by its excessive uptake from soil by crop plants or the irrigation of plants with As-contaminated water. The toxic metalloid has been reported to be carcinogenic even at low levels. Consumption of the As-contaminated food or groundwater over a long period leads to the As poisoning or the arsenicosis, which has become a major threat to the public health. Arsenic exposure affects different morphophysiological processes in plants leading to decrease in plant height, leaf number, biomass, photosynthetic activities, and productivity (Farooq et al. 2016). The As-induced ROS accumulation causes oxidation of lipids and proteins resulting in the cell death (Chen et al. 2017). Arsenic toxicity in the soil leads to straight head

Table 19.2 Different heavy metal hyperaccumulating plant species

Heavy metals	Concentration	Hyperaccumulators	Responses	Reference
Aluminum (Al)	400 μm	*Neolamarckia cadamba*	Affect plant growth	Dai et al. (2020)
Arsenate [As(V)]	12.5, 25, 50, and 100 mg kg^{-1}	*Zea mays* L.	Low As levels improved plant growth, and grain nutrition quality, high As levels reduced ear length, kernel number per row, kernel weight, and grain yield	Ci et al. (2012)
	5, 10, 50 mg kg^{-1}	*Trifolium pretense*	Increase in SOD, POD activities, increased polyamine accumulation, decreased glutathione content, reduction in chlorophyll and carotenoid concentrations	Mascher et al. (2002)
	100 mg l^{-1}	*Cicer arietinum* L.	Decreased seed germination, reduced plant height, and dry weight, reduced seed-setting, decrease in mineral nutrients and amino acid contents in seeds, induction in non-protein thiols, decreased in antioxidant enzymes (SOD, CAT, APX, GPX, and GR) activities	Tripathi et al. (2017)
	100, and 200 μM	*Vigna mungo*	Reduced chlorophyll and carotenoid contents, increased lipid peroxidation, increased SOD, POD, and APX activities, decreased CAT activity	Srivastava et al. (2017)
Arsenite [As(III)]	50 μM	*Oryza sativa* L.	Reduction in seed germination, decreased plant growth, biomass production, relative water content, and chlorophyll content, increased electrolyte leakage, and lipid peroxidation	Kumar et al. (2021a)

(continued)

Table 19.2 (continued)

Heavy metals	Concentration	Hyperaccumulators	Responses	Reference
	150 μM	*Zea mays* L.	Reduced plant growth, and yield, decreased photosynthetic rate, transpiration rate, and stomatal conductance, reduced chlorophyll content	Anjum et al. (2017)
Cadmium (Cd)	20 mg kg^{-1}	Mediterranean saltbush (*Atriplex halimus* L.)	Increase in the amount of photosynthetic pigments	Manousaki and Kalogerakis (2009)
	1, 2, 4, 8, and 16 mg l^{-1}	Castor bean (*Ricinus communis*)	Decrease the production of root and shoot, severe visual symptoms of toxicity both in the roots and in the shoots	de Souza Costa et al. (2012)
	30, 60, 90, 120, 150, and 180 mg kg^{-1} (in soil) 5, 10, 15, 20, 30, and 40 mg l^{-1} (in hydroponics)	*Amaranthus hybridus*	Increased POD, and CAT activities	Zhang et al. (2010)
Chromium (Cr)	10, 20, 50, and 100 μM	*Ocimum tenuiflorum*	Leaves showed increased proline level	Rai et al. (2004)
	50 mg kg^{-1}	*Phaseolus vulgaris*	Decreased carotenoids	Karthik et al. (2016)
	50, 100, 200, and 300 μM	*Zea mays*	Increased SOD and GPX activities, increased lipid peroxidation and H_2O_2 content	Maiti et al. (2012)
	300, 400, 500, and 600 mg kg^{-1}	*Camellia sinensis*	Increased SOD and CAT activities	Tang et al. (2012)
	25, 50, 100, and 200 μM	*Oryza sativa* L.	Increased ethylene synthesis, enhanced SOD, POD, and CAT activities	Ma et al. (2016)
	500 μmol l^{-1}	*Pterogyne nitens*	Polyamines were decreased in leaves and increased in roots; ethylene was increased in the whole plant and NO was increased in the roots	Paiva et al. (2014)
	1.2 mM	*Raphanus sativus*	Enhanced ROS scavenging capacities	Choudhary et al. (2012)

(continued)

Table 19.2 (continued)

Heavy metals	Concentration	Hyperaccumulators	Responses	Reference
	30 mg kg^{-1}	*Vigna radiata, Zea mays*	Increase in superoxide dismutase (SOD), CAT, and POD activities	Dheeba et al. (2014)
	3, 60, and 120 μM	*Matricaria chamomilla L.*	Elevation of nitric oxide, increased phenol, and lignin content, enhanced POD activity, increase in mineral nutrients (Ca, Fe, Cu, Zn) in roots	Kovacik et al. (2014)
Copper (Cu)	100, and 500 μM	*Zea mays* L.	Decreased growth traits, photosynthetic pigments, soluble sugars, phosphorous (P) and potassium (K) contents, and CAT activity increased, proline, MDA content, POD activity, and Cu ion concentration at root and shoot level increased	Abdel Latef et al. (2020)
	50, 100, 200, and 500 mg kg^{-1}	Soybean (*Glycine max)*	Decreased root length	Yusefi-Tanha et al. (2020)
Lead (Pb)	40, 80, and 160 mg kg^{-1}	Mesquite (*Prosopis juliflora-velutina)*	Increased total amylase activity	Arias et al. (2010)
Lithium (Li)	99.6–226.4 g kg^{-1}	*Cirsium arvense, Solanum dulcamara, Holoschoenus vulgaris, Nicotiana tabacum*	Exhibited necrotic spots and reduced growth associated with altered rhythmic movements, abnormal pollen germination	Shahzad et al. (2016)
Mercury (Hg)	2500 μg g^{-1}	Tomato (*Lycopersicon esculentum*)	Reduces their rate of germination, stem height, fruit yield, and chlorosis	Basri et al. (2020)
		O. sativa L.	Decreases tiller, panicle formation, stem height, and yield	

(continued)

Table 19.2 (continued)

Heavy metals	Concentration	Hyperaccumulators	Responses	Reference
Zinc (Zn)	>200 mg kg^{-1}	Spinach (*Spinacia oleracea*), radish (*Raphinus sativus*), and clover (*Trifolium repens*)	Plants had stunted growth of shoots, curling and rolling of young leaves, death of leaf tips, and chlorosis	Mishra et al. (2020)

disease in rice plants (Rahman et al. 2008). It is therefore, a serious concern to take necessary steps to counteract the problem of As toxicity in plants.

Aquatic macrophytes, including *Azolla pinnata*, *Hydrilla verticillata*, and *Lemna minor,* have been shown to have competencies for removal of As from contaminated water (Mishra et al. 2008). Srivastava et al. (2014) have also revealed the aquatic macrophytes *H. verticillata*, *L. minor*, *Ceratophyllum demersum*, *Wolffia globosa*, and *Eichhornia crassipes* to be capable in As accumulation from the contaminated water. In this study, the combination of *C. demersum* and *L. minor* was found to have the highest potential for As removal than the plants used in other combinations or the single plants. Among ferns, *Pteris longifolia*, *P. cretica*, (Zhao et al. 2002) *P. ryukyuensis*, *P. biaurita*, *P. quadriaurita* (Srivastava et al. 2006), and *Pityrogramma calomelanos* (Francesconi et al. 2002) have been found to have greater potential for As hyperaccumulation from contaminated soil. Higher plants including *Eclipta alba* (Dwivedi et al. 2008), *Isatis cappadocia* (Karimi et al. 2009), and *Sesuvium portulacastrum* (Lokhande et al. 2011) have also been identified to be potential As hyperaccumulators.

19.5.3 Cadmium

Cadmium (Cd) is a tremendously toxic environmental pollutant causing lethal effects to animals and plants. It is classified as a human class I carcinogen. Chronic Cd poisoning due to prolonged oral Cd ingestion causes itai-itai disease in human (Genchi et al. 2020). In plants, Cd stress reduces growth, leaf area, dry matter, and yield (Shanying et al. 2017). It also affects photosynthesis and respiration, induces oxidative damage, and decreases nutrient uptake ability in plants. It is therefore, a serious concern to take necessary steps to remove Cd toxicity from the environment (Kapoor et al. 2021).

The process of phytoextraction of Cd has been mediated by several potential hyperaccumulating plant species, including *Cassia alata*, *Celosia argentea*, *Kummerowia striata*, *Nicotiana tabacum*, *Momordica charantia*, *Solanum melonaena*, *Swietenia macrophylla*, *Salix mucronata*, and *Vigna unguiculata* (Raza et al. 2020). Several other plant species like *Abelmoschus manihot*, *Atriplex halimus*, *Brassica chinensis*, *B. juncea*, *B. napus*, *Glycine max*, *Lolium perenne*, *Macleaya cordata*, *Oryza sativa*, *Paspalum scrobiculatum*, *Petroselinum hortense*, *Quercus robur*, *Sedum alfredii*, *Solanum lycopersicum*, *S. tuberosum*, and *Triticum aestivum* have also been shown the potential of Cd tolerance and phytoremediation through their enhanced antioxidant defense system. Few microorganisms like *Aspergillus niger* (fungus) (Ren et al. 2009), *Aspergillus versicolor* (Fazli et al. 2015), *Pleurotus ostreatus* (Kapahi and Sachdeva 2017), *Pseudomonas aeruginosa*, *Streptomyces* sp., *Fomitopsis pinicola*, and *Bacillus* sp. (Bagot et al. 2006) have also been reported to play a crucial role in removal of Cd from the contaminated soil. Salinity has been shown to be a key factor in the translocation of Cd from the roots to the shoots in *Aster tripolium*, *Temnothorax smyrnensis*, and *Potamogeton pectinatus* (Manousaki et al. 2008). Salinity increased the Cd concentration of the shoots as well as that of the whole plants.

19.5.4 Chromium

Chromium (Cr) is the most widespread toxic trace elements that adversely affect crop productivity throughout the world. The Cr contamination is caused by natural (weathering of rocks) or anthropogenic activities (used in various industries like chrome plating, alloys, paints, use of excess fertilizers). The Cr is found in many forms but Cr(0), Cr(III), and Cr(VI) are the most stable and common forms. Among these forms, the hexavalent chromium Cr(VI) is highly toxic to animals and plants. The Cr (VI) and its compounds have carcinogenic effects when inhaled or ingested. Being more soluble in water, Cr(VI) has greater availability for plants and it has higher ability of penetrating plant roots (Shanker et al. 2005).

Uptake of Cr(VI) in plants occurs through sulfur transporters (Kovacik et al. 2013). Toxic levels of Cr in the soil decrease plant height, roots and shoot biomass, chlorophyll content, transpiration rate, stomatal conductance, net photosynthesis, and water use efficiency of plants (Sharma et al. 2020). It also induces the ROS overproduction causing oxidative stress and damage to lipids and proteins. The Cr (VI) also causes alterations in nutrient assimilation, hormonal homeostasis, and genotoxicity in plants, thereby inhibiting plant growth and development. Furthermore, Cr(VI) causes reduced tissue density of roots due to enhanced cellular accumulation of Cr which leads to the cell death.

Plants develop antioxidant defense system to alleviate Cr toxicity through effective ROS detoxification (Maiti et al. 2012). In a recent study, Cr toxicity has been revealed to be ameliorated by increased superoxide dismutase (SOD), peroxidase (POD), catalase (CAT), and ascorbate peroxidase (APX) activities in *Brassica napus* L. (Zaheer et al. 2020). Increased glutathione production has also been reported to detoxify Cr toxicity in *Oryza sativa*, *Actinidia deliciosa*, *Pistia stratiotes*, *B. napus*, *Salvinia natans*, *S. rotundifolia*, and *S. minima* (Shahid et al. 2017).

19.5.5 Copper

Copper (Cu) is one of the important nutrients for plants and humans. The Cu toxicity has detrimental effect on human health leading to liver disorder and Alzheimer's disease. In plants, Cu toxicity causes chlorosis and necrosis, leaf discoloration stunting, and root growth inhibition (Kumar et al. 2021b). The Cu toxicity also has detrimental effects on morphophysiology and nutrient uptake in plants. High Cu concentration induces DNA damage, decreased photosynthetic rate, loss of cell membrane integrity, decreased enzyme activity, and respiration leading to growth reduction in plants.

Several wild plant species having Cu phytoremediation potential growing in the mine polluted areas include *Hypericum perforatum*, *Phleum pretense*, *Thymus kotschyanus*, and *Teucrium orientale* (Ghazaryan et al. 2019). In a study, Lu et al. (2018) have shown aquatic plant species *Acorus calamus*, *Arundina graminifolia*, *Eichhornia crassipes*, *Echinodorus major*, *Juncus effusus*, *Nymphaea tetragona*, *Pistia stratiotes*, and *Sagittaria sagittifolia* to exhibit exceptional potential for remediation of Cu pollution. Another recent study by Covre et al. (2020) revealed *Cedrela fissilis* and *Khaya ivorensis* to have good potential for Cu accumulation. Another plant species *Vetiveria zizanioides* has been reported to have potential for Cu phytostabilization in Cu-mine tailing (Chu et al. 2020).

19.5.6 Lead

Lead (Pb) is a persistent toxic pollutant of concern that is produced due to increasing anthropogenic pressure on the environment. The Pb is absorbed by the plant roots via the Ca^{2+}-permeable channels or the apoplastic pathway. Excessive Pb accumulation in plants directly or indirectly impairs the morphophysiological and biochemical functions, thereby inducing various deleterious effects. The Pb toxicity causes swollen, bent, short, and stubby roots with increased number of secondary roots per unit root length. Severe Pb toxicity in plants results in growth inhibition with fewer, smaller, and more brittle leaves having dark purplish abaxial surfaces. Plant growth retardation from Pb exposure may be attributed to nutrient metabolic disturbances and disturbed photosynthesis. Exposure of *Allium sativum* roots to toxic concentration of Pb leads to mitochondrial swelling, loss of cristae, vacuolization of endoplasmic reticulum and dictyosomes, and injured plasma membrane. In most cases, the toxic effect of Pb on plant growth is time- and dose-dependent. Moreover, the effects of Pb toxicity vary with plant species, i.e., hyperaccumulators naturally tolerate more Pb toxicity than the sensitive plants (Pourrut et al. 2011).

The Pb adsorption onto roots has been documented to occur in several plant species like *Vigna unguiculata*, *Festuca rubra*, *Brassica juncea*, *Lactuca sativa*, and *Funaria hygrometrica*. For most plant species, the majority of absorbed Pb (approximately 95% or more) is accumulated in the roots, and only a small fraction is translocated to the aerial plant parts, as reported in *Avicennia marina*, *Phaseolus vulgaris*, *Pisum sativum*, *Vicia faba*, *Vigna unguiculata*, *Lathyrus sativus*, *Nicotiana tabacum*, and *Zea mays*. Several hyperaccumulator plant species, such as *Brassica pekinensis* and *Pelargonium* sp., are capable of translocating higher concentrations of Pb to aerial plant parts, without incurring damage to their basic metabolic functions.

19.5.7 Lithium

The essentiality and toxicity of lithium (Li) on higher plants are not clear till date. Previous studies indicated Li salts to be highly toxic inducing the formation of necrosis in plants. It also causes considerable reduction in plant growth. Other Li toxic effects include altered rhythmic movement of petals and disrupted pollen development. Plant root is the first organ that comes in contact with Li in soil, and Li in excess alters gravitropic growth of maize roots. Furthermore, Li toxicity affects cold-induced dephosphorylation of microtubules in mesophyll cells of spinach. In the Li-rich soils, damage of root tips and chlorotic and necrotic spots on leaves have been observed in corn. Nonetheless, different plant species showed plastic behavior in sensitivity and tolerance to and noted that plants belonging to the families *Asteraceae* and *Solanaceae* show tolerance against Li toxicity and sustain normal plant growth.

Some plants, notably *Cirsium arvense* and *Solanum dulcamara,* accumulate higher concentration of Li. Halophilic plants like *Apocynum pictum*, *Carduus arvense*, and *Holoschoenus vulgaris* may reach up to 99.6–226.4 g kg^{-1} Li contents. *Apocynum venetum* is a potentially rich target of Li biofortification owing to its ability to accumulate Li in natural habitat. However, the medicinal effects (existence of various flavonoid compounds in leaves) of *A. venetum* might also be attributed to the existence of high level of Li and evaluate the feasibility of *A. venetum* for the Li bio-enrichment (Jiang et al. 2019).

19.5.8 Mercury

Mercury (Hg) is a naturally occurring persistent environmental pollutant generated from minings, petrochemicals, paintings, industries and agricultural sources like fertilizers, fungicidal sprays, and bioaccumulated in fish, animals, and human beings. Severe Hg poisoning (methylmercury) in human causes neurological disease known as Minamata or Chisso-Minamata (Yorifuji and Tsuda 2014). The Hg also affects growth and productivity of different plant species. Sahu et al. (2012) have shown the Hg stress to limit plant growth and nutrition and cause oxidative damage in wheat. High concentration soil Hg affects the roots of *Aeschynomene fluminensis* and *Polygonum acuminatum* (Mariano et al. 2020). The Hg contamination reduces the germination rate, stem height, and fruit yield and causes chlorosis in tomatoes (Shekar et al. 2011). Likewise, in rice, Hg stress impedes the tiller and panicle formation, leading to the decrease in stem height and yield (Basri et al. 2020). Chen and Yang (2012) showed the Hg to interfere with the electron transport chain in the chloroplasts and mitochondria consequently, affecting the photosynthesis and oxidative metabolism in plants. It also reduces water uptake in plants by hindering the activities of the aquaporins.

A number of plant species, including *Zea mays, Ceratophyllum demersum, Anodonta grandix, Victoria amazonica, Sphagnum girgensohnii, Convolvulus* sp., *Cyrtomium macrophyllum,* and *Eichhornia crassipes* have been reported to be the hyperaccumulators of Hg (Kumar et al. 2017). The leaf tissue of *Cyrtomium macrophyllum* shows high resistance to Hg stress (Xun et al. 2017). In a field study conducted by Fernandez et al. (2017), some native plant species (*Festuca rubra* L., *Leontodon taraxacoides, Equisetum telmateia*) were used for the phytoextraction of Hg from the mining area, where higher concentrations Hg were accumulated mainly in the leaves of the plants.

19.5.9 Zinc

Zinc (Zn) is an important component of thousands of proteins and is essential for mineral nutrition. However, increased concentration of Zn can become toxic in both plants and animals. Chronic ingestion of Zn leads to sideroblastic anemia, granulocytopenia, and myelodysplastic syndrome in human (Irving et al. 2003). In plants, Zn toxicity causes stunted plant growth, defective chlorophyll biosynthesis, chloroplast degradation, reduced Mg, Mn, and P uptake, and reduced yields (Broadley et al. 2007). Crops differ in their susceptibility to Zn toxicity. Dicots are more sensitive to the Zn toxicity in acidic soils; whereas, Graminaceous plants show more sensitivity in alkaline soil.

Numerous metallophytes belong to the families *Amaranthaceae*, *Brassicaceae*, *Caryophyllaceae*, *Lamiaceae*, *Rubiaceae*, *Polygonaceae,* and *Poaceae* are known to be the Zn hyperaccumulators. The Zn hypertolerance has been reported in *Agrostis stolonifera*, *A. capillaris*, *Arabidopsis halleri*, *A. arenosa*, *Arenaria patula*, *Avicenna marina*, *Betula pendula*, *Mimulus guttatus*, *Mirabilis jalapa*, *Silene vulgaris*, *S. dioica*, *Thlaspi alpestre*, *T. caerulescens,* and *Thlaspiceras oxyceras*. Some of the Zn hyperaccumulator species including *Acer pseudoplatanus*, *Biscutella laevigata*, *Dianthus* sp., *Festuca rubra*, *Galium mollugo*, *Minuartia verna*, *Polycarpaea synandra*, and *Rumex acetosa* accumulate up to 3000 μg Zn g^{-1} DW in their shoots.

19.6 Improvement of Phytoremediation Ability of Plants

Effective phytoremediation requires the improvement of certain traits and minimization of limitations to enhance the plants' ability for the HM removal from the environment. To improve the growth rate and biomass of the hyperaccumulator plant species or to introduce the hyperaccumulation traits, traditional plant breeding or genetic engineering may be employed (DalCorso et al. 2019).

Traditional plant breeding includes somatic hybridization technique to transfer the HM hyperaccumulation trait to the plants having high biomass. Protoplasts isolated from the Zn hyperaccumulator *T. caerulescens* and higher biomass producing *B. napus* were fused using the electrofusion (Brewer et al. 1999). The somatic hybrids showed ability to accumulate enhanced Cd and Zn. Likewise, sunflower giant mutant has been developed by using chemical mutagen ethyl methanesulfonate (EMS). The mutants showed increased ability for extraction of Cd, Pb, and Zn (7.5,

8.2, and 9.2 times more accumulation, respectively, than control plants) (Nehnevajova et al. 2007).

Genetic engineering has also been proved to be a prospective method for enhancing the phytoremediation abilities of the plants (Sarma et al. 2021; Sonowal et al. 2022). This technique takes less time to introduce the desirable traits for phytoremediation in plants than the traditional breeding. Sexually incompatible plant species can also be improved through the genetic engineering by transferring the desirable genes from the HM hyperaccumulators. Improvement of HM removal capacity in plants can be achieved by overexpressing the candidate genes for the HM uptake (*ZIP*, *HMA*, *MATE*, *MTP*), translocation, and sequestration in the hyperaccumulators (Das et al. 2016). Chelators improve the bioavailability of HMs by acting as metal-binding ligands. Therefore, genes encoding the natural chelators can also be overexpressed for the improvement of the HM uptake and translocation in plants (Yan et al. 2020).

19.7 Conclusion

HMs are environmental and food chain contaminants posing serious threat to sustainable agricultural production. Therefore, alleviation of the HM contamination from the soil and water has become an essential requirement to encounter the food security. Phytoremediation has been established as a novel and promising technique to clean-up the HM-contaminated soil and water. Application of HM hyperaccumulating metallophytes is the most candid tactic for the phytoremediation. Additionally, application of genetic engineering in improvement of the metallophytes' performance may further be advantageous for the efficacious phytoremediation. Comprehensive understanding of the physiological as well as the molecular mechanisms of the HM uptake, translocation, and detoxification in plants may be beneficial in improving the phytoremediation potential of the metallophytes through genetic engineering.

References

Abdel Latef AAH, Zaid A, Abo-Baker AA, Salem W, Abu Alhmad MF (2020) Mitigation of copper stress in maize by inoculation with *Paenibacillus polymyxa* and *Bacillus circulans*. Plants (Basel) 9(11):1513

Aborode FA, Raab A, Voigt M, Costa LM, Krupp EM, Feldmann J (2016) The importance of glutathione and phytochelatins on the selenite and arsenate detoxification in *Arabidopsis thaliana*. J Environ Sci 49:150–161

Ali H, Khan E, Sajad MA (2013) Phytoremediation of heavy metals—concepts and applications. Chemosphere 91:869–881

Anjum SA, Tanveer M, Hussain S, Ashraf U, Khan I, Wang L (2017) Alteration in growth, leaf gas exchange, and photosynthetic pigments of maize plants under combined cadmium and arsenic stress. Water Air Soil Pollut 228:13

Arao T, Makino T, Kawasaki A, Akahane I, Kiho N (2018) Effect of air temperature after heading of rice on the arsenic concentration of grain. Soil Sci Plant Nutr 64(3):433–437

Arias JA, Peralta-Videa JR, Ellzey JT, Viveros MN, Ren M, Mokgalaka-Matlala NS, Castillo-Michel H, Gardea-Torresdey JL (2010) Plant growth and metal distribution in tissues of *Prosopis juliflora-velutina* grown on chromium contaminated soil in the presence of *Glomus deserticola*. Environ Sci Technol 44(19):7272–7279

Arif N, Yadav V, Singh S, Singh S, Ahmad P, Mishra RK, Sharma S, Tripathi DK, Dubey NK, Chauhan DK (2016) Influence of high and low levels of plant-beneficial heavy metal ions on plant growth and development. Front Environ Sci 4:69

Arora P, Singh G, Tiwari A (2017) Effect of microbial inoculation in combating the aluminium toxicity effect on growth of *Zea mays*. Cell Mol Biol (Noisy-le-Grand) 63:79–82

Assuncao AG, Da Costa MP, De Folter S, Vooijs R, Schat H, Aarts MGM (2001) Elevated expression of metal transporter genes in three accessions of the metal hyperaccumulator *Thlaspi caerulescens*. Plant Cell Environ 24:217–226

Bagot D, Lebeau T, Jezequel K (2006) Microorganisms for remediation of cadmium-contaminated soils. Environ Chem Lett 4:207–211

Banuelos G, Meek D (1990) Accumulation of selenium in plants grown on selenium-treated soil. J Environ Qual 19:772–777

Basri, Sakakibara M, Sera K (2020) Mercury in soil and forage plants from artisanal and small-scale gold mining in the bombana area, Indonesia. Toxics 8:15

Bastow EL, Garcia De La Torre VS, Maclean AE, Green RT, Merlot S, Thomine S, Balk J (2018) Vacuolar iron stores gated by NRAMP3 and NRAMP4 are the primary source of iron in germinating seeds. Plant Physiol 177:1267–1276

Basu S, Kumar G (2020a) Stress signalling in the phytomicrobiome: breadth and potential. In: Kumar M, Kumar V, Prasad R (eds) Phyto-Microbiome in stress regulation. environmental and microbial biotechnology. Springer, Singapore, pp 245–268

Basu S, Kumar G (2020b) Nitrogen fixation in a legume-rhizobium symbiosis: the roots of a success story. In: Varma A, Tripathi S, Prasad R (eds) Plant microbe symbiosis. Springer, Singapore, pp 35–53

Basu S, Kumar G (2021a) Plant microbe interaction for changing endophytic colonization to improve plant productivity. In: Verma JP, Macdonald CA, Gupta VK, Podile AR (eds) New and future developments in microbial biotechnology and bioengineering. Elsevier, Amsterdam, pp 137–147

Basu S, Kumar G (2021b) Exploring the significant contribution of silicon in regulation of cellular redox homeostasis for conferring stress tolerance in plants. Plant Physiol Biochem 166:393–404

Basu S, Giri RK, Benazir I, Kumar S, Rajwanshi R, Dwivedi SK, Kumar G (2017) Comprehensive physiological analyses and reactive oxygen species profiling in drought tolerant rice genotypes under salinity stress. Physiol Mol Biol Plants 23:837–850

Basu S, Kumar G, Kumari N, Kumari S, Shekhar S, Kumar S, Rajwanshi R (2020) Reactive oxygen species and reactive nitrogen species induce lysigenous aerenchyma formation through programmed cell death in rice roots under submergence. Environ Exp Bot 177:104118

Basu S, Kumari S, Kumar P, Kumar G, Rajwanshi R (2021a) Redox imbalance impedes photosynthetic activity in rice by disrupting cellular membrane integrity and induces programmed cell death under submergence. Physiol Plant 172(3):1764–1778

Basu S, Kumari S, Kumar A, Shahid R, Kumar S, Kumar G (2021b) Nitro-oxidative stress induces the formation of roots' cortical aerenchyma in rice under osmotic stress. Physiol Plant 172(2): 963–975

Belimov AA, Shaposhnikov AI, Syrova DS, Kichko AA, Guro PV, Yuzikhin OS, Azarova TS, Sazanova AL, Sekste EA, Litvinskiy VA, Nosikov VV, Zavalin AA, Andronov EE, Safronova VI (2020) The role of symbiotic microorganisms, nutrient uptake and rhizosphere bacterial community in response of pea (*Pisum sativum* L.) genotypes to elevated Al concentrations in soil. Plan Theory 9(12):1–25

Bizily SP, Rugh CL, Meagher RB (2000) Phytodetoxification of hazardous organomercurials by genetically engineered plants. Nat Biotechnol 18:213–217

Brewer EP, Saunders JA, Angle JS, Chaney RL, Mcintosh MS (1999) Somatic hybridization between the zinc accumulator *Thlaspi caerulescens* and *Brassica napus*. Theor Appl Genet 99:761–771

Broadley MR, White PJ, Hammond JP, Zelko I, Lux A (2007) Zinc in plants. New Phytol 173(4): 677–702

Burges A, Alkorta I, Epelde L, Garbisu C (2018) From phytoremediation of soil contaminants to phytomanagement of ecosystem services in metal contaminated sites. Int J Phytoremediation 20: 384–397

Cailliatte R, Lapeyre B, Briat JF, Mari S, Curie C (2009) The NRAMP6 metal transporter contributes to cadmium toxicity. Biochem J 422:217–228

Cailliatte R, Schikora A, Briat JF, Mari S, Curie C (2010) High-affinity manganese uptake by the metal transporter NRAMP1 is essential for *Arabidopsis* growth in low manganese conditions. Plant Cell 22:904–917

Chen J, Yang ZM (2012) Mercury toxicity, molecular response and tolerance in higher plants. Biometals 25(5):847–857

Chen Y, Han Y-H, Cao Y, Zhu Y-G, Rathinasabapathi B, Ma LQ (2017) Arsenic transport in rice and biological solutions to reduce arsenic risk from rice. Front Plant Sci 8:268

Choudhary SP, Kanwar M, Bhardwaj R, Yu J-Q, Tran L-SP (2012) Chromium stress mitigation by polyaminebrassinosteroid application involves phytohormonal and physiological strategies in *Raphanus sativus* L. PLoS One 7(3):e33210

Chu Z, Wang X, Wang Y, Zha F, Dong Z, Fan T, Xu X (2020) Influence of coal gangue aided phytostabilization on metal availability and mobility in copper mine tailings. Environ Earth Sci 79(3):68

Ci X-K, Liu H-L, Hao Y-B, Zhang J-W, Liu P, Dong S-T (2012) Arsenic distribution, species, and its effect on maize growth treated with arsenate. J Integr Agric 11(3):416–423

Cieslinski G, Van Rees K, Szmigielska A, Krishnamurti G, Huang P (1998) Low-molecular-weight organic acids in rhizosphere soils of durum wheat and their effect on cadmium bioaccumulation. Plant and Soil 203:109–117

Covre WP, da Silveira Pereira WV, Goncalves DAM, Teixeira OMM, do Amarante CB, Fernandes AR (2020) Phytoremediation potential of *Khaya ivorensis* and *Cedrela fissilis* in copper contaminated soil. J Environ Manage 268:110733

Dai B, Chen C, Liu Y, Liu L, Qaseem MF, Wang J, Li H, Wu A-M (2020) Physiological, biochemical, and transcriptomic responses of *Neolamarckia cadamba* to aluminum stress. Int J Mol Sci 21(24):9624

DalCorso G, Fasani E, Manara A, Visioli G, Furini A (2019) Heavy metal pollutions: state of the art and innovation in phytoremediation. Int J Mol Sci 20:1–17

Dalvi AA, Bhalerao SA (2013) Response of plants towards heavy metal toxicity: an overview of avoidance, tolerance and uptake mechanism. Ann Plant Sci 2:362–368

Das N, Bhattacharya S, Maiti MK (2016) Enhanced cadmium accumulation and tolerance in transgenic tobacco overexpressing rice metal tolerance protein gene OsMTP1 is promising for phytoremediation. Plant Physiol Biochem 105:297–309

de Souza Costa ET, Guilherme LR, de Melo EE, Ribeiro BT, Dos Santos B, Inacio E, da Costa Severiano E, Faquin V, Hale BA (2012) Assessing the tolerance of castor bean to Cd and Pb for phytoremediation purposes. Biol Trace Elem Res 145(1):93–100

Dheeba B, Sampathkumar P, Kannan K (2014) growth, chromium accumulation potential and biochemical changes of *Vigna radiata* and *Zea mays* grown with effective microbes. Proc Natl Acad Sci, India, Sect B Biol Sci 84:381–387

Dong B, Niu L, Meng D, Song Z, Wang L, Jian Y, Fan X, Dong M, Yang Q, Fu Y (2019) Genome-wide analysis of MATE transporters and response to metal stress in *Cajanus cajan*. J Plant Interact 14(1):265–275

Dwivedi S, Srivastava S, Mishra S, Dixit B, Kumar A, Tripathi RD (2008) Screening of native plants and algae growing on fly-ash affected areas near National Thermal Power Corporation, Tanda, Uttar Pradesh, India for accumulation of toxic heavy metals. J Hazard Mater 158:359–365

Etim EE (2012) Phytoremediation and its mechanisms: a review. Int J Environ Bioenergy 2:120–136

Exley C (2016) The toxicity of aluminium in humans. Morphologie 100(329):51–55

Farooq MA, Islam F, Ali B, Najeeb U, Mao B, Gill RA, Yan G, Siddique KHM, Zhou W (2016) Arsenic toxicity in plants: cellular and molecular mechanisms of its transport and metabolism. Environ Exp Bot 132:42–52

Fazli MM, Soleimani N, Mehrasbi M, Darabian S, Mohammadi J, Ramazani A (2015) Highly cadmium tolerant fungi: their tolerance and removal potential. J Environ Health Sci Eng 13:19

Fernandez S, Poschenrieder C, Marceno C, Gallego JR, Jimenez-Gamez D, Bueno A, Afif E (2017) Phytoremediation capability of native plant species living on Pb-Zn and hg-as mining wastes in the Cantabrian range, north of Spain. J Geochem Explor 174:10–20

Francesconi K, Visoottiviseth P, Sridokchan W, Goessler W (2002) Arsenic species in an arsenic hyperaccumulating fern, *Pityrogramma calomelanos*: a potential phytoremediator of arsenic-contaminated soils. Sci Total Environ 284:27–35

Genchi G, Sinicropi MS, Lauria G, Carocci A, Catalano A (2020) The effects of cadmium toxicity. Int J Environ Res Publ Health 17:3782

Gerhardt KE, Gerwing PD, Greenberg BM (2017) Opinion: taking phytoremediation from proven technology to accepted practice. Plant Sci 256:170–185

Ghazaryan K, Movsesyan H, Ghazaryan N, Watts BA (2019) Copper phytoremediation potential of wild plant species growing in the mine polluted areas of Armenia. Environ Pollut 249:491–501

Guerinot ML (2000) The ZIP family of metal transporters. Biochim Biophys Acta 1465:190–198

Hanikenne M, Baurain D (2014) Origin and evolution of metal P-type ATPases in Plantae (Archaeplastida). Front Plant Sci 4:544

Irving JA, Mattman A, Lockitch G, Farrell K, Wadsworth LD (2003) Element of caution: a case of reversible cytopenias associated with excessive zinc supplementation. CMAJ 169(2):129–131

Ismael MA, Elyamine AM, Moussa MG, Cai M, Zhao X, Hu C (2019) Cadmium in plants: uptake, toxicity, and its interactions with selenium fertilizers. Metallomics 11(2):255–277

Jaiswal SK, Naamala J, Dakora FD (2018) Nature and mechanisms of aluminium toxicity, tolerance and amelioration in symbiotic legumes and rhizobia. Biol Fertil Soils 54:309–318

Jiang L, Wang L, Tanveer M, Tian C (2019) Lithium biofortification of medicinal tea *Apocynum venetum*. Sci Rep 9:8182

Kamal M, Ghaly AE, Mahmoud N, CoteCote R (2004) Phytoaccumulation of heavy metals by aquatic plants. Environ Int 29:1029–1039

Kamal S, Prasad R, Varma A (2010) Soil microbial diversity in relation to heavy metals. In: Sherameti I, Varma A (eds) Soil heavy metals, vol 19. Springer-Verlag, Berlin, Heidelberg, pp 31–64

Kapahi M, Sachdeva S (2017) Mycoremediation potential of *Pleurotus* species for heavy metals: a review. Bioresour Bioprocess 4:32

Kapoor D, Singh S, Ramamurthy PC, Jan S, Bhardwaj S, Gill SS, Prasad R, Singh J (2021) Molecular consequences of cadmium toxicity and its regulatory networks in plants. Plant Gene. https://doi.org/10.1016/j.plgene.2021.100342

Karimi N, Ghaderian SM, Raab A, Feldmann J, Meharg AA (2009) An arsenic-accumulating, hypertolerant *Brassica*, *Isatis capadoica*. New Phytol 184:41–47

Karthik C, Oves M, Thangabalu R, Sharma R, Santhosh SB, Arulselvi PI (2016) *Cellulosimicrobium funkei*-like enhances the growth of *Phaseolus vulgaris* by modulating oxidative damage under Chromium(VI) toxicity. J Adv Res 7(6):839–850

Kidwai M, Dhar YV, Gautam N, Tiwari M, Ahmad IZ, Asif MH, Chakrabarty D (2019) *Oryza sativa* class III peroxidase (OsPRX38) overexpression in *Arabidopsis thaliana* reduces arsenic accumulation due to apoplastic lignification. J Hazard Mater 362:383–393

Kochian LV, Pineros MA, Liu J, Magalhaes JV (2015) Plant adaptation to acid soils: the molecular basis for crop aluminum resistance. Annu Rev Plant Biol 66:571–598

Komal T, Mustafa M, Ali Z, Kazi AG (2015) Heavy metal uptake and transport in plants. In: Sherameti I, Varma A (eds) Heavy metal contamination of soils. Springer, Cham

Kovacik J, Babula P, Klejdus B (2013) Chromium uptake and consequences for metabolism and oxidative stress in chamomile plants. J Agric Food Chem 61:7864–7873

Kovacik J, Babula P, Hedbavny J, Klejdus B (2014) Hexavalent chromium damages chamomile plants by alteration of antioxidants and its uptake is prevented by calcium. J Hazard Mater 273:110–117

Kumar G, Kushwaha HR, Panjabi-Sabharwal V, Kumari S, Joshi R, Karan R, Mittal S, Singla-Pareek SL, Pareek A (2012) Clustered metallothionein genes are co-regulated in rice and ectopic expression of OsMT1e-P confers multiple abiotic stress tolerance in tobacco via ROS scavenging. BMC Plant Biol 12:107

Kumar B, Smita K, Flores LC (2017) Plant mediated detoxification of mercury and lead. Arab J Chem 10(2):S2335–S2342

Kumar A, Basu S, Kumar G (2021a) Evaluating the effect of seed-priming for improving arsenic tolerance in rice. J Plant Biochem Biotechnol. https://doi.org/10.1007/s13562-021-00666-0

Kumar V, Pandit S, Sidhu GPS, Sharma A, Khanna K, Kaur P, Bali AS, Setia R (2021b) Copper bioavailability, uptake, toxicity and tolerance in plants: a comprehensive review. Chemosphere 262:127810

LeDuc DL, Terry N (2005) Phytoremediation of toxic trace elements in soil and water. J Ind Microbiol Biotechnol 32:514–520

Liu H, Zhao H, Wu L, Liu A, Zhao F-J, Xu W (2017) Heavy metal ATPase 3 (HMA3) confers cadmium hypertolerance on the cadmium/zinc hyperaccumulator *Sedum plumbizincicola*. New Phytol 215(2):687–698

Lokhande VH, Srivastava AK, Srivastava S, Nikam TD, Suprasanna P (2011) Regulated alterations in redox and energetic status are the key mediators of salinity tolerance in the halophyte *Sesuvium portulacastrum* (L.) L. Plant Growth Regul 65:287–298

Lorestani B, Yousefi N, Cheraghi M, Farmany A (2013) Phytoextraction and phytostabilization potential of plants grown in the vicinity of heavy metal-contaminated soils: a case study at an industrial town site. Environ Monit Assess 185:10217–10223

Lu D, Huang Q, Deng C, Zheng Y (2018) Phytoremediation of copper pollution by eight aquatic plants. Pol J Environ Stud 2(1):175–181

Ma Y, Prasad M, Rajkumar M, Freitas H (2011) Plant growth promoting rhizobacteria and endophytes accelerate phytoremediation of metalliferous soils. Biotechnol Adv 29:248–258

Ma J, Lv C, Xu M, Chen G, Lv C, Gao Z (2016) Photosynthesis performance, antioxidant enzymes, and ultrastructural analyses of rice seedlings under chromium stress. Environ Sci Pollut Res Int 23(2):1768–1778

Ma Q, Yi R, Li L, Liang Z, Zeng T, Zhang Y, Huang H, Zhang X, Yin X, Cai Z, Mu Y, Cheng Y, Zeng Q, Li X, Nian H (2018) GsMATE encoding a multidrug and toxic compound extrusion transporter enhances aluminum tolerance in *Arabidopsis thaliana*. BMC Plant Biol 18:212

Madhaiyan M, Poonguzhali S, Sa T (2007) Metal tolerating methylotrophic bacteria reduces nickel and cadmium toxicity and promotes plant growth of tomato (*Lycopersicon esculentum* L.). Chemosphere 69:220–228

Mahar A, Wang P, Ali A, Awasthi MK, Lahori AH, Wang Q, Li R, Zhang Z (2016) Challenges and opportunities in the phytoremediation of heavy metals contaminated soils: a review. Ecotox Environ Safe 126:111–121

Maiti S, Ghosh N, Mandal C, Das K, Dey N, Adak MK (2012) Responses of the maize plant to chromium stress with reference to antioxidation activity. Braz J Plant Phsiol 24(3):203–212

Manousaki E, Kalogerakis N (2009) Phytoextraction of Pb and Cd by the Mediterranean saltbush (*Atriplex halimus* L.): metal uptake in relation to salinity. Environ Sci Pollut Res Int 16(7):844–854

Manousaki E, Kadukova J, Papadantonakis N, Kalogerakis N (2008) Phytoextraction and phytoexcretion of cd by the leaves of *Tamarix smyrnensis* growing on contaminated non-saline and saline soils. Environ Res 106(3):326–332

Mariano C, Mello IS, Barros BM, da Silva GF, Terezo AJ, Soares MA (2020) Mercury alters the rhizobacterial community in Brazilian wetlands and it can be bioremediated by the plant-bacteria association. Environ Sci Pollut Res 27:13550–13564

Marques APGC, Rangel AOSS, Castro PML (2009) Remediation of heavy metal contaminated soils: phytoremediation as a potentially promising clean-up technology. Crit Rev Environ Sci Technol 39(8):622–654

Mascher R, Lippmann B, Holzinger S, Bergmann H (2002) Arsenate toxicity: effects on oxidative stress response molecules and enzymes in red clover plants. Plant Sci 163(5):961–969

Mendez MO, Maier RM (2008) Phytostabilization of mine tailings in arid and semiarid environments—an emerging remediation technology. Environ Health Perspect 116:278–283

Migocka M, Kosieradzka A, Papierniak A, Maciaszczyk-Dziubinska E, Posyniak E, Garbiec A, Filleur S (2014) Two metal-tolerance proteins, MTP1 and MTP4, are involved in Zn homeostasis and cd sequestration in cucumber cells. J Exp Bot 66(3):1001–1015

Mishra B, McDonald LM, Roy M, Lanzirotti A, Myneni SCB (2020) Uptake and speciation of zinc in edible plants grown in smelter contaminated soils. PLoS One 15(4):e0226180

Mishra VK, Upadhyay AR, Pathak V, Tripathi BD (2008) Phytoremediation of mercury and arsenic from tropical opencast coalmine effluent through naturally occurring aquatic macrophytes. Water Air Soil Pollut 192:303–314

Mokhtar H, Morad N, Fizri FFA (2011) Phytoaccumulation of copper from aqueous solutions using *Eichhornia crassipes* and *Centella asiatica*. Int J Environ Sci Dev 2:205–210

Moreno FN, Anderson CWN, Stewart RB, Robinson BH (2004) Mercury phytoextraction and phytovolatilisation from hg-contaminated artisanal mine sites. In: Phytoremediation of mercury-contaminated mine wastes, pp 147–159

Muthusaravanan S, Sivarajasekar N, Vivek JS, Paramasivan T, Naushad M, Prakashmaran J, Gayathri V, Al-Duaij OK (2018) Phytoremediation of heavy metals: mechanisms, methods and enhancements. Environ Chem Lett 16:1339–1359

Nagajyoti PC, Lee KD, Sreekanth TVM (2010) Heavy metals, occurrence and toxicity for plants: a review. Environ Chem Lett 8:199–216

Nehnevajova E, Herzig R, Federer G, Erismann K-H, Schwitzguebel J-P (2007) Chemical mutagenesis—a promising technique to increase metal concentration and extraction in sunflowers. Int J Phytoremediation 9:149–165

Nevo Y, Nelson N (2006) The NRAMP family of metal-ion transporters. BBA-Mol Cell Res 1763(7):609–620

Ojuederie OM, Babalola OO (2017) Microbial and plant-assisted bioremediation of heavy metal polluted environments: a review. Int J Env Res Publ Health 14:1–26

Paiva LB, Correa SF, Santa Catarina C, Floh EIS, Silva MGD, Vitoria AP (2014) Ecophysiological and biochemical parameters for assessing Cr^{+6} stress conditions in *Pterogyne nitens* Tul.: New and usual methods for the management and restoration of degraded areas. Environ Eng Manag J 13:3073–3081

Parrotta L, Guerriero G, Sergeant K, Cai G, Hausman J-F (2015) Target or barrier? The cell wall of early- and later-diverging plants vs cadmium toxicity: differences in the response mechanisms. Front Plant Sci 6:133

Peng HY, Yang XE, Jiang LY (2005) Copper phytoavailability and uptake by *Elsholtzi asplendens* from contaminated soil as affected by soil amendments. J Environ Sci Health 40:839–856

Pourrut B, Shahid M, Dumat C, Winterton P, Pinelli E (2011) Lead uptake, toxicity, and detoxification in plants. Rev Environ Contam Toxicol 213:113–136

Rahman MA, Hasegawa H, Rahman MM, Miah MAM, Tasmin A (2008) Straight head disease of rice (*Oryza sativa* L.) induced by arsenic toxicity. Environ Exp Bot 62:54–59

Rai V, Vajpayee P, Singh SN, Mehrotra S (2004) Effect of chromium accumulation on photosynthetic pigments, oxidative stress defense system, nitrate reduction, proline level and eugenol content of *Ocimum tenuiflorum* L. Plant Sci 167(5):1159–1169

Rajakaruna N, Tompkins KM, Pavicevic PG (2006) Phytoremediation: an affordable green technology for the clean-up of metalcontaminated sites in Sri Lanka. Cey J Sci (Bio Sci) 35:25–39

Rashid A, Mahmood T, Mehmood F, Khalid A, Saba B, Batool A, Riaz A (2014) Phytoaccumulation, competitive adsorption and evaluation of chelators-metal interaction in lettuce plant. Environ Eng Manag J 13:2583–2592

Raza A, Habib M, Kakavand SN, Zahid Z, Zahra N, Sharif R, Hasanuzzaman M (2020) Phytoremediation of cadmium: physiological, biochemical, and molecular mechanisms. Biology (Basel) 9(7):177

Ren W-X, Li P-J, Geng Y, Li X-J (2009) Biological leaching of heavy metals from a contaminated soil by *aspergillus Niger*. J Hazard Mater 167:164–169

Rezania S, Taib SM, Md Din MF, Dahalan FA, Kamyab H (2016) Comprehensive review on phytotechnology: heavy metals removal by diverse aquatic plants species from wastewater. J Hazard Mater 318:587–599

Ricachenevsky FK, Menguer PK, Sperotto RA, Williams LE, Fett JP (2013) Roles of plant metal tolerance proteins (MTP) in metal storage and potential use in biofortification strategies. Front Plant Sci 4:144

Sahu GK, Upadhyay S, Sahoo BB (2012) Mercury induced phytotoxicity and oxidative stress in wheat (*Triticum aestivum* L.) plants. Physiol Mol Biol Plants 18(1):21–31

Sarma H, Forid N, Prasad R, Prasad MNV, Ma LQ, Rinklebe J (2021) Enhancing phytoremediation of hazardous metal(loid)s using genome engineering CRISPR–Cas9 technology. J Hazard Mater. https://doi.org/10.1016/j.jhazmat.2021.125493

Seguel A, Cumming JR, Klugh-Stewart K, Cornejo_P, Borie F (2013) The role of arbuscular mycorrhizas in decreasing aluminium phytotoxicity in acidic soils: a review. Mycorrhiza 23: 167–183

Shahandeh H, Hossner LR (2002) Role of soil properties in phytoaccumulation of uranium. Water Air Soil Pollut 141:165–180

Shahzad B, Tanveer M, Hassan W, Shah AN, Anjum SA, Cheema SA, Ali I (2016) Lithium toxicity in plants: reasons, mechanisms and remediation possibilities—a review. Plant Physiol Biochem 107:104–115

Shahid M, Shamshad S, Rafiq M, Khalid S, Bibi I, Niazi NK, Dumat C, Rashid MI (2017) Chromium speciation, bioavailability, uptake, toxicity and detoxification in soil-plant system: a review. Chemosphere 178:513–533

Shanker AK, Cervantes C, Loza-Tavera H, Avudainayagam S (2005) Chromium toxicity in plants. Environ Int 31(5):739–753

Shanying H, Xiaoe Y, Zhenli H, Baligar VC (2017) Morphological and physiological responses of plants to cadmium toxicity: a review. Pedosphere 27:421–438

Sharma A, Kapoor D, Wang J, Shahzad B, Kumar V, Bali AS, Jasrotia S, Zheng B, Yuan H, Yan D (2020) Chromium bioaccumulation and its impacts on plants: an overview. Plants (Basel) 9(1): 100

Shekar CC, Sammaiah D, Shasthree T, Reddy KJ (2011) Effect of mercury on tomato growth and yield attributes. Int J Pharma Biol Sci 2(2):B358–B364

Shi G, Ma H, Chen Y, Liu H, Song G, Cai Q, Lou L, Rengel Z (2019) Low arsenate influx rate and high phosphorus concentration in wheat (*Triticum aestivum* L.): a mechanism for arsenate tolerance in wheat plants. Chemosphere 214:94–102

Singh S, Parihar P, Singh R, Singh VP, Prasad SM (2016) Heavy metal tolerance in plants: role of transcriptomics, proteomics, metabolomics, and ionomics. Front Plant Sci 6:1143

Sonowal S, Nava AR, Joshi SJ, Borah SN, Islam NF, Pandit S, Prasad R, Sarma H (2022) Biosurfactants assisted heavy metals phytoremediation: green technology for the United Nations sustainable development goals. Pedosphere 2(1):198–210. https://doi.org/10.1016/S1002-0160(21)60067-X

Srivastava M, Ma LQ, Santos JAG (2006) Three new arsenic hyperaccumulating ferns. Sci Total Environ 364:1–3

Srivastava S, Sinha P, Sharma YK (2017) Status of photosynthetic pigments, lipid peroxidation and anti-oxidative enzymes in *Vigna mungo* in presence of arsenic. J Plant Nutr 40:298–306

Srivastava S, Sounderajan S, Udas A, Suprasanna P (2014) Effect of combinations of aquatic plants (*Hydrilla*, *Ceratophyllum*, *Eichhornia*, *Lemna* and *Wolffia*) on arsenic removal in field conditions. Ecol Eng 73:297–301

Sujkowska-Rybkowska M, Borucki W (2015) Pectins esterification in the apoplast of aluminum-treated pea root nodules. J Plant Physiol 184:1–7

Suman J, Uhlik O, Viktorova J, Macek T (2018) Phytoextraction of heavy metals: a promising tool for clean-up of polluted environment? Front Plant Sci 9:1476

Tamburini E, Sergi S, Serreli L, Bacchetta G, Milia S, Cappai G, Carucci A (2017) Bioaugmentation-assisted phytostabilisation of abandoned mine sites in south West Sardinia. Bull Environ Contam Toxicol 98:310–316

Tang J, Xu J, Wu Y, Li Y, Tang Q (2012) Effects of high concentration of chromium stress on physiological and biochemical characters and accumulation of chromium in tea plant (*Camellia sinensis* L.). Afr J Biotechnol 11:2248–2255

Thakare M, Sarma H, Datar S, Roy A, Pawar P, Gupta K, Pandit S, Prasad R (2021) Understanding the holistic approach to plant-microbe remediation technologies for removing heavy metals and radionuclides from soil. Curr Res Biotechnol 3:84–98. https://doi.org/10.1016/j.crbiot.2021.02.004

Tiwari M, Sharma D, Dwivedi S, Singh M, Tripathi RD, Trivedi PK (2014) Expression in *Arabidopsis* and cellular localization reveal involvement of rice NRAMP, OsNRAMP1, in arsenic transport and tolerance. Plant Cell Environ 37:140–152

Tome FV, Rodriguez PB, Lozano J (2008) Elimination of natural uranium and 226Ra from contaminated waters by rhizofiltration using *Helianthus annuus* L. Sci Total Environ 393: 351–357

Tripathi P, Singh PC, Mishra A, Srivastava S, Chauhan R, Awasthi S, Mishra S, Dwivedi S, Tripathi P, Kalra A, Tripathi RD, Nautiyal CS (2017) Arsenic tolerant *Trichoderma* sp. reduces arsenic induced stress in chickpea (*Cicer arietinum*). Environ Pollut 223:137–145

van de Mortel JE, Villanueva LA, Schat H, Kwekkeboom J, Coughlan S, Moerland PD et al (2006) Large expression differences in genes for iron and zinc homeostasis, stress response, and lignin biosynthesis distinguish roots of *Arabidopsis thaliana* and the related metal hyperaccumulator *Thlaspi caerulescens*. Plant Physiol 142:1127–1134

Varun M, D'Souza RD, Pratas J, Paul MS (2011) Evaluation of phytostabilization, a green technology to remove heavy metals from industrial sludge using *Typha latifolia* L. experimental design. Biotechnol Bioinformatics Bioeng 1:137–145

Viehweger K (2014) How plants cope with heavy metals. Bot Stud 55:35

Wang X, Peng B, Tan C, Ma L, Rathinasabapathi B (2015) Recent advances in arsenic bioavailability, transport, and speciation in rice. Environ Sci Pollut Res Int 22:5742–5750

Wang J, Liang S, Xiang W, Dai H, Duan Y, Kang F, Chai T (2019) A repeat region from the *Brassica juncea* HMA4 gene *BjHMA4R* is specifically involved in Cd^{2+} binding in the cytosol under low heavy metal concentrations. BMC Plant Biol 19:89

Whiting SN, Leake JR, McGrath SP, Baker AJM (2000) Positive responses to zinc and cadmium by roots of the hyperaccumulator *Thlaspi caerulescens*. New Phytol 145:199–210

Xun Y, Feng L, Li Y, Dong H (2017) Mercury accumulation plant *Cyrtomium macrophyllum* and its potential for phytoremediation of mercury polluted sites. Chemosphere 189:161–170

Yan A, Wang Y, Tan SN, Mohd Yusof ML, Ghosh S, Chen Z (2020) Phytoremediation: a promising approach for revegetation of heavy metal-polluted land. Front Plant Sci 11:359

Yorifuji T, Tsuda T (2014) Minamata. In: Wexler P (ed) Encyclopedia of toxicology, 3rd edn. Academic Press, London, pp 340–344

Yusefi-Tanha E, Fallah S, Rostamnejadi A, Pokhrel LR (2020) Root system architecture, copper uptake and tissue distribution in soybean (*Glycine max* (L.) Merr.) grown in copper oxide nanoparticle (cuonp)-amended soil and implications for human nutrition. Plan Theory 9 (10):1326

Zaheer IE, Ali S, Saleem MH, Imran M, Alnusairi GSH, Alharbi MB, Riaz M, Rizwan M, Soliman MH (2020) Role of iron-lysine on morpho-physiological traits and combating chromium toxicity in rapeseed (*Brassica napus* L.) plants irrigated with different levels of tannery wastewater. Plant Physiol Biochem 155:70–84

Zhang X, Zhang S, Xua X, Li T, Gong G, Jiaa Y, Li Y, Denga L (2010) Tolerance and accumulation characteristics of cadmium in *Amaranthus hybridus* L. J Hazard Mater 180:303–308

Zhao FJ, Dunham SJ, McGrath SP (2002) Arsenic hyperaccumulation by different fern species. New Phytol 15:27–31

Zhao FJ, McGrath SP, Meharg AA (2010) Arsenic as a food chain contaminant: mechanisms of plant uptake and metabolism and mitigation strategies. Annu Rev Plant Biol 61:535–559

The manufacturer's authorised representative in the EU is Springer Nature Customer Service Centre GmbH, Europaplatz 3, 69115 Heidelberg, Germany. If you have any concerns regarding our products, please contact ProductSafety@springernature.com

Printed and bound by CPI Group (UK) Ltd, Croydon, CR0 4YY
30/09/2026
02225332-0017